S0-CAS-638

Specialist Periodical Reports Online

Application for Free Access to Electronic Chapters

Amino Acids, Peptides and Proteins
Volume 35

Customers purchasing a print volume are now entitled to free site-wide access to the electronic version of that title, including back volumes from 1998 - 2005 inclusive.
(For the definition of a site, please consult **www.rsc.org/subagree**)

To apply for free access, please complete and return this form.

Contact Name:______________________________

Job Title:______________________________

Organisation:______________________________

Address:______________________________

____

____

Town:______________________________

Country:______________________________

Post/Zip Code:______________________________

E-mail:______________________________

Telephone:______________________________

RSC Publishing

Continued over

Please give IP addresses of the site overleaf in the format w.x.y.z.
Wildcards and ranges are allowed

e.g. for Class B 128.128.*.*
for Class C 192.192.192.*
for a range 123.456.1-99.*

Please put each entry on a new line:

____

____

____

____

(please use a continuation page if necessary)

I have read and agree to the terms of the Electronic Information Licence Agreement at **www.rsc.org/subagree**

Signed: ____________________ **Date:** _______________

☐ Please contact me about access for additional sites outside the site definition in the Electronic Information Licence Agreement

☐ Please arrange username/password access and not IP address control

Please return this form by non-electronic means to the Royal Society of Chemistry at the address below.

Photocopies or facsimiles are not acceptable.

Sales & Customer Care Dept
Royal Society of Chemistry · Thomas Graham House
Science Park · Milton Road · Cambridge · CB4 0WF · UK
T +44(0)1223 432360 · **F** +44(0)1223 426017
E sales@rsc.org

Amino Acids, Peptides and Proteins

Volume 35

A Specialist Periodical Report

Amino Acids, Peptides and Proteins

Volume 35

A Review of the Literature Published during 2002.

Editor
J.S. Davies, *University of Wales, Swansea, UK*

Authors
W.C. Chan, *University of Nottingham, Nottingham, UK*
D.T. Elmore, *Oxfordshire, UK*
E. Farkas, *University of Debrecen, Debrecen, Hungary*
A. Higton, *Nottingham Trent University, Nottingham, UK*
B. Penke, *University Szeged, Hungary*
I. Sóvágó, *University of Debrecen, Debrecen, Hungary*
G. Tóth, *University Szeged, Hungary*
G. Váradi, *University Szeged, Hungary*

RSC Publishing

If you buy this title on standing order, you will be given FREE access to the chapters online. Please contact sales@rsc.org with proof of purchase to arrange access to be set up.

Thank you

ISBN-10: 0-85404-247-4
ISBN-13: 978-0-85404-247-0
ISSN 1361-5904

A catalogue record for this book is available from the British Library

Published by The Royal Society of Chemistry,
Thomas Graham House, Science Park, Milton Road,
Cambridge CB4 0WF, UK

Registered Charity Number 207890

For further information see our web site at www.rsc.org

Typeset by Macmillan India Ltd, Bangalore, India
Printed by Henry Ling Ltd, Dorchester, Dorset, UK

Preface

This 35th Volume in the series has not appeared within a time frame that we would all be proud of, but it is often difficult to recover from critical delays brought about mainly through pressure of work. However it is our sincere hope that the hard work of our Chapter authors will form a real workhorse of a source book, representing the relevant and important work associated with the period up to the end of 2002. With the expansion of the literature all our Reporters have increasingly had to approach their task with a more selective brief than was the case in earlier volumes of the series.

In the absence of a chapter on Amino Acids in volume 34, Chapter 1 in this Volume covers the two-year period 2001–2, and this time also we welcome back the biennial Chapter on 'Metal Complexes' authored by Etelka Farkas and Imre Sóvágó covering the same period. Colleagues from the Medical University of Szeged (Botond Penke, Gábor Tóth and Györgyi Váradi) have continued with their wide-ranging Chapter 3 reviewing Analogue and Conformational Studies on Peptides. The continuing involvement of Don Elmore, Weng Chan, Avril Higton and John Davies from the UK has taken care of the other core Chapters. Sadly due to family commitments Graham Barrett had to relinquish the role of Joint Senior Reporter, but we are grateful to Graham for his contributions as an author within these Reports over many years. So the current volume has been co-ordinated by one Senior Reporter, who appreciates very much the genuine and unstinting efforts of everyone that has brought this Volume to fruition. This includes the RSC personnel, who have patiently waited for manuscripts to appear and ensured that everyone's efforts have been transcribed to the document you are now reading.

Again to preserve good practice in presentation in peptide science, we are grateful for the permission to reproduce John Jones's most recent plea for conformity of nomenclature in such a wide-ranging discipline. Over recent volumes of these Reports we have reproduced his 'Short Guide to Abbreviations' published in 1999, but this time we have the opportunity of re-printing the most recent version which appeared in J. Peptide Science 2006, **11**, pp. 1–12. During the production of this volume 'Proceedings of the 18th American Symposium at Boston' in 2003 (eds. M. Chorev and T. K. Sawyer) appeared. From the title of the book from these proceedings – *Peptide Revolution: Genomics, Proteomics and Therapeutics* – comes a hint as to where our subject might be heading.

John S. Davies
University of Wales, Swansea

Contents

Cover
The crystal structure of particulate methane monooxygenase (pMMO) reveals many unexpected features including, a trimeric oligomerization state and three distinct metal centers. Image reproduced by permission of Amy Rosenzweig from *Dalton Transactions*, 2005.

A Short Guide to Abbreviations and Their Use in Peptide Science

Abbreviations, acronyms and symbolic representations are very much part of the language of peptide science – in conversational communication as much as in its literature. They are not only a convenience, either – they enable the necessary but distracting complexities of long chemical names and technical terms to be pushed into the background so the wood can be seen among the trees. Many of the abbreviations in use are so much in currency that they need no explanation. The main purpose of this editorial is to identify them and free authors from the hitherto tiresome requirement to define them in every paper. Those in the tables that follow – which will be updated from time to time – may in future be used in this Journal without explanation.

All other abbreviations should be defined. Previously published usage should be followed unless it is manifestly clumsy or inappropriate. Where it is necessary to devise new abbreviations and symbols, the general principles behind established examples should be followed. Thus, new amino-acid symbols should be of form Abc, with due thought for possible ambiguities (Dap might be obvious for diaminoproprionic acid, for example, but what about diaminopimelic acid?).

Where alternatives are indicated below, the first is preferred.

Amino Acids

Proteinogenic Amino Acids

Ala	Alanine	A
Arg	Arginine	R
Asn	Asparagine	N
Asp	Asp artic acid	D
Asx	Asn *or* Asp	
Cys	Cysteine	C
Gln	Glutamine	Q
Glu	Glutamic acid	E
Glx	Gln *or* Glu	
Gly	Glycine	G
His	Histidine	H

Ile	Isoleucine	I
Leu	Leucine	L
Lys	Lysine	K
Met	Methionine	M
Phe	Phenylalanine	F
Pro	Proline	P
Ser	Serine	S
Thr	Threonine	T
Trp	Tryptophan	W
Tyr	Tyrosine	Y
Val	Valine	V

Other Amino Acids

Aad	α-Aminoadipic acid
βAad	β-Aminoadipic acid
Abu	α-Aminobutyric acid
Aib	α-Aminoisobutyric acid; α-methylalanine
βAla	β-Alanine; 3-aminopropionic acid (avoid Bal)
Asu	α-Aminosuberic acid
Aze	Azetidine-2-carboxylic acid
Cha	β-cyclohexylalanine
Cit	Citrulline; 2-amino-5-ureidovaleric acid
Dha	Dehydroalanine (also ΔAla)
Gla	γ-Carboxyglutamic acid
Glp	pyroglutamic acid; 5-oxoproline (also pGlu)
Hph	Homophenylalanine (Hse = homoserine, and so on). Caution is necessary over the use of the use of the prefix homo in relation to α-amino-acid names and the symbols for homo-analogues. When the term first became current, it was applied to analogues in which a side-chain CH_2 extension had been introduced. Thus homoserine has a side-chain CH_2CH_2OH, homoarginine $CH_2CH_2CH_2NHC(=NH)NH_2$, and so on. In such cases, the convention is that a new three-letter symbol for the analogue is derived from the parent, by taking H for homo and combining it with the first two characters of the parental symbol – hence, Hse, Har and so on. Now, however, there is a considerable literature on α-amino acids which are analogues of α-amino acids in which a CH_2 group has been inserted between the α-carbon and carboxyl group. These analogues have also been called homo-analogues, and there are instances for example not only of 'homophenylalanine', $NH_2CH(CH_2CH_2Ph)CO_2H$, abbreviated Hph, but also 'homophenylalanine', $NH_2CH(CH_2Ph)CH_2CO_2H$ abbreviated Hph. Further, members of the analogue class with CH_2 interpolated between the α-carbon and the carboxyl group of the parent α-amino acid structure have been

	called both 'α-homo'- and β-homo'. Clearly great care is essential, and abbreviations for 'homo' analogues ought to be fully defined on every occasion. The term 'β-homo' seems preferable for backbone extension (emphasizing as it does that the residue has become a β-amino acid residue), with abbreviated symbolism as illustrated by βHph for $NH_2CH(CH_2Ph)CH_2CO_2H$.
Hyl	δ-Hydroxylysine
Hyp	4-Hydroxyproline
αIle	*allo*-Isoleucine; 2*S*, 3*R* in the L-series
Lan	Lanthionine; *S*-(2-amino-2-carboxyethyl)cysteine
MeAla	*N*-Methylalanine (MeVal=*N*-methylvaline, and so on). This style should not be used for α-methyl residues, for which either a separate unique symbol (such as Aib for α-methylalanine) should be used, or the position of the methyl group should be made explicit as in αMeTyr for α-methyltyrosine.
Nle	Norleucine; α-aminocaproic acid
Orn	Ornithine; 2,5-diaminopentanoic acid
Phg	Phenylglycine; 2-aminophenylacetic acid
Pip	Pipecolic acid; piperidine-s-carboxylic acid
Sar	Sarcosine; *N*-methylglycine
Sta	Statine; (3*S*, 4*S*)-4-amino-3-hydroxy-6-methyl-heptanoic acid
Thi	β-Thienylalanine
Tic	1,2,3,4-Tetrahydroisoquinoline-3-carboxylic acid
αThr	*allo*-Threonine; 2*S*, 3*S* in the L-series
Thz	Thiazolidine-4-carboxylic acid, thiaproline
Xaa	Unknown or unspecified (also Aaa)

The three-letter symbols should be used in accord with the IUPAC-IUB conventions, which have been published in many places (e.g. *European J. Biochem*. 1984; 138: 9–37), and which are (May 1999) also available with other relevant documents at: http://www.chem.qnw.ac.uk/iubmb/iubmb.html#03

It would be superfluous to attempt to repeat all the detail which can be found at the above address, and the ramifications are extensive, but a few remarks focussing on common misuses and confusions may assist. The three-letter symbol standing alone represents the unmodified intact amino acid, of the L-configuration unless otherwise stated (but the L-configuration may be indicated if desired for emphasis: e.g. L-Ala). The same three-letter symbol, however, also stands for the corresponding amino acid *residue*. The symbols can thus be used to represent peptides (e.g. AlaAla or Ala-Ala = alanylalanine). When nothing is shown attached to either side of the three-letter symbol it is meant to be understood that the amino group(always understood to be on the left) or carboxyl group is unmodified, but this can be emphasized, so AlaAla = H-AlaAla-OH. Note however that indicating free termini by presenting the terminal group in full is wrong;$NH_2AlaAlaCO_2H$ implies a hydrazino group at one end and an α-keto acid derivative at the other. Representation of a free terminal carboxyl group by writing H on the right is also wrong because that implies a terminal aldehyde.

Side chains are understood to be unsubstituted if nothing is shown, but a substituent can be indicated by use of brackets or attachment by a vertical bond up or down. Thus an *O*-methylserine residue could be shown as **1**, **2**, or **3**.

—Ser(Me)— **1**

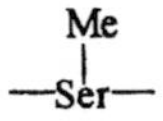

```
 Me
  |      2
—Ser—

—Ser—
  |      3
 Me
```

Note that the oxygen atom is not shown: it is contained in the three-letter symbol – showing it, as in Ser(OMe), would imply that a peroxy group was present. Bonds up or down should be used only for indicating side-chain substitution. Confusions may creep in if the three-letter symbols are used thoughtlessly in representations of cyclic peptides. Consider by way of example the hypothetical cyclopeptide threonylalanylalanylglutamic acid. It might be thought that this compound could be economically represented **4**.

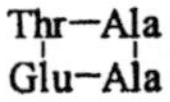

```
Thr—Ala
 |    |    4
Glu—Ala
```

But this is wrong because the left hand vertical bond implies an ester link between the two side chains, and strictly speaking if the right hand vertical bond means anything it means that the two Ala α-carbons are linked by a CH_2CH_2 bridge. This objection could be circumvented by writing the structure as in **5**.

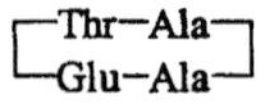

```
┌Thr—Ala┐
└Glu—Ala┘   5
```

But this is now ambiguous because the convention that the symbols are to be read as having the amino nitrogen to the left cannot be imposed on both lines. The direction of the peptide bond needs to be shown with an arrow pointing from CO to N, as in **6**.

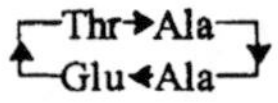

```
┌Thr→Ala┐
↑       ↓   6
└Glu←Ala┘
```

Actually the simplest representation is on one line, as in **7**.

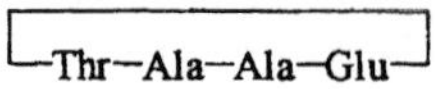

```
┌───────────────────┐
└Thr—Ala—Ala—Glu┘   7
```

Substituents and Protecting Groups

Ac	Acetyl
Acm	Acetamidomethyl
Adoc	1-Adamantyloxycarbonyl
Alloc	Allyloxycarbonyl
Boc	*t*-Butoxycarbonyl
Bom	π-Benzyloxymethyl
Bpoc	2-(4-Biphenylyl)isopropoxycarbonyl
Btm	Benzylthiomethyl
Bum	π-*t*-Butoxymethyl
Bu^i	*i*-Butyl
Bu^n	*n*-Butyl
Bu^t	*t*-Butyl
Bz	Benzoyl
Bzl	Benzyl (also Bn); Bzl(OMe)=4-methoxybenzyl and so on
Cha	Cyclohexylammonium salt
Clt	2-Chlorotrityl
Dcha	Dicyclohexylammonium salt
Dde	1-(4,4-Dimethyl-2,6-dioxocyclohex-1-ylidene)ethyl
Ddz	2-(3,5-Dimethoxyphenyl)-isopropoxycarbonyl
Dnp	2,4-Dinitrophenyl
Dpp	Diphenylphosphinyl
Et	Ethyl
Fmoc	9-Fluorenylmethoxycarbonyl
For	Formyl
Mbh	4,4′-Dimethoxydiphenylmethyl, 4,4′-Dimethoxybenzhydryl
Mbs	4-Methoxybenzenesulphonyl
Me	Methyl
Mob	4-Methoxybenzyl
Mtr	2,3,6-Trimethyl,4-methoxybenzenesulphonyl
Nps	2-Nitrophenylsulphenyl
OAll	Allyl ester
OBt	1-Benzotriazolyl ester
OcHx	Cyclohexyl ester
Onp	4-Nitrophenyl ester
Opcp	Pentachlorophenyl ester
OPfp	Pentafluorop henyl ester
OSu	Succinimido ester
OTce	2,2,2-Trichloroethyl ester
Otcp	2,4,5-Trichlorophenyl ester
Tmob	2,4,5-Trimethoxybenzyl
Mtt	4-Methyltrityl
Pac	Phenacyl, $PhCOCH_2$ (care! Pac also = $PhCH_2CO$)
Ph	Phenyl
Pht	Phthaloyl
Scm	Methoxycarbonylsulphenyl

Pmc	2,2,5,7,8-Pentamethylchroman-6-sulphonyl
Pr^i	*i*-Propyl
Pr^n	*n*-Propyl
Tfa	Trifluoroacetyl
Tos	4-Toluenesulphonyl (also Ts)
Troc	2,2,2-Trichloroethoxycarbonyl
Trt	Trityl, triphenylmethyl
Xan	9-Xanthydryl
Z	Benzyloxycarbonyl (also Cbz). Z(2Cl)=2-chlorobenzyloxycarbonyl and so on

Amino Acid Derivatives

DKP	Diketopiperazine
NCA	*N*-Carboxyanhydride
PTH	Phenylthiohydantoin
UNCA	Urethane *N*-carboxyanhydride

Reagents and Solvents

BOP	1-Benzotriazolyloxy-tris-dimethylamino-phosphonium hexafluorophosphate
CDI	Carbonyldiimidazole
DBU	Diazabicyclo[5.4.0]-undec-7-ene
DCCI	Dicyclohexylcarbodiimide (also DCC)
DCHU	Dicyclohexylurea (also DCU)
DCM	Dichloromethane
DEAD	Diethyl azodicarboxylate (DMAD = the dimethyl analogue)
DIPCI	Diisopropylcarbodiimide (also DIC)
DIPEA	Diisopropylethylamine (also DIEA)
DMA	Dimethylacetamide
DMAP	4-Dimethylaminopyridine
DMF	Dimethylformamide
DMS	Dimethylsulphide
DMSO	Dimethylsulphoxide
DPAA	Diphenylphosphoryl azide
EEDQ	2-Ethoxy-1-ethoxycarbonyl-1,2-dihydroquinoline
HATU	This is the acronym for the 'uronium' coupling reagent derived from HOAt, which was originally thought to have the structure **8**, the Hexafluorophosphate salt of the *O*-(7-Azabenzotriazol-lyl)-Tetramethyl Uronium cation.

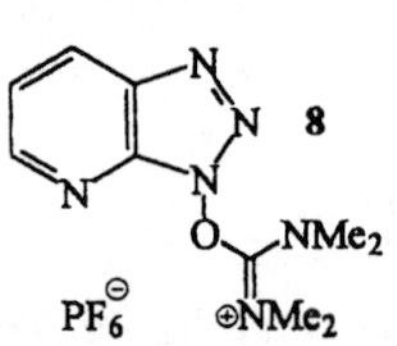

In fact this reagent has the isomeric *N*-oxide structure **9** in the crystalline state, the unwieldy correct name of which does not conform logically with the acronym, but the acronym continues in use.

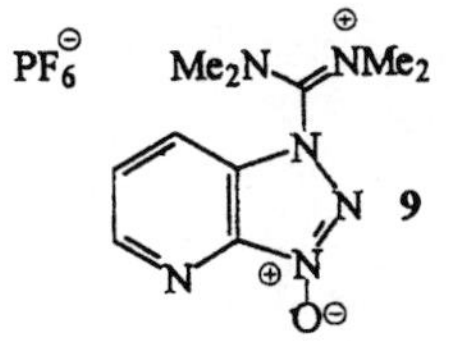

$PF_6^{\ominus}$ Me_2N $\overset{\oplus}{N}Me_2$ N N N N $\overset{\oplus}{N}$ $O^{\ominus}$ **9**

Similarly, the corresponding reagent derived from HOBt has the firmly attached label HBTU (the tetrafluoroborate salt is also used: TBTU), despite the fact that it is not actually a uronium salt.

HMP Hexamethylphosphoric triamide (also HMPA, HMPTA)
HOAt 1-Hydroxy-7-azabenzotriazole
HOBt 1-Hydroxybenzotriazole
HOCt 1-Hydroxy-4-ethoxycarbonyl-1,2,3-triazole
NDMBA *N,N′*-Dimethylbarbituric acid
NMM *N*-Methylmorpholine
PAM Phenylacetamidomethyl resin
PEG Polyethylene glycol
PtBOP 1-Benzotriazolyloxy-tris-pyrrolidinophosphonium hexafluorophosphate
SDS Sodium dodecyl sulphate
TBAF Tetrabutylammonium fluoride
TBTU See remarks under HATU above
TEA Triethylamine
TFA Trifluoroacetic acid
TFE Trifluoroethanol
TFMSA Trifluoromethanesulphonic acid
THF Tetrahydrofuran
WSCI Water soluble carbodiimide: 1-ethyl-3-(3′-dimethylaminopropyl)-carbodiimide hydrochloride (also EDC)

Techniques

CD Circular dichroism
COSY Correlated spectroscopy
CZE Capillary zone electrophoresis
ELISA Enzyme-linked immunosorbent assay
ESI Electrospray ionization
ESR Electron spin resonance
FAB Fast atom bombardment
FT Fourier transform
GLC Gas liquid chromatography
hplc High performance liquid chromatography
IR Infra red
MALDI Matrix-assisted laser desorption ionization
MS Mass spectrometry

NMR	Nuclear magnetic resonance
nOe	Nuclear Overhauser effect
NOESY	Nuclear Overhauser enhanced spectroscopy
ORD	Optical rotatory dispersion
PAGE	Polyacrylamide gel electrophoresis
RIA	Radioimmunoassay
ROESY	Rotating frame nuclear Overhauser enhanced spectroscopy
RP	Reversed phase
SPPS	Solid phase peptide synthesis
TLC	Thin layer chromatography
TOCSY	Total correlation spectroscopy
TOF	Time of flight
UV	Ultraviolet

Miscellaneous

Ab	Antibody
ACE	Angiotensin-converting enzyme
ACTH	Adrenocorticotropic hormone
Ag	Antigen
AIDS	Acquired immunodeficiency syndrome
ANP	Atrial natriuretic polypeptide
ATP	Adenosine triphosphate
BK	Bradykinin
BSA	Bovine serum albumin
CCK	Cholecystokinin
DNA	Deoxyribonucleic acid
FSH	Follicle stimulating hormone
GH	Growth hormone
HIV	Human immunodeficiency virus
LHRH	Luteinizing hormone releasing hormone
MAP	Multiple antigen peptide
NPY	Neuropeptide Y
OT	Oxytocin
PTH	Parathyroid hormone
QSAR	Quantitative structure–activity relationship
RNA	Ribonucleic acid
TASP	Template-assembled synthetic protein
TRH	Thyrotropin releasing hormone
VIP	Vasoactive intestinal peptide
VP	Vasopressin

J. H. Jones

Amino Acids

BY WENG C. CHAN,[a] AVRIL HIGTON[b] AND JOHN S. DAVIES[c]
[a] *Department Pharmaceutical Sciences, University of Nottingham, NG7 2RD, Nottingham, UK*
[b] *Nottingham Trent University, 224 Rutland RoadWest Bridgeford, NG2 5EB, Nottingham, UK*
[c] *Department of Chemistry, University of Wales Swansea, Singleton Park, SA2, 8PP, Swansea, UK*

1 Introduction

As the coverage of Amino Acids did not make it into Volume 34 of these Reports,[1] this Chapter covers the years 2001 and 2002. This inevitably means that the authors this time needed to be a little more selective in the papers reviewed, due to space limitations. The main source of the citations was again Chemical Abstracts (Vols 134–136), CA Selects on Amino Acids Peptides and Proteins[2] and the Web of Knowledge.[3] No references to conference proceedings have been included, and the patent literature has not been scanned. With the addition of an extra author, the style of the Chapter might show minor changes, but in order to preserve continuity for those taking 'year on year' surveys within a field, the pattern of sub-headings have been retained.

2 Reviews

The main aim of this Report is to review the original refereed papers in this subject area, so reporting on reviews covering similar areas is included in 'title-only' format as a token of respect for all those that have similarly laboured through the literature to bring us highlights from specific areas of endeavour. Reviews cited during 2001–2002 were:-

New Strecker Synthesis-Asymmetric Synthesis and Chiral Catalysts[4]
Methods for the Synthesis of Unnatural Amino Acids[5]
Chiral Oxazinones and Pyrazinones as α-Amino Acid Templates[6]
Amino Acid Derivatives by Multicomponent Reactions[7]
Progress on the Asymmetric Synthesis of α–Amino Acids[8]
Comparison of different Chemoenzymatic Process Routes to Enantiomerically Amino Acids[9]

Amino Acids, Peptides and Proteins, Volume 35

α-Imino Esters: Versatile Substrates for the Catalytic, Asymmetric Synthesis of α- and β-Amino Acids and Lactones[10]

The Asymmetric Synthesis of Unnatural α–Amino Acids as building blocks for Complex Synthesis[11]

Asymmetric Hydrogenation and other Methods for the Synthesis of Unnatural Amino Acids[12]

Metabolic Engineering of Glutamate Production[13]

Amino Acids Production Processes[14]

Biotechnological Manufacture of Lysine[15]

The Threonine Story[16]

The Economic Aspects of Amino Acid Production[17]

Synthesis of Enantiometrically pure Pipecolic Acid Derivatives via Bio- and Transition Metal Catalysis[18]

A Journey from Unsaturated Amino Acid Synthesis to Cyclic Peptides[19]

Side-chain modifications and Applications of Aliphatic Unsaturated α-Amino Acids[20]

Fullerene-based Amino Acids and Peptides[21]

Selenocysteine Derivatives for Chemoselective Ligations[22]

Study on Resolution of Chiral Amino Acid Enantiomers[23]

Highly Diastereoselective Michael Reactions between Nucleophilic Glycine Equivalents and β-Substituted α,β-Unsaturated Acids: A General Approach to chi-Constrained Amino Acids[24]

3 Naturally-Occurring Amino Acids

While interest continues in the synthesis of known naturally-occurring amino acids, it has not been a productive period in papers highlighting the existence of amino acids from new sources.

3.1 New Naturally Occurring Amino Acids. – Amongst the new nortropane alkaloids isolated from the fruit of *Morus alba* LINNE in Turkey, six new amino acids have been characterised.[25] They have been allocated pyrrolidinyl dodecanoic and piperidinyl dodecanoic structures: (3*R*)-3-hydroxy-12-{(1*S*,4*S*)-4-[(1-hydroxyethyl-pyrrolidin-1-yl}-dodecanoic acid-3-*O*-β-D-glucopyranoside; its free acid; (3*R*)-3-hydroxy-12-[(1*R*, 4*R*, 5*S*)-4-hydroxy-5-methyl-piperidin-1-yl]-dodecanoic acid-3-*O*-β-D-glucopyranoside; its free acid; (3*R*)-3-hydroxy-12-[(1*R*, 4*R*, 5*S*)-4-hydroxy-5-hydroxymethyl-piperidin-1-yl]-dodecanoic acid-3-*O*-β-D-glucopyranoside and (3*R*)-3-hydroxy-12-[(1*R*, 4*S*, 5*S*)-4-hydroxy-5-methyl-piperidin-1-yl]-dodecanoic acid.

4 Chemical Synthesis and Resolution of Amino Acids

As in past years this remains the core section of the activity in the field, with some aspects already covered by the reviews listed in sub-section 2. Some subject matter also overlaps with material in subsequent sections of this Report.

4.1 General Methods for the Synthesis of α-Amino Acids, including Enantioselective Synthesis. – The development of benzophenone imines of glycine derivatives for the synthesis of α-amino acids has been outlined.[26] Substituted phenylalanines have been prepared[27] using UV photolysis of protected glycines in the presence of di-*t*-butylperoxide, substituted toluenes and the photosensitiser, benzophenone. After removal of the chiral auxiliary by lithium hydroperoxide, *N*-acetyl-D-serine methyl ester was the final product of the rearrangement summarised[28] in Scheme 1.

Reaction of 5-oxazolidinones such as (**1**), with alcohols in bicarbonate solution,[29] and with PLE and HLE[30] has led to good yields of amino acid derivatives. A chiral copper catalyst can catalyse[31] an enantioselective Mannich type reaction as summarised in Scheme 2. Catalysis by L-proline has enabled a general reaction[32] between ketones and PMP-protected α-imino ethyl glyoxylate (Scheme 3) to be made highly stereoselective. Its simplicity would make it an attractive proposal regarding prebiotic synthesis. Either enantiomer of both α- or β-amino acids have been made available[33] if this reaction includes an aldehyde instead of a ketone.

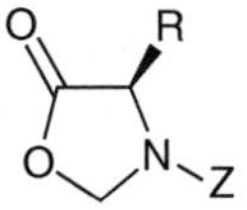

(**1**)

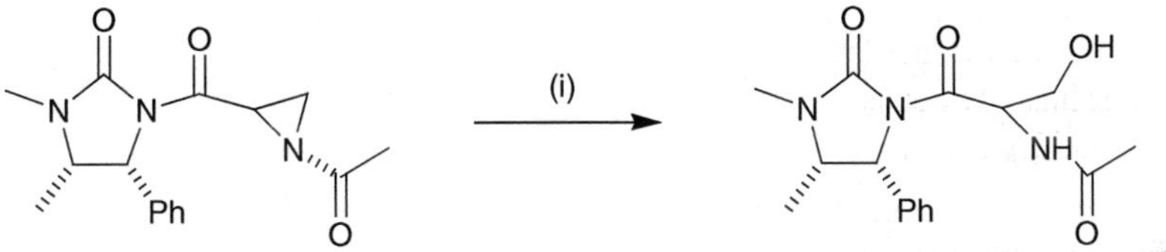

Scheme 1 *Reagents: $CeCl_3$ $6H_2O$, microwaves 240W 50–60°.*

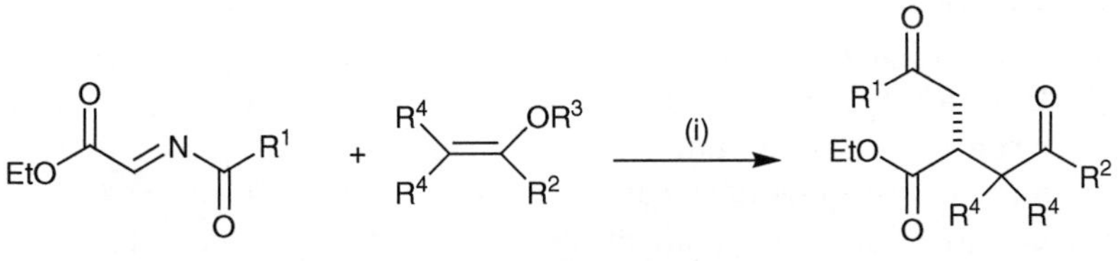

Scheme 2 *Reagents: (i) Cu catalyst.*

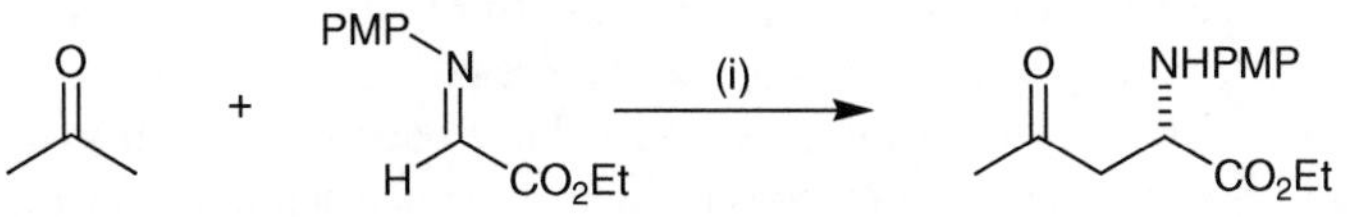

Scheme 3 *Reagents: (i) L-Pro (cat), DMSO.*

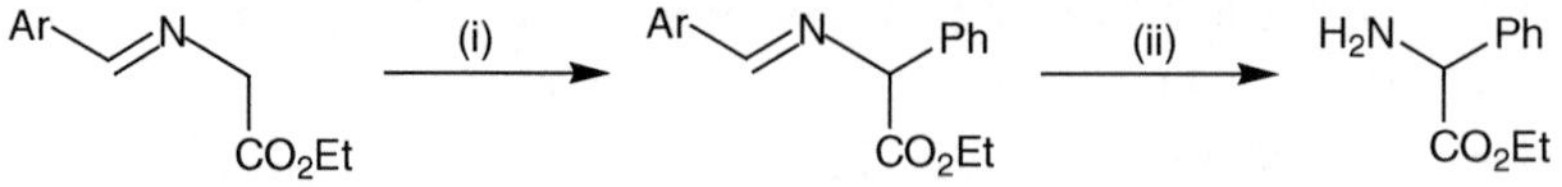

Scheme 4 *Reagents: (i) PhBr/Pd, t-Bu_3P/K_3PO_4 100°C (ii) H^+*

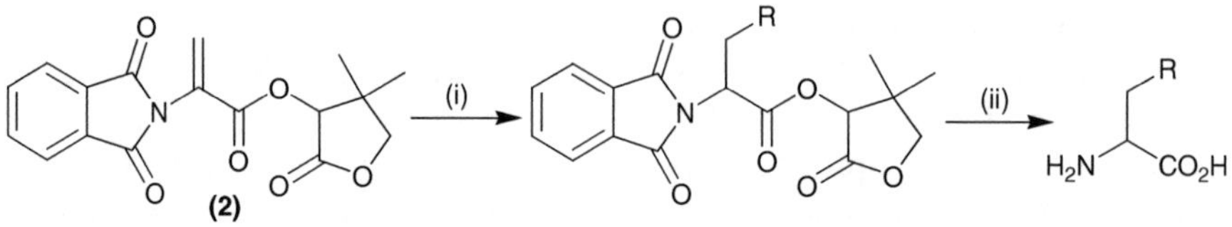

Scheme 5 *Reagents: (i) e.g. RX, Bu_3SnH Et_3B, Lewis acid (ii) 6M HCl/glacial acetic acid.*

A broad-based methodology[34] for the synthesis of non-natural amino acids has used catalytic enantioselective alkylation of α-imino esters and acetals with enol silanes, allyl silanes and olefins using chiral Cu(I) phosphine complexes. With suitable substitution (using 4-methoxybenzylidene) the pathway of Pd-catalysed acrylation can be guided[35] to amino acid esters as summarised in Scheme 4. A novel method, via a radical pathway for the asymmetric synthesis of α-amino acids has been reported.[36] Starting from the pantolactone (**2**) and using the conditions summarised in Scheme 5, it was shown that the absolute configuration of the stereogenic centre was dependent on the nature of the added radical. Rhodium-catalysed conjugated addition of α-aminoacrylates, with organotin and organobismuth reagents have yielded[37] amino acids under ambient conditions of air and water. Similar conditions have been used by the same authors[38] for the zinc-mediated conjugate addition of alkyl halides to α-phthalimidoacrylate.

α- and β-Substituted alanine derivatives have been efficiently produced[39] by α-amidoacrylation or Michael addition reactions using microwave irradiation and catalysis by silica-supported Lewis acids. Diverse functionalities, such as chlorides, nitriles, azides, acetates, thioacetates, thioethers and amines have been inserted[40] at varying chain lengths away from the α-centre, if amino acids (Ala or Phe), attached to a Wang resin, and derivatised with 3,4-dichlorobenzaldehyde were subjected to alkylation by α-bromo-ω-chloroelectrophiles. The one-step conversion[41] of azetidine-2,3-diones (**3**) to amino acids in the presence of cadmium/wet methanol has been explained by chelation of the ketone and amide groups to the metal, which allows for attack of the keto group by methanol followed by CO extrusion. The demands of the chemical libraries fraternity for a fast throughput of synthons, has brought the Ugi 4-component condensation into mainstream activity, and the formation of chiral products has now been made easier through better access[42] to chiral 1-amino carbohydrates. The reductive amination of ketones[43] has been adapted for the synthesis of racemic amino acids from α-keto acids, as shown by the synthesis of phenylglycine, and its 3-indolyl or 2-thienyl analogues.

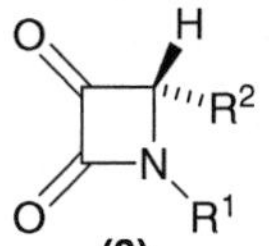

(3)

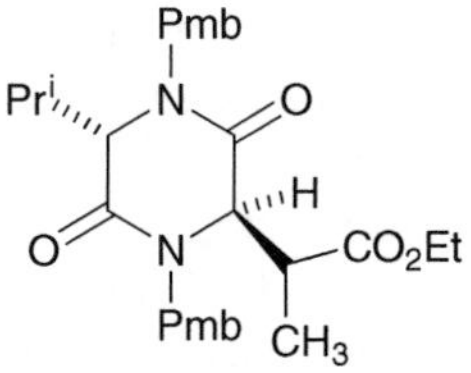

(4)

4.1.1 Use of Chiral Synthons in Amino Acid Synthesis. This still represents a booming area of interest and justifies its own sub-section. The Oxford school continues its productivity in this area as exemplified by the chiral glycine enolate, (*S*)-*N*,*N′*-bis-(*p*-methoxybenzyl)-3-isopropylpiperazine-2,5-dione.[44] This was able to discriminate between enantiomers of 2-bromopropionate esters in forming (**4**), which on further manipulation resulted in the synthesis of chiral 3-methylaspartates. A more detailed examination of the same chiral synthesis has also been carried out.[45] Diastereoselective conjugate addition of lithium (*R*)-*N*-benzyl-*N*-α-methylbenzylamide to α,β-unsaturated esters followed by enolate hydroxylation, reduction and oxidative cleavage has been shown[46] to be a route to α-amino acids in high enantiomeric excess. (*S*)-α-Amino acids in high chiral yields have been obtained[47] from (*S*)-*N*,*N′*-bis(*p*-methoxybenzyl)-3-methylene-6-isopropylpiperazine by reaction with a range of organocuprates.

The imine moiety of (5*S*)-3-([2-methoxycarbonyl]ethyl)-5-phenyl-5,6-dihydro-2H-1,4-oxazin-2-one has been shown[48] to undergo highly diastereocontrolled reduction followed by Lewis-acid-mediated nucleophilic addition of Grignard reagents to give enantiomerically pure glutamic acid analogues. The same authors[49] have also shown that iminium ions derived from (*S*)-5-phenylmorpholine-2-one undergo diastereoselective Strecker reactions using copper(I) cyanide/anhydrous hydrochloric acid, which lead eventually to D-α-amino acids. An inexpensive chiral auxiliary, the imino lactone (**5**), from (1*R*)-(+)-camphor on alkylation afforded[50,51] good yields of monosubstituted products at the H_{endo} position, which on hydrolysis yielded D-α-amino acids. Using camphor of the opposite configuration, or by switching the OH group of the auxiliary from C_2 to C_3, gave the L-enantiomer. *N*-Methyl pseudoephedrine has also been used[52] as a chiral auxiliary, by mediating a dynamic resolution of α-bromo-α-alkyl esters in nucleophilic substitution. Enantiomeric ratios of 98: 2 were achieved in the α-amino acids finally produced.

A variety of α-amino acids have been produced[53] diastereoselectively using indium-mediated allylation and alkylation of the Oppolzer camphorsultam derivative of glyoxylic oxime ether. A new chiral auxiliary (*S*)-*N*-(2-benzoylphenyl)-1-(3,4-dichlorobenzyl)-D-pyrrolidine-2-carboxamide, and related halogen-containing auxiliaries as their Ni(II) complexes have been shown[54] to give an increased chiral bias in the formation of (*S*)-α-amino acids, due possibly to the halogen substitution in the *N*-benzyl group. A 5-step asymmetric synthesis[55] of the (2*R*, 4*R*) 4-hydroxy-D-pyroglutamic acid involved the 1,3-dipolar cycloaddition of a chiral nitrone (from glyoxylic and protected D-ribosyl hydroxylamine) with the acrylamide of Oppolzer's sultam. The scope of this reaction[56] using other analogues has also been studied experimentally and theoretically for the formation of the (2*S*,4*S*)-isomer.

By association with substrates, chiral catalysts can also be considered under the heading 'chiral auxiliaries', and recently together with phase-transfer reagents constitute a popular means of asymmetric synthesis. A novel substrate/catalytic pair under phase-transfer conditions turned out[57] to be based on complex (**6**) which reacted quickly with 2-amino-2′-hydroxy-1,1-binaphthyl (NOBIN) as the phase transfer catalyst and could then be alkylated asymmetrically to give purifiable complexes for further processing to amino acids.

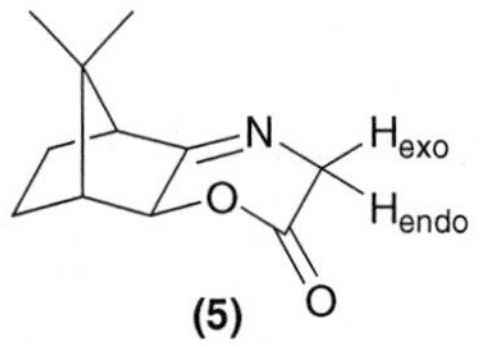

(**5**) 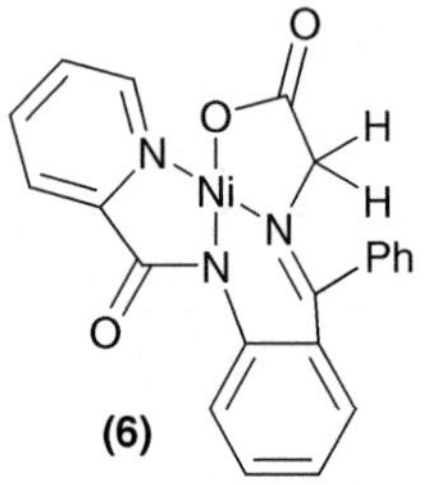 (**6**)

Stereoselective alkylation reactions of *N*′[(*S*)-1′-phenylethyl]-*N*-(diphenylmethylene)glycinamide using 18-crown-6 as catalyst, gave[58] a series of enantioenriched (83:17 ratio) unnatural amino acids. Structures based on the chirality of cinchona alkaloids have a noble track record in this area, and when polymer-supported cinchona alkaloid salts with different spacers were used[59] as catalysts in the *C*-alkylation of *N*-diphenylmethylene glycine *t*-butyl esters, it was found that the best result (81% ee) came from the polymer bearing a 4-carbon spacer. When dimeric cinchonidine- and cinchonine-derived ammonium salts incorporating a dimethylanthracenyl bridge were studied in the same type of asymmetric alkylations, 90% ee was achieved.[60] In the enantioselective synthesis[61] summarised in Scheme 6 cinchonidine is used to control the stereochemistry of the α-carbon when side-chains are introduced using β-alkyl-9-BBN organoborane reagents. In 12 examples studied 54–95% ee's were recorded. A new class of naphthalene-based dimeric cinchona alkaloids have been developed[62] which show excellent enantioselectivity in the alkylation of glycine derivatives and seem good prospects for adoption by industry. Cinchonidinium salts bearing a 3,5-dialkoxybenzyl have been shown[63] to be efficient catalysts for the alkylation of *N*-(diphenylmethylene)glycine Pr^i ester with benzyl bromide, but surprisingly give the (*S*)-enantiomer when KOH is used as base and the (*R*)-enantiomer when NaOH was used. A cinchonidine-based phase-transfer catalyst in the

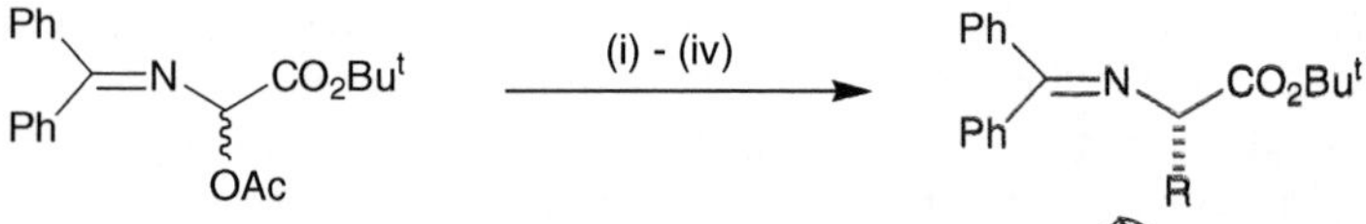

Scheme 6 *Reagents: (i) cinchonidine/ LiCl (ii) nBuLi (iii)* 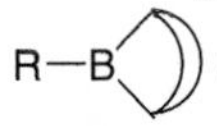 R—B *(iv) 0°C THF*

presence of KOD/D_2O, provided[64] the means to incorporate deuterium into the α-position of benzophenone-derived glycine imine to produce α-deuterated α-amino acids.

4.1.2 Synthesis via Rearrangements. γ-Amino acids were produced[65] via the hydrogen-mediated ring-opening of (**7**) to the carboxy nitrile, followed by hydrogenation of the nitrile group. *N*-Benzyl-4-acetylproline has been prepared[66] from *N*-(2-hydroxy-2-methyl)but-2-enyl-*N*-benzylamine and glyoxylic acid via a tandem cationic aza-Cope rearrangement and Mannich reaction under mild conditions. A TEMPO-mediated ring expansion to the α-keto-β-lactam (**8**) resulted[67] in the formation of *N*-carboxyanhydrides which could be hydrolysed without loss of chirality to the α-amino acid derivatives as summarised in Scheme 7. *N*-Protected allylic amines produced from allylic alcohols via Overman's [3,3] sigmatropic rearrangement of trichloroacetimidates have been converted[68] to *N*-protected amino acids by using $NaIO_4$ with catalytic amounts of $RuCl_3.3H_2O$ or by ozonolysis, without loss of chirality.

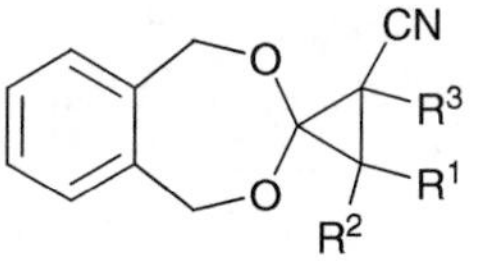

(**7**)

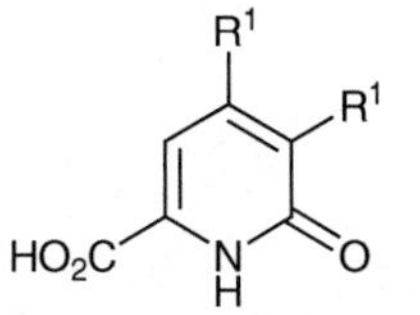

(**9**) R^1= $CH_2CH(NH_2)CO_2H$, R^2 = H
(**10**) R^1= H, R^2 = $CH_2CH(NH_2)CO_2H$

4.1.3 Synthesis from Dehydroamino Acids and by Carbohydroamination. Secondary phosphanes have proved[69] to be useful ligands for the asymmetric hydrogenation of acetamidocinnamic and itaconic acids using [rhodium (cycloocta-1,5-diene)$_2$]BF_4 as catalyst. Enantiomeric excesses of up to 97% were found for both the bidentate and monodentate ligands. The asymmetric hydrogenation of dehydroaminoacid precursors was the key step[70] in the synthesis of *S*-(−)-acromelobic acid (**9**) and *S*-(−)-acromelobonic acid (**10**). An ee of >98% was achieved at the key stage in the synthesis of (**9**) through the use of (*R*,*R*)-Rh(DIPAAMP)(COD)]BF_4, while (S,S)-[Rh(Et-DuPHOS)(COD)]BF_4 was used for (**10**) and gave >96% ee. In the hydrogenation of (*Z*)-acetamido-3-arylacrylic acid methyl esters it has been discovered[71] that the introduction of N for O into the binaphthyl chiral ligands allows catalysts such

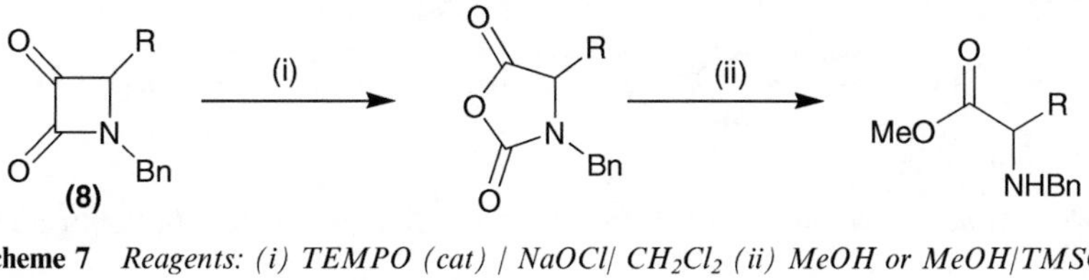

(**8**)

Scheme 7 *Reagents: (i) TEMPO (cat) / NaOCl/ CH_2Cl_2 (ii) MeOH or MeOH/TMSCl*

as [Rh(H_8-BINAPO)] and [Rh(H_8-BDPAB)] to give improved ee values. (2*S*, 6*S*)- and *meso*-Diaminopimelic acids have been synthesised[72] by the asymmetric hydrogenation (95% ee) of their dehydroamino acid precursor under catalysis by [Rh(I))COD)-(*S,S*) or (*R,R*)-Et-DuPHOS]OTf.

No asymmetric bias, but good yields (up to 86%) have been claimed[73] for the interesting formation of PhCH(NHR)CONHR, when various iodoarenes, primary amines and carbon monoxide were condensed together in a Pd-catalysed one-pot double carbonylation reaction.

4.2 Synthesis of Protein Amino Acids and Other Naturally Occurring Amino Acids. – The use of enzymes to carry out key conversions in the synthesis of amino acids has been a useful part of the armoury for many years. During this period, examples come from: the formation[74] of L-[4-^{11}C]-aspartate and L-[5-^{11}C]-glutamate by enzymatic catalysis of ^{11}C-hydrogen cyanide into *O*-acetyl-serine and –homoserine respectively; phenylalanine ammonia lyase was used[75] to produce [1-^{14}C]- and [2-^{14}C]-phenylalanine from corresponding cinnamic acids, and these isotopomers converted further to [1-^{14}C]- and [2-^{14}C]-tyrosine using L-phenylalanine hydroxylase; tyrosine phenol lyase from *Citrobacter freundii* can catalyse[76] conversion of 2-aza-1- and 3-aza-1-tyrosine from 3-hydroxy- and 2-hydroxypyridine respectively and ammonium pyruvate; four tritium-labelled isotopomers of L-phenylalanine (2-^{3}H-, 2′,6′-^{3}H, 3*R*-^{3}H and 3*S*-^{3}H-phenylalanine) have been made[77] and converted into [2-^{3}H]-, [2′,6′-^{3}H]-, [3R-^{3}H]- and [3S-^{3}H]-tyrosine using phenylalanine-4′-monooxygenase; reductive amination[78] of pyruvate to form L-alanine using alanine dehydrogenase from the hyperthermophilic archeon, *Archaeoglobus fulgidus*; the biotransformation[79] of *p*-hydroxyphenylpyruvic acid to L-tyrosine using L-aspartate amino transferase from *E. coli.*

The chemical synthesis of well known amino acids and derivatives still commands a great deal of attention. Thus glutamic acid analogues have been prepared[80] from PEG-supported intermediates using a Heck reaction as summarised in Scheme 8, followed by further processing to the amino acid derivatives. Starting from the same Schiff base, conjugative addition[81] of Michael acceptors, either in solution or on solid phase, in the presence of quaternary salts from cinchona alkaloids have also given glutamic acid derivatives. Asymmetric synthesis[82] of β-substituted aspartic acid derivatives has been secured via a catalysed [2+2] cycloaddition of ketenes and imines to form acyl-β-lactams,

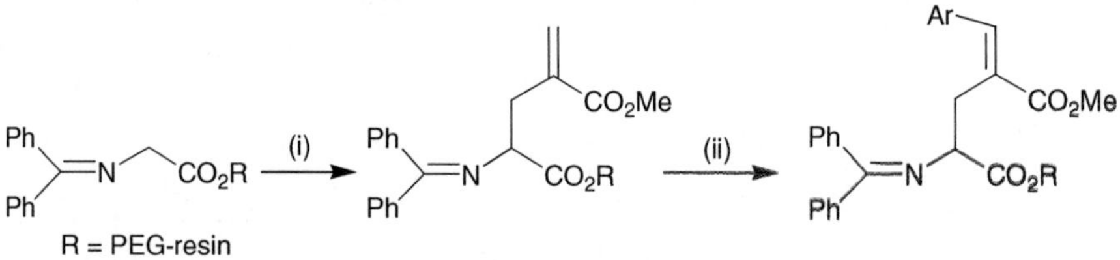

Scheme 8 *Reagents: (i) $BrCH_2(C{=}CH_2)CO_2Me$/ Cs_2CO_3/ MeCN (ii) ArX /$Pd(OAc)_2$/10% PPh_3/Cs_2CO_3.*

which ring-open under catalysis by benzoylquinine with high enantio- and diastereo-selectivity. Aziridines, such as (**11**) and cyclic sulfates based on (**12**) were the basis[83] of the asymmetric synthesis of *syn* and *anti* forms of β-substituted cysteines and serines, and reductive amination[84] of phenylpyruvic acid over Pd/C catalyst yielded DL-phenylalanine. Oxazolidinones (**13**) and (**14**) have been shown[85] to be efficient synthons on the pathway to α-amino-aldehydes and α-amino acids respectively.

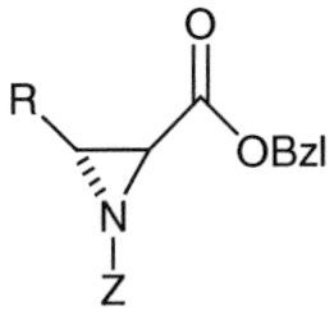

(**11**) 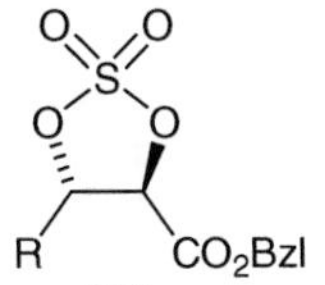 (**12**) 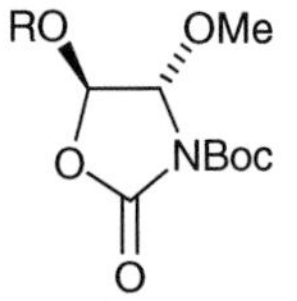 (**13**) R = Bn (**14**) R = Me

The Schollkoff chiral auxiliary (**15**) formed the basis,[86] which on conversion to the alkyne (**16**), and reaction with *o*-iodoanilines gave a series of tryptophan analogues (Scheme 9). Enantiopure 4-substituted prolines have been prepared[87] via intramolecular radical cyclisation of *N*-arylsulfonyl-*N*-allyl-3-bromo-L-alanines, while a practical and convenient enzymatic[88] enantioselective hydrolysis of DL-glycine nitriles with nitrile hydratase from *Rhodococcus* sp. AJ270 cells have yielded amino acids and their amides. Both solution and solid phase versions of the Ugi multicomponent reaction[89] have produced a library of arginine derivatives. Selectively protected L-DOPA derivatives were the products[90] from the hydroformylation of 3-iodo-L-tyrosine followed by Baeyer-Villiger oxidation of the derived 3-formyl group.

Further investigations[91] have optimised the oxidation conditions required to produce (*R*)-glycine-d-^{15}N from *N*-(*p*-methoxyphenyl methylamine)-2,2,2-trichloroethyl carbamate to be the use of periodic acid.

N-Enriched-L-histidine (99% enrichment in each position) were the products[92] of introduction of the labels into the precursor 1-benzyl-5-hydroxymethylimidazole, and leucine labelled with ^{13}C-carbon and deuterium as shown in (**17**) was the product[93] of a multistep synthesis starting from pyroglutamic acid derivatives. A chiral centre located[94] on a *N*-phthaloyl protecting group has secured control of stereochemistry in the formation of (2-^{2}H)- and (2,3-^{2}H)-phenylalanine, while a chirally deuterated 3-aminopropanol derivative

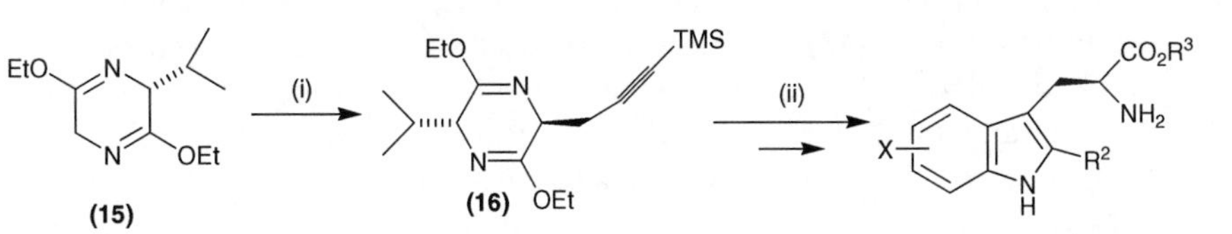

(**15**) (**16**)

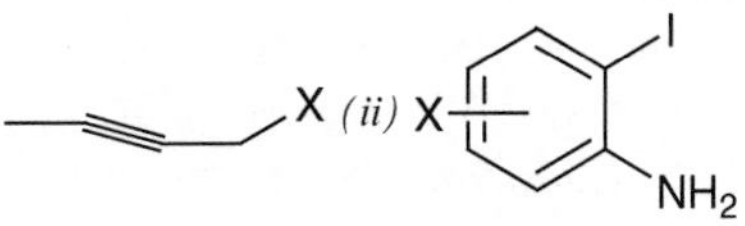

Scheme 9 *Reagents: (i) BuLi/THF/TMS*—≡—X *(ii)* X-substituted 2-iodoaniline

proved[95] to be key to the synthesis of L-[2,3,4,5-^{2}H]-ornithine. Synthesis[96,97] of [2*S*, 3*R*]-[3-^{2}H,^{15}N]-phenylalanine involved a key alkylation of the chiral glycine template ^{15}N-labelled 8-phenylmenthyl hippurate with (*S*)-(+)-benzyl-α-d-mesylate, giving 92% de at the 2-position and 74% de at the 3-position.

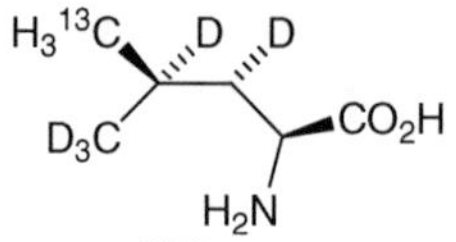

$H_3{}^{13}C$, D, D, D_3C, CO_2H, H_2N

(17)

Starting from a diprotected L-aspartic acid, (2*S*, 4*R*)-4-hydroxyornithine has been synthesised, and from L-aspartic acid itself after esterification, azo-hydrolysis, aminolysis and a Hofmann rearrangement, (*S*)-isoserine has been synthesised.[98] The partially hydrogenated aromatic ring of phenylalanine (1,4-dihydro-L-phenylalanine), produced as a minor product in the Birch reduction of the amino acid, has been shown[99] to be a moderate competitive inhibitor of phenylalanine ammonia lyase rather than a substrate. In an enzyme-assisted[100] preparation of D-*tert*-leucine, it was (±)-*N*-acetyl-*tert*-leucine chloroethyl ester that exhibited the highest rate of hydrolysis.

Amino acid anhydride hydrochlorides have been used[101] for the first time as acylating agents in Friedel-Crafts reactions, resulting in the synthesis of L-homophenylalanine from aspartic acid. Reduction of cystine with Fe/HCl yielded[102] cysteine, while di- and tri-nuclear Cu(II) complexes catalysed[103] the condensation of glycine with HCHO to yield serine. A process (40g-scale) for making enantiomerically pure (*S*)- and (*R*)-valine *t*-butyl esters has been developed[104] from *N*-TFA-valine and 2-methylpropene, but *N*-Boc-4-hydroxy-methyl-oxazolidin-2-ones from L-serine[105] undergo rearrangement and race-misation making these unsuitable polymer-mounted auxiliaries. Homolytic free radical alkylations via silver-catalysed oxidative decarboxylation with ammo-nium persulfate, has been used[106] to convert L-histidine methyl ester to 2,3-dialkylated-histidines. *N*-Boc-Phenylglycine *t*-butyl esters were the result[107] of a 1,2 Boc-migration on treating *N*, *N*-di-Boc-benzylamines with KDA/*t*-BuOLi. Iodine powder/hydrogen peroxide mixtures have furnished[108] L-thyroxine from L-tyrosine.

Non-proteinogenic amino acids now encompass a wide range of structures and are sometimes difficult to retrieve from the literature as they are often referred to only by their 'trivial' name. However statines, due to their phar-macological importance, are well-documented. Thus, all four 2,3 stereoisomers 2-substituted statines (**18**) have been synthesised.[109] Both the *anti* and *syn* forms were obtained from precursor β-ketoesters via reduction and aldol reactions. Both enantiomers of statine (**18**, R^1=Bu^i, R^2=H, R^3=H, without Boc) have been synthesised[110] by exploiting an α-lithiated alkyl sulfoxide as a chiral α-hydroxyalkyl carbanion equivalent, while another method[111] utilised orthogon-ally protected *syn*-2-amino-1, 3, 4-butantriol as a general *syn*-aminoalcohol building block. A *N*-hydroxymethyl group tethered to the amino group of *N*-Boc-L-leucinal has been shown[112] to undergo intramolecular conjugate addition to an α,β-unsaturated ester formed by condensation with the aldehyde group of the leucinal. The resulting adduct hydrolysed to (–)-statine.

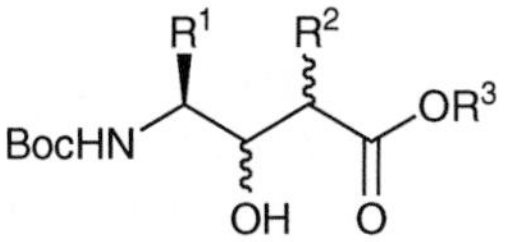

(18) 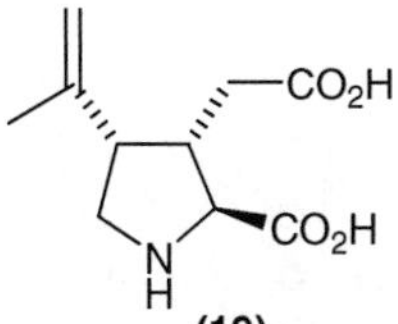 **(19)**

Interest in the polysubstituted proline family of kainoids for their anthelmintic and insecticidal properties has provided interesting synthetic challenges, since natural sources have been interrupted. Thus (–) kainic acid (**19**) has been synthesised[113] from chiral lithium amide bases which are able to deprotonate *N*-benzyl-*N*-cumylanisamide enantioselectivity to yield enantiomerically enriched benzylic organolithiums. These spontaneously undergo dearomatising cyclisation to yield partially saturated isoindolones, which are processed further in nine steps to (**19**). A potent neuroexcitatory kainoid analogue MFPA (2-methoxyphenyl group in position 4 in **19**) has been synthesised[114] with the proline ring built using a photoinduced benzyl radical cyclisation which had excellent stereoselectivity. Model studies[115] on another photochemical approach to the kainoid ring system have been reported, as well as a formal synthesis[116] of the kainoid amino acid FPA, using a ketyl radical cyclisation as a key step. Some reflections on the synthesis of (–)-kainic acid (**19**) have been recorded in a short review.[117] Full details[118] have now emerged for the synthesis of 4-arylsulfonyl-substituted kainoid analogues starting from 4-hydroxy-L-proline.

(*S*)-(+)-Carnitine (**20**) and analogues have been produced[119] by sequential mono-addition of organometallic reagents to the lactone of (5*R*, 6*S*)-4-(benzyloxycarbonyl)-5,6-diphenyl-2,3,5,6-tetrahydro-4H-1,4-oxazin-2-one followed by Lewis acid-promoted stereoselective allylation of the resulting hemiacetals. The *R*-(–)-carnitine has also become available[120] from the same oxazinone template but via its reaction in a $TiCl_4$ – promoted Mukaiyama-type aldol reaction of the ketenesilyl acetal of ethyl acetate. A key step in another stereoselective synthesis of *S*-(+)-lycoperdic acid (**21**), was achieved[121] by the stereoselective hydroxylation of the enolate of a bicyclic lactam using oxodiperoxymolybdenum(pyridine)hexamethyl phosphoric triamide as oxidising agent.

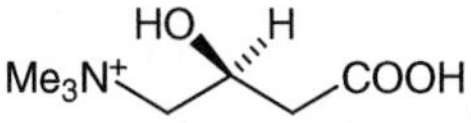

(20) 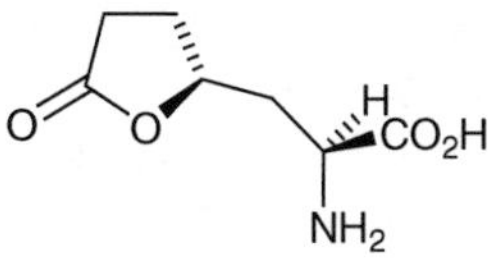 **(21)** 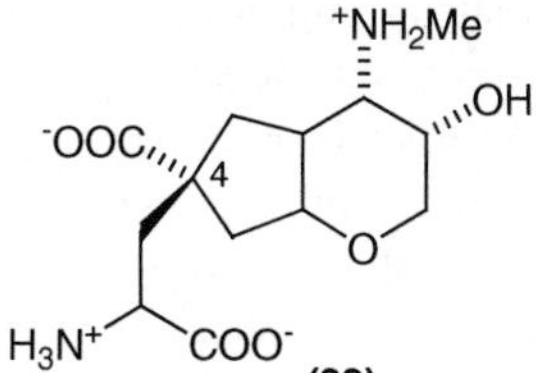 **(22)**

The side-chain component of paclitaxel (taxol), (2*R*, 3*S*)-*N*-benzoyl-3-phenylisoserine, has been synthesised[122] utilising dihydroxylation and regio-and diastereo-selective iminocarbonate rearrangement. *N, N*-Dichlorinated derivatives of taurine, homotaurine, GABA and leucine have been shown[123] to be more lipophilic than their parent compounds, and an asymmetric synthesis[124] of *cis*-α,β-propanoleucine has used a Strecker synthesis as a key step. Another total synthesis[125] of the selective glutamate receptor agonist dysiherbaine (**22**) has been reported, this time in a one-pot halogenation-ring-contraction to prepare the bicyclic ring system with excellent stereochemical control at the C-4 centre. *N*-Methylhydroxyleucine and another three unusual components (**23–25**) of cyclomarin A have been synthesised[126] in protected form ready for further processing to the cyclic peptide. Amino acid (**23**) was derived from diastereoselective methylation of an aspartic acid lactone, while (**24**) was formed via aldol reaction with Schollkopf's chiral glycine enolate, and (**25**) was achieved by the AQN ligand-promoted Sharpless aminohydroxylation protocol. New derivatives of L-canavanine have been produced[127] in order to study the effect of oxygen in the S-position, on their efficiency as nitric oxide synthase inhibitors.

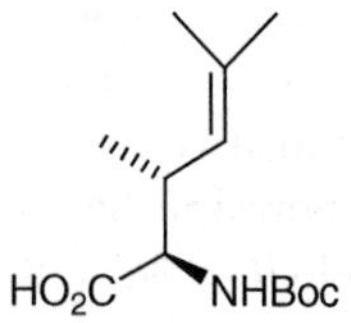

HO_2C NHBoc **(23)** OMe CO_2Me NHZ **(24)** HO CO_2Et NHZ N O **(25)**

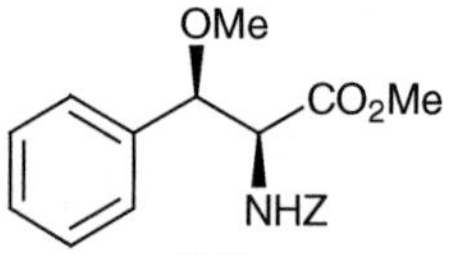

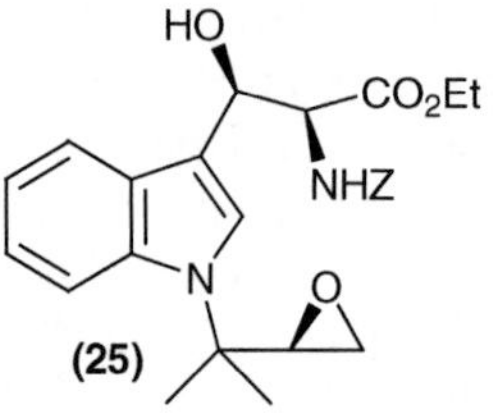

All eight stereoisomers of puleherrimine (**26**), the bitter principle from a sea urchin ovary have been synthesized,[128] and has resulted in the re-designation of the stereochemistry of the natural pulcherrimine as (2′*S*, 2*R*, 4*S*). The stereoselective synthesis[129] of (+)-myriocin (**27**) from D-mannose has been briefly reported, using an Overman rearrangement as a key step.

5-[4-^{13}C, ^{15}N]- and 5-[5-^{13}C, ^{15}N]-Aminolevulinic acids have been synthesised[130] in 4 steps from labelled glycine, and *N*-Boc-aziridine-2-carboxylates treated with ^{11}C-cyanide with no carrier added[131] produced DL-2,4-diamino[4-^{11}C]butanoic acid, 4-^{11}C-asparagine and 4-^{11}C-aspartic acid. [l-^{11}C]-γ-Vinyl γ-aminobutyric acid (Vigabatrin®), a suicide inhibitor of GABA-transaminase has been synthesised[132] in order to better understand its pharmokinetics using positron emission tomography. A 7-step synthesis[133] to (*S*, *S*)-dysibetaine (**28**) from a marine sponge, established its natural stereochemistry as well as providing other isomers for testing. A first asymmetric synthesis[134] has been recorded for (2*S*)- and (2*R*)-amino-3,3-dimethoxy propanoic acid through quenching of a chiral glycine titanium enolate with trimethyl orthoformate, and (*S*)-α-amino-oleic acid has been formed[135] from Me (2*S*)-2-[bis-(Boc)amino]-5-oxopentanoate.

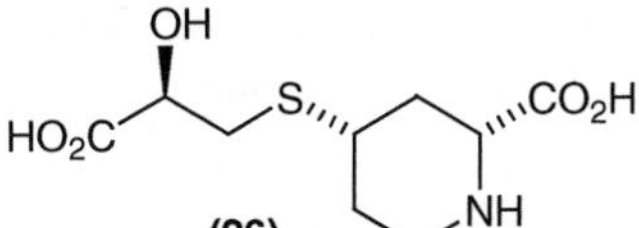

4.3 Synthesis of α-Alkyl α-Amino Acids. – The chromatographically isolatable diastereoisomers of 2, 2-disubstituted 2H-azirin-3-amine (**29**) provided[136] useful synthons for the synthesis of (*R*)- and (*S*)-isomers of isovaline, 2-methylvaline, 2-cyclopentylalanine, 2-methylleucine and 2-methylphenylalanine, and in an extension of the work[137] 2-methyltyrosine and 2-methylDOPA were produced. The latter compound has also been used,[138] to create via its catechol hydroxyls new crown ether carriers. Both enantiomers of α-methyl serine were synthesised[139] with the use of Ni(II) complexes of (*S*)-*N*(2-benzoylphenyl-l-benzylpyrrolidine)-2-carboxamide, and can also be obtained[140] on a multigram scale from the Weinreb amide of 2-methyl-2-propenoic acid via a Sharpless asymmetric di-hydroxylation. The products were also converted to (*S*) and (*R*)-*N*-Boc-*N*, *O*-isopropylidene-α-methylserinals in new approaches to the synthesis of quaternary α-methyl amino acids. (*S*, *S*)-and (*R*, *R*)-Cyclohexane-l,2-diols have been used[141] as chiral auxiliaries in the asymmetric synthesis of (*S*)-butylethylglycine and (*S*)-ethylleucine, while chiral synthons containing metal ions gave α,α-amino acids in a 'one-pot' reaction.[142]

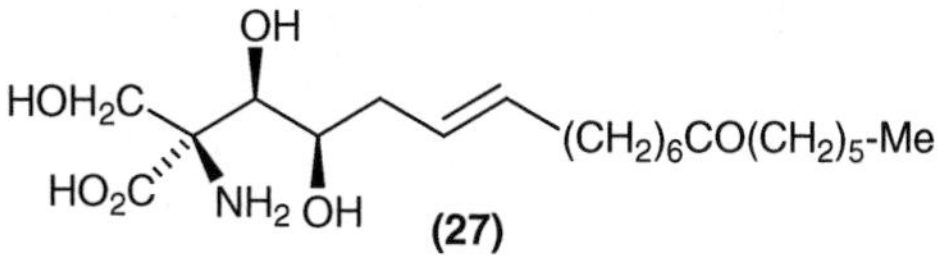

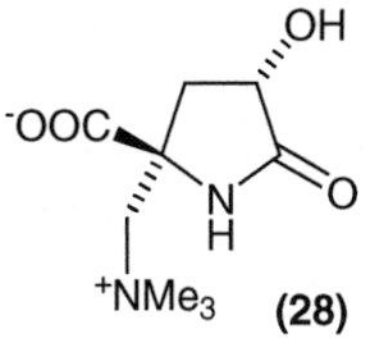

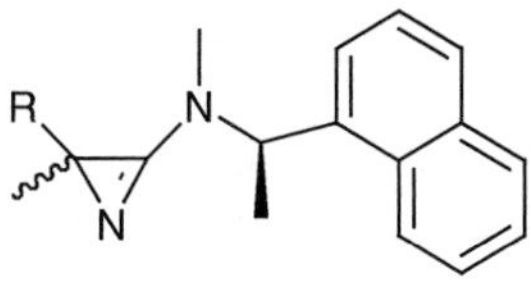

Copper salen complexes have been found[143] to catalyse the asymmetric alkylation of enolates from a variety of amino acids, and after a wide survey of conditions it was concluded that there was a clear relationship between size of the substrate side chain and the enantioselectivity of the process. A chiral nitrone from L-erythrulose has been subjected[144] to reaction by various Grignard reagents, to give protected α,α-disubstituted amino acids and their corresponding *N*-hydroxy derivatives. Using α,β-didehydroglutamates as starting material,[145] α-methyl pyroglutamates have been synthesised via α-methyl-6-oxoperhydropyridazme-3-carboxylates with ring contraction using LiHMDS. A range of chiral α-alkyl-α-phenylglycine derivatives were prepared[146] by alkylation of (3*R*)-phenylpyrazine, which was obtained from the arylation of (*S*)-2, 5-dihydro-3,6-dimethoxy-2-isopropylpiperazine with benzene-$Mn(CO)_3$ complex. Treatment of ethyl nitroacetate with *N*,*N*-diisopropylethylarnine/tetraalkylammonium salt followed by addition of an alkyl halide or Michael acceptor gave the doubly C-alkylated product in good yield which gave the corresponding amido esters on selective nitro group reduction.[147] On *N*-protection, a series of these $C^{\alpha,\alpha}$ -disubstituted amino acids were incorporated into peptides, while in a separate publication[148] esters of Boc- and

Z-α,α,-dialkylamino acids have been prepared via the mixed anhydride method. Large scale syntheses[149] of C^{α}-tetrasubstituted α-amino acids such as (**30**), important for ring closing metathesis have been carried out using phase-transfer catalysed alkylation of *N*-benzylidene-DL-alaninamide using two amidases for the resolution of the amino acid into chirally pure forms. Another synthesis[150] of APCD (**31**), a selective agonist of metabotropic glutamate receptors has appeared which is based upon an alkylidene carbene 1,5-CH insertion reaction as a key step. A Wittig homologation of Garner's aldehyde, with subsequent catalytic hydrogenation gave a precursor ketone which with lithio(trimethylsilyl)diazomethane resulted in the cyclopentene-CH insertion product.

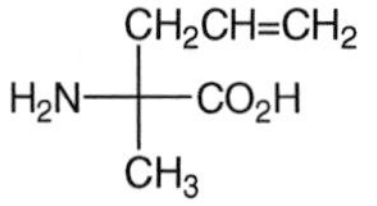

$CH_2CH{=}CH_2$ / H_2N—C—CO_2H / CH_3 **(30)** H_2N, HO_2C, CO_2H **(31)**   Bn, R, N, Ph, 3, O, O **(32)**

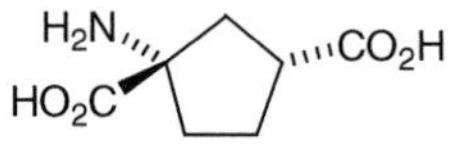

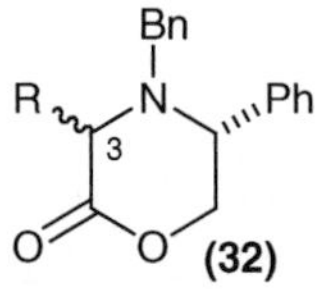

Chiral α,α-disubstituted amino acid derivatives possessing a vinyl silane synthetic handle have been obtained[151] from aza-[2,3]-Wittig rearrangement precursors derived from Ala, Val, Phe and PhGly. Upon irradiation[152] with suitable α-alkoxy carbon radical precursors plus a sensitiser, substitution occurred onto C=N bonds of ketoxime ethers to form β-oxygenated quaternary α-amino acid derivatives.

Protected α-methyl-α-phenylglycine and α-methylisoleucine have been prepared[153] by oxidative cleavage of *N*-Boc-3-amino-l, 2-diols which had been formed from 3-azido-l, 2-diols. Treatment[154] of enolates of (**32**) with alkyl halides or aldehydes, gave a quaternary C at 3 with *S*-configuration, but with methyl bromoacetate the *R*-configuration predominated. The products were de-protected to form enantiopure α, α-dialkyl amino acids. In the chiral phase transfer-catalysed alkylation of protected amino acids, anaerobic conditions offer advantages[155] of yield and enantioselectivity. 6-Benzyl-piperazine-2, 3, 5-trione has been selectively alkylated[156] at the C-6 position, which is equivalent to the C^{α} position of phenylalanine.

4.4 Synthesis of α-Amino Acids with Alicyclic and Long Aliphatic Side Chains. – There seems to be an increase in activity under this Section during this period, with the 3-membered cyclopropyl ring amongst the most popular. Options for the synthesis of a large class of chiral 2-*S*-alkyl-l-aminocyclopropane carboxylic acids have been made available[157] through the synthesis of (*Z*)-l-benzoylamino-2-tritylsulfanylcyclopropane carboxylic acid, formed from (−) or (+)-menthyl-2-benzoylamino-3-tritylsulfenyl acrylates and diazomethane. (1*S*, 2*R*)- and (1*R*, 2*S*)-Allocoronamic acids (**33**) have been made[158] using Belokon's complex {Ni(II) complex of glycine-(*S*)-2-[*N*′-(*N*-benzylprolyl)amino] benzophenone Schiff base ligands}, and chiral sulfate cyclopropane amino acids have been reported[159] as products from ylide insertion on the exocyclic double bond of a chiral 5(4H)- oxazolone from D-glyceraldehyde. There is a first report[160] of an enzymatic (pig liver esterase) asymmetrisation of a prochiral precursor in the form of (1*S*)-l-amino-2, 2-dimethylcyclopropane-l-carboxylic acid, and (2*R*, 1′*S*, 2′*S*)-2-(carboxycyclopropyl)glycine has been formed[161] via an extension of Taguchi's protocol for Simmons-Smith cyclopropanation to a chiral amino allyl alcohol. *S*-Cleonin (**34**) from the anti-tumour cleomycin has been prepared[162] from *R*-serine via a Kulinkovich cyclopropanation of the methyl ester of Z-serine acetonide. Of the two novel antagonists of group 2 metabotropic glutamate receptors, synthesised[163] as (2*S*, 1′*S*, 2′*S*, 3′*R*)-2-(3′-xanthenyl, methyl-2′-carboxycyclopropyl)glycine and its xanthenylethyl analogue, it was the latter which had submicromolar activity. *exo*-Nucleophilic addition[164] to (bicyclo [5.1.0] octadienyl)iron (1+) established the stereochemistry of the ring and the α-stereocentre in the synthesis of *cis*-2-(2′-carboxycyclopropyl)glycine (**35**), while the synthesis[165] of (2*S*, 1′*S*, 2′*S*, 3′*R*)-2-(2′-carboxy-3′-methylyclopropyl)glycine and its epimer at C-3′ has shown the former to be a potent and selective metabotropic group 2 receptor agonist. Cyclopropanation of dehydroamino acid derivatives[166] with alkyl diazoacetates, catalysed by dirhodium tetraacetate gave cyclopropane analogues of aspartic and adipic acids. Full details[167] have appeared for the titanium-mediated cyclopropanation of homoallyl alk-2-enoates to give (Z)-vinylcyclopropanols, which can be processed via azidation and oxidative cleavage to give alkyl 2,3-methanoamino acids such as (**36**). *Syn*- and *anti*- forms of 3,4-cyclopropylarginitie have been produced[168] using diazomethane addition to Z-dehydroglutamate, 4-methyl-2,6,7-trioxabicyclo [2,2,2] ortho ester followed by irradiation of the resulting pyrazoline.

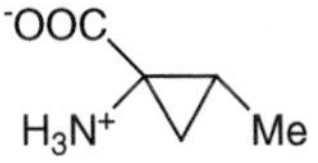

(33) 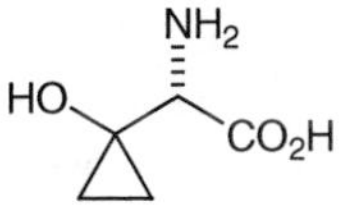 **(34)** 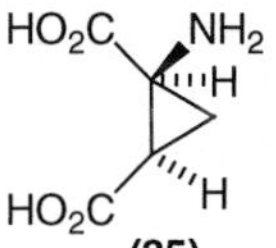 **(35)** 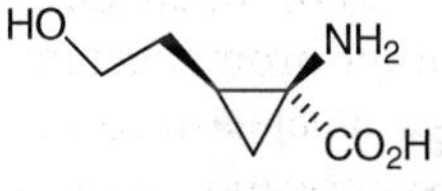 **(36)**

Also generating continuing interest in this sub-section is the conformational constraints offered by the 5-membered ring in proline, so a number of syntheses of substituted prolines have been reported. A number of 3-substituted prolines have been synthesised enantioselectively[169] starting from 3-OH-(*S*)-2-Pro using the enol triflate derived from *N*-trityl-3-oxo-(*S*)-2-proline methyl ester, followed by hydrolysis/hydrogenation. A chirally stabilised azomethine ylide[170] from 5-(*S*)-5-phenylmorpholin-2-one and 2,2-dimethoxypropane has been trapped diastereoselectively with singly and doubly-activated dipolarophiles to give cycloadducts dismantled in one-pot to enantiomerically pure 5,5-dimethylproline derivatives. A reductive cyanation[171] of 2-pyrrolidones with Schwartz's reagent has also given the same disubstituted proline. Two methods have been described for the synthesis of (2*S*, 5*S*)-5-*t*-butylprolin. One involved[172] converting 2(*S*)-l-*t*-butyldimethylsiloxy-2-*N*-(PhF)amino-5-oxo-6, 6-dimethyl heptane into its imino alcohol followed by reduction of the imine function with >96% enantiomeric purity, while another study[173] made the same 5-substituted proline via the addition of *t*-butylcuprate to 2(*S*)-Boc-Δ^5-dehydroproline ethyl ester with 78:22 diastereoselectivity. Starting from 2,3-disubstituted pentenoic acid derivatives it has been shown[174] that a hydroboration-Swern oxidation sequence created a *N*-acyliminium precursor which could be transformed into 3-phenyl-5-vinylproline in 70% yield. *N*-Alkylated glycine esters[175] with excess acrolein in presence of acid, followed by treatment with Et_3N have provided 4-formyl-5-vinylproline carboxylates with good regio- and stereo-selectivity.

Trans-4-Cyclohexylproline has been obtained[176] by stereoselective alkylation of *N*-benzyl-pyroglutamic acid with 3-bromocyclohexene, which after hydrogenation afforded *trans*-4-cyclohexylpyroglutamic acid and then processed via a thioester (Lawesson's reagent) and Raney Ni to give the proline derivative with 93% ee. A library of 4-alkoxyprolines has been produced[177] using solid phase techniques, and a detailed survey[178] of the phase transfer catalysis conditions required for the cycloaddition of imino esters derived from alanine and glycine with alkenes to form substituted prolines has been carried out. A number of heteroaromatic acromelic acid analogues have been synthesised[179] from (−)-α-kainic acid and (2*S*, 3*R*)-3-hydroxy-3-methylproline a component of the polyoxypeptins has been synthesised[180] via a SmI_2-mediated cyelisation of an iodoketone. Both *syn, exo*, and *anti, exo*-3, 4, 5-trisubstitiued-prolines can separately be prepared[181] from 2, 4, 5-pyrrolidinyl carbene complexes formed from glycine ester aldimines and chiral alkoxyalkenylcarbene complexes of chromium, oxidation producing the former, acid hydrolysis producing the latter. The azabicycle (**39**) with sec-butyllithium/TMEDA at 0°C afforded[182] the C bridgehead anion which could be quenched with e.g. CO_2 to form the naturally-occurring 2, 4-methanoproline (**38**), but when conditions were changed to −78°C the hitherto unknown 3, 5-methanoproline (**37**) was amongst the products. The 2, 4-isomer (**38**) was also synthesised[183] in five steps from allyl benzyl ether. Further bridged analogues of proline have come in the form of (**40**) and (**41**), which were produced[184] from transformations of an azabicyclic intermediate obtained from the asymmetric Diels-Alder reaction of a chiral oxazolone derived from *R*-glyceraldehyde with Danishefsky's diene. Cyclopenta[c]proline derivatives (**42**) were formed[185] from enyne amino acid derivatives in a stereocontrolled manner using catalytic Pauson-Khand reactions. A facile synthesis[186] of protected 3*R*, 5*R*-dihydroxyhomoproline has been achieved using L-threonine aldolase as an enzyme catalyst. The potent metabotropic glutamate receptor agonist (1*S*, 3*R*)-l-aminocyclopentane-l, 3-dicarboxylic acid (**43**) has been prepared[187] from serine via C–H insertion of an alkylidene carbene, and a stereoselective route[188] has been used to form 3-substituted 2-amino cyclopentane carboxylic acid derivatives.

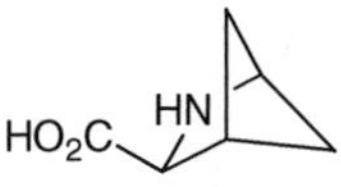

(37) 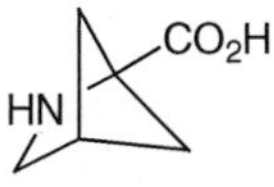 **(38)** 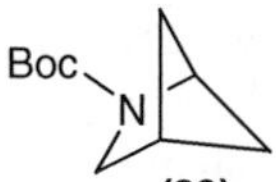 **(39)**

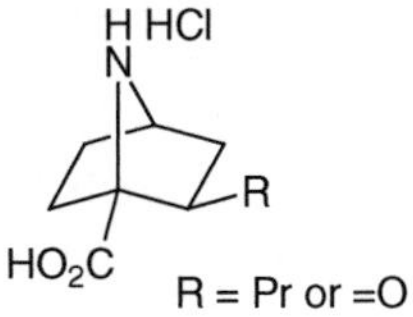

R = Pr or =O

(40) 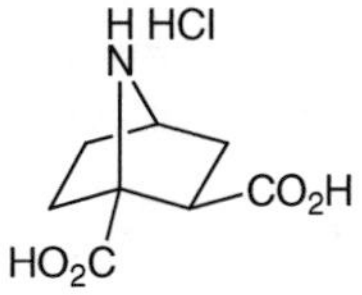 **(41)**

Pyroglutamic acid has been the starting point for the synthesis[189] of (3*S*, 4*R*)-3,4-dimethyl-L-pyroglutamic acid and (3*S*, 4*R*)-3,4-dimethylglutamine, with the methyl groups being introduced via a cuprate and enolate addition. *N*-Ts and *N*-Boc derivatives of (2*S*, 4*S*)-4-phenylamino-5-oxoprolines have also been synthesised.[190] A common strategy has been devised[191] for the synthesis of pipecolic acid derivatives (**44**) and (**45**) and involves intramolecular eneiminium cyclisation. Analogue (**45**) has also been prepared[192] using the stereoselective addition of allyltrimethylsilane with *N*-tosyl- and *N*-phenyl-iminoglyoxylates of (*R*)-8-phenylmenthol. Several pipecolic acid derivatives have been synthesised[193] from 2, 3-epoxy-5-hexen-l-ol, followed by regio- and stereo-selective ring opening with allylamine, while the constrained phenylalanine analogue (2*S*, 3*R*)-3-phenylpipecoiic acid has been obtained[194] from the Evans chiral auxiliary (4*S*)-4-benzyl-l, 3-oxazolidin-2-one. (2*R*, 3*R*)-3-Hydroxypipecolic acid has been obtained[195] from the *O*-protected methyl mandelate, *via* the nucleophilic substitution of an azide epoxide.

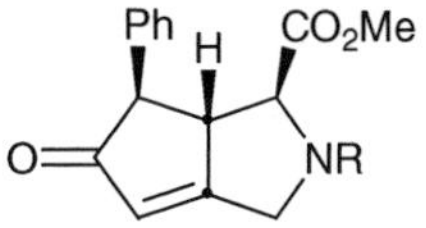

(42) 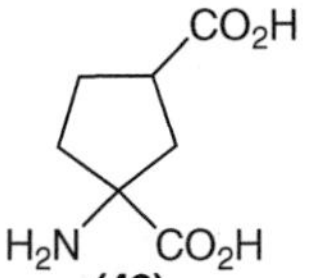 **(43)** 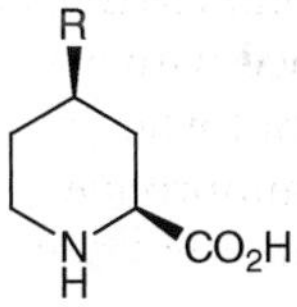 **(44)** R = Me

(45) R = OH

Both enantiomers of 4-oxo-pipecolic acid have been synthesized[196] via 1, 3-dipolar cycloaddition of C-ethoxycarbonyl-*N*-(1*R*)-phenylethylnitrone to but-3-en-l-ol, and a range of different disubstituted pipecolic acid derivatives have been made[197] via an oxidative cleavage of azabicycloalkene synthesised from an aza Diels-Alder reaction. *Trans*-3,4-Piperidindicarboxylic acid derivative (**46**) and a *trans*-3, 4-analogue have been synthesised[198] asymmetrically in 5 steps starting from aspartic acid *t*-butyl ester and Z-(*S*)-alanine respectively via intermediates which on ozonolysis and reductive animation provided the cyclic structures. Five and six-membered cyclic amino acids can be obtained[199] in a one-pot protocol by a rhodium-catalysed (Rh-DuPHOS) hydroformylation/

cyclisation sequence. Benzocyclic α, α-dialkyl amino acids (**47**) have been constructed[200] via an asymmetric Strecker reaction using *S*-α-methylbenzylamine and *R*-phenylglycinol as chiral auxiliaries, while the benzene ring in the benzocycloheptene α-amino acid derivatives (**48**) was built up[201] from a Diels-Alder reaction of a seven-membered ring diene with various dienophiles. Rigidified bicyclic α-amino acids have also been obtained[202] from appropriate 1,6-heptenynes and the reactions[203] of acyclic and cyclic dehydroalanines with 1,3-dienylcobaloxime complexes have yielded functionalised carbocyclic amino acids. Crown ether macro-rings such as (**49**) have been built up[204] from masked tris(hydroxymethyl)amino methane, and shown to be capable of stacking in the presence and in the absence of alkali metal ions.

Compounds (**50**) with *cis*- and *trans*-relationships between the 2,5-dihydroxy groups have been made[205] starting from Diels-Alder reactions between oxazolones and dienes. Cyclohexylglycine scaffolds have been synthesised[206] and tested for potency as matrix metalloproteinase inhibitors, and *cis*- and *trans* forms of 4-Boc-cyclohexylglycine have been obtained[207] from aminohydroxylation of styrene. Rhodium-catalysed[208] hydrogenation of enamide (**51**) gave (*R*)-4-piperidinylglycine in good yield. The cyclopentyl glutamate analogue (**52**) formed[209] using a [3 + 2] cycloaddition reaction of dehydroamino acids turned out to be a potent agonist of the mGlu5 and mGlu2 receptors. Methyl substituted cyclohexyl-l-amino-3-hydroxy-l-carboxylic acids have been prepared[210] from 5,5-tethered dienes of (2*R*)-2, 5-dihydro-2-isopropyl-3, 6-dimethylpyrazine.

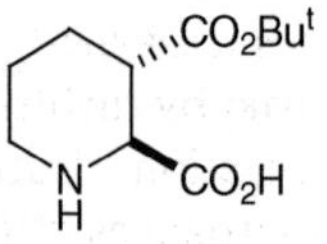

CO_2Bu^t, CO_2H, NH

(46)

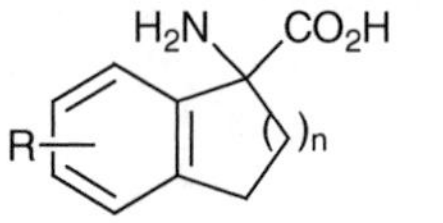

H_2N, CO_2H, R, n

(47) n = 1, 2 or 3

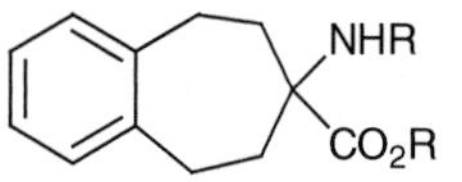

NHR, CO_2R

(48)

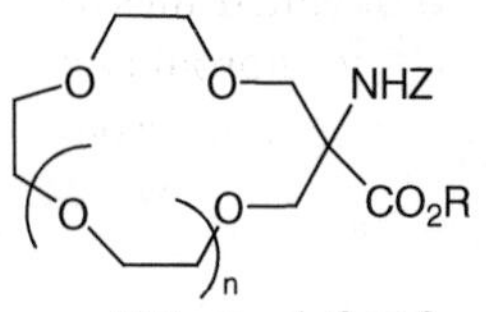

O, O, O, O, NHZ, CO_2R, n

(49) n = 1, 2 or 3

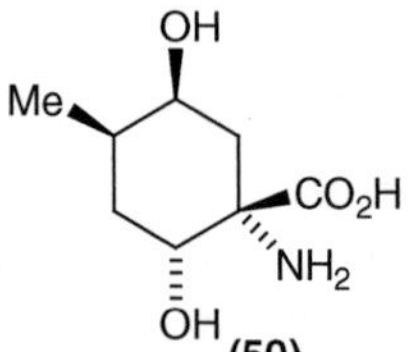

OH, Me, CO_2H, NH_2, OH

(50)

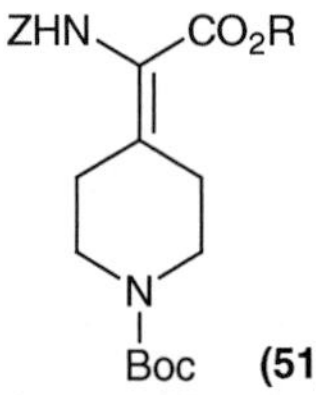

ZHN, CO_2R, N, Boc

(51)

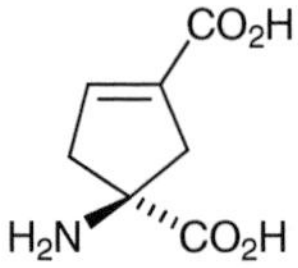

CO_2H, H_2N, CO_2H

(52)

A new class of cyclic amino acids based on (**53**) (6-oxoperhydropyridazine-3-carboxylic acid derivative) has been created[211] by diastereoselective transformation of α, β-didehydroglutamates. An improved synthesis[212] of (−) CIP-AS (**54**), an analogue of glutamic acid has been reported, which involves cycloaddition of ethoxycarbonyl formonitrile oxide to a *N*-(4-methoxybenzyl)α-ethoxycarbonyl nitrone. Further clarification[213] has been given of the conditions that favour cyclopropane *vis à vis* cyclopentene ring formation from a Schiff base derivative of glycine and bis-alkylating alkenes.

The synthesis[214] of 2-cyclopentadienylglycine (**55**) from α-bromohippuric acid with nickelocene, or other cyclopentadienyl complexes has provided a means of studying its dimerisation in Diels-Alder reactions and in ferrocenylene formation. By coupling ethyl isocyanoacetate with 1,2-bis (4-bromoethylphenyl)ethane under phase transfer conditions a macrocyclic cyclophane α-amino acid was produced,[215] and in a Diels-Alder reaction[216] between methyl 2-acetamidoacrylate and anthracene a highly constrained α-amino acid derivative was produced.

Fmoc-Protected lipophilic amino acids with alkyl side chains varying from C_{12}–C_{16} have originated[217] from the alkylation of Schiff bases obtained from 2-hydroxypinan-3-one with Gly-OBu^t. (*S*)-2-Amino-8-oxodecanoic acid a constituent of the cyclic tetrapeptides, the apicidins, has been synthesised[218] from an iodo-ester of glutamic acid which was subjected to photolytic condensation with ethyl vinyl ketone in the presence of tri-*N*-buryltin hydride. (*S*)-2-Amino-oleic acid was obtained[219] from a chiral aldehyde, Bu-(2*S*)-2[bis(Boc-amino)]-5-oxopentanoate, derived from glutamic acid, which underwent a Wittig reaction with appropriate ylides.

4.5 Models for Prebiotic Synthesis of Amino Acids. – Only one publication was found that fitted into this category, and was a report[220] of low energy nitrogen ions being implanted into carboxylic salt solutions to give hplc evidence for the production of amino acids.

4.6 Synthesis of Halogenoalkyl α-Amino Acids. – The only member of the halogen series with the correct stability profile is fluorine, so it commands a monopoly of the papers. The Reformatsky reaction between Garner's aldehyde (from D-serine) and ethyl bromodifluoroacetate[221] followed by further processing yielded L-4, 4-difluoroglutamine, and the same reaction[222] after slight modification gave L-4, 4-difluoroglutamic acid. New approaches[223] for incorporating fluorine stereoselectively have included, alkylation of chiral glycine Schiff bases, intramolecular cyclisation of chiral cyanohydrins and catalytic hydrogenation of fluorinated imino esters. Chiral α-fluoroalkylated mesylates with Boc-Gly-OH in the presence of Pd-catalyst have given[224] γ-fluoroalkylated allyl esters, which after an Ireland-Claisen rearrangement gave α-fluoroalkylated-β,γ-unsaturated amino acids. β, β-Difluoro amino acids have been made[225,226] via the alkylation of a hydroxypinanone glycinate from ethyl trifluoropyruvate, or via carboxamides or substituted ureas.[227] 5, 5, 5, 5′, 5′, 5′-Hexafluoroleucine has been obtained[228] from addition of an organozincate (from Z-L-serine) to hexafluoroacetone followed by a radical-mediated deoxygenation. (2*S*, 3*R*)-Difluorothreonine's synthesis was evolved[229] from 3, 3-difluoroacetaldehyde, an alkenyl or arylboronic acid and an amine in high yield and ee. Hydride reduction stereocontrolled[230] by intramolecular π-stacking of 1-naphthylsulfinyl and *N*-aryl groups, non-oxidative Pummerer rearrangement and ring-closing metathesis, have given both cyclic and acyclic fluorinated α-amino acid derivatives. Ring-closing metathesis using a ruthenium catalyst, also formed[231] the last stage in the synthesis of 5-, 6- and 7-membered cyclic amino esters such as (**56**), and used similarly in another publication.[232]

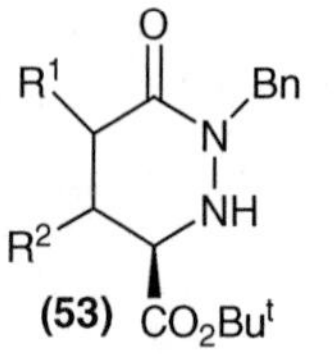

A number of fluorinated proline analogues have been made, such as *cis*- and *trans*-4-trifluoromethyl D-proline from serine[233] involving the reaction between Garner's aldehyde and an ylide, followed by trifluoromethylation. From the same authors[234] comes a report that *N*-Boc-*cis*-4-trifluoromethyl (and difluoromethyl)-L-proline could be made from *N*-Boc-4-oxo-L-proline using either CF_3SiMe_3 or CF_2Br_2/Zn respectively, and in another approach,[235] the 4-oxoproline was again alkylated with CF_3SiMe_3, whose adduct was allowed to dehydrate to a cyclic alkene which was hydrogenated stereoselectively using $Ir(cod)(py)PCy_3$ (Crabtree's catalyst). Fmoc-(4*R*) and (4*S*)-fluoroprolines have been synthesised[236] from 4(*R*)-hydroxyproline using a Mitsunobu reaction.

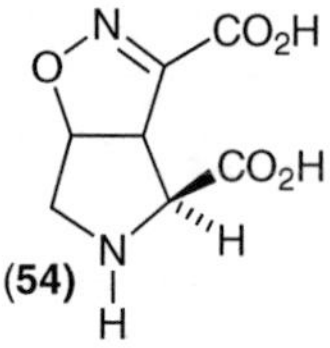

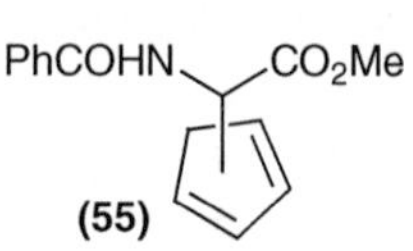

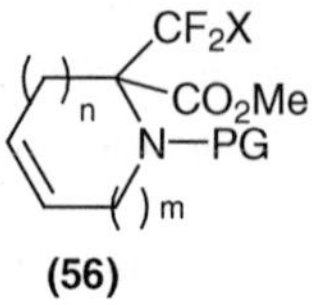

4-Fluoro- and 4,4-difluoropipecolic acid have been stereoselectively synthesised[237] from Z-protected 4-OH- and 4-oxopipecolates via fluorodehydroxylation and fluorodeoxygenation. For the preparation[238] of α-fluoromethyltryptophans, highly electrophilic imines derived from methyl bromodifluoro- and trifluoropyruvate were reacted with l-sulfonyl-3-methyleneindolines, and for the synthesis[239] of α,α-difluoro-β-amino acids, starting from aldehydes and ethyl bromodifluoroacetate, a Mitsunobu reaction was again the key step. Similar products were obtained[240] from the solid phase condensation of amines, aldehydes, benzotriazole and a Reformatsky reagent prepared *in situ* from ethyl bromodifluoroacetate, trimethylsilyl chloride and zinc. Using the recently discovered electrophilic *N*-fluoro-cinchona alkaloid reagent, it has been proven[241] that it will carry out α-deprotection/fluorination with enantiomeric excess up to 94% in the synthesis of α-fluoro-*N*-phthaloylphenylglycinonitrile. Taking the pathway summarised in Scheme 10, α-trifluoromethyl α-amino acids have been obtained[242] from sulfinimes and trifluoropyruvates using Grignard reagents with the source of chirality being menthylsulfinate which can be recycled.

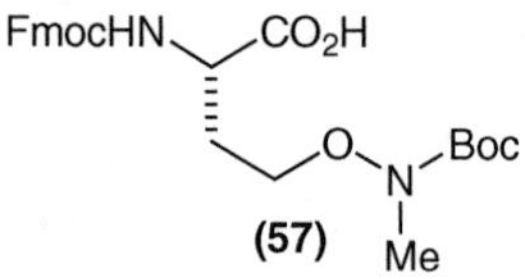

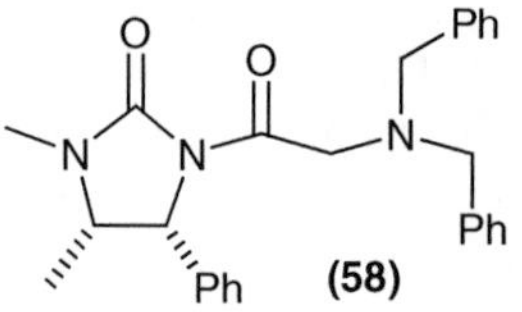

4.7 Synthesis of Hydroxyalkyl α-Amino Acids. – A general method[243] whereby 2-benzyloxyaziridine-2-carboxylates undergo regioselective hydrogenolysis has been shown to yield α-substituted serines. A stereoselective synthesis[244] of Z-D-Ser-OMe involved an intermediate tetrahydrooxazin-4-one generated from a hetero Diels-Alder reaction between 2-aza-3-trimethylsilyloxy-1,3-diene and

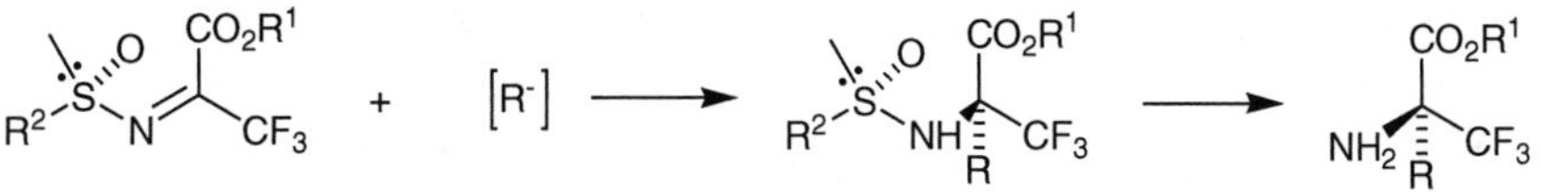

Scheme 10

gaseous formaldehyde. Z- and Boc-(*S*)-Isoserines have been obtained[245] from the appropriate (*S*)-malic acid monoester via an oxazolidin-2-one, and a protected form (**57**) of *N*-methylated amino-homoserine suitable for Fmoc-protocols has been produced.[246]

There have been a number of examples of OH groups generated in the side chains of proteinogenic acids, thus 2*S*, 3*R*, 4*S*-4-hydroxyisoleucine, a potent insulinotropic amino acid from the seeds of fenugreek, has been made[247] by biotransformation of ethyl, 2-methylacetoacetate to (2*S*, 3*S*)-2-methyl-3-hydroxybutanoate followed by an asymmetric Strecker synthesis. Protected β-hydroxyvalines were obtained[248] from *N*-protected serine methyl esters via a Grignard addition (MeMgBr) to the ester group followed by selective oxidation of the diol formed with a hypochlorite cocktail. (2*S*, 3*S*)-3-Hydroxyleucine was the model compound synthesised[249] to test out a stereodivergent approach starting from a symmetrical alk-2-yne-l, 4-diol and using a Pd(0)-catalysed process to select the stereochemistry of the α-carbon in the amino acid produced. Four stereoisomers of 3,4-dihydroxyglutamic acid have been formed[250] as a result of a stereoselective cyanation of an *N*-acyliminium intermediate derived from L- or D-tartaric acid. (2*S*)-2-Hydroxymethylglutamic acid (HMG), a potent agonist of metebotropic glutamate receptor mGlu R3 has been obtained[251] from *S*-pyroglutaminol, via a bicyclic siloxypyrrole. Another route to HMG[252] involved starting from D-serine, followed by a tandem Michael addition, a ring-closure protocol, followed by a stereoselective alkylation reaction from the convex face of the bi-cycle. The imidazolidinone (**58**) has been shown[253] to give excellent stereocontrol and can be considered a chiral enolate equivalent, which can undergo diastereoselective aldol reactions which give rise to β-hydroxy-α-amino acids. *Syn*-(*S*)-β-Hydroxy α-amino acids have also been synthesised[254] via asymmetric aldol reactions of aldehydes with a chiral Ni(II) BPB/glycine Schiff base complex in the presence of NaH. A multigram scale preparation of *syn*-(*S*)-β-hydroxyleueine was possible using this method. L-2-Amino-4, 5-dihydroxypentanoic acid has been prepared[255] from L-allylglycine and using appropriate protection it is a good precursor for conversion to an aldehyde side chain in peptides. Enantiopure ω-hydroxy-α-amino acids were key[256] to the synthesis of a number of C-15 α-amino carboxylates and were produced as a result of the Wittig reaction of methyl (2*S*)-2-[bis(Boc-amino)]-5-oxopentanoate with ω-trityloxyalkylidene triphenylphosphoranes. A general strategy[257] has been developed to access α, β-dihydroxy-α-amino acid via N-carboxyanhydrides produced by ring expansion of 3-hydroxy-β-lactams. Enantiomerically and diastereomerically pure 2(*S*)-amino-6(*R*)-hydroxy-l, 7-heptanedioic acid dimethyl ester has been derived[258] from cycloheptadiene using an acylnitroso Diels-Alder reaction as the key step. *Anti*-β-Hydroxy-α-amino acid esters were obtained[259] as the major diastereoisomer with moderate enantiomeric excess using a direct aldol reaction of glycinate Schiff bases with aldehydes, using heterobimetallic asymmetric complexes as catalyst.

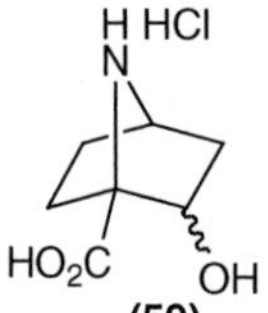

H HCl N HO2C OH **(59)**

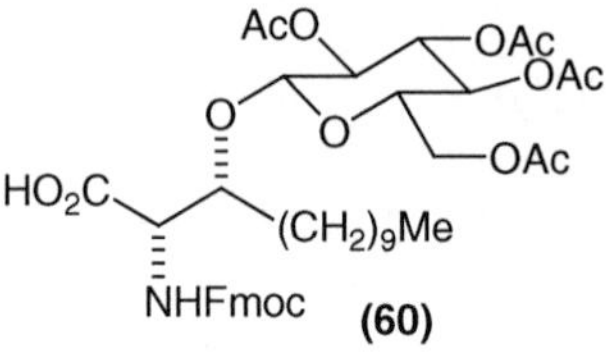

AcO OAc OAc OAc O O O HO2C (CH2)9Me NHFmoc **(60)**

Two building blocks as part of the structure of the cyclic depsipeptide, callipeltin A, have been synthesised. (2*R*, 3*R*, 4*S*)-3-Hydroxy-2, 4, 6-trimethylpentanoic acid has been obtained[260] from L-valine using the Heathcock variant of the Evans aldol reaction, while a fully protected (2*R*, 3*R*, 4*S*)-4, 7-diamino-2,3-dihydroxyheptanoic acid has been produced[261] by a multistage process from D-glucose. The protease inhibitor statine, continues to be thoroughly researched as in the synthesis of all four 2,3-stereoisomers of 2-substituted statines.[262] The 2,3 *syn*- and *anti*-isomers were synthesised via β-ketoester and aldol reactions. Since Mitsunobu reactions of *syn*-2,3-dihydroxy esters exhibited[263] complete regioselectivity for the β-hydroxyl to give *anti* α-hydroxy β-Nu ester (Nu = OBz, OTs, N_3), which could be processed to form natural *syn*-statine and its *anti* diastereoisomer. Fmoc-(2*S*, 3*S*)-2-hydroxy-3-amino acids have been synthesised[264] starting from 2-furaldehyde, cyanation being catalysed by R-oxynitrilase, and final removal of the furan ring by ozonolysis.

Cyclic amino acids bearing hydroxyl groups have also been the focus of interest. *N*-Boc-(2*S*, 5*R*)-5-(1′-hydroxy-1′-methylethyl)proline was chosen[265] as a *cis*-conformation inducer in Xaa-Pro amide bonds and was synthesised from enantiopure 2, 5-disubstituted pyrrolidine. With the availability of chiral 2-amino-3-hydroxy-4-pentanoate from sugars,[266] cyclisation stages produced both *trans*-3-hydroxy-L-proline and *cis*-3-hydroxy-D-proline. (2*S*, 3*R*)-3-Hydroxy-3-methylproline, found in the polyoxypeptins, has been synthesised[267] via the Sharpless regioselective opening of a cyclic sulfate by NaN_3 and an intramolecular ring closing reaction. By starting with *trans*-4-hydroxy-L-proline and using acetic anhydride, the intermediate lactone contains an inverted stereochemistry at C-4 so that acidic cleavage of the lactone gave *cis*-hydroxy-D-proline.[268] Two independently-reported routes to 3, 4-dihydroxyprolines, involved either a strategy[269] starting with a pentose sugar γ-lactone, or a protocol[270] based on ring-closing metathesis of unsaturated chiral allyl amines. These allyl amines were synthesised from unsaturated epoxy alcohols or from 2,3-epoxy-3-phenylpropanol. All four conformationally constrained analogues (**59**) of 3-hydroxyproline have been synthesised[271] starting from a Diels-Alder reaction between methyl benzamidoacrylate and Danishefsky's diene (1-methoxy-1,3-butadiene) followed by an internal nucleophilic displacement of methanesulfonyl group in the cyclohexane ring. Crystallisation techniques afforded the final resolution to chiral purity.

$RuO_2/NaIO_4$ oxidation[272] of *N*-Boc-4-silyloxy- and 4-acetoxyproline methyl esters under biphasic conditions and after deprotection, have given both *cis* and *trans* methyl, *N*-Boc-4-hydroxypyroglutamates. Constrained serine analogues[273] (1*S*, 2*S*)-, (1*R*, 2*R*), and (1*R*, 2*S*)-1-amino-hydroxy cyclohexane carboxylic acids were synthesised in racemic forms using the Diels-Alder reaction of 2-benzamidoacrylate with Danishefsky's diene, followed by resolution methods. Hydroxylation of 6-substituted piperidine-2-ones has provided[274] an efficient synthesis of (2*S*, 5*R*)-5-hydroxylysine and related amino acids. Lipo-β-hydroxy amino acids and their glycoside derivatives such as (**60**) have been formed[275] by selective oxidation of protected amino diols, formed from a D-serine Schiff base. In order to clarify the configuration of the natural β-methoxytyrosine in the cyclodepsipeptide, the papuamides all four stereoisomers have been synthesised[276] via their β-hydroxy counterparts. The latter were synthesised from Garner's aldehyde [from (*S*)-serine] with 4-benzyloxyphenyllithium at −78°C in presence of LiBr. Ethyl isothiocyanatoacetate and a range of aromatic aldehydes, in the presence of triethylamine, magnesium(II) perchlorate and bipyridine, reacted together[277] to form β-hydroxy-α-amino acids.

4.8 Synthesis of *N*-Substituted α-Amino Acids. – While the *N*-methyl analogues of most of the proteinogenic amino acids are known, their synthesis still demands chemical rigour, so a methodology[278] based on reductive alkylation of Schiff bases both in solution and on solid phase is welcome. Similarly, the unified approach for the methylation of the 20 common acids through 5-oxazolidinones has been researched,[279] and it is concluded that the side-chains of Ser, Thr, Tyr, Cys, Met, Trp, Asn, His and Arg needed protecting groups during the methylation process, but these can be chosen to be compatible with applications for peptide synthesis, A novel solid phase method for monomethylation has also been develop,[280] whereby amino acids supported on Wang or Sasrin resin can be methylated with pinacol chloromethylboronic ester followed by rearrangement of the resulting aminomethylboronate and subsequent cleavage. Both *N*-*Z* and *N*-Fmoc protected MeSer and MeThr have been synthesised[281] via their oxazolidinones, but using *t*-butyldimethylsilyl as the transient side chain protecting group.

A series of benzylamides of *N*-alkylated and *N*-acylated cyclic and linear amino acids have been synthesised[282] for the testing of their anticonvulsant activity, and *N*-formyl-L-aspartic anhydride has been prepared[283] using $HCOOH/Ac_2O/MgO$. N-Acylated amino acids have been synthesised[284] from *N*-acylamino esters using a polymer-supported amine and scandium triflate, and *N*-phthaloyl derivatives have been obtained[285] in a rapid one-pot procedure involving monomethyl phthalate, BOP and $Pr^i{}_2Net$ to give an intermediate which then cyclises in aqueous sodium carbonate. A selection of *N*-acyl homoserine lactone analogues have been synthesised[286] and tested for their ability to inhibit bioluminescence, while *N*-acetyl L-glutamic acid was formed[287] efficiently using Ac_2O in alkaline solution. A new reagent 2-[phenyl(methyl)sulfonio]ethyl-4-nitrophenylcarbonate tetrafluoroborate[288] has been introduced for attachment of a water soluble protecting group onto sulfur-containing amino acids. Scalable syntheses of Z- and Fmoc-Orn have been noted,[289] p-Toluenesulfonamido glutaramides have been prepared[290] from Tos-Cl, Glu and amines, and a novel method[291] has been described for the synthesis of *N*-sulfonyl protected amino acids. Orthogonally protected amino acid building blocks for combinatorial *N*-backbone cyclic peptides have been produced[292] for all amino acids except proline. N^{α}–Aminoallyloxycarbonyl- and carboxyallyl derivatives were first produced, via alkylation (alkyl halides), reductive amination (aldehydes), and Michael addition (α,β-unsaturated carbonyl compound), followed by Fmoc protection of the remaining N—H bond.

3-*N*-Phthaloyl homophenylalanine lactone has been synthesized,[293] and sodium salts of valine, leucine and phenylalanine when treated[294] with acid chloride of trichlorovinylacetic acid and 3, 3, 4, 4, 4-pentachlorobutanoic gave *N*-chloroacyl derivatives. *N*-Stearylleucine was formed[295] from stearyl chloride under Schotten-Baumann conditions and a number of 2, 6-disubstituted benzoyl derivatives have been prepared[296] in the series N-benzoyl-4-[(2,6-dichlorobenzoylamino)]-L-phenylalanine to test for VCAM/VLA-4 antagonist activity. N-Boc- and N-benzoyl-(*S*)-phenylglycinals have been prepared[297] by the oxidation of the respective alcohols (racemisation free) with Dess-Martin periodinane. *N*-(2-Boc-aminoethyl)glycine esters have been produced[298] from Boc-aminoacetaldehyde and glycine esters, and when a H_2/O_2 flame was blown against an aqueous solution of urea and maleic acid, *N*-carbamoylaspartic acid was generated.[299]

A library of *N*-substituted amino acid esters have been produced[300] by a novel 4-component synthesis using a polymer-bound isocyanide, and when the N^{α} - group is protected[301] by a hydrolysable protecting group (e.g. trifluoroacetyl or an enamine), methyl esters could still be made using Me_2SO_4 with the tetramethylguanidinium salts of the acids. *Threo*-*N*-Benzoyl-3-phenylisoserine has been prepared,[302] and *N*-arachidonyl derivatives of both *O*-phospho-L-serine and *O*-phospho-L-tyrosine have been formed[303] through firstly forming the *N*-arachidonyl derivative followed by phosphorylation with cyanoethylphosphate. A series of *N*-aryl-2, 6-dimethoxybiphenylalanine analogues were prepared[304] for inhibition studies on integrin VLA-4, using aryl halides/ $NaOBu^t$/BINAP/Pd_2dba_3 in toluene at 75°C. A 1,4-benzoquinone and three 1,4-naphthaquinones have been directly reacted[305] (2 equivalents) with a series of ω-amino acids to form adducts which have been subjected to electrochemical studies. In model studies[306] for scyphostatin *N*-palmitoylation of the amino group was necessary to produce (*S*)-*N*-(l-benzyl-2-hydroxyethyl)hexadecanamide and *N*-pyrazolidinyl amino acids resulted[307] from reaction of l-acetyl-2-phenyl-5-hydroxypyrazolidine with amino acid esters. Details have been disclosed[308] for the synthesis of N^{α}[4-(2-(2,4-diaminoquinazolin-6-yl)ethyl)]benzoyl-N^{δ}-hemiphthaloyl-L-ornithine and a 4-amino-deoxypteroyl analogue.

4.9 Synthesis of α-Amino Acids carrying Unsaturated Aliphatic Side Chains. – Allylic alkylations have been carried out in different modes, e.g as in Scheme 11, where the alkylations[309] with simple allylic substrates were catalysed by chiral

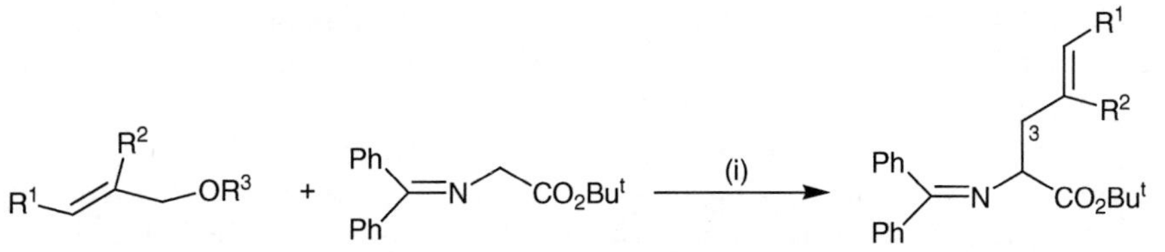

Scheme 11 *Reagents: (i) Pd/phosphine/PTC/CsOH H_2O/Solvent.*

quaternary ammonium salts giving ee's of up to 61%. In a variation of this approach, the glycine esters were added[310] as zinc enolates. Producing allylic isomers with substitution at position 3 (3 on product in Scheme 11) is more synthetically demanding, but has been achieved[311] by molybdenum-catalysed asymmetric allylic alkylation using azlactones as the 'glycine' element. Good stereoselectivity (*anti*-configured products) has been obtained[312] due to suppression of π-σ-π isomerisation, if zinc enolates of TFA-protected glycine esters were reacted in the presence of $[allylPdCl]_2$.

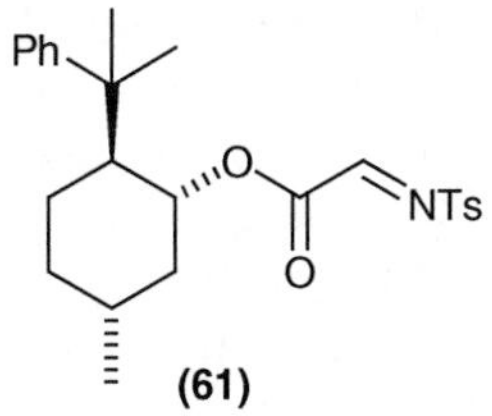

(61)

Asymmetric addition[313] of allyltrimethylsilane to a chiral *N*-tosylimine (**61**) derived from 8-phenylmethyl glyoxylate in the presence of various Lewis acids has been monitored. Good diasteroselectivity was found for $ZnBr_2$, $ZnCl_2$, $TiCl_4$ and $SnCl_4$, but poor selectivity using $BF_3.Et_2O$.

Chiral (*Z*)-α, β-didehydro amino acids have been generated[314] from a chiral iminic cyclic glycine template with 1, 2, 3, 6-tetrahydropyrazin-2-one by condensation with Eschenmoser's salt and Bredereck's reagent. These can be further arylated in the presence of $Pd(OAc)_2$. Enantioselective hydrolysis of amino acid amides using an amidase has been applied[315] to a series of unsaturated amino acids, which can be up-graded to multigram scale. α-Methylene-β-amino acid derivatives have been obtained[316] from aldehydes, sulfonamides and α, β -unsaturated carbonyl compounds using the aza-version of the Baylis-Hillman reaction, with DABCO as base and $La(OTf)_3$ as a Lewis acid. An asymmetric version[317] of the same reaction between *N*-*p*-toluenesulfinimines and methyl acrylate in the presence of $In(OTf)_3$ has also been successfully developed for β-amino-α-methylene esters. The effect of base and solvents on the formation of dehydroalanine through elimination of *p*-Ts from *p*-tosylserine derivatives has been monitored.[318] *N*-Boc- and *N*-Z-α-tosyl ethylglycinates have been reacted[319] with aldehydes in the presence of tributylphosphine and a base under Wittig conditions to give the corresponding α, β-didehydro-amino acid derivatives with high *Z*-selectivity and in good yields.

A concise, scalable route to both isomers of *Z*-2-Boc-amino-6-hydroxyhex-4-enoic acid, has been carried out[320] starting from 2-butyne-l, 4-diol, featuring acylase enzymes in the resolution step. Introduction of allyl groups into the side chain without loss of chirality has been facilitated[321] via the intermediate formation of a terminal aldehyde group in the side chain (generated from the Weinreb amides of aspartic or glutamic acids), which was then subjected to methyl triphenylphosphonium bromide in a Wittig reaction. Starting from L-vinylglycine, enantiomeric purity has been preserved[322] in the making of D- or L-quaternary α-(2-stannylvinyl) amino acids. A series of reaction steps has enabled[323] L-vinylglycine to be synthesised in a novel way from D-xylose. An enolate-Claisen rearrangement of α-acyloxysilane has led[324] to an enantioselective synthesis of compounds such as (**62**) possessing two consecutive chiral centres. A vinylidene-^{3}H analogue of (−)-α-kainic acid has been synthesised[325] via an intermediate keto group substituted in the side-chain of the proline compound, and the 4-keto group in the proline ring of kainic acid was the starting point for transformations[326] leading to (+)-α-allokainic acid. Isoxazolyl and pyrazolyl moieties have been added[327] via 2H-pyran-2-ones to α, β-didehydroamino acid derivatives.

Reaction of *N*-butylsulfonyl, *N*-trimethylsilylethanesulfonyl and *N*-Ts α-imino esters with bis (allyl) titanium complexes, in the presence of $Ti(OPr^i)_4$ and $ClTi(OPr^i)_3$ has given, with high regio- and stereo-selectivity, δ-sulfonimidoyl functionalised β-alkyl-γ,δ-unsaturated amino acids.[328] Dehydrophenylalanine cyclophanes bearing structures such as (**63**) have been synthesised,[329] and the influence of the length of the tethers between the amino acid residues assessed.

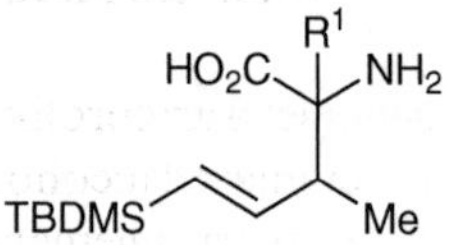

(62)

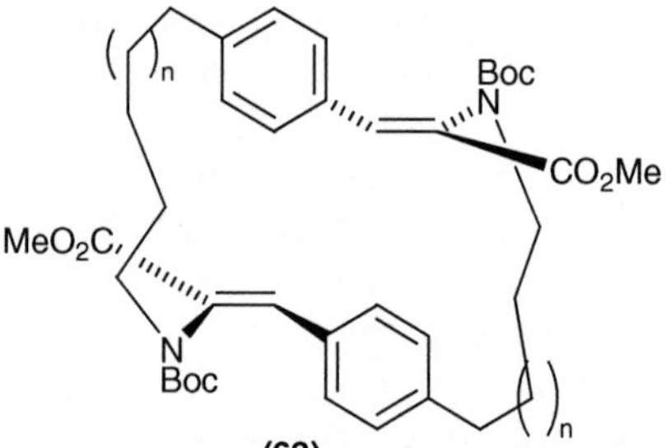

(63)

In synthetic transformations leading to dihydrofurans from amino acids intermediate synthons involved the synthesis[330] of allenic α-amino acid derivatives by 1, 6-addition of the cyano-Gilman reagent, *t*-But-CuLi-LiCN to substituted enynoates, followed by deprotection stages, A convenient route[331] has been found to incorporate alkyne groups into the γ, δ-position of the side chain by treating an aziridine, 1-[(*S*)-l-(2-nitrobenzensulfonyl)-aziridin-2-yl]-4-methyl-2, 6, 7-trioxabicyclo[2,2,2]-octane, with a variety of lithium acetylides. The alkyne-containing amino acids were further transformed to C-glycosyl amino acids. A Heck reaction[332] of (*S*)-*Z*-*N*-allylglycine Bu^t ester with aromatic halides has yielded a series of arylallylglycine derivatives.

Although not strictly side–chain unsaturation, it is appropriate to record the preparation of *N*-allyl amino esters[333] by either reaction of allylamine with $Cl(CH_2)_nCOOEt$ or with ethyl acrylate, and the addition of a range of (hetero)-aryl bromides to the alkyne function in *N*-propargyl alanine[334] using a cross-coupling reaction catalysed by 10% Pd/C.

4.10 Synthesis of α-Amino Acids with Aromatic or Heterocyclic Side-Chains. – There has been a flurry of activity in incorporating fluorine into aromatic side-chains, for use as tracers in medical imaging (PET). It has been shown[335] that direct fluorination ($^{18}F_2$) of L-α-methyltyrosine in TFA or in HF, produced 3-^{18}F-α-methyltyrosine in up to 30% radiochemical yield. 3, 5-Difluoro- and α-methyltyrosine could also be produced. Alkylation under phase transfer conditions of the Oppolzer chiral sultam with fluorinated analogues of 3, 4-$(OMe)_2$-benzyl chloride, proceeded[336] diasteroselectively, to produce after deprotection 2-, 5-, 6-fluoro- and 2,6-difluoro-DOPA. Fluorinated DL-phenyl-alanines were synthesised[337] from fluorinated aromatic aldehydes by a 'one-pot' procedure involving Erlenmeyer reactions and subsequent reduction (P/HI). 6-^{18}F-Fluoro-L-DOPA has been synthesised in less than 2 hr, using chiral catalytic transfer alkylation techniques.[338] The Erlenmeyer azlactone strategy has also been employed[339] for the synthesis of 6-fluoro-*meta-tyrosine* from 2-fluoro-5-hydroxybenzaldehyde, and *O*-(2-[^{18}F]-fluoroethyl)-L-tyrosine was made[340] in a remote-controlled 'no carrier added' synthesis. 2-Fluoro- and 6-fluoro-(2*S*, 3*R*)-(3, 4-dihydroxy-phenyl)serines, were also produced[341] via oxazolidine intermediates formed from 3, 4- (and 4, 5)dibenzyloxy-2-fluoro-benzaldehyde respectively. Cell-free extracts from a number of bacterial strains were found to catalyse[342] the transamination of 4-fluorophenylglyoxylic acid to 4-(*S*)- fluorophenylglycine. Aniline derivatives have been the source[343] of halogenated-substituted phenylalanines and radio-iodination techniques for aromatic amino acids have been reviewed.[344]

An efficient 5-step sequence to synthesise optically active 3-arylprolines has been developed[345] starting from aromatic aldehydes and cinnamyl alcohol, with L-proline derivatives as chiral auxiliaries. An aza-Claisen rearrangement with azidoacetyl fluoride became a key stage in the synthesis. 4-*cis*-Phenyl-L-proline has been synthesised[346] starting from naturally-occurring 4-*trans*-HO-L-Pro, via the formation of the 4-oxo derivative which underwent a diastereoselective Grignard reaction with PhMgBr. A novel family of chiral *N*-4-pyridinylproline derivatives have been developed[347] as potential stereoselective catalysts, using nucleophilic displacement of 4-chloropyridine by a proline imino group.

Constraining the flexibility of the aromatic side-chains of amino acids has been actively developed in designing pharmacophores. Over this period examples come from:- the benzazepinone derivative (**64**) which was potent in a VCAM/VLA-4 ELISA assay;[348] indane-based constraints as in (**65**) introduced via a [4+2] cycloaddition, a dialkyne strategy[349] or by starting with phenyl-glycine;[350] the novel endo-12-aminotricyclo[6.3.2.0(2,7)]trideca-2(7), 3, 5-tri-ene-12-exo-carboxylic acid made from cycloheptadiene;[351] *o*-aryl substituted phenylalanines, naphthylalanines and tryptophan analogues synthesised[352] via asymmetric hydrogenation of α-enamides using Burke's DuPHOS-based Rh(I) catalyst, followed by Suzuki cross-coupling with boronic acid derivatives. L-2-Naphthylalanine has also been produced[353] from 2-naphthylpyruvate and L-glutamate catalysed by a thermostable aminotransferase. Two separate reports[354,355] summarise different approaches to the synthesis of 1, 2, 3, 4,-tetrahydroisoquinoline-carboxylic acid (Tic). A first synthesis[356] of 6-hydroxyoctahydroindole-2-carboxylic acid, a key strategic element in the structure of aeruginosins 298 has been reported. The stereochemical route started with L-tyrosine which underwent a Birch reduction followed by aminocyclisation. Octahydroanthracene amino acids affording conformationally-constrained lysine analogues have also been synthesised (18 steps).[357]

Arylglycines, as components of a number of natural products, have found popularity over recent years. A 'one-pot' novel procedure[358] for their synthesis involved treatment of the side chain OH group of serine with (diacetoxyiodo)benzene and iodine, with the resulting radical being oxidised by a cationic glycine equivalent which can be trapped by nucleophiles (aryl and a number of other types). *N*, *N'*-Di-Boc-protected benzylamines have been shown to undergo[359] 1,2-Boc migration on treatment with KDA/Bu^{t}OLi to give *N*-Boc-phenylglycine Bu^t esters. Mannich-type reaction of phenols with an iminolactone from phenylglycine has given[360] highly stereoselective yields of α-arylglycines. The highly fluorescent L-3-(l-pyrenyl)alanine has been obtained[361] by asymmetric hydrogenaton of 1-acetyl-3-pyrenemethylidene-6-methylpiperazine-2, 5-dione, and novel pyrazolylglycines (**66**) have been produced,[362] including l-hydroxypyrazoleglycine formed[363] by addition of organomagnesium and lithium intermediates to diEt-*N*-Boc-iminomalonate. *N*-Protected arylglycines have been diastereoselectively synthesised[364] via $TiCl_4$ promoted Friedel-Crafts reaction of phenols with chiral *N*, O-hemiacetals. α-Alkyl-α-phenylglycines were obtained[365] by asymmetric synthesis, via phenylation/alkylation of (2*R*, 3*S*)-*N*-Boc-6-oxo-2, 3-diphenylmorpholine followed by hydrolytic ring opening. α-Arylation in high yields has been reported[366] for the reaction of aryl bromides with protected glycinate esters in the presence of Li or Na hexamethyldisilazide and $Pd(dba)_2$/ligand.

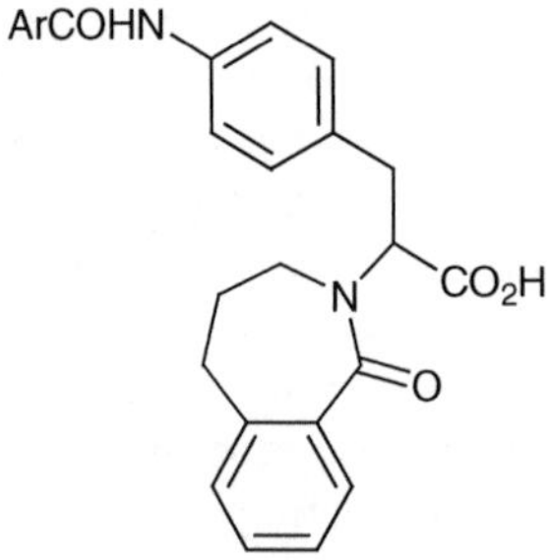

(64) 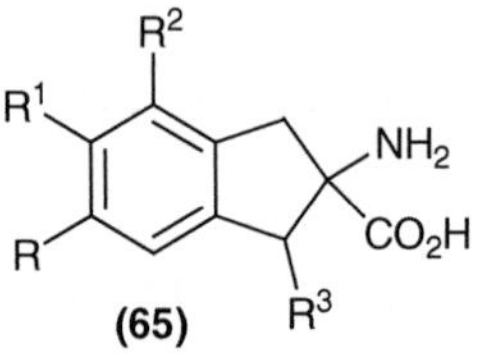 **(65)** 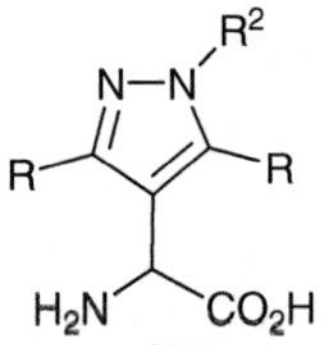 **(66)**

Fmoc-L-*p*-Azidotetrafluorophenylalanine has been prepared[367] from acetamidomalonate, followed by enzymic resolution. 4-Aminophenylalanine has been synthesised[368] from 4-hydroxymethyl anilines via aza quinone methide intermediates, while Boc-L-*p*-aminophenylalanine has been utilised[369] to make FRET cassettes. A practical 'one-pot' catalytic procedure[370] has been developed for both aromatic and heteroaromatic amino acids, utilising a chiral BINAP-Cu catalyst for the addition of aromatic/heteroaromatics to *N*-alkoxycarbonyl α-iminoesters. A cross-enyne metathesis reaction has been the foundation[371] of a synthetic route to many highly substituted phenylalanines. In order to access β-hydroxyphenylalanine, a component of ustiloxin D, use has been made[372] of the Sharpless asymmetric aminohydroxylation of substituted cinnamic acids, using 2-trimethylsilylethyl carbamate with Os(VIII)/ $(DHQD)_2AQN$ as catalyst. Homophenylalanine derivatives have been made using cycloaddition[373] of a cyclic nitrone glycine template with styrene derivatives or from L-malic acid.[374] The synthesis of (*S*)-α-methylphenylalanine has featured[375] in the development of better chiral catalysts for C-alkylation of aldimine Schiff bases of alanine esters. (*S*)-3, 3′-Bis[(diethylamino)-methyl]-2, 2′dihydroxy-l, 1′-binaphthalene gave the best ee's. Several α-benzylphenylalanines have been prepared[376] by alkylation of ethyl isocyanoacetate with different benzyl bromides, the products being further derivatised via the Suzuki-Miyaura coupling reaction. The two enantiomers of α-methyldiphenylalanine have been resolved[377] using chiral hplc separation. New photoactivatable phenylalanine analogues have been synthesised[378] via the asymmetric synthesis of (*S*)-Boc-*p*-(propanoyl)phenylalanine from the alkylation of sultam *N*-(diphenylmethylene)glycinate, which was further transformed to Boc-*p*-[3′-(phenylselenenyl)propanoylphenylalanine.

Recent advances in the chemistry of fullerenes (C_{60} buckyballs) have included the synthesis of fullerene-based amino acids, and the developments (especially incorporating proline) have been reviewed.[379] A C_{60}-fullerene unit has been directly attached to the α-position of glycine via the addition[380] of N-(diphenylmethylene)glycine esters to [60] fullerene under Bingel cyclopropanation conditions (C_{60}, DBU, CBr_4, C_6H_5Cl). The sterically constrained 4, 5-diazafluorene amino acid (**67**) has been obtained[381] via the acylation of the anion of *N*-benzyl 4, 5-diazafluorene-9-methylene amine, but incorporation of this building block into peptides could only be done via the azide method due to decarboxylation.

(2*S*, 3*R*)-*N*-(1, 1′-Dimethyl-2′-propenyl)-3-hydroxytryptophan, a constituent of cyclomarin C, has been stereoselectively synthesised[382] from L-tryptophan, and as part of the synthesis of ergot alkaloids the intermediate *N*-Boc-4-bromo-*N*-methyl-1-tosyl-D-tryptophan methyl ester has been synthesized.[383] The Pmc group (2, 2, 5, 7, 8-pentamethylchroman-6-sulfonyl), usually associated with the side chain protection of Arg residues, has been substituted into the 2-position of tryptophan.[384] L-Tryptophanol (COOH → CH_2OH) and indole-substituted analogues have been made[385] from 4(*R*)-iodomethyl-2-oxazolidinone and indolyl magnesium bromide. β-Substituted tryptophans can be made[386] from an enantiomerically pure 3-phenylaziridine-2-carboxylic ester (made via Sharpless dihydroxylation) followed by ring opening with indole.

There has been quite an interest in amino acids bearing heteroaromatic side chains such as oxazole/thiazole amino acids[387] which have been found widely in marine organisms. The isoaxazole amino acids (**68**) have been prepared[388] on-resin, starting with alkynyl groups attached to tritylchloride resin, then treated with aldoximes (RCH=NOH)/N-chlorosuccinimide. Homologues (**69**) of ibotenic acid have been synthesised[389] with different aryl substituents at R^1 using previously published strategies, and thioanalogues of ibotenic acid have also been made[390] using regioselective lithiation and functionalisation of 3-benzyloxy isothiazole. Both isomers[391] of the neuroexcitant (**70**) have been synthesised asymmetrically. A route[392] to spiroisoxazolino-Pro (**71**) started with nitrile oxides.

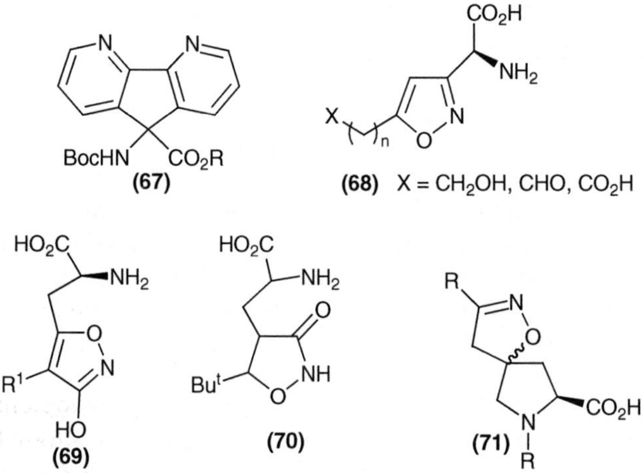

(67)

(68) X = CH_2OH, CHO, CO_2H

(69) **(70)** **(71)**

Selective agonists at group II metabotropic glutamate receptors have been found amongst stereoselectively synthesised[393] (*S*)- and (*R*)-2-amino-4-(4-hydroxy[l, 2 5]thiadiazol-3-yl)butyric acids. A soluble polyethylene glycol-supported protecting group related to silylethylsulfonyl has been used[394] to make unsaturated compounds which could be transformed to pyridyl amino acids using ring closing metathesis. Included is a wider study[395] of the application of aza Diels-Alder reactions was a protocol to make anthracenylglycine derivatives. A range of heterocyclic amino acid systems[396] including quinoxalines, pyrazines and 1, 2, 4-triazines have been obtained from the reaction of diamines and amidrazones with α-amino acid vicinal tricarbonyls. *N*-(Iodoethyl)- and *N*-(3-iodopropyl)pyrimidines and purines have shown[397] to undergo conjugate radical addition to chiral oxazolidinone acceptors to give purine and pyrimidine amino acids. A (thymin-l-yl)methyl function at the α-position[398] has been inserted via 2-(N^3-benzoylthymin-l-yl) methyl 1, 3-propanediol using enzymic desymmetrization catalysed by lipase PS. Amongst a series of nucleophiles added[399] to dehydroalanine derivatives were nitrogen heterocycles. The 'organic' stages in the synthesis[400] of ruthenium trisbipyridyl amino acid (**72**) were achieved using chiral phase transfer catalysis of the reaction between a bromomethylbipyridine and *N*-(diphenylmethylene) glycine t-butyl ester.

Different benzocycloalkane amino acids (**73**) have been synthesised[401] using (*S*)-α-methylbenzylamine and (*R*)-phenylglycerol as chiral auxiliaries, and which highlighted significant solvent effects in the use of TMSCN. 5, 5-Diaryl-2-amino-4-pentenoates (**74**) have been made[402] using Oppolzer's sultam and a Pd-catalysed stereoselective hydrostannylation. L-2-Amino-3-(6, 7-dimethoxy-4-coumaryl) propionic acid, a fluorescent molecule, has been synthesized,[403] and an analogue, L-2-amino-3-(7-methoxy-4-coumaryl) propionic acid,[404] also a fluorogen, has been made from an oxazinone alkylated with the fluorogenic group. A novel fluorescent amino acid has been synthesised[405] in the form of *N*-Boc-3-[2-(1H-indol-3-yl)benzoxazol-5-yl]-L-alanyl methyl ester from 3-nitro-L-tyrosine and 1H-indole-3-carboxaldehyde.

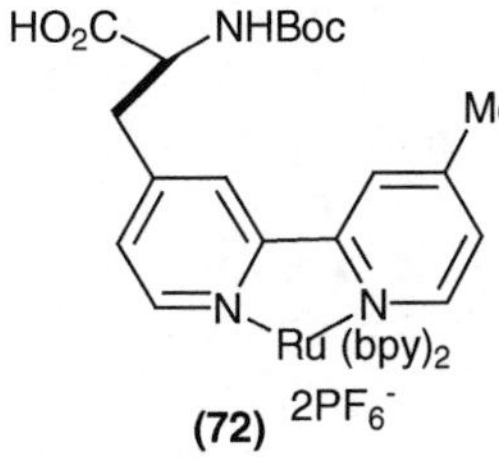

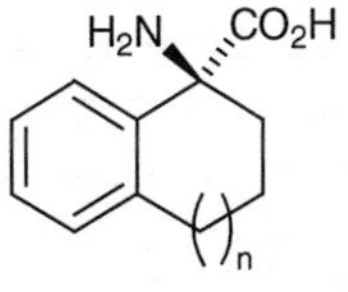

(*S*)-2-Fmoc amino-3-(5-phenyl-8-hydroxyquinoline-2-yl) propionic acid has been made[406] for its fluorescence and ability to act as a sensor for divalent zinc, and 6-(2-methylaminonaphthyl)alanine (DANA) has also been synthesised[407] and used to monitor protein-protein interactions. 6-Aminoquinoline amino acids have been made[408] for antibacterial testing and *N*-(5-hydroxy-3′, 4′-ethylenedioxy-7-isoflavonyloxyacetyl) amino acids have also been synthesized.[409] (S)-Acromelobinic acid (**75**) has been characterised through synthesis.[410] 6-Amino-l, 4, 6, 7-tetrahydroimidazo[4, 5, b]-pyridin-5-ones have been prepared[411] from acetyl-4-nitrohistidine as confromationally restricted His analogues. (1-Benzimidazolonyl) alanine (**76**) has been synthesised[412] as a potential tryptophan mimetic, starting with Z-L-diaminopropanoic acid and building up the benzylimidazolonyl ring from 2-fluoro-l-nitrobenzene. The *cis* isomer of oxazolidinone (**77**), obtained in high purity using $ZnCl_2/SOCl_2$ in the cyclisation step, was alkylated with l-bromobenzylbromide to provide chiral α-(4-bromobenzyl) alanine ethyl ester.[413] (*R*) and (*S*)-α-Hydroxymethylnaphthyl alanine has been made[414] and 3-(3-hydroxy-4-isoamyloxybutyl [and 3-(3-hydroxy)propyl]-4-phenyl-5-mercapto-1, 2, 4-triazoles have been added[415] to dehydroalanine using Ni(II) complexes to form the corresponding phenyltriazolylcysteines. The triazinyl unit itself has also been incorporated into pseudopeptide structures.[416]

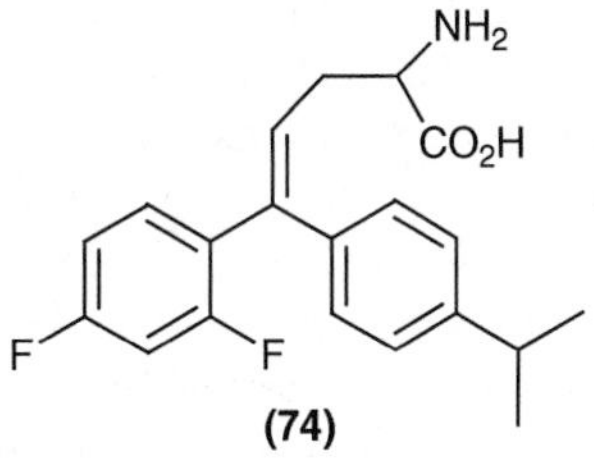

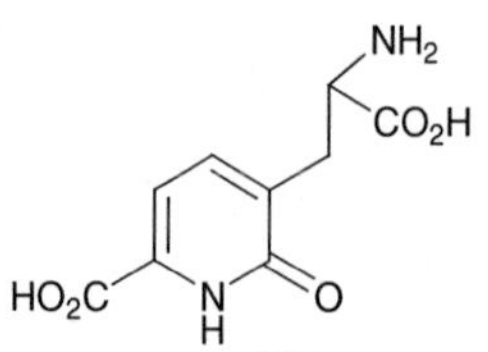

4.11 Synthesis of α-Amino Acids carrying Amino Group and related Nitrogen Functional Groups in Aliphatic Side Chains. – This is the sub-section where interest in diamino alkanoic acids can be accommodated. Thus glycine-derived chiral synthons of type (**78**) have been the source of an enantioselective synthesis [417] of (1*R*, 4*R*)- and (1*S*, 4*S*)-forms of 2, 6-diaminopimelic acid, while differentially protected *meso*-2, 6-diaminopimelic acid has been obtained[418] from both aspartic and glutamic acids. The second chiral centre was established by the asymmetric reduction of a pyruvate moiety with Alpine-Borane. All four stereoisomers of *N*, *N′*-protected 2, 3-diaminobutanoic acid have been synthesised[419] using an asymmetric Rh(I)-phosphine-catalysed hydrogenation of isomeric enamides. Desymmetrisation of dimethyl 3-benzylaminoglutarate through enzymic ammonolysis[420] has given enantiopure (*R*)-3, 4-diaminobutanoic acid. ^{14}C-Labelled (*S*, *S*)-2, 7-di-Boc-diamino[l, 8 $^{14}C_2$] suberic acid,[421] as well as (*R*, *R*) and (*S*, *S*) isomers[422] have been synthesized, the former by inserting ^{14}CN into 1, 6-hexandial via a thermodynamically-controlled asymmetric Strecker synthesis using (*R*)-2-phenylglycinol as chiral auxiliary. All four stereoisomers of 4-aminoglutamic acid have also been obtained[423] using Ni complex-catalysed Michael-type condensation on to dehydroalanine. (2*R*, 4*S*)-and (2*R*, 4*R*)-Diaminoglutamic acids were obtained[424] in three steps from 6-oxo, 2, 3-diphenyl-4-morpholine carboxylate using a radical reaction of a selenide with methyl 2-acetamidoacrylate.

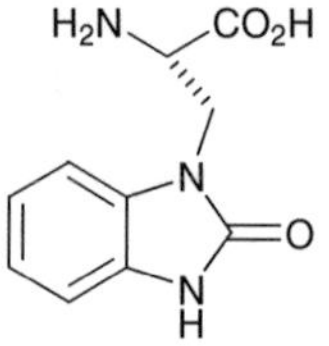

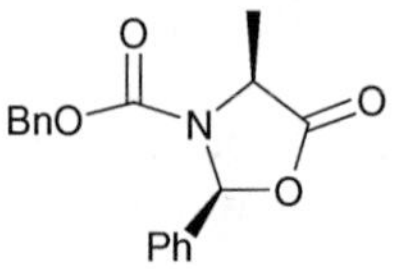

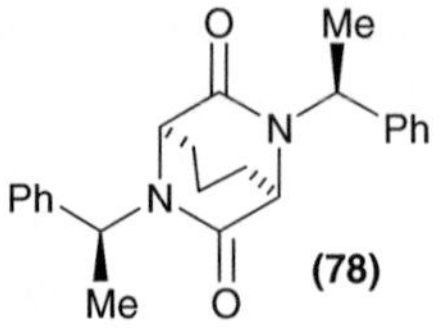

Quinazolinone and pyrazolopyrimidone derivatives of *cis*-4-aminoproline have been made,[425] and an improved synthesis[426] of protected *cis* and *trans* 3(and 4)-azido-L(and D)-prolines has been reported. The latter evolved from carrying out Mitsunobu reactions with diphenylphosphoryl azide on hydroxyprolines. Homochiral building blocks of 4-azalysine (2, 6-diamino-4-aza-hexanoic acid) have been made[427] by exploiting the reductive amination of aldehydes with amines. One route started from L-serine (to form D-isomers) and another route from L-asparagine gave orthogonally protected 4-azalysine derivatives. An enantiomerically pure β-lactone (4-trichloromethyl-2-oxetanone) has been shown[428] to be a versatile synthon leading to a variety of γ-substituted α-amino acids. A synthesis[429] of optically active β-nitro-α-amino esters has provided an entry into α, β-diamino acid derivatives. The former were produced via a copper-bisoxazoline-catalysed aza-Henry (nitroaldol) reaction between silyl nitronates and α-imino esters. 2-Nitromethyl-ornithine has been obtained[430] from ornithine, mediated by cobalt (III), and *S*-nitroso-L-cysteine ethyl ester, known to be involved in *trans* -*S*-nitrosation of thiol proteins has been made[431] by direct nitrosation of L-cysteine ethyl ester hydrochloride with ethyl nitrite.

Reaction[432] of unmodified aldehydes with *N*-Pmp-protected α-imino ethyl glyoxylate in the presence of catalytic amounts of L-proline followed by addition of Et_2AlCN has provided enantiometrically pure β-cyanohydroxymethyl α-amino acid derivatives. A good yield of (*S*)-(+)-2-amino-6-(aminoxy)-hexanoic acid was obtained[433] from (*S*)-(−)-6-amino-2-{[(benzyloxy)carbonyl]amino}-hexanoic acid.

4.12 Synthesis of α-Amino Acids with Side-Chains carrying Boron Functional Groups. – A previously published synthesis in 1993 of L-valylpyrrolidine-(2*R*)-boronic acid has been significantly improved[434] by developing efficient recycling of chiral auxiliary (+)-pinanediol. α-Chymotrypsin has been used[435] to resolve p-boronophenylalaninol, while a conference review[436] concentrated on the concise synthesis of enantiomerically pure L-borophenylalanine from L-tyrosine. Both ^{18}F and ^{11}C-labelled *p*-boronophenylalanine[437] and l-amino-3-[2-[7-(6-deoxy-β-galactopyranos-6-yl)-1, 7-dicabododecarboran(12)-l-yl]ethyl cyclobutane carboxylic acid[438] have been synthesised for boron neutron capture therapy.

4.13 Synthesis of α-Amino Acids with Side Chains Carrying Silicon Functional Groups. – Synthesis[439] of racemic and (*R*)-$Me_2Si(CH_2R)CH_2CH(NH_2)CO_2H$, with R=$NH_2$, OH and SH was effected from 3, 6-diethoxy-2, 5-dihydropyrazine and (*R*)-3, 6-diethoxy-2-isopropyl-2, 5-dihydropyrazine respectively. Racemic and non-racemic $NH_2CH(CH_2ElR_3)COOH$ where El = Si or Ge, have been produced[440] starting also from 3, 6-diethoxy-2, 5-dihydropyrazine, using chiral hplc for resolution. Sodium salts of many amino acids have been reacted[441] with dichlorodimethylsilane to form a series of complexes. Esters of three types of silylated amino acids have been prepared[442] from zircona aziridines.

4.14 Synthesis of α-Amino Acids with Side Chains carrying Phosphorus Functional Groups. – Electrophilic fluorinations[443] of lithiated bis-lactim ethers allowed direct access to monofluorinated phosphonate mimetics of naturally-occurring phospho-serine and -threonine, suitable for solid phase peptide synthesis. Facile synthesis[444] and X-ray structural analysis, has been carried out on $Ar_2PCH_2CH(NHBoc)CO_2Me$, the synthesis requiring a nucleophilic phosphination of *N*-(Boc)-3-iodoalanine methyl ester using potassium carbonate as base. Phosphino amino acids were products[445] from the reaction of phosphine with formaldehyde and amino acids. 2*S*-2-(4-Phosphonophenylmethyl)-3-aminopropanoic acid [D-β-2(4-phosphono)-phenylalanine] has been made[446] by the diastereoselective alkylation of a chiral enolate. Access to 4-substituted 2-amino-4-phosphonobutanoic acid was possible[447] either from conjugated addition of the lithiated bis-lactim ether from cyclo (Gly-D-Val) to α-substituted vinyl phosphonates or electrophilic substitution on the lithiated bis-lactim ether from cyclo(2-NH_2-4-phosphonobutanoyl-D-Val). Preliminary results[448] on the synthesis of constrained analogues of phospho-isostere of glutamic acid (AP4) have been reported. Analogues constructed were (**79**) and (**80**) and their synthesis started from dibromo cycloalkanes. Although stretching the heading of this sub-section the synthesis[449] of proline and pipecolic acid phosphorus analogues such as (**81**) via diastereoselective carboxylation appears to be an interesting concept.

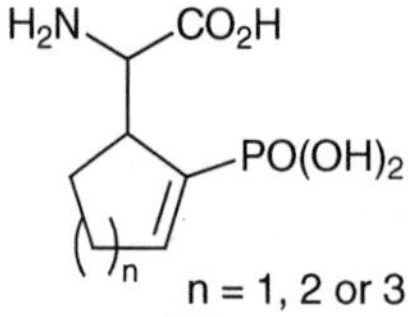

H_2N CO_2H $PO(OH)_2$ n = 1, 2 or 3 **(79)**

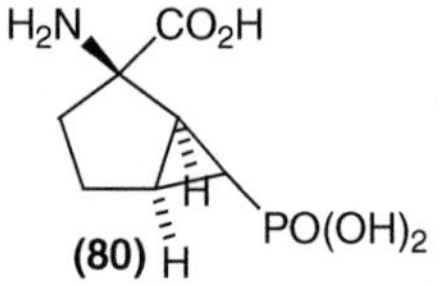

H_2N CO_2H H $PO(OH)_2$ **(80)** H

CO_2H P Ph BH_3 **(81)**

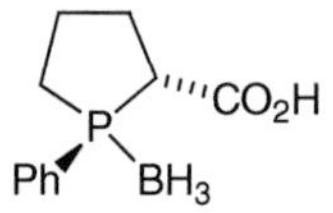

4.15 Synthesis of α-Amino Acids carrying Sulfur and Selenium Containing Side Chains. – With an increasing number of proteins known to contain selenocysteine (Sec) and the amino acid itself a good source of dehydroamino acids, new methods for its synthesis have been developed. So high yielding upgrades to the synthesis[450] of Fmoc-Sec(PMB) and Fmoc-Sec(Ph) have been reported. The former was synthesised from Fmoc Ser(Ts)-allyl ester, which is treated with p-methoxybenzylselenol, while the latter followed a similar route but using diphenyl methyl ester for carboxyl protection. (Ac-Gly-SecOH)$_2$ has been used[451] to test selenocysteine in native chemical ligation. Aromatic selenoamino acids have been made from 4-aminophenylalanine[452] via diazotisation of the 4-amino group and replacement with SCN (polar neutral), SeO_2^- hydrophilic anionic) or SeR(hydrophobic). Selenazolidines, cyclic analogues (**82**) of selenocysteine have been made[453] as masks for the chemically reactive groups. One analogue (**82**, X = C=O) was made from selenocysteine using 1, 1-carbonyldiimidazole, the other (**82**, X = CHMe) using acetaldehyde as the carbonyl donor.

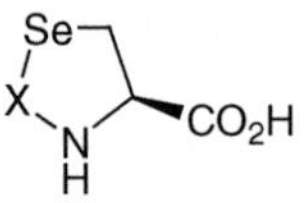

Se X N H CO_2H

(82) X = C=O or CHMe

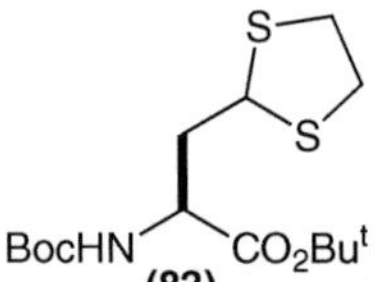

S S BocHN CO_2Bu^t **(83)**

Asymmetric synthesis[454] of *S*-alkylated cysteines has made use of the nucleophilic addition of alkane thiols to dehydroalanine derivatives using Ni(II) complexes as chiral catalysts, and 4-methoxytrityl-mercapto acids have been trialled[455] in solid phase construction of libraries of mercapto-acyl peptides. The *S*-2-amino-3-(l, 2-dithiolan-4-yl) propanoic acid (**83**) and its dithiolic form have been made[456] using a parallel route to a previously synthesised dihydroxyleucine starting from t-butyl(*S*)-Boc-pGlu. Synthesis of novel N$^{\alpha}$-(ω-thioalkyl) amino acid building units have been reported[457] and *N*-[2-(indan-l-yl)-3-mercaptopropionyl) amino acids have been made[458] as highly potent inhibitors of NEP, ACE and ECE. (+)-Biotin was synthesised[459] in 11 steps (25% overall yield) from cysteine using a Lewis base-catalysed cyanosilylation of (2*R*, 4*R*)-*N*-Boc-2-phenylthiazolidine-4-carbaldehyde followed by a Pd-catalysed allylic amination. New routes[460] from homoserine and methionine have been successful for the synthesis of α-amino acid, β, γ-thioenol ether. Suzuki coupling reactions have been applied[461] to the synthesis of several S analogues of dehydrotryptophan e.g. benzo[b]thiophenes.

4.16 Synthesis of β-Amino Acids and Higher Homologous Amino Acids. – Peptides based on the β-amino acid building blocks have acquired renewed interest in the 21st century, and therefore require efficient means of synthesising the building blocks. Thus enantioselective methanolysis[462] of cyclic *meso* anhydrides mediated by cinchona alkaloids provided the monomethyl esters required for further Curtius degradation of acyl azides, to yield *N*-protected β-amino esters. Enantiopure β-amino acids were the result[463] of hydrogenolysis of β-enamino esters, and also the result[464] of the addition of 2 or 3 equivalents of lithium (*S*)-*N*-benzyl-*N*-α-methylbenzamide to α, β-unsaturated ester fragments followed by hydrogenolytic deprotection. Conjugated addition[465] of heteroaromatic amides to ethyl acrylate in a CsF-Si(OEt)$_4$ system gave *N*-substituted β-amino acid ethyl esters, while a mild 2-step synthesis of racemic β-amino acids from Z-alkyl-Δ^2-oxazolines has been shown[466] to take place in high yields. Stereoselective conjugate addition[467] of homochiral lithium *N*-benzyl *N*-α-methyl-4-methoxybenzylamide to α, β-unsaturated esters gave, after mono deprotection either under oxidative or acid-promoted reaction, β-amino acids or β-lactams.

As summarised in Scheme 12 the 1, 3-dipolar cyloaddition of nitrones with ynolates have yielded[468] 5-isoxazolidinones which readily yield β-amino acids. Twenty one 3-amino-3-arylpropanoic acids have been synthesised[469] in a 'one-pot' reaction by refluxing a substituted benzaldehyde derivative with malonic acid and 2 eq. of ammonium acetate in ethanol. Electron-donating groups on the benzaldehyde ring favoured the reaction, as well as solubility in the solvent used. 3-Substituted- and 3, 3-disubstituted aziridine-2-carboxylate esters were a source[470] of β-amino acid and quaternary β-amino acids. The aziridine intermediates were sourced from the aza-Darzens reaction of α-bromoenolates with *N*-sulfinyl imines, and by addition of Grignard reagents to 2H-azirine-2-carboxylate esters. Enantioselective hydrogenation of α, β-unsaturated nitriles, using Rh-DuPHOS as catalyst has given β amino acid precursor, but with only 48% ee.[471] Higher diastereoselectivity (91% ee) was achieved[472] in the synthesis of (*S*)-β-aminophenylpropanoic acid by addition of lithium enolate of

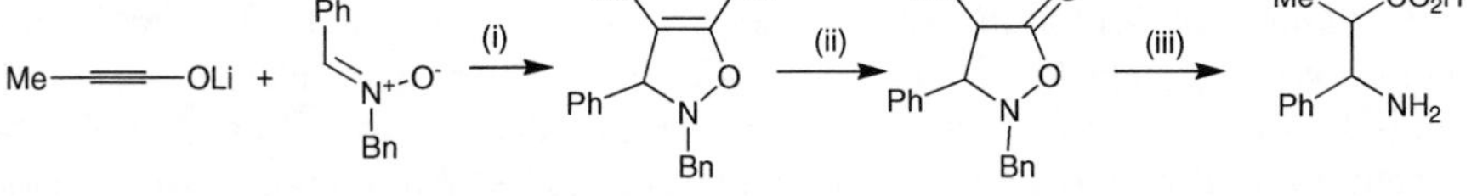

Me—≡—OLi + Ph–CH=N⁺(Bn)–O⁻ —(i)→ [Me, OLi, Ph, O, N, Bn] —(ii)→ [Me, O, Ph, O, N, Bn] —(iii)→ Me–CH(CO$_2$H)–CH(Ph)–NH$_2$

Scheme 12 *Reagents: (i) 0°C 1 hr (ii) HX (iii) H$_2$/Pd/C 60°C.*

Bu^t-(+)-(*R*)-*p*-toluenesulfinylacetate to *N*-(benzylidene)toluene-4-sulfonamides, followed by reductive cleavage, ester hydrolysis and detosylation. Cyclic β-amino acids, (*S*)-homoproline and (*S*)-homopipecolic acid have been made[473] via the diastereoselective conjugate addition of lithium (*S*)-*N*-allyl-N$^{\alpha}$-methyl benzamide to α, β-unsaturated esters followed by ring-closing metathesis.

There have been a few examples of conversion of α-amino- to β-amino acids. Thus the Wolff rearrangement of α-aminodiazoketones derived from *N*-urethane protected α-amino acids in the presence of *o*-nitrophenol produced[474] protected β-amino acids. Similarly the Arndt-Eistert homologation method[475] has been applied to protected α-amino acids using *p*-toluenesulfonyl chloride for carboxyl activation, and using Boc_2O as a coupling agent[476] allows for scale up of the homologation process. The β-amino homologue of histidinol has been synthesised[477] utilising the *N*-Mts group for the protection of both the imidazole and amino nitrogens during the transformation of histidinol via its methansulfonate ester and cyanation to its higher homologue.

Several examples of α-substitution of β-amino acids have been reported. Conjugate radical additions[478] and Heck reactions on conjugated double-bonded synthons have yielded a number of α-substituted diverse structures. *N*-Protected α-phenylethyl amides of α-amino acids can be alkylated[479] diastereoselectively (up to 89% ds) in the α-position (via lithium enolates). Following Arndt-Eistert homologation of α-amino acid esters, diastereoselective α-alkylation has been reported,[480] while *anti*-α-methyl-β-amino acid derivatives[481] have emerged from highly enantio- and diastereo-selective Mannich reactions using a zirconium complex of (*R*)-6,6′-bis(pentafluoromethyl)-l, l′-bi-2-naphthol. Asymmetric synthesis[482] of α-phenyl-β-alanine was carried out via intermolecular catalytic C–H insertion of carbenoids derived from aryldiazoacetates. As key compounds in a number of medicinally important compounds, asymmetric synthesis of α-hydroxy-β-amino acids has been a source of increasing interest. One synthesis[483] incorporated the Lewis acid-mediated addition of Z-α-methoxyketene methyltrimethylsilyl acetal to chiral amines, *N*-alkylidene (*S* or *R*)-α-methylbenzyl-amine, followed by demethylation, hydrogenolysis and hydrolysis. In the formation[484] of α-hydroxy-α-methyl-β-amino acids the key steps were a catalytic asymmetric aldol reaction, a modified Curtius reaction to form oxazolidinones, which could be chemoselectively opened.

Ring opening of (2*R*, 3*S*)-2-benzyl-3-vinyl-1-[(*R*)-1-phenylethyl]-aziridine with acetic acid yielded[485] 3-acetyloxy-4-[(*R*)-l-phenylethyl]amino-5-phenyl-pent-l-ene which was transformed using known procedures to (2*S*, 3*R*)-3-amino-2-hydroxy-4-phenylbutanoic acid, a key component of bestatin. As part of the demand for aminopeptidase and HIV protease inhibitors as anti-cancer agents, 3-amino-2-hydroxy-4-phenylbutanoates have been synthesised[486] using $H–C(CN)_2OSiR_3$ as a key reagent with protected α-amino aromatic aldehydes. Epoxidation of l-tolylthio-l-nitroalkenes containing an allylic Boc-protected amino group has yielded[487] *cis*-oxazolidinones, which, can be transformed to *anti*-α-hydroxy-β-amino acids. Asymmetric α-hydroxylation[488] of *N*, *N*-diprotected homo-β-amino acid methyl esters has afforded β-amino-α-hydroxy acids with full orthogonal protection using KHMDS/2-[(4-methylphenyl)sulfonyl]-3-phenyloxaziridine for introduction of the α-OH group.

Among the fluorinated β-amino acids produced were, CF_3-containing amino acids, synthesised[489] via *cis*-3-CF_3 aziridine-2-carboxylate (**84**), β-fluoroalkyl-β-amino acid derivatives starting from 2-alkyl-Δ^2-oxazolines and fluorinated imidoyl chloride.[490] Trifluoromethylated dehydro β-amino acids[491] such as (**85**) were the product of [α-(alkoxycarbonyl)vinyl]diisobutylaluminium with *N*-acylimines of hexafluoroacetone and methyl trifluoropyruvate. (+)-4, 4, 4-Trifluoro-3-aminobutanoic acid has been made[492] from a β-amino ester derived from chiral 2-trifluoromethyl-l, 3-oxazolidine. *p*-Toluene sulfinimines proved to be efficient[493] chiral amine equivalents in the high temperature Reformatsky-type additions with $BrZnCF_2CO_2Et$ which gave enantiomerically pure α, α-difluoro-β-amino acids. Using a similar set of reactions[494] α, α-difluoro-β-amino acid derivatives have been made from *N*-*t*-butylsulfinimines. α-Substituted-β, β-bis(trifluoromethyl)-β-amino acids have been made[495] via a Morita-Baylis-Hillman reaction with the double-bond containing adducts formed, either being hydrogenated or subjected to cuprate addition. Although low yielding, a route[496] to enantiopure conformationally constrained fluorine-containing β-amino acids has been worked out starting from D-glucose. Glycosylated β-amino acid derivatives such as (**86**) have been made via Wittig olefination of xylofuranos-5-ulose and found to be good antitubercular agents.[497] Homochiral β-haloaryl β-amino esters have been obtained[498] by conjugate addition of lithium *N*-benzyl-*N*-α-methyl-4-methoxybenzyl-amide to cinnamates.

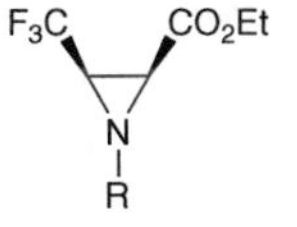

F_3C CO_2Et N R **(84)** — O F_3C CF_3 CO_2Me R N H **(85)** — EtO_2C NR^1R^2 O O RO O **(86)** — O O O NHBn CO_2Me O O **(87)**

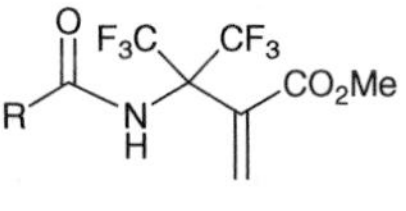

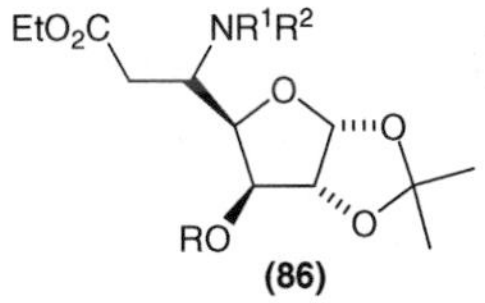

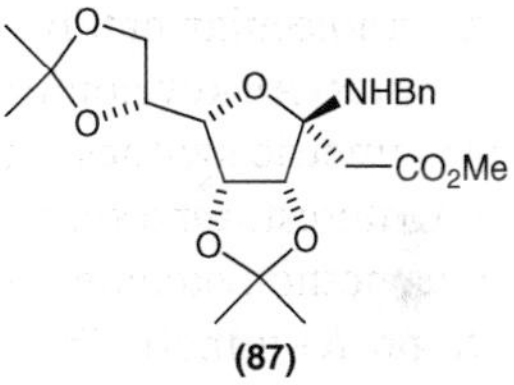

Quite an interest has been shown in cyclic β-amino acids e.g. β-aminocyclopropane carboxylic acids with the side chain functionality of asparagines, arginine, cysteine or serine and have been synthesised[499] via an established route involving the cyclopropanation of *N*-Boc-pyrrole. A photochemical [2 + 2] cycloaddition between ethylene and uracil has furnished[500] *cis*-cyclobutane-β-amino acid, and a chemoenzymatic approach[501] has given both enantiomers of the methyl esters of β-proline by enzymatic resolution of intermediate β-methoxycarbonyl-γ-lactams using α-chymotrypsin. Facile syntheses[502] of fused furanosyl β-amino acids e.g. (**87**) started from protected sugar lactones, and enantiomerically pure *trans*-2-aminocyclohexane acid has been obtained[503] from a 'one-pot' procedure from *trans*-cyclohexane 1, 2-dicarboxylic acid.

Aromatic β-amino acids have been prepared[504] by the Radionow reaction and used as Asp-Phg mimics in VLA-4 antagonism. Nucleophilic addition[505] of lithium enolates of (*S*)-(−)-4-benzyl-2-oxazolidinones to *N*-tosyl aldimines have given β-aryl-β-amino acid derivatives with excellent diastereoselectivity. Commencing with 4-vinylbenzyl chloride, preparation of suitably protected β-aminophosphotyrosine has been carried out[506,507] for the first time, while use of an acid and aldehydes in 'ring switching' reactions with hydrazines have given β-(l-aminopyrrole)-amino acid and β-(pyrazine)-amino acids respectively.

β-Amino acids have been produced[508] via the opening of *N*-nosyl aziridine rings with cyanide ions, followed by hydrolysis, while base-promoted[509] isomerisation of aziridinyl ethers offer an entry into β-amino acids. All four stereoisomers of 2-alkyl-3, 4-iminobutanoic acid (aziridine ring in β-position) have been synthesised[510] starting from aspartic acid, which underwent alkylation at the β-position, followed by reduction of the α-COOH to the alcohol, which was then subjected to cyclisation and aziridine formation. β-Lactone rings open up quite readily with amine nucleophiles, azides or sulfonamide anions[511] to yield β-amino acid derivatives.

The rhodium complex of an imidazolidinone with bisphosphine ligands was an effective catalyst[512] for hydrogenation of both (*E*) and (*Z*)-forms of 3-acylaminoacrylates, and cyclisation of sulfamate esters using a Rh-catalysed C–H bond oxidation/insertion to form oxathiazinones has provided[513] the pathway to form β-amino acids such as (*R*)-Z-β-Ile in 81% yield. 4-Spiro-3-lactams when treated[514] with KCN/MeOH, gave α, α-disubstituted β-amino esters, and when *trans*- and *cis*-oxazoline-5-carboxylates were reacted[515] with thiol acetic acid *syn* and *anti* forms of *S*-acetyl-*N*-benzoyl-3-phenylisocysteine (2′-sulfur analogues of taxol C-13 side chain) were obtained. Methodology has been developed[516] for the enatitioselective synthesis of differentially protected *erythro* α, β-diamino acids from *N*-tosyloxylactams, made from β-keto esters and enantioselective addition of *t*- butyldimethylsilylketone to nitrones in the presence of isopropoxide and phenols have yielded[517] β amino acid esters. Novel sila-substituted β-amino acids (**88**) have been synthesised[518] via a nucleophilic opening of an intermediate aziridine which generated the *trans* amino/carboxyl relationship. Tritium-labelled β-alanine was formed[519] after solid-phase catalytic hydrogenation of uracil with gaseous tritium to form [5, 5, 6, 6-$^3H_3$]5, 6-dihydrouracil, which was cleaved to the labelled β-alanine. An efficient synthesis[520] of α-dehydro-β-amino esters has been achieved through regioselective palladium(0)-catalysed reactions of primary amines with acetates derived from Baylis-Hillman adducts.

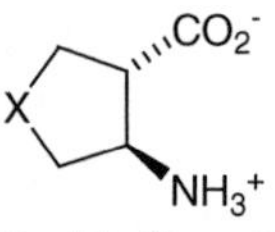

(88) X = Me_2Si or Ph_2Si

4.17 Resolution of DL-Amino Acids. – With this having been one of the most fundamental necessities of amino acid chemistry over the decades, its role recently has received competition from direct methods of asymmetric synthesis. However, while enzymatic resolution remains a staple method, new developments in chiral recognition for liquid chromatography continue to flourish. In the context of enzymic techniques the following resolutions were reported:

D- and L-forms from *N*-acetyltryptophan (using acylase);[521] L-Phe from *N*-acetyl-Phe-OEt (using aminoacylase);[522] de-racemisation of DL-amino acids (using L-amino acid oxidase[523] or D-amino acid oxidase[524]) followed by amino-borane or hydride reduction of the imino acids; all four diastereoisomers of 4, 4, 4-trifluorovaline and 5, 5, 5-trifluoroleucine by flash chromatography of derivatives followed by enzymic (porcine kidney acylase)[525] deacetylation of *N*-acetylated derivatives; resolution of *N*-Ac-DL-methionine (*Aspergillus*-derived amino acylase);[526,527] enantiopure α-methyl-β-alanine esters by lipase-catalysed (CAL-A and CAL-B) kinetic resolution. Chiral recognition of individual enantiomers forms the basis of a number of chiral hplc separations as listed in Table 1.

Chiral recognition of individual enantiomers is not restricted to CSP's for hplc however. Chiral recognition of D- and L-amino acids was seen[539] in the tandem mass spectrometry of Ni(II)-bound trimeric complexes. Some spontaneous chiral separation[540] of non-covalently bound clusters of amino acids occurred using soft-sampling electrospray ionisation. Amino acids such as hSer, 4OH-Pro, aThr and aIle, underwent guest-exchange reactions when complexed to permethylated β-cyclodextrin in the gas phase.[541] Neutral H-bonding receptors based on a *trans*-benzoxanthene skeleton have shown good stereoselective association towards carbamates of amino acids,[542] and *cis*-tetrahydrobenzoxanthene discriminated[543] between enantiomers of Z-amino

Table 1

Racemate	*Chiral Stationary Phase (CSP) or complex added to eluant*	*Refs.*
DL-amino acids	Cu(II) complexes of tetradentate diaminodiamido ligands in eluant	528
N-DNB-Leu-deriv.	Secondary and tertiary amide linked CSP's	529
DNP-amino acids	Carboxyethyl-β-cyclodextrin-coated-zirconia(CSP)	530
DNP-amino acid methyl esters	Silica-based CSP-derived from L-Ala and piperidyl cyanuric chloride	531
Aromatic amino acids	Amylose column (CSP)	532
1-Boc-amino 2, 3-diphenyl-l-cyclopropane COOH	Polysaccharide –derived CSP	533
DL-Amino acids	Diphenyl-disubstituted l, l′-binaphthyl crown ether (CSP)	534
DL-Amino acids	Chiral monolithic column (CSP)	535
Boc/Z/Fmoc aminoacid anilides	Modified commercial (*S*)-leucine (CSP)	536
DNP-aminoacid hexyl amide	Macrocycle from (R, R)-l, 2-diphenyl ethylamine and 5-allyloxyphthaloyl chloride on silica	537
DL-Amino acids	Pre-derivatisation by Marfey's reagent-a review	538

acid derivatives. Critical differences in chiral recognition of Z-DL-Asp and -Glu by mono- and bis(trimethylammonio)-β-dextrins have been reported.[544] Chiral molecular tweezers derived[545] from deoxycholic acid have been shown to form 1:1 inclusion complexes with methyl esters of D- and L-Phe and –Leu due to H-bonding and Van der Waals interaction. Cholic acid guanidines and carbamates have been found[546] to extract enantioselectively *N*-acetylated amino acids form aq. buffer into chloroform. Sixty references on enantiomeric amino acid recognition have been compiled in a review.[547] A series of acyclic thiourea derivatives[548] having four H-bond donors showed a moderate enantioselectivity towards *N*-protected amino acid carboxylate salts. (+)-(1*S*)-1, 1′-Binaphthalene-2, 2′-diyl hydrogen phosphate was known[549] to recognise L-α-amino acids and separating them by fractional crystallisation, and X-ray data now confirms the chiral space required for association.

Ultrafiltration through immobilised DNA membranes (channel-type) separated D- from L-phenylalanine,[550] and the same amino acid featured[551] in ultrafiltration using non-ionic micelles containing cholesteryl-L-glutamate at pH 7. Affinity ultrafiltration using bovine serum albumin as a stereoselective ligand has been studied[552] for the separation of D- and L-tryptophan. The D-form of alanine showed[553] preferential flux through a liquid membrane supported polypropylene hollow-fibre molecule using N-3, 5-dinitrobenzoyl-L-alanine octyl ester in toluene as a chiral selector. Enantiomeric discrimination between D- and L-amino acids has been shown[554] to occur using potential changes of optically active membranes. An enantioselective surface-imprinted poly(vinylbenzene)matrix using benzyldimethyl-*n*-tetradecyl ammonium chloride, could recognise[555] the chirality of *N*-protected glutamic acid, and a polypyrrole colloid on over oxidation[556] was able to show higher affinity for L- than for D-alanine.

(*RS*)-2-Benzoylamino-2-benzyl-3-hydroxypropanoic acid has been resolved[557] both by cinchonidine as resolving agent or by preferential crystallisation, while the optical purity[558] of enantioselective reactions can be improved through recrystallisation of Fmoc-α-amino acid *t*-butyl esters. In a dynamic kinetic resolution, (4*S*)-3-(2′-pyrrolidinyl)-3-oxo-2-methylpropanoate hydrochloride underwent[559] asymmetric hydrogenation in the presence of Ru[(*S*)-MeO-BIPHEP$_2$ to yield *anti*-β-hydroxy-α-methyl ester quantitatively. Using micellar electrokinetic chromatography,[560] eleven pairs of DL-amino acids derivatised with *o*-phthalaldehyde and *N*-acylcysteine could be separated using β-cyclodextrin. A kilogram of racemic Z-*t*-leucine could be resolved[561] by continuous chromatography on cellulose-based 'Chiralcel OD' while 0.5 kg Boc-*t*-leucine benzyl ester was resolved using 'Chiralpak AD'. Enantiopure (*R*)- and (*S*)-2-hydroxy-2-methyl-l-tetralone has been efficiently used[562] to epimerise/de-racemise amino acids. A first efficient and general non-enzymatic catalytic method[563] of synthesis involved the asymmetric alcoholysis of urethane *N*-carboxy anhydrides such as (**89**) catalysed by cinchona alkaloids. After its synthesis[564] RS-$\beta^{2,2}$-Bin (**90**) could be obtained in enantiomeric forms by benzoylation, coupling with L-Phe-cyclohexamide to give diastereoisomeric dipeptides which could be separated by chromatography. Further transformation led to the pure enantiomers, which are individually chirally stable due to axial dissymmetry.

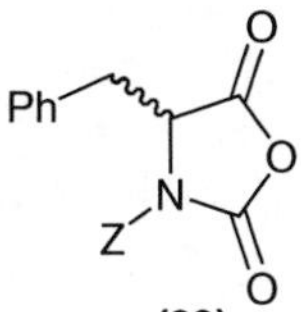

(89) 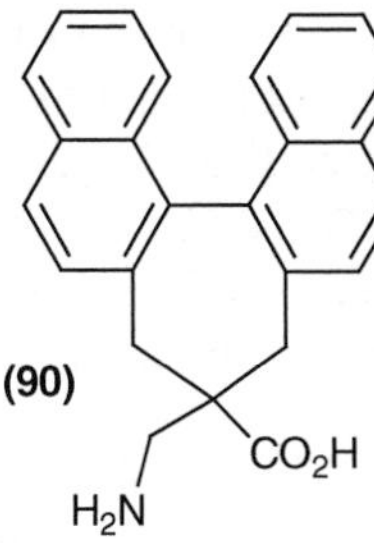 **(90)**

5 Physico-Chemical Studies of Amino Acids

5.1 X-Ray Crystal Analysis of Amino Acids and Their Derivatives. – A keyword scan of the literature for this sub-section brought in less than the usual harvest of papers, and the ones collected do show an overlap with discussions elsewhere on molecular recognition in this Chapter. Two nearly isomorphous complexes of *N*-acetyl-L-Phe-OMe and its amide with β-cyclodextrin (2:2) have been analysed[565] by X-ray crystallography, which found the former guest molecule showing dynamic disorder, while the latter showed evidence of having shifted position within the complex. No H-bonding existed between host dimer and the guest molecules but hydrophobic interactions involving the phenyl rings were seen[566] in both cases. A perturbation of the aromatic side chain of the guest molecule was seen in a room temperature crystal determination of the 2:2 *N*-acetyl *p*-methoxy-L-Phe-NH^2/β-cyclodextrin complex. Amongst the many physical techniques used to study[567] the selective recognition of amino acids by a novel crown ether (containing pyrillium, thiopyrillium and pyridinium moieties) was X-ray crystallography, which showed the conformation of the aromatic rings to be almost planar, but allowed selective binding to a selection of amino acids. The interaction[568] between a corrugated Langmuir film of cholesteryl-L-glutamate and various α-amino acids at the air-aq. solution interface has been examined by grazing incidence X-ray diffraction (GKD), and the incorporation of the 'guest molecule' within the host monolayer depended on hydrophobicity, shape and chirality of the solute molecules. Silver complexes of Asp, Gly and Asn have also been analysed by X-ray crystallography.[569]

5.2 Nuclear Magnetic Resonance Spectroscopy. – This technique is used widely in the general elucidation of structure, and finds itself as a major study vehicle in only a few papers which can be included in this sub-section. Host-guest NMR studies[570] between β-cyclodextrin and tryptophan enantiomers have revealed that the D-Trp is more strongly bound than its L-enantiomer, The chiral selector (18-crown-6)-tetracarboxylic acid, employed for resolution of amino acids also showed[571] chemical shift differences for D- and L-phenylglycine and the nature of the association between chiral selector and the 'guest molecules' was due to (i) 3 H-bonds in a tripod arrangement between the crown ether and the ammonium moiety (ii) hydrophobic interaction between polyether ring and Ph ring of the enantiomer and (iii) H-bonding between COOH of the crown ether and the CO oxygen of the D-enantiomer. A three-versus-two point attachment[572] of (*R*)- and (*S*)-amino acid methyl esters to chloro cobalt (III) tetramethyl chiroporphyrin has been used to explain the large diastereomeric dispersions observed with this novel chiral shift reagent. NMR spectroscopy in deuterated mesitylene solution has been used[573] to study the binding between a self-assembled cylindrical capsule and the 'guests', Boc-L-Ala and Boc-β-Ala esters. The most strongly bound to the capsule was Boc-β-Ala-OEt. A study[574] on modified β-cylodextrin,. 6 (A) deoxy 6(AH1) 4, 1, l0-tetreazacyclododecan-10-yl) β-cyclodextrin showed that it formed host-guest complexes with a number of amino acids but only showed enantioselectivity with tryptophan. The structure of liquid-crystalline phases formed from *N*-acetyl-L-glutamic oligopeptide benzyl esters have been examined[575] by NMR, the results indicating that the molecules were in an alignment as in a nematic mesophase.

Solid state NMR has been used[576] to study Lys, Arg and His intercalated into layered zirconium phosphates, and found the ε-NH_2 of Lys and the α-NH_2 of Arg and His interacted with P-OH of the phosphates. H-Bonding in amino acids and peptides as a function of temperature has been monitored[577] by solid state ^2H NMR.

5.3 Circular Dichroism. – CD and FT-IR have been used[578] to assess the secondary structure of poly(L-glutamic acid) segments and the binding of α-amino acids onto mixed monolayers. Induced CD signals have been observed[579] in synergistic binding of zwitterionic amino acids to lanthanide porphyrinate crown ethers.

5.4 Mass Spectrometry. – Electrospray mass spectrometry has been applied[580] to the study of Pd(II) complexes of amino acid-substituted calix[4]arenes. ESI-TOF-MS was used[581] in the chiral recognition of D- and L-Tyr and –Trp by cyclodextrins. It was found that D-isomers were more strongly bound than were the L-forms.

5.5 Other Spectroscopic Studies on Amino Acids. – The adsorption of *S*-proline vacuum-deposited on clean Cu(l 1 0) has been investigated[582] using reflective absorption IR and low energy electron diffraction (LEED). Throughout the adsorption, proline bonds to the Cu surface via its COO and N–H functionalities.

5.6 Measurements on Amino Acids in Solution. – Thermodynamic properties of a number of amino acid situations in solution have been reported and include: dissociation constants of Gly in ethanol/water[583]; stability constants for complexation of dioxovanadium with Glu in methanol/water[584]; mixing enthalpies of Gly, Ala, Ser, Pro, Thr and Val with 1,3-dimethylurea[585] and with monomethylurea[586] in aq. solution; densities and sound velocities of Gly in aq. nickel sulphate;[587] enthalpies of solution for α-aminobutyric acid in aq. alkali metal halide solutions;[588] free energies, enthalpies and entropies of transfer of amino acids from water into sodium sulphate;[589] stability constants for complex formation of Cd-amino acid-Vit B systems;[590] enthalpies of dilution of Gly, Ala and Ser in aq. ethylene glycol solution;[591] Gibbs Free energies of Gly zwitterions in hydro-organic solvents (DMSO, EtOH, dioxan and acetonitrile);[592] enthalpies of aq. solutions of Cys, His, Asn, Arg, Trp and Glu in order to determine homogeneous pair interaction coefficients;[593] diffusion coefficients of Pro, Thr and Arg in water at 25°C;[594] activity coefficients of Gly, Ser and Val in aq, sodium and potassium nitrate.[595]

Amongst other parameters reported were; solubility polytherm for Phe-NH_4HPO_4-H_2O systems;[596] extraction equilibrium constants of leucine with di(2-ethylhexyl)phosphonic acid;[597] constant volume combustion energies for Zn-L-threonate and Ca-L-threonate[598]; partial molar volumes of Gly, Ala, Val, Leu and Phe in aq. glycerol solution;[599] stability constants of Cu (II)-glycine complex in mixed solvents;[600] density, viscosity, solubility and diffusivity of N_2O in aq. amino acid solutions;[601] viscosities of amino acid-urea –water systems;[602] activity coefficients of amino acids in variable dielectric permeability media;[603] partial molar volumes of Gly-Gly and Ser at elevated temperatures and pressure[604] and *N*-acetyl-*N*-methylamino acid amides in aq. solution;[605] effect of pH on the diffusion coefficient of Cu(H) ions in Gly and β-Ala aq. Solutions;[606] apparent molar volume and compressibility of Gly in aq. vanadyl sulphate;[607] dissociation functions of Gly and β-Ala in 2-propanol/water mixtures;[608] apparent and partial molar heat capacities and volumes of L-Lys and L-Arg hydrochlorides in aq. solution;[609] water activity, pH and density of aq. amino acid solution;[610] partition coefficients of amino acids in PEG-4000/sodium sulphate 2-phase systems.[611]

Lysine has been extracted with a combination of ion-exchange and membrane ultrafiltration,[612] while N-cholyl amino acid alkyl esters were found[613] to act as potent organogelators for aromatic solvents and cyclohexene. Special surface and aggregation behaviour of the amphiphile *N*-(decyloxy-2-hydroxybenzylidene)glycine has been investigated,[614] and the reactivity of radical dications of protonated amino acids in microsolutions has been investigated.[615] The hydration of amino acids has been found[616] to depend on ion form, the anions of neutral amino acids being the most hydrated. The spatial effect of hydrophobic groups in amino acids on the volume phase transitions of hydrogels has been investigated[617] and new lysine derivatives[618] with positively charged terminal groups could gel water below 1 wt.%. In a survey of chiral aggregation[619] acyl amino acids were found to be present as monomers in acetonitrile, and in a molecular dynamics calculation[620] of solvation properties of non-polar amino acids, the results show that the solvation structure around the amino acids is richer for methanol than for water. Protonation constants and solvation of some α-amino acid methyl and ethyl esters in ethanol-water mixtures have been determined[621] using potentiometry, and the effect of amino acids on the dynamics of water has been studied.[622] Hydration characteristics of aromatic amino acids have been investigated by an isopiestic method[623] and the effect of additives (ammonium sulfate and dextrose) on the transformation behaviour of phenylalanine has been studied using powder X-ray diffraction.[624] The properties of hydroxy-glycine in aq. solution have been explored[625] and Gly and Lys in a mixed aq. solution[626] demonstrated buffering action, and there was proton transfer from Gly to the Lys zwitterions.

Cystine, being naturally the least soluble of the amino acids, has been subjected[627] to a constant ionic medium of NaCl at different concentrations to study its protonation equilibria, and solvent effects on the conformational behaviour[628] of acetylated amino acids revealed significant differences dependent on the solvent used (DMSO, D_2O, hexafluoroisopropanol or CH_2Cl_2), as determined by IR techniques. The pH dependence of the anisotropy factors (g) for the essential amino acids has been assessed[629] and it was found that g factors at pH 1 were 2–3 times those at pH 7. The results[630] of subjecting Gly to super- and sub-critical water conditions to simulate submarine hydrothermal conditions showed that $(Gly)_2$, $(Gly)_3$ and dioxopiperazine were found in the reaction mixtures. Glycine also decomposed[631] under high temperature and pressure water, by decarboxylation and methylamine formation or by production of ammonia and an organic acid.

The Stark-effect of spectral holes burnt into the long wavelength absorption of phenylalanine in glycerol-water glass showed[632] two protonation-deprotonation transitions. Phenylalanine and aspartic acid can be separated in aq. solution using nanofiltration.[633] Using glycine methyl ester, it has been shown[634] that on forming an iminium adduct with acetone the α-carbon experienced a 7 pK unit increase in acidity, and NMR evidence showed[635] that in α-(benzotriazol-l-yl)-*N*-acylglycines the benzotriazole ring can mop up protons from the ionisation of the carboxyl groups.

5.7 Measurements on Amino Acids in the Solid State. – X-Ray, cyclic voltammetry and IR spectroscopic evidence concurred[636] that the carboxyl group in phthalimido acetic acid preorientates prior to photodecarboxylation by photo induced electron transfer. The crystallisation rates of L-glutamic acid were retarded by the addition of Val, Leu, Ile and Nle, with the effect being larger for Val.[637] Tryptophan was selected[638] for measurement of its permanent electrical dipole in a molecular beam, using a MALDI source coupled to an electron beam deflection setup, and the experimental value agreed with the lowest energy conformation found by calculation. In a study of precipitating agents[639] 3, 4-dimethylbenzene sulfonic acid was highly selective for leucine, and flavianic acid was good for arginine. A novel batch crystallizer[640] has successfully produced large crystals of aspartic acid, while growing glycine crystals at an air-aq. solution interface[641] initiated the occlusion of one α-amino acid enantiomer from an added racemate resulting in enrichment in solution of the other isomer. The latter spontaneous separation of enantiomers has been termed ‘chemistry in 2D’.

IR and Inelastic Neutron Vibrational Spectroscopy have been used[642] to examine both the internal and external vibration in crystalline-alanine, and for the first time a splitting of the NH_3^+ torsional band has been seen at a temperature below 220 K. Atomic force microscopy has been applied[643] to study the blocking behaviour of step motion on the (1 0 0) face of an Asp crystal when doped with other amino acids (Ala, Lys, Phe, Asn and GIu). Scanning tunnel microscopy (STM) data[644] on lysine adsorbed on graphite showed that some of its CH groups were located between the hexagonal lattice of the graphite, while the other groups rise above the surface. STM investigations have also been carried out on lipid amino acids on pyrolytic graphite,[645] and on the adsorption and assembly of L-tryptophan on to the Cu (0 0 1) surface.[646]

5.8 Amino Acid Adsorption and Transport Phenomena. – A short review[647] has been noted entitled, 'Adsorption of amino acids at solid/liquid interfaces', and the effect of pH and aluminium content of zeolites on their adsorption and separation of amino acids from aq. solution has been surveyed.[648] Selective adsorption of groups of amino acids occurred from aq. solution using MCM-41 mesoporous molevular sieves.[649] 'Acidic' amino acids were hardly adsorbed, 'basic' amino acids showed high affinity, while adsorption of 'neutral' amino acids increased with side-chain length. The effect of solvophobic interactions[650] on the molecular sorption of four amino acids from water/non-electrolyte binary systems onto mono carboxycellulose has been investigated, and amino acids tested[651] for binding onto calcium materials found in human kidney stones (calcium oxalate, $CaHPO_4.2H_2O$ and Ca_3PO_4), showed maximum binding at pH 5, with the 'acidic' amino acids showing the strongest adsorption.

An investigation[652] for optimising the ion-exchange extraction of amino acids from an L-proline culture liquid has been carried out. Mixed-ligand Cu (II) and Ni (II) chelates have been shown[653] to extract phenylglycine, phenylalanine and tryptophan into dichloroethane from aq. solution. Phenylalanine has been used[654] as test molecule for studying the effect of physical factors on its extraction by Aerosol-OT reverse micelles, and calix[6]arene carboxylic acid derivative was found[655] to be the strongest extractant of the target tryptophan ester in an aliphatic organic solvent. The micellar extraction of tryptophan and tyrosine, using *N*-*n*-dodecyl-L-proline and *trans* *N*-*n*-dodecyl-4-hydroxyproline as chiral selectors with Cu(II) ions has been evaluated,[656] theoretically and experimentally, and a very similar approach has been applied to study[657] the conditions needed for enantiomeric separations by dense permeation-selective membranes, using the diffusion of phenylalanine through polypropylene beads coated with *N*-dodecyl-L-hydroxyproline:Cu(II), which showed a selectivity value of 1.25 (D/L). The transport of Phe, Tyr and Trp in a buffer solution through Aerosol-OT reverse micelles, has been studied[658] and the conclusion drawn that Tyr does not cross the membrane, but transport rates for Trp and Phe are of the same order. The selectivity of an activated composite membrane containing bis-(2-ethyl-hexyl) phosphoric acid for aromatic amino acids has been shown[659] to be Trp > Phe > Tyr, and when Langmuir monolayers of N-stearoyl glutamic acid have been investigated by π-A measurements and Atomic Force Microscopy, layers built up from aq $CdCl_2$ solution could aeccomodate the L-*N*-stearoyl glutamic acid better than the racemate.

Distribution coefficients of isomorphic amino acids between a crystal phase and aq. solution have been worked out[660] and indicated that the extent of impurity in a crystal is related to the ratio of the pure compound solubility of the primary solute to that of the impurity in the same solvent Diffusion measurements using NMR[661] have enabled the association constants for weak interactions between cyclodextrin and guest molecules (Phe, Leu and Val) to be worked out, while chiral discrimination between DL-dansyl amino acids and immobilised teicoplanin[662] rely on hydrophobic interactions and H-bond formation.

5.9 Host-Guest Studies with Amino Acids. – Aspects of the complexation of amino acids with macrocyclic receptors such as crown ethers, cryptands, calixerenes have been reviewed,[663] and recent progress on the synthesis/molecular recognition of amino acids/peptides has been reported.[664] The cyclodextrins and crown ether derivatives vie for the top popularity spot as host over this review period and will be reported on first in this sub-section.

Examples having cyclodextrin as host receptor include, calorimetric and NMR studies[665] which showed a direct correlation between complex structure and the thermodynamics of Z-L-Asp and Z-L-Glu inclusion complexes with mono- and bis-(trimethylammonio)-β-cyclodextrins. A β-cyclodextrin derivative bearing a pyridinio on the primary side was synthesised[666] and its complexation stability constants with several aliphatic amino acids determined, with the highest enantioselectivity shown for serine (D/L 5.4). *R*-(−)-2-Phenylglycinol–modified β-cyclodextrin was synthesised[667] under microwave irradiation and showed high chiral discrimination when complexed with amino acids. β-Cyclodextrin bearing nicotinic or isonicotinic moieties have been synthesised[668] and their inclusion complexes with L/D tryptophan, analysed fluorometrically, showed a preference for the D-enantiomer. A kinetic study[669] using ultrasonic relaxational methods on a β-cyclodextrin/L-isoleucine system implied that the departure of the guest molecule from the host cavity was influenced by the amino acid structure. A theoretical study[670] using a fast annealing evolutionary algorithm (FAEA) on the interactions of amino acids with α-cyclodextrin revealed that of the four pairs of L/D-amino acid compared, interaction energies were lower for the L-amino acids than the D-forms which is in agreement with experimental results. Studies have been reported[671] on the separation of the enantiomers of Leu, Val, Tyr and Phe on thin layer chromatography plates modified by β-cyclodextrin, and fluorescence enhancement was recorded[672] in all the examples of coumarin-6-sulfonyl amino acid derivatives on forming inclusion complexes with cyclodextrin. In the latter technique glycine showed the greatest enhancement, tyrosine the least. The association of dansyl amino acids with permethylated β-cyclodextrin has been further investigated[673] using Na^+ as a RPLC retention marker.

Chiral 15-metallacrown-5-complexes, based on L-phenylalaninehydroxamic acid, copper diacetate and lanthanum or gadolinium trinitrate have been prepared and characterized.[674] They adopt a dimeric structure in the solid state, and show selective binding of carboxylate ions. Chiral recognition[675] of carboxylic acids was a feature found in bis-crown ether peptides when complexed with Z-Phe-OH or Z-Ala-OH. The thermodynamics of the complexing of L-Ala-OMe.HCl, L-Phe-OMe.HCl and L-Val-OMe with a series of crown ethers has been reported[676] and the complexation of similar protonated amino acid methyl esters with 18-crown-6 and benzo-18-crown-6 has been studied using calorimetric titrations.[677] 18-Crown-6-tetracarboxylic acid, when used as a chiral additive, has provided[678] simultaneous separation of *o*-, *m*-, *p*-enantiomers of tyrosine and fluorophenylalanine by capillary electrophoresis. Synergistic binding and chiral recognition of unprotected amino acids were properties found[679] for a 18-crown-6 ring connected with a carboxyl group via a ferrocene spacer.

Interactions between calix[4]resorcinarene and amino acids have been studied[680] in Langmuir films and chiral recognition turned out to be a bit patchy depending upon sub-phase conditions. However gas-phase proton bound complexes between chiral resorcin[4]ene and enantiomers of Ala and Ser underwent[681] exchange with the enantiomers of 2-butylamine with significant enantioselectivity. *p*-Tetra *tert*-butyl calix[4]arene derivatives bearing chiral bicyclic guanidinium moieties have served[682] as receptors of amino acid zwitterions and showed selectivity for L-aromatic amino acids. Complexes of tetra-*p*-sulfonated calix[4]arene with racemic Ala, His and Phe, and (*S*)-forms of Ala, His and Tyr have been analysed by X-ray diffraction.[683] The racemates were found in capsules within the host in a bilayer arrangement, while the enantiomers formed a 1:1 complex within the bilayer.

A new kind of threonine-modified porphyrinato zinc(II) host has been shown to form[684] 1:1 and 1:2 adducts with amino acid esters, while a commercially available Zn (II) protoporphyrin has had its binding constants with a series of amino acids analysed by UV-VIS titrations.[685] Three novel chiral zinc porphyrins with protected chiral amino acid substituents, showed[686] enantioselectivity towards amino acid methyl esters, and spectroscopic studies[687] have been carried out on the interactions of Co (II), *N*, *N′*, *N″*, *N‴*-tetramethyltetra-3, 4-tetrapyridinoporphyrazine with amino acids and nitrogen oxides.

Some further selectivity has been incorporated into a pyridyl host molecule with four phosphonate groups[688] attached, which now binds to basic amino acid esters in water. A UV-VIS spectral investigation[689] of chiral SalenCo towards four pairs of enantiomeric amino acid esters showed that 1:1 complexes were formed and the associative constants for the molecular recognition processes were in the order KD > KL > K(LeuOMe) > K(AlaOMe) > K(SerOMe) > K(TyrOMe). The thermodynamic stereoselectivity involved in the chiral recognition of amino acids by the Cu (II) complex of 6-deoxy-6-[4-(2-aminoethyl)imidazolyl]cyclomaltoheptaose has been investigated[690] by a number of techniques. Aromatic amino acids showed stronger binding characteristics with the L-form being favoured. An EPR and molecular mechanics study[691] of the influence of amino acid side chains on water binding to Cu (II) bis (*N*, *N*-dimethyl-L-isoleucinato) has been carried out, and a 12-reference review has highlighted[692] the intercalation of amino acids and sugars in anionic clays.

A series of unsymmetrical tris-amide receptors, which show a particular affinity for *N*-acetylglycine have been synthesized,[693] and adsorption isotherms for amino acids in BEA zeolites have been analysed.[694] Good chiral recognition properties for enantiomeric amino acid derivatives have been shown[695] by newly synthesised chiral imidazole cyclophane receptors, and enantiomeric recognition[696] could be visualised through development of a purple colour when chiral phenolphthalein derivatives were explored. A synthesised poly (L-glutamic acid) segment grafted on the third generation poly (amidoamine) dendrimer[697] allowed amino acids into its inner core, and showed a preference for the D-forms of Trp, Phe or Tyr.

The thermodynamics of binding of (*R*)- and (*S*)-DNB-Ieueine to cinchona alkaloids and their t-butylcarbamates have been assessed using a number of techniques.[698] The direct intercalation of amino acids into layered double hydroxides of Mg-Al, Mn-Al, Ni-Al and Zn-Al and Zn-Cr has been studied[699] and shown to be dependent on pH and the structure of the amino acid side chains. Reaction of glycine with ninhydrin has been catalysed[700] by cationic micelles, of cetyltrimethylammonium bromide and cetylpyridinium bromide, and the chromatographic performanee[701] in chiral recognition by polymeric amino acid based surfactants has been monitored.

5.10 Theoretical Calculations involving Amino Acids. – Some subject matter covered here will also have found its way into other areas of the Chapter, as theoretical interpretations nowadays are published side by side with experimental deductions. Improved agreement[702] between calculated values and experimental data for the partial molar volume of the 20 amino acids, has been achieved by combining the bridge-corrected ID reference interaction site model (1D-RISM) with the Kirkwood-Buff theory, and even further improvement was gained using a three-dimensional 3D-RISM. The ability of the GROMOS96 force field to reproduce partition constants between water and cyclohexane/chloroform for analogues of 18 of the amino acids has been investigated,[703] and found to work well for non-polar analogues, but overestimates the free energies of polar analogues in water.

Interactions in the contact region of the trypsin-pancreatic trypsin inhibitor complex have been evaluated using simulation methods and thermodynamic cycles,[704] and electrodiffusion of amino acids through fixed charge membranes has been modelled[705] using Nernst-Planck flux equations in the (Goldman) constant electric field assumptions. The formation of various cation radical structures in the irradiated L-Ala crystal has been simulated[706] using a 208-atom cluster, and the relative total energies and equilibrium geometries of various radical conformations were obtained at the PM3 level. The geometries of the 20 genetically encoded amino acids have been optimised[707] at the restricted Hartree-Fock level of theory using 6-31+G* basis set, which turned out to be in excellent agreement with those determined by X-ray crystallography.

6. Reactions and Analysis of Amino Acids

6.1 General and Specific Reactions of Amino Acids. – Many of the items associated with this topic are really dealt with in other Chapters of this Volume, e.g functional group protection for peptide synthesis (Chapter 2) and metal complexes of amino acids (Chapter 5). So the few papers that remain unattached to other 'sub-sections' are discussed here. The processes involving amino acids, which are catalysed by pyridoxal in nature have been mimicked by pyridoxal analogues such as (**91**) and (**92**) in an attempt to design catalysts that can be applied to organic reactions.[708] Compounds (**91**) and (**92**) steered the reactions of amino acids away from transamination or racemisation towards decarboxylation, retroaldol reaction and aldol reaction. Some enantioselectivity was seen during the protonation of the newly created amino acid carbon chiral centre, but this aspect needs enhancing. Six α-amino acids (Gly, Ala, Abu, iAbu, Val and norVal), β-Ala, and β-aminobutyric acid and γ-amino butyric acid have been subject to nitrosation conditions in aq. media,[709] mimicking the conditions of the stomach lumen. Conclusions there were: dinitrogen trioxide was the main nitrosating agent in aq. media; the reactivity order was α- > β- > γ-amino acids. The kinetics of the oxidation of amino acids by chloramines T in the presence of Fe(II) ion has been studied[710] in aq. sulphuric acid, and show almost identical behaviour with all simple amino acids. Several amino acids were transformed[711] into alkyl esters using triphosgene (trichloromethyl carbonate), and acylation of aromatic amino acids with furancarboxylic acid chlorides was carried out effectively[712] in acetone/water at pH 8–9. In the latter process aliphatic amino acids had to be acylated as methyl esters and then hydrolysed.

To clarify the differences between the reactivity of the γ- and α-COOH's in *N*-phosphorylated glutamic acid, MNDO calculations have been carried out[713] to mimic the results of mixed anhydride activation. Theory and practice agree that it is the α-COOH, via the intermediate 5-membered phosphoric-carboxylic mixed anhydride, that is the most readily activated. A new Fmoc-protected dipeptide isostere, named BTS, (**93**) has been synthesised[714] in 9 steps (11% yield) from *R*, *R*-tartaric acid and *O*-benzyl-L-serine, for use as a peptidomimetic. Chlorination of *N*-acetyl-L-tyrosine with NaOCl and the myeloperoxidase chlorinating system has been investigated[715] and the position of chlorination in the aromatic ring (position 3 and 5) found to be dependent on the reactant concentration ratio and on the pH. FT-IR Spectroscopy has been used to study[716] the decarboxylation of alanine at 280–330°C, as a function of pH, where the rate of decomposition of the zwitterions was found to be 3 × greater than that of the cationic and anionic forms. Pyrolysis of Asn, Pro and Trp has been carried out[717] and the relative formation of polycyclic aromatic amino acids in the products assessed. Seven amino acids have been assessed[718] as substrates in Belousov-Zhabotinskii oscillation reactions, with mixed results. Asp exhibited typical oscillations, Tyr exhibited oscillations where the metal ion catalyst was not necessary, Cys-Cys, Ala, Gly and Glu gave sustained oscillations only after addition of acetone, while serine showed oscillations even in the absence of acetone. Selective hydroxylation[719] of amino acids in water has been possible with the aid of a catalytic system made up of 5mol % K_2PtCL_4 with 7 eq. $CuCl_2$. The results suggested that the functionalisation of α-amino acids is via a chelate-directed (involving α-NH_2 and COOH groups) C–H bond activation.

CHO SR HO N

(91) R =

(92) R = $-(CH_2)_2-$

HO_2C Fmoc—N O O CO_2Me BTS **(93)**

6.2 Analysis of Amino Acids. – Keyword scanning for papers in this area have brought in very few within this period of reporting, which is out of line with previous periods. This situation could be interpreted in many ways, with the possibility that most analytical methods have over the years been applied to amino acid analysis. The final judgement will have to await the next Volume of these reports to find out what the success rate was during 2003–04.

Better resolution of amino acids by reversed-phase HPLC has been possible using C-18 reverse-phase packing, dynamically coated with 2-aminotetraphenyl porphyrin and in the presence of Zn (II) ion. Twelve of the 20 amino acids could be resolved under isocratic conditions.[720] Ethyl chloroformate in an aq. medium was the derivatisation method[721] chosen in a GC-MS analysis of amino acids and other acids used in artistic paintings. Microchip devices[722] integrating electrophoretic separations, enzymic reactions (amino acid peroxidase), and amperometric detection (of hydrogen peroxide) have been developed.

References

1. *Amino Acids Peptides and Proteins*, eds. G.C. Barrett and J.S. Davies, Royal Society of Chemistry, 2003, Vol. 34.
2. C.A. Selects on *Amino Acids Peptides and Proteins*, published by the American Chemical Society and Chemical Abstracts Service, Columbus, Ohio.
3. '*The ISI Web of Knowledge Service for UK Education*' on http://wok.mimas.ac.uk.
4. T. Lindel, *Nachr. Chem.*, 2000, **48**, 790.
5. D.J. Ager and I.G. Fotheringham, *Curr. Opin. Drug Discovery Dev.*, 2001, **4**, 800.
6. T. Abellan, R. Chinchilla, N. Gallindo, G. Guillena, C. Najera and J. M. Sansano, *Targets in Het. Systems*, 2000, **4**, 57.

7. G. Dyker, *Org. Synth. Highlights IV*, ed. H-G.Schmaltz, Wiley-VCH Verlag, 2000, 53.
8. Y.-h. Cheng, X.-m. Zou, C. Wu and H.-z. Yang, *Jingxi Huagong Zhongjianti*, 2001, **31**, 1.
9. A.S. Bommarius, M. Schwarm and K. Drauz, *Chimia*, 2001, **55**, 50.
10. A. Taggi, A.M. Hafez and T. Lectka, *Acc. Chem. Res.*, 2002, **36**, 10.
11. M.G. Natchus and X. Tian, *Org. Synth: Theory and Applications*, 2001, **5**, 89.
12. D.J. Ager, *Curr. Opin. Drug Discovery Dev.*, 2002, **5**, 892.
13. E. Kimura, *Adv. Biochem.Eng/Biotech.*, 2003, **79**, 37.
14. M. Ikeda, *Adv. Biochem.Eng/Biotech.*, 2003, **79**, 1.
15. W. Pfefferle, B. Mockel, B. Bathe and A. Marx, *Adv. Biochem.Eng/Biotech.*, 2003, **79**, 59.
16. V.G. Debabov, *Adv. Biochem.Eng/Biotech.*, 2003, **79**, 113.
17. U. Mueller and S. Huebner, *Adv. Biochem.Eng/Biotech.*, 2003, **79**, 137.
18. S.S. Kinderman, J.W. van Beijma, L.B. Wolf, H.E Schoemaker, H.R. Hiemstra and F.P.J.T. Rutjes, *Proc. ECSOC-4*, ed E. Pombo-Villar, Mol. Diversity Preservation International, 2000, 709.
19. T. Kimachi, *Farumashia*, 2001, **37**, 920.
20. F.P.J.T. Rutjes, L. B. Wolf and H. E. Schoemaker, *J. Chem. Soc. Trans 1*, 2000, 4197.
21. A. Bianco, T. DaRos, M. Prato and C. Toniolo, *J. Pept. Sci.*, 2001, **7**, 346.
22. M.D. Gieselman, Y. Zhu, H. Zhou, D. Galonic and W.A. Van der Donk, *ChemBiochem.*, 2002, **3**, 709.
23. Y.-X. Fang, Y.-L. Wong, X.-J. Xiong and K. Zhan, *Guangdon Gongye Daxue*, 2002, **19**, 7.
24. V.A. Soloshonok, *Curr. Org. Chem.*, 2002, **6**, 341.
25. G. Kusano, S. Orihara, D. Tsukamoto, M. Shibano, M. Coskun, A. Guvenc and C. S. Erdurak, *Chem. Pharm. Bull.*, 2002, **50**, 185.
26. M.J. O'Donnell, *Aldrichim. Acta*, 2001, **34**, 3.
27. H.S. Knowles, K. Hunt and A.F. Parsons, *Tetrahedron*, 2001, **57**, 8115.
28. G. Cardillo, L. Gentilucci, M. Gianotti and A. Tolemelli, *Tetrahedron*, 2001, **57**, 2807.
29. P. Allevi, G. Cighetti and M. Anastasia, *Tetrahedron Lett.*, 2001, **42**, 5319.
30. C. Tanyeli and B. Sezen, *Enantiomer*, 2001, **6**, 229.
31. S. Kobayashi, R. Matsubara and H. Kitagawa, *Org. Lett.*, 2002, **4**, 143.
32. A. Cordova, W. Notz, G. Zhong, J.M. Betancort and C.F. Barbas III, *J. Am. Chem. Soc.*, 2002, **124**, 1842.
33. A. Cordova, S. Watanabe, F. Tanaka, W. Notz and C.F. Barbas III, *J. Am. Chem. Soc.*, 2002, **124**, 1866.
34. D. Ferraris, B. Young, C. Cox, T. Dudding, W.J. Drury III, L. Ryzhkov, A.E. Taggi and T. Lectka, *J. Am. Chem. Soc.*, 2002, **124**, 67.
35. G.C. Lloyd-Jones, *Angew. Chem. Int. Ed.*, 2002, **41**, 953.
36. A.-M. Yim, Y. Vidal, P. Viallefont and J. Martinez, *Tetrahedron Asymmetry*, 2002, **13**, 503.
37. T.-S. Huang and C.-J. Li, *Org. Lett.*, 2001, **3**, 2037.
38. T.-S. Huang, C.C.K. Keh and C.-J. Li, *Chem. Commun. (Cambridge)*, 2002, 2440.
39. A. de la Hoz, A. Diaz-Ortiz, M.V. Gomez, J.A. Mayoral, M.A. Sanchez-Migallon and A.M. Vazquez, *Tetrahedron*, 2001, **57**, 6421.
40. W.L. Scott, M.J. O'Donnell, F. Delgado and J. Alsina, *J. Org. Chem.*, 2002, **67**, 2960.

41. B. Alcaide, P. Almendros and C. Aragonolo, *Chem.-A Eur. J.*, 2002, **8**, 3646.
42. (*a*) J.M. Drabik, J. Achatz and I. Ugi, *Proc. Estonian Acad. Sci.*, 2002, **51**, 156; (*b*) G. Ross and I. Ugi, *Can. J. Chem.*, 2001, **79**, 1934.
43. M. Kitamura, D. Lee, S. Hayashi, S. Tanaka and M. Yoshimura, *J. Org. Chem.*, 2002, **67**, 8685.
44. S.D. Bull, S.G. Davies, A.C. Garner and N. Mujtaba, *Synlett*, 2001, 781.
45. S.D. Bull, S.G. Davies, M.D. O'Shea, E.D. Savoury and E. Snow, *J. Chem. Soc. Perkin Trans 1*, 2002, 2442.
46. S.G. Davies, S.W. Epstein, O. Ichihara and A.D. Smith, *Synlett*, 2001, 1599.
47. S.D. Bull, S.G. Davies, M.D. O'Shea and A.C. Garner, *J. Chem. Soc. Perkin Trans 1*, 2002, 2442.
48. L.M. Harwood, S.N.G. Tyier, M.G.B. Drew, A. Jahans and I.D. MacGilp, *ARKJVOC*, 2000, **1**, 820.
49. L.M. Harwood, M.G.B. Drew, D.J. Hughes and R.J. Vickers, *J. Chem. Soc. Perkin Trans 1*, 2001, 1581.
50. P.-F. Xu, Y.-S. Chen, S.-I. Lin and T.-J. Lu, *J. Org. Chem.*, 2002, **67**, 2309.
51. P.-F. Xu and T.-J. Lu, *J. Org. Chem.*, 2002, **68**, 658.
52. S.K. Lee, J. Nam and Y.S. Park, *Synlett*, 2002, 790.
53. H. Miyabe, A. Nishimura, M. Ueda and T. Naito, *Chem. Commun. (Cambridge)*, 2002, 1454.
54. (*a*) A.S. Sagiyan, A.A. Petrosyan, A.A. Ambartsumyan, V.I. Maleev and Y.N. Belokon, *Hayastani Kimiak. Handes*, 2002, **55**, 150; (*b*) Y. N. Belokon, V.I. Maleev, A.A. Petrosyan, T.F. Savel'eva, N.S. Ikonnikov, A.S. Peregudov, V.N. Khrustalev and A.S. Saghiyan, *Russ. Chem. Bull.*, 2002, **51**, 1593.
55. P. Merino, J. Revuelta, T. Tejero, U. Chiacchio, A. Rescifina, A. Piperno and G. Romeo, *Tetrahedron Asymmetry*, 2002, **13**, 167.
56. P. Merino, J.A. Mates, J. Revuelta, T. Tejero, U. Ciacchio, G. Romeo, D. Lannazzo and R. Romeo, *Tetrahedron Asymmetry*, 2002, **13**, 173.
57. Y.N. Belokon, K.A. Kochetov, T.D. Churkiana, N.S. Ikonnikov, O.V. Larionov, S.R. Harutyunyan, S. Vyskocil, M. North and H.B. Kagan, *Angew. Chem. Int. Ed.*, 2001, **40**, 1948.
58. H.J. Kim, S.-K. Lee and Y.S. Park, *Synlett*, 2001, 613.
59. B. Thierry, J.C. Plaquevent and D. Cahard, *Tetrahedron: Asymmetry*, 2001, **12**, 983.
60. R. Chinchilla, P. Mazon and C. Najera, *Tetrahedron: Asymmetry*, 2002, **13**, 927.
61. M.J. O'Donnell, M.D. Drew, J.T. Cooper, F. Delgado and C. Zou, *J. Am. Chem. Soc.*, 2002, **124**, 9348.
62. H.-g. Park, B.-S. Jeong, M.-S. Yoo, J.-H. Lee, M.-k. Park, Y.-J. Lee, M.J. Kim and S.-s. Jew, *Angew. Chem. Int. Ed.*, 2002, **41**, 3036.
63. P. Mazon, R. Chinchilla, C. Najera, G. Guillena, R. Kreiter, G. Klein, J.M. Robertus and G. van Koten, *Tetrahedron Asymmetry*, 2002, **13**, 2181.
64. B. Lygo and L.D. Humphreys, *Tetrahedron Lett.*, 2002, **43**, 6677.
65. F. Royer, F.-X. Felpin and E. Doris, *J. Org. Chem.*, 2001, **66**, 6487.
66. A. Cooke, J. Bennett and E. McDaid, *Tetrahedron Lett.*, 2002, **43**, 903.
67. T.B. Durham and M.J. Miller, *J. Org. Chem.*, 2002, **68**, 27.
68. Y.K. Chen, A.E. Lurain and P.J. Walsh, *J. Am. Chem. Soc.*, 2002, **124**, 12225.
69. M. Ostermeier, J. Priess and G. Helmchen, *Angew. Chem. Int. Ed.*, 2002, **41**, 612.
70. M. Adamczyk, S.R. Akireddy and R.E. Reddy, *Tetrahedron*, 2002, **58**, 6951.
71. F.-Y. Zhang, W.H. Kwok and A.S.C. Chan, *Tetrahedron: Asymmetry*, 2001, **12**, 2337.

72. W. Wang, C.Y. Xiong and V.J. Hruby, *Synthesis-Stuttgart*, 2002, **94**.
73. Y.-S. Lin and H. Alper, *Angew. Chem. Int. Ed.*, 2001, **40**, 779.
74. G. Antoni, H. Omura, M. Ikemoto, R. Moulder, Y. Watanabe and B. Langstrom, *J. Labelled Compd. and Radiopharm.*, 2001, **44**, 287.
75. W. Augustyniak, R. Kanski and M. Kanska, *J. Labelled Compd. and Radiopharm.*, 2001, **44**, 553.
76. E.B. Watkins and R.S. Phillips, *Biorg. Med. Chem.Lett.*, 2001, **11**, 2085.
77. (*a*) W. Augustyniak, P. Suchecki, J. Jemielity, R. Kanski and M. Kanska, *J. Labelled Compd. and Radiopharm.*, 2002, **45**, 559; (*b*) J. Jemielity, R. Kanski and M. Kanska, *ibid.*, 2001, **44**, 295.
78. A.J.H. Vadas, I.M. Schroeder and H.G. Monbouquette, *Biotech. Prog.*, 2002, **18**, 909.
79. M. Wu, P. Wei and H. Zhou, *Huaxue Shijie*, 2002, **43**, 476.
80. B. Sauvagnat, F. Lamaty, R. Lazaro and J. Martinez, *Tetrahedron*, 2001, **57**, 9711.
81. M.J. O'Donnell, F. Delgado, E. Dominguez, J. de Blas and W.L. Scott, *Tetrahedron Asymmetry*, 2001, **12**, 821.
82. T. Dudding, A.M. Hafez, A.E. Taggi, T.R. Wagerle and T. Lectka, *Org. Lett.*, 2002, **4**, 387.
83. C. Xiong, W. Wang and V.J. Hruby, *J. Org. Chem.*, 2002, **67**, 3514.
84. G.-x. Li, F.-m. Mei and Y. Juan, *Jingxi Huagong*, 2001, **18**, 579.
85. T. Morita, Y. Nagasawa, S. Yahiro, H. Matsunga and T. Kunieda, *Org. Lett.*, 2001, **3**, 897.
86. C. Ma, X. Liu, X. Li, J. Flippen-Anderson, S. Yu and J. M. Cook, *J. Org. Chem.*, 2001, **66**, 4525.
87. A. Basak, S.S. Bag, K.R. Rudra, J. Barman and S. Dutta, *Chem. Lett.*, 2002, 710.
88. M.-X. Wang and S.-J. Lin, *J. Org. Chem.*, 2002, **67**, 6542.
89. S. Guery, M. Schmitt and J.-J. Bourguignon, *Synlett*, 2002, 2003.
90. E. Morero and G. Ortar, *Synth. Commun.*, 2001, **31**, 2215.
91. J.R. Walker and R.W. Curley, *Tetrahedron*, 2001, **57**, 6695.
92. C. Soede-Huijbregts, M. Van Laren, F.B. Hulsbergen, J. Raap and J. Lugtenburg, *J. Labelled Compd and Radiopharm.*, 2001, **44**, 831.
93. M. Oba, M. Kobayashi, F. Oikawa, K. Nishiyama and M. Kainasho, *J. Org. Chem.*, 2001, **66**, 5919.
94. C.J. Easton, N.L. Fryer, J.B. Kelly and K. Kociuba, *ARKIVOC*, 2001, **2**, U48.
95. M. Oba, T. Ishihara, H. Satake and K. Nishiyama, *J. Labelled Compd. and Radiopharm.*, 2002, **45**, 619.
96. D.W. Barrett, M.J. Panigot and R.W. Curley, *Tetrahedron: Asymmetry*, 2002, **13**, 1893.
97. J. Rudolph, F. Hannig, H. Theis and R. Wischnat, *Org. Lett.*, 2001, **3**, 3153.
98. F.-G. Pan, L.-G. Chen and M. Lu, *Guangzhou Huaxue*, 2001, **26**, 18.
99. A. Skolaut and J. Retey, *Arch. Biochem. Biophys.*, 2001, **393**, 187.
100. K. Laumen, O. Ghisalba and K. Auer, *Biosc. Biotech. Biochem.*, 2001, **65**, 1977.
101. W. Lin, Z. He, H. Zhang, X. Zhang, A. Mi and Y. Jing, *Synthesis*, 2001, 1007.
102. Y. Zhang, X. Liang and Y. Lin, *Huagong Shikan*, 2002, **16**, 46.
103. M. Yashiro, *Bull. Chem. Soc. Jap.*, 2002, **75**, 1383.
104. V.P. Krasnov, G.L. Levit, I.M. Bukrina and A.M. Demin, *Tetrahedron Asymmetry*, 2002, **13**, 1911.
105. S.P. Bew, S.D. Bull, S.G. Davies, E.D. Savory and D.J. Watkin, *Tetrahedron*, 2002, **58**, 9387.
106. S. Narayanan, S. Vangapandu and R. Jain, *Chem. Lett.*, 2001, **11**, 1133.

107. N. Kise, H. Ozaki, H. Terui, K. Ohya and N. Ueda, *Tetrahedron Lett.*, 2001, **42**, 7637.
108. Q. Yin, B. Jiang, Z. Mao, X. Sun and Y. Wang, *Huaxue Shijie*, 2001, **42**, 29.
109. J.M. Travins, M.G. Bursavich, D.F. Veber and D.H. Rich, *Org. Lett.*, 2001, **3**, 2725.
110. C. Pesenti, P. Bravo, E. Corradi, M. Frigerio, S.V. Meille, W. Panzeri, F. Viani and M. Zanda, *J. Org. Chem.*, 2001, **66**, 5637.
111. S.J. Kwon and S.Y. Koo, *Tetrahedron Lett.*, 2002, **43**, 639.
112. D. Yoo, J.S. Oh and Y.G. Kim, *Org. Lett.*, 2002, **4**, 1213.
113. J. Clayden, C.J. Menet and K. Tchabanenko, *Tetrahedron*, 2002, **58**, 4727.
114. S.-T.S. Itadani, C. Tanigawa, K. Hashimoto and M. Shirahama, *Tetrahedron Lett.*, 2002, **43**, 7777.
115. E.S. Greenwood and P.J. Parsons, *Synlett.*, 2002, 167.
116. M. Kamabe, T. Miyazaki, K. Hashimoto and H. Shirahama, *Heterocycles*, 2002, **56**, 105.
117. G. Rosini, *Chim. Ind.*, 2001, **83**, 75.
118. J.E. Baldwin, G.J. Pritchard and D.S. Williamson, *Tetrahedron*, 2001, **57**, 7991.
119. R.P. Jain and R.M. Williams, *Tetrahedron*, 2001, **57**, 6505.
120. R.P. Jain and R.M. Williams, *Tetrahedron Lett.*, 2001, **42**, 4437.
121. K. Makino, K. Shintani, T. Yamatake, O. Hara, K. Hatano and Y. Hamada, *Tetrahedron*, 2002, **58**, 9737.
122. G.Y. Cho, K.M. An and S.Y. Ko, *Bull. Korean Chem. Soc.*, 2001, **22**, 432.
123. N.M. Van Gelder and R.J. Bowers, *Neurochem. Res.*, 2001, **26**, 575.
124. P. Bisel, E. Breitling, M. Schlauch, F.-J. Volk and A.W. Frahm, *Pharmazie*, 2001, **56**, 770.
125. D. Phillips and A.R. Chamberlain, *J. Org. Chem.*, 2002, **67**, 3194.
126. H. Sugiyama, T. Shioiri and F. Yokokawa, *Tetrahedron Lett.*, 2002, **43**, 3489.
127. X. Li, R.N. Atkinson and S.B. King, *Tetrahedron*, 2001, **57**, 6557.
128. N.U. Sata, R. Kuwahara and Y. Murata, *Tetrahedron Lett*, 2002, **43**, 115.
129. T. Oishi, K. Ando and N. Chida, *Chem. Commun. (Cambridge)*, 2001, 1932.
130. K. Lida and M. Kajiwara, *J. Labelled Compd. and Radiopharm.*, 2002, **45**, 139.
131. N.M. Gillings and A.D. Gee, *J. Labelled Compd. and Radiopharm.*, 2001, **44**, 909.
132. Z. Zhang, Y.-S. Ding, A.R. Studenov, M.R. Gerasimov and R.A. Ferrieri, *J. Labelled Compd. and Radiopharm.*, 2002, **45**, 199.
133. B.B. Snider and Y. Gu, *Org. Lett.*, 2001, **3**, 1761.
134. D.E. DeMong and R.M. Williams, *Tetrahedron Lett.*, 2002, **43**, 2355.
135. V. Magrioti and V. Constantinou-Kokotou, *Lipids*, 2002, **37**, 223.
136. K.A. Brun, A. Linden and H. Heimgartner, *Helv. Chim. Acta*, 2001, **84**, 1756.
137. K.A. Brun, A. Linden and H. Heimgartner, *Helv. Chim. Acta*, 2002, **85**, 3422.
138. K. Wright, F. Melandri, C. Cannizzo, M. Wakselman and J.-P. Mazaleyrat, *Tetrahedron*, 2002, **58**, 5811.
139. Y.N. Belekon, V.I. Maleev, T.F. Saveleva and N.S. Ikonnikov, *Russ. Chem. Bull.*, 2001, **50**, 1037.
140. A. Avenoza, C. Cativiela, F. Corzana, J.-M. Peregrina, H.D. Sucunza and M.M. Zurbano, *Tetrahedron Asymmetry*, 2001, **12**, 949.
141. M. Tanaka, M. Oba, M. Kurihara, Y. Demizu, S. Nishimura, K. Hayashida and H. Suemune, *Pept. Sci.*, 2001, **38**, 263.
142. A.G. Hu, L.-Y. Zhang, S.-W. Ing and J.-T. Wang, *Synth. Commun.*, 2002, **32**, 2143.

143. (*a*) Y.N. Belokon, D. Bhave, D. D'Addario, E. Groaz, V. Maleev, M. North and A. Pertrosyan, *Tetrahedron Lett.*, 2003, **44**, 2045; (*b*) Y.N. Belokon, R.G. Davies, J.A. Fuentes, M. North and T. Parsons, *ibid.*, 2001, **42**, 8093.
144. R. Portoles, J. Murga, E. Falomir, M. Carda, S. Uriel and J. A. Marco, *Synlett*, 2002, **711**.
145. C. Alvarez-Ibarra, A.G. Csaky and C.G. De la Oliva, *Eur. J. Org. Chem.*, 2002, 4190.
146. S.-H. Lee and E.-K. Lee, *Bull. Korean Chem. Soc.*, 2001, **22**, 551.
147. Y. Fu, L.G.J. Hammarstroem, T.J. Miller, F.R. Fronczek, M.L. McLaughlin and R.P. Hammer, *J. Org. Chem.*, 2001, **66**, 7118.
148. V.V.S. Babu, K. Ananda and G.-R. Vasanthakumar, *Protein and Peptide Lett.*, 2002, **9**, 345.
149. B. Kaptein, Q.B. Broxterman, H.E. Schoemaker, F.P.J.T. Rutjes, J.J.N. Veerman, J. Kamphuis, C. Peggion, F. Formaggio and C. Toniolo, *Tetrahedron*, 2001, **57**, 6567.
150. D.M. Bradley, R. Mapiste, N.M. Thomson and C.J. Hayes, *J. Org. Chem.*, 2002, **67**, 7613.
151. J.C. Anderson and S. Skerratt, *J. Chem. Soc Perkin Trans 1*, 2002, 2871.
152. S. Torrente and R. Alonso, *Org. Lett.*, 2001, **3**, 1985.
153. R. Martin, G. Islas, A. Moyana, M.A. Pericas and A. Riera, *Tetrahedron*, 2001, **57**, 6367.
154. K. Ding and D. Ma, *Tetrahedron*, 2001, **57**, 6361.
155. T. Ooi, M. Takeuchi, D. Ohara and K. Maruoka, *Synlett.*, 2001, 1185.
156. P.D. Bailey, N. Bannister, M. Bemad, S. Blanchard and N.A. Boa, *J. Chem. Soc. Perkin Trans 1*, 2001, 3245.
157. F. Clerici, M.L. Gelmi, D. Pocar and T. Pilati, *Tetrahedron Asymmetry*, 2001, **12**, 2663.
158. A. Debache, S. Collet, P. Bauchat, D. Danion, L. Euzenat, A. Hercouet and B. Carboni, *Tetrahedron Asymmetry*, 2001, **12**, 761.
159. E. Bunuel, A.I. Jimenez, M.D. Diazde-Villegas and C. Cativiela, *Targets Het. Systems*, 2001, **5**, 79.
160. A. Salgado, T. Huybrechts, A. Eeckhaut, J. Van der Eycken, Z. Szakonyi, F. Fulop, A. Tkachev and N. DeKimpe, *Tetrahedron*, 2001, **57**, 2781.
161. D.K. Mohapatra, *J. Chem. Soc. Perkin Trans 1*, 2001, **1851**.
162. A. Esposito, P.P. Piras, D. Ramazzotti and M. Taddei, *Org. Lett.*, 2001, **3**, 3273.
163. R. Pellicciari, G. Constantino, M. Marinozzi, A. Macchiarulo, L. Amori, P. Josef Flor, F. Gasparini, R. Kuhn and S. Urwyler, *Bioorg. Med. Chem. Lett.*, 2001, **11**, 3179.
164. N.J. Wallock and W.A. Donaldson, *Tetrahedron Lett.*, 2002, **43**, 4541.
165. I. Collado, C. Pedregal, A. Mazon, J. Felix Espinosa, J. Blanco-Urgoiti, D.D. Schoepp, R.A. Wright, B.G. Johnson and A.E. Kingston, *J. Med. Chem.*, 2002, **45**, 3619.
166. N.A. Anisimova, G.A. Berkova, T.Y. Paperno and L.I. Deiko, *Russ. J. Gen. Chem.*, 2002, **72**, 272.
167. S. Racouchot, I. Sylvestre, J. Ollivier, Y.Y. Kozyrkov, A. Pukin, O.G. Kulinkov and J. Salaun, *Eur. J. Org. Chem.*, 2002, 2160.
168. D. Fishlock, J.G. Guillemette and G.A. Lajoie, *J. Org. Chem.*, 2002, **67**, 2352.
169. T.M. Kamenecka, Y.-J. Park, L.S. Lin, T. Lanza and W.K. Hagmann, *Tetrahedron Lett.*, 2001, **42**, 8571.

170. D.J. Aldous, M.G.B. Drew, E.M.-N. Hamelin, L.M. Harwood, A.B. Jahans and S. Thurairatnam, *Synlett*, 2001, 1836.
171. Q. Xia and B. Ganem, *Tetrahedron Lett.*, 2002, **43**, 1597.
172. L. Halab, L. Belec and W.D. Lubell, *Tetrahedron*, 2001, **57**, 6439.
173. E.A.A. Wallen, J.A.M. Christiaans, J. Gynther and J. Vepsalainen, *Tetrahedron Lett.*, 2003, **44**, 2081.
174. S. Duan and K.D. Moeller, *Tetrahedron*, 2001, **57**, 6407.
175. Y.G. Gu, Y. Xu, A.C. Krueger, D. Madigan and H.L. Sham, *Tetrahedron Lett.*, 2002, **43**, 955.
176. X. Chen, D.-M. Du and W.-T. Hua, *Tetrahedron Asymmetry*, 2002, **13**, 43.
177. A.M. Boldi, J.M. Oener and T.P. Hopkins, *J. Comb. Chem.*, 2001, **3**, 367.
178. J. Casas, R. Grigg, C. Najera and J.M. Sansano, *Eur. J. Org. Chem.*, 2001, 1971.
179. J.E. Baldwin, A.M. Fryer and G.J. Pritchard, *J. Org. Chem.*, 2001, **66**, 2588.
180. K. Makino, A. Kondoh and Y. Hamada, *Tetrahedron Lett.*, 2002, **43**, 4695.
181. I. Merino, Y.R. Santosh Laxmi, J. Florez, J. Barluenga, J. Ezquerra and C. Pedregal, *J. Org. Chem.*, 2002, **67**, 648.
182. G.R. Crow, S.B. Herzon, G. Lin, F. Qui and P.E. Sonnet, *Org. Lett.*, 2002, **4**, 3151.
183. T. Rammeloo and C.V. Stevens, *Chem. Commun. (Cambridge)*, 2002, 250.
184. (*a*) E. Bunuel, A.M. Gil, M.D. Diaz de Villegas and C. Cativiela, *Tetrahedron*, 2001, **57**, 6417; (*b*) A. Avenoza, J.I. Barriobero, J.H. Busto, C. Cativiela and J.S. Peregrina, *Tetrahedron Asymmetry*, 2002, **13**, 625; (*c*) A. Avenoza, C. Cativiela, J.H. Busto, M.A. Fernandez-Recio, J.M. Peregrina and F. Rodriguez, *Tetrahedron*, 2001, **57**, 545.
185. B. Jiang and M. Xu, *Org. Lett.*, 2002, **4**, 4077.
186. T. Miura and T. Kajimoto, *Chirality*, 2001, **13**, 577.
187. S. Ohira, M. Akiyama, K. Kamihara, Y. Isoda and A. Kuboki, *Biosci. Biotech. and Biochem.*, 2002, **66**, 887.
188. M.G. Woll, J.D. Fisk, P.R. LePlae and S.H. Gellman, *J. Am. Chem. Soc.*, 2002, **124**, 12447.
189. C.M. Acevedo, E.F. Kogul and M.A. Lipton, *Tetrahedron*, 2001, **57**, 6353.
190. I.A. Nizova, V.P. Krasnov, G.L. Levit and M.I. Kodess, *Amino Acids*, 2002, **22**, 179.
191. C. Agami, F. Bisaro, S. Comesse, S. Guesne, C. Kadouri-Puchot and R. Morgentin, *Eur. J. Org. Chem.*, 2001, 2385.
192. A. Kulesza, A. Mieczkowski and J. Jurczak, *Tetrahedron Asymmetry*, 2002, **13**, 2061.
193. X. Ginesta, M.A. Pericas, A. Riera and F. Marti, *Tetrahedron Lett.*, 2002, **43**, 779.
194. D.-G. Liu, Y. Gao, X. Wang, J.A. Kelley and T.R. Burke Jr., *J. Org. Chem.*, 2002, **67**, 1448.
195. M. Haddad and M. Larcheveque, *Tetrahedron Lett.*, 2001, **42**, 5223.
196. F. Machetti, F.M. Cordero, F. de Sarlo and A. Brandi, *Tetrahedron*, 2001, **57**, 4995.
197. W. Maison and G. Adiwidjaja, *Tetrahedron Lett.*, 2002, **43**, 5957.
198. C.-B. Xue, X. He, J. Roderick, R.L. Corbett and C.P. Decicco, *J. Org. Chem.*, 2002, **67**, 865.
199. E. Teoh, E.M. Campi, W.R. Jackson and A.J. Robinson, *New J. Chem.*, 2003, **27**, 387.
200. R. Warmuth, T.E. Munsch, R.A. Stalker, B. Li and A. Beatty, *Tetrahedron*, 2001, **57**, 6383.
201. S. Kotha, N. Sreenivasachary and E. Brahmachary, *Tetrahedron*, 2001, **57**, 6261.

202. J. Wang, M.L. Falck-Pederson, C. Romming and K. Undheim, *Synth. Commun.*, 2001, **31**, 1141.
203. C.L.L. Chai, R.C. Johnson and J. Koh, *Tetrahedron*, 2002, **58**, 975.
204. M. Belohradsky, I. Cisarova, P. Holy, J. Pastor and J. Zavada, *Tetrahedron*, 2002, **58**, 8811.
205. F. Clerici, M.L. Gelmi, A. Gambini and D. Nava, *Tetrahedron*, 2001, **57**, 6429.
206. J.S. Tulus, M.J. Laufersweiler, J.C. VanRens, M.G. Natchus, R.G. Bookland, N.G. Almstead, S. Pikul, B. De, L.C. Hsieh, M.J. Janusz, T.M. Branch, S.X. Peng, Y.Y. Jin, T. Hudlicky and K. Oppong, *Bioorg. Med. Chem. Lett.*, 2001, **11**, 1975.
207. S. Venkatraman, F.G. Njoroge, V. Girijavallabhan and A.T. McPhail, *J. Org. Chem.*, 2002, **67**, 2686.
208. W.-C. Shieh, S. Xue, N. Reel, R. Wu, J. Fitt and O. Repic, *Tetrahedron Asymmetry*, 2001, **12**, 2421.
209. A.T. Ung, K. Schafer, K.B. Lindsay, S.G. Pyne, K. Amornraksa, R. Wouters, U. Van der Linden, U. Biesmans, A.S.J. Lesage, B.W. Skelton and A.H. White, *J. Org. Chem.*, 2001, **67**, 227.
210. S. Krikstolaityte, A. Sackus, C. Roinmilng and K. Undheim, *Tetrahedron Asymmetry*, 2001, **12**, 393.
211. C. Alvarez-Ibarra, A.G. Csaky and C. Gomez de la Oliva, *J. Org. Chem.*, 2002, **67**, 2789.
212. P. Conti, G. Roda and P.F. Barberia Negra, *Tetrahedron Asymmetry*, 2001, **12**, 1363.
213. K.-H. Park, T.M. Kurth, M.M. Olmstead and M.J. Kurth, *Tetrahedron Lett.*, 2001, **42**, 991.
214. H. Dialer, W. Steglich and W. Beck, *Tetrahedron*, 2001, **57**, 4855.
215. S. Kotha, S. Halder, L. Damodharan and V. Pattabhi, *Bioorg. Med. Chem. Lett.*, 2002, **12**, 1113.
216. S. Kotha, A.K. Ghosh and M. Behera, *Ind. J. Chem. Section B*, 2002, **41B**, 2330.
217. A.M. Papini, E. Nardi, F. Nuti, I. Uziel, M. Ginanneschi, M. Chelli and A. Brandi, *Eur. J. Org. Chem.*, 2002, 2736.
218. L. Mou and G. Singh, *Tetrahedron Lett.*, 2001, **42**, 6603.
219. V. Constantitinou-Kokotou, V. Magrioti, T. Markidis and G. Kokotos, *J. Pept. Res.*, 2001, **58**, 325.
220. H.-b. Shi, C.-l. Shao and Z.-l. Yu, *Huaxue Yanjiu*, 2001, **12**, 1.
221. P. Meffre, R.H. Dave, J. Leroy and B. Badet, *Tetrahedron Lett.*, 2001, **42**, 8625.
222. Y. Ding, A.J. Wang, K.A. Abboud, Y. Xu, W.R. Dolbier Jr. and N.G.J. Richards, *J. Org. Chem.*, 2001, **66**, 6381.
223. K. Uneyama, T. Katagiri and H. Amii, *Yuki Gosei, Kagaku Kyokaishi*, 2002, **60**, 1069.
224. T. Konno, T.I. Daitoh, T. Ishihara and H. Yamanaka, *Tetrahedron Asymmetry*, 2001, **12**, 2743.
225. T. Katagiri, M. Handa, Y. Matsukawa, J.S. Dileep Kumar and K. Uneyama, *Tetrahedron Asymmetry*, 2001, **12**, 1303.
226. M. Crucianelli, N. Battista, P. Bravo, A. Volonterio and M. Zanda, *Elec.J. Geotech. Eng.*, 2000, **5**, 1251.
227. A.Y. Aksinenko, A.N. Pushin and V.B. Sokolov, *Russ. Chem. Bull.*, 2002, **51**, 2136.
228. J.T. Anderson, P.L. Toogood and E.N.G. Marsh, *Org. Lett*, 2002, **4**, 4281.
229. G.K.S. Prakash, M. Mandal, S. Schweizer, N.A. Petasis and G.A. Olah, *J. Org. Chem.*, 2002, **67**, 3718–6286.

230. S. Fustero, A. Navarro, B. Pina, J.G. Soler, A. Bartolome, A. Asensio, A. Simon, P. Bravo, G. Fronza, A. Volonterio and M. Zanda, *Org. Lett.*, 2001, **3**, 2621.
231. S.G. Osipov, O.I. Artyushin, A.F. Kolomiets, C. Brunaeau, M. Picquet and P.H. Dixneuf, *Eur. J. Org. Chem.*, 2001, 3891.
232. S.N. Osipov, N.M. Kobelikova, G.T. Shchetnikov, A.F. Kolomeits, C. Bruneau and P.H. Dixneuf, *Synlett*, 2001, 621.
233. X.-l. Qui and F.-l. Qing, *J. Chem. Soc. Perkin Trans.*, 2002, 2052.
234. X.-l. Qui and F.-l. Qing, *J. Org. Chem.*, 2002, **67**, 7162.
235. J.R. Del Valle and M. Goodman, *Angew. Chem. Int. Ed.*, 2002, **41**, 1600.
236. M. Doi, Y. Nishi, N. Kiritoshi, T. Iwata, M. Nago, H. Nakano, S. Uchiyama, T. Nakazawa, T. Wakamiya and Y. Kobayashi, *Tetrahedron*, 2002, **58**, 8453.
237. A.S. Gobulev, H. Schedel, G. Radios, J. Sieler and K. Burger, *Tetrahedron Lett.*, 2001, **42**, 7941.
238. S.N. Osipov, N.M. Kobel Kova, A.F. Kolomiets, K. Pumpor, B. Kotsch and K. Burger, *Synlett*, 2001, 1287.
239. N.A. Fokina, A.M. Komilov and V.P. Kukhar, *J. Fluorine Chem.*, 2001, **111**, 69.
240. A. Vidal, A. Nefzi and R.A. Houghten, *J. Org. Chem.*, 2001, **66**, 8268.
241. B. Mohar, J. Badoux, J.-C. Plaquevent and D. Cahard, *Angew. Chem. Int. Ed.*, 2001, **40**, 4214.
242. A. Asensio, P. Bravo, M. Crucianelli, A. Farina, S. Fuslero, J.G. Soler, S.V. Meille, W. Fanzcri, F. Viani. A. Vobnterio and M. Kanda, *Eur. J. Org. Chem.*, 2001, 1449.
243. F.A. Davis, Y. Zhang, A. Rao and Z. Zhang, *Tetrahedron*, 2001, **57**, 6345.
244. M. Panunzio, E. Bandini, E. Campana and P. Vicennati, *Tetrahedron Asymmetry*, 2002, **13**, 2113.
245. R. Andruszkiewicz and M. Wyszogrodzka, *Synlett*, 2002, 2101.
246. M.R. Carrasco, R.T. Brown, Y.M. Serafinova and O. Silva, *J. Org. Chem.*, 2002, **68**, 195.
247. Q. Wang, J. Ouazzani, N.A. Sasaki and P. Potier, *Eur. J. Org. Chem.*, 2002, 834.
248. J.E. Dettwiler and W.D. Lubell, *J. Org. Chem.*, 2003, **68**, 177.
249. M. Amador, X. Ariza, J. Garcia and S. Sevilla, *Org. Lett.*, 2002, **4**, 4511.
250. M. Oba, S. Koguchi and K. Nishiyama, *Tetrahedron*, 2002, **58**, 9359.
251. P.X. Choudhury, D.X. Le Nguyen and N. Langlois, *Tetrahedron Lett.*, 2002, **43**, 463.
252. J. Zhang, J.L. Flippen-Anderson and A.P. Kozikowski, *J. Org. Chem.*, 2001, **66**, 7555.
253. S. Caddick, N.J. Parr and M.C. Pritchard, *Tetrahedron*, 2001, **57**, 6615.
254. Y.N. Belokon, K.A. Kochetkov, N.S. Ikannikov, T.V. Strelkova, S.R. Hartyunyan and A.S. Saghiyan, *Tetrahedron Asymmetry*, 2001, **12**, 481.
255. J. Spetzler and T. Hoeg-Jensen, *J. Pept. Sci.*, 2001, **7**, 537.
256. T. Markidis and G. Kokotos, *J. Org. Chem.*, 2002, **67**, 1685.
257. C. Palomo, M. Oiarhide, A. Landn, A. Esnal and A. Lindtin, *J. Org. Chem.*, 2001, **66**, 4180.
258. B.T. Shireman and M.J. Miller, *J. Org. Chem.*, 2001, **66**, 4809.
259. N. Yoshikawa and M. Shibasaki, *Tetrahedron*, 2002, **58**, 8289.
260. V. Guerlavais, P.J. Carroll and M.M. Joullie, *Tetrahedron Asymmetry*, 2002, **13**, 675.
261. S. Chandrasekhar, T. Ramachandar, B. Rao and B. Venkateswara, *Tetrahedron Asymmetry*, 2001, **12**, 2315.
262. J.M. Travins, M.G. Bursavich, D.F. Veber and D.H. Rich, *Org. Lett.*, 2001, **3**, 2725.

263. S.Y. Ko, *J. Org. Chem.*, 2002, **67**, 2689.
264. R.A. Tromp, M. van der Hoeven, A. Amore, J. Brussee, M. Overhand and G.A. van der Gen, *Tetrahedron Asymmetry*, 2001, **12**, 1109.
265. Q. Wang, M.E. Tran Huu Dau, N. Andre Sasaki and P. Potier, *Tetrahedron*, 2001, **57**, 6455.
266. J.H. Lee, J.E. Kang, M.S. Yang, K.Y. Kang and K.-H. Park, *Tetrahedron*, 2001, **57**, 10071.
267. D.G. Qin, H.-Y. Zha and Z.-J. Yao, *J. Org. Chem.*, 2002, **67**, 1038.
268. P. Dalla Croce and C. La Rosa, *Tetrahedron Asymmetry*, 2002, **13**, 197.
269. C.M. Taylor, W.D. Barker, C.A. Weir and J.H. Park, *J. Org. Chem.*, 2001, **67**, 4466.
270. R. Martin, M. Alcon, M.A. Pericas and A. Riera, *J. Org. Chem.*, 2002, **67**, 6896.
271. A. Avenoza, J. Barriobero, J.H. Busto, C. Cativiela and J.M. Peregrina, *Tetrahedron Asymmetry*, 2002, **13**, 625.
272. X. Zhang, A.C. Schmitt and W. Jiang, *Tetrahedron Lett.*, 2001, **42**, 5335.
273. A. Avenoza, J.I. Barriobero, C. Cativiela, M.A. Fernandez-Recio, J.M. Peregrina and F. Rodriquez, *Tetrahedron*, 2001, **57**, 2745.
274. J. Marin, C. Didierjean, A. Aubry, J.-P. Briand and G. Guichard, *J. Org. Chem.*, 2002, **67**, 8440.
275. M.M. Palian and R. Polt, *J. Org. Chem.*, 2001, **66**, 7178.
276. N. Okamoto, O. Hara, K. Makino and Y. Hamada, *J. Org. Chem.*, 2002, **67**, 9210.
277. M.C. Willis and V.J.-D. Piccio, *Synlett*, 2002, 1625.
278. F. Mutilis, I. Mutule and J.E.S. Wikberg, *Bioorg. Med. Chem. Lett.*, 2001, **12**, 1039.
279. L. Aurelio, J.S. Box, R.T.C. Brownlee, A.B. Hughes and M.M. Sleebs, *J. Org. Chem.*, 2003, **68**, 2652.
280. C. Laplante and D.G. Hall, *Org. Lett.*, 2001, **3**, 1487.
281. Y. Luo, G. Evindar, D. Fishlock and G.A. Lajoie, *Tetrahedron Lett.*, 2001, **42**, 3807.
282. R. Paruszewski, M. Strupinska, J.P. Stables, M. Swiader, S. Czuczwar, Z. Kleinrok and W. turski, *Chem. Pharm. Bull.*, 2001, **49**, 629.
283. Y.-x. Leng, X.-s. Rui, J.-q. Ma, H.-y. Sun and H.-b. Zhou, *Jingxi Huagong*, 2001, **18**, 50.
284. S. Kobayashi, H. Kitagawa and R. Matsubara, *J. Comb. Chem.*, 2001, **3**, 401.
285. J.R. Casimir, G. Guichard, D. Tourwe and J.-P. Briand, *Synthesis-Stuttgart*, 2001, 1985.
286. S. Reverchon, B. Chantegrel, C. Deshayes, A. Doutheau and N. Cotte-Pattat, *Bioorg. Med. Chem. Lett.*, 2002, **12**, 1153.
287. Y. Zhang, Q. Yang and Y. Sun, *Huaxue Shijie*, 2002, **43**, 363.
288. K. Hojo, M. Maeda, Y. Takahara, S. Yamamoto and K. Kawasaki, *Tetrahedron Lett.*, 2002, **44**, 2849.
289. E. Masiukiewicz, S. Eiejak and B. Rzeszotarska, *Org. Prep. & Proc. Int.*, 2002, **34**, 521.
290. B.C. Shekar, K. Roy and A.U. De, *J. Het. Chem.*, 2001, **10**, 237.
291. Y. Xu and B. Zhu, *Synthesis*, 2001, **9**, 690.
292. G. Gellenuaii, A. Elgavi, Y. Salitra and M. Kramer, *J. Pept. Res.*, 2001, **57**, 277.
293. H. Shen, A.P. Li, H. Wang, T.X. Wu and X.F. Pan, *Chinese Chem. Lett.*, 2002, **13**, 117.
294. (*a*) V.A. Knizhnikov, V.I. Potkin, S.K. Petkevich, A.S. Skripchenko and A.V. Mikulich, *Vest. Nats. Akad. Navuk. Belarusi*, 2002, 82; (*b*) V.A. Knizhnikov, V.I.

Potkin, K.A. Zhavnerko, L.S. Yakubovich, S.K. Petkevich and S.P. Kacherskaya, *Russ. J. Org. Chem.*, 2002, **38**, 915.
295. X.-j. Wang, *Guangzhou Huaxue*, 2001, **26**, 27.
296. A. Sidduri, J.W. Tilley, J.P. Lou, L. Chen, G. Kaplan, F. Mennona, R. Campbell, R. Guthrie, T.-N. Huang, K. Rowan, V. Schwinge and L.M. Renzetti, *Bioorg. Med. Chem. Lett.*, 2002, **12**, 2479.
297. A.E. Wroblewski and D.G. Piotrowska, *Tetrahedron Asymmetry*, 2002, **13**, 2509.
298. W.-h. Li, *Hecheng Huaxue*, 2001, **9**, 563.
299. M. Terasaki, S. Nomoto, H. Mita and A. Shimoyama, *Bull. Chem. Soc. Jap.*, 2002, **75**, 855.
300. B. Henkel and L. Weber, *Synlett*, 2002, 1877.
301. S. McGhie, *Synth. Commun.*, 2002, **32**, 1275.
302. Z.-Q. Zhu and X.-G. Mei, *Synth. Commun.*, 2001, **31**, 3609.
303. D.V. Arsenov, M.A. Kisel and O.A. Strel'chenok, *Dokl. Nat. Akad. Nauk. Belarousi*, 2001, **45**, 71.
304. G.A. Doherty, T. Kamenecka, E. McCauley, G. Van Riper, R.A. Mumford, S. Tong and W.K. Hagmann, *Bioorg. Med. Chem. Lett.*, 2002, **12**, 729.
305. S. Bittner, S. Gorovsky, O. Paz-Tal and J.Y. Becker, *Amino Acids*, 2002, **22**, 71.
306. T. Izuhara, W. Yokota, M. Inoue and T. Katoh, *Heterocycles*, 2002, **56**, 553.
307. L.A. Sviridova, I.F. Leschova and G.K. Vertelov, *Khim. Geterotsikl Soedinenii*, 2000, 1335.
308. C.M. Vaidya, J.E. Wright and A. Rosowsky, *J. Med. Chem.*, 2002, **45**, 1690.
309. G. Chen, Y. Deng, L. Gong, A. Mi, X. Cui, Y. Jiang, M.C. K. Choi and A.S.C. Chan, *Tetrahedron Asymmetry*, 2001, **12**, 1567.
310. T.D. Weiss, G. Helmchen and U. Kazmaier, *Chem. Commun. (Cambridge)*, 2002, 1270.
311. B.M. Trost and K. Dogra, *J. Am. Chem. Soc.*, 2002, **124**, 7256.
312. U. Kazmeier and F.L. Zumpe, *Eur. J. Org. Chem.*, 2001, 4067.
313. A. Kulesza and J. Jurczak, *Chirality*, 2001, **13**, 634.
314. T. Abellan, B. Mancheno, C. Najera and J.M. Sansano, *Tetrahedron*, 2001, **57**, 6627.
315. L.B. Wolf, T. Sonke, K.C.M.F. Tjen, B. Kaptein, Q.B. Broxterman, H.E. Schoemaker and F.P.J.T. Rutjes, *Adv. Synth. Catal.*, 2001, **343**, 662.
316. D. Balan and H. Adolfsson, *J. Org. Chem.*, 2001, **66**, 6498.
317. V.K. Aggarwal, A.M.M. Castro, A. Mereu and H. Adams, *Tetrahedron Lett.*, 2002, **43**, 1577.
318. S. Suezen, *Ankara Univ. Eczacilik Fak. Derg.*, 2001, **30**, 17.
319. R. Kimura, T. Nagano and H. Kinoshita, *Bull. Chem. Soc. Jpn.*, 2002, **75**, 2517.
320. K.E. Holt, J.P. Swift, M.E.B. Smith, S.J.C. Taylor and R. McCague, *Tetrahedron Lett.*, 2002, **43**, 1545.
321. C. Douat, A. Heitz, J. Martinez and J.-A. Fehrentz, *Tetrahedron Lett.*, 2001, **42**, 3319.
322. D.B. Berkowitz, E. Chisowa and J.M. McFadden, *Tetrahedron*, 2001, **57**, 6329.
323. S. Chandrasekhar, A. Raza and M. Takhi, *Tetrahedron Asymmetry*, 2002, **13**, 423.
324. K. Sakaguchi, H. Suzuki and Y. Ohfune, *Chirality*, 2001, **13**, 357.
325. D. G. Ahem, R. Seguin and C.N. Filer, *J. Labelled Comp. Radiopharm.*, 2002, **45**, 401.
326. D. Ma, W. Wu and P. Deng, *Tetrahedron Lett.*, 2001, **42**, 6929.
327. L. Vranicar, A. Meden, S. Polanc and M. Kocevar, *J. Chem. Soc. Perkin Trans 1*, 2002, **675**.

328. M. Schleusner, H.-J. Gais, S. Koep and G. Raabe, *J. Am. Chem. Soc.*, 2002, **124**, 7789.
329. S.E. Gibson, J. Jones, S.B. Kalindjian, J.D. Knight, J.W. Steed and M.J. Tozer, *Chem. Commun. (Cambridge)*, 2002, 1938.
330. N. Krause, A. Hoffmann-Roder and J. Canisius, *Synthesis-Stuttgart*, 2002, 1759.
331. J.J. Turner, M.A. Leeuwenburgh, G.A. van der Marel and J.H. van Boom, *Tetrahedron Lett.*, 2001, **42**, 8713.
332. M.K. Gurjar and A. Talukdar, *Synthesis-Stuttgart*, 2002, 315.
333. S.F. Khalilova, Zh.N. Kirbagakova and K.B. Erzhanov, *Izv. Minist. Obraz. Nauki. Resp. Kaz.*, 2000, 94.
334. M.P. Lopez-Deber, L. Castedo and J.R. Granja, *Org. Lett.*, 2001, **3**, 2813.
335. N. Vasdev, R. Chirakal, G.J. Schrobilgen and C. Nahmias, *J. Fluorine Chem.*, 2001, **11**, 17.
336. W.P. Deng, K.A. Wong and K.L. Kirk, *Tetrahedron Asymmetry*, 2002, **13**, 1135.
337. A.V. Samet, D.J. Coughlin, A.C. Buchanan III and A.A. Gakh, *Synth. Commun.*, 2002, **32**, 941.
338. (*a*) G. Tang, L. Zhang, X.-l. Tang, Y.-X. Wang and D. Yin, *He Huaxue Yu Fangshe Huaxue*, 2001, **23**, 211; (*b*) G. Tang, L. Zhang, X.-l. Tang, Y.-X. Wang and D. Yin, *Hejishu*, 2002, **25**, 1019; (*c*) G. Tang, L. Zhang, X.-l. Tang, Y.-X. Wang and D. Yin, *Appl. Rad. Isotop.*, 2002, **57**, 145; (*d*) G. Tang, L. Zhang, X.-l. Tang, Y.-X. Wang and D. Yin, *Zwngguo Yaoke Daxue Xuebao*, 2001, **32**, 166.
339. J.T. Konkel, J. Fan, B. Jayachandran and K.L. Kirk, *J. Fluorine Chem.*, 2002, **115**, 27.
340. K. Hamacher and H.H. Coenen, *Appl. Rad. Isotop.*, 2002, **57**, 853.
341. B. Herbert, I.H. Kim and K.L. Kirk, *J. Org. Chem.*, 2001, **66**, 4892.
342. M. Cameron, D. Cohen, I.F. Cottrell, D.J. Kennedy, C. Roberge and M. Chartrain, *J. Mol. Cat. B: Enz.*, 2001, 1.
343. W.-S. Yu, Y.-J. Liang, K.-L. Liu and Y.-F. Zhao, *Gaodeng Xuexiao Huaxue Xuebao*, 2002, **23**, 1314.
344. J. Vahatalo, M. Kulvik, S. Savolamen and S.-L. Karonen, *Frontiers in Neutron Capture Therapy*, eds. M. F. Hawthorne, K. Shelly and R. J. Wiersema, publ. Kluwer/Plenum, 2001, **2**, 835.
345. S. Laabs, W. Munch, J.W. Bats and U. Nubbemeyer, *Tetrahedron*, 2002, **58**, 1317.
346. M. Tamaki, G. Han and V.J. Hruby, *J. Org. Chem.*, 2001, **66**, 3593.
347. G. Priem, M.S. Anson, S.J.F. MacDonald, B. Pelotier and I.B. Campbell, *Tetrahedron Lett.*, 2002, **43**, 6001.
348. A. Sidduri, J.P. Lou, R. Campbell, K. Rowan and J.W. Tilley, *Tetrahedron Lett.*, 2001, **42**, 8757.
349. (*a*) S. Kotha, N. Bieenivanachary and E. Brahmachary, *Eur. J. Org. Chem.*, 2001, 787; (*b*) S. Kotha and E. Brahmachary, *Bioorg. Med. Chem.*, 2002, **10**, 2291; (*c*) S. Kotha, S. Halder and K. Lahiri, *Synthesis-Stuttgart*, 2002, 339.
350. K. Ding, X.-R. Zhang, D.-W. Ma and B.-M. Wang, *Chin. J. Chem.*, 2001, **19**, 1232.
351. P.A. Crooks and J. Matheru, *Synth. Commun.*, 2002, **32**, 3813.
352. (*a*) W. Wang, C. Xiong, J. Yang and V.J. Hruby, *Tetrahedron Lett.*, 2001, **42**, 7717; (*b*) W. Wang, M. Cai, C. Xiong, J. Zhang, D. Trivedi and V.J. Hruby, *Tetrahedron*, 2002, **58**, 7365; (*c*) W. Wang, M. Cai, C. Xiong, J. Zhang, D. Trivedi and V.J. Hruby, *Tetrahedron Lett.*, 2002, **43**, 2137; (*d*) W. Wang, C. Xiong, J. Zhang and V.J. Hruby, *Tetrahedron*, 2001, **58**, 3101.
353. S. Hanzawa, S. Oe, K. Tokuhisa, K. Kawano, H. Kakidani and T. Ishiguro, *Toso Kenkyu Gijutsu Hokoku*, 2001, **45**, 11.

354. T. Ooi, M. Takeuchi and K. Maruoka, *Synthesis-Stuttgart*, 2001, 1716.
355. J. Spengler, H. Schedel, J. Sieler, P.J.L.M. Quaedser, Q.B. Broxterman, A.L.T. Duchateau and K. Burger, *Synthesis-Stuttgart*, 2001, 1513.
356. N. Vails, M. Lopez-Canet, M. Vallribera and J. Bonjoch, *Chem. -Eur. J.*, 2001, **7**, 3446.
357. R.A. Stalker, T.E. Munsch, J.D. Tran, X. Nie, R. Warmuth, A. Beatty and C.B. Aakeroy, *Tetrahedron*, 2002, **58**, 4837.
358. A. Boto, R. Hernandez, A. Montoya and E. Suarez, *Tetrahedron Lett.*, 2002, **43**, 8269.
359. N. Kise, H. Ozaki, H. Terui, K. Ohya and N. Ueda, *Tetrahedron Lett.*, 2001, **42**, 7637.
360. S. Tohma, A. Endo, T. Kan and T. Fukuyama, *Synlett.*, 2001, 1179.
361. A. Szymanska, W. Wiczk and L. Lankiewicz, *Amino Acids*, 2001, **21**, 265.
362. M. Zia-Ul-Hao, M. Arshad and Saeed-Ur-Reliman, *J. Chin. Chem. Soc.*, 2001, **48**, 45.
363. P. Call and M. Begtrup, *Tetrahedron*, 2002, **58**, 1595.
364. C.-S. Ge, Y.-J. Chen and D. Wang, *Synlett.*, 2002, 37.
365. S.-h. Lee, E.-K. Lee and S.-M. Jeun, *Bull. Korean Chem. Soc.*, 2002, **23**, 931.
366. S. Lee, N.A. Beare and J.F. Hartwig, *J. Am. Chem. Soc.*, 2001, **123**, 8410.
367. J.E. Redman and M.R. Ghadiri, *Org. Lett.*, 2002, **4**, 4467.
368. H. Takahishi, N. Kashiwa, H. Kobayashi, Y. Hashimoto and K. Nagasawa, *Tetrahedron Lett.*, 2002, **43**, 6751.
369. S. Nampalli, W. Zhang, R.T. Sudhakar, H. Xiao, L.P. Kotra and S. Kumar, *Tetrahedron Lett.*, 2002, **43**, 1999.
370. S. Saaby, P. Bayon, P.S. Aburel and K.A. Jorgensen, *J. Org. Chem.*, 2002, **67**, 4352.
371. S. Kotha, S. Halder and E. Brahmachary, *Tetrahedron*, 2002, **58**, 9203.
372. H. Park, B. Cao and M.M. Joullie, *J. Org. Chem.*, 2001, **66**, 7223.
373. A. Long and S.W. Baldwin, *Tetrahedron Lett.*, 2001, **42**, 5343.
374. W.-Q. Lin, Z. He, Y. Jing, X. Cui, H. Liu and A.-Q. Mi, *Tetrahedron Asymmetry*, 2001, **12**, 1583.
375. J. Casas, C. Najera, J.M. Sansano, J. Gonzalez, J.M. Saa and M. Vega, *Tetrahedron Asymmetry*, 2001, **12**, 699.
376. S. Kotha, M. Behera and R.V. Kumar, *Bioorg. Med. Chem. Lett.*, 2002, **12**, 105.
377. S. Royo, P. Lopez, A.I. Jimenez, L. Oliveros and C. Cativiela, *Chirality*, 2002, **14**, 39.
378. V. Jullian, V. Monjardet-Bas, C. Fosse, S. Lavielle and G. Chassaing, *Eur. J. Org. Chem.*, 2002, 1677.
379. A. Bianco, T. DaRos, M. Prato and C. Toniolo, *J. Pept. Sci.*, 2001, **7**, 208.
380. G.A. Burley, P.A. Keller, S.G. Pyne and G.E. Ball, *J. Org. Chem.*, 2002, **67**, 8316.
381. J.-P. Mazaleyrat, K. Wright, M. Wakselman, F. Formaggio, M. Crisma and C. Toniolo, *Eur. J. Org. Chem.*, 2001, 1821.
382. S.-J. Wen, H.W. Zhang and Z.-J. Yao, *Tetrahedron Lett.*, 2002, **43**, 5291.
383. Y. Yokoyama, K. Osanai, M. Mitsuhashi, K. Kondo and Y. Murakami, *Heterocycles*, 2001, **55**, 653.
384. B.E. Haug, J. Andersen, O. Rekdal and J.S. Svendsen, *J. Pept. Sci.*, 2002, **8**, 307.
385. D.K. Pyun, C.H. Lee, H.-J. Ha, C.S. Park, J.-W. Chang and W.K. Lee, *Org. Lett.*, 2001, **3**, 4197.
386. C. Xiong, W. Wang, C. Cai and V.J. Hruby, *J. Org. Chem.*, 2002, **67**, 1399.
387. A. Bertram and G. Pattenden, *Synlett*, 2001, 1873.

388. L. De Luca, G. Giacomelli and A. Riu, *J. Org. Chem.*, 2001, **66**, 6823.
389. H. Kromann, F.A. Slok, T.B. Stensbol, H. Braeuner-Osborne, U. Madsen and P. Krogsgaard-Larsen, *J. Med. Chem.*, 2002, **45**, 988.
390. L. Bunch, P. Krogsgaard-Larsen and U. Madsen, *J. Org. Chem.*, 2002, **67**, 2375.
391. H. Pajouhesli, M. Hosaaini-Meresht, S.H. Pajouhesta and K. Cunry, *Tetrahedron Asymmetry*, 2000, **11**, 4955.
392. W.-C. Cheng, Y. Liu, M. Wong, M.M. Olmstead, K.S. Lam and M.J. Kurth, *J. Org. Chem.*, 2002, **67**, 5673.
393. R.P. Clausen, H. Braeuner-Osborne, J.R. Greenwood, M.B. Hermit, T.B. Stensbol, B. Nielsen and P. Krogsgaard-Larsen, *J. Med. Chem.*, 2002, **45**, 4240.
394. S. Van-ay, R. Lazaro, J. Martinez and F. Lamaty, *Eur. J. Org. Chem.*, 2002, 2308.
395. S.K. Bertilsson, J.K. Ekegren, S.A. Modin and P.G. Andersson, *Tetrahedron*, 2001, **57**, 6399.
396. R.M. Adlington, J.E. Baldwin, D. Catterick and G.J. Pritchard, *J. Chem. Soc. Perkin Trans 1*, 2001, 668.
397. R.C.F. Jones, D.J.C. Berthelot and J.N. Iley, *Tetrahedron*, 2001, **57**, 6539.
398. T. Yokomatsu, K. Takada, Y. Yuasa and S. Shibuya, *Heterocycles*, 2002, **56**, 545.
399. P.M. T. Ferreira, H.L.S. Maia, L.S. Monteiro and J. Sacramento, *J. Chem. Soc. Perkin Trans 1*, 2001, 3167.
400. K.J. Kise Jr. and B.E. Bowler, *Inorg. Chem.*, 2002, **41**, 379.
401. R. Warmuth, T.E. Munsch, R.A. Stalker, B. Li and A. Beatty, *Tetrahedron*, 2001, **57**, 6383.
402. M. Isaac, A. Slassi, K. Da Silva and T. Xin, *Tetrahedron Lett*, 2001, **42**, 2957.
403. G. Sui, P. Kele and J. Orbulescu, *Lett. Pept. Sci.*, 2001, **8**, 47.
404. P. Kele, G. Sui, Q. Huo and R.M. Leblanc, *Tetrahedron Asymmetry*, 2000, **11**, 4959.
405. K. Guzow, M. Szabelski, J. Malicka and W. Wiczk, *Helv. Chim. Acta*, 2001, **84**, 1086.
406. N. Jotterand, D.A. Pearce and B. Imperiali, *J. Org. Chem.*, 2001, **66**, 3224.
407. A.R. Nitz Mark-Mezo, M.H. Ali and B. Imperiali, *Chem. Commun. (Cambridge)*, 2002, 1912.
408. S. Jiranusomkul, B. Sirithun, H. Nemoto and H. Takahata, *Heterocycles*, 2002, **56**, 487.
409. T.V. Shokol, O.S. Ogorodnuchuk, V.V. Shilin, V.B. Milevskaya and V.P. Khilya, *Chem. Het. Comp.*, 2002, **38**, 151.
410. M. Adamczyk, S.R. Akireddy and R.E. Reddy, *Tetrahedron Asymmetry*, 2001, **12**, 2385.
411. C. Escolano, M. Rubiralta and A. Diez, *Tetrahedron Lett.*, 2002, **43**, 4343.
412. V.J. Huber, T.W. Arroll, C. Lum, B.A. Goodman and H. Nakanishi, *Tetrahedron Lett.*, 2002, **43**, 6729.
413. S.K. Kapadia, D.M. Spero and M. Eriksson, *J. Org. Chem.*, 2001, **66**, 1903.
414. A. Olma, A. Gniadzik, A.W. Lipkowski and M. Lachwa, *Acta Biochim. Pol.*, 2001, **48**, 1165.
415. A.S. Sagiyan, A.V. Geolchanyan, N.R. Martiosyan, S.A. Dadayan, V.I. Tararov, Yu.N. Belokon, T.V. Kochikyan, V.S. Arutyunyan and A.A. Avetisyan, *Hayastani Kimiakan Handes*, 2002, **55**, 84.
416. J.A. Zerkowski, L.M. Hensley and D. Abramowitz, *Synlett.*, 2002, 557.
417. F. Paradisi, F. Piccinelli, G. Porzi and S. Sandri, *Tetrahedron Asymmetry*, 2002, **13**, 497.
418. J.L. Roberts and C.K. Chan, *Tetrahedron Lett.*, 2002, **43**, 7679.

419. A. J. Robinson, P. Stanislawski, D. Mulholland, L. He and H.-Y. Li, *J. Org. Chem.*, 2001, **66**, 4148.
420. M. Lobez-Garcia, I. Alfanso and V. Gotor, *J. Org. Chem.*, 2003, **68**, 648.
421. A.J. Villani, D. Saunders, A.Y.L. Shu and R.J. Heys, *J. Labell. Comp. Radiopharm.*, 2002, **45**, 49.
422. F. Paradisi, G. Porzi, S. Rinaldi and S. Sandri, *Tetrahedron Asymmetry*, 2000, **11**, 4617.
423. A.S. Segiyan, A.v. Geolchanyan, L.G. Minasayan, L.L. Manasayan, R.V. Ovspeyan and Yu.N. Belokon, *Hayastani Kimiakan Handes*, 2002, **55**, 103.
424. M.M. Kabat, *Tetrahedron Lett.*, 2001, **42**, 7521.
425. S.K. Das, V.L. Narishima Rao Krowidi, H. Jagadheshan and J. Iqbal, *Bioorg. Med. Chem. Lett.*, 2002, **12**, 3579.
426. J.A. Gomez-Vidal and R.B. Silverman, *Org. Lett.*, 2001, **3**, 2481.
427. S.R. Chhabra, A. Mahajan and W.C. Chan, *J. Org. Chem.*, 2002, **67**, 4017.
428. R.L. Tennyson, G.S. Cortez, H.J. Galicia, C.R. Kreiman, C.M. Thomson and D. Romo, *Org. Lett.*, 2002, **4**, 533.
429. (*a*) K.R. Knudsen, T. Risgaard, N. Nishiwaki, K.V. Gothelf and K.A. Jorgenson, *J. Am. Chem. Soc.*, 2001, **123**, 5843; (*b*) K.R. Knudsen, N. Nishiwaki, K.V. Gothelf and K.A. Jorgenson, *Angew. Chem. Int. Ed.*, 2001, **40**, 2992.
430. P.A. Butler, C.G. Crane, B.T. Golding, A. Hammershoi, D.C. Hockless, T.B. Petersen, A.M. Sargeson and D.C. Ware, *Inorg. Chimica Acta*, 2002, **331**, 318.
431. R. Clancy, A.I. Cederbaum and D.A. Stoyanovsky, *J. Med Chem.*, 2001, **44**, 2035.
432. S. Watanabe, A. Cordova, F. Tanaka and C.F. Carlos III, *Org. Lett.*, 2002, **4**, 4519.
433. M. Adamczyk and R.E. Reddy, *Synth. Commun.*, 2001, **31**, 579.
434. F.S. Gibson, A.K. Singh, M.C. Sourmeillant, P.S. Manchant, M. Humora and D.R. Kronenthal, *Org. Proc. Res, Dev.*, 2002, **6**, 814.
435. A. Kumanishi, N. Oaaki, S. Tanfanori, T. Tsuda, M. Tagaguki, K. Ono and M. Kirihata, *KURKI KR*, 2000, **541**, 315.
436. H. Nakamura, M. Figiwara and Y. Yamamoto, *Frontiers in Neutron Capture Therapy*, 2001, **2**, 765.
437. G.W. Kabalka, T.L. Nichols, M. Akula, C.P.D. Longford and L. Miller, '*Synth. and Appl. Isotop.-labelled Compds.*' Proc.7th Int. Symp. Dresden, eds. U. Pless and R. Voges, J. Wiley & Sons Ltd., 2001, 329.
438. B.C. Das, S. Das, G. Li, W. Bao and G.W. Kabalka, *Synlett*, 2001, 1419.
439. R. Tacke and V.I. Handmann, *Organometallics*, 2002, **21**, 2619.
440. M. Merget, K. Gunther, M. Bernd, E. Gunther and R. Tacke, *J. Organomet. Chem.*, 2001, **628**, 183.
441. M. Nath and S. Goyal, *Phosph. Sulfur, Silicon and Rel. Elements*, 2002, **177**, 841.
442. J.-X. Chen, J.A. Tonge and J.R. Norton, *J. Org. Chem.*, 2002, **67**, 4366.
443. M. Ruiz, V. Ojea, J.M. Quintela and J.J. Guillin, *Chem. Commun. (Cambridge)*, 2002, 1600.
444. D.J. Brauer, K.W. Kottsieper, S. Schenk and O. Stelzer, *Z. Anorg. Allg. Chem.*, 2001, **627**, 1151.
445. A.A. Karasik, O.G. Sinyashin, J. Heinicke and E. Hey-Hawkins, *Phosph. Sulfur, Silicon and Rel. Elements*, 2002, **177**, 1469.
446. W.-Q. Liu, C. Olszowy, L. Bischoff and G. Garbay, *Tetrahedron Lett.*, 2002, **43**, 1417.
447. M.C. Feraandez, J.M. Yumtela, M. Ruiz and V. Ojea, *Tetrahedron Asymmetry*, 2001, **13**, 233.

448. B. Bessieres, A. Schoenfelder, C. Verrat, A. Mann, P. Ornstein and C. Pedregal, *Tetrahedron Lett.*, 2002, **43**, 7659.
449. S. Kobayashi, N. Shiraishi, W.W.L. Lam and K. Manabe, *Tetrahedron Lett.*, 2001, **42**, 7303.
450. M.U. Gieaelman, L. Xie and W.A. van der Donk, *Org. Lett.*, 2001, **3**, 1331.
451. R.J. Hondal, B.L. Nilsson and R.T. Raines, *J. Am. Chem. Soc.*, 2001, **123**, 5140.
452. H.B. Ganther, *Bioorg. Med. Chem.*, 2001, **9**, 1459.
453. Y. Xie, M.D. Short, P.B. Cassidy and J.C. Roberts, *Bioorg. Med. Chem. Lett.*, 2001, **11**, 2911.
454. A.S. Sagiyan, A.V. Geolchanyan, S.V. Vardapetvan, A.A. Avetisyan, V.I. Tararov, N.A. Kuzmina, Yu.N. Belakon and M. North, *Russian Chem. Bull.*, 2000, **49**, 1460.
455. S. Mourtas, D. Gatos, V. Kalaitzi, C. Katakalou and K. Barlos, *Tetrahedron Lett.*, 2001, **42**, 6965.
456. E. Morera, F. Pinnen and G. Lucente, *Org. Lett.*, 2002, **4**, 1139.
457. S. Gazal, G. Gellerman, E. Glukhov and C. Gilon, *J. Pept. Res.*, 2001, **43**, 527.
458. N. Inguimbert, H. Poras, F. Teffo, F. Beslot, M. Selkti, A. Tomas, E. Scalbert, C. Bennejean, P. Renard, M.-C. Fournie-Zaluski and B.-P. Roques, *Bioorg. Med. Chem. Lett.*, 2002, **12**, 2001.
459. M. Seki, M. Hatsuda, Y. Mori and S.-i. Yamada, *Tetrahedron Lett.*, 2002, **43**, 3269.
460. P. Meffre and P. Durand, *Synth. Commun.*, 2002, **32**, 287.
461. N.O. Silva, A.S. Abreu, P.M.T. Ferreira, L.S. Monteiro and M.-J.R.P. Queiroz, *Eur. J. Org. Chem.*, 2002, 2524.
462. C. Bolm, I. Schiffers, C.L. Dinter, L. Defrere, A. Gerlach and G. Raabe, *Synthesis-Stuttgart*, 2001, 1719.
463. C. Cimarelli, G. Palmieri and E. Volpini, *Synth. Commun.*, 2001, **31**, 2943.
464. S.D. Bull, S.G. Davies and A.D. Smith, *J. Chem. Soc. Perkin Trans 1*, 2001, 2931.
465. I.L. Iovel, J. Golomba, A. Popelis and E.L. Gaukhmann, *Appl. Organomet. Chem.*, 2001, **15**, 67.
466. S. Fuatero, M.D. Diaz, A. Navarro, E. Salavert and E. Aguilar, *Tetrahedron*, 2001, **57**, 703.
467. S.D. Bull, S.G. Davies, P.M. Kelly, M. Gianotti and A.D. Smith, *J. Chem. Soc. Perkin Trans. 1*, 2001, 3106.
468. M. Lindo, K. Itoh, C. Tsuchiya and K. Shishido, *Org. Lett.*, 2002, **4**, 3119.
469. C.Y.K. Tan and D.F. Weaver, *Tetrahedron*, 2002, **58**, 7449.
470. F.A. Davis, J. Deng, Y. Zhang and R.C. Haltiwanger, *Tetrahedron*, 2001, **58**, 7135.
471. D. Saylik, E.M. Campi, A.C. Donohue, W.R. Jackson and A.J. Robinson, *Tetrahedron Asymmetry*, 2001, **12**, 657.
472. A.V. Sivakumar, G.S. Babu and S.V. Bhat, *Tetrahedron Asymmetry*, 2001, **12**, 1095.
473. S.G. Davies, K. Iwamoto, C.A.P. Smethurst, A.D. Smith and H. Rodriguez-Solla, *Synlett*, 2002, 1146.
474. K. Ananda, H.N. Gopi and V.V. S. Babu, *Ind. J. Chem. Section B: Org. Chem. Med. Chem.*, 2001, **40B**, 790.
475. G.-R. Vasanthakumar and V.V.S. Babu, *Synth. Commun.*, 2002, **32**, 651.
476. G.-R. Vasanthakumar, B.S. Patil and V.V.S. Babu, *J. Chem. Soc. Perkin Trans. 1*, 2002, 2087.
477. A. Kumar, S. Ghilagaber, J. Knight and P.B. Wyatt, *Tetrahedron Lett.*, 2001, **43**, 6991.

478. J. Huck, J.-M. Receveur, M.-L. Roumestant and J. Martinez, *Synlett*, 2001, 1467.
479. (*a*) V.M. Gutierrez-Garcia, H. Lopez-Ruiz, G. Reyes-Rangel and E. Juaristi, *Tetrahedron*, 2001, **57**, 6487; (*b*) V. M. Gutierrez-Garcia, H. Lopez-Ruiz, G. Reyes-Rangel, O. Munoz-Muniz and E. Juaristi, *J. Braz. Chem. Soc.*, 2001, **12**, 652.
480. Z.H. Ma, C. Liu, Y.H. Zhao, W. Li and J.B. Wang, *Chin. Chem. Lett.*, 2002, **13**, 721.
481. S. Kobayashi, J. Kobayashi, H. Ishiani and M. Ueno, *Chem. –A Eur. J.*, 2002, **8**, 4185.
482. H.M.L. Davies and C. Venkataramani, *Angew. Chem. Int. Ed.*, 2001, **41**, 2197.
483. H.-J. Ha, Y.-G. Ahn, J.-S. Woo, G.S. Lee and W.K. Lee, *Bull. Chem. Soc. Jpn.*, 2001, **74**, 1667.
484. R.V. Roers, *Tetrahedron Lett.*, 2001, **42**, 3563.
485. K.-D. Lee, J.M. Suh, J.H. Park, H.J. Ha, H.G. Choi, C.S. Park, J.W. Chang, W.K. Lee, Y. Dong and H. Yun, *Tetrahedron*, 2001, **57**, 8267.
486. H. Nemoto, R. Ma, X. Li, I. Suzuki and M. Shibuya, *Tetrahedron Lett.*, 2001, **42**, 2145.
487. L. Ambroise, E. Dumez, A. Szeki and R.F.W. Jackson, *Synthesis-Stuttgart*, 2001, 2296.
488. R. Caputo, G. Cecere, A. Guaragna, G. Palumbo and S. Pedatella, *Eur. J. Org. Chem.*, 2002, 3050.
489. B. Crousse, S. Nanzuka, D. Bonnet-Delpon and J.-P. Begue, *Synlett*, 2001, 679.
490. (*a*) S. Fustero, E. Salavert, B. Pina, M.C. Ramirez de Arallano and A. Asensio, *Tetrahedron*, 2001, **57**, 6475; (*b*) S. Fustero, E. Salavert, B. Pina, A. Navarro, M.C. Ramirez de Arallano and A.S. Fuentes, *J. Org. Chem.*, 2002, **67**, 4667.
491. N.N. Sergeeva, A.S. Golubev, L. Hennig, M. Findeisen, E. Paetzold, G. Oehme and K. Burger, *J. Fluorine Chem.*, 2001, **111**, 41.
492. N. Lebouvier, C. Laroche, F. Heguenot and T. Brigaud, *Tetrahedron Lett.*, 2002, **43**, 2827.
493. V.A. Soloshonok, H. Ohkura, A. Sorochinsky, N. Voloshin, A. Markovsky, M. Belik and T. Yamasaki, *Tetrahedron Lett.*, 2002, **43**, 5445.
494. D.D. Staas, K.L. Savage, C.F. Homnik, N.N. Tsou and R.G. Ball, *J. Org. Chem.*, 2001, **67**, 8276.
495. N.N. Sergeeva, A.S. Golubev and K. Burger, *Synthesis-Stuttgart*, 2001, 281.
496. Y. Vera-Ayoso, P. Borrachero, F. Cabrera-Escribano, M.J. Dianez, M.D. Estrada, M. Gomez-Guillen, A. Lopez-Castro and S. Perrez-Garrido, *Tetrahedron Asymmetry*, 2001, **12**, 2031.
497. R.P. Tripathi, R. Tripathi, V.K. Tiwari, L. Bala, S. Sinha, A. Srivastava, R. Srivastava and B.S. Srivastava, *Eur. J. Med. Chem.*, 2002, **37**, 773.
498. S.D. Bull, S.G. Davies, S. Delgado-Ballester, P.M. Kelly, L.J. Kotchie, M. Gianotti, M. Laderas and A.D. Smith, *J. Chem. Soc. Perkin Trans. 1*, 2001, 3112.
499. R. Beumer and O. Reiser, *Tetrahedron*, 2001, **57**, 6497.
500. D.J. Aitken, C. Gauzy and E. Pereira, *Tetrahedron Lett.*, 2002, **43**, 6177.
501. F. Felluga, G. Pitacco, M. Prodan, S. Prici, M. Visintin and E. Valentin, *Tetrahedron Asymmetry*, 2001, **12**, 3241.
502. C. Taillefumier, Y. Lakhrissi and M. Lakhrissi, and Y. Chapleur, *Tetrahedron Asymmetry*, 2002, **13**, 1707.
503. A. Berkessel, K. Glaubitz and J. Lex, *Eur. J. Org. Chem.*, 2002, 2948.
504. V. Wehner, H. Blum, M. Kurz and H.U. Stilz, *Synthesis-Stuttgart*, 2002, 2023.
505. Z. Ma, Y.-h. Zhao, N. Jiang, X. Jin and J. Wang, *Tetrahedron Lett.*, 2002, **43**, 3209.

506. K. Lee, M. Zhang, D. Yang and T.R. Burke, *Bioorg. Med. Chem. Lett.*, 2002, **12**, 3399.
507. A. Dinsmore, P.M. Doyle, M. Steger and D.W. Young, *J. Chem. Soc. Perkin Trans. 1*, 2002, 613.
508. J. Fan-as, X. Ginesta, P.W. Sutton, J. Taltavull, F. Egeler, P. Romea, F. Urpi and J. Vilarrasa, *Tetrahedron*, 2001, **57**, 7665.
509. A. Mordini, L. Sbaragli, M. Valacchi, F. Russo and G. Reginato, *Chem. Commun. (Cambridge)*, 2002, **778**.
510. J.-i. Park, G.M. Tian and D.H. Kim, *J. Org. Chem.*, 2001, **66**, 3696.
511. S.G. Nelson, K.L. Spencer, W.S. Cheung and S.J. Mamie, *Tetrahedron*, 2002, **58**, 7081.
512. S. Lee and Y.J. Zhang, *Org. Lett.*, 2002, **4**, 2429.
513. C.G. Espino, P.M. When, J. Chow and J. DuBois, *J. Am. Chem. Soc.*, 2001, **123**, 6935.
514. E. Alonso, C. del Pozo and J. Gonzalez, *Synlett*, 2002, 69.
515. S.-H. Lee, X. Qi, J. Yoon, K. Nakamura and Y.-S. Lee, *Tetrahedron*, 2002, **58**, 2777.
516. T.B. Durham and M.J. Miller, *J. Org. Chem.*, 2003, **68**, 35.
517. S. Murahashi, Y. Imada, T. Kawakami, K. Harada, Y. Yonemushi and N. Tomita, *J. Am. Chem. Soc.*, 2002, **124**, 2888.
518. J.L. Matthews, D.R. McArthur and K.W. Muir, *Tetrahedron Lett.*, 2002, **43**, 5401.
519. G.V. Sidorov and N.F. Myasoedov, *Radiochemistry*, 2002, **44**, 295.
520. S. Rajesh, B. Banerji and J. Iqbal, *J. Org. Chem.*, 2002, **67**, 7852.
521. B.-r. Liao, Y.-y. Hong, Zh.-p. Chen and B. Liu, *Uprhing Huaxue*, 2001, 48.
522. Z.-Y. Jiang and H.-F. Chen, *Ymgyong Huaxue*, 2001, **18**, 231.
523. F.-R. Alexandre, D.P. Panteleone, P.P. Taylor, I.G. Fotheringham, D.J. Ager and N.J. Turner, *Tetrahedron Lett.*, 2002, **43**, 707.
524. T.M. Beard and N.J. Turner, *Chem. Commun. (Cambridge)*, 2002, 246.
525. X. Xing, A. Fichera and K. Kumar, *J. Org. Chem.*, 2002, **67**, 1722.
526. Y. Liu, J.-w. Cao and Z. Chao, *Ziran Kexwban*, 2000, **46**, 769.
527. M. Solymar, A. Liljeblad, L. Lazar, F. Fulop and L.T. Kanerva, *Tetrahedron Asymmetry*, 2002, **13**, 1923.
528. G. Galavema, R. Corradini, F. Dallavalle, G. Folesani, A. Dossena and R. Marchelli, *J. Chromatogr. A*, 2001, **922**, 151.
529. W. Lee, *Anal. Lett.*, 2001, **34**, 2785.
530. S.Y. Park, J.K. Park, J.H. Park, C.V. McNeff and P.W. Carr, *Microchem. J.*, 2001, **70**, 179.
531. C.-H. Lin, C.-E. Lin, C.-C. Chen and L.-F. Liao, *J. Chin. Chem. Soc.*, 2001, **48**, 1069.
532. Y.K. Ye, B.S. Lord, L. Yin and R.W. Stringham, *J. Chromatogr. A*, 2002, **945**, 147.
533. A.I. Jimenez, P. Lopez, L. Ollveroe and C. Cativiela, *Tetrahedron*, 2001, **57**, 6019.
534. M.H. Hyun, S.C. Han, B.H. Lipshutz, Y.-J. Shin and C.J. Welch, *J. Chromatogr. A*, 2001, **910**, 359.
535. K. Hamase, *Farumashia*, 2002, **38**, 437.
536. M.H. Huyn, Y.J. Cho and I.K. Baik, *Bull. Korean. Chem. Soc.*, 2002, **23**, 1291.
537. F. Gasparrini, D. Misiti, M. Pierini and C. Villani, *Org. Lett.*, 2002, **4**, 3993.
538. C. B'Hymer, M. Montes-Bayon and J.A. Caruso, *J. Sep. Sci.*, 2003, **26**, 7.
539. D. Zhang, W.A. Tao and R.G. Cooks, *Int. J. Mass Spectrom.*, 2001, **204**, 159.
540. R. Hodyss, R.R. Julian and J.L. Beauchamp, *Chirality*, 2001, **13**, 703.
541. J.F. Gal, M. Stone and C.B. Lebrilla, *Int. J. Mass Spectrom.*, 2002, **222**, 259.

542. E.M. Perez, A.I. Oliva, J.V. Hernandez, L. Simon, J.R. Moran and F. Sanz, *Tetrahedron Lett.*, 2001, **42**, 5853.
543. A.I. Oliva, L. Simon, J.V. Hernandez, F.M. Muniz, A. Lithgow, A. Jimenez and J.R. Moran, *J. Chem. Soc. Perkin Trans. 2*, 2002, 1050.
544. M. Rekharsky, H. Yamamura, M. Kawai and Y. Inoue, *J. Am. Chem. Soc.*, 2001, **123**, 5360.
545. (*a*) H. Zhao, C.-H. Xue, Q.M. Mu, L. Li and S.-H. Chen, *Yingyong Huaxue*, 2001, **18**, 614; (*b*) C.-H. Xue, Q.-M. Mu and S.-H. Chen, *Huaxue Xuebao*, 2002, **60**, 355.
546. L.L. Lawless, A.G. Blackburn, A.J. Ayling, M.N. Perez-Payan and A.P. Davis, *J. Chem. Soc. Perkin Trans. 1*, 2001, 1329.
547. Q. Mu, C. Xie and S. Chen, *Huaxue Yanjiu Yu Yingyong*, 2001, **13**, 473.
548. G.M. Kyne, M.E. Light, M.B. Hursthouse, J. de Mendoza and J. D. Kilburn, *J. Chem. Soc. Perkin Trans. 1*, 2001, 1258.
549. I. Fujii and N. Hirayama, *Helv. Chim. Acta*, 2002, **85**, 2946.
550. A. Higuchi, K. Furuta, H. Yomogita, B.O. Yoon, M. Hara, S. Maniwa and M. Saitoh, *Desalination*, 2002, **148**, 155.
551. P.E.M. Overdevest, T.J.M. de Bruin, E.J.R. Sudhoelter, K. van't Riet, J.T.F. Keurentjes and A. van der Padt, *Ind. Eng. Chem. Res.*, 2001, **40**, 5991.
552. J. Romero and A.L. Zydney, *Desalination*, 2002, **148**, 159.
553. (*a*) P. Hadik, L.-P. Szabo and E. Nagy, *Desalination*, 2002, **148**, 193; (*b*) P. Hadik, L.-P. Szabo and E. Nagy, *Bulgarian Chem. Commun.*, 2001, **33**, 389.
554. H. Chibvongodze, K. Hayashi and K. Toko, *Sens. Mater.*, 2001, **13**, 99.
555. K. Araki, M. Goto and S. Furusaki, *Anal. Chim. Acta*, 2002, **469**, 173.
556. H. Okuno, T. Kitano, H. Yakabe, M. Kishimoto, B.A. Deore, H. Siigi and T. Nagaoka, *Anal. Chem.*, 2002, **74**, 4184.
557. T. Shiraiwa, M. Suzuki, Y. Sakai, H. Nagasawa, K. Takatani, D. Noshi and K. Yamanashi, *Chem. Pharm. Bull.*, 2002, **50**, 1362.
558. M.J. O'Donnell and F. Delgado, *Tetrahedron*, 2001, **57**, 6641.
559. D. Lavergne, C. Mordant, V. Ratovelomanana-Vidal and J.-P. Genet, *Org. Lett.*, 2001, **3**, 1909.
560. L.-l. Yang, D.-q. Zhang and Z.-b. Yuan, *Anal. Chim. Acta*, 2001, **433**, 23.
561. E. Francotte, T. Leutert, L. La Vecchia, F. Ossala, P. Richert and A. Schmidt, *Chirality*, 2002, **14**, 313.
562. A. Solladie-Cavallo, O. Sedy, M. Salisova and M. Schmitt, *Eur. J. Org. Chem.*, 2002, 3042.
563. J.-f. Hang, S.-K. Tian, L. Tang and L. Deng, *J. Am. Chem. Soc.*, 2001, **123**, 12696.
564. A. Gaucher, Y. Zuliani, D. Cabaret, M. Wakselman and J.-P. Mazaleyrat, *Tetrahedron Asymmetry*, 2001, **12**, 2571.
565. J.L. Clark and J.J. Stezowski, *J. Am. Chem. Soc.*, 2001, **123**, 9880.
566. J.L. Clark, B.R. Booth and J.J. Stezowski, *J. Am. Chem. Soc.*, 2001, **123**, 9889.
567. M.F. Rastegar, M. Ghandi, M. Taghizadeh, A. Yari, M. Shamsipur, G.P.A. Yap and H. Rahbamoohi, *J. Org. Chem.*, 2002, **67**, 2065.
568. C. Alonso, R. Eliash, T.R. Jensen, K. Kjaer, M. Lahav and L. Leiserowitz, *J. Am. Chem. Soc.*, 2001, **123**, 10105.
569. K. Nomiya and H. Yokoyama, *J. Chem. Soc. Dalton Trans.*, 2002, 2483.
570. Q.-h. Zhu, W.-y. Shao, J.-f. He and Q.-y. Deng, *Bopuxue Zazhi*, 2001, **18**, 377.
571. E. Bang, J.-W. Jung, W. Lee, D.W. Lee and W. Lee, *J. Chem.. Soc. Perkin Trans. 2*, 2001, 1685.
572. M. Claeys-Bruno, D. Toronto, J. Pecaut, M. Bardet and J.-C. Marchon, *J. Am. Chem. Soc.*, 2001, **123**, 11067.

573. O. Hayashida, L. Sebo and J. Rebek Jr., *J. Org. Chem.*, 2002, **67**, 8291.
574. S.D. Keane, C.J. Easton, S.F. Lincoln and D. Parker, *Aus. J. Chem.*, 2001, **54**, 535.
575. A. Yoshino, M. Ishida, H. Yuki, H. Okabayashi, H. Masuda and C.J. O'Connor, *Coll. Poly. Sci.*, 2001, **279**, 1144.
576. A. Hayashi, S. Saito, Y. Nakatani, A. Nishiyama, Y. Matsumura, H. Nakayama and M. Tsuhako, *Phosphorus Res. Bull.*, 2001, **12**, 129.
577. S. Ono, T. Taguma, S. Kuroki, I. Ando, H. Kimura and K. Yamauchi, *J. Mol. Struct.*, 2001, **603-;3**, 49.
578. N. Higashi, T. Koga, Y. Fujii and M. Niwa, *Langmuir*, 2001, **17**, 4061.
579. H. Tsukube, M. Wada, S. Shinoda and H. Tamiaki, *J. Alloys Compd.*, 2001, **323-4**, 133.
580. W. He, F. Liu, Y. Mei, Z. Guo and L. Zhu, *New J. Chem.*, 2001, **25**, 1330.
581. Y. Cheng and D.M. Hercules, *J. Mass Spectrom.*, 2001, **36**, 834.
582. M.E. Mateo, S.M. Barlow, S. Haq and R. Raval, *Surface Sci.*, 2002, **501**, 191.
583. Q.-P. Wang, D.-Z. Lu, L. Shen and J.-Z. Yang, *Wuli Huaxue Xuebao*, 2001, **17**, 952.
584. M. Monajjemi, E. Moniri and H.A. Panahi, *J. Chem. Eng. Data.*, 2001, **46**, 1249.
585. S. Shao and R.-S. Lin, *Zhejiang Daxue Xuebao*, 2001, **28**, 269.
586. S. Shao, R.-S. Lin, X.-G. Hu, W.-J. Fang and X.-H. Ying, *Wuli Huaxue Xuebao*, 2001, **17**, 645.
587. P.G. Rohankar and A.S. Aswar, *Indian J. Chem. A, Inorg. Bio-inorg. Phys. Theor. Anal. Chem.*, 2001, **40A**, 1086.
588. Y. Lu, W. Xie and J. Lu, *Thermochim. Acta*, 2002, **385**, 1.
589. M.N. Islam and R.K. Wadi, *Phys. Chem. Liq.*, 2001, **30**, 77.
590. F. Khan and P.L. Sahu, *J. Inst. Chem.*, 2000, **72**, 127.
591. Q. Lau, X. Hu, R. Lin, S. Li and W. Sang, *Thermochim. Acta*, 2001, **369**, 31.
592. A. Al-Khouly, *Roi. Soc. Quim. Peru*, 2000, **66**, 9.
593. B. Palecz, *J. Am. Chem. Soc.*, 2002, **124**, 6003.
594. Y. Wu, P. Ma, Y. Liu and S. Li, *Fluid Phase Equil.*, 2001, **186**, 27.
595. C. Gao and J.H. Vera, *J. Chem. Eng.*, 2001, **79**, 392.
596. B.M. Beglov, B.S. Zakirov and K.N. Karimova, *Uzb. Khim. Zh.*, 2001, **8**.
597. S.-Y. Tan, H. Yang and B. Tang, *Yingywg Huaxue*, 2001, **18**, 252.
598. S. Chen, X. Yang, Z. Ju, H. Li and S. Gao, *Chem. Pap.*, 2001, **55**, 239.
599. T.S. Banipal, G. Singh and B.S. Lark, *J. Solution Chem.*, 2001, **30**, 657.
600. J. Fan, X. Shen and J. Wang, *Electroanalysis*, 2001, **13**, 1115.
601. P.S. Kumar, J.A. Hogendoon, P.H.M. Feron and G.F. Versteeg, *J. Chem. Eng. Data*, 2001, **46**, 1357.
602. B. N. Waris, U. Hassan and N. Shrivastava, *Indian J. Chem. A, Inorg. Bio-inorg. Phys. Theor. Anal. Chem.*, 2001, **40A**, 1218.
603. N.Y. Mokshina, V.F. Selemenev and G.Y. Oros, *Izv. Vyssh. Khimiya Khimicheskaya Technolog.*, 2001, **44**, 17.
604. R.A. Marriott, *J. Chem. Thermodynamics*, 2001, **33**, 959.
605. J.L. Liu, A.W. Hakin and G.R. Hedwig, *J. Solution Chem.*, 2001, **30**, 861.
606. T.I. Lezhava, N.Sh. Ananiashvili, M.P. Kikabidze and N.O. Berdzenishvili, *Russian J. Electrochem.*, 2001, **37**, 1395.
607. P.G. Rohankar and A.S. Aswar, *Indian J. Chem. A, Inorg. Bio-inorg. Phys. Theor. Anal. Chem.*, 2002, **41A**, 312.
608. E.N. Tsurko, T.M. Shihova and N.V. Bondarev, *J. Mol. Liq.*, 2002, **96-7**, 425.
609. A.W. Hakin and G.R. Hedwig, *J. Chem. Thermodynamics*, 2001, **33**, 1709.
610. L. Ninni and A.J.A. Meirelles, *Biotechnol. Progr.*, 2001, **17**, 703.

611. A. Shono, T. Okabe and K. Satoh, *Solvent Ext. Res.Dev. Jpn.*, 2001, **8**, 120.
612. S.-x. Zhang, Y. Yao, X.-k. Yan, Y.-q. Yang and S. Zhang, *Huadong Ugong Daxue Xitebao*, 2000, **26**, 678.
613. H.M. Willemen, T. Vermonden, A.T.M. Marcelis and E.J.R. Sudhölter, *Eur. J. Org. Chem.*, 2001, 2329.
614. C. Wang, J. Huang, S. Tang and B. Zhu, *Langmuir*, 2001, **17**, 6389.
615. M. Sorensen, J.S. Forster, P. Hvelplund, T.J.D. Jorgensen, S.B. Nielsen and S. Tomita, *Chem-Eur. J.*, 2001, **7**, 3214.
616. A.N. Zyablov, T.V. Eliseeva, V.F. Selemenev and N.N. Samoilova, *Zh. Biz. Khim.*, 2001, **75**, 545.
617. J. Hendri, A. Hiroki, Y. Maekawa, M. Yoshida and R. Katakai, *Radiat. Phys. Chem.*, 2001, **61**, 155.
618. M. Suzuki, M. Yumoto, M. Kimura, H. Shirai and K. Hanabusa, *Chem. Commun. (Cambridge)*, 2002, 884.
619. M. Matsuzawa, H. Minami, T. Yano, T. Wakabayashi, M. Iwahashi, K. Sakamoto and D. Kaneko, *Stud. Surface Sci. and Cat.*, 2001, **132**, 137.
620. D. Renzi, C.M. Carlevaro, C. Stoico and F. Vericat, *Mol. Phys.*, 2001, **99**, 913.
621. A. Dogan, F. Koseoglu and E. Kilic, *Indian J. Chem. A, Inorg. Bio-inorg. Phys. Theor. Anal. Chem.*, 2002, **41A**, 960.
622. I.N. Kochnev, A.I. Khaloimov, E.I. Grigor'ev, L.V. Shurpova and V.Kh. Khavinson, *Biofizika*, 2002, **47**, 12.
623. D.L. Kotova, D.S. Beilina, V.F. Selemenev and A. Shepeleva, *Pharm. Chem. J.*, 2001, **35**, 221.
624. R. Mohan, K.-K. Koo, C. Strege and A.S. Myerson, *Ind. Eng. Chem. Res.*, 2001, **40**, 6111.
625. F.-G. Pan, D.-H. Wang, F. Fang, L. Cao, X.-L. Yan and L.-G. Chen, *Youji Huaxue*, 2002, **22**, 341.
626. I.V. Aristov, O.V. Bobreshova and O.Y. Strel'nikova, *Russian J. Electrochem.*, 2002, **38**, 567.
627. F. Apruzzese, E. Bottari and M.R. Festa, *Talanta*, 2002, **56**, 459.
628. M. Plass, C. Griehl and A. Kolbe, *J. Mol. Struct.*, 2001, **570**, 203.
629. H. Nishino, A. Kosaka, G.A. Hembury, K. Matsushima and Y. Inoue, *J. Chem. Soc. Perkin Trans. 2*, 2002, 582.
630. D.K. Alargov, S. Deguchi, K. Tsujii and K. Horikoshi, *Origins of Life Evol. of Biosphere*, 2002, **32**, 1.
631. N. Sato, H. Daimon and K. Fujie, *Kagaku Kogaku Ronbunshu*, 2002, **28**, 113.
632. M. Stuebner, E. Schneider and J. Friedrich, *Phys. Chem. Chem. Phys.*, 2001, **3**, 5369.
633. A.-l. Ying and X. Wang, *Huaihai Gongxueyan Xuebao*, 2001, **10**, 49.
634. A. Rios, J. Crugeiras, T.L. Amyes and R.P. John, *J. Am Chem. Soc.*, 2001, **123**, 7949.
635. A. Khalaj, M. Pirali and R. Dowlatabadi, *J. Chem. Res. Synop.*, 2001, 412.
636. M. Oelgemoeller, A.G. Griesbeck, J. Lex, A. Haeuseler, M. Schmittel, M. Niki, D. Hesek and Y. Inoe, *Org. Lett.*, 2001, **3**, 1593.
637. M. Kitamura and T. Nakamura, *Powder Tech.*, 2001, **121**, 39.
638. I. Compagnon, F.C. Hagemeister, R. Antoine, D. Rayane, M. Broyer, P. Dugourd, R.R. Hudgins and M.F. Jarrold, *J. Am. Chem. Soc.*, 2001, **123**, 8440.
639. G.-c. Yang and X.-r. Chen, *Sichuan Daxue Xuebao*, 2001, **33**, 51.
640. G. Shan, K. Igarashi, H. Noda and H. Ooshima, *Chem. Eng.*, 2002, **85**, 161–169.
641. I. Weissbuch, L. Leiserowitz and M. Lahav, *ACS Symp. Series*, 2002, **810**, 242.

642. M. Barthes, A.F. Vik, A. Spire, H.N. Bordallo and J. Eckert, *J. Phys. Chem.*, 2001, **106**, 5230.
643. M. Yokota, K. Kawaguchi, S. Sakaki and N. Kubota, *Nippon Kaisui Gakkaishi*, 2002, **56**, 261.
644. A.N. Zyablov, D.S. Dolgikh, T.V. Eliseeva, V.F. Selemenev, L.A. Bityutskaya and I.S. Surovtsev, *J. Struct. Chem.*, 2001, **42**, 503.
645. S. Hoeppener, L.F. Chi, J. Wonnemann, G. Erker and H. Fuchs, *Surf. Sci.*, 2001, **487**, 9.
646. X. Zhao, R.G. Zhao and W.S. Yang, *Langmuir*, 2001, **18**, 433.
647. Z. Zhao, *Huaxue Yanjiu Yingyong*, 2001, **13**, 599.
648. S. Munsch, M. Hartmann and S. Ernst, *Chem. Commun. (Cambridge)*, 2001, 1978.
649. S. Ernst, M. Hartmann and S. Munsch, *Stud. Surf. Sci. Catal.*, 2001, **135**, 4566.
650. G.L. Starobinets, T.L. Yurkshtovich, P.M. Bychkovskii and F.N. Kaputskii, *Zhurnal Fizicheskoi Khimii*, 2001, **75**, 1702.
651. D.E. Fleming, W. Van Bronswijk and R.L. Ryall, *Clin. Sci.*, 2001, **101**, 159.
652. A.E. Aghajanyan, K.I. Ygian, A.S. Saghiyan and G.J. Oghanisyan, *Khim. Zh. Arm.*, 2001, **54**, 112.
653. Y. Ihara, S. Kurose and T. Koyama, *Monatsh. für Chem.*, 2001, **132**, 1433.
654. T. Nishiki, K. Nakamura, M. Hisatsune and D. Kato, *Solv. Ext. Res. Dev.*, 2002, **9**, 99.
655. K. Nakashima, T. Oshima and M. Goto, *Solv. Ext. Res. Dev.*, 2002, **9**, 69.
656. M. Hebrant, P. Burgoss, X. Assfeld and J.-P. Joly, *J. Chem. Soc. Perkin Trans. 2*, 2001, 998.
657. E.M. van der Ent, K. van't Riet, J.T.P. Keurentjes and A. van dar Padt, *J. Membr. Sci.*, 2001, **185**, 207.
658. A.M. Antunes, M.M.C. Ferreira and P.L.O. Voipe, *J. Chemometr.*, 2002, **16**, 111.
659. Y.J. Zhang, Y. Song, Y. Zhao, T.J. Li, L. Jiang and D. Zhu, *Langmuir*, 2001, **17**, 1317.
660. J. Givand, B.-K. Chang, A.S. Teja and R.W. Rousseau, *Ind Eng. Chem. Res.*, 2002, **41**, 1873.
661. R. Wimmer, F.L. Aachmann, K.L. Larsen and S.B. Petersen, *Carbohydr. Res.*, 2002, **337**, 841.
662. E. Peyrin, A. Ravel, C. Grosset, A. Villet, C. Ravelet, E. Nicolle and J. Alary, *Chromatographia*, 2001, **53**, 645.
663. H.-J. Buschmann, L. Mutihac and K. Jansen, *J. Incl. Phenom. Macrocycl. Chem.*, 2001, **39**, 1.
664. L.-x. Song and Z.-j. Guo, *Wuji Huaxue Xuebao*, 2001, **17**, 457.
665. H. Yamamura, M. Rekharsky, A. Akasaki, S. Araki, M. Kawai and Y. Inoue, *J. Phys. Org. Chem.*, 2001, **14**, 416.
666. Y. Liu and S. Kang, *Sci. China Ser. B: Chem.*, 2001, **44**, 260.
667. C. Ye, Y. Zhao, J. Chang and W. Liu, *J. Chem. Res. Synopsis*, 2001, 330.
668. Y. Liu, B. Lin, T. Wada and Y. Inoue, *Bioorg. Chem.*, 2001, **29**, 19.
669. T. Ugawa and S. Nishikawa, *J. Phys. Chem. A*, 2001, **105**, 4248.
670. B.-Y. Xia, W.-S. Cai, X.-G. Shao, Q.-X. Guo. B. Maigret and Z.-X. Pan, *THEOCHEM.*, 2001, **546**, 33.
671. S.-C. Xiang, Y. Zheng, J.-b. Weng and Z.-s. Zhu, *Hecheng Huaxue*, 2001, **9**, 499.
672. S.M.Z. Al-Kindy, F.F.O. Suliinan and A.A. Al-Hamadi, *Anal. Sci.*, 2001, **17**, 639.
673. Y.C. Guillaume, E. Peynn, A. Villet, A. Nicolaa, C. Guinchard, J. Millet and J.F. Robert, *Chromatographia*, 2001, **52**, 753.

674. A.D. Cutland, J.A. Halfen, J.W. Kampf and V.L. Pecoraro, *J. Am. Chem. Soc.*, 2001, **123**, 6211.
675. N. Voyer, S. Cote, E. Biron, M. Beaumont, M. Chaput and S. Levac, *J. Supramol. Chem.*, 2001, **1**, 1.
676. J. Al-Mustafa, S. Hamzah and D. Marji, *J. Solution Chem.*, 2001, **30**, 681.
677. H.-J. Buschmann, E. Schollmeyer and L. Mutihac, *J. Incl. Phenom. Macrocycl. Chem.*, 2001, **40**, 199.
678. Z. Chen, K. Uchiyana and T. Hobo, *Enantiomer*, 2001, **6**, 19.
679. H. Tsukube, H. Fukui and S. Shinoda, *Tetrahedron Lett.*, 2001, **42**, 7583.
680. P. Prus, M. Pietraszkiewicz and R. Bilewicz, *Mater. Sci. Eng. C*, 2001, **18**, 157.
681. B. Botta, M. Botta, A. Pilippi, A. Tafi, G. Delle Monache and M. Speranza, *J. Am. Chem. Soc.*, 2002, **124**, 7658.
682. F. Liu, G.-Y. Lu, W.-J. He, Z.S. Wang and L.-G. Zhu, *Chin. J. Chem.*, 2001, **19**, 317.
683. J.L. Atwood, T. Ness, P.J. Nichols and C.L. Raston, *Cryst. Growth and Design*, 2002, **2**, 171.
684. X. Peng, L. Liang, G. Yuan and S. Liu, *Huaxue Tongbao*, 2001, **65**, 126.
685. A. Tung, Q. Yang, H. Dong, L. Li and C.W. Huie, *Anal. Sci.*, 2001, **17**, a207.
686. C.Z. Wang, Z.A. Zhu, Y. Li, Y.T. Chen, F.M. Miao, W.L. Chan, X. Wien and A.S.C. Chan, *New J. Chem.*, 2001, **25**, 801.
687. M. Thamae and T. Nyokong, *J. Porphyrins Phthalocyanins*, 2001, **5**, 839.
688. T. Grawe, T. Schrader, P. Finocchario, G. Consiglio and S. Failla, *Org. Lett.*, 2001, **3**, 1597.
689. T. Liu, W.-J. Ruan, Y. Li, D.-Q. Yang, Z.-A. Zhu, Y.-T. Chen and A.S.C. Chan, *Gaodeng Xuexiao Huaxue Xuebao*, 2001, **22**, 159.
690. R.P. Bonomo, V. Cucinotta, G. Maccarrone, E. Rizzarelli and G. Vecchio, *J. Chem. Soc. Dalton Trans.*, 2001, 1366.
691. J. Sabolovic and V. Noethig-Laslo, *Cell. Mol Biol. Lett.*, 2002, **7**, 151.
692. E. Naritu, *Nendo Kogaku*, 2001, **40**, 173.
693. S. Goswami and R. Mukherjee, *Indian J. Chem. B: Org. Chem. Med. Chem.*, 2001, **40B**, 960.
694. C. Buttersack and A. Perlberg, *Stud. Surf. Sci. Catal.*, 2001, **135**, 2944.
695. J.-S. You, X.-Q. Yu, G.-L. Zhang, Q.-X. Xiang, J.-B. Lan and R.-G. Xie, *Chem. Commun. (Cambridge)*, 2001, 1816.
696. K. Tsubaki, M. Nuruzzaman, T. Kusumoto, N. Hayashi, B.-G. Wang and K. Fuji, *Org. Lett.*, 2001, **3**, 4071.
697. N. Higasi, T. Koga and M. Niwa, *ChemBioChem.*, 2001, **3**, 448.
698. M. Jurij, N.M. Maier, W. Lindner and G. Vesnaver, *J. Phyt. Chem. B*, 2001, **105**, 1070.
699. S. Aisawa, S. Takahishi, W. Ogasawara, Y. Umetsu and E. Narita, *J. Solid State Chem.*, 2001, **162**, 52.
700. Kabir-Ud-Din, J.K.J. Salem, S. Kumar and Z. Khan, *Indian J. Chem. B: Org. Chem. Med. Chem.*, 2001, **40B**, 1196.
701. F. Billiot, E.J. Billiot and I.M. Wamer, *J. Chromat. A*, 2001, **922**, 329.
702. (*a*) M. Kinoshita, T. Imai, A. Kovalenko and F. Hirata, *Chem. Phys. Lett.*, 2001, **348**, 337; (*b*) Y. Harano, T. Imai, A. Kovalenko, M. Kinoshita and F. Hirata, *J. Chem. Phys.*, 2001, **114**, 9606.
703. A. Villa and A.E. Mark, *J. Comput. Chem.*, 2002, **23**, 548.
704. A. Melo and M.J. Ramos, *THEOCHEM*, 2002, **580**, 251.
705. P. Ramirez, A. Alcaraz and S. Mate, *J. Colloid Interface Sci.*, 2001, **242**, 164.

706. T.L. Petrenko, *J. Phys Chem. A*, 2001, **106**, 149.
707. C.F. Matta and R.F.W. Bader, *Proteins-Structure Funct. Genetics*, 2002, **48**, 519.
708. L. Liu, M. Rozenman and R. Breslow, *Bioorg. Med. Chem.*, 2002, **10**, 3973.
709. M.D. Garcia-Santos, S. Gonzalez-Mancebo, J. Hernandez-Benito, E. Calle and J. Casado, *J. Am. Chem. Soc.*, 2002, **124**, 2177.
710. S.D. Quine and B.T. Gowda, *Oxid. Commun.*, 2001, **24**, 450.
711. I.A. Rivero, S. Heredia and A. Ochao, *Synth. Commun.*, 2001, **31**, 2169.
712. I.M. Lapina and L.M. Pevzner, *Russ. J. Gen. Chem.*, 2001, **71**, 1479.
713. Z.Z. Chen, C.-M. Chen and Y.-F. Zhao, *THEOCHEM*, 2001, **574**, 163.
714. N. Cini, F. Machetti, G. Menchi, E.G. Occhiato and A. Guarna, *Eur. J. Org. Chem.*, 2002, 873.
715. G. Drabik and J.W. Nashalski, *Acta Biochim. Pol.*, 2001, **48**, 271.
716. J. Li, X. Wang, M.T. Klein and T.B. Brill, *Int. J. Chem. Kinetics*, 2002, **34**, 271.
717. R.K. Sharma, W.G. Chan, J.I. Seeman and M.R. Hajaligol, *Preprints of Symposia ACS.*, 2002, **47**, 398.
718. H.X. Li, Y.P. Xu and M.H. Wang, *Int. J. Chem. Kinetics*, 2002, **34**, 405.
719. B.D. Dangel, J.A. Johnson and D. Sames, *J. Am. Chem. Soc.*, 2001, **123**, 8149.
720. M. Soleimani and M.N. Sarbolouki, *Chromatographia*, 2002, **56**, 505.
721. R. Mateo-Castro, J.V. Gimeno-Adelantado, F. Bosch-Reig, A. Domenech-Carbo, M.J. Casas-Catalan, L. Osete-Cortina, J. De La Cruz-Canizares and M.T. Domenech-Carbo, *Fresnius J. Anal. Chem.*, 2001, **369**, 642.
722. J. Wang, A.P. Chatrathi, A. Ibanez and A. Escarpa, *Electroanalysis*, 2002, **14**, 400.

Peptide Synthesis

BY DONALD T ELMORE
Eynsham, Witney, Oxfordshire

1 Introduction

As in the previous Report,[1] many reviews have been published. Some[2–11] relate to several sections or none in particular whereas others are cognate to particular sections as follows: Section 2.1,[12] Section 2.5,[13] Section 2.6,[14–19] Section 2.7,[20–25] Section 3.1,[26–30] Section 3.3,[31–34] Section 3.4,[35,36] Section 3.5,[37–46] Section 3.7,[47] Section 3.8, [48–52] Section 3.9[53] and Section 3.10.[54]

2 Methods

2.1 Amino-group Protection. – Introduction of the *N*-phthaloyl group using a reactive ester of monomethyl phthalate has been reported (Scheme 1).[55] As suggested earlier[1], however, *N*-phthaloyl protection is unlikely to find much application unless milder cleavage conditions are available such as those used with tetrachlorophthaloyl derivatives are available. Z-Protection can be effected with a new reagent, benzyl-4,6-dimethoxy-1,3,5-triazinyl carbonate.[56] Z groups

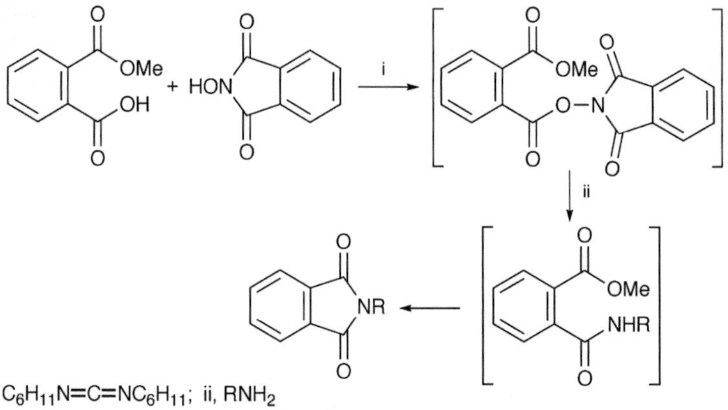

Reagents: i, $C_6H_{11}N{=}C{=}NC_6H_{11}$; ii, RNH_2

Scheme 1

can be hydrogenolysed in the presence of hydrazinium formate using either Mg or Pd-C as catalyst;[57–60] Bu^t and base-labile groups are stable to this treatment. Boc groups can be removed from amino, hydroxy or thiol functions by heating with $Ce(NH_4)_2(NO_3)_6$ in MeCN under reflux or with SiO_2 impregnated with the same reagent. Product yields are excellent.[61] The same method can be used to detach Boc groups from β-amino groups.[62] The use of Boc_2O as coupling agent in the homologation of α-amino acids by the Arndt-Eistert method proceeds with good yields and without loss of chiral purity.[63] *N*-Bis-Boc derivatives of amino acids are mono-deprotected by In or Zn metal in refluxing MeOH.[64] Boc derivatives of cis-4-trifluoromethyl- and cis-4-difluoro-methyl-L-prolines are accessible from Boc-4-oxoproline.[65] During *N*-protection of α-methyl DOPA methyl ester with Boc_2O, a *N,O*-bis-Boc derivative was formed.[66] The *O*-Boc group could be removed using pyrrolidine as nucleophile. t-Amyloxycarbonyl (Aoc) derivatives can be obtained[67] by treatment of 1,1-dimethylprop-2-yn-1-ol with triphosgene to give $CH{\equiv}CCMe_2OCOCl$. This reacts normally with amino groups and subsequent hydrogenation over Rh/Al_2O_3 catalyst in Et_2O gives the Aoc derivative. This final step was reported not to be reliable and depended on the batch of catalyst, so the Aoc group may not be widely used unless this final step can be improved. *N*-Propargyloxycarbonyl (Poc) amino acids have been synthesized from $CH{\equiv}CCH_2OCOOC_6H_5$.[68] The presence of the triple bond might be a useful site for introducing coloured or radioactive labels into peptide derivatives. More importantly, Poc amino acid chlorides are new coupling agents (Section 2.5).[69] The Poc group is stable under conditions that remove the Boc group but is detached by resin-bound tetrathiomolybdate in MeOH at room temperature in 1–1.5 h. The tetrathiomolybdate was attached to Amberlite IRC-400 resin by stirring the ammonium salt with pre-swollen resin in water.

Turning now to Fmoc chemistry, in the synthesis of peptide thioesters, the *C*-terminus survives treatment with DBU and HOBt and this removes Fmoc groups during SPPS.[70] The inappropriate detachment of Fmoc groups has been reported in SPPS when a peptide attached to the resin has unprotected ε-amino groups.[71] This side reaction is not caused by free α- or β-amino groups but is caused by the free ω-amino groups of Dab and Orn. The 2,7-di-Bu^t-Fmoc group has been linked to polymeric *N*-hydroxysuccinimide and the product has been used to protect α-amino groups.[72] The derivatives of amino acids are reported to have favourable solubilities. The use of 2-fluoro-Fmoc amino acids offers a different advantage.[73] The synthesis of base-sensitive peptide derivatives including peptide thioesters is possible without damage occurring to the *C*-terminus. An interesting new approach to peptide synthesis uses the *N*-carbamoyl group for amino-group protection.[74] The *N*-carbamoyl group is easily introduced on an amino group by reaction with KCNO in aqueous solution at pH 8.0–8.5 and 40–50°C for a few hours. No loss of chiral purity occurred. Coupling to an amino acid ester was effected by the unsymmetrical anhydride method using isobutyl chloroformate. Unfortunately, loss of chiral purity was not negligible especially if the *C*-terminal residue had a small side chain. Strangely this complication was diminished if the chloroformate was

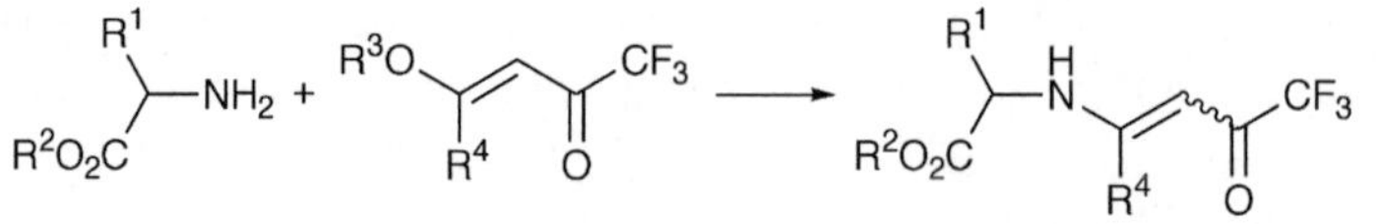

Scheme 2

added after the amino ester. The *N*-carbamoyl group was removed by nitrosation with a mixture of NO and O_2 (2.5:1). The authors stress the need to ensure that all traces of nitrosoureas are destroyed after deprotection of the product. 4-Alkoxy-1,1,1-trifluoro[chloro]alk-3-ene-2-ones can be used for amino-group protection (Scheme 2).[75] Deprotection is effected by treatment with 6M-HCl in MeOH at 70°C for 20 h. Both of these last two methods need further examination. PhAcOZ can be used for amino protection during peptide bond formation in the presence of papain or thermolysin.[76] Here the risk is of peptide-bond hydrolysis, but this can be avoided by several methods (see Section 2.7). Another interesting method for amino-group protection uses the 2,2-dimethyl-2-(2′-nitrophenyl)acetyl group which can be introduced using either the acyl chloride or a typical coupling method.[77] Hydrogenation over Pd-C (5%) reduces the nitro group and cyclisation with detachment of the protecting group occurs under acidic conditions. The 2-[phenyl(methyl)-sulfonio]ethyloxycarbonyl group, which is labile to base, has been used for the synthesis of Leu enkephalin.[78] This group resembles others and is worth a more demanding examination. The final method of amino-group protection discussed in this section uses 9-bora-bicyclo[3.3.1]nonane (Aldrich).[79] The protecting group (BBN) can be removed either by exchange with $NH_2CH_2CH_2NH_2$ or with dilute HCl in MeOH. The BBN derivatives are soluble in a range of polar solvents. Certain blocking groups for amino functions can be converted into coloured or fluorescent derivatives,[80,81] but these find no use to date in peptide synthesis.

2.2 Carboxy-group Protection. – Esters of Boc or Z-α,α-dialkyl-amino acids can be prepared by the reaction of an alcohol with an unsymmetrical acid anhydride obtained by the interaction of the amino-acid derivative and an aryl or alkyl chloroformate in the presence of DMAP.[82] Good yields of product are obtained. Bu^t esters can be selectively deprotected in the presence of other esters using ytterbium triflate as catalyst (5 mol per cent of catalyst) in $MeNO_2$ at 45–50°C.[83] A useful carboxy ester derivative is the (2-phenyl-2-trimethylsilyl)ethyl ester which is prepared as in Scheme 3.[84] These esters are stable during the hydrogenolysis of Z and Bzl groups, the Pd(0)-catalysed removal of Alloc and the acidolytic cleavage of Boc groups. The protecting group is removed by $Bu_4BF_4.3H_2O$ in CH_2Cl_2. A carboxy group can be protected by converting it into a phenylhydrazide with the aid of a carbodiimide.[85] Construction of a peptide can then be undertaken after the liberation of the α-amino group. After that, deprotection of the carboxy group can be effected using tyrosinase from *Agaricus bisporus* at pH 7 and room temperature with O_2 bubbling through the solution (Scheme 4). The method is highly chemo- and

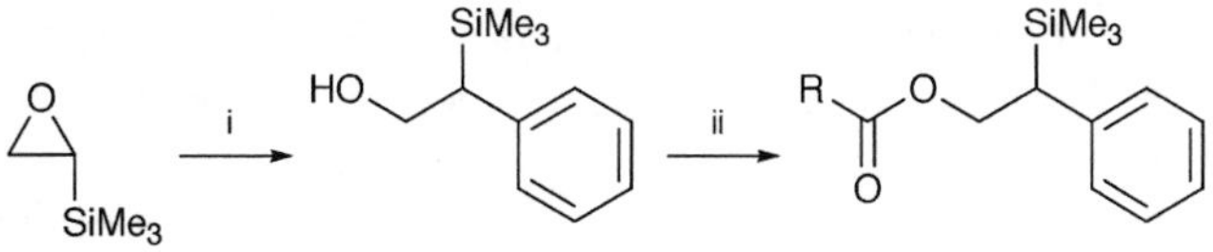

Reagents: i, Ph_2CuLi, Et_2O, −50 °C; ii, RCO_2H, DCCI, DMAP, CH_2Cl_2, 0 °C

Scheme 3

—CONHNHPh $\xrightarrow{i}$ —CON=NPh $\xrightarrow{ii}$ $—CO_2H + C_6H_6 + N_2$

Reagents: i, O_2, tyrosinase, pH7; ii, H_2O

Scheme 4

regio-selective. The side chain of any Met residues is unaffected by oxidation. Information about the fate of the side chains of Tyr and Trp, however, is desirable.

2.3 Side-chain Protection. – There is little new knowledge in this area. In the SPPS of Lys-vasopressin, the side chains of Tyr and Cys residues were protected by Bzl groups, the amide side chains of Asn and Gln residues were blocked by 1-tetralinyl and benzhydryl groups and the ε-amino group bore a Z-group.[86] A prospective study of possible acyl protecting groups for the imidazole nucleus of His has been carried out. Their removal by 3% CF_3CO_2H and 2% 1,8-diazabicyclo[5.4.0]undec-7-ene in $CHONMe_2$ was included in the study.[87] A detailed study in peptide synthesis is required to determine if any group is superior to the Bum group. The sulfation of Tyr hydroxy groups is not strictly for protective purposes because it is not intended to remove it. Its introduction, however, is attended by a minor problem. The hydroxy group of Tyr must be protected during peptide coupling and this is effected with the azidomethyl group.[88] This is sensitive to acid, so if Boc chemistry is being used, amino group deprotection must be achieved using TMSOTf. The liberated hydroxy group can then be esterified with $CHONMe_2$–SO_3.

2.4 Disulfide Bond Formation. – Work in this area has consisted of consolidation and exploitation rather than chemical exploration. Aromatic thiols react with Fmoc-Cys(Npys)-OH under acidic conditions to give stable heterodimers.[89] Amino acids of the type Acm-$S(CH_2)_nNHCH_2CO_2H$ (n=1–3) have been synthesised and incorporated into small peptides.[90] After removal of the Acm group, the products were converted into disulfides using iodine or thallic trifluoroacetate. The heat-stable bacterial enterotoxin ST peptide was prepared by SPPS and three disulfide bridges were formed regio-selectively using Trt, Bu^t and 4-methylbenzyl groups for protection of appropriate thiol groups.[91,92] Detachment of the peptide from the resin followed by oxidation with Me_2SO in MeCN formed one disulfide bridge. The second disulfide bridge was formed by removal of Bu^t groups and oxidation with $CF_3CO_2H/Me_2SO/CHONH_2$. Subsequent heating cleaved the 4-methylbenzyl groups with concomitant oxidation. The other research group used a combination of Trt, Acm and Bu^t groups with similar results. Single-chain peptides or dimers containing disulfide bridges both related to collagen tend to self-associate into triple helical homotrimers.[93] When bis-Boc-cystine and cystine dimethyl ester were coupled using DCCI and HOSu, a mixture of macrocycles containing either 32 atoms (4 cystine units) or 48 atoms (6 cystine units) were formed.[94] If Boc groups were removed, the products might be useful scaffolds for the attachment of epitopic peptides in studies of production of vaccines. Cystine derivatives can act as catalysts for *in vitro* oxidation with correct refolding.[95]

2.5 Peptide Bond Formation. – A new synthesis of *N*-carboxy-anhydrides involves exposing *N*-carbamoylamino acids to reaction with NO_x in the absence of solvent.[96] Fmoc-amino acid azides have been used for peptide synthesis and are favoured because of their stability at room temperature.[97] They are obtained from the reaction of Fmoc amino acid chlorides and NaN_3. When hydrazides are converted into azides using $MeCO_2H$ or HCO_2H, some *N*-acylation occurs as a side reaction.[98] Removal of *N*-formyl group was studied using N_2H_4 or NH_2OH and it was recommended that HCO_2H should not be used in the preparation of azide. Amides can be obtained from unprotected amino acids using dichlorodialkyl silanes (Scheme 5).[99] *N*-Protected amino acid bromides are recommmended for difficult coupling reactions.[100] The required reagents are generated by treatment of the acids with 1-bromo-*N,N*-2-trimethylpropenylamine (1). Peptide coupling is achievable using 4-nitrophenyl esters and unprotected amino acids such as Cys, His or Ser in the presence of Et_3N (1 equiv.) and DMAP (0.1 equiv.) in MeCN.[101] Other polar aprotic solvents can also be used. Unfortunately, no systematic study of the chiral purity of products was reported, although some hplc and nmr data indicate chiral purity. Additionally, coupling involving lysine tended to occur at the ε-amino group. The *N*-oxysuccinimido ester of α-*N*-tosyl-α-aminoisobutyric acid has been isolated in the solid state and its crystalline structure has been determined.[102] The trifluoroacetyl ester of *N*-hydroxysuccinimide (2) reacts with amino acids to give the reactive ester of the *N*-trifluoroacetyl- amino acid which can then be used in a coupling reaction to give peptides.[103] The reagent (2) has been used previously to prepare reactive esters of *N*-protected amino acids.

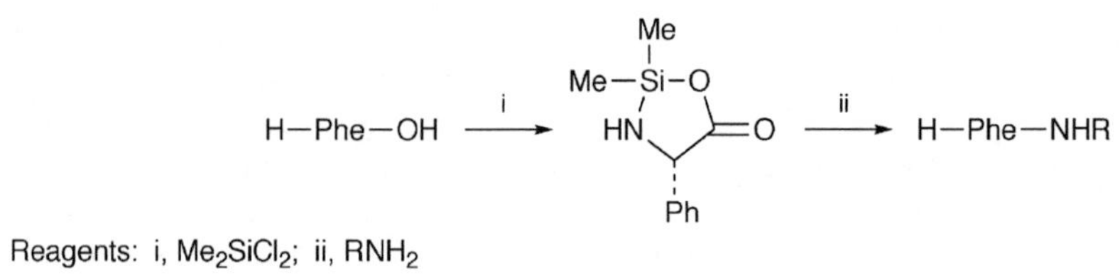

Scheme 5

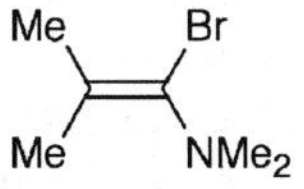

(1) 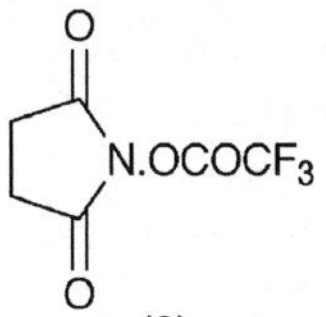 (2) 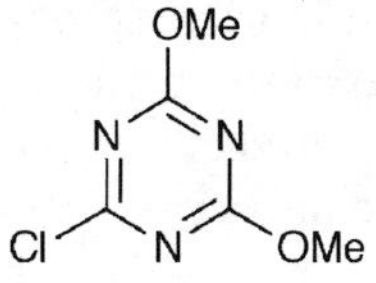 (3)

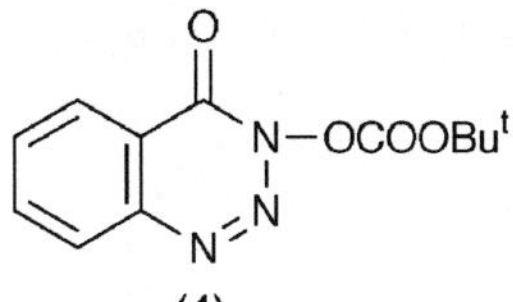

(4) 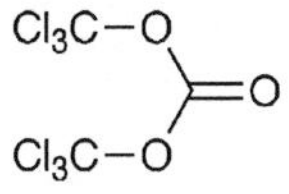 (5) 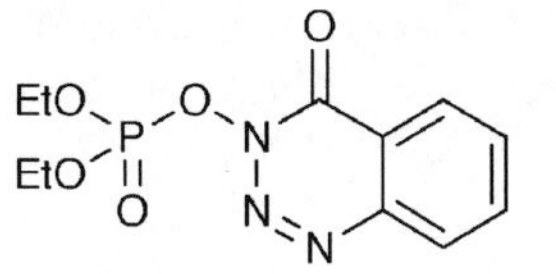 (6)

Frequently, the nucleophile used in a coupling reaction is added in the form of a salt and liberated by addition of a tertiary base. This can cause loss of chiral purity and techniques that avoid this danger are valuable. One method uses Zn dust to deprotonate hydrochloride.[104] In the synthesis of dipeptides from amino acids, the latter can be coupled in MeCN when treated with phosphazene bases.[105] Another problem has been reported when Et_3N was used in coupling reactions involving Boc amino acids and phenacyl esters.[106] Byproducts were formed and this problem was avoided by using Na_2CO_3 supported on Al_2O_3.

Reactive esters continue to be popular reagents for peptide synthesis. Addition of catechol has been found to catalyse this type of reaction.[107] A one-pot, one-step procedure using 2-chloro-4,6-dimethoxy-1,3,5-triazine (3) in EtOAc or similar solvent is recommended to favour high yields of chirally pure product.[108] When a two-step procedure was used, an azlactone intermediate was detected and loss of chiral purity resulted. Another triazine derivative (4) reacted with *N*-protected amino acids to give benzotriazinonyl esters that reacted with primary and secondary amines to give the corresponding amides in good yield.[109] For example, a good yield of a dipeptide of Aib was obtained. The kinetics of interaction of Boc-Xaa-OC_6H_4F and H-Leu-NH_2 were measured.[110] The kinetics were not simple and obeyed an equation of the type: $v=k[amide]^a[ester]^b$ and a and b were not always integral depending on the ester used. The reagent, bis(trichloromethyl)carbonate but perhaps better known by the trivial name of triphosgene (5), is claimed to be the best coupling agent for the sterically hindered *N*-alkylamino acids.[111] No loss of chiral purity has been detected using a combination of Pr^i_2NEt and collidine as base. In coupling Fmoc-MeVal-OH and H-MeVal-MeIle-MeGly-OH, only (5) gave any of the desired product. Moreover, cyclosporin O, an immunosuppressant cycloundecapeptide containing 7 *N*-methylamino acids, was successfully synthesised. In addition, omphalotin A, a cyclododecapeptide that contains 6 *N*-methylamino acids was also synthesised by the same group.[112]

Another very satisfactory coupling reagent is 3-diethoxy-phosphoryloxy-1,2,3-benzotriazin-4(3H)-one (6).[113,114] It has been used for the synthesis of peptide alcohols and a glycopeptide as well as various peptides. The authors claim that hydroxy groups do not have to be protected. Instead of using soluble active esters and assembling a peptide on an insoluble support, the converse had been studied. Three immobilised hydroxy compounds (7–9) have been converted into active esters and allowed to react with the growing peptide chain that is in the soluble phase.[115] The immobilised esters were synthesised on TentaGel resin, which is composed of a low cross-linked polystyrene grafted on PEG. Esters constructed on (7) tended to hydrolyse rather quickly and the best results were obtained with (9).

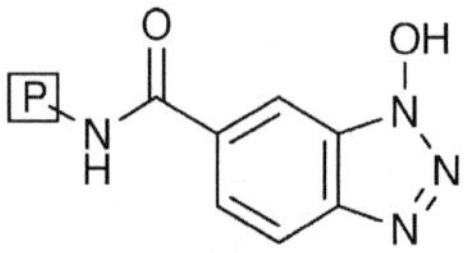

(7) 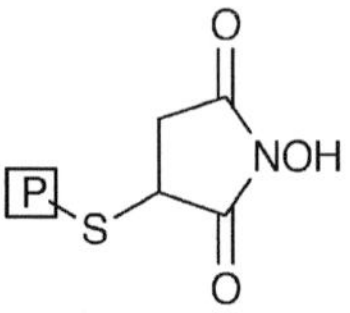 (8) 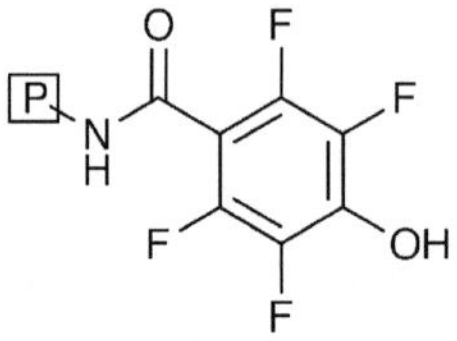 (9)

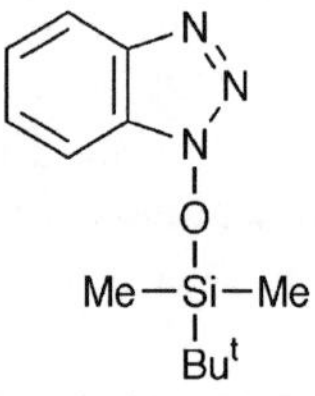

(10) 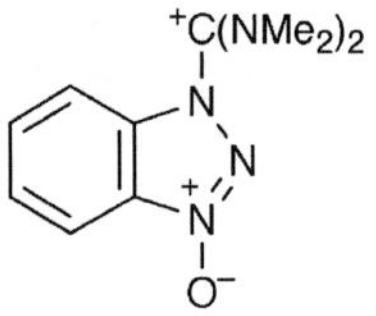 (11) 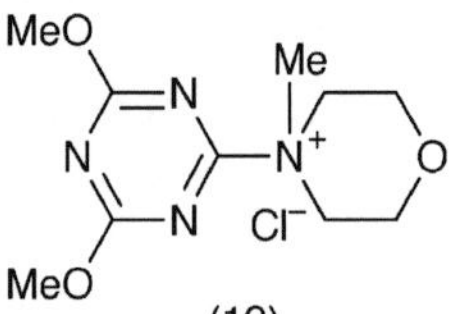 (12)

In coupling reactions involving Fmoc-amino acid chlorides, a new derivative of benztriazole (10) has been designed.[116] It requires no addition of base, effects rapid reaction and no loss of chiral purity was detected.

Carpino's group has been studying the structure of reagents used in peptide coupling reactions; special attention has been directed to reagents that might be expected to undergo structural rearrangement before or during coupling.[117] For example, HATU (11) and HBTU are not normally uronium compounds, but guanidinium derivatives as revealed by X-ray crystallography. By substituting KOBt or KOAt for the unprotonated compounds, it has been shown that the uronium forms can be isolated by using a rapid working up procedure. $O \rightarrow N$ isomerisation occurs rapidly in the presence of organic bases (e.g. Et_3N). The converse rearrangement has never been observed. The *O*-compounds are more efficient coupling reagents. O-HATU converts Z-Aib-OH into the methyl ester rapidly ($t_{1/2}$<2 min.) in presence of 1 equivalent of collidine whereas N-HATU requires much longer. The reagent isomers are easily distinguished by IR spectra. The *O*-isomers have bands at 1709–1711 cm^{-1} whereas the *N*-isomers absorb at 1664–1675 cm^{-1}. The isomeric form of the reagent and the reaction conditions determine the extent of loss of chiral purity. Although coupling occurs faster with the uronium isomer, so does loss of chiral purity. The last word on peptide coupling has yet to be written. Uronium salts have been synthesised from 1,3-dimethylpyrimid-2-one and *N*-hydroxysuccinimide (Scheme 6).[118]

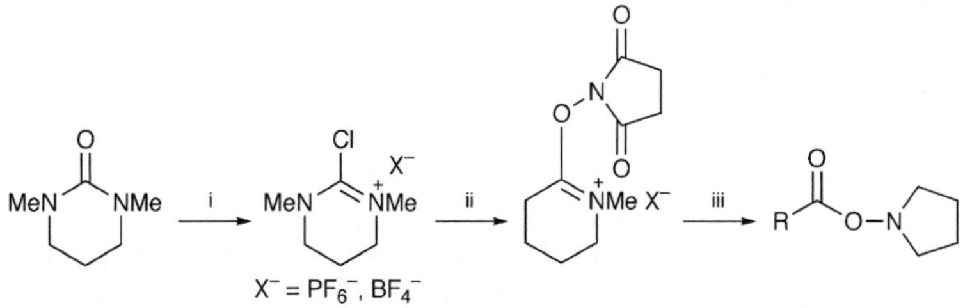

$X^- = PF_6^-, BF_4^-$

Reagents: i, $COCl_2$, DMF, CH_2Cl_2 then KPF_6 or $NaBF_4$; ii, *N*-hydroxysuccinimide, Et_3N; iii, RCO_2H, Et_3N

Scheme 6

The initial product reacts with *N*-protected amino acids to give esters of *N*-hydroxysuccinimde which can then couple in the usual way. Reactions are rapid and loss of chiral purity is not detectable. Treatment of cyanuric chloride with chiral amines or esters of amino esters give modest yields of 2,4-dichloro-6-alkylamino triazines which are useful coupling agents, but the usual battery of tests is required for unreserved recommendation.[119] A new test for potential loss of chiral purity has been described.[120] Using 4-(4,6-dimethoxy-1,3,5-triazin-2-yl)-4-methylmorpholinium chloride (12), the coupling of Z(OMe)-Gly-Ala-OH and H-Phe-OBzl was carried out in a range of solvents. There was very little loss of chiral purity in AcOEt or THF but more in MeSOMe, EtOH or MeOH, a predictable result.

Although acyl azides are part of the classical history of peptide synthesis, an azido group at the other end of the molecule may not seem to be very relevant, but the ingenuity of peptide chemists should not be underestimated. The α-amino group can be converted into an azide by diazo transfer using triflyl azide in the presence of Cu^{2+} ions in the form of $CuSO_4$ in a mixture of CH_2Cl_2, MeOH and water.[121] The reaction can be carried out in solution or on a solid phase using the Wang linker resin. The azide group can be viewed as a protected amino group and the carboxy group can converted into a reactive derivative for coupling. Azidopeptides are potential intermediates for 1,3-dipolar cycloaddition reactions or they can participate in a Staudinger ligation reaction. Scheme 7 outlines this method of coupling two peptides and the stereochemistry is preserved.[122] Since a fairly free choice of amino-acid residues appears to be possible at the ligation site, this promises to be an important new avenue to the synthesis of large peptides. Azetidine-2,3-diones react with amines to give amino acid amides and with amino acid esters to give dipeptide products.[123]

Modifications to experimental protocol have permitted the synthesis of amyloid β-peptides in solution using $CHCl_3$–PhOH as solvent.[124] The coupling reagents are $EtN{=}C{=}N(CH_2)_3NMe_2$ and 3,4-dihydro-3-hydroxy-4-oxo-1,2,3-benzotriazine (HOOBt). No significant loss of chiral purity occurred. Aminoxypeptoids have been synthesised using the 2-nitrobenzenesulfonyl protecting group for the aminoxy moiety.[125]

An interesting new technique involves the use of a thermomorphic solvent system.[126] This is defined as a system that consists of two phases at 5°C but is

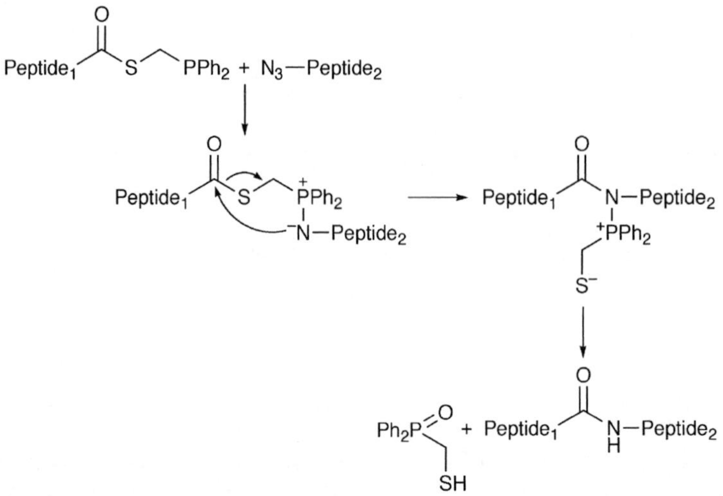

Scheme 7

homogenous at 35°C. (3,4,5-trioctadecyloxy)benzyl alcohol is used as a substitute for a solid support and confers on the system the ability to separate into two phases at low temperature. The coupling step is carried out at the higher temperature. Cooling permits separation of product and unused starting reagents. One could visualise several problems and a wide series of test syntheses is required for it to be generally attractive, but numerous modifications are possible and it avoids the use of expensive equipment. A thermostatically controlled water bath and some ice should suffice.

The native chemical ligation method has been further developed permitting the linkage of four unprotected peptides using three chemoselective reactions to give cysteine, pseudoproline and pseudoglycine at the three ligation sites.[127] A further development of the native chemical ligation technique has made it possible to join unprotected peptides at sites that do not include cysteine.[128] The key substance was the auxiliary substance 3,4,5-trimethoxy-6-(4-methylbenzylthio)benzylamine (13). The 3- and 5-methoxy groups are present to increase acid lability when the auxiliary group is detached. The 4-methoxy group is present to increase the nucleophilicity of the sulfur atom and to accelerate thioester exchange. The amino group of (13) carries temporarily the *C*-terminal fragment of the ultimate product. The *N*-terminal fragment of the final product is presented as a thioester and the mechanism of chemical ligation is summarised in Scheme 8. Although the main objective of the chemical ligation method is to join together two or more peptide chains, there is no obvious reason why the method can not be used to link one end of a linear peptide to the other. This has now been effected with a 16 residue peptide containing a *C*-terminal thioester and an *N*-terminal selenocysteine (Scheme 9).[129]

To end this section, two technical publications are cited, both of which concern the use of microreactors. One[130] describes the solution phase synthesis of β-peptides while the other[131] demonstrated that by using more dilute

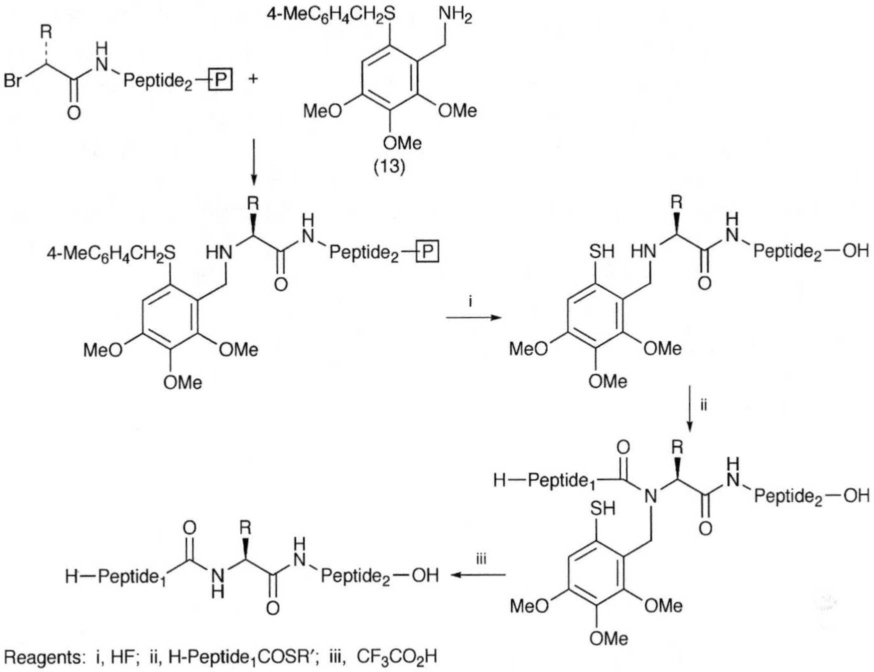
4-MeC6H4CH2S
NH2
R
H
N
Br
Peptide2
P
O
MeO
OMe
OMe
(13)
HN
i
SH
Peptide2—OH
ii
H—Peptide1
iii
Reagents: i, HF; ii, H-Peptide1COSR′; iii, CF3CO2H

Scheme 8

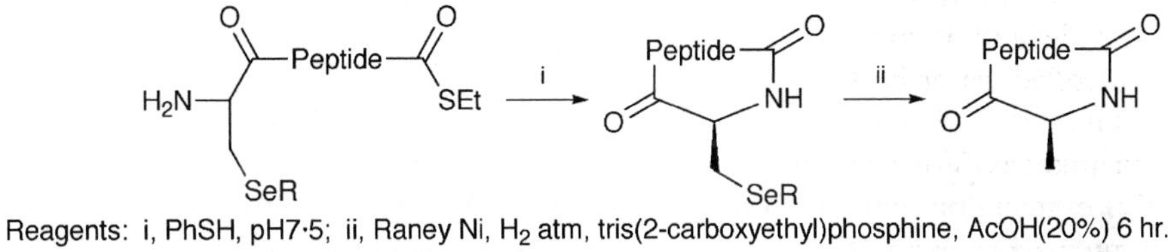
Peptide
H2N
SEt
SeR
i
ii
NH
O
Reagents: i, PhSH, pH7·5; ii, Raney Ni, H2 atm, tris(2-carboxyethyl)phosphine, AcOH(20%) 6 hr.

Scheme 9

solutions of reagents for shorter reaction times, the loss of chiral purity is less than in bulk syntheses.

2.6 Peptide Synthesis on Macromolecular Supports and Methods of Combinatorial Synthesis. – Commercially available JandaJel has been claimed to be better than Merrifield resin for the synthesis of ACP(65-74) because it has increased swelling power that allows solvent to disrupt secondary structures on the resin.[132] Searches for new and improved supports for SPPS have continued. A flexible amphiphilic resin was made by suspension polymerisation of styrene and 1,4-butanediol dimethacrylate. The product was converted into a benzhydryl resin and this was tested for the synthesis of delta sleep inducing peptide using Fmoc/Bu^t chemistry. Yield and purity were very good.[133] This type of resin gave better yields of purer product than when an earlier rigid and hydrophobic polystyrene-divinylbenzene was used for the synthesis of difficult sequences.[134] In addition, the better resin is stable to neat CF_3CO_2H, 6N HCl and 6N KOH.[135] The same resin grafted to PEG was equipped with a 4-(hydroxymethyl)-phenoxyacetic acid and used to synthesise two fragments of Alzheimer's β-amyloid peptide.[136] The results were more satisfactory than when Tentagel resin was employed under identical conditions. A new resin prepared by crosslinking polystyrene with tri(propyleneglycol)-glycerolate (14) has free hydroxy groups without the need for a special linker.[137] The resin was used for the successful synthesis of a fragment of the Alzheimer's β-amyloid peptide and a 23 residue substrate of Ca^{2+}/calmodulin binding peptide. A copolymer of polystyrene and PEG is reported to be better than Tentagel in SPPS, photolytic and enzymatic cleavages.[138] The capacity of the resin was increased by grafting Tris as a dendrimer.

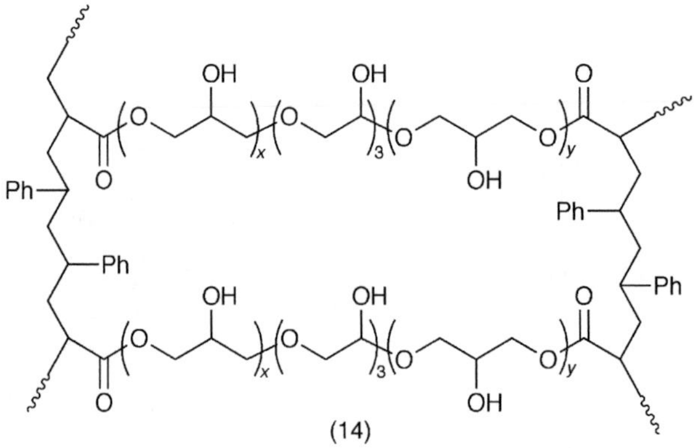

(14)

Research on the design of new linkers for use in SPPS continues apace, but Kenner's sulphonamide linker as modified by Backes and Ellman is still popular. For example, it has been used for the assembly of a hairpin shaped polyamide-peptide conjugate.[139] The attachment of the *C*-terminal residue to a sulphonamide linker can be difficult and the problem seems to have been satisfactorily solved by using the fluoride of the Fmoc derivative.[140] The acyl fluoride is accessible using cyanuric fluoride as described by Carpino. The introduction of the 4-alkoxy-2-hydroxybenzaldehyde linker mentioned in the last Report has been followed by the 4,6-dimethoxy-2-hydroxybenzaldehyde linker.[141] Detachment of the assembled peptide is achieved under mild conditions with CF_3CO_2H (either 5% aqueous or 1% in CH_2Cl_2). The synthesis of the linker is straightforward. Even milder conditions suffice for the detachment of a peptide if a silylated linker is used. The (2-phenyl-2-trimethylsilyl)ethyl linker (15) functions well in SPPS involving Fmoc chemistry and the peptide product can be detached using $Bu_4N^+F^- \cdot 3H_2O$ in CH_2Cl_2.[142] Such mild conditions make this linker very suitable for the synthesis of glycopeptides. This linker was also used to synthesise a tripeptide containing the -Pro-Gly- sequence which often causes the formation of a diketopiperazine. A different approach has been made to avoid the formation of diketopiperazines.[143] It is argued that depending on the nature of the *C*-terminal two amino acids, when the *N*-protection group is removed from the second amino acid, the molecular road is open for intramolecular amide formation to compete with the desired acylation by the reactive derivative of the third amino acid. A suitable strategy involved the use of the acyl fluoride for the acylation by the third residue from the *C*-terminus and a measured liberation of the amino group of the second residue. The protecting group on the *N*-terminus of the *C*-terminal dipeptide of the peptide to be synthesised is blocked by a $Pr^i_3SiO_2$-group. This is removed steadily by F^- catalysis and acylation by the acyl fluoride rapidly ensues (Scheme 10). A silyl linker has also been used to synthesise peptide thioesters.[144] A disulfide linker has been developed with the view that single-bead libraries could be rapidly screened (Scheme 11). The disulfide bond is easy to synthesise

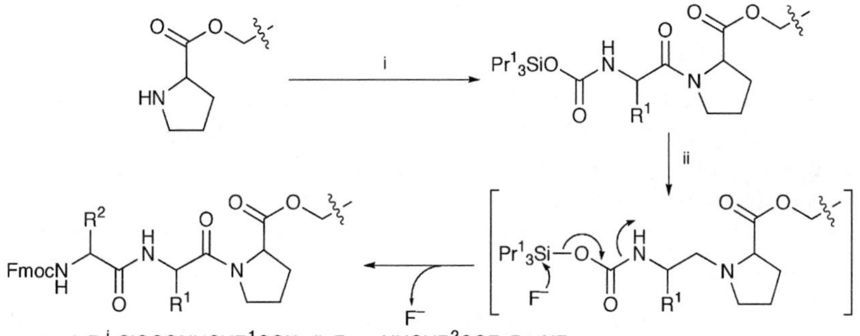

Reagents: i, $Pr^i_3SiOCONHCHR^1COX$; ii, $FmocNHCHR^2COF$, Bu_4NF

Scheme 10

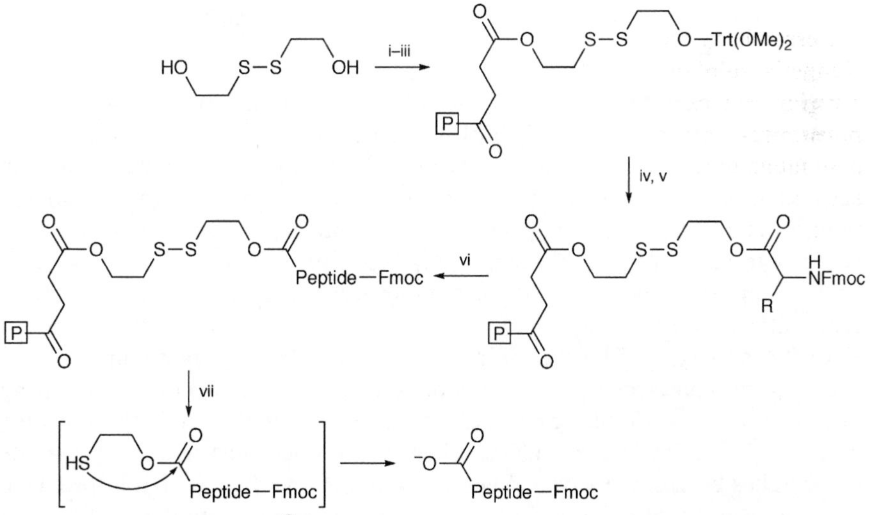

Reagents: i, $(MeO)_2TrtCl$; ii, $\overline{CO(CH_2)_2CO}$(–O–); iii, Resin, HATU, HOAt; iv, CF_3CO_2H, H_2O, $Pr^i_3SiH_4$; v, Fmoc–Xaa–OH, DCCI, DMAP, Pyridine; vi, SPPS as for v; vii, RSH, NH_4OH(pH10)

Scheme 11

and easy to cleave, so it is an obvious candidate for incorporation into a linker moiety in SPPS. A symmetrical disulfide is an attractive candidate.[145] The monosuccinyl derivative can easily be attached to a resin bearing an amino group. After SPPS is complete and the disulfide bond has been cleaved by reduction, the peptide is isolated as a 2-mercapto ethyl ester which is easily hydrolysed by dilute aqueous ammonia or can afford the amide by treatment with 15% ammonia in water.

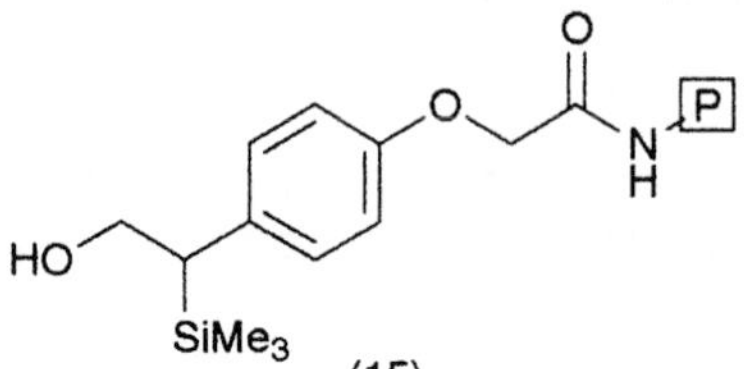

(15)

The final linker to be considered is (16).[146] This has the important property that when peptide assembly is complete the product can be detached under mild basic conditions with no need for a scavenger. There is, however, a possible disadvantage; the hydrophobic character of the fluorene ring in the linker could interact with a group of hydrophobic amino acid residues anywhere in the chain causing some coupling steps to be troublesome.

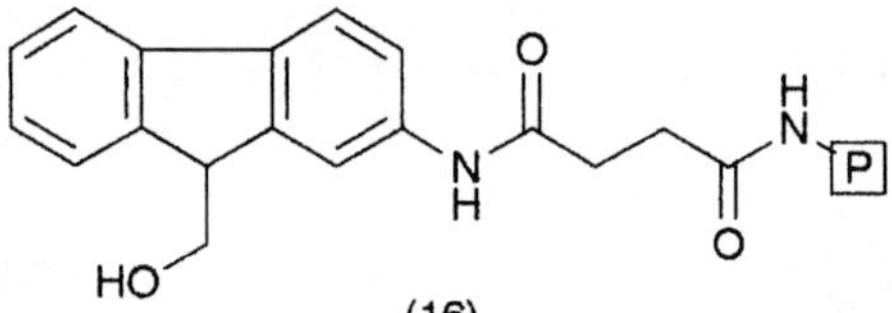

(16)

SPPS has tended to be dependent on either Boc or Fmoc for protection of the *N*-terminus, but there is a revival of interest in the use of the Trt group with base labile groups for side-chain proection.[147] A stable $Me_2O\text{-}(HF)_n$ complex has been found to be effective for detaching completed peptides from Merrifield resin.[148] The covalent capture purification of peptides bearing either Cys or Thr at the *N*-terminus can be effected with a resin bearing a free –CHO group.[149] Such a support can be produced by coupling acrylic acid at a hydroxy or amino group already on the support followed by oxidation with periodate to give a glyoxylyl moiety.[150] The use of microwave radiation to speed up sluggish coupling steps has previously been reported and there are two more supporting publications.[151,152] Coupling times can be as low as 1 min. even in the assembly of peptoids. A simple, manual synthesiser for the construction of peptide libraries has been described.[153] A high-throughput synthesiser that is geared to working on a μmol scale in a 96-well format gave products that were about 90% pure.[154]

The remainder of this section is less concerned with equipment and more with giving a flavour of successes achieved. A number of noncoded amino acids have been incorporated into peptides using SPPS. One of the more disappointing results was obtained when attempting to synthesise peptides of fulleropyrrolidino-Glu.[155] An attempt to synthesise an analogue of enkephalin gave poor yields possibly due to strong binding of the fullerene system by the resin. Peptides can be labelled on resin with ^{18}F 4-fluorobenzoic acid.[156] The products can be used clinically for PET scans. Peptides containing the tripeptide sequence RGD are likely to be of considerable interest since it is present at the binding site of integrins. Orthogonally protected derivatives of norlanthionine[157] and lanthionine[158] have been synthesised for incorporation into peptides. Likewise, 4-azalysine has been synthesised.[159] Derivatives were not tested as substrates or inhibitors of trypsin-like enzymes such as thrombin and similar blood-clotting enzymes.

Although the synthesis of cyclic peptides is not the main thrust of this Report, a few publications are cited here. A corticotropin-releasing factor antagonist, astressin, its retro-, inverso-, and retro-inverso isomers have been synthesised using Fmoc/Bu^t chemistry. Some side chains were protected by allyl groups and a requisite pair were deprotected on the resin and a lactam was formed between Lys and Glu.[160] A rather similar approach involved incorporating α-azido-γ-9-fluorenylmethyl-Glu and α-azido-ε-Fmoc-Lys into the chain at an appropiate distance apart and the side chains were deprotected for lactam formation.[161] Attempts were made to effect a ring-closing metathesis reaction using Grubbs catalyst on peptides containing unsaturated side chains.[162] The experiments were not always successful but incorporation of Mutter's pseudoproline Ser($\psi^{Me,Me}$Pro) overcame this problem. The remaining double bond in the cyclised peptide could be reduced with H_2 over Pd/C catalyst.

Peptides containing several residues of cysteine with the thiol group protected with the Acm group can have this blocking group removed on resin and the liberated thiol group can be acylated with aliphatic acids.[163] The metabolism and biological properties of such peptide derivatives have not been widely studied hitherto. Perhaps this simple chemistry will encourage the filling of this gap. The construction of peptide libraries have been mentioned elsewhere in this Report, but an extensive study meriting special mention has been made of peptides that are antagonists against insulin-like growth factor[164] A four-dimensional orthogonal SPPS of new scaffolds based on cyclic tetra-β-peptides has been described.[165] The scaffolds contain four free amino groups as sites for assembling different peptides. A fragment of transmembrane bradykinin receptor comprising 34 residues has been assembled by SPPS and the process was followed by EPR spectroscopy.[166] A marked reduction in yield and chain mobility was evident during the assembly of the 12–16 region when $CHONMe_2$ was used as solvent. Higher velocity of coupling and chain mobility were observed when the solvent was an equimolar mixture of an electron donor and an electron acceptor. This may be a satisfactory solution for chemical syntheses, but perhaps not if any stage is enzymatically catalysed. Peptide synthesis on macromolecular supports usually involves construction of the desired product on a solid support. Just occasionally all the chemistry occurs in the liquid phase and there is such a process described in the year under review.[167] A typical support is (17) and the peptide is assembled on the hydroxy group of the trityl moiety. Various linkers and protecting groups are possible and it is hoped that data will become available on the synthesis of peptides of different sizes and amino-acid composition. Soluble supports may be very suitable for syntheses catalysed by enzymes.

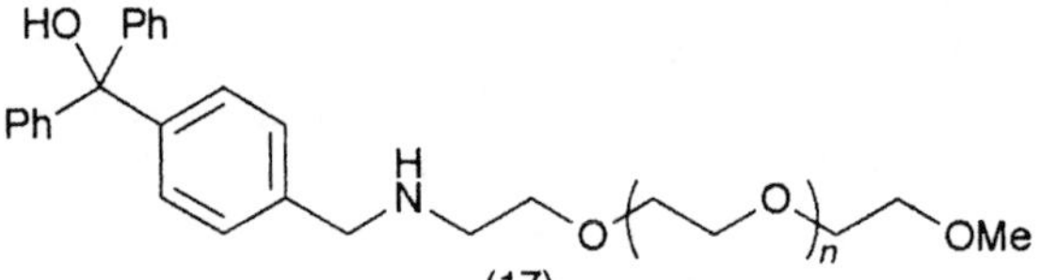

HO Ph

Ph

H
N

O O n OMe

(17)

Finally, in this section, there are two noteworthy syntheses to report. A 31-residue peptide has been synthesised that has a structure based on that of bovine pancreatic polypeptide.[168] It has a stable conformation in solution with helical domains. It catalyses the decarboxylation of oxaloacetate following Michaelis-Menten kinetics. Although the kinetic constants are not spectacular (K_m=64.8 mM, k_{cat}=0.229 s^{-1}) by comparison with native enzyme, it is four orders of magnitude more effective than a simple amine. The authors attribute the results to the flexibility of Lys side chains and doubtless a more effective catalyst will be forthcoming. Perhaps more spectacular is the synthesis of a human group IIa secretory phospholipase, a protein containing 124 amino acid residues and 7 disulfide bonds.[169] The enzyme was found to be identical to the native enzyme. This was not the first synthetic enzyme, but the number of disulfide bonds was a potential difficult hurdle. Interestingly, the authors maintain that the chemical synthesis is as quickly achieved as the biological approach from the isolated gene. This is important because the synthetic approach has greater flexibility in selection of amino acids and their sequence.

2.7 Enzyme-mediated Synthesis and Semi-synthesis. – Work continues on various reactions involving amino acids and their simple derivatives. A mutant of thiolsubtilisin functions as an acetyltransferase with ethyl acetate as the acyl donor and amino acids as acyl acceptors.[170] Chymotrypsin catalyses trans-esterification reactions such as the formation of Ac-Phe-OPr^n from Ac-Phe-OEt and Pr^{n}OH in ionic solvents.[171] Supercooled liquid CO_2 also serves as a suitable solvent. The value of this solvent has been reported before. Ac-Tyr-OPr^n has also been synthesised in the presence of organic salts.[172] As the size of the organic cation or anion increased, synthetic activity decreased, but the stability of the enzyme increased under these conditions. Amino acids can be esterified by hexoses, the reaction occurring at the primary hydroxyl group. DL-α-Amino-nitriles are hydrolysed by *Rhodococcus sp. AJ270 hydratase* in phosphate buffer at 30°C to give a mixture of D-amino acid amide and free L-amino acid.[173] Diesters of Asp can undergo amidification in presence of *Candida antarctica* lipase.[174] Amino acid esters of sugars can be synthesised, preferably in pyridine, using a complex of subtilisin and a surfactant.[175] The amino acid is introduced at C_6 with glucose and at C'_6 with maltose.

Three dipeptide amides were obtained from Ac-Phe-OEt and Xaa-NH_2 (Xaa=Leu, Ala, Val) in the presence of lipase.[176] It is possible that synthesis was favoured by the hydrophobic character of the substrates. Dipeptide *N*-phenyl-hydrazides were formed using thermoase in organic solvents.[177] A new cysteine proteinase, morrenain b II, has been isolated from the latex of a S. American climbing plant, *Morrenia brachystephana.*[178] The enzyme displays a high specificity for the synthesis of Cys-Phe-OMe in a mixture of 0.1M Tris-HCl buffer, pH 8.5, and $CHCl_3$. *N*-Protected peptide alcohols can be synthesised using either α-chymotrypsin or subtilisin as catalyst in organic solvents and water mixtures.[179] Subtilisin and thermolysin have been immobilised on polyvinyl alcohol and used to synthesise *N*-protected small peptides.[180] Z-$(Ala)_2$-Leu-pNA was obtained in 90% yield in 1 hr. The immobilised enzyme was stable in aqueous buffer for 6 months. L-H-Glu$(OEt)_2$ was converted into H-(Glu-γ-OH)$_n$-OH by papain in 50 mM phosphate buffer, pH 8.0, containing 0–2 M NaCl at 25°C.[181] The presence of NaCl promoted the early stages of reaction but tended to give a lower final yield. Hardly any H-Glu(OEt)-OH was formed.

In a solid-phase synthesis, Fmoc-Phe-OH was coupled to $PEGA_{1900}$ via a Wang linker. The Fmoc group was removed with piperidine and an *N*-protected amino acid was added in excess in the presence of thermolysin. The product was detached with 95% CF_3CO_2H.[182] Good yields were obtained; the shift in position of equilibrium for peptide formation was often sufficient to give complete conversion. The solid-to-solid synthetic technique favoured by the Edinburgh group has been further studied.[183] No product precipitated in CH_2Cl_2 and the yield was low. In contrast, when precipitation occurred, the product accumulated in both solution and as solid. Yields were highest when both substrate and product solubilities were lowest. Other workers have found that good yields can arise when organic solvents containing very low concentrations of water are present. For example, α-chymotrypsin gave a good yield of pentapeptide when Boc-Tyr-Gly-Gly-Phe-$OCOCH_2CONH_2$ was coupled to H-Leu-NH_2 in a mixture of MeCN (2 ml) and Tris buffer (pH 7.8) at 30°C.[184] Similar results were obtained using *B. licheniformis* proteinase in MeCN.[185] A protected pentapeptide fragment of osteogenic growth peptide, Z-Tyr-Gly-Phe-Gly-Gly-OEt, has been synthesized enzymatically by both a 3 + 2 and a 2 + 3 route.[186,187] In one stage, a microporous molecular sieve was used to immobilise α-chymotrypsin as a catalyst. The use of β-trifluoroethyl esters gave higher yields than ethyl esters or carboxylic acids. A stepwise $N \rightarrow C$ strategy was used for the synthesis of some small peptides using either chymopapain or subtilisin immobilised on Celite in aqueous ethanol.[188] A protected tetrapeptide was obtained in an overall yield of 34% in a one-pot, three-step procedure. The *N*-phenylacetoxy-4-benzyloxy carbonyl group (PhAcOZ) has been used in the synthesis of nucleopeptides.[189] The special attraction is the possibility of detaching the phenylacetyl moiety with penicillin G acylase, when the remainder of the group is spontaneously cleaved. CCK-8 has been synthesised as its *N*-phenylacetyl derivative using immobilised enzymes in organic solvents containing low concentrations of water.[190] An overall yield of 15% was obtained.

The use of inverse substrates of trypsin and the use of enzymes from different biological sources enabled the synthesis of Boc-[Leu^5]-enkephalin amide.[191] The use of enzymes from different biological sources leads naturally to the use of artificially produced mutant enzymes. Mutants of the serine proteinase from *B. lentus* with polar residues at position 166 inside the primary specificity site catalyses syntheses with both L- and D-residues at the S_1 position of the substrate.[192] Several mutant forms of bovine trypsinogen have been obtained and studied.[193] Replacement of Lys^{15} by Ala gave a protein (K15A) that could not be activated. A double mutant (D189S, D194N) involving residues at the active site afforded a synthetic catalyst of broad specificity, but with virtually no proteolytic activity. A triple mutant (K60E, D189S, D194N) gave a catalyst with even higher synthetic activity. The catalysed reaction between a protected virtual substrate Boc-Phe-Gly-Gly-OGp and H-Ala-Ala-Arg-Ala-Gly-OH afforded the octapeptide derivative and no hydrolysis occurred at the Arg residue. Microscale semisynthetic techniques are sometimes feasible to produce analogues of biologically active peptides. For example, the *C*-terminal dipeptide of epidermal growth factor (EGF) can be removed with trypsin and analogues can be obtained by reversing this procedure.[194] The enzyme-catalysed synthesis of a peptide can be improved if the amino-acid sequence around the site of synthesis favours the formation of an α-helical region.[195] The presence of a suitable dipeptide sequence (e.g. Phe-Lys) between a monoclonal antibody and doxorubicin in a covalent conjugate of the two molecules can be cleaved, for example, by cathepsin B.[196] Finally, enzymic synthesis is not limited to forming peptide bonds. Some examples are to be found in the appendix. It is sufficient here to cite the use of Ser and Gly mutants of *Agrobacterium sp. β-glucosidase*. Acceptors were linked to PEGA resin through a backbone amide linker. A galactose residue was transferred from α-D-galactosyl fluoride to a polypeptide substrate in high yield.[197]

2.8 Miscellaneous Reactions Related to Peptide Synthesis. – This section is not intended to be exhaustive in scope. Some side reactions can be an unexpected nuisance and an example arose in the synthesis of human parathyroid hormone (1–34) in which a tosyl group was used to protect the side chain of His.[198] A byproduct containing a His residue bearing a Me group was identified. This was found to be due to the presence of Tos-O-Me which was generated during reprecipitation of the product using MeOH.

The remaining examples give rise to products that offer further synthetic possibilities. For example, treatment of nitriles with H_2S gas in aqueous methanol or ethanol and in the presence of an anion exchange resin (Dowex 1X8, SH^- form) gave thioamides.[199] Further reaction with e.g. a bromoacetyl derivative of an amino acid or peptide followed by reaction with NH_3 should give products with a protonated amidine group as a pseudopeptide bond. These might bind to enzymes possessing a carboxy group at the catalytic site. Hydroxylysine has been converted into α-Fmoc-Hyl(Boc-oxazolidine).[200] Peptides can be synthesized from this, the oxazolidine group can be detached with CF_3CO_2H and finally treatment with periodate generates an aldehyde group.

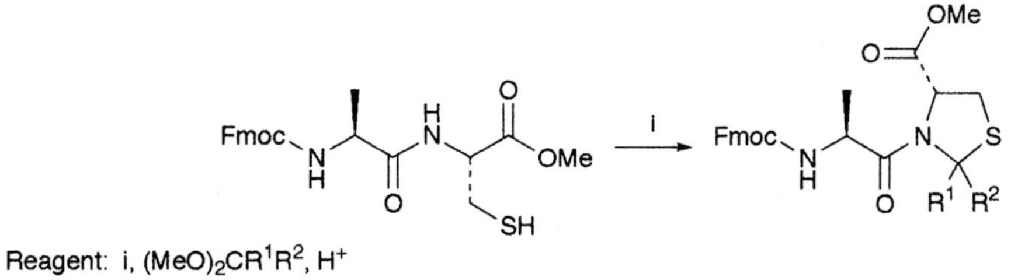

Reagent: i, $(MeO)_2CR^1R^2$, H^+

Scheme 12

This offers numerous opportunities for further synthetic development. Intra-residual acetalisation of a peptide containing an esterified Cys residue at the *C*-terminus (Scheme 12) gives diastereoisomeric ψ-proline.[201] The authors synthesised an immunosuppressive derivative of cyclosporin A. With a peptide containing 4-NH_2-Phe and either Tyr or His at a suitable distance, diazotisation and coupling of the diazonium salt to the side chain of Tyr or His gives rise to a novel type of cyclopeptide.[202] This section ends with a couple of examples of enzyme action on synthetic peptides. The first applies the 1,6-elimination reaction of 4-aminobenzyl ethers to the design of potential anticancer drugs.[203] 4-Hydroxymethyl-anilides of Z-Val-Cit are hydrolysed by cathepsin B, an enzyme that is present in metastatic cancer. If a phenolic anticancer drug such as etoposide or combretastatin A-4 is linked to this peptide derivative through an ether link, hydrolysis of the amide bond by cathepsin B results in the liberation of the free drug as a result of the ensuing fragmentation after hydrolysis. Further examples of prodrugs are cited in the appendix. The second example of enzyme action on a synthetic peptide uses mushroom tyrosinase at high [E]/[S] ratios to convert Tyr residues to DOPA. Borate is incorporated into the medium to prevent the formation of 3,4,5-trihydroxytyrosine residues.[204] The formation of vicinal hydroxy groups on Tyr side chains provides sites for chelation of transition metal ions. This might constitute *inter alia* a route to artificial redox enzymes.

3 Appendix: A List of Syntheses in 2002

Peptide/Protein	*Ref.*
3.1 Natural Peptides, Proteins and Partial Sequences	
Amelogenin	
27-mer repeat peptide	205
Androgenic gland hormone	
Nonglycosylated hormone from *Armadillidium vulgare*	206

3 Appendix (*continued*)

Peptide/Protein	*Ref.*
Angiotensin	
Spin-labelled analogues	207
Cyclic analogues	208
Antibiotics	
Ampullosporin A	209
Chloroorienticin B derivatives	210, 211
Distamycin analogues	212
Berninamycin A	213
α-Defensin HNP-1 and analogues	214
β-Defensin	215
Derivatives of eremomycin	216
Analogues of drosocin and apidaecin	217
Vancomycin analogues	218
Glycobleomycin	219, 220
Gramicidin	221
Lactoferricin analogue	222
Pyloricidins	223
Statherins	224
Antibiotic peptides from haemolytic lectin	225
Antibiotic peptides rich in dialkylglycines	226
Polymyxin B peptides	227
Cationic and steroid antibiotics	228
Antibacterial peptides containing Aib	229
Anti-HIV peptides	230–236
Bradykinin	
Analogues	207, 237, 238
Antagonists	239
Calcitonin	
Human calcitonin	240
Glycosylated derivatives	241
Calcium channel blockers	
SAR studies	242
Cancer	
Proapoptotic peptide for treatment of solid tumours	243
Antiangiogenic/heparin mimic of endostatin	244
Catenane	
Catenane from tumour suppressor protein p53	245
Chaperonines	
Chaperonine 10	246
Chemokines	
Human CC chemokine	247
Chemotactic peptides	

3 Appendix (*continued*)

Peptide/Protein	*Ref.*
Analogues of fMLP	248–252
Cholecystokinin and gastrin	
CCK-8 derivative	253
Analogues	254
Coleophomone	
Coleophomone D	255
Collagen	
Triple helical collagen mimetics	256
Phosphorylated model	257
Collagen mimetic dendrimers	258
Heterotrimeric peptides with integrin recognition site	259
Cyclin A	
Inhibitors of CDK2 cyclin A	260
Elastin	
Fragments of bovine elastin	261
Endothelin	
Antagonists	262
Erythropoietin	
Cis-trans isomers of a small fragment	263
Fibrinogen	
Thrombolytic activity of fragment analogues	264
Galanin	
Analogues	265
GnRH/LHRH	
Antagonists	266, 267
Analogues	268, 269
Glutathione	
Analogues	270
Immunosuppressants	
Sanglifehrin A	271
Insulin	
Analogue	272
Integrins	
Antagonists	273–280
Kaitocephalin	
Revision of stereochemistry	281
Keratin	
Fragment analogues	282
Laminin	
Laminin fragment	283
Leptin	
Synthesis of human protein	284

3 Appendix (*continued*)

Peptide/Protein	*Ref.*
Loloatins	
SPPS of loloatins A, B and C	285
Magainin	
Analogues	286
Dimer of magainin 2	287
Marine organisms	
Fragment (94–97) of sea cucumber globin	288
Melanotropins	
Analogues	289–293
Melanocortin analogues	294, 295
Metastin	
Bioorganic synthesis	296
Motilin	
Agonists	297
Natriuretic peptides	
Analogues	298
Neuropeptides	
Neurotensin analogues	299, 300
Proctolin analogues	301, 302
Chimera polypeptide of o-Ctx MVIIA and HWTX-I	303
Octreotide	
Modification of termini of Tyr^3-octreotate	304
Opioids, antinociceptive peptides and receptors	
Metal complexes of enkephalin	305
Enkephalin analogues	306–309
Deltorphin analogues	310
Dermorphin analogues	311–313
Nociceptin analogues	314
Opioid mimetics	315–318
Analogues of opioid peptide YkFA	319
Endomorphin-1 analogues	320
Affinity label for δ-opioid receptors	321
Opioid receptor antagonist	322
Osteogenic growth factor	
C-terminal fragment	323
Posterior pituitary hormones	
Oxytocin analogues	324–330
Vasopressin analogues	331, 332
Isotocin	333
RGD peptides	
Liquid-phase synthesis of tripeptide	334
Analogues	335–343

3 Appendix (*continued*)

Peptide/Protein	*Ref.*
Antithrombotic activity of carboline sequence	344
Synthetic receptor for RGD sequence	345
Peptide containing RGD and synergistic site (PHSRN) of fibronectin linked to PEG	346
Somatostatin	
Analogues	347–349
Substance P	
Tc-labelled substance P	350
T-cell ligands	
Antagonists of the $p56^{ick}$SH2 domain	351
Tachykinins	
Human tachykinin NK-2 receptor antagonists	352
Thymic humoral factor	
Thymic humoral factor-γ-2 analogues	353, 354
Thyroliberin	
Antagonist	355
Analogues	356–361
Toxic peptides	
Derivative of conotoxin MII	362
AK-toxins	363
Tentoxin	364
Transcription factors	
Factor IIIa of *Xenopus laevis*	365
Urotensin	
Analogues of hU-II (4-11)	366
Viral proteins	
Hepatitis C fragments	367
Human T-cell leukaemia virus type-1 proteinase	368, 369
Fragment (513-522) of hepatitis G virus	370

3.2 Sequential Oligo- and Poly-peptides

Boc-Homo-oligopeptides	371
Homooligo-pyroglutamic acid	372
Aggregation of Boc-(L-Glu)$_n$-OBzl (n=4,6,8)	373
Binding of amino acids in helical poly(L-Glu) shelled dendrimer in water	374
Cross-linked poly(ethylene glycol)peptides and glycopeptides as carriers for gene delivery	375
Copolypeptides having Pro as the most common residue	376

3 Appendix (*continued*)

Peptide/Protein	*Ref.*
Oligomer of Ser and Pro has polyproline II structure	377
Poly(Xaa-Pro)	378
Poly(Xaa-Pro-Pro)	379
Poly(Gly-Pro-Pro-Pro)	380
Poly *N*-substituted glycines	381
Poly(Orn-Gly-Gly-Orn-Gly); elastin-based models	382
Poly(Val-Pro-Gly-Val-Gly); elastin-based models	383
Sulfated homo-oligomers of Tyr	384

3.3 Enzyme Substrates and Inhibitors

SPPS of fluorogenic proteinase substrates	385
Trypsin inhibitors	386–388
Proteinase inhibitors from *N*-Boc-α-aminoaldehydes	389
Azapeptidomimetic β-lactams as proteinase inhibitors	390
Substrate for factor Xa	391
Inhibitors of factor Xa	392
Thrombin inhibitors	393–399
SAR of plasmin and kallikrein inhibitors	400
Inhibitors of tissue kallikrein	401
Urokinase inhibitors	402
Inhibitors of neutrophil elastase	403
Peptide boronic inhibitors of HCV NS3 proteinase	404
Inhibitor of plasminogen activator inhibitor	405
Mimetics of PA receptor-binding domain	406
Inhibitors of recombinant *L. mexicana* proteinase B	407
Peptidyltrifluoromethyl ketones as proteinase inhibitors	408
Azapeptides as inhibitors of cysteine proteinases	409
Calpain inhibitors	410–412
Caspase inhibitors	413, 414
Cathepsin K inhibitors	415, 416
Cathepsin L inhibitors	417
Cathepsin S inhibitors	418
Aspartyl proteinase inhibitors	419, 420
Inhibitor of human β-secretase	421
HIV proteinase inhibitors	422–430
Synthesis of dipeptide 4-nitroanilides	431
Aminopeptidase inhibitors	432–434
Carboxypeptidase A inhibitors	435, 436
Thionopeptides as inhibitors of aminopeptidase of *Acromonas proteolytica*	437

3 Appendix (*continued*)

Peptide/Protein	*Ref.*
Carboxypeptidase B substrate	438
Metalloproteinase inhibitors	439–447
ACE inhibitors	448–450
Macrocyclic potential inhibitors of proteinases	451
Inhibitors of D-alanyl-D-alanine dipeptidase	452
Inhibitors of tumour necrosis factor converting factor	453, 454
Inhibitors of peptidylproline isomerase	455
Rhinovirus 3C proteinase inhibitors	456, 457
Hepatitis C virus proteinase	458, 459
Total synthesis of TMC-95A/B proteasome inhibitors	460
Potential irreversible inhibitors of transglutaminase	461
Inhibitors of human cyclophilin hCyp-18	462
Protein phosphatase inhibitors	463, 464
β-*C*-Mannosides as selectin inhibitors	465
Mannosyl peptide as acceptor substrate for *N*-acetylglucosaminyl transferase	466
S-Alkyl GSH analogue as glutathione *S*-transferase inhibitor	467
Disulfide substrates for trypanothione reductase	468
Inhibitors of glutathionyl spermidine synthetase	469
Inhibitors of 17β-hydroxysteroid dehydrogenase	470
Potential substrate of Syk kinase	471
Prodrug of inducible nitric oxide synthase	472
SAR of peptide deformylase inhibitors	473
Histone deacetylase inhibitors	474, 475
Inhibitors of protein: geranylgeranyl-transferase I	476

3.4 Conformations of Synthetic Peptides

3- and 4-helix bundle TASP molecule	477
Scaffolds for helical peptide mimics	478
Cyclic pseudo 3_{10}-helical structure	479
R.H. helix containing central two D-amino acids	480
Ferrocenylglycylcystamine in helical structure	481
Screw sense of Aib-Xaa-(Aib-Δ^2Phe)$_x$-Aib-OMe	482
α-Helical peptides on self-assembled monolayers	483
Helical peptides containing a thioamide bond	484
SAR of 14-helical antimicrobial β-peptides	485
14-Helical peptides stabilised by side-chain branching adjacent to β-carbon atom	486
Helical tripeptide containing a non-coded amino acid	487

3 Appendix (*continued*)

Peptide/Protein	*Ref.*
Self-assembly of a tetrapeptide in the solid state	488
Side-chain arrays in 12-helical peptides	489
Rôle of *N*-t-Boc group in helix initiation	490
Zwitter-ionic α-cyclodextrin self-assemblies into nanotubes	491
Helicity and hydrophobicity relationships of membrane-spanning model peptides	492
3_{10}-Helical peptides containing ΔPhe residues	493
Cytokine receptor antagonists from helix-loop-helix peptide library	494
Assembly of triple-stranded β-sheet peptides	495
Fibril-forming peptides containing 3-aminophenyl-acetic acid	496
Fibrillar β-sheets from peptide-grafted polyamine	497
6-Amino-5-oxo-1,2,3,5-tetrahydro-3-indolizine-carboxylic acids as β-sheet peptidomimetics	498
β-Sheet polypeptide stabilised by a disulfide bridge	499
Parallel β-sheet assembly of a model dipeptide	500
β-Sheet folding induced by an unnatural amino acid	501
Peptide polymerised on self assembled monolayer	502
Conformation of oligomers of cis-2-aminocyclopentane-carboxylic acid	503
β-Barrels with alternating L and H at inner and K and E at outer surfaces	504
3-Stranded mixed artificial β-sheets	505
Amyloid grafts of poly-Glu(OMe) and polyallylamine	506
Complementary assembly of heterogeneous peptides into amyloid fibrils	507
Peptoid residues and β-turn formation	508
AAGDYY forms β-turn as in β-lactamase inhibitor	509
Turns and helices from 3-aminoxy-2,2-dimethyl-propionic acid	510
Dipeptide β-turn mimetics	511–513
Retro-inverso tetrapeptide β-turn mimetic	514
Oxidative induction of β-turn conformation in cyclic peptidomimetics	515
Scaffold for β-sheet capping turns	516
β-Turns involving SO to HN hydrogen bonding	517
Cross-coupling of dehydroamino acids to form β-turns	518
N-Terminal Abu in tri- or tetra-peptides is involved in 12-membered H-bonded rings	519
Constrained β-turn mimic derived from α-dehydro-β-amino esters	520
Ring-closure of Δ-Phe peptides to give β-turns	521

3 Appendix (*continued*)

Peptide/Protein	*Ref.*
Conformationally restricted glycoamino acids are potent β-turn mimics	522
7-Azocyclo[2.2.1]heptan-1-carboxylic acid in model dipeptides can participate in β-turn formation	523
Cyclic $\alpha_2\beta$-tripeptides as β-turn mimetics	524
β-Turn involving (1S,2R)-1-amino-2-hydroxycyclohexane-carboxylic acid	525
β-Turn thiazolidine mimetic	526
Turn formation initiated by a bis-sulfoximine motif	527
Conformation of peptides of 5-t-butylproline	528
Dodecapeptide containing D-Pro and L-Ser	529
β-Turns in peptides containing α and β-amino acids	530
β-Turn in peptide containing a pyrrole amino acid	531
Diamidopyridine derived tweezer receptors for peptides	532
Assembly of single-peptide nanotubes	533
Channel-forming peptides	534
Spirolactams: conformationally restricted peptides	535
Conformation of dipeptide isosteres derived from L-leucine and mesotartaric acid	536
Copolymers of PEG and coiled-coil peptides	537
Construction of heterotrimeric α-helical coiled-coil	538
Coiled-coil peptides containing nucleobase interactions and their catalytic function on self-replication	539
Pro tripeptides constrained with aziridine and dehydro-Phe residues	540
A fully extended tetrapeptide	541
Peptides containing α-ethylated amino acid	542
Conformation of Pheψ[CH_2O] pseudopeptides	543
Gel-like structures of peptidomimetics containing furanoid sugar amino acids	544
Conformation of peptides containing α,α-disubstituted amino acids	545
Self-replication of coiled-coil peptides	546
Effect of added His residues on structure of ASYQDL	547

3.5 Glycopeptides

N-Linked glycosylasparagine derivatives	548
Conformation of glycosylated Asn oligopeptides	549
Glycosylamino acids: components of neoglycoconjugates	550
Core-class 2 type glycopeptides	551, 552

3 Appendix (*continued*)

Peptide/Protein	*Ref.*
Automated high-throughput synthesis of glycopeptides	553
T-helper cell stimulation by *O*-linked glycopeptide	554
Glycopeptidomimetics containing *O*- and *N*-glycosylated α-aminoxy acids	555
C-Neoglycopeptides	556
Neoglycopeptides containing an *N*′-methylaminoxy amino acid	557
Cyclic analogue of muramyl dipeptide	558
Muramyl peptides containing mesodiaminopimelic acid	559
N-Acetylmuramyl-L-alanyl-D-isoglutamine aryl β-glycosides	560
C-Glycosyl tyrosine analogues	561
C-Glycoside analogue of Fmoc-Ser-β-*N*-acetylglucos-aminide by Ramberg-Baeckland reaarangement	562
C-Linked glyco β-amino acids	563
Fused furanosyl-β-amino acids: derived peptides	564
Oligomers containing furanoid sugar amino acids	565
N-Linked glycoamino acids: SPPS of glycopeptides	566
Structure of glycopeptides in solution and membranes	567
Mannosyl and oligomannosyl threonine derivatives	568
Silyl linker used in SPPS of Fmoc glycopeptide thioesters	569
Mucin-type glycopeptides	570
56 component library of sugar β-peptides	571
Neoglycoproteins containing *O*-methylated trisaccharides related to antigens of *Toxocara* larvae	572
S-Linked glycosyl amino acids	573
Thioether-linked glycopeptide mimics	574, 575
N-Linked glycopeptide containing 6-thioGlcNAc	576
Synthetic inhibitors of cell adhesion	577
β-Glycan-type tetraosyl hexapeptides	578
Sugar amino acid based scaffolds	579
Proteinase-catalysed synthesis of carbohydrate-peptide conjugates at peptide carboxy groups	580
Thioester analogues of peptidoglycans	581
Chemoenzymatic synthesis of *O*-linked sialyl oligo-saccharides	582
Chemoenzymatic synthesis of GlcNAc-Z-Asn	583
Chemoenzymatic synthesis of *S. cerevisiae* glycosylated mating factor	584
Chemoenzymatic synthesis of a glycopeptide using microbial endoglycosidase	585
Enzymatic linkage of LSQ(or N)VHR and various sugars	586

3 Appendix (*continued*)

Peptide/Protein	*Ref.*
Enzymatic attachment of glycosaminoglycans to peptides	587
Chemoenzymatic synthesis of a mucooligosaccharide	588
Major histocompatibility peptide	589

3.6 Phosphopeptides and Related Compounds

Atherton-Todd phosphorylation of Tyr	590
β-Aminophosphotyrosyl mimetic	591
Phosphoangiotensin II	592
SPPS of a peptide containing phosphotyrosine	593
Phosphopeptide ligands of Grb2 SH2 domain	594, 595
Caged phosphopeptides	596
Phosphopeptide mimics of p53 domain	597
Oxathiaphospholane approach to phosphoamino acids	598
SPPS of phosphopeptide thioesters	599
SPPS of *O*-boranophospho- and *O*-dithiophosphopeptides	600
Peptides of 2-(4-phosphonophenylmethyl)-3-aminopropanoic acid	601
SPPS of α-aminoalkylphosphonopeptides	602
Phosphinic acid analogues of Gly-Gly	603
Phosphinodipeptides	604, 605
Fmoc-3-(4-(di-Bu^t-phosphonomethyl)phenyl)pipecolic acid	606

3.7 Immunogenic and Immunosuppressant Peptides

Immunogenicity of group A streptococcal peptide epitopes	607
Antihexapeptide antibodies that recognise ribosomal proteins of *E. coli*	608
Antigenic activity of fragment (87–99) of myelin basic protein	609
Immunosuppressants	610, 611
Argyrin B, an immunosuppressive cyclic peptide	612
Antigenic peptides from the enolase sequence	613

3.8 Nucleopeptides, PNAs

General methods of synthesis	614–625
PNAs rich in particular bases	626–629
PNA oligomers modified with Pt (II) complex	630
PNA libraries	631

3 Appendix (*continued*)

Peptide/Protein	*Ref.*
Chromium tricarbonyl labelled thymine PNA monomers	632
Peptide-PNA-peptide conjugation via native ligation	633
3′-Oligonucleotide conjugation via chemoselective oxime bond formation	634
Prolyl carbamate nucleic acids	635
PNAs with single and multiple peptides attached to 2′-aldehydes through various linkages	636
Phosphorothioate DNA-peptide conjugate	637
Fluoroaromatic universal bases in PNAs	638
F^{18}-Labelling of PNAs	639
Thiazane and thiazolidine PNA monomers	640
Synthesis and properties of thiazolidine PNAs	641
SPPS and disulfide cyclisation of peptide-PNA-peptides	642
Attachment of fluorescent reporter groups to PNAs	643
Liquid-phase synthesis of diPNA-arginine conjugates	644
Liquid-phase synthesis of oligonucleotide-PEG-peptide conjugates	645
N-Nonpolar nucleobase dipeptides	646

3.9 Miscellaneous Peptides

SPPS of peptidomimetics based on 4-imidazolidinones	647
Bis(cyclobutane) β-dipeptides	648
An optically active bicyclic dipeptide mimetic	649
Peptidyl α-keto heterocycles	650
β-Peptidosulfonamide/β-peptide hybrids	651
β-Peptide synthesis from β-lactones	652
β- and mixed α/β-peptides	653
α,α-Disubstituted β-amino esters and peptides	654
γ-Peptides	655
A helical peptide receptor for [60] fullerene	656
Tripeptide that binds LDL in plasma	657
Peptides that bind DNAs	658–660
Nitrosylpeptides as catalysts of enantioselective oxidations	661
Peptides containing diethylene triamine pentaacetic acid	662
Structural comparison of Ac-ΔAla-NMe_2 and Ac-Ala-Nme_2	663
Peptides of dehydrobutyrine	664
N-Z-ΔLeu-L-Ala-L-Leu-OMe	665
A constrained surrogate of Gly-ΔAla	666
Peptide-lipid conjugates	667–669
Fluorescent labels for peptides and proteins	670–672

3 Appendix (*continued*)

Peptide/Protein	*Ref.*
Tag reagent for separation of deletion peptides	673
Labelling with ferrocene derivatives	674–676
Protoporphyrin complexes	677
Polypeptide complexes of transition elements of periods 5 and 6	678–680
Peptides based on a rigid receptor for transition metal ions	681
Peptides containing Ru(II) and Pd(II) bipyridine complexes	682
Folding of Zn finger peptides	683
Metal-binding and catalytic peptoids labelled with ^{19}F	684
Patellamide A derivatives and Cu(II) complexes	685
Conversion of Tyr peptides into metal-binding peptides	686
SPPS of polyamides containing pyrrole amino acid	687
Homoazacalix[4]arenes containing amino acids that bind quaternary ammonium ions	688
Aspartame analogues	689
Antifungal nonapeptide libraries	690
Synthetic tripeptide as an organogelator	691
C-Backbone branched peptides	692
Reduction of ethanethiol esters to aldehydes	693
Peptide aldehydes	694
Peptide-arene hybrids from tetrachlorophthaloyl amino acids	695
Interaction of amphiphilic α-helical peptide with phospholipid	696
Retro-and retro-inverso ψ[NHCH(CF_3)]-Gly peptides	697
Dipeptides containing 2-methyl-Tyr and 2-methyl-DOPA	698
Peptide-linked dicatechol derivatives	699
Orthogonally-protected lanthionines	700
L-Valyl-pyrrolidine(2R)-boronic acid	701
Pilot-scale synthesis of frakefamide analgaesic	702
Solid-phase synthesis of spirobicyclic peptides	703
(3-Trifluoromethyl)phenyldiazirine based Phe compounds	704
Peptide-small molecule hybrids on scaffolds	705
Photoactive peptide of 4-azidotetrafluorophenylalanine	706
Hexapeptides of ε-*N*-(1,2-dihydro-1-hydroxy-2-oxo-pyrimidin-4-yl)-L-Lys-β-Ala-OH	707
Peptides containing α-difluoromethyl-α-amino acids	708
Peptide-2,2′-biphenyl hybrids	709
Antitumour imidazotetrazines conjugated to peptides	710
Molecular tweezers from deoxycholic acid for amino-acid methyl esters	711
Constrained Phe and naphthylalanine derivatives	712
3-[2-(Pyridyl)-benzoxazol-5-yl]-alanine derivatives	713
Derivatives of mercaptoacetyltriglycine	714, 715

3 Appendix (*continued*)

Peptide/Protein	*Ref.*
Triazinyl-amino acids, components of pseudopeptides	716
Guanidinium-derived receptor libraries	717
N-Z-*S*-Acm-cysteine peptides	718
SPPS of phosphine-oxazoline peptides	719
Antitumour drug-peptide conjugates	720, 721
N-Aminoamide pseudopeptides	722
(2S)-4,4-Difluoroglutamyl-γ-peptides	723
Proteinase activity of peptidosteroids containing His and Ser residues	724
Peptidomimetics of hydroxylated D-Pro-Gly motifs	725
Polycationic dendrimers on lipophilic peptides for oligonucleotide transport	726
Dendritic-graft polypeptides	727
N-[α-*N*-(4-amino-4-deoxypteroyl)-δ-*N*-hemiphthaloyl]-L-Orn-L-Phe-OH, prodrug activated by carboxypeptidase	728

3.10 Purification Methods

Capillary liquid chromatography and electrospray ionisation in MS analysis of peptides	729, 730
Separation of peptides by normal phase chromatography	731
Peptide purification on preparative scale by RP-HPLC	732
Peptide separation by RP and ion-exchange chromatography on a monolithic capillary column	733
Affinity chromatography of coagulation factor VIII	734
Analysis of phosphinic pseudopeptides by capillary zone electrophoresis	735

References

1. D.T. Elmore, Specialist Periodical Report, *Amino Acids, Peptides and Proteins*, 2003, **34**, 1.
2. C. Najera, *Synlett*, 2002, 1388.
3. G.A. Grant, *Synthetic Peptides (2nd ed.)* Oxford Univ. Press, New York, 2002.
4. M.L. Moore and G.A. Grant, ref. 3, p. 10.
5. V.J. Hruby and T.O. Matsunaga, ref. 3, p. 292.
6. M. Yashiro, *Kagaku to Kyoiku*, 2002, **50**, 177.
7. K. Sadler and J.P. Tam, *Rev. Mol. Biotechnol.*, 2002, **90**, 195.
8. G.S. Shaw, *Meth. Mol. Biol.*, 2002, **173**, 175.

9. J.-M. Ahn, N.A. Boyle, M.J. MacDonald and K.D. Janda, *Mini-Rev. Med. Chem.*, 2002, **2**, 463.
10. A. Zega and U. Urleb, *Acta Chem. Slovenica*, 2002, **49**, 649.
11. R.G.S. Berlinck, *Nat. Prod. Rep.*, 2002, **19**, 617.
12. S.A. Lawrence, *Pharma Chem.*, 2002, 32.
13. J.P. Tam, Q. Yu and Y.-A. Lu, *Biologicals*, 2001, **29**, 189.
14. G.B. Fields, J.-L. Lauer-Fields, R.-q. Liu and G. Barany, ref. 3, p. 93.
15. G.C. Fromont, V. Pomel and M. Bradley, in *Integrated Drug Discovery Technologies*, eds. H.-Y. Mei and A.W. Czarnik, Marcel Dekker Inc., New York, 2002, p. 463.
16. G. Petersen, in *Peptide Arrays on Membrane Supports*, eds. J. Koch and M. Mahler, Springer-Verlag, Berlin, 2002, p. 41.
17. H. Gausepohl and C. Behn, ref. 16, p. 55.
18. N. Zander and H. Gausepohl, ref. 16, p. 23.
19. J. Koch, M. Mahler, M. Bluthner and S. Dubel, in *Protein-Protein Interactions*, ed. E. Golemis, Cold Spring Harbor Laboratory Press, Cold Spring Harbor, N.Y., 2002, p. 569.
20. B. Schulze and E. De Vroom, in *Enzyme Catalysis in Organic Synthesis (2nd ed.)*, eds. K. Drauz and H. Waldmann, Wiley-VCH Verlag GmbH, Weinheim, 2002, p. 716.
21. H.-D. Jakubke, ref. 20, p. 800.
22. A. Yan, G. Tian and Y. Ye, *Huaxue Jinzhan*, 2001, **13**, 203.
23. K. Aso, *Kagaku to Kogyo*, 2002, **76**, 18.
24. F. Bordusa, *Chem. Rev.*, 2002, **102**, 4817.
25. F. Bordusa, *Curr. Protein Pept. Sci.*, 2002, **3**, 159.
26. M. Takayanagi, *Kagaku Kyokaishi*, 2002, **60**, 240.
27. E. Illyes, S. Bosze, O. Lang, F. Sebestyen, L. Kohidai and F. Hudecz, *Chem. Oggi.*, 2002, **20**, 55.
28. Y. Park, D.G. Lee, H.N. Kim, H.K. Kim, E.-R. Woo, C.-H. Choi and K.-S. Hahm, *Biotechnol. Lett.*, 2002, **24**, 1209.
29. A.-C. Thierry, S. Pinaud, N. Bigler, G. Perrenoud, B. Denis, M.A. Roggero, N. Fasel, C. Moulon and S. Demotz, *Biologicals*, 2001, **29**, 259.
30. K. Yamada, *Kagaku to Kogyo*, 2002, **55**, 54.
31. V.J. Hruby, R.S. Agnes and C. Cai, *Methods Emzymol.*, 2001, **343**, 73.
32. D.L. Steer, R.A. Lew, P. Perlmutter, A.J. Smith and M.-I. Aguilar, *Lett. Pept. Sci.*, 2001, **8**, 241.
33. R.D. Tung, D.J. Livingston, B.G. Rao, E.E. Kim, C.T. Baker, J.S. Boyer, S.P. Chambers, D.D. Deininger, M. Dwyer, L. Elsayed, J. Fulghum, B. Li, M.A. Murcko, M.A. Navia, P. Novak, S. Pazhanisamy, C. Stuver and J.A. Thomson, *Infect. Disease Ther.*, 2002, **25**, 101.
34. T. Monteil, D. Danvy, M. Sihel, R. Leroux and J.-C. Plaquevent, *Mini-Rev. Med. Chem.*, 2002, **2**, 209.
35. M.P. Glenn and D.P. Fairlie, *Mini-Rev. Med. Chem.*, 2002, **2**, 433.
36. I. Karle, *Biopolymers*, 2002, **60**, 351.
37. H. Hojo, *Kagaku*, 2001, **56**, 66.
38. R.J. Ferrier, R. Blattner, R.A. Field, R.H. Furneaux, J.M. Gardiner, J.O. Hoberg, K.P.R. Kartha, D.M.G. Tilbrook, P.C. Tyler and R.H. Wightman, *Carbohydr. Chem.*, 2002, **33**, 144.
39. L.A. Marcaurelle and C.R. Bertozzi, *Glycobiol.*, 2002, **12**, 69R.
40. N. Bezay and H. Kunz, in *Solid Support Oligosaccharide Synthesis and Combinatorial Carbohydrate Libraries*, ed. P.H. Seeberger, John Wiley & Sons Inc., New York, 2001, p. 257.

41. P.M. St. Hilaire, K.M. Halkes and M. Meldal, ref. 40, p. 283.
42. C. Brocke and H. Kunz, *Bioorg. Med. Chem.*, 2002, **10**, 3085.
43. P. Sears, T. Tolbert and C.-H. Wong, *Genetic Eng.*, 2001, **23**, 45.
44. R. Roy and M. Baek, *Rev. Mol. Biotechnol.*, 2002, **90**, 291.
45. P.M.St. Hilaire, M. Meldal and K. Bock, *Spec. Publ. Roy. Soc. Chem.*, 2001 **264**, 293.
46. S. Manabe, *Yakugaku Zasshi*, 2002, **122**, 295.
47. C. Granier, *J. Immunol. Methods*, 2002, 267.
48. E.M. Zubin, E.A. Romanova and T.S. Oretskaya, *Russ.Chem. Rev.*, 2002, **71**, 239.
49. T. Zhang and P. Xu, *Zhongguo Yaowu Huaxue Zazhi*, 2002, **12**, 172.
50. R. Iwase and A. Murakami, *Yuki Gosei Kagaku Kyokaishi*, 2002, **60**, 1179.
51. Y. Zhang and F. He, *Shengli Kexue Jinzhan*, 2002, **33**, 59.
52. S.I. Antsypovitch, *Russ. Chem. Rev.*, 2002, **71**, 71.
53. R.A. Breitenmoser, A. Linden and H. Heimgartner, *Helv. Chim. Acta*, 2002, **85**, 990.
54. G.A. Grant, ref. 3, p. 220.
55. J.R. Casimir, G. Guichard and J.-P. Briand, *J. Org. Chem.*, 2002, **67**, 3764.
56. K. Hioki, M. Fujiwara, S. Tani and M. Kunishima, *Chem.Lett.*, 2002, 66.
57. D.C. Gowda, K. Abiraj and P. Augustine, *Lett. Pept. Sci.*, 2002, **9**, 43.
58. D.C. Gowda, *Tetrahedron Lett.*, 2002, **43**, 311.
59. D.C. Gowda and B. Mahesh, *Protein Pept. Lett.*, 2002, **9**, 225.
60. D.C. Gowda, *Indian J. Chem.*, 2002, **41B**, 1064.
61. J.R. Hwu, M.L. Jain, F.-Y. Tsai, A. Balakumar, G.H. Hakimelahi and S.-C. Tsay, *ARKIVOC*, 2002, 29.
62. S.D. Bull, S.G. Davies, P.M. Kelly, M. Gianotti and A.D. Smith, *J. Chem. Soc., Perkin Trans. 1*, 2001, 3106.
63. G.-R. Vasanthakumar, B.S. Patil and V.V.S. Babu, *J. Chem. Soc., Perkin Trans. 1*, 2002, 2087.
64. J.S. Yadav, B.V.S. Reddy, K. Srinivasa and K.B. Reddy, *Tetrahedron Lett.*, 2002, **43**, 1549.
65. X.-l. Qiu and F.-l. Qing, *J. Org. Chem.*, 2002, **67**, 7162.
66. K. Wright, F. Melandri, C. Cannizzo, M. Wakselman and J.-P. Mazaleyrat, *Tetrahedron*, 2002, **58**, 5811.
67. G. Davies and A.T. Russell, *Tetrahedron Lett.*, 2002, **43**, 8519.
68. R.G. Bhat, E. Kerouredan, E. Porhiel and S. Chandrasekaran, *Tetrahedron Lett.*, 2002, **43**, 2467.
69. R.G. Bhat, S. Sinha and S. Chandrasekaran, *Chem. Commun.*, 2002, 812.
70. X. Bu, G. Xie, C.W. Law and Z. Guo, *Tetrahedron Lett.*, 2002, **43**, 2419.
71. J. Farrera-Sinfreu, M. Royo and F. Albericio, *Tetrahedron Lett.*, 2002, **43**, 7813.
72. R. Chinchilla, D.J. Dodsworth, C. Najera and J.M. Soriano, *Bioorg. Med. Chem. Lett.*, 2002, **12**, 1817.
73. K. Hasegawa, J.-K. Bang, T. Kawakami, K. Akaji and S. Aimoto, *Pept. Sci.*, 2002, 65.
74. O. Lagrille, J. Taillades, L. Boiteau and A. Commeyras, *Eur. J. Org. Chem.*, 2002, 1026.
75. N. Zanatta, A.M.C. Squizani, L. Fantinel, F.M. Nachtigall, H.G. Bonacorso and M.A.P. Martins, *Synthesis*, 2002, 2409.
76. K. Braun and P. Kuhl, *Pharmazie*, 2002, **57**, 310.
77. Y. Jiang, J. Zhao and L. Hu, *Tetrahedron Lett.*, 2002, **43**, 4589.
78. K. Hojo, M. Maeda and K. Kawasaki, *Pept. Sci.*, 2002, 41.

79. W.H. Dent, W.R. Erickson, S.C. Fields, M.H. Parker and E.G. Tromiczak, *Org. Lett.*, 2002, **4**, 1249.
80. O.M. Kustenko, S.Y. Dmitrieva, O.I. Tolmachev and S.M. Yarmoluk, *J. Fluoresc.*, 2002, **12**, 173.
81. A.D. Abell, D.C. Martyn, C.H.B. May and B.K. Nabbs, *Tetrahedron Lett.*, 2002, **43**, 3673.
82. V.V.S. Babu, K. Ananda and G.-R. Vasanthakumar, *Protein Pept. Lett.*, 2002, **9**, 345.
83. P.R. Sridhar, S. Sinha and S. Chandrasekaran, *Indian J. Chem.*, 2002, **41B**, 157.
84. M. Wagner and H. Kunz, *Z. Naturforsch. B: Chem. Sci.*, 2002, **57**, 928.
85. M. Völkert, S. Koul, G.H. Müller, M. Lehnig and H. Waldmann, *J. Org. Chem.*, 2002, **67**, 6902.
86. A.O. Yusuf, B.M. Bhatt and P.M. Gitu, *S. African J. Chem.*, 2002, **55**, 87.
87. S. Zaramella, R. Stromberg and E. Yeheskiely, *Eur. J. Org. Chem.*, 2002, 2633.
88. T. Young and L.L. Kiessling, *Angew. Chem., Int. Ed.*, 2002, **41**, 3449.
89. E. Cros, M. Planas and E. Bardaji, *Lett. Pept. Sci.*, 2002, **9**, 1.
90. S. Gazal, G. Gellerman, E. Glukhov and C. Gilon, *J. Pept. Res.*, 2002, **58**, 527.
91. H. Gali, G.L. Sieckman, T.J. Hoffman, N.K. Owen, D.G. Mazuru, L.R. Forte and W.A. Wolkert, *Bioconjugate Chem.*, 2002, **13**, 224.
92. A. Cuthbertson, E. Jarnaess and S. Pullan, *Tetrahedron Lett.*, 2001, **42**, 9257.
93. B. Sacca, D. Barth, H.-J. Musiol and L. Moroder, *J. Pept. Sci.*, 2002, **8**, 205.
94. S. Ranganathan, K.M. Muraleedharan, M. Vairamani, A.C. Kunwar and A. Ravi Sankar, *Chem. Commun.*, 2002, 314.
95. C. Cabrele, S. Fiori, S. Pegoraro and L. Moroder, *Chem. Biol.*, 2002, **9**, 731.
96. L. Boiteau, H. Collet, O. Lagrille, J. Taillades, W. Vayaboury, O. Giani, F. Schue and A. Commeyras, *Polym. Intern.*, 2002, **51**, 1037.
97. G.-R. Vasanthakumar, K. Ananda and V.V.S. Babu, *Indian J. Chem.*, 2002, **41B**, 1733.
98. K. Hojo, M. Maeda, T.J. Smith and K. Kawasaki, *Chem. Pharm. Bull.*, 2002, **50**, 140.
99. S.H. van Leeuwen, P.J.L.M. Quaedflieg, Q.B. Broxterman and R.M.J. Liskamp, *Tetrahedron Lett.*, 2002, **43**, 9203.
100. A. DalPozzo, M. Ni, L. Muzi, A. Caporale, R. de Castiglione, B. Kaptein, Q.B. Broxterman and F. Formaggio, *J. Org. Chem.*, 2002, **67**, 6372.
101. P. Gagnon, X. Huang, E. Therrien and J.W. Keillor, *Tetrahedron Lett.*, 2002, **43**, 7717.
102. M. Crisma and C. Toniolo, *Acta Cryst.*, 2002, **C58**, o275.
103. T.S. Rao, S. Nampalli, P. Sekher and S. Kumar, *Tetrahedron Lett.*, 2002, **43**, 7793.
104. S.J. Tantry, K. Ananda and V.V.S. Babu, *Indian J. Chem.*, 2002, **41B**, 1028.
105. C. Palomo, A.L. Palomo, F. Palomo and A.L. Mielgo, *Org. Lett.*, 2002, **4**, 4005.
106. G. Ivanova, E. Bratovanova and D. Petkov, *J. Pept. Sci.*, 2002, **8**, 8.
107. A.D. Corbett and J.L. Gleason, *Tetrahedron Lett.*, 2002, **43**, 1369.
108. C.E. Garrett, X. Jiang, K. Prasad and O. Repič, *Tetrahedron Lett.*, 2002, **43**, 4161.
109. Y. Basel and A. Hassner, *Tetrahedron Lett.*, 2002, **43**, 2529.
110. E.A. Permyakov, S.E. Permyakov and V.N. Medvedkin, *Russ. J. Bioorg. Chem.*, 2002, **28**, 9.
111. B. Thern, J. Rudolph and G. Jung, *Tetrahedron Lett.*, 2002, **43**, 5013.
112. B. Thern, J. Rudolph and G. Jung, *Angew. Chem., Int. Ed.*, 2002, **41**, 2307.
113. H. Liu, L. Xia and Y.H. Ye, *Chin. Chem. Lett.*, 2002, **13**, 601.

114. P. Liu, B.-Y. Sun, X.-H. Chen, G.-L. Tian and Y.-H. Ye, *Synth. Commun.*, 2002, **32**, 473.
115. A.D. Corbett and J.L. Gleason, *Tetrahedron Lett.*, 2002, **43**, 1369.
116. S.J. Tantry and V.V.S. Babu, *Lett. Pept. Sci.*, 2002, **9**, 35.
117. L.A. Carpino, H. Imazumi, A. El-Faham, F.J. Ferrer, C. Zhang, Y. Lee, B.M. Foxman, P. Henklein, C. Hanay, C. Mugge, H. Wenschuh, J. Klose, M. Beyermann and M. Bienert, *Angew. Chem., Int. Ed.*, 2002, **41**, 441.
118. M.A. Bailén, R. Chinchilla, D.J. Dodsworth and C. Nájera, *Tetrahedron Lett.*, 2002, **43**, 1661.
119. Z.J. Kaminski, K.J. Zajac and K. Jastrzabek, *Acta Biochim. Pol.*, 2001, **48**, 1143.
120. M. Kunishima, A. Kitao, C. Kawachi, Y. Watanabe, S. Iguchi, K. Hioki and S. Tani, *Chem. Pharm. Bull.*, 2002, **50**, 549.
121. D.T.S. Rijkers, H.H.R. van Vugt, H.J.F. Jacobs and R.M.J. Liskamp, *Tetrahedron Lett.*, 2002, **43**, 3657.
122. M.B. Soellner, B.L. Nilsson and R.T. Raines, *J. Org. Chem.*, 2002, **67**, 4993.
123. B. Alcaide, P. Almendros and C. Aragoncillo, *Chem.-Eur. J.*, 2002, **8**, 3646.
124. T. Inui, J. Bodi, H. Nishio, Y. Nishiuchi and T. Kimura, *Lett. Pept. Sci.*, 2001, **8**, 319.
125. I. Shin and K. Park, *Org. Lett.*, 2002, **4**, 869.
126. K. Chiba, Y. Kono, S. Kim, K. Nishimoto, Y. Kitano and M. Tada, *Chem. Commun.*, 2002, 1766.
127. J.P. Tam, K.D. Eom and J. Xu, *Pept. Sci.*, 2002, 13.
128. J. Offer, C.N.C. Boddy and P.E. Dawson, *J. Amer. Chem. Soc.*, 2002, 4642.
129. R. Quaderer and D. Hilvert, *Chem. Commun.*, 2002, 2620.
130. P. Watts, C. Wiles, S.J. Haswell and E. Pombo-Villar, *Tetrahedron*, 2002, **58**, 5427.
131. P. Watts, C. Wiles, S.J. Haswell and S. Pombo-Villar, *Lab. Chip*, 2002, **2**, 141.
132. J.A. Moss, T.J. Dickerson and K.D. Janda, *Tetrahedron Lett.*, 2002, **43**, 37.
133. (*a*) I.M.K. Kumar and B. Mathew, *Protein Pept. Lett.*, 2002, **9**, 167; (*b*) I.M.K. Kumar and B. Mathew, *J. Appl. Polym. Sci.*, 2002, **86**, 1717.
134. I.M. Krishnakumar and B. Mathew, *Eur. Polym. J.*, 2002, **38**, 1745.
135. I.M.K. Kumar, V.N.R. Pillai and B. Mathew, *J. Pept. Sci.*, 2002, **8**, 183.
136. I.M. Krishnakumar and B. Mathew, *Lett. Pept. Sci.*, 2001, **8**, 339.
137. P.G. Sasikumar, K.S. Kumar, C. Arunan and V.N.R. Pillai, *J. Chem. Soc., Perkin Trans 1*, 2002, 2886.
138. D.-W. Kim, J. Namgung, J.-K. Cho and Y.-S. Lee, *Pept. Sci.*, 2002, 17.
139. D. Fattori, O. Kinzel, P. Ingallinella, E. Bianchi and A. Pessi, *Bioorg. Med. Chem. Lett.*, 2002, **12**, 1143.
140. R. Ingenito, D. Drežnjak, S. Guffler and H. Wenschuh, *Org. Lett.*, 2002, **4**, 1187.
141. U. Boas, J. Brask, J.B. Christensen and K.J. Jensen, *J. Comb. Chem.*, 2002, **4**, 223.
142. M. Wagner and H. Kunz, *Angew. Chem., Int. Ed.*, 2002, **41**, 317.
143. K. Sakamoto, Y. Nakahara and Y. Ito, *Tetrahedron Lett.*, 2002, **43**, 1515.
144. A. Ishii, H. Hojo, Y. Nakahara, Y. Nakahara and Y. Ito, *Pept. Sci.*, 2002, 59.
145. O. Lack, H. Zbinden and W.-D. Woggon, *Helv. Chim. Acta*, 2002, **85**, 495.
146. J.J. Pastor, I. Fernández, F. Rabanal and E. Giralt, *Org. Lett.*, 2002, **4**, 3831.
147. B.G. de la Torre, M.A. Marcos, R. Eritja and F. Albericio, *Lett. Pept. Sci.*, 2001 **8**, 331.
148. B. Török, I. Bucsi, G.K. Prakash and G.A. Olah, *Chem. Commun.*, 2002, 2882.
149. J. Vizzavona, M. Villain and K. Rose, *Tetrahedron Lett.*, 2002, **43**, 8693.
150. Q. Xu and K.S. Lam, *Tetrahedron Lett.*, 2002, **43**, 4435.

151. H.J. Olivos, P.G. Alluri, M.M. Reddy, D. Salony and T. Kodadek, *Org. Lett.*, 2002, **4**, 4057.
152. M. Erdélyi and A. Gogoll, *Synthesis*, 2002, 1592.
153. K. Nokihara, S. Yamamoto, C. Toda and J. Wang, *Pept. Sci.*, 2001, 61.
154. R. Pipkorn, C. Boenke, M. Gehrke and R. Hoffman, *J. Pept. Res.*, 2002, **59**, 105.
155. D. Pantarotto, A. Bianco, F. Pellarini, A. Tossi, A. Giangaspero, I. Zelezetsky, J.-P. Briand and M. Prato, *J. Amer. Chem. Soc.*, 2002, **124**, 12543.
156. J.-F. Sutcliffe-Goulden, M.J. O'Doherty, P.K. Marsden, I.R. Hart, J.F. Marshall and S.S. Bansal, *Eur. J. Nucl. Med. Mol. Imag.*, 2002, **29**, 754.
157. M.F. Mohd Mustapa, R. Harris, J. Mould, N.A.L. Chubb, D. Schultz, P.C. Driscoll and A.B. Tabor, *Tetrahedron Lett.*, 2002, **43**, 8363.
158. V. Swali, M. Matteucci, R. Elliot and M. Bradley, *Tetrahedron*, 2002, **58**, 9101.
159. S.R. Chhabra, A. Mahajan and W.C. Chan, *J. Org. Chem.*, 2002, **67**, 4017.
160. D.T.S. Rijkers, J.A. Den Hartog and R.M.J. Liskamp, *Biopolymers*, 2002, **63**, 141.
161. J.T. Lundquist and J.C. Pelletier, *Org. Lett.*, 2002, **4**, 3219.
162. N. Schmiedeberg and H. Kessler, *Org. Lett.*, 2002, **4**, 59.
163. A. Harishchandran, B. Pallavi and R. Nagaraj, *Protein Pept. Lett.*, 2002, **9**, 411.
164. K. Deshayes, M.L. Schaffer, N.J. Shelton, G.R. Nakamura, S. Kadkhodayan and S.S. Sidhu, *Chem. Biol.*, 2002, **9**, 495.
165. M. Royo, J. Farrera-Sinfreu, L. Solé and F. Albericio, *Tetrahedron Lett.*, 2002, **43**, 2029.
166. E. Oliveira, E.M. Cilli, A. Miranda, G.N. Jubilut, F. Albericio, D. Andreu, A.C.M. Paiva, S. Schreier, M. Tominaga and C.R. Nakaie, *Eur. J. Org. Chem.*, 2002, 3686.
167. P.M. Fischer and D.I. Zheleva, *J. Pept. Sci.*, 2002, **8**, 529.
168. S.E. Taylor, T.J. Rutherford and R.K. Allemann, *J. Chem. Soc. Perkin Trans. 2*, 2002, 751.
169. C.-Z. Dong, A. Romieu, C.M. Mounier, F. Heymans, B.P. Roques and J.-J. Godfroid, *Biochem. J.*, 2002, **365**, 505.
170. D.-F. Tai and W.-C. Liaw, *FEBS Lett.*, 2002, **517**, 24.
171. J.A. Laszlo and D.L. Compton, *A.C.S. Symp. Ser.*, 2002, **818**, 387.
172. P. Lozano, T. De Diego, J.-P. Guegan, M. Vaultier and J.L. Iborra, *Biotechnol. Bioeng.*, 2001, **75**, 563.
173. M.-X. Wang and S.-J. Lin, *J. Org. Chem.*, 2002, **67**, 6542.
174. S. Conde and P. Lopez-Serrano, *Eur. J. Org. Chem.*, 2002, 922.
175. T. Maruyama, S.-I. Nagasawa and M. Goto, *J. Biosci. Bioeng.*, 2002, **94**, 357.
176. Z.-Y. Jiang and H.-F. Chen, *Youji Huaxue*, 2002, **22**, 1050.
177. H. Yang, C. Zhou, G.-L. Tian and Y.-H. Ye, *Chin. J. Chem.*, 2002, **20**, 1354.
178. S. Barberis, E. Quiroga, M.C. Arribere and N. Priolo, *J. Mol. Catal. B: Enzymatic*, 2002, **17**, 39.
179. P. Liu, G.-I. Tian, W.-H. Lo, K.-S. Lee and Y.-H. Ye, *Prep. Biochem. Biotechnol.*, 2002, **32**, 29.
180. I.Yu. Filippova, A.V. Bacheva, O.V. Baibak, F.M. Plieva, E.N. Lysogorskaya, E.S. Oksenoit and V.I. Lozinsky, *Russ. Chem. Bull.*, 2001, **50**, 1896.
181. K. Aso, A. Narai, N. Endo and R. Uchida, *Nippon Jui Chikusan Daigaku Kenkyu Hokoku*, 2001, **50**, 38.
182. R.V. Ulijn, B. Baragaña, P.J. Halling and S.L. Flitsch, *J. Amer. Chem. Soc.*, 2002, **124**, 10988.
183. R.V. Ulijn, L. De Martin, L. Gardossi, A.E.M. Janssen, B.D. Moore and P.J. Halling, *Biotechnol. Bioeng.*, 2002, **80**, 509.

184. T. Miyazawa, E. Ensatsu, M. Hiramatsu, R. Yanagihara and T. Yamada, *J. Chem. Soc., Perkin Trans. 1*, 2002, 396.
185. T. Miyazawa, M. Hiramatsu, R. Yanagihara and T. Yamada, *Pept. Sci.*, 2002, 57.
186. P. Liu, G.-l. Tian, K.-S. Lee, M.-S. Wong and Y.-h. Ye, *Tetrahedron Lett.*, 2002, **43**, 2423.
187. P. Liu, Y.-H. Ye, G.-L. Tian, K.-S. Lee, M.-S. Wong and W.-H. Lo, *Synthesis*, 2002, 726.
188. I. Gill and R. Valivety, *Org. Proc. Res. Develop.*, 2002, **6**, 684.
189. D.A. Jeyaraj, H. Prinz and H. Waldmann, *Chem.-Eur. J.*, 2002, **8**, 1879.
190. M. Fité, P. Clapés, J. López-Santin, M.D. Benaiges and G. Caminal, *Biotechnol. Prog.*, 2002, **18**, 1214.
191. H. Sekizaki, K. Itoh, E. Toyata and K. Tanizawa, *J. Pept. Sci.*, 2002, **8**, 521.
192. K. Matsumoto, B.G. Davis and J.J. Bryan, *Chem.-Eur. J.*, 2002, **8**, 4129.
193. K. Rall and F. Bordusa, *J. Org. Chem.*, 2002, **67**, 9103.
194. E.C. Nice, T. Domagala, L. Fabri, M. Nerrie, F. Walker, R.N. Jorissen, A.W. Burgess, D.-F. Cui and Y.-S. Zhang, *Growth Factors*, 2002, **20**, 71.
195. S. Srinivasulu and A.S. Acharya, *Protein Sci.*, 2002, **11**, 384.
196. G.M. Dubowchik, R.A. Firestone, L. Padilla, D. Willner, S.J. Hofstead, K. Mosure, J.O. Knipe, S.J. Lasch and P.A. Trail, *Bioconjugate Chem.*, 2002, **13**, 855.
197. J.F. Tolborg, L. Petersen, K.J. Jensen, C. Mayer, D.L. Jakeman, R.A.J. Warren and S.G. Withers, *J. Org. Chem.*, 2002, **67**, 4143.
198. H. Sato, S. Kubo and T. Kimura, *Pept. Sci.*, 2002, 67.
199. R. Liboska, D. Zyka and M. Bobek, *Synthesis*, 2002, 1649.
200. J.C. Spetzler and T. Hoeg-Jensen, *Tetrahedron Lett.*, 2002, **43**, 2303.
201. J.-F. Guichou, L. Patiny and M. Mutter, *Tetrahedron Lett.*, 2002, **43**, 4389.
202. G. Fridkin and C. Gilon, *J. Pept. Res.*, 2002, **60**, 104.
203. B.E. Toki, C.G. Cerveny, A.F. Wahl and P.D. Senter, *J. Org. Chem.*, 2002, **67**, 1866.
204. S.W. Taylor, *Analyt. Biochem.*, 2002, **302**, 70.
205. V. Renugopalakrishnan, *J. Pept. Sci.*, 2002, **8**, 139.
206. K. Goto, T. Nazaki, H. Hojo, H. Nagasawa and Y. Nakahara, *Pept. Sci.*, 2002, 77.
207. C.R. Nakaie, E.G. Silva, E.M. Cilli, R. Marchetto, S. Schreier, T.B. Paiva and A.C.M. Paiva, *Peptides*, 2002, **23**, 65.
208. R. Roumelioti, L. Polevaya, P. Zoumpoulakis, N. Giatas, I. Mutule, T. Keivish, A. Zoga, D. Vlahakos, E. Iliodromitis, E. Kremastinos, S.G. Grdadolnik, T. Mavromoustakos and J. Matsoukas, *Bioorg. Med. Chem. Lett.*, 2002, **12**, 2627.
209. H.-H. Nguyen, D. Imhof, M. Kronen, B. Schlegel, A. Härtl, U. Graefe, L. Gera and S. Reissmann, *J. Med. Chem.*, 2002, **45**, 2781.
210. O. Yoshida, T. Yasukata, Y. Sumino, T. Munekage, Y. Narukawa and Y. Nishitani, *Bioorg. Med. Chem. Lett.*, 2002, **12**, 3027.
211. T. Yasukata, H. Shindo, O. Yoshida, Y. Sumino, T. Munekage, Y. Narukawa and Y. Nishitani, *Bioorg. Med. Chem. Lett.*, 2002, **12**, 3033.
212. M. Thomas, U. Varshney and S. Bhattacharya, *Eur. J. Org. Chem.*, 2002, 3604.
213. H. Saito, T. Yamada, K. Okumura, Y. Yonezawa and C.-G. Shin, *Chem. Lett.*, 2002, 1098.
214. M. Mandal and R. Nagaraj, *J. Pept. Res.*, 2002, **59**, 95.
215. E. Kluver, A. Schulz, W.-G. Forssmann and K. Adermann, *J. Pept. Res.*, 2002 **59**, 241.
216. S.S. Printsevskaya, A.Y. Pavlov, E.N. Olsufyeva, E.P. Mirchink, E.B. Isakova, M.I. Reznikova, R.C. Goldman, A.A. Branstrom, E.R. Baizman, C.B. Longley,

F. Sztaricskai, G. Batta and M.N. Preobrazhenskaya, *J. Med. Chem.*, 2002, **45**, 1340.
217. M. Gobbo, L. Biondi, F. Filira, R. Gennaro, M. Benincasa, B. Scolaro and R. Rocchi, *J. Med. Chem.*, 2002, **45**, 4494.
218. T.A. Blizzard, R.M. Kim, J.D. Morgan, J. Chang, J. Kohler, R. Kilburn, K. Chapman and M.L. Hammond, *Bioorg. Med. Chem. Lett.*, 2002, **12**, 849.
219. C.J. Thomas, A.O. Chizhov, C.J. Leitheiser, M.J. Rishel, K. Konishi, Z.-F. Tao and S.M. Hecht, *J. Amer. Chem. Soc.*, 2002, **124**, 12926.
220. K.L. Smith, Z.-F. Tao, S. Hashimoto, C.J. Leitheiser, X. Wu and S.M. Hecht, *Org. Lett.*, 2002, **4**, 1079.
221. H.-D. Arndt, A. Vascovi, A. Schrey, J.R. Pfeifer and U. Koert, *Tetrahedron*, 2002, **58**, 2789.
222. B.E. Haug, J. Andersen, O. Rekdal and J.S. Svendsen, *J. Pept. Sci.*, 2002, **8**, 307.
223. A. Hasuoka, Y. Nishikimi, Y. Nakayama, K. Kamiyama, M. Nakao, K.-I. Miyagawa, O. Nishimura and M. Fujino, *J. Antibiotics*, 2002, **55**, 191.
224. W. Kamysz, B. Kochanska, J. Ochocinska, Z. Mackiewicz and G. Kupryszewski, *Pol. J. Chem.*, 2002, **76**, 801.
225. T. Suenaga, T. Hatakeyama, T. Niidome and H. Aoyagi, *Pept. Sci.*, 2002, 199.
226. T.S. Yokum, R.P. Hammer, M.L. McLaughlin and P.H. Elzer, *J. Pept. Res.*, 2002, **59**, 9.
227. K. Ohki, H. Sawanishi, S. Nagai, K. Okimura, Y. Uchida and S. Naoki, *Pept. Sci.*, 2002, 189.
228. P.B. Savage, *Eur. J. Org. Chem.*, 2002, 759.
229. H. Yamaguchi, H. Kodama, S. Osada, M. Jelokhani-Niaraki, F. Kato and M. Kondo, *Bull. Chem. Soc. Japan*, 2002, **75**, 1563.
230. C. Boussard, V.E. Doyle, N. Mahmood, T. Klimkait, M. Pritchard and I.H. Gilbert, *Eur. J. Med. Chem.*, 2002, **37**, 883.
231. H. Tamamura, K. Hiramatsu, K. Mujamoto, A. Omagari, S. Oiskhi, H. Nakashima, N. Yamamoto, Y. Kuroda, T. Nakagawa, A. Otaka and N. Fujii, *Bioorg. Med. Chem. Lett.*, 2002, **12**, 923.
232. A. Litovchick, A. Lapidot, M. Eisenstein, A. Kalinkovich and G. Borkow, *Biochemistry*, 2001, **40**, 15612.
233. A. Otaka, M. Nakamura, D. Nameki, E. Kodama, S. Uchiyama, S. Nakamura, H. Nakano, H. Nobutaka, H. Tamamura, Y. Kobayashi, M. Matsuoka and N. Fujii, *Angew. Chem., Int. Ed.*, 2002, **41**, 2938.
234. D.M. Eckert and P.S. Kim, *Proc. Natl. Acad. Sci., U.S.A.*, 2001, **98**, 11187.
235. M. Nakamura, A. Otaka, E. Kodama, M. Matsuoka, S. Uchiyama, S. Nakamura, Y. Kobayashi, H. Tamamura and N. Fujii, *Pept. Sci.*, 2002, 73.
236. K. Hiramatsu, H. Tamamura, A. Omagari, H. Nakashima, Y. Xu, M. Matsuoka, A. Otaka and N. Fujii, *Pept. Sci.*, 2002, 175.
237. A. Prahl, T. Wierzba, I. Derdowska, W. Juzwa, K. Neubert, B. Hartrodt and B. Lammek, *Pol. J. Chem.*, 2002, **76**, 713.
238. A. Prahl, I. Derdowska, K. Neubert, B. Hartrodt, T. Wierzba, W. Jugwa and B. Lammek, *Pol. J. Chem.*, 2002, **76**, 1435.
239. D. Chan, L. Gera, J. Stewart, B. Helfrich, M. Verella-Garcia, G. Johnson, A. Baron, J. Yang, T. Puck and P. Bunn, *Proc. Nat. Acad. Sci., U.S.A.*, 2002, **99**, 4608.
240. T. Shi and D.L. Rabenstein, *Bioorg. Med. Chem. Lett.*, 2002, **12**, 2237.
241. K. Haneda, M. Takeuchi, T. Inazu, K. Toma, M. Tagashira, K. Kobayashi, K. Yamamoto and K. Takegawa, *Pept. Sci.*, 2002, 89.

242. T. Seko, M. Kato, H. Kohno, S. Ono, K. Hashimura, H. Takimizu, K. Nakai, H. Maegawa, N. Katsube and M. Toda, *Bioorg. Med. Chem. Lett.*, 2002, **12**, 915.
243. J.C. Mai, Z. Mi, S.-H. Kim, B. Ng and P.D. Robins, *Cancer Res.*, 2001, **61**, 7709.
244. S. Kasai, H. Nagasawa, M. Shimamura, Y. Uto and H. Hori, *Bioorg. Med. Chem. Lett.*, 2002, **12**, 951.
245. L.Z. Yan and P.E. Dawson, *Angew. Chem. Int. Ed.*, 2001, **40**, 3625.
246. D.R. Englebretsen, B. Garnham and P.F. Alewood, *J. Org. Chem.*, 2002, **67**, 5883.
247. S.E. Escher, E. Kluever and K. Adermann, *Lett. Pept. Sci.*, 2001, **8**, 349.
248. G.P. Zecchini, M. Nalli, A. Mollica, G. Lucente, M.P. Paradisi and S. Spisani, *J. Pept. Res.*, 2002, **59**, 283.
249. M. Miyazaki, M. Shibue, M. Yoshiki, S. Osada, H. Kodama, I. Fujita, Y. Hamasaki, H. Maeda and M. Kondo, *Pept. Sci.*, 2002, 153.
250. S. Pooyan, B. Qiu, M.M. Chan, D. Fong, P.J. Sinko, M.J. Leibowitz and S. Stein, *Bioconjugate Chem.*, 2002, **13**, 216.
251. G. Pagani Zechini, E. Morera, M. Nalli, M. Paglialunga Paradisi, G. Lucente and S. Spisani, *Farmaco*, 2001, **56**, 851.
252. G. Cavicchioni, M. Turchetti and S. Spisani, *J. Pept. Res.*, 2002, **60**, 223.
253. G. Morelli, S. De Luca, D. Tesauro, M. Saviano, C. Pedone, A. Dolmella, R. Visentin and U. Mazzi, *J. Pept. Sci.*, 2002, **8**, 373.
254. S. Herrero, M.T. Garcia-López, M. Latorre, E. Cenarruzabietia, J. Del Río and R. Herranz, *J. Org. Chem.*, 2002, **67**, 3866.
255. K.C. Nicolaou, T. Nontagnon and G. Vassilikogiannakis, *Chem. Commun.*, 2002, 2478.
256. J. Kwak, A. De Capua, E. Locardi and M. Goodman, *J. Amer. Chem. Soc.*, 2002, **124**, 14085.
257. F. Hubalek, D.E. Edmondson and J. Pohl, *Anal. Biochem.*, 2002, **306**, 124.
258. G.A. Kinberger, W. Cai and M. Goodman, *J. Amer. Chem. Soc.*, 2002, **124**, 15162.
259. B. Sacca and L. Moroder, *J. Pept. Sci.*, 2002, **8**, 192.
260. G.E. Atkinson, A. Cowan, C. McInnes, D.I. Zheleva, P.M. Fischer and W.C. Chan, *Bioorg. Med. Chem. Lett.*, 2002, **12**, 2501.
261. C. Spezzacatena, T. Perri, V. Guantieri, L.B. Sandberg, T.F. Mitts and A.M. Tamburro, *Eur. J. Org. Chem.*, 2002, 95.
262. W. Yu, Y. Liang, K. Liu, Y. Zhao, G. Fei and H. Wang, *J. Pept. Res.*, 2002, **59**, 134.
263. R.D. Husain, J. McCandless, P.J. Stevenson, T. Large, D.J.S. Guthrie and B. Walker, *J. Chromatogr. Sci.*, 2002, **40**, 1.
264. Y. Wu, M. Zhao, C. Wang and S. Peng, *Bioorg. Med. Chem. Lett.*, 2002, **12**, 2331.
265. J. Ruczynski, A. Sawula, Z. Konstanski, R. Korolkiewicz and P. Rekowski, *Pol. J. Chem.*, 2002, **76**, 57.
266. G. Digilio, C. Bracco, L. Barbero, D. Chicco, M.D. Del Curto, P. Esposito, S. Traversa and S. Aime, *J. Amer. Chem. Soc.*, 2002, **124**, 3431.
267. X.-M. Gao, Y.-H. Ye, M. Bernd and B. Kutscher, *J. Pept. Sci.*, 2002, **8**, 418.
268. J. Izdebski, E. Witkowska, D. Kunce, A. Orlowska, B. Baranowska, M. Radzikowska and M. Smoluch, *J. Pept. Sci.*, 2002, **8**, 289.
269. O. Langer, H. Kählig, K. Zierler-Gould, J.W. Bats and J. Mulzer, *J. Org. Chem.*, 2002, **67**, 6878.
270. K. Amssoms, S.L. Oza, E. Ravaschino, A. Yamani, A.-M. Lambeir, P. Rajan, G. Bal, J.B. Rodriguez, A.H. Fairlamb, K. Augustyns and A. Haemers, *Bioorg. Med. Chem. Lett.*, 2002, **12**, 2553.

271. L.A. Paquette, M. Duan, I. Konetzki and C. Kempmann, *J. Amer. Chem. Soc.*, 2002, **124**, 4257.
272. M. Li, D. Cui and Y. Zhang, *IUBMB Life*, 2002, **53**, 57.
273. L. Chen, J. Tilley, R.V. Trilles, W. Yun, D. Fry, C. Cook, K. Rowan, V. Schwinge and R. Campbell, *Bioorg. Med. Chem. Lett.*, 2002, **12**, 137.
274. G.A. Doherty, T. Kamenecka, E. McCauley, G. Van Riper, R.A. Mumford, S. Tong and W.K. Hagmann, *Bioorg. Med. Chem. Lett.*, 2002, **12**, 729.
275. I.E. Kopka, D.N. Young, L.S. Lin, R.A. Mumford, P.A. Magriotis, M. MacCoss, S.G. Mills, G. Van Riper, E. McCauley, L.E. Egger, U. Kidambi, J.A. Schmidt, K. Lyons, R. Stearns, S. Vincent, A. Coletti, Z. Wang, S. Tong, J. Wang, S. Zheng, K. Owens, D. Levorse and W.K. Hagmann, *Bioorg. Med. Chem. Lett.*, 2002, **12**, 637.
276. I. Lewis, B. Rohde, M. Mengus, M. Weetall, S. Maeda, R. Hugo and P. Lake, *Molec. Divers.*, 2000, 61.
277. T.M. Kamenecka, T. Lanza, S.E. de Laszlo, B. Li, E.D. McCauley, G. Van Riper, L.A. Egger, U. Kidambi, R.A. Mumford, S. Tong, M. MacCoss, J.A. Schmidt and W.K. Hagmann, *Bioorg. Med. Chem. Lett.*, 2002, **12**, 2205.
278. G.M. Castanedo, F.C. Sailes, N.J.P. Dubree, J.B. Nicholas, L. Caris, K. Clark, S.M. Keating, M.H. Beresini, H. Chiu, S. Fong, J.C. Marsters, D.Y. Jackson and D.P. Sutherlin, *Bioorg. Med. Chem. Lett.*, 2002, **12**, 2913.
279. V. Wehner, H. Blum, M. Kurz and H.U. Stilz, *Synthesis*, 2002, 2023.
280. D. Gottschling, J. Boer, L. Marinelli, G. Voll, M. Haupt, A. Schuster, B. Holzmann and H. Kessler, *ChemBioChem.*, 2002, **3**, 575.
281. M. Okue, H. Kobayashi, K. Shin-ya, K. Furihata, Y. Hayakawa, H. Seto, H. Watanabe and T. Kitahara, *Tetrahedron Lett.*, 2002, **43**, 857.
282. T.-G. Nam, R. Sangaiah, A. Gold, G.D. Lacks, L.-A. Nylander-French and J.E. French, *Polycycl. Arom. Comp.*, 2002, **22**, 239.
283. G.J. Tian, S.J. Li and D.X. Wang, *Chin. Chem. Lett.*, 2002, **13**, 1154.
284. T. Inui, H. Nishio, Y. Nishiuchi and T. Kimura, *Pept. Sci.*, 2002, 21.
285. J. Scherkenbeck, H. Chen and R.K. Haynes, *Eur. J. Org. Chem.*, 2002, 2350.
286. T. Tachi, R.F. Epand, R.M. Epand and K. Matsuzaki, *Biochemistry*, 2002, **41**, 10723.
287. Y. Matsushita, T. Niidome, T. Hatakeyama and H. Aoyagi, *Pept. Sci.*, 2002, 195.
288. P.N.S. Kumar, K.S. Devaky, C. Sadasivan and M. Haridas, *Protein Pept. Lett.*, 2002, **9**, 403.
289. P. Grieco, P. Srinivasan, G. Han, D. Weinberg, T. MacNeil, L.H.T. Van der Ploeg and V.J. Hruby, *J. Pept. Res.*, 2002, **59**, 203.
290. A.W.-H. Cheung, W. Danho, J. Swistok, L. Qi, G. Kurylko, L. Franco, K. Yagaloff and L. Chen, *Bioorg. Med. Chem. Lett.*, 2002, **12**, 2407.
291. M.J. Kavarana, D. Trivedi, M. Cai, J. Ying, M. Hammer, C. Cabello, P. Grieco, G. Han and V.J. Hruby, *J. Med. Chem.*, 2002, **45**, 2644.
292. W. Wang, M. Cai, C. Xiong, J. Zhang, D. Trivedi and V.J. Hruby, *Tetrahedron*, 2002, **58**, 7365.
293. J.R. Holder, Z. Xiang, R.M. Bauzo and C. Haskell-Luevano, *J. Med. Chem.*, 2002, **45**, 5736.
294. J.R. Holder, R.M. Bauzo, Z. Xiang and C. Haskell-Luevano, *J. Med. Chem.*, 2002, **45**, 2801.
295. J. Bondebjerg, Z. Xiang, R.M. Bauzo, C. Haskell-Luevano and M. Meldal, *J. Amer. Chem. Soc.*, 2002, **124**, 11046.
296. M. Suenaga, T. Yamada, T. Itoh, N. Koyama, M. Wakimasu, T. Ohtaki, C. Kitada, O. Nishimura and M. Fujino, *Pept. Sci.*, 2002, 395.

297. M. Haramura, K. Tsuzuki, A. Okamachi, K. Yogo, M. Ikuta, T. Kozono, H. Takanashi and E. Murayama, *Bioorg. Med. Chem.*, 2002, **10**, 1805.
298. K. Sugase, Y. Oyama, K. Kitano, T. Iwashita, T. Fujiwara, H. Akutsu and M. Ishiguro, *J. Med. Chem.*, 2002, **45**, 881.
299. F. Cavelier, B. Vivet, J. Martinez, A. Aubry, C. Didierjean, A. Vicherat and M. Marraut, *J. Amer. Chem. Soc.*, 2002, **124**, 2917.
300. F. Hong, J. Jaidi, B. Cusack and E. Richelson, *Bioorg. Med. Chem.*, 2002, **10**, 3849.
301. W. Szeszel-Fedorowicz, G. Rosinski, J. Issberner, R. Osborne, J. Sliwowska and D. Konopinska, *Pol. J. Pharmacol.*, 2001, **53**, 31.
302. I. Woznica, G. Rosinski and D. Konopinska, *Pol. J. Chem.*, 2002, **76**, 1425.
303. H. Tang, X.-c. Wang and S.-p. Liang, *Hunan Shifan Daxue Ziran Kexue Xuebao*, 2002, **25**, 68.
304. K.A.N. Graham, Q. Wang, M. Eisenhut, U. Aberkorn and W. Mier, *Tetrahedron Lett.*, 2002, **43**, 5021.
305. D.R. van Staveren and N. Metzler-Nolte, *Chem. Commun.*, 2002, 1406.
306. N.V. Sumbatyan, A.N. Topin, M.V. Taranenko and G.A. Korshunova, *Vestnik Moskovskogo Universiteta, Seriya 2*, 2001, **42**, 287.
307. S. Maricic, U. Berg and T. Frejd, *Tetrahedron*, 2002, **58**, 3085.
308. Y. Rew and M. Goodman, *J. Org. Chem.*, 2002, **67**, 8820.
309. K. Kaczmarek, M. Kaleta, N.N. Chung, P.W. Schiller and J. Zabrocki, *Acta Biochim. Pol.*, 2001, **48**, 1159.
310. A. Olma, N. Gniadzik, A.W. Lipkowski and M. Lachwa, *Acta Biochim. Pol.*, 2001, **48**, 1165.
311. T. Ogawa, M. Araki, T. Miyamae, T. Okayama, M. Hagiwara, S. Sakurada and T. Morikawa, *Pept. Sci.*, 2002, 101.
312. T. Ogawa, T. Miyamae, K. Murayama, K. Okuyama, T. Okayama, M. Hagiwara, S. Sakurada and T. Morikawa, *J. Med. Chem.*, 2002, **45**, 5081.
313. T. Ogawa, T. Miyamae, T. Okayama, M. Hagiwara, S. Sakurada and T. Morikawa, *Chem. Pharm. Bull.*, 2002, **50**, 771.
314. A. Ambo, N. Hamasaki and Y. Sasaki, *J. Tohoku Pharm. Univ.*, 2000, **47**, 101.
315. Y. Fujisawa, K. Shiotani, A. Miyazaki, Y. Tsuda, T. Yokoi, S.D. Bryant, L.H. Lazarus, A. Anbo, Y. Sasaki and Y. Okada, *Pept. Sci.*, 2002, 143.
316. Y. Fujita, Y. Shimizu, M. Takahashi, T. Yokoi, Y. Tsuda, L.H. Lazarus, S.D. Bryant, A. Anbo, Y. Sasaki and Y. Okada, *Pept. Sci.*, 2002, 105.
317. Y. Rew, S. Malkmus, C. Svensson, T.L. Yaksh, N.N. Chung, P.W. Schiller, J.A. Cassel, R.N. DeHaven and M. Goodman, *J. Med. Chem.*, 2002, **45**, 3746.
318. A. Ambo, T. Akihiro and Y. Sasaki, *Chem. Pharm. Bull.*, 2002, **50**, 1401.
319. J.E. Burden, P. Davis, F. Porreca and A.F. Spatola, *Bioorg. Med. Chem. Lett.*, 2002, **12**, 213.
320. G. Cardillo, L. Gentilucci, A.R. Qasem, F. Sgarzi and S. Spampinato, *J. Med. Chem.*, 2002, **45**, 2571.
321. V. Kumar, T.F. Murray and J.V. Aldrich, *J. Med. Chem.*, 2002, **45**, 3820.
322. L. Halab, J.A.J. Becker, Z. Darula, D. Tourwe, B.L. Kieffer, F. Simonin and W.D. Lubell, *J. Med. Chem.*, 2002, **45**, 5353.
323. Y.-C. Chen, A. Muhlrad, A. Shteyer, M. Vidson, I. Bab and M. Chofrev, *J. Med. Chem.*, 2002, **45**, 1624.
324. K. Bakos, J. Havass, F. Fulop, L. Gera, J.M. Stewart, G. Falkay and G.K. Tóth, *Lett. Pept. Sci.*, 2002, **8**, 35.
325. M. Jelinski, K. Hamacher and H.H. Coenen, *J. Labelled Comp. Radiopharm.*, 2002, **45**, 217.

326. Z. Yuan, D. Blomberg, I. Sethson, K. Brickmann, K. Ekholm, B. Johansson, A. Nilsson and J. Kihlberg, *J. Med. Chem.*, 2002, **45**, 2512.
327. J. Marik, M. Budesinsky, J. Slaninova and J. Hlavacek, *Coll. Czech. Chem. Comm.*, 2002, **67**, 373.
328. S. Terrillon, L.L. Cheng, S. Stoev, B. Mouillac, C. Barberis, M. Manning and T. Durroux, *J. Med. Chem.*, 2002, **45**, 2579.
329. G. Flouret, T. Majewski, L. Balaspiri, W. Brieher, K. Mahan, O. Chaloin, L. Wilson and J. Slaninova, *J. Pept. Sci.*, 2002, **8**, 314.
330. J.L. Stymiest, B.F. Mitchell, S. Wong and J.C. Vederas, *Org. Lett.*, 2003, **5**, 47.
331. M. Hedenström, Z. Yuan, K. Brickmann, J. Carlsson, K. Ekholm, B. Johansson, E. Kreutz, A. Nilsson, I. Sethson and J. Kihlberg, *J. Med. Chem.*, 2002, **45**, 2501.
332. B. Jastrzebska, I. Derdowska, P. Kuncarowa, J. Slaninova, B. Lammek, B. Olejniczak and J. Zabrocki, *Pol. J. Chem.*, 2002, **76**, 823.
333. A.O. Yusuf, B.M. Bhatt and P.M. Gitu, *Bull. Chem. Soc. Ethiopia*, 2001, **15**, 143.
334. Q. Zhang, M.-z. Zhang, C.-h. Zhang and W. Ding, *Sichuan Daxue Xuebao, Gongcheng Kexuaban*, 2001, **39**, 78.
335. C.F. McCusker, P.J. Kocienski, F.T. Boyle and A.G. Schätzlein, *Bioorg. Med. Chem. Lett.*, 2002, **12**, 547.
336. S. Jasseron, C. Contino-Pepin, J.C. Maurizis, M. Rapp and B. Pucci, *Bioorg. Med. Chem. Lett.*, 2002, **12**, 1067.
337. A. Massaguer, I. Haro, M.A. Alsina and F. Reig, *J. Liposome Res.*, 2001, **11**, 103.
338. S. Oishi, T. Kamano, A. Niida, H. Takamura, A. Otaka and N. Fujii, *Pept. Sci.*, 2002, 241.
339. S. Schabbert, M.D. Pierschbacher, R.-H. Mattern and M. Goodman, *Bioorg. Med. Chem.*, 2002, **10**, 3331.
340. S. Oishi, T. Kamano, A. Niida, Y. Odagaki, N. Hamanaka, M. Yamamoto, K. Ajito, H. Tamamura, A. Otaka and N. Fujii, *J. Org. Chem.*, 2002, **67**, 6162.
341. J. Jiang, W. Wang, D.C. Sane and B. Wang, *Bioorg. Chem.*, 2001, **29**, 357.
342. K. Belekoukias, J. Sarigiannis, G. Stavropoulos and T. Mavromoustakos, *Pharmakeut.*, 2002, **15**, 2.
343. M. Zhao, C. Wang, X. Jiang and S. Peng, *Prep. Biochem. Biotechnol.*, 2002, **32**, 363.
344. N. Lin, M. Zhao, C. Wang and S. Peng, *Bioorg. Med. Chem. Lett.*, 2002, **12**, 585.
345. S. Rensing and T. Schrader, *Org. Lett.*, 2002, **4**, 2161.
346. Y. Susuki, K. Hojo, I. Okazaki, H. Kamata, M. Sasaki, M. Maeda, M. Nomizu, Y. Yamamoto, S. Nakagawa, T. Mayumi and K. Kawasaki, *Chem. Pharm. Bull.*, 2002, **50**, 1229.
347. S. Gazal, G. Gelerman, O. Ziv, O. Karpov, P. Litman, M. Bracha, M. Afargan and C. Gilon, *J. Med. Chem.*, 2002, **45**, 1665.
348. M. Morpurgo, C. Monfardini, L.J. Hofland, M. Sergi, P. Orsolini, J.M. Dumont and F.M. Veronese, *Bioconjugate. Chem.*, 2002, **13**, 1238.
349. S.A. Potekhin, T.N. Melnik, V. Popov, N.F. Lanina, A.A. Vazina, P. Rigler, A.S. Verdini, G. Corradin and A.V. Kajava, *Chem. Biol.*, 2001, **8**, 1025.
350. S.K. Ozker, R.S. Hellman and A.Z. Krasnow, *Appl. Rad. Isotopes*, 2002, **57**, 729.
351. C.J. Hobbs, R.A. Bit, A.D. Cansfield, B. Harris, C.H. Hill, K.L. Hilyard, I.R. Kilford, E. Kitas, A. Kroehn, P. Lovell, D. Pole, P. Rugman, B.S. Sherborne, I.E.D. Smith, D.R. Vesey, D.L. Walmsley, D. Whittaker, G. Williams, F. Wilson, D. Banner, A. Surgenor and N. Borkakoti, *Bioorg. Med. Chem. Lett.*, 2002, **12**, 1365.
352. A. Giolitti, M. Altamura, F. Bellucci, D. Giannotti, S. Meini, R. Patacchini, L. Rotondaro, S. Zappitelli and C.A. Maggi, *J. Med. Chem.*, 2002, **45**, 3418.

353. T. Abiko and R. Ogawa, *Prep. Biochem. Biotechnol.*, 2002, **32**, 117.
354. T. Abiko and R. Ogawa, *Prep. Biochem. Biotechnol.*, 2002, **32**, 269.
355. N.-H. Nguyen, J.W. Apriletti, S.T.C. Lima, P. Webb, J.D. Baxter and T.S. Scanlan, *J. Med. Chem.*, 2002, **45**, 3310.
356. G. Chiellini, N.H. Nguyen, J.W. Apriletti, J.D. Baxter and T.S. Scanlan, *Bioorg. Med. Chem.*, 2002, **10**, 333.
357. J.C. Simpson, C. Ho, S.E.F. Berkeley, M.C. Gerhengorn, G.R. Marshall and K.D. Moeller, *Bioorg. Med. Chem.*, 2002, **10**, 291.
358. A.P. Smirnova, S.P. Krasnoschekova, A.M. Nikitina and Yu.P. Shvachkin, *Pharm. Chem. J.*, 2001, **35**, 393.
359. K. Prokai-Tatrai, P. Perjesi, A.D. Zharikova, X. Li and L. Prokai, *Bioorg. Med. Chem. Lett.*, 2002, **12**, 2171.
360. A.P. Smirnova, S.M. Funtova, V.V. Knyazeva and Yu.P. Shvachkin, *Pharm. Chem. J.*, 2001, **35**, 437.
361. S.-H. Yoon, J. Wu and N. Bodor, *Bull. Korean Chem. Soc.*, 2002, **23**, 761.
362. J. Blanchfield, J. Dutton, R. Hogg, D. Craik, D. Adams, R. Lewis, P. Alewood and I. Toth, *Lett. Pept. Sci.*, 2001, **8**, 235.
363. I. Uemura, H. Miyagawa and T. Ueno, *Tetrahedron*, 2002, **58**, 2351.
364. N. Loiseau, F. Cavelier, J.-P. Noel and J.M. Gomis, *J. Pept. Sci.*, 2002 **8**, 335.
365. M. Giel-Pietraszuk, P. Mucha, P. Rekowski, G. Kupryszewski and M.Z. Barciszewska, *Pol. J. Chem.*, 2002, **76**, 815.
366. P. Grieco, A. Carotenuto, R. Patacchini, C.A. Maggi, E. Novellino and P. Rovero, *Bioorg. Med. Chem.*, 2002, **10**, 3731.
367. M. Roice, K.S. Kumar and V.N.R. Pillai, *Protein Pept. Lett.*, 2002, **9**, 245.
368. K. Teruya, T. Kawakami, K. Akaji and S. Aimoto, *Pept. Sci.*, 2002, 25.
369. K. Teruya, T. Kawakami, K. Okaji and S. Aimoto, *Tetrahedron Lett.*, 2002, **43**, 1487.
370. K.J. Weronski, M.A. Busquets, M. Munoz, I. Haro, J. Prat and V. Girona, *Mater. Sci. Eng., C: Biomim. Supramol. Sys.*, 2002, **22**, 279.
371. H. Takiguchi, *Fukuoka Daigaku Rigaku Shuho*, 2002, **32**, 69.
372. F. Bernardi, M. Garavelli, M. Scatizzi, C. Tomasini, V. Trigari, M. Crisma, F. Formaggio, C. Peggion and C. Toniolo, *Chem.-Eur. J.*, 2002, **8**, 2516.
373. M. Ishida, H. Tamaoki, H. Masuda, H. Okabayashi and C.J. O'Connor, *Colloid Polym. Sci.*, 2002, **280**, 1157.
374. N. Higashi, T. Koga and M. Niwa, *ChemBioChem.*, 2002, **3**, 448.
375. Y. Park, K.Y. Kwok, C. Boukarim and K.G. Rice, *Bioconjugate Chem.*, 2002 **13**, 232.
376. M. Kitamura, M. Oka and T. Hayashi, *Pept. Sci.*, 2002, 297.
377. P. Tremmel and A. Geyer, *J. Amer. Chem. Soc.*, 2002, **124**, 8548.
378. Y. Hirano, M. Shimoda, S. Tokuyama, M. Morita, M. Hattori, M. Oka and T. Hayashi, *Pept. Sci.*, 2002, 273.
379. M. Wakahara, T. Nakamura, M. Oka, T. Hayashi, M. Hatori and Y. Hirano, *Pept. Sci.*, 2002, 289.
380. Y. Onoda, M. Oka, T. Hayashi and Y. Hirano, *Pept. Sci.*, 2002, 293.
381. T.J. Sanborn, C.W. Wu, R.N. Zuckermann and A.E. Barron, *Biopolymers*, 2002, **63**, 12.
382. M. Martino, T. Perri and A.M. Tamburro, *Biomacromolecules*, 2002, **3**, 297.
383. B.A. Ramesha and D. C. Gowda, *Lett. Pept. Sci.*, 2001, **8**, 309.
384. M. Ueki, H. Komiya, A. Tajima and H. Nakashima, *Pept. Sci.*, 2002, 81.

385. D.J. Maly, F. Leonetti, B.J. Backes, D.S. Dauber, J.L. Harris, C.S. Craik and J.S. Ellman, *J. Org. Chem.*, 2002, **67**, 910.
386. P. Fraszczak, K. Rolka and Kupryszewski, *Pol. J. Chem.*, 2002, **76**, 519.
387. P. Fraszczak, K. Kazmierczak, M. Stawikowski, A. Jaskiewicz, G. Kupryszewski and K. Rolka, *Pol. J. Chem.*, 2002, **76**, 1441.
388. H. Noguchi, T. Aoyama and T. Shioiri, *Heterocycles*, 2002, **58**, 471.
389. L. Banfi, G. Guanti, R. Riva, A. Basso and E. Calcagno, *Tetrahedron Lett.*, 2002, **43**, 4067.
390. W.P. Malachowski, C. Tie, K. Wang and R.L. Broadrup, *J. Org. Chem.*, 2002, **67**, 8962.
391. C. Jiang, Q. Miao, Y. Ke and T. You, *Zhongguo Yiyao Gongye Zazhi*, 2002, **33**, 55.
392. M. Czekaj, S.I. Klein, K.R. Guertin, C.J. Gardner, A.L. Zulli, H.W. Pauls, A.P. Spada, D.L. Cheney, K.D. Brown, D.J. Colussi, V. Chu, R.J. Leadley and C.T. Dunwiddie, *Bioorg. Med. Chem. Lett.*, 2002, **12**, 1667.
393. J.Z. Ho, T.S. Gibson and J.E. Semple, *Bioorg. Med. Chem. Lett.*, 2002, **12**, 743.
394. A. Dahlgren, P.-O. Johansson, I. Kvarnstrom, D. Musil, I. Nilsson and B. Samuelsson, *Bioorg. Med. Chem.*, 2002, **10**, 1829.
395. A.A.K. Hasan, M. Warnock, S. Srikanth and A.H. Schmaier, *Thromb. Res.*, 2001, **104**, 451.
396. U.E.W. Lange and C. Zechel, *Bioorg. Med. Chem. Lett.*, 2002, **12**, 1571.
397. S. Hanessian, R. Margarita, A. Hall, S. Johnstone, M. Tremblay and L. Parlanti, *J. Amer. Chem. Soc.*, 2002, **124**, 13342.
398. A.E.P. Adang, A.P.A. de Man, G.M.T. Vogel, P.D.J. Grootenhuis, M.J. Smit, C.A.M. Peters, A. Visser, J.B.M. Rewinkel, T. van Dinther, H. Lucas, J. Kelder, S. van Aelst, D.G. Meuleman and C.A.A. vam Boeckel, *J. Med. Chem.*, 2002, **45**, 4419.
399. J.J. Cui, G.-L. Araldi, J.E. Reiner, K.M. Reddy, S.J. Kemp, J.Z. Ho, D.V. Siev, L. Mamedova, T.S. Gibson, J.A. Gaudette, N.K. Minami, S.M. Anderson, A.E. Bradbury, T.G. Nolan and J.E. Semple, *Bioorg. Med. Chem. Lett.*, 2002, **12**, 2925.
400. Y. Tsuda, M. Tada, K. Wanaka, U. Okamoto, A. Hijikata-Okunomiya, S. Okamoto and Y. Okada, *Chem. Pharm. Bull.*, 2001, **49**, 1457.
401. D.C. Pimenta, R.L. Melo, G. Caliendo, V. Santagada, F. Fiorino, B. Severino, G. De Nucci, L. Juliano and M.A. Juliano, *Biol. Chem.*, 2002, **383**, 853.
402. S. Künzel, A. Schweinitz, S. Reissmann, T. Stürzebecher and T. Steinmetzer, *Bioorg. Med. Chem. Lett.*, 2002, **12**, 645.
403. F. Sato, Y. Inoue, T. Omodani, K. Imano, H. Okazaki, T. Takemura and M. Komiya, *Bioorg. Med. Chem. Lett.*, 2002, **12**, 551.
404. E.S. Priestley, I. De Lucca, B. Ghavimi, S. Erickson-Viitanen and C.P. Decicco, *Bioorg. Med. Chem. Lett.*, 2002, **12**, 3199.
405. P.R. Guzzo, M.P. Trova, T. Inghardt and M. Linschoten, *Tetrahedron Lett.*, 2002, **43**, 41.
406. N. Schmiedeberg, M. Schmitt, C. Roelz, V. Truffault, M. Sukopp, M. Buergle, O.G. Wilhelm, W. Schmalix, V. Magdolen and H. Kessler, *J. Med. Chem.*, 2002, **45**, 4984.
407. P.M. St.Hilaire, L.C. Alves, F. Herrera, M. Renil, S.J. Sanderson, J.C. Mottram, G.H. Coombs, M.A. Juliano, L. Juliano, J. Arevalo and M. Meldal, *J. Med. Chem.*, 2002, **45**, 1971.
408. W.G. Gutheil and Q. Xu, *Chem. Pharm. Bull.*, 2002, **50**, 688.
409. E. Wieczerzak, P. Drabik, L. Lankiewicz, S. Oldziej, Z. Grzonka, M. Abrahamson, A. Grubb and D. Broemme, *J. Med. Chem.*, 2002, **45**, 4202.

410. W. Lubisch and A. Möller, *Bioorg. Med. Chem. Lett.*, 2002, **12**, 1335.
411. M. Nakamura and J. Inoue, *Bioorg. Med. Chem. Lett.*, 2002, **12**, 1603.
412. E. Mann, A. Chana, F. Sanchez-Sancho, C. Puerta, A. Garcia-Marino and B. Herradon, *Adv. Synth. Catal.*, 2002, **344**, 855.
413. D.J. Lauffer and M.D. Mullican, *Bioorg. Med. Chem. Lett.*, 2002, **12**, 1225.
414. S.D. Linton, D.S. Karanewsky, R.J. Ternansky, J.C. Wu, B. Pham, L. Konandapani, R. Smidt, J.-L. Diaz, L.C. Fritz and K.J. Tomaselli, *Bioorg. Med. Chem. Lett.*, 2002, **12**, 2969.
415. E. Altmann, J. Renaud, J. Green, D. Farley, B. Cutting and W. Jahnke, *J. Med. Chem.*, 2002, **45**, 2352.
416. R.V. Mendonca, S. Vankatraman and J.T. Palmer, *Bioorg. Med. Chem. Lett.*, 2002, **12**, 2887.
417. S.F. Chowdhury, J. Sivaraman, J. Wang, G. Devanathan, P. Lachance, H. Qi, R. Menard, J. Lefebvre, Y. Konishi. M. Cygler, T. Sulea and E.O. Purisima, *J. Med. Chem.*, 2002, **45**, 5321.
418. Y.D. Ward, D.S. Thomson, L.L. Frye, C.L. Cywin, T. Morwick, M.J. Emmanuel, R. Zindell, D. McNeil, Y. Bekkali, M. Giradot, M. Hrapchak, M. DeTuri, K. Crane, D. White, S. Pav, Y. Wang, M.-H. Hao, C.A. Grygon, M.E. Labadia, D.M. Freeman, W. Davidson, J.L. Hopkins, M.L. Brown and D.M. Spero, *J. Med. Chem.*, 2002, **45**, 5471.
419. M. Chino, M. Wakao and J.A. Ellman, *Tetrahedron*, 2002, **58**, 6305.
420. L.C. Dias, A.A. Ferreira and G. Diaz, *Synlett*, 2002, 1845.
421. J.S. Tung, D.L. Davis, J.P. Anderson, D.E. Walker, S. Mamo, N. Jewett, R.K. Hom, S. Sinha, E.D. Thorsett and V. John, *J. Med. Chem.*, 2002, **45**, 259.
422. S. Rajesh, J. Srivastava, B. Banergi and J. Iqbal, *Ind. J. Chem. B*, 2001, **40B**, 1029.
423. Z. Xu, J. Singh, M.D. Schwinden, B. Zheng, T.P. Kissick, B. Patel, M.J. Humora, F. Quiroz, L. Dong, D.-M. Hsieh, J.E. Heikes, M. Pudipeddi, M.D. Lindrud, S.K. Srivastava, D.R. Kronenthal and R.H. Mueller, *Org. Proc. Res. Dev.*, 2002, **6**, 203.
424. G. Quéléver, M. Bouygues and J.L. Kraus, *J. Chem. Soc., Perkin Trans. I*, 2002, 1181.
425. B. Saha, J.P. Nandy, S. Shukla, I. Siddiqui and J. Iqbal, *J. Org. Chem.*, 2002, **67**, 7858.
426. H.L. Sham, C. Zhao, L. Li, D.A. Betebenner, A. Saldivar, S. Vasavanonda, D.J. Kempf, J.J. Plattner and D.W.S. Norbeck, *Bioorg. Med. Chem. Lett.*, 2002, **12**, 3101.
427. S. Rajesh, E. Ami, T. Kotake, T. Kimura, Y. Hayashi and Y. Kiso, *Bioorg. Med. Chem. Lett.*, 2002, **12**, 3615.
428. P.V. Murphy, J.L. O'Brien, L.J. Gorey-Feret and A.B. Smith, *Bioorg. Med. Chem. Lett.*, 2002, **12**, 1763.
429. K. Hidaka, T. Kimura, Y. Hayashi, K.F. MaDaniel, T. Dekhtyar, L. Colletti and Y. Kiso, *Bioorg. Med. Chem. Lett.*, 2003, **13**, 93.
430. H.L. Sham, D.A. Betebenner, X. Chen, A. Saldivar, S. Vasavanonda, D.J. Kempf, J.J. Plattner and D.W. Norbeck, *Bioorg. Med. Chem. Lett.*, 2002, **12**, 1185.
431. K. Senten, P. Van der Veken, G. Bal, A. Haemers and K. Augustyns, *Tetrahedron Lett.*, 2001, **42**, 9135.
432. E.N. Prabhakaran, S. Rajesh, M.M. Reddy and J. Iqbal, *J. Indian Inst. Sci.*, 2001, **81**, 143.
433. J. Grembecka, A. Mucha, T. Cierpicki and P. Kafarski, *Phosphorus, Sulfur, Silicon Relat. Elem.*, 2002, **177**, 1739.
434. M. Drag, J. Grembecka and P. Kafarski, *Phosphorus, Sulfur, Silicon Relat. Elem.*, 2002, **177**, 1591.

435. J.D. Park and D.H. Kim, *J. Med. Chem.*, 2002, **45**, 911.
436. J.D. Park, D.H. Kim, S.-J. Kim, J.-R. Woo and S.E. Ryu, *J. Med. Chem.*, 2002, **45**, 5295.
437. D.L. Bienvenue, D. Gilner and R.C. Holz, *Biochemistry*, 2002, **41**, 3712.
438. W.L. Mock and D.J. Stanford, *Bioorg. Med. Chem. Lett.*, 2002, **12**, 1193.
439. M. Yamamoto, S. Ikeda, H. Kondo and S. Inoue, *Bioorg. Med. Chem. Lett.*, 2002, **12**, 375.
440. A. Volonterio, S. Bellosta, P. Bravo, M. Canavesi, E. Corradi, S.V. Meille, N. Monetti, N. Moussier and M. Zanda, *Eur. J. Org. Chem.*, 2002, 428.
441. D. Krumme and H. Tschesche, *Bioorg. Med. Chem. Lett.*, 2002, **12**, 933.
442. T. Fujisawa, K. Igeta, S. Odake, Y. Morita, J. Yasuda and T. Morikawa, *Bioorg. Med. Chem.*, 2002, **10**, 2569.
443. A. Reichelt, C. Gaul, R.R. Frey, A. Kennedy and S.F. Martin, *J. Org. Chem.*, 2002, **67**, 4062.
444. T. Fujisawa, S. Odake, Y. Ogawa, J. Yasuda and T. Morikawa, *Pept. Sci.*, 2002, 171.
445. K. Peters, G. Jahreis and E.-M. Kotters, *J. Enzyme Inhib.*, 2001, **16**, 339.
446. M.W. Mutahi, T. Nittoli, L. Guo and S.M. Sieburth, *J. Amer. Chem. Soc.*, 2002, **124**, 7363.
447. T. Fujisawa, S. Odake, Y. Ogawa, J. Yasuda, Y. Morita and T. Morikawa, *Chem. Pharm. Bull.*, 2002, **50**, 239.
448. A.P. Coll and S.S. Morte, *Afinidad*, 2001, **58**, 391.
449. N. Inguimbert, H. Poras, F. Teffo, F. Beslot, M. Selkti, A. Tomas, E. Scalbert, C. Bennejean, P. Renard, M.C. Fournié-Zaluski and B.-P. Roques, *Bioorg. Med. Chem. Lett.*, 2002, **12**, 2001.
450. C. Palomo, I. Ganboa, M. Oiarbide, G.T. Sciano and J.I. Miranda, *ARKIVOC*, 2002, 8.
451. E. Dumez, J.S. Snaith, R.F.W. Jackson, A.B. McElroy, J. Overington, M.J. Wythes, J.M. Withka and T.J. McLellan, *J. Org. Chem.*, 2002, **67**, 4882.
452. L. Yaouancq, M. Anissimova, M.-A. Badet-Denisot and B. Badet, *Eur. J. Org. Chem.*, 2002, 3573.
453. J. Holms, K. Mast, P. Marcotte, I. Elmore, J. Li, L. Pease, K. Glaser, D. Morgan, M. Michaelides and S. Davidsen, *Bioorg. Med. Chem. Lett.*, 2001, **11**, 2907.
454. M.A. Letavic, M.Z. Axt, J.T. Barberia, T.J. Carty, D.E. Danley, K.F. Geoghegan, N.S. Halim, L.R. Hoth, A.V. Kamath, E.R. Laird, L.L. Lopresti-Morrow, K.F. McClure, P.G. Mitchell, V. Natarajan, M.C. Noe, J. Pandit, L. Reeves, G.K. Schulte, S.L. Snow, F.J. Sweeney, D.H. Tan and C.H. Yu, *Bioorg. Med. Chem. Lett.*, 2002, **12**, 1387.
455. L. Wei, Y.-Q. Wu, D.E. Wilkinson, Y. Chen, R. Soni, C. Scott, D.T. Ross, H. Guo, P. Howorth, H. Valentine, S. Liang, D. Spicer, M. Fuller, J. Steiner and G.S. Hamilton, *Bioorg. Med. Chem. Lett.*, 2002, **12**, 1429.
456. P.S. Dragovich, T.J. Prins, R. Zhou, T.O. Johnson, E.L. Brown, F.C. Maldonado, S.A. Fuhrman, L.A. Zalman, A.K. Patick, D.A. Matthews, X. Hou, J.W. Meador, R.A. Ferre and S.T. Worland, *Bioorg. Med. Chem. Lett.*, 2002, **12**, 733.
457. P.S. Dragovich, T.J. Prins, R. Zhou, E.L. Brown, F.C. Maldonado, S.A. Fuhrman, L.S. Zakman, T. Tuntland, C.A. Lee, A.K. Patick, D.A. Matthews, T.F. Hendrickson, M.B. Kosa, B. Liu, M.R. Batugo, J.-P.R. Gleeson, S.K. Sakata, L. Chen, M.C. Guzman, J.W. Meador, R.A. Ferre and S.T. Worland, *J. Med. Chem.*, 2002, **45**, 1607.
458. R. Beevers, M.G. Carr, P.S. Jones, S. Jordan, P.B. Kay, R.C. Lazell and T.M. Raynham, *Bioorg. Med. Chem. Lett.*, 2002, **12**, 641.

459. S. Colarusso, B. Gerlach, U. Koch, E. Muraglia, I. Conte, I. Stansfield, V.G. Matassa and F. Narjes, *Bioorg. Med. Chem. Lett.*, 2002, **12**, 705.
460. B.K. Albrecht and R.M. Williams, *Org. Lett.*, 2002, **4**, 3711.
461. P. de Macedo, C. Marrano and J.W. Keillor, *Bioorg. Med. Chem.*, 2002, **10**, 355.
462. L. Demange, M. Moutiez and C. Dugave, *J. Med. Chem.*, 2002, **45**, 3928.
463. S.D. Larsen, T. Barf, C. Liljebris, P.D. May, D. Ogg, T.J. O'Sullivan, B.J. Palazuk, H.J. Schostarez, F.C. Stevens and J.E. Bleasdale, *J. Med. Chem.*, 2002, **45**, 598.
464. T. Hu and J.S. Panek, *J. Amer. Chem. Soc.*, 2002, **124**, 11368.
465. N. Kaila, L. Chen, B.E. Thomas, D. Tsao, S. Tam, P.W. Bedard, R.T. Camphausen, J.C. Alvarez and G. Ullas, *J. Med. Chem.*, 2002, **45**, 1563.
466. M. Mizuno, H.-k. Ishida, F. Ito, T. Endo and T. Inazu, *Pept. Sci.*, 2002, 85.
467. D. Burg, L. Hameetman, D.V. Filippov, G.A. van der Marel and G.J. Mulder, *Bioorg. Med. Chem. Lett.*, 2002, **12**, 1579.
468. K. Chibale, A. Chipelame and S. Warren, *Tetrahedron Lett.*, 2002, **43**, 1587.
469. K. Amssoms, S.L. Oza, K. Augustyns, A. Yamani, A.-M. Lambeir, G. Bal, P. Van der Veken, A.H. Fairlamb and A. Haemers, *Bioorg. Med. Chem. Lett.*, 2002, **12**, 2703.
470. R. Maltais, V. Luu-The and D. Poirier, *Bioorg. Med. Chem.*, 2001, **9**, 3101.
471. P. Ruzza, A. Calderan, A. Osler, S. Elardo and G. Borin, *Tetrahedron Lett.*, 2002, **43**, 3769.
472. E.A. Hallinan, S. Tsymbalov, C.R. Dorn, P.S. Pitzele, D.W. Hansen, W.M. Moore, G.M. Jerome, J.R. Connor, L.F. Branson, D.L. Widomski, Y. Zhang, M.G. Currie and P.T. Manning, *J. Med. Chem.*, 2002, **45**, 1686.
473. H.K. Smith, R.P. Beckett, J.M. Clements, S. Doel, S.P. East, S.P. Launchbury, L.M. Pratt, Z.M. Spavold, W. Thomas, R.S. Todd and M. Whitaker, *Bioorg. Med. Chem. Lett.*, 2002, **12**, 3595.
474. S.B. Singh, D.L. Zink, J.M. Liesch, R.T. Mosley, A.W. Dombrowski, G.F. Bills, S.J. Darkin-Rattray, D.M. Schmatz and M.A. Goetz, *J. Org. Chem.*, 2002, **67**, 815.
475. S. Wittich, H. Scherf, C. Xie, G. Brosch, P. Loidl, C. Gerhaeuser and M. Jung, *J. Med. Chem.*, 2002, **45**, 3296.
476. F. El Oualid, L. Bruining, I.M. Leroy, L.H. Cohen, J.H. Van Boom, G.A. Van der Marel, H.S. Overkleeft and M. Overhand, *Helv. Chim. Acta*, 2002, **85**, 3455.
477. A.S. Causton and J.C. Sherman, *J. Pept. Sci.*, 2002, **8**, 275.
478. R. Bannerjee, G. Basu, P. Chene and S. Roy, *J. Pept. Res.*, 2002, **60**, 88.
479. B. Banerji, B. Mallesham, S. Kiran Kumar, A.C. Kunwar and J. Iqbal, *Tetrahedron Lett.*, 2002, **43**, 6479.
480. S. Aravinda, N. Shamala, S. Desiraju and P. Balaram, *Chem. Commun.*, 2002, 2454.
481. I. Bediako-Amoa, R. Silerova and H.B. Kraatz, *Chem.Commun.*, 2002, 2430.
482. Y. Inai, Y. Kurokawa and N. Kojima, *J. Chem. Soc., Perkin Trans. 2*, 2002, 1850.
483. M. Higuchi, T. Koga, K. Taguchi and T. Kinoshita, *Trans. Mat. Res. Soc. Jpn.*, 2002, **27**, 489.
484. J.H. Miwa, L. Pallivathucal, S. Gowda and K.E. Lee, *Org. Lett.*, 2002, **4**, 4655.
485. T.L. Raguse, E.A. Porter, B. Weisblum and S.H. Gellman, *J. Amer. Chem. Soc.*, 2002, **124**, 12774.
486. T.L. Raguse, J.R. Lai and S.H. Gellman, *Helv. Chim. Acta*, 2002, **85**, 4154.
487. D. Haldar, S.K. Maji, W.S. Sheldrick and A. Banerjee, *Tetrahedron Lett.*, 2002, **43**, 2653.
488. S.K. Maji, A. Banerjee, M.G.B. Drew, D. Haldar and A. Banerjee, *Tetrahedron Lett.*, 2002, **43**, 6759.

489. P.R. LePlae, J.D. Fisk, E.A. Porter, B. Weisblum and S.H. Gellman, *J. Amer. Chem. Soc.*, 2002, **124**, 6820.
490. S. Ganesh and R. Jayakumar, *J. Pept. Res.*, 2002, **59**, 249.
491. T. Kraus, M. Buděšinský, I. Císařová and J. Závada, *Angew. Chem., Int. Ed.*, 2002, **41**, 1715.
492. M. Maeda, L. Liu and C.M. Deber, *Pept. Sci.*, 2002, 371.
493. U.A. Ramagopal, S. Ramakumar, P. Mathur, R. Joshi and V.S. Chauhan, *Protein Eng.*, 2002, **15**, 331.
494. Y. Takaoka, T. Mizukoshi, H. Shimizu and I. Fujii, *Pept. Sci.*, 2002, 309.
495. H. Rapaport, G. Moeller, C.M. Knobler, T.R. Jensen, K. Kjaer, L. Leiserowitz and D.A. Tirrell, *J. Amer. Chem. Soc.*, 2002, **124**, 9342.
496. S.K. Maji, D. Haldar, A. Banerjee and A. Banerjee, *Tetrahedron*, 2002, **58**, 8695.
497. T. Koga, K. Taguchi, M. Higuchi and T. Kinoshita, *Trans. Mat. Res. Soc. Jpn.*, 2002, **27**, 493.
498. X. Zhang, A.C. Schmitt and C.P. Decicco, *Tetrahedron Lett.*, 2002, **43**, 9663.
499. J. Venkatraman, G.A.N. Gowda and P. Balaram, *J. Amer. Chem. Soc.*, 2002, **124**, 4987.
500. S.K. Kundu, P.A. Mazumdar, A.K. Das, V. Bertolasi and A. Pramanik, *J. Chem. Soc., Perkin Trans. 2*, 2002, 1602.
501. J.S. Nowick, K.S. Lam, T.V. Khasanova, W.E. Kemnitzer, S. Maitra, H.T. Mee and R. Liu, *J. Amer. Chem. Soc.*, 2002, **124**, 4972.
502. M. Higuchi, T. Koga, K. Taguchi and T. Kinoshita, *Chem. Commun.*, 2002, 1126.
503. T.A. Martinek, G.B. Toth, E. Vass, M. Hollosi and F. Fulop, *Angew. Chem., Int. Ed.*, 2002, **41**, 1718.
504. G. Das, N. Sakaiand and S. Matile, *Chirality*, 2002, **14**, 18.
505. J.S. Nowick, E.M. Smith, J.W. Ziller and A.J. Shaka, *Tetrahedron*, 2002, **58**, 727.
506. T. Koga, K. Taguchi, T. Kinoshita and M. Higuchi, *Chem. Commun.*, 2002, 242.
507. Y. Takahashi, T. Yamashita, A. Ueno and H. Mihara, *Pept. Sci.*, 2002, 353.
508. M. Rainaldi, V. Moretto, M. Crisma, E. Peggion, S. Mammi, C. Toniolo and G. Cavicchioni, *J. Pept. Sci.*, 2002, **8**, 241.
509. F. Gao, Y. Wang, Y. Qiu, Y. Li, Y. Sha, L. Lai and H. Wu, *J. Pept. Res.*, 2002, **60**, 75.
510. D. Yang, Y.-H. Zhang and N.-Y. Zhu, *J. Amer. Chem. Soc.*, 2002, **124**, 9966.
511. W. Wang, J. Yang, J. Ying, C. Xiong, J. Zhang, C. Cai and V.J. Hruby, *J. Org. Chem.*, 2002, **67**, 6353.
512. X. Gu, X. Tang, S. Cowell, J. Ying and V.J. Hruby, *Tetrahedron Lett.*, 2002, **43**, 6669.
513. P. Grieco, P. Campiglia, I. Gomez-Monterrey and E. Novellino, *Tetrahedron Lett.*, 2002, **43**, 6297.
514. Y. Han, C. Giragossian, D.F. Mierke and M. Chorev, *J. Org. Chem.*, 2002, **67**, 5085.
515. L. Jiang and K. Burgess, *J. Amer. Chem. Soc.*, 2002, **124**, 9028.
516. A. Basak, K.R. Rudra, S.S. Bag and A. Basak, *J. Chem. Soc., Perkin Trans. 1*, 2002, 1805.
517. L. Jiang and K. Burgess, *Tetrahedron*, 2002, **58**, 8743.
518. J. Zhang, C. Xiong, W. Wang, J. Ying and V.J. Hruby, *Org. Lett.*, 2002, **4**, 4029.
519. S.K. Maji, R. Bannerjee, D. Velmurugan, A. Razak, H.K. Fun and A. Bannerjee, *J. Org. Chem.*, 2002, **67**, 633.
520. S. Rajesh, B. Banerji and J. Iqbal, *J. Org. Chem.*, 2002, **67**, 7852.
521. T.V.R.S. Sastry, B. Banerji, S. Kiran Kumar, A.C. Kunwar, J. Das, J.P. Nandy and J. Iqbal, *Tetrahedron Lett.*, 2002, **43**, 7621.

522. A. Nguyen Van Nhien, H. Ducatel, C. Len and D. Postel, *Tetrahedron Lett.*, 2002, **43**, 3805.
523. A. Avenoza, J.H. Busto, J.M. Peregrina and F. Rodriguez, *J. Org. Chem.*, 2002, **67**, 4241.
524. B. Wels, J.A.W. Kruijtzer and R.M.J. Liskamp, *Org. Lett.*, 2002, **4**, 2173.
525. A. Avenoza, J.H. Busto, C. Cativiela, J.M. Peregrina and F. Rodriguez, *Tetrahedron Lett.*, 2002, **43**, 1429.
526. P. Grieco, P. Campiglia, I. Gomez-Monterey and E. Novellino, *Tetrahedron Lett.*, 2002, **43**, 1197.
527. C. Bolm, D. Müller and C.P.R. Hackenberger, *Org. Lett.*, 2002, **4**, 893.
528. L. Halab and W.D. Lubell, *J. Amer. Chem. Soc.*, 2002, **124**, 2474.
529. Y.-l. Sha, Y.-l. Li, Y. Qiu, Q. Wang, L.-h. Lai and Y.-q. Tang, *Wuli Huaxue Xuebao*, 2002, **18**, 907.
530. H.N. Gopi, R.S. Roy, S.R. Raghothama, I.L. Karle and P. Balaram, *Helv. Chim. Acta*, 2002, **85**, 3313.
531. T.K. Chakraborty, B.K. Mohan, S.K. Kumar and A.C. Kunwar, *Tetrahedron Lett.*, 2003, **44**, 471.
532. R. Arienzo and J.D. Kilburn, *Tetrahedron*, 2002, **58**, 711.
533. T. Nakanishi, H. Okamoto, Y. Nagai, T. Takeda, I. Obataya, H. Mihara, H. Azehara, Y. Suzuki, W. Mizutani, K. Furukawa and K. Torimitsu, *Phys. Rev. B: Cond. Matter Mater. Phys.*, 2002, **66**, 165417/1.
534. J.M. Sanderson and S. Yazdani, *Chem. Commun.*, 2002, 1154.
535. M.M. Fernández, A. Diez, M. Rubiralta, E. Montenegro, M. Casamitjana, M.J. Kogan and E. Giralt, *J. Org. Chem.*, 2002, **67**, 7587.
536. A. Trabocchi, E.G. Occhiato, D. Potenza and A. Guarna, *J. Org. Chem.*, 2002, **67**, 7483.
537. M. Pechar, P. Kopeckova, L. Joss and J. Kopecek, *Macromol. Biosci.*, 2002, **2**, 199.
538. T. Kiyokawa, K. Kanaori, K. Tajima, M. Kawaguchi, J.-i. Oku and T. Tanaka, *Pept. Sci.*, 2002, 347.
539. S. Matsumura, A. Ueno and H. Mihara, *Pept. Sci.*, 2002, 307.
540. E.N. Prabhakaran, J.P. Nandy, S. Shukla, A. Tewari, S. Kumar Das and J. Iqbal, *Tetrahedron Lett.*, 2002, **43**, 6461.
541. H. Birkedal, D. Schwarzenbach and P. Pattison, *Chem. Commun.*, 2002, 2812.
542. M. Oba, M. Tanaka, M. Kurihara and H. Suemune, *Helv. Chim. Acta*, 2002, **85**, 3197.
543. M. Hedenström, L. Holm, Z. Yuan, H. Emtenäs and T. Kihlberg, *Bioorg. Med. Chem. Lett.*, 2002, **12**, 841.
544. T.K. Chakraborty, S. Jayaprakash, P. Srinivasu, S.S. Madhavendra, A. Ravi Sankar and A.C. Kunwar, *Tetrahedron*, 2002, **58**, 2853.
545. M. Tanaka, *Yuki Gosei Kagaku Kyokaishi*, 2002, **60**, 125.
546. R. Isaac and J. Chmielewski, *J. Amer. Chem. Soc.*, 2002, **124**, 6808.
547. O. Spiga, A. Bernini, M. Scarselli, L. Giovannoni, F. Laschi, P. Neri, L. Bracci, L. Lozzi and N. Niccolai, *Spectroscopy. Lett.*, 2002, **35**, 111.
548. A. Ishiwata, M. Takatani, Y. Nakahara and Y. Ito, *Synlett*, 2002, 634.
549. L. Biondi, F. Filira, M. Gobbo and R. Rocchi, *J. Pept. Sci.*, 2002, **8**, 80.
550. T. Ziegler, D. Röseling and L.R. Subramanian, *Tetrahedron Asymm.*, 2002, **13**, 911.
551. J. Watabe, L. Singh, Y. Nakahara, Y. Ito, H. Hojo and Y. Nakahara, *Biosci., Biotechnol.Biochem.*, 2002, **66**, 1904.
552. Y. Takano, M. Habiro, M. Someya, H. Hojo and Y. Nakahara, *Tetrahedron Lett.*, 2002, **43**, 8395.

553. P. Arya, A. Barkley and K.D. Randell, *J. Comb. Chem.*, 2002, **4**, 193.
554. M. Cudic, H.C.J. Ertl and L. Otvos, *Bioorg. Med. Chem.*, 2002, **10**, 3859.
555. M.-R. Lee, J. Lee and I. Shin, *Synlett*, 2002, 1463.
556. G.J. McGarvey, T.E. Benedum and F.W. Schmidtmann, *Org. Lett.*, 2002, **4**, 3591.
557. M.R. Carrasco, M.J. Nguyen, D.R. Burnell, M.D. MacLaren and M.S. Hengel, *Tetrahedron Lett.*, 2002, **43**, 5727.
558. S.D. Zhang, G. Liu, S.Q. Xia and P. Wu, *Chin. Chem. Lett.*, 2002, **13**, 17.
559. N. Kubasch and R.R. Schmidt, *Eur. J. Org. Chem.*, 2002, 2710.
560. A.E. Zemlyakov, V.V. Tsikalov, V.O. Kur'yanov, V.Ya. Chirva and N.V. Bovin, *Russ. J. Bioorg. Chem.*, 2001, **27**, 390.
561. E. Brenna, C. Fuganti, P. Grasselli, S. Serra and S. Zambotti, *Chem.-Eur. J.*, 2002, **8**, 1872.
562. Y. Ohnishi and Y. Ichikawa, *Bioorg. Med. Chem. Lett.*, 2002, **12**, 997.
563. C. Palomo, M. Oiarbide, A. Landa, M.C. Gonzalez-Rego, J.M. Garcia, A. Gonzalez, J.M. Odriozola, M. Martin-Pastor and A. Linden, *J. Amer. Chem. Soc.*, 2002, **124**, 8637.
564. C. Taillefumier, Y. Lakhrissi, M. Lakhrissi and Y. Chapleur, *Tetrahedron Asymm.*, 2002, **13**, 1707.
565. S.A.W. Gruner, V. Truffault, G. Voll, E. Locardi, M. Stockle and H. Kessler, *Chem.-Eur. J.*, 2002, **8**, 4365.
566. J. van Ameijde, H.B. Albada and R.M.J. Liskamp, *J. Chem. Soc., Perkin Trans. 1*, 2002, 1042.
567. L. Stella, M. Venanzi, M. Carafa, E. Maccaroni, M.E. Straccamore, G. Zanotti, A. Palleschi and B. Pispisa, *Biopolymers*, 2002, **64**, 44.
568. D. Varon, E. Lioy, M.E. Patarroyo, X. Schratt and C. Unvertzagt, *Austral. J. Chem.*, 2002, **55**, 161.
569. A. Ishii, H. Hojo, Y. Nakahara, Y. Ito and Y. Nakahara, *Biosci. Biotech. Biochem.*, 2002, **66**, 225.
570. T. Ichiyanagi, M. Takatani, K. Sakamoto, Y. Nakahara, Y. Ito, H. Hojo and Y. Nakahara, *Tetrahedron Lett.*, 2002, **43**, 3297.
571. A. Lohse, F. Schweizer and O. Hindsgaul, *Comb. Chem. High Throughput Screening*, 2002, **5**, 389.
572. H. Amer, A. Hofinger and P. Kosma, *Carbohydr. Res.*, 2002, **338**, 35.
573. S.B. Cohen and R.L. Halcomb, *J. Amer. Chem. Soc.*, 2002, **124**, 2534.
574. D. Macmillan, A.M. Daines, M. Bayrhuber and S.L. Flitsch, *Org. Lett.*, 2002, **4**, 1467.
575. J.P. Malkinson and R.A. Falconer, *Tetrahedron Lett.*, 2002, **43**, 9549.
576. I. Shin, S. Park, K. Sungjin, K.S. Kim, J.W. Cho and D. Lim, *Bull. Korean Chem. Soc.*, 2002, **23**, 15.
577. M. Rösch, H. Herzner, W. Dippold, M. Wild, D. Vestweber and H. Kunz, *Angew. Chem., Int. Ed.*, 2002, **40**, 3836.
578. J.-i. Tamura, A. Yamaguchi and J. Tanaka, *Bioorg. Med. Chem. Lett.*, 2002, **12**, 1901.
579. T.K. Chakraborty, S. Jayaprakash and S. Ghosh, *Comb. Chem. High Throughput Screening*, 2002, **5**, 373.
580. N. Wehofsky, R. Loser, A. Buchynskyy, P. Welzel and F. Bordusa, *Angew. Chem., Int. Ed.*, 2002, **41**, 2735.
581. R.L. Harding, J. Henshaw, J. Tilling and T.D.H. Bugg, *J. Chem. Soc., Perkin Trans. 1*, 2002, 1714.

582. O. Blixt, K. Allin, L. Pereira, A. Datta and J.C. Paulson, *J. Amer. Chem. Soc.*, 2002, **124**, 5739.
583. Y. Tokuda, Y. Takahashi, K. Matoishi, Y. Ito and T. Sugai, *Synlett*, 2002, 57.
584. I. Saskiawan, M. Mizuno, T. Inazu, K. Haneda, S. Harashima, H. Kumagai and K. Yamamoto, *Arch. Biochem. Biophys.*, 2002, **406**, 127.
585. K. Yamamoto, *J. Biosci. Bioeng.*, 2001, **92**, 493.
586. K. Ajisaka, M. Miyasato, C. Ito, Y. Fujita, Y. Yamazaki and S. Oka, *Glyconjug. J.*, 2001, **18**, 301.
587. K. Takagaki, K. Ishido, M. Iwafune and M. Endo, *Biochem. Biophys. Res. Commun.*, 2002, **293**, 220.
588. K. Suzuki, I. Matsuo, M. Isomura and K. Ajisaka, *J. Carbohyd. Chem.*, 2002, **21**, 99.
589. S. Rawale, L.M. Hrihorczuk, W.-Z. Wei and J. Zemlicka, *J. Med. Chem.*, 2002, **45**, 937.
590. G. Zhao, Y.-m. Li, S.-z. Luo, B. Han and Y.-f. Zhao, *Gaodeng Xuexiao Huaxue Xuebao*, 2001, **22**, 2034.
591. K. Lee, M. Zhang, D. Yang and T.R. Burke, *Bioorg. Med. Chem. Lett.*, 2002, **12**, 3399.
592. S.-Z. Luo, G. Zhao, Y.-M. Li, L. Xu and Y.-F. Zhao, *Gaodeng Xuexiao Huaxue Xuebao*, 2002, **23**, 852.
593. K. Tamura, *Seirigaku Gijutsu Kenkyukai Hokoku*, 2001, **23**, 32.
594. J.-Y. Tsai and F.-D.T. Lung, *Chin. Pharm. J.*, 2002, **54**, 141.
595. C.-Q. Wei, B. Li, R. Guo, D. Yang and T.R. Burke, *Bioorg. Med. Chem. Lett.*, 2002, **12**, 2781.
596. D.M. Rothman, M.E. Vazquez, E.M. Vogel and B. Imperiali, *Org. Lett.*, 2002, **4**, 2865.
597. G. Varadi and L. Otvos, *J. Pept. Sci.*, 2002, **8**, 621.
598. J. Baraniak, R. Kaczmarek, D. Korczyński and E. Wasilewska, *J. Org. Chem.*, 2002, **67**, 7267.
599. K. Hasegawa, Y.L. Sha, J.K. Bang, T. Kawakami, K. Akaji and S. Aimoto, *Lett. Pept. Sci.*, 2001, **8**, 277.
600. K.E. Jenkins, A.P. Higson, P.H. Seeberger and M.H. Caruthers, *J. Amer. Chem. Soc.*, 2002, **124**, 6584.
601. W.-Q. Liu, C. Olszowy, L. Bischoff and C. Garbay, *Tetrahedron Lett.*, 2002, **43**, 1417.
602. M. Rinnová, A. Nefzi and R.A. Houghten, *Tetrahedron Lett.*, 2002, **43**, 4103.
603. M. Lukas, P. Vojtisek, P. Hermann, J. Rohovec and I. Lukes, *Synth. Commun.*, 2002, **32**, 79.
604. H.-J. Cristau, A. Genevois-Borella, F. Sanchez and J.-L. Pirat, *J. Organomet. Chem.*, 2002, **643–644**, 381.
605. J.-L. Pirat, A. Coulombeau, A. Genevois-Borella and H.-J. Cristau, *Phosphorus, Sulfur, Silicon Relat Elem.*, 2002, **177**, 1793.
606. D.-G. Liu, X.-Z. Wang, Y. Gao, B. Li, D. Yang and T.R. Burke, *Tetrahedron*, 2002, **58**, 10423.
607. A. Horváth, C. Olive, A. Wong, T. Clair, P. Yarwood, M. Good and I. Toth, *J. Med. Chem.*, 2002, **45**, 1387.
608. A. Metaxas, S. Tzartos and M. Liakopoulou-Kyriakides, *J. Pept. Sci.*, 2002, **8**, 118.
609. T. Tselios, V. Apostolopoulos, I. Daliani, S. Deraos, S. Grdadolnik, T. Mavromoustakos, M. Melachrinou, S. Thymianou, L. Probert, A. Mouzaki and J. Matsoukas, *J. Med. Chem.*, 2002, **45**, 275.

610. P. Zubrzak, K. Kociolek, M. Smoluch, J. Silberring, M. Kowalksi, B. Szudlinska and J. Zabrocki, *Acta Biochim. Pol.*, 2001, **48**, 1151.
611. Z. Szewczuk, A. Wilczynski, I. Petry, I.Z. Siemion and Z. Wieczorek, *Acta Biochim. Pol.*, 2001, **48**, 1147.
612. S.V. Ley and A. Priour, *Eur. J. Org. Chem.*, 2002, 3995.
613. M. Kokubo, H. Oku, K. Yamada, K. Sato, S. Kano, M. Suzuki and R. Katakai, *Pept. Sci.*, 2002, 331.
614. M. Antopolsky, E. Azhayeva, U. Tengvall and A. Azhayev, *Tetrahedron Lett.*, 2002, **43**, 527.
615. M. Ollivier, C. Olivier, C. Gouyette, T. Huynk-Dink, H. Gras-Mosse and O. Melnyk, *Tetrahedron Lett.*, 2002, **43**, 997.
616. S.A. Kuznetsova, N.V. Sumbatyan, C. Malvy, J.-R. Bertrand, A. Harel-Bellan, G.A. Korshunova and F.P. Svinarchuk, *Vestnik Moskovskogo Universiteta 2: Khimiya*, 2001, **42**, 281.
617. E. Millo, R. Nicolai, S. Scarfi', C. Scapolla, B. Biasotti, U. Benatti and G. Diamonte, *Tetrahedron Lett.*, 2002, **43**, 2961.
618. G. Kovács, Z. Timár, Z. Kupihár. Z. Kele and L. Kovács, *J. Chem. Soc., Perkin Trans. 1*, 2002, 1266.
619. J. Kaneno, T. Kubo and M. Fujii, *Pept. Sci.*, 2002, 339.
620. K. Yokoyama, T. Kubo and M. Fujii, *Pept. Sci.*, 2002, 313.
621. L. Debethune, V. Marchan, G. Fabregas, E. Pedroso and A. Grandas, *Tetrahedron*, 2002, **58**, 6965.
622. T. Tedeschi, R. Corradini, R. Marchelli, A. Pushl and P.E. Nielsen, *Tetrahedron Asymm.*, 2002, **13**, 1629.
623. B. Falkiewicz, *Chem. Preprint Server, Org. Chem.*, 2002, 1.
624. A.M. Bruckner, H.W. Schmitt and U. Diederichsen, *Helv. Chim. Acta*, 2002, **85**, 3855.
625. B. Falkiewicz, *Nucleosides, Nucleotides, Nucleic Acids*, 2002, **21**, 883.
626. S.M. Viladkar, *Tetrahedron*, 2002, **58**, 495.
627. R.H.E. Hudson, G. Li and J. Tse, *Tetrahedron Lett.*, 2002, **43**, 1381.
628. H. Ikeda, Y. Nakamura and I. Saito, *Tetrahedron Lett.*, 2002, **43**, 5525.
629. T. Vilaivan and G. Lowe, *J. Amer. Chem. Soc.*, 2002, **124**, 9326.
630. K.S. Schmidt, M. Boudvillain, A. Schwartz, G.A. Van der Marel, J.H. Van Boom, J. Reedijk and B. Lippert, *Chem--Eur. J.*, 2002, **8**, 5566.
631. S.K. Awasthi and P.E. Nielsen, *Comb. Chem. High Throughput Screening*, 2002 **5**, 253.
632. C. Baldoli, S. Maiaorana, E. Licandro, G. Zinzalla and D. Perdicchia, *Org. Lett.*, 2002, **4**, 4341.
633. M.C. de Koning, D.V. Filippov, N. Meeuwenoord, M. Overhand, G.A. van der Marel and J.H. van Boom, *Tetrahedron Lett.*, 2002, **43**, 8173.
634. D. Forget, O. Renaudet, D. Boturyn, E. Defrancq and P. Dumy, *Tetrahedron Lett.*, 2001, **42**, 9171.
635. K.V.A. Meena and K.N. Ganesh, *Nucleosides, Nucleotides, Nucleic Acids*, 2001, **20**, 1193.
636. T.S. Zatsepin, D.A. Stetsenko, A.A. Arzumanov, E.A. Romanova, M.J. Gait and T.S. Oretskaya, *Bioconjugate Chem.*, 2002, **13**, 822.
637. M. Yano, T. Kubo and M. Fujii, *Pept. Sci.*, 2002, 335.
638. K.A. Frey and S.A. Woski, *Chem. Commun.*, 2002, 2206.
639. B. Kuhnast, F. Dolla and B. Tavitian, *J. Labelled Compds. Radiopharm.*, 2002, **45**, 1.
640. S. Bregant, F. Burlina and G. Chassaing, *Bioorg. Med. Chem. Lett.*, 2002, **12**, 1047.

641. S. Bregant, F. Burlina, J. Vaissermann and G. Chassaing, *Eur. J. Org. Chem.*, 2001, 3285.
642. X. Tian and E. Wickstrom, *Org. Lett.*, 2002, **4**, 4013.
643. O. Seitz and O. Köhler, *Chem.-Eur. J.*, 2001, **7**, 3911.
644. M.-L. Leroux, C. Di Giorgio, N. Patino and R. Condom, *Tetrahedron Lett.*, 2002, **43**, 1641.
645. S. Drioli, I. Adamo, M. Ballico, F. Morvan and G.M. Bonora, *Eur. J. Org. Chem.*, 2002, 3473.
646. B. Kumar Das, N. Shibata and Y. Takeuchi, *J. Chem. Soc., Perkin Trans. 1*, 2002, 197.
647. M. Rinnová, A. Nefzi and R.A. Houghten, *Tetrahedron Lett.*, 2002, **43**, 2343.
648. S. Izquierdo, M. Martin-Vilà, A.G. Moglioni, V. Branchadell and R.M. Ortuño, *Tetrahedron: Asymmetry*, 2002, **13**, 2403.
649. P.S. Dragovich, R. Zhou and T.J. Prins, *J. Org. Chem.*, 2002, **67**, 741.
650. C. Subramanyam and S.P. Chang, *Tetrahedron Lett.*, 2002, **43**, 6313.
651. R. de Jong, D.T.S. Rijkers and R.M.J. Liskamp, *Helv. Chim. Acta*, 2002, **85**, 4230.
652. S.G. Nelson, K.L. Spencer, W.S. Cheung and S.J. Mamie, *Tetrahedron*, 2002, **58**, 7081.
653. T. Kimmerlin, D. Seebach and D. Hilvert, *Helv. Chim. Acta*, 2002, **85**, 1812.
654. E. Alonso, C. del Pezo and J. González, *Synlett*, 2002, 69.
655. D. Seebach, M. Brenner, M. Rueping and B. Jaun, *Chem.-Eur. J.*, 2002, **8**, 573.
656. A. Bianco, C. Corvaja, M. Crisma, D.M. Guldi, M. Maggini, E. Sartori and C. Toniolo, *Chem.-Eur. J.*, 2002, **8**, 1544.
657. Y.-x. Xu, C.-x. Wan, H.-w. Yao, X.-h. Huang and Y.-l. Yue, *Sichuan Daxue Xuebao, Gongcheng Kexueban*, 2002, **34**, 65.
658. J.M. Belitsky, D.H. Nguyen, N.R. Wurtz and P.B. Dervan, *Bioorg. Med. Chem.*, 2002, **10**, 2767.
659. P. Chaltin, E. Lescrinier, T. Lescrinier, J. Rozenski, C. Hendrix, H. Rosemeyer, R. Busson, A. Van Aerschot and P. Herdewijn, *Helv. Chim. Acta*, 2002, **85**, 2258.
660. Z. Zhang, P. Chaltin, A. Van Aerschot, J. Lacey, J. Rozenski, R. Busson and P. Herdewijn, *Bioorg. Med. Chem.*, 2002, **10**, 3401.
661. F. Formaggio, M. Bonchio, M. Crisma, C. Peggion, S. Mezzato, A. Polese, A. Barazza, S. Antonello, F. Maran, Q.B. Broxterman, B. Kaptein, J. Kamphuis, R.M. Vitale, M. Saviano, E. Benedetti and C. Toniolo, *Chem.-Eur. J.*, 2002, **8**, 84.
662. J.S. Davies and L. Al-Jamri, *J. Pept. Sci.*, 2002, **8**, 663.
663. B. Rzeszotarska, D. Siodlak, M.A. Broda, I. Dybala and A.E. Koziol, *J. Pept. Res.*, 2002, **59**, 79.
664. H. Zhou and W.S.A. van der Donk, *Org. Lett.*, 2002, **4**, 1335.
665. J. Makker, S. Dey, P. Kumar and T.P. Singh, *Acta Cryst.*, 2002, **C58**, o212.
666. T.K. Chakraborty, B. Krishna Mohan, S. Kiran Kumar and A.C. Kunwar, *Tetrahedron Lett.*, 2002, **43**, 2589.
667. B. Ludolph, F. Eisele and H. Waldmann, *J. Amer. Chem. Soc.*, 2002, **124**, 5954.
668. H. Kawakami, K. Toma, M. Takagi, T. Yoshida and H. Tamiaki, *Pept. Sci.*, 2002, 381.
669. B. Ludolph, F. Eisele and H. Waldmann, *ChemBioChem*, 2002, **3**, 901.
670. S. Sato, K. Ishikawa, H. Kajita and S. Ohuchi, *Pept. Sci.*, 2002, 407.
671. K. Usui, M. Takahashi, A. Ueno, K. Nokihara and H. Mihara, *Pept. Sci.*, 2002, 405.
672. D.V. Filippov, D.J. van Zoelen, S.P. Oldfield, G.A. van der Marel, H.S. Overkleeft, J.W. Drijfhout and J.H. van Boom, *Tetrahedron Lett.*, 2002, **43**, 7809.

673. N. Vavourakis, L. Leondiadis and N. Ferderigos, *Tetrahedron Lett.*, 2002, **43**, 8343.
674. H. Dialer, W. Steglich and W. Beck, *Z. Naturforsch B.*, 2001, **56**, 1084.
675. T. Moriuchi, K. Yoshida and T. Hirao, *J. Organomet. Chem.*, 2001, **637–639**, 75.
676. K.K.-W. Lo, J.S.-Y. Lau, D.C.-M. Ng and N. Zhu, *J. Chem. Soc., Dalton Trans.*, 2002, 1753.
677. M.A. Razzaque, G.A. Lord and C.K. Lim, *Rapid Commun. Mass Spectrom.*, 2002, **16**, 1675.
678. K.J. Kise and B.E. Bowler, *Inorg. Chem.*, 2002, **41**, 379.
679. Z. Cheng, J. Chen, Y. Miao, N.K. Owen, T.P. Quinn and S.S. Jurisson, *J. Med. Chem.*, 2002, **45**, 3048.
680. K. Haas and W. Beck, *Z. Anorg. Allgem. Chem.*, 2002, **628**, 788.
681. C. Peggion, M. Crisma, F. Formaggio, C. Toniolo, K. Wright, M. Wakselman and J.P. Mazaleyrat, *Biolymers*, 2002, **63**, 314.
682. T.-A. Okamura, T. Iwamura, K. Nozaki, T. Ohno, H. Yamamoto and N. Ueyama, *Mol. Cryst. Liq. Cryst. Sci., Sect. A*, 2002, **379**, 431.
683. A. Nomura and Y. Sugiura, *Inorg. Chem.*, 2002, **41**, 3693.
684. M.C. Pirung, K. Park and N.L. Tumey, *J. Comb. Chem.*, 2002, **4**, 329.
685. P.V. Berhardt, P. Comba, D.P. Fairlie, L.R. Gahan, G.R. Hanson and L. Lötzbeyer, *Chem.-Eur. J.*, 2002, **8**, 1527.
686. S. Ranganathan and N. Tamilerasu, *Ind. J. Chem. B*, 2001, **40B**, 1081.
687. J.S. Choi, H.-S. Lee, Y. Lee, N. Jeong, H.-J. Kim, Y.-D. Kim and H. Han, *Tetrahedron Lett.*, 2002, **43**, 4295.
688. K. Ito, M. Noike, A. Kida and Y. Ohba, *J. Org. Chem.*, 2002, **67**, 7519.
689. A. Avenoza, M. Paris, J.M. Peregrina, M. Alias, M.P. Lopez, J.I. Garcia and C. Cativiela, *Tetrahedron*, 2002, **58**, 4899.
690. B. Kundu, T. Srinivasan, A.P. Kesarwani, A. Kavishwar, S.K. Raghuwanshi, S. Batra, S. Batra and P.K. Shukla, *Bioorg. Med. Chem. Lett.*, 2002, **12**, 1473.
691. S. Malik, S.K. Maji, A. Banerjee and A.K. Nandi, *J. Chem. Soc., Perkin Trans. 2*, 2002, 1177.
692. S. Herrero, M.L. Suarez-Gea, M.T. García-López and R. Herranz, *Tetrahedron Lett.*, 2002, **43**, 1421.
693. H. Tokuyama, S. Yokoshima, S.-C. Lin, L. Li and T. Fukuyama, *Synthesis*, 2002, 1121.
694. C. Gros, C. Boulegue, N. Galeotti, G. Niel and P. Jouin, *Tetrahedron*, 2002, **58** 2673.
695. M. Planas, E. Cros, R.-A. Rodriguez, R. Ferre and E. Bardaji, *Tetrahedron Lett.*, 2002, **43**, 4431.
696. T. Furuya, S. Lee, K. Murata and N. Takami, *Pept. Sci.*, 2002, 317.
697. M. Sani, P. Bravo, A. Volonterio and M. Zanda, *Coll. Czech. Chem. Commun.*, 2002, **67**, 1305.
698. K.A. Brun, A. Linden and H. Heimgartner, *Helv. Chim. Acta*, 2002, **85**, 3422.
699. M. Albrecht, O. Spiess and M. Schneider, *Synthesis*, 2002, 126.
700. M.F. Mohd Mustapa, R. Harris, J. Mould, N.A.L. Chubb, D. Schultz, P.C. Driscoll and A.B. Tabor, *Tetrahedron Lett.*, 2002, **43**, 8359.
701. F.S. Gibson, A.K. Singh, M.C. Soumeillant, P.S. Manchand, M. Hunora and D.R. Kronenthal, *Org. Proc. Res. Develop.*, 2002, **6**, 814.
702. H.M. Franzen, G. Bessidskaia, V. Abedi, A. Nilsson, M. Nilsson and L. Olsson, *Org. Proc. Res. Develop.*, 2002, **6**, 788.
703. P. Virta, J. Sinkkonen and H. Lonnberg, *Eur. J. Org. Chem.*, 2002, 3616.

704. M. Hashimoto, Y. Hatanaka, Y. Sadakane and K. Nabeta, *Bioorg. Med. Chem. Lett.*, 2002, **12**, 2507.
705. A.B. Smith, S.N. Savinov, U.V. Manjappara and I.M. Chaiken, *Org. Lett.*, 2002, **4**, 4041.
706. J.E. Redman and M.R. Ghadiri, *Org. Lett.*, 2002, **4**, 4467.
707. A. Katoh, Y. Inoue, H. Nagashima, Y. Nikita, J. Ohkanda and R. Saito, *Heterocycles*, 2002, **58**, 371.
708. T. Michel, B. Koksch, S.N. Osipov, A.S. Golubev, J. Sieler and K. Burger, *Coll. Czech. Chem. Commun.*, 2002, **67**, 1533.
709. E. Mann, A. Montero, M.A. Maestro and B. Herradón, *Helv. Chim. Acta*, 2002, **85**, 3624.
710. J. Arrowsmith, S.A. Jennings, A.S. Clark and M.F.G. Stevens, *J. Med. Chem.*, 2002, **45**, 5458.
711. C.-H. Xue, Q.-M. Mu and S.-H. Chen, *Huaxue Xuebao*, 2002, **60**, 355.
712. W. Wang, J. Zhang, C. Xiong and V.J. Hruby, *Tetrahedron Lett.*, 2002, **43**, 2137.
713. K. Guzow, M. Szabelski, J. Malicka, J. Karolczak and W. Wiczk, *Tetrahedron*, 2002, **58**, 2201.
714. S.M. Okarvi, K. Verbeke, P. Adriaens and A.M. Verbruggen, *J. Labelled Comp. Radiopharm.*, 2002, **45**, 115.
715. S.M. Okavi, P. Torfs, P. Adriaens and A.M. Verbruggen, *J. Labelled Comp. Radiopharm.*, 2002, **45**, 407.
716. J.A. Zerkowski, L.M. Hensley and D. Abramowitz, *Synlett*, 2002, 557.
717. K.B. Jensen, T.M. Braxmeier, M. Demarcus, J.G. Frey and J.D. Kilburn, *Chem.-Eur. J.*, 2002, **8**, 1300.
718. N. Sofroniev and S. Minchev, *Bulgarian Chem. Ind.*, 2002, **73**, 45.
719. S.R. Gilbertson and L. Ping, *Tetrahedron Lett.*, 2002, **43**, 6961.
720. R.W. Kinas, A. Opolski, B. Kolesinska and J. Kaminski, *Acta Pol. Pharm.*, 2000, **57**(Suppl.), 36.
721. R.W. Kinas, A. Opolski, B. Kolesinska and J. Kaminski, *Acta Pol. Pharm.*, 2000, **57**(Suppl.), 143.
722. N. Brosse, A. Grandeury and B. Jamart-Grégoire, *Tetrahedron Lett.*, 2002, **43**, 2009.
723. D.W. Konas, J.J. Pankuch and J.K. Coward, *Synthesis*, 2002, 2616.
724. A. Madder, L. Li, H. De Muynck, N. Farcy, D. Van Haver, F. Fant, G. Vanhoenacker, P. Sandra, A.P. Davis and P.J. De Clercq, *J. Comb. Chem.*, 2002, **4**, 552.
725. T.K. Chakraborty, P. Srinivasu, S.K. Kumar and A.C. Kunwar, *J. Org. Chem.*, 2002, **67**, 2093.
726. N. Wimmer, R.J. Marano, P.S. Kearns, E.P. Rakoczy and I. Toth, *Bioorg. Med. Chem. Lett.*, 2002, **12**, 2635.
727. H.-A. Klok and J. Rodríguez-Hernández, *Macromolecules*, 2002, **35**, 8718.
728. J.E. Wright and A. Rosowsky, *Bioorg. Med. Chem.*, 2002, **10**, 493.
729. L.M. Matz, H.M. Dion and H.H. Hill, *J. Chromatogr, A*, 2002, **946**, 59.
730. J. Rozenski, P. Chaltin, A. Van Aerschot and P. Herdewijn, *Rapid Commun. Mass Spectrom.*, 2002, **16**, 982.
731. T. Yoshida and S. Kasahara, *Chromatography*, 2002, **23**(Suppl.), 33.
732. Q. Bai, X. Ge and X. Geng, *Fenxi Huaxue*, 2002, **30**, 1126.
733. R. Wu, H. Zou, H. Fu, W. Jin and M. Ye, *Electrophoresis*, 2002, **23**, 1239.
734. K. Pflegerl, R. Hahn, E. Berger and A. Jungbauer, *J. Pept. Res.*, 2002, **59**, 174.
735. D. Koval, V. Kasicka, J. Jiracek, M. Collinsova and T.A. Garrow, *Electrophoresis*, 2002, **23**, 215.

Analogue and Conformational Studies on Peptides, Hormones and Other Biologically Active Peptides

BY BOTOND PENKE, GÁBOR TÓTH AND GYÖRGYI VÁRADI
Dept. Medical Chemistry, University of Szeged, Dom ter 8, H-6720, Szeged, Hungary

1 Introduction

The sub-divisions and the basic structure of this Chapter remains unchanged compared to previous volumes.[1] Core references for this Chapter were obtained from the Chemical Abstract Service (CA Selects on Amino Acids, Peptides and Proteins, up to issue 10, 2003).[2] The expansion in the availability of scientific journals in electronic format and computer scanning of published titles has become much easier and gave us a big help for compiling the material for this Chapter. An increasing number of Web of Science databases[3,4] was used. Several proceedings of international conferences were published in year 2002: first of all the scientific material of the 27th European Peptide Symposium at Sorrento ("Peptides 2002" a 1057 page volume)[5] and the Proceeding of the 17th American Peptide Symposium[6] as well as the Proceedings of the 39th Japanese Peptide Symposium.[7] However, as in the past, conference reports are not reviewed in this Periodical Report. Another important event was the publication of a comprehensive work in the series "Houben-Weyl Methods of Organic Chemistry", 4th Edition ("Synthesis of Peptides and Peptidomimetics").[8] It is a great aid for peptide chemists in their everyday work. No patents have been used as source material.

2 Peptide Backbone Modifications and Peptide Mimetics

A review[9] of different sequence-specific peptidomimetic oligomers, including β- and hydrazino peptides as antibacterial agents, oligourea, oligocarbamate or β-peptide mimics of HIV-Tat protein, azapeptide ligands for MHC-II, novel somatostatin analogues, and helical peptoid mimics of lung surfactant protein C has appeared. Types of peptide isosteres, the recent developments in their synthesis, design and inhibitory activities were the subject of a 299-reference

Amino Acids, Peptides and Proteins, Volume 35

review.[10] Strategies for producing modified peptides and mimetics in order to convert a peptide into a potential drug have been reviewed.[11] A review[12] highlighting molecular modeling techniques used to reduce the time of designing of a potential lead from a peptide has been published. Comparison of conformer profiles with molecular recognition templates which allows estimates of molecular recognition as bioactive ligands, was illustrated for pseudopeptides, isosteres and conformationally-constrained peptide analogues in a 119-reference review.[13] A review[14] discussing the applications of peptide mimetics for cancer immunotherapy has appeared.

2.1 Aza, Oxazole, Oxazoline, Triazole, Triazine and Tetrazole Peptides. – Azapeptide based inhibitors of the Hepatitis C Virus NS3 Serine Protease were designed and synthesized.[15] Some of the azanorvalyl residue-containing inhibitors (**1, 2**) that had an ester group at the C-terminus with a good leaving group were found to be the most potent and had a K_i in the 0.2–2 μM range. Azapeptide epoxides (**3**), a new class of irreversible protease inhibitors were found to be specific for the clan CD cysteine proteases.[16]

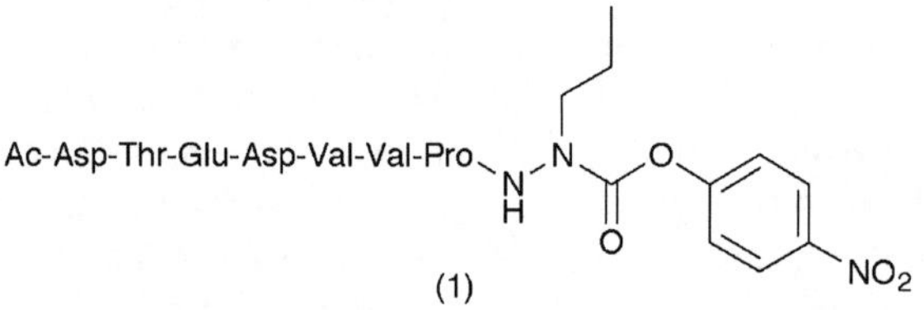

Ac-Asp-Thr-Glu-Asp-Val-Val-Pro–NH–N(propyl)–C(=O)–O–C$_6$H$_4$–NO$_2$

(1)

Ac-Asp-Thr-Glu-Asp-Val-Val-Pro–NH–N(propyl)–C(=O)–O–CH(Cl)–CCl$_3$

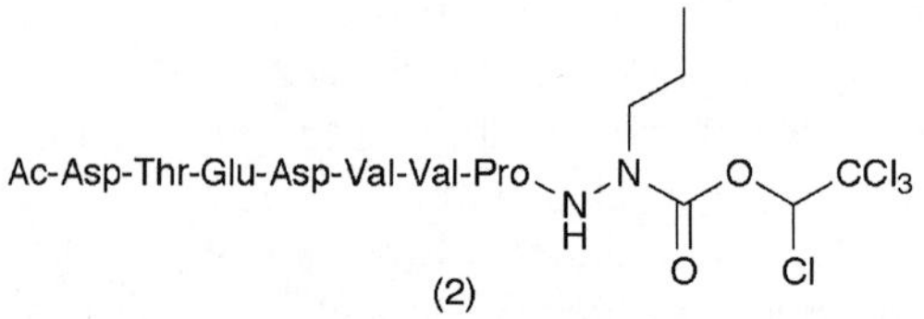

(2)

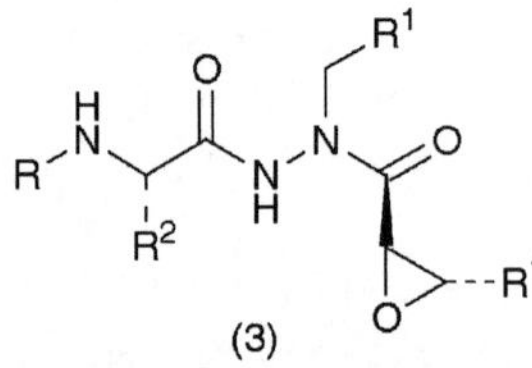

(3)

Systematic examination of the conformational properties of N-methyl azapeptide derivatives (Figure 1) provided information for use in the design of secondary structure for peptides and proteins, and in the development of new drugs.[17] Compound (**4**), a conformationally constrained analogue of azaproline or azapipecolic acid was synthesized applying a novel template, 2, 3-diazabicyclo [2.2.1]heptane.[18] New approaches to the synthesis of possible building blocks of peptide mimetics have been reported: 2,4-disubstituted

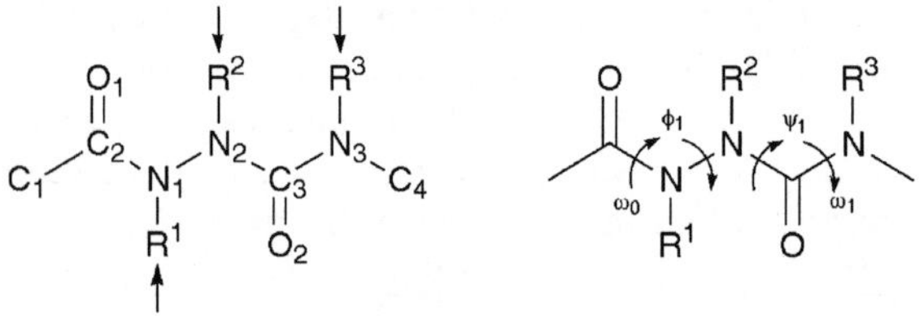

Figure 1

oxazoles[19] and substituted oxazoline amines[20] (**5**) were prepared from α-amino acids. Incorporation of 1,2,4-triazole into peptides resulted in the formation of a novel *cis* amine bond surrogate (**6**).[21]

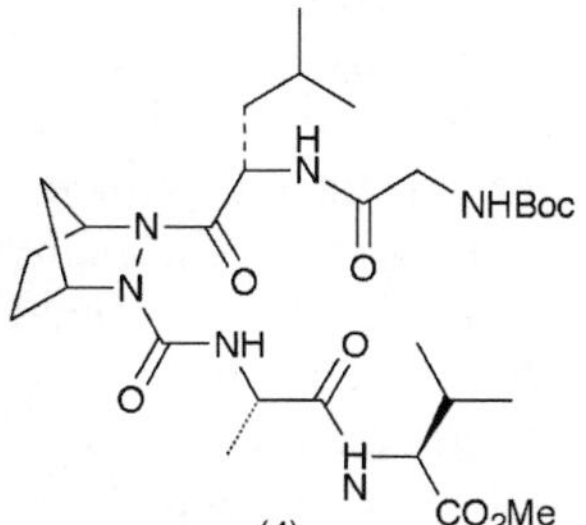

(4) 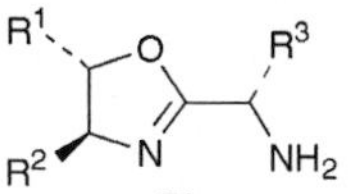 (5) 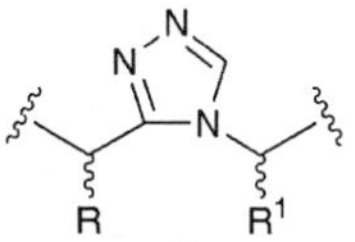 (6)

Synthesis of some new 1,2,3-triazole derived potential peptidomimetics was described.[22] Tetrazole analog of the highly conserved C-terminal pentapeptide of the insect kinins (Phe-Phe-Ψ[CN_4]Ala-Trp-Gly-NH_2) which mimics a *cis*-oriented peptide bond could discriminate between the two possible conformations known from previous studies.[23] The incorporation of a tetrazole moiety with D stereochemistry on the α-carbon of Ala into the active core could provide a lead candidate for an antagonist of the insect kinins. An α-methylene tetrazole-based dipeptidomimetic has been prepared and incorporated into peptides (Figure 2) which were assayed against HIV protease.[24] Linear and cyclic tetrazole analogues of hymenistatin I did not affect lymphocyte proliferation in the lymphocyte proliferation test.[25]

2.2 Ψ[CH=CH], Ψ[Z-CF=CH], Ψ[CH(OH)–CH_2], Ψ[CH(OH)–CH_2–NH], retro- and retro-inverso-Ψ[NHCH(CF_3)], Ψ[CH_2O], retro-Ψ[CONR], Ψ[CO-N(NPht)], Ψ[O–CO–N], Ψ[CO–NR–O], Ψ[CH_2NH], Ψ[CO–CH_2–cyclopropyl-NH], Ψ[PO_2R–N], Ψ[PO_2R], Ψ[NHCO]. – Regio- and stereoselective ring-opening of chiral 1,3-oxazolidin-2-one derivatives by organocopper reagents provided a new synthetic route to Ψ[(*E*)-CH=CH], Ψ[(*E*)-CMe=CH] and Ψ[(*E*)-CH=CMe] type alkene dipeptide isosteres (Figure 3).[26,27] Alkylation

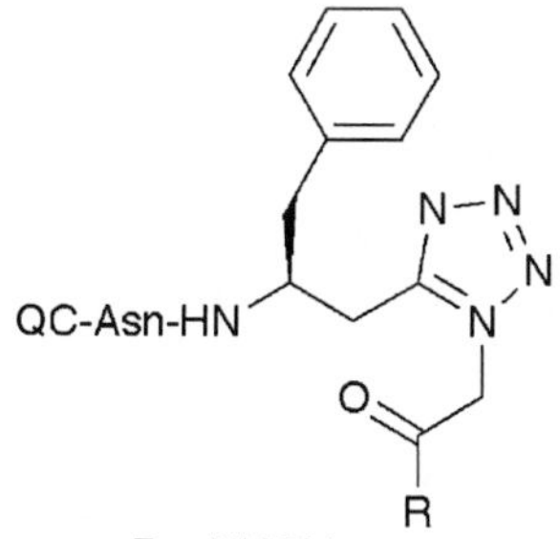

QC-Asn-HN

R = $NHCMe_3$
R = Ile-OMe
R = Ile-Val-OMe

Figure 2

of γ-mesyloxy-β-methyl-α,β-unsaturated esters with organocyanocuprates afforded the Ψ[(*E*)-CH=CMe]- or Ψ[(*Z*)-CH=CMe]-isomer (Figure 4).[28] Application of the two latter isosteres to conformationally restricted cyclic RGD analogs has also been published by the same group.[29] Anti-S(N)2′ reactions shown in Figure 3 were used for regio- and stereoselective synthesis of (*E*)-alkene trans-Xaa-Pro dipeptide mimetics.[30] The turn region of a specific CXCR4 inhibitor, T140, could be replaced by an (*E*)-alkene dipeptide isostere with the maintenance of strong anti-HIV activity.[31] Enantiopure (*Z*)-ethylenic pseudopeptides were synthesized using an intramolecular alkene metathesis (ring-closing metathesis) on diethylenic amides followed by hydrolytic ring opening of the dihydropyridone (Figure 5).[32] A completely stereoselective synthesis of functionalized (*Z*)-fluoroolefins (**7**) has been reported.[33] The key step of the preparation is a Cu(I)-mediated alkyl-transfer reaction of trialkyl-aluminum with (*E*)-4,4-difluoro-5-hydroxyallylic alcohol derivatives which provides 2-alkylated 4-fluorohomoallylic alcohol derivatives in a highly stereoselective manner.

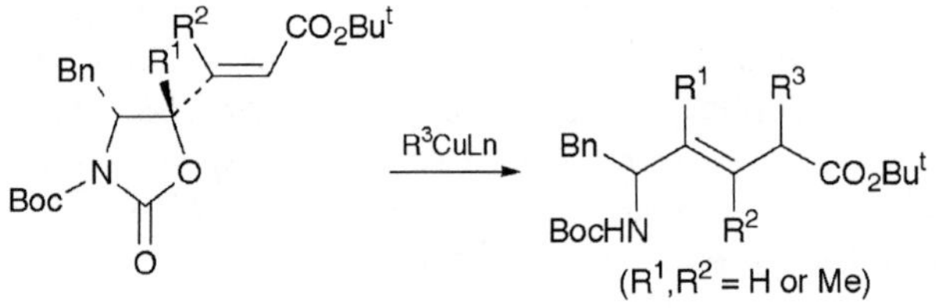

Bn R^2 R^1 CO_2Bu^t Boc N O O

R^3CuLn

Bn R^1 R^3 CO_2Bu^t BocHN R^2

(R^1,R^2 = H or Me)

Figure 3

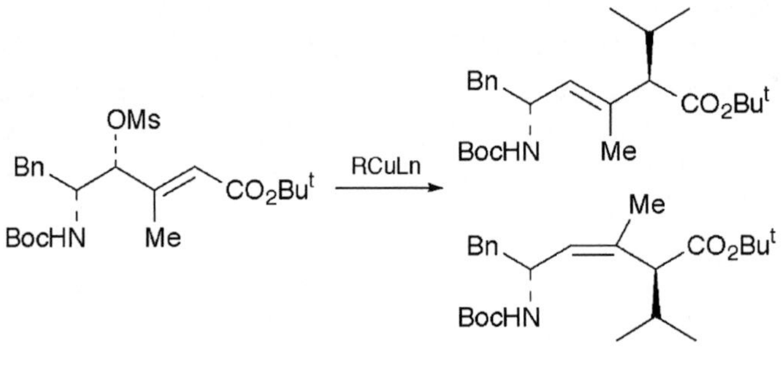

OMs Bn CO_2Bu^t BocHN Me

RCuLn

Bn CO_2Bu^t BocHN Me

Me Bn CO_2Bu^t BocHN

Figure 4

Aux N R^1 O R^2

RCM

Aux N R^1 O R^2

H_2N R^1 HO_2C R^2

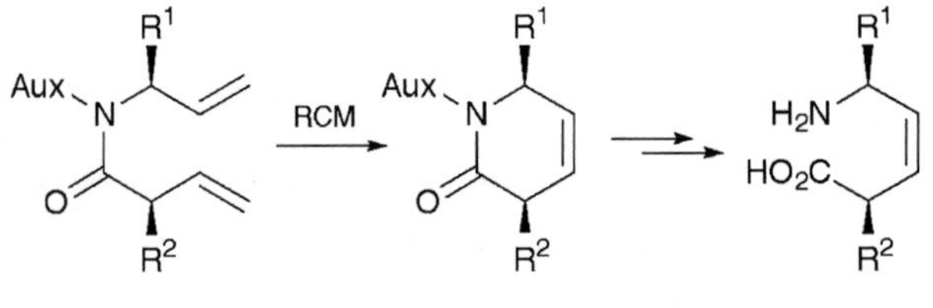

Figure 5

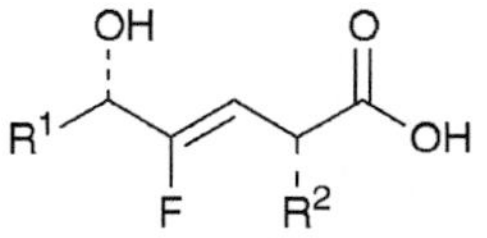

(7) 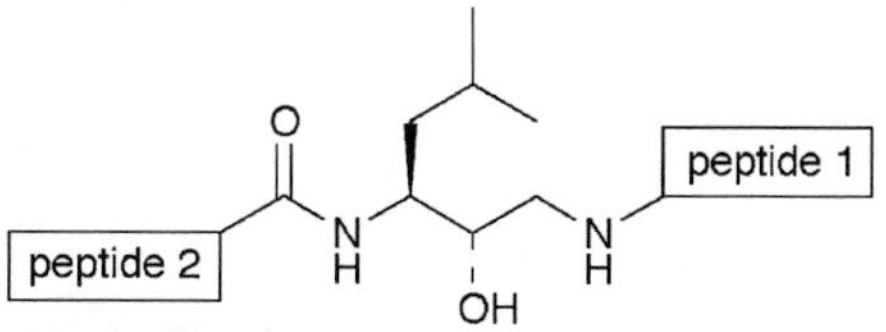 (8)

A stereoselective synthesis of hydroxyethylene dipeptide isosteres has been described and applied for a series of compounds, including the diamino alcohol core of a potent HIV-protease inhibitor, ritonavir and its C-2 epimer.[34] Olefination of α-amino acid-derived phosphonates, stereoselective reduction of the resulting enones, and epoxydation of the formed allylic alcohols led to 1-amino-2-hydroxy-3,4-epoxybutanes, key intermediates of the synthesis. The aza-Payne rearrangement and *O, N*-acyl transfer reactions were utilized in a novel stereoselective synthesis of hydroxyethylamine dipeptide isostere containing peptidomimetics (**8**).[35] As a continuation of last-year's project, a new series of retro- and retro-inverso-peptidyl hydroxamates have been prepared with good to excellent purity.[36] X-ray diffraction of one of the peptides showed an interesting turn-like conformation. Three retro-peptidyl hydroxamates (**9–11**) reduced MMP-9 gelatinolytic activity.[37] The introduction of pseudopeptides H-TyrΨ[CH_2O]Ile-OH and H-TyrΨ[CH_2O]Phe-OH into positions 2 and 3 of oxytocin and vasopressin caused total absence of all biological activities.[38]

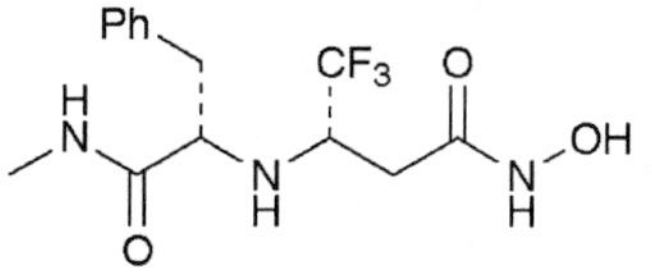

(9) 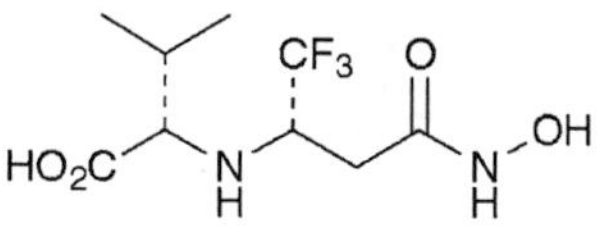 (10)

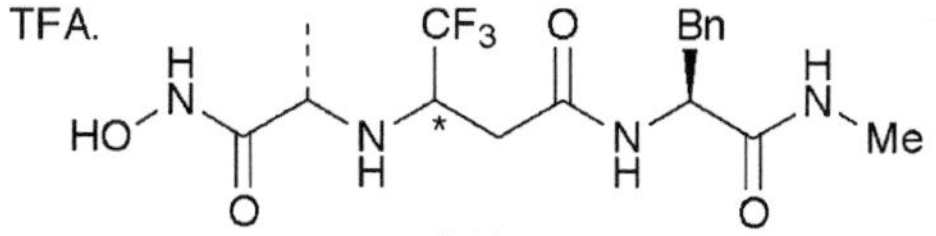

(11)

These results and NMR studies confirmed the importance of the H-bond between the backbone carbonyl of the Tyr(2) and NH proton of the Asn(5) residues, stabilizing the β-turn for their biological activity. Stereoselective synthesis of XaaΨ[CH_2O]Ala/Gly pseudopeptides and insertion of one of them in the peptide drug desmopressin have been published.[39] Design, synthesis and *in vitro* activities of oligopeptoid amide and ester analogues (Figure 6) which bind TAR RNA with high affinities were reported.[40] *N*-Phthalimidoamide pseudopeptides could be easily obtained by condensing α-hydroxyesters via the Mitsunobu protocol onto amino acid phthaloyl hydrazide derivatives (Scheme 1).[41] The glutathione-ethacrynic acid (GS-EA) conjugate is a potent inhibitor of most glutathione-*S*-transferase (GST) isoenzymes. A series of novel peptidomimetic GS-EA analogues (Figure 7) that are stabilized against peptidase-mediated breakdown were synthesized by replacing the peptide

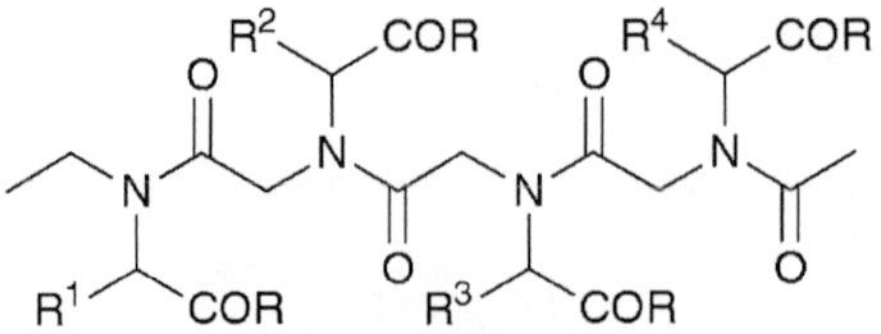

Peptoid ester R = OMe
Peptoid amide R = NH_2

Figure 6

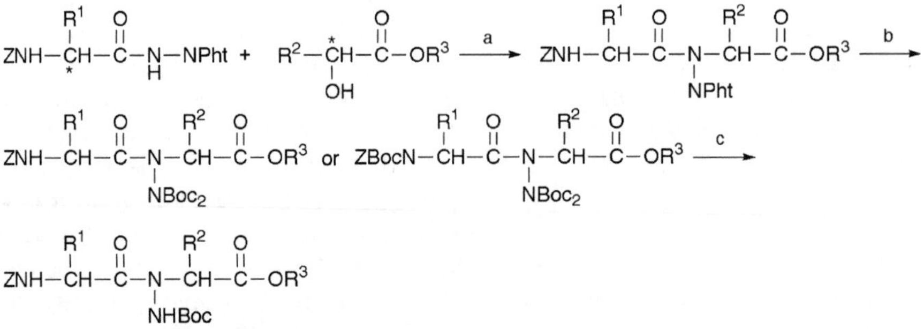

Reagents and conditions: a, DBAD, PPh_3, THF;
b, $MeNH_2$, THF, rt, 3h, evaporation, $(Boc)_2O$, DMAP cat., THF, rt;
c, $Mg(ClO_4)_2$, cat., CH_3CN

Scheme 1

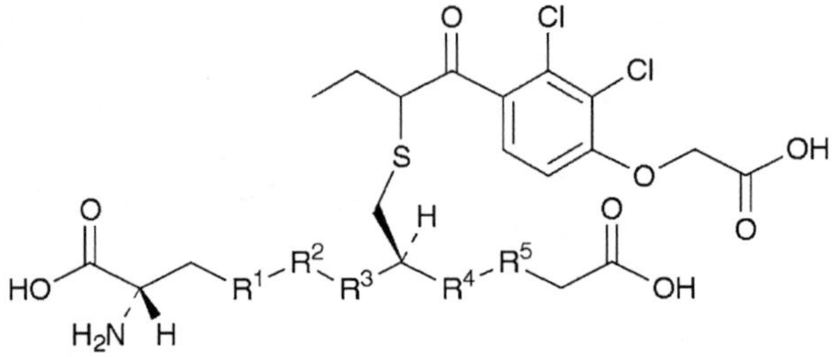

$R^1 = CH_2$, $R^2 = C{=}O$, $R^3 = NH$, $R^4 = C{=}O$, $R^5 = N{-}Me$
$R^1 = CH_2$, $R^2 = C{=}O$, $R^3 = N{-}Me$, $R^4 = C{=}O$, $R^5 = NH$
$R^1 = CH_2$, $R^2 = C{=}O$, $R^3 = NH$, $R^4 = CH_2$, $R^5 = CH_2$
$R^1 = CH_2$, $R^2 = CH_2$, $R^3 = NH$, $R^4 = C{=}O$, $R^5 = NH$
$R^1 = O$, $R^2 = C{=}O$, $R^3 = NH$, $R^4 = C{=}O$, $R^5 = NH$

Figure 7

bonds in glutathione by isosteres such as reduced amide, N-methylamide, urethane, and methylene linkages.[42] In addition to their improved stability towards γ-glutamyl-transpeptidase, all novel analogues were shown to inhibit rat liver cytosolic GSTs. Since the GS-EA conjugate is a good substrate of multidrug resistance protein 1 (MRP1), GS-EA derivatives are expected to be good inhibitors of MRP1. An effective competitive MRP1 inhibitor has been obtained by replacing the γ-glutamyl-cysteine peptide bond by a urethane

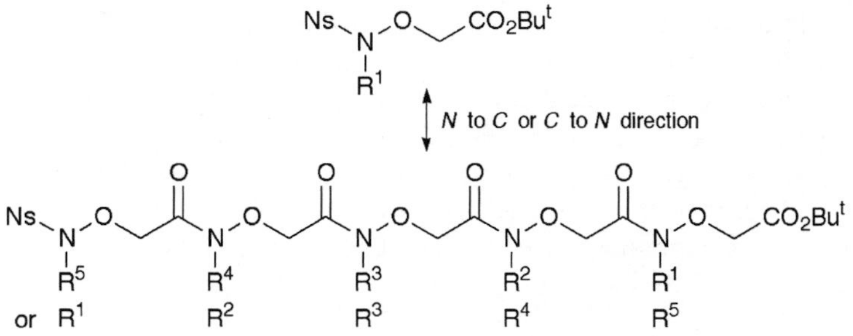

Figure 8

isostere.[43] Synthesis and assembly of monomers in both the C to N and N to C directions resulted in aminooxy peptoids (Figure 8) have been described.[44] Solid-phase synthesis of 1,4,5-substituted-2-oxopiperazins (**12**), cyclic pseudo-dipeptides containing a reduced peptide bond was published.[45] The strategy is based on reductive alkylation of resin-bound amino acids using N-protected α-amino aldehydes, followed by acylation with α-chloroacetyl chloride and intramolecular cyclization. Xenin 6 is a shorter fragment of a 25-amino acid peptide which interacts with the neurotensin receptor subtype 1 of intestinal muscles of the guinea pig. Replacement of the C-terminal peptide bond by a reduced peptide bond led to the pseudopeptide K-Ψ(CH_2NH)-R-P-W-I-L which showed increased binding affinity to isolated jejunal and colonic muscle membranes.[46] Compounds (**13**) and (**14**), cyclopropane-derived peptide mimics of two known inhibitors of matrix metalloproteinases and their flexible analogues (**15, 16**) were prepared.[47] Pseudopeptides (**13**) and (**15**) were found to be weak competitive inhibitors of a series of MMPs, suggesting that in compound (**13**) the loss of hydrogen bond could not be compensated with any favorable conformational constraints.

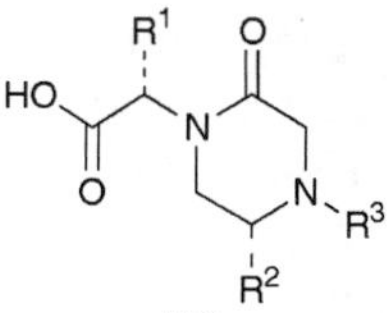

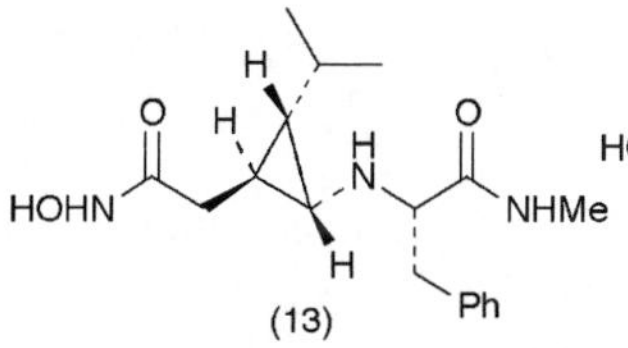

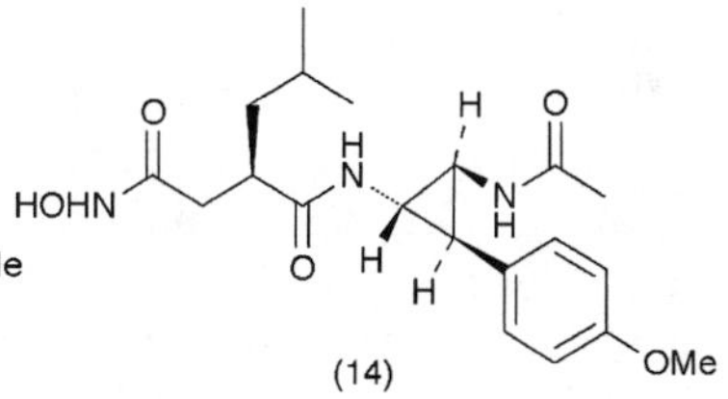

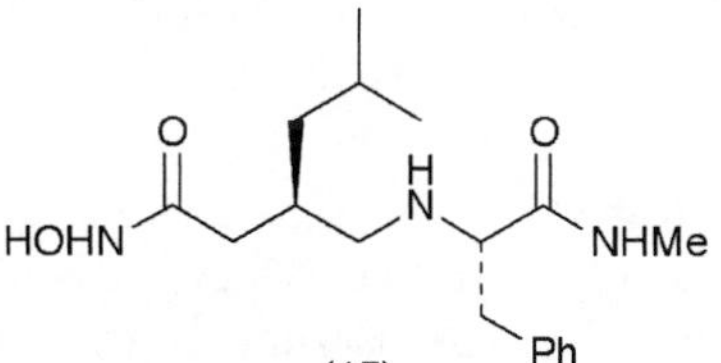

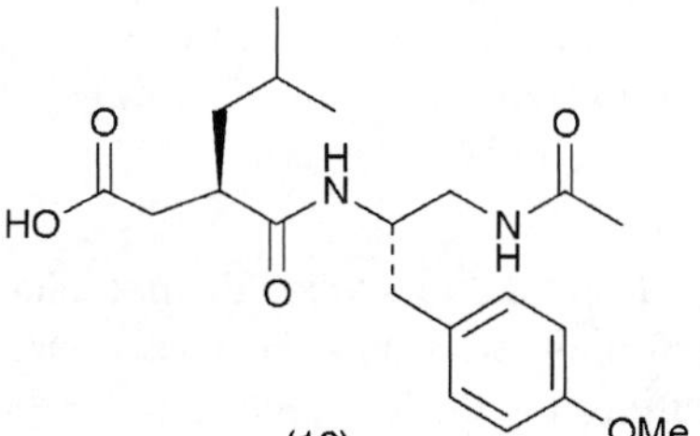

However, compounds (**14**) and (**16**), the retro-amide bond containing mimetics were modest competitive inhibitors of some MMPs with the constrained pseudopeptide more potent than the flexible analogue. GlyΨ[PO_2R–N]Pro-containing pseudopeptides as novel inhibitors of the human cyclophilin, hCyp-18 have been designed and synthesized.[48] Compound (**17**) bound to hCyp-18 with a K_d of 20 μM and inhibited selectively its peptidyl-prolyl cis-trans isomerase activity ($IC_{50} = 15$ μM). Deprotection of the phosphonamidate moiety (**18**) resulted in a complete lack of inhibition. A simple one-pot preparation of phosphinodipeptides (**19**) has been developed.[49]

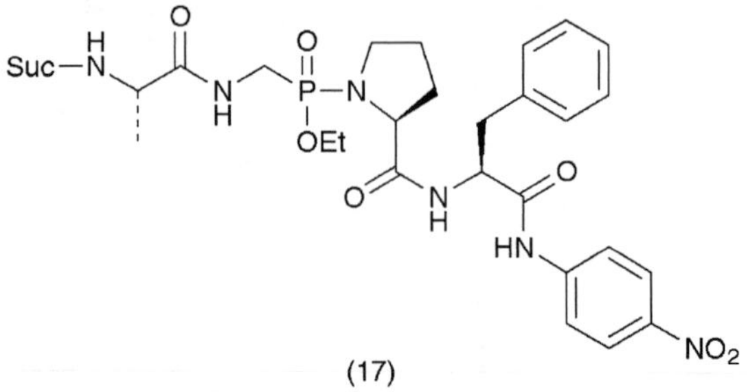

(17)

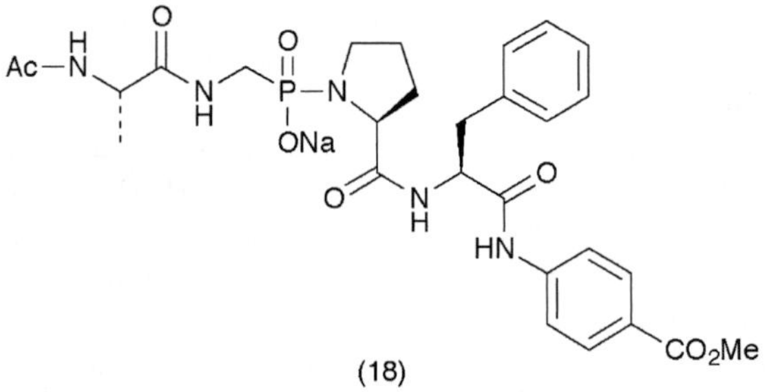

(18)

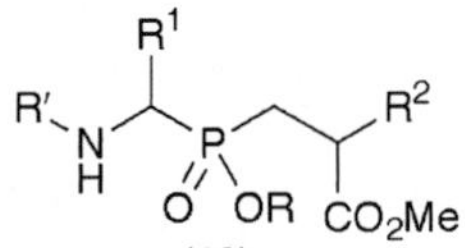

(19)

To show the value of these analogues as synthetic building blocks for combinatorial or parallel synthesis, selective or complete deprotection of the various protective groups has also been performed. Retro-inverso analogue of hemagglutinin (HA) 91–108 from influenza virus HA conjugated with ovalbumin was used to immunize mice intranasally.[50] Strong systemic and mucosal antibody responses have been produced, and the immunized mice became protected against a lethal dose of influenza virus demonstrating the possible application of this type of peptidomimetic in vaccine development. Replacing the peptide bonds in osteogenic growth peptide (OGP) with surrogates such as Ψ[CH_2Me], Ψ[CONMe], and Ψ[CH_2CH_2], showed OGP(10–14), to be the minimal sequence that retains the full osteogenic growth peptide activity.[51] The cyclic retro-inverso analogue of OGP(10–14) found to be at least as potent as the parent peptide.

2.3 Rigid Amino Acid, Peptide and Turn Mimetics. – Design of non-peptide peptidomimetics has been discussed in a 137-reference review.[52] Michael addition of nucleophilic glycine derivatives to α,β-unsaturated carboxylic acid derivatives results in the formation of χ-constrained five-carbon-atom amino acids used in *de novo* peptide design. A review[53] summarizes the

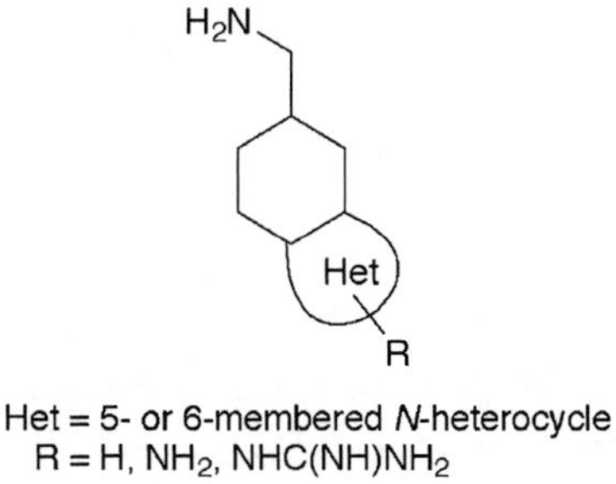

Het = 5- or 6-membered *N*-heterocycle
R = H, NH_2, NHC(NH)NH_2

Figure 9

literature methods and the author's own results on the syntheses of these compounds. An efficient protocol has been developed for the stereoselective synthesis of polyfunctionalized pipecolic acid derivatives (**20**) that can serve as rigid analogues of the *cis*- and *trans*-prolyl amide portion in peptides.[54]

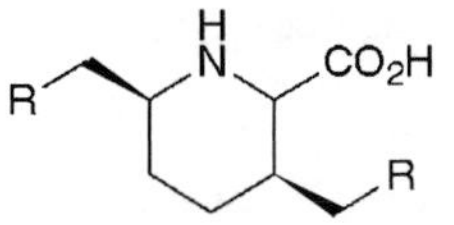

(20)

A convenient synthesis of the 1,2-dihydro-3(6*H*)-pyridinone unit, a cyclic amino acid surrogate designed to mimic the extended, β-sheet conformation of peptides has been carried out using solution and solid-phase methods.[55] A general synthesis of novel partially saturated, conformationally restricted arginine side chain mimetics containing a five- or six-membered *N*-heterocyclic ring (Figure 9) was described.[56] These compounds could possess reduced basicity in trypsin-like serine protease inhibitors. Sulphono(difluoromethyl)phenylalanine (F_2Smp, (**21**) was built into PTP1B inhibitors.[57] The most effective inhibitor (ELEF(F_2Smp)MDYE-NH_2) exhibited a K_i of 360 nM. Electrophilic fluorinations of lithiated bis-lactim ethers derived from readily available *cyclo*-[*L*-AP4-*D*-Val] led to the preparation of α-monofluorinated phosphonate mimetics of phosphoserine (**22, 23**) and phosphothreonine (**24, 25**).[58] The first general synthesis of racemic, orthogonally protected α-alkylamino glycines, having the general formula of (**26**), as new building blocks for the synthesis of peptidomimetics has been described.[59]

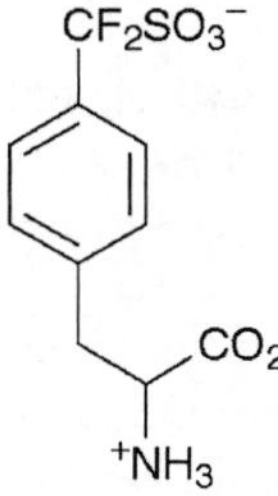

(21) 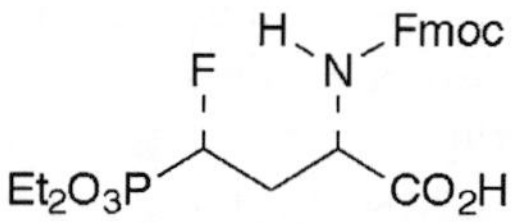 (22) 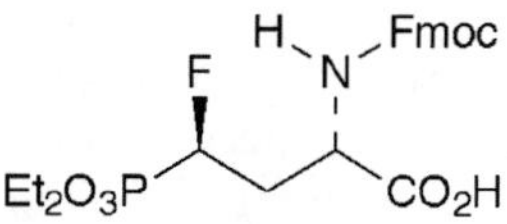 (23)

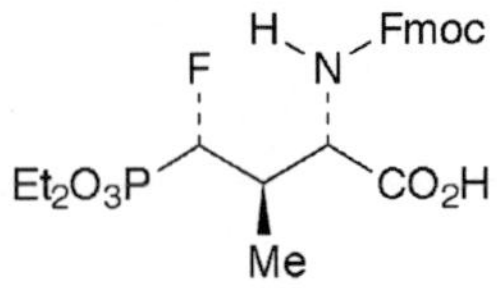

(24) 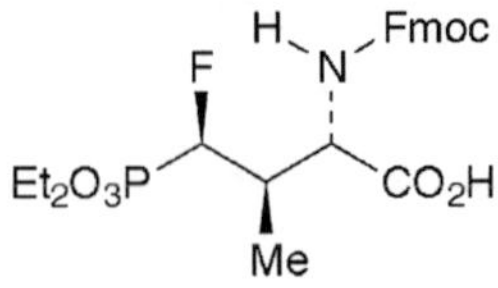 (25) 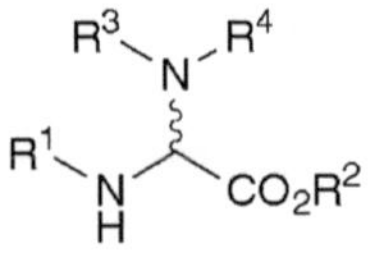 (26)

Compounds (**27**) and (**28**), conformationally constrained *L*- and *D*-lysines which belong to the α,α-disubstituted glycine family have been designed.[60] The asymmetric synthesis of their selectively protected derivatives (**29, 30**) was also described. Homochiral 4-azalysine building blocks were synthesized via two routes.[61] In the first strategy, naturally occurring *L*-serine has been used as starting material and as the synthesis changed the chirality, the *D*-isomer (**31**) was prepared. The alternative synthetic strategy started from *L*-asparagine and the chirality of the target molecule (**32**) remained the same as that of the starting material. Novel hydrophobic, bulky χ^2-constrained phenylalanine and naphthylalanine derivatives were designed and synthesized.[62]

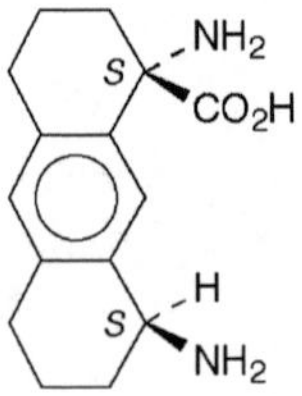

(27) 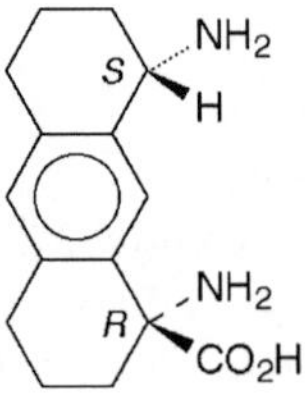 (28) 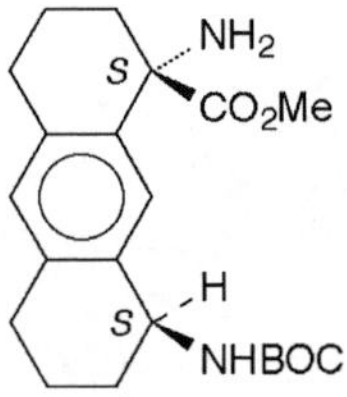 (29) 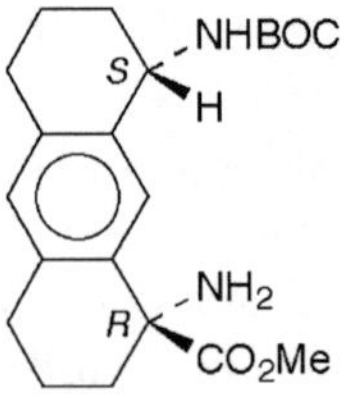 (30)

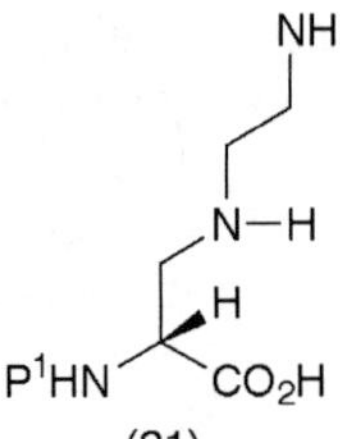

(31) 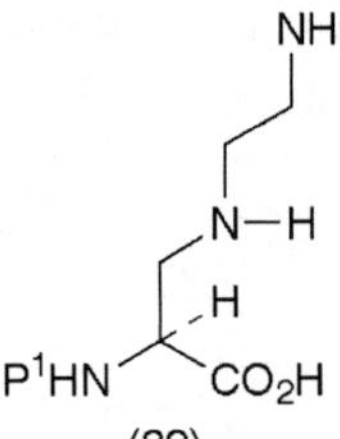 (32)

The amino acids were prepared through asymmetric hydrogenations using Burk's DuPHOS-based catalysts, followed by Suzuki cross couplings with boronic acid derivatives in high yields and with high enantioselectivity. Novel χ^2-constrained tryptophan analogues were also designed and synthesized by the same research group.[63] Two analogues of the super agonist MT-II were prepared incorporating the phenylalanine analogues (**33**) and (**34**) and the preliminary biological studies showed high selectivity for hMCR3, a human melanocortin receptor with good activity (IC_{50}: 71.0 and 28.0 nM, respectively).

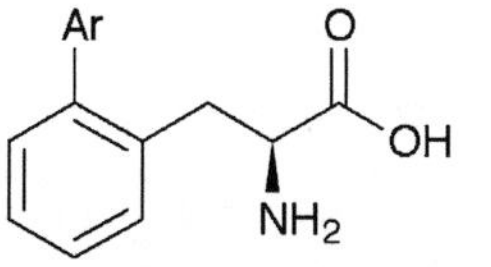

(33) Ar = Phenyl or Naphthyl 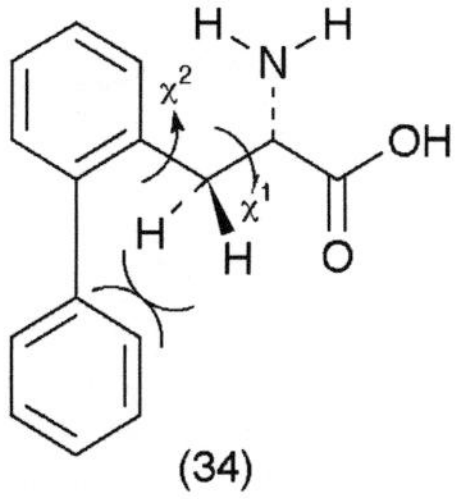 (34)

Highly functionalized phenylalanine derivatives were synthesized by cross-enyne metathesis reactions (Figure 10) followed by Diels-Alder reactions of the dienes with suitable dienophiles (Figure 11).[64] The same research group applied successfully the 'building block approach' for the synthesis of several 5- and 5,6-disubstituted indan-based α-amino acid derivatives via a [2 + 2 + 2] cycloaddition mediated by Wilkinson's and Vollhardt catalysts (Figure 12).[65] The first enantioselective synthesis of N^{α}-Fmoc protected (2*S*,3*R*)-3-phenylpipecolic acid (**35**) as a conformationally constrained phenylalanine analogue

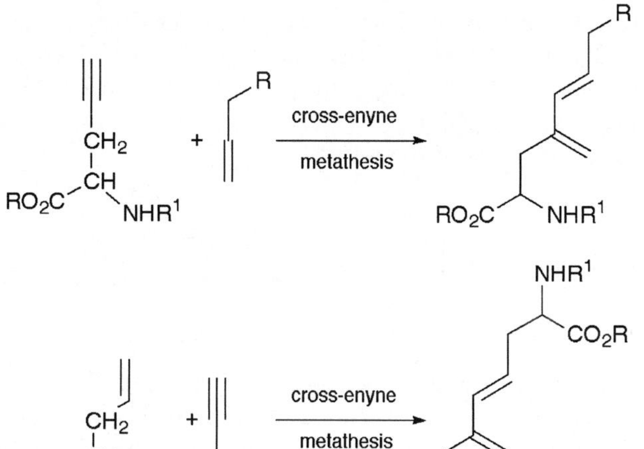

Figure 10

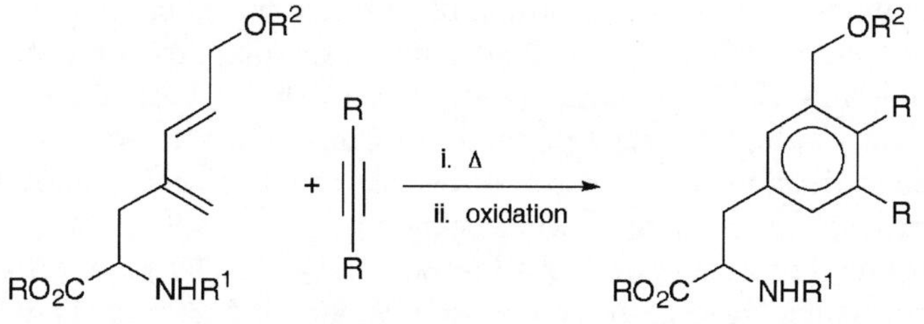

Figure 11

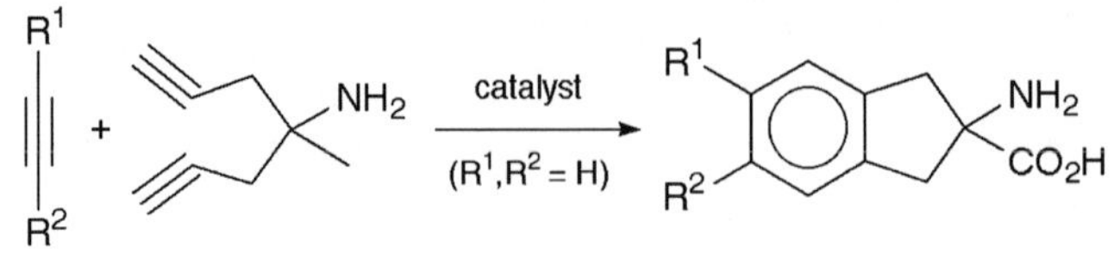

Figure 12

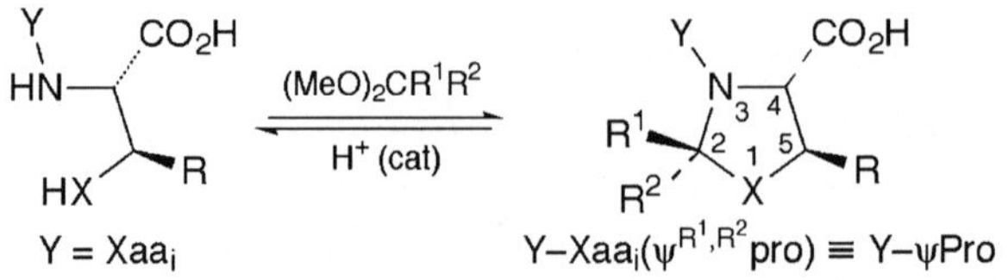

Figure 13

has been reported.[66] A racemic precursor of α-methyldiphenylalanine [(αMe)-dip] was prepared in high yield from easily available starting materials and subjected to HPLC resolution.[67] The enantiomers were then converted into the appropriate isomers of (αMe)dip. Computer assisted conformational analysis was performed on eight stereoisomers of 2-amino-3-phenylnorbornane-2-carboxylic acid (**36**), a constrained phenylalanine analogue.[68] Substituted prolines such as compounds (**37**) and (**38**) were prepared via selective reduction of 2-pyrrolidinones using Cp_2ZrHCl (Schwartz's reagent) followed by cyanation of the imines and hydrolysis of the nitriles.[69]

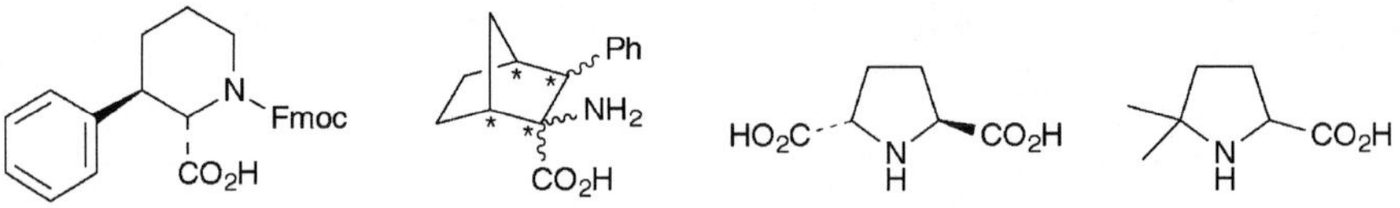

(35) (36) (37) (38)

The direct conversion of cysteine containing peptides into pseudo-proline derivatives by intraresidual *N*,*S*-acetalisation (Figure 13) resulted in conformationally constrained derivatives which represented a versatile tool in structure-activity studies of bioactive peptides.[70] The Alloc-protected derivative of (1-benzimidazolonyl)alanine, a new tryptophan mimetic was synthesized from commercially available N^{α}-Z-*L*-diaminopropionic acid as shown in Scheme 2.[71] A truncated p53 analogue showed nearly identical binding to HDM2 protein as the native sequence upon replacement of a Trp residue with this amino acid. A new hydrophobic tyrosine-mimetic, 3-[1-hydroxy-1,12-dicarba-*closo*-dodecaboran (12)-12-yl]alanine, was prepared and incorporated into biologically active peptides or boron-rich delivery tools for experimental Boron Rich Capture Therapy of cancer.[72] Constrained analogues of tyrosine were used to develop peptide substrates which could be readily phosphorylated by Syk protein tyrosine kinase, while being unaffected by Src family enzymes.[73] The tyrosine analogues were as follows: 3(*S*)- or 3(*R*)-7-hydroxy-1,2,3,4-etrahydroisoquinoline-3-carboxylic acid (**39, 40**) and 8-hydroxy-4(*S*)-amino-1,3,4,5-tetrahydro-3-oxo-2*H*-benzazepine-2-acetic acid (**41**). Novel classes of

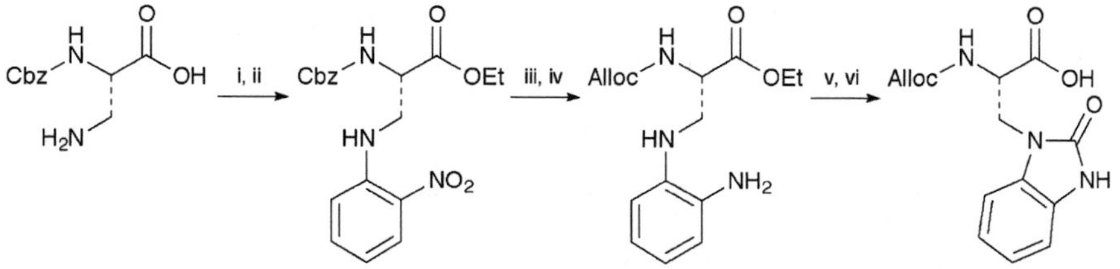

Reagents and conditions: i, EtOH, MsOH, reflux (81%); ii, 2-fluoro-1-nitrobenzene, TEA, MeCN, reflux (87%); iii, H_2, Pd/C, THF, MeOH; iv, Alloc-Cl, NHS, TEA, THF (85%, steps iii and iv); v, $COCl_2$, pyr, DCM, 0°C (70%); vi, NaOH, MeOH, H_2O, HCl (93%)

Scheme 2

phosphopeptide mimetics have been designed and synthesized.[74] *O*-Boranophosphopeptides and *O*-dithiophosphopeptides having the general formulae of (**42**) and (**43**), respectively, derivatives of tyrosine, serine and threonine, and an *O*-thiophosphopeptide (**44**) were successfully prepared on the solid-phase using either *H*-phosphonate chemistry or an *H*-phosphonothioate synthon.

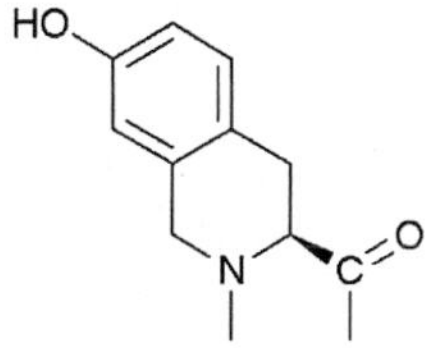

(39) L-Htc 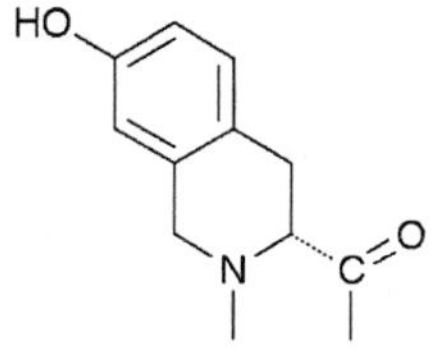 (40) D-Htc 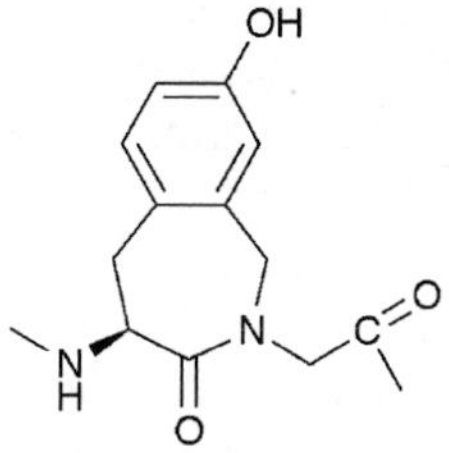 (41) Hba–Gly

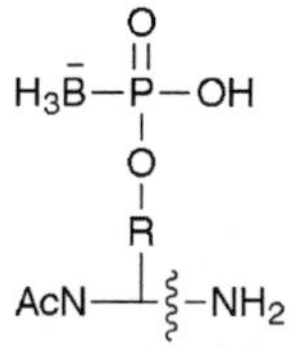

(42) R = tyrosine, R = threonine, R = serine (43) R = tyrosine, R = threonine, R = serine (44) 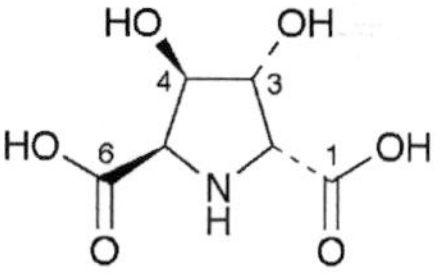 (45)

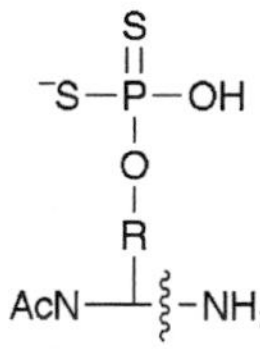

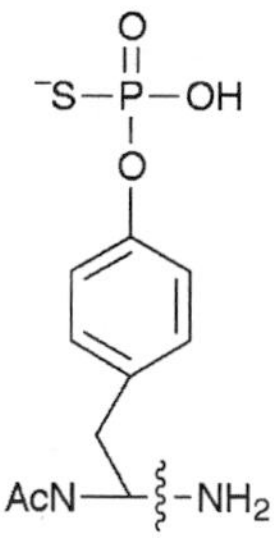

A 100-reference review highlights the development of sugar amino acids as a novel class of peptidomimetic building blocks and their applications in creating structurally diverse peptide-based molecules.[75] The use of sugar amino acid based scaffolds in combinatorial chemistry has also been published.[76] A rigid pyrrolidine-based scaffold, 2,5-dideoxy-2,5-imino-*D*-idaric acid (**45**) has been developed and synthesized.[77] The diacid moiety of (**45**) was bi-directionally elongated with identical peptide chains (Figure 14) and subjected to conformational analysis by NMR. Furanoid sugar amino acid (6-amino-2,5-anhydro-6-deoxy-*D*-gluconic acid) based peptidomimetics, (**46**) and its dimer (**47**)

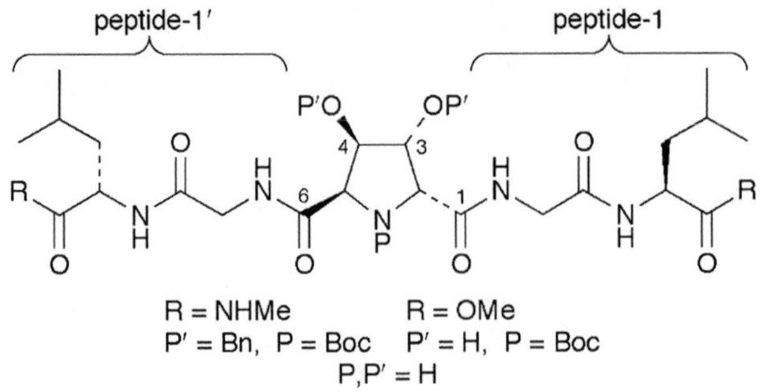

peptide-1′ peptide-1

R = NHMe R = OMe
P′ = Bn, P = Boc P′ = H, P = Boc
P,P′ = H

Figure 14

displayed a repeating β-turn-type secondary structure at lower concentrations in $CDCl_3$, but started to form aggregates that turned into excellent organogels as the concentrations were increased.[78]

Two classes of heterocyclic-based peptidomimetics have been reviewed.[79] In one of these classes the conformation of a peptide is conformed into a particular geometry, while chemical reactivity is masked and released in a controller manner in the other. *Cis-trans* isomerization and puckering of pseudoproline dipeptides such as *N*-acetyl-*N′*-methylamides of oxazolidine (**48**) and thiazolidine (**49**) were calculated using the *ab initio* and density functional computations with the reaction field theory.[80] An efficient solid-phase method for the synthesis of 4-imidazolidinone-peptide derivatives, having the general formula of (**50**), expected to have similar properties to pseudoprolines, was described.[81]

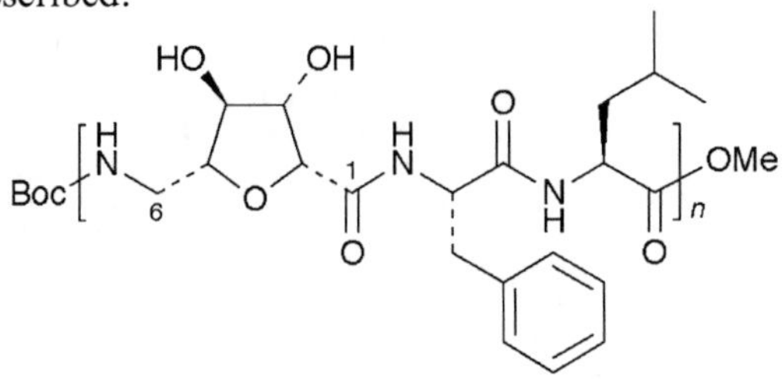

(46) $n = 1$
(47) $n = 2$

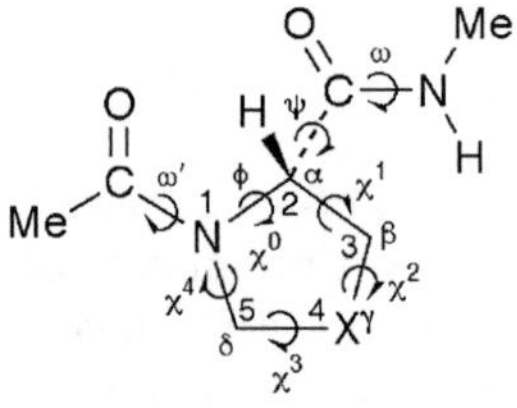

(48) X = O
(49) X = S

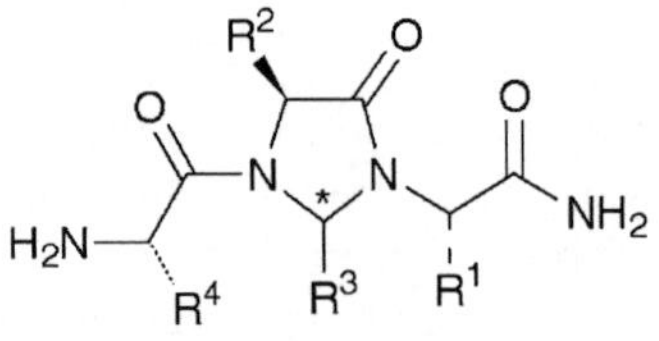

(50)

A study reported a refined synthesis of a homologous series of 1,2,5-hexahydro-3-one-1*H*-1,4-diazepine (Figure 15), a novel class of dipeptidomimetics.[82] This new synthetic procedure was utilized to expand the structural diversity by

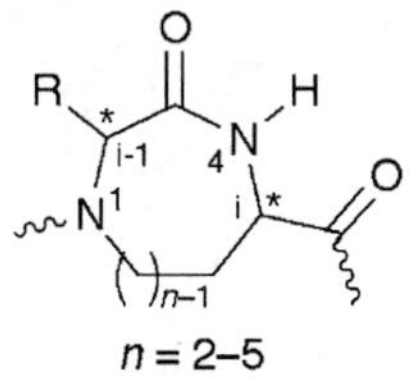

Figure 15

introducing functionalized side-chains at the *i*-1 position. Benzyl (6*S*)-1,3-dichloro-4-oxo-4,6,7,8-tetrahydro-pyrrolo[1,2-*a*]pyrazine-6-carboxylic ester (**51**, a new conformationally constrained peptidomimetic was prepared in seven steps from (*S*)-pyroglutamic acid.[83]

A constrained threonine-valine dipeptide mimetic (**52** has been synthesized in a stereoselective manner and incorporated into a tetrapeptide.[84] The resulting compound (**53**) possessed an order of magnitude less affinity for plasminogen activator inhibitor-1 than the lead peptide (TVAS). Intermolecular olefin cross-metathesis was used for the synthesis of a Pro-Gly dipeptide isostere (**54**).[85]

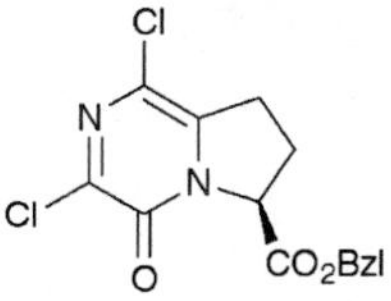

(51) 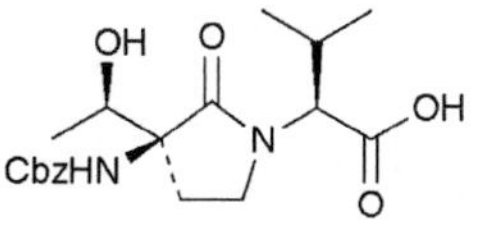 (52) 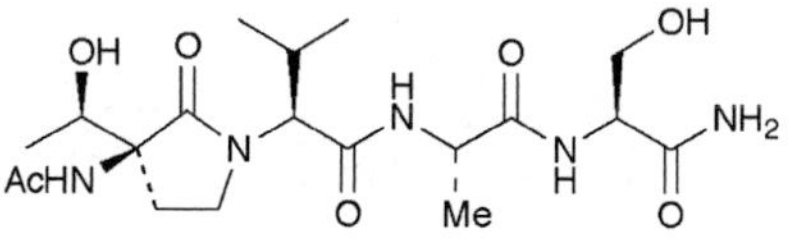 (53)

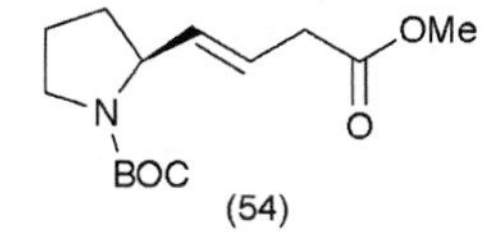

(54)

Hba-Gly (**55**), a conformationally constrained Tyr-Gly dipeptide mimetic has been synthesized and used for the replacement of the potentially phosphorylated tyrosine residue in the peptide Glu-Asp-Asp-Glu-Tyr-Glu-Glu-Val, a modified sequence of the phosphorylation domain of the haematopoietic lineage-specific protein HS1.[86] The obtained peptides possessed similar phosphorylation efficacy to the parent peptide. 5-(Aminomethyl)pyrrole-2-carboxylic acid, a new building block was developed and applied in peptidomimetic studies as a constrained surrogate of the Gly-ΔAla dipeptide.[87] Conformational analysis of the resulting peptides (**56** and **57**) was performed using various NMR techniques.

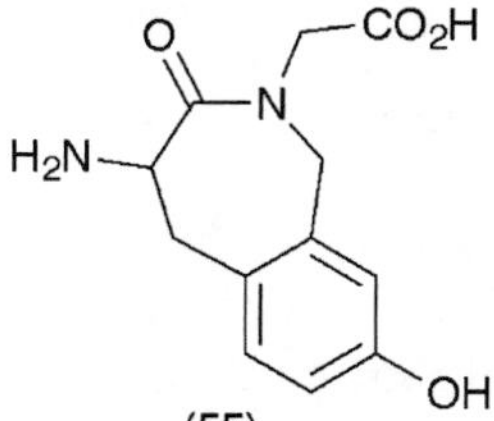

(55) 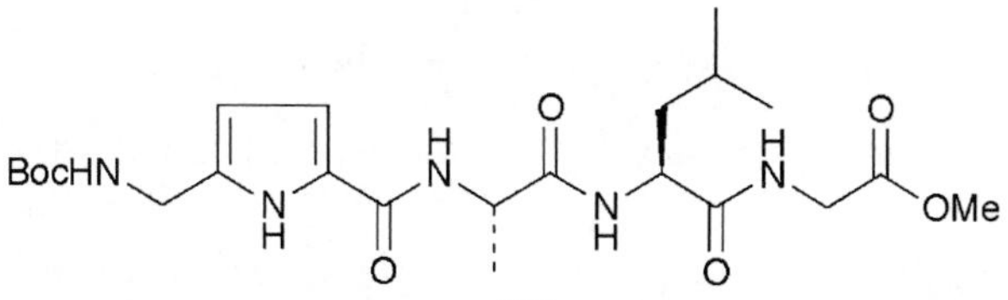 (56)

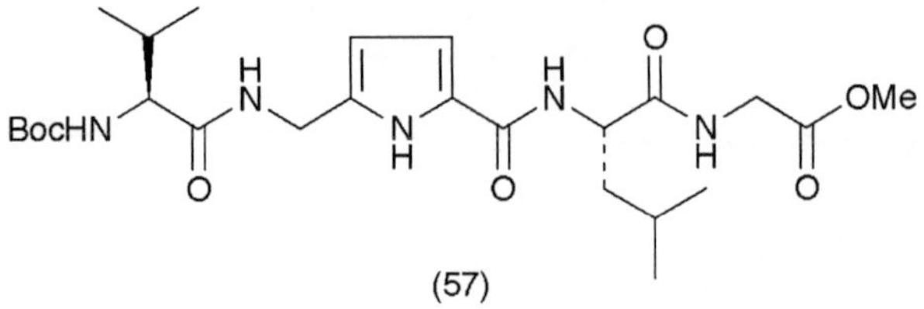

(57)

A short synthesis of the rigid dipeptide mimetic (3*S*,6*S*)-6-[(benzyloxy)carbonyl]amino-5-oxo-1,2,3,5,6,7-hexahydro-3-indolizine-carboxylic acid and of its (3*S*,6*R*) diastereomer was presented (Scheme 3).[88] Compounds with the 1-azabicyclo[4.3.0]nonan-2-one nucleus having methyl and isobutyl groups at C-6 (Figure 16) were also prepared.[89] A new synthetic route to enantiomerically pure diazabicycloalkane dipeptide mimetics has been reported.[90] A stereoselective synthesis of 1-(*tert*-butoxycarbonyl)-7-[1-(*tert*-butoxycarbonyl)-3-methylbutyl]-6-oxo-1,7-diazaspiro[4.5]decanes (**58, 59**) as constrained surrogates

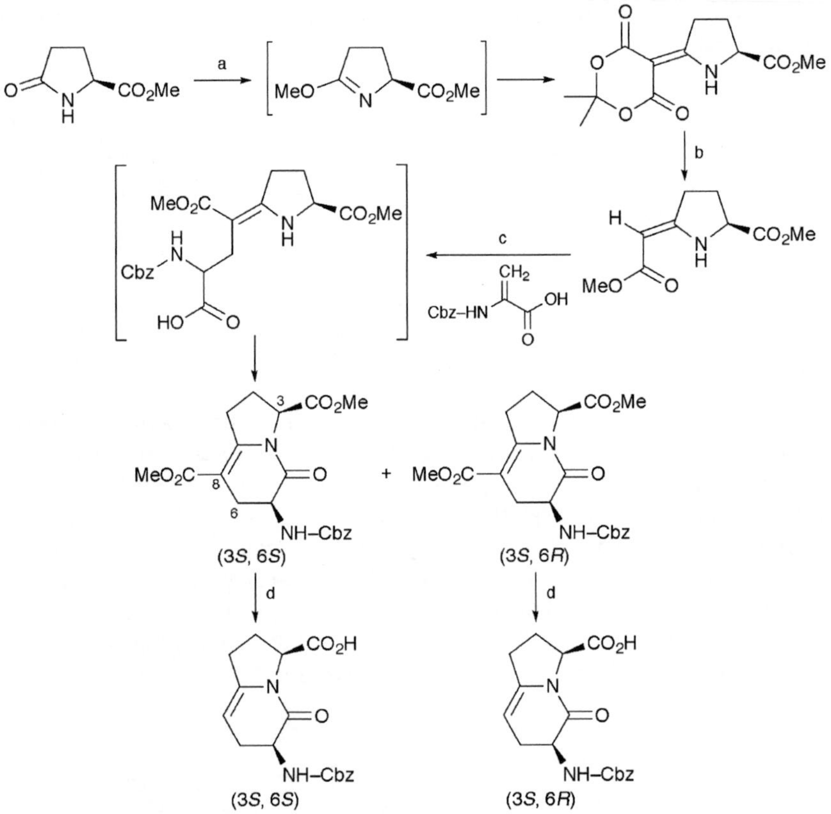

Reagents and conditions: (a) i, Me_2SO_4, NEt_3, 60 °C, 12 h; ii, Meldrum's acid, 24 h, rt, 83%;
(b) $BF_3{\cdot}Et_2O$, benzene, MeOH, reflux, 24 h, 60%;
(c) WSC, HOBt, CH_2Cl_2, rt, 48 h;
(d) i, LiOH (2N), dioxane/H_2O, rt, 4 h; ii, HCl (2N) 100%

Scheme 3

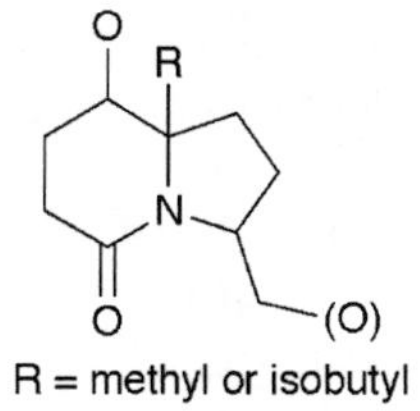

R = methyl or isobutyl

Figure 16

of the Pro-Leu and Gly-Leu dipeptides was described.[91] Incorporation of these 5,6-spirolactams into the Gly-Leu-Met-NH_2 C-terminal region of tachykinins led to tripeptide analogues which showed γ-turn/distorted type II β-turn structure by NMR experiments. A paper described the preparation of the optically active, bicyclic 2-pyridone dipeptide mimetic [(3*S*)-6-(benzyloxycarbonylamino)-5-oxo-1,2,3,5-tetrahydroindolizine-3-carboxylic acid] (**60**) in eleven steps from commercially available 2-hydroxy-6-methylnicotinonitrile.[92]

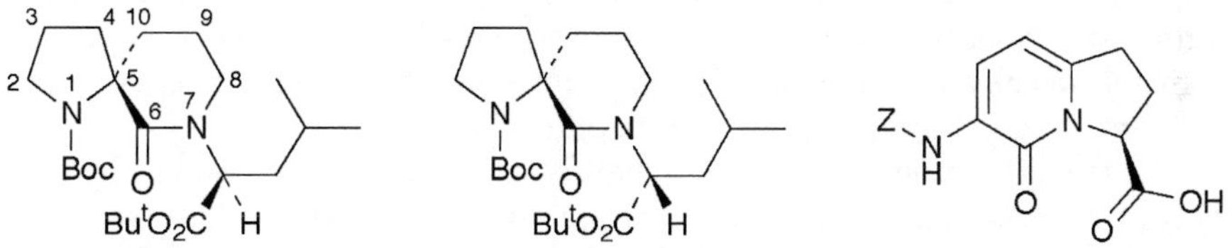

(58) (59) (60)

Diastereoselective synthesis of 6-oxoperhydropyridazine-3-carboxylic acid derivatives (**61** and **62**) as new constrained dipeptide mimics was reported.[93] A tartaric acid derivative was the starting material of the preparation of a new enantiopure bicyclic γ/δ-amino acid (BTKa, **63**) and an Fmoc-protected dipeptide isostere (BTAas, **64**).[94,95] Different synthetic strategies for the preparation of a range of 3-(phenylsulfonimidoyl)propanoate derivatives having the general formula of (**65**) were discussed.[96]

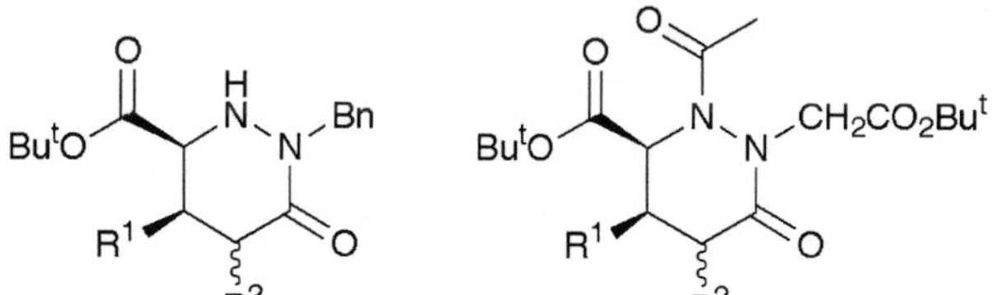

(61) (62)

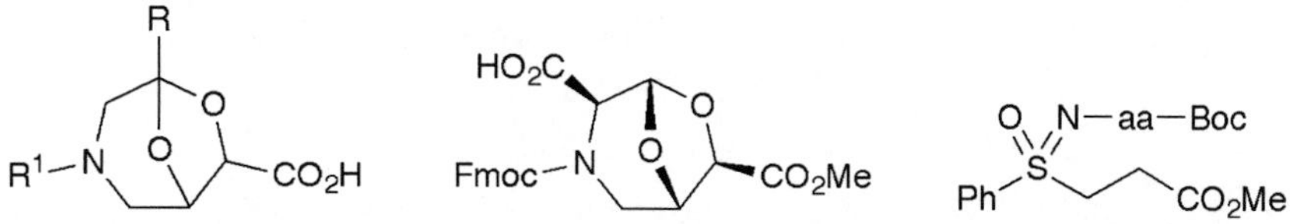

(63) (64) (65)

NMR studies showed interesting conformational properties of the pseudodipeptides formed, giving evidence of intramolecular H-bonds in all cases, except for the proline derivative. P–N bond – containing phosphonamidate dipeptide analogues (Figure 17) and P–C bond – containing phosphinate dipeptide analogues (Figure 18) which act as new inhibitors of leucine aminopeptidase were designed and synthesized.[97] A facile, high yielding method for the preparation of chiral non-racemic 3-(dioxo-piperazine-2-yl)propionates (**66–68**) was presented.[98]

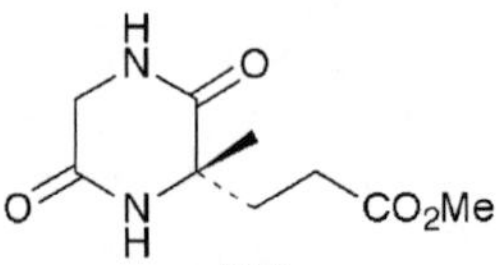

(66) 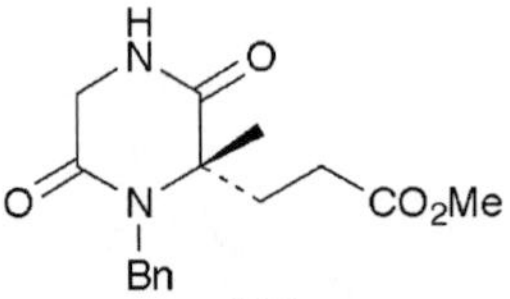 (67) 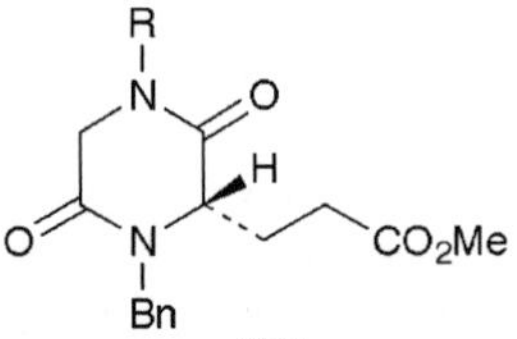 (68)

A study discussed the possibility of evaluating peptidomimetics for therapeutic applications by comparing virtual backbone ψ, ω and ϕ torsions in dipeptides with those of backbone modified pseudopeptides.[99] The energy surfaces of a stereochemically complete set of peptide analogues based on a *cis*-enediol unit (**69**) have been examined by different approaches.[100]

The reductive amination of cyanomethyleneamino pseudopeptides, by catalytic hydrogenation in the presence of amino acid derivatives, led to C-backbone branched peptides (**70**).[101] When the amino acid was a glycine derivative, these branched peptides could lactamize to give 2-oxopiperazines (**71**) as conformationally restricted pseudopeptides. Chiral 1,6,8-trioxohydropyrazino[1,2-*c*]pyrimidines (**72**) as novel highly functionalized scaffolds for peptidomimetics have been synthesized.[102] Novel building blocks for library generation of peptidomimetic compounds, with densily functionalized pyrrolidinone templates (**73** have been prepared by an intramolecular oxo-Diels-Alder reaction.[103] Several classes of peptidomimetics of the efflux pump inhibitor *D*-ornithine-*D*-homophenylalanine-3-aminoquinoline (MC-02,595)

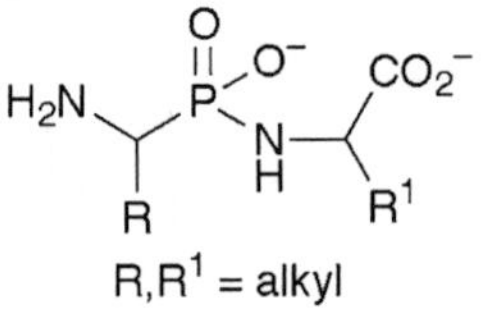

R,R[1] = alkyl

Figure 17

R,R[1] = alkyl

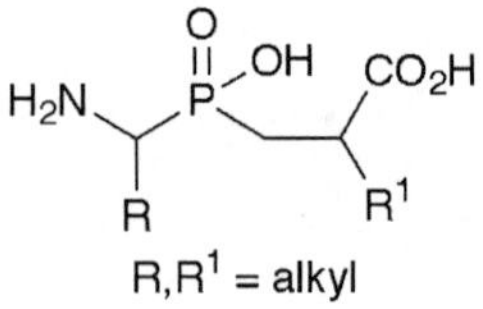

Figure 18

such as ethers, thioethers, tertiary amides, oxazole and other derivatives were prepared and evaluated for their ability to potentiate the activity of the fluoroquinolone levofloxacin (LVFX).[104] The results demonstrated that the peptide bond was not necessary for the potentiation activity of this class of compounds. A biosynthetic method was successfully used to incorporate (aminooxy)acetic acid into positions 10 and 27 of *Escherichia coli* dihydrofolate reductase opening the way for the preparation of proteins containing conformationally biased peptidomimetic motifs at given sites.[105] Silylated amino acids have been incorporated into peptides and then converted into *N*-acyliminium ions with the use of an anodic oxidation reaction allowing either the synthesis of constrained peptidomimetics or introduction of external nucleophiles into peptides.[106] Compound (**74**), an 'orthogonally' protected lanthionine was synthesized in an efficient stereo-specific route.[107]

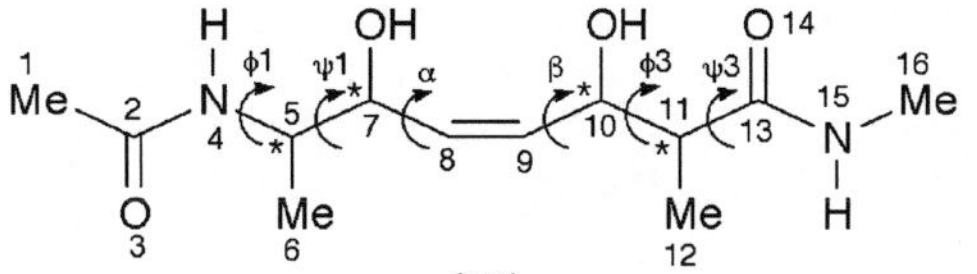

(69)

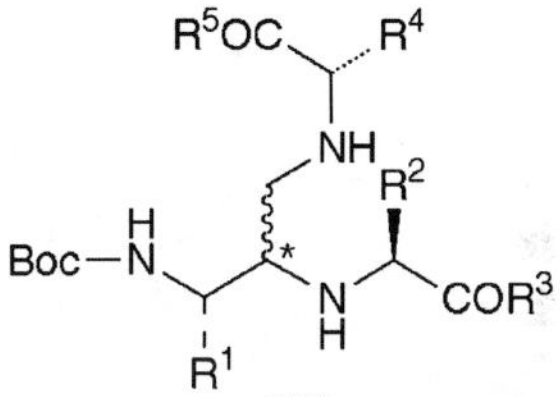

(70)

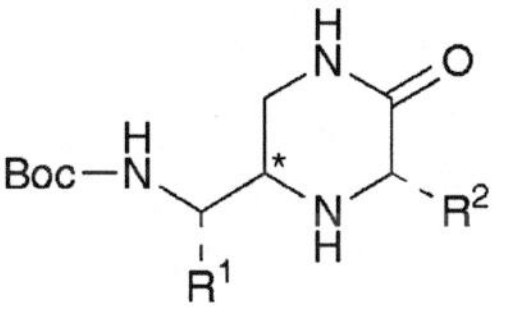

(71)

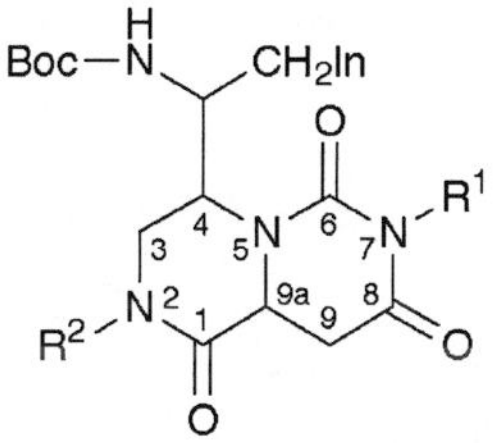

(72)

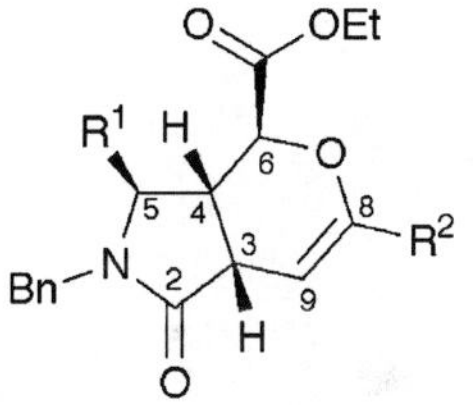

(73)

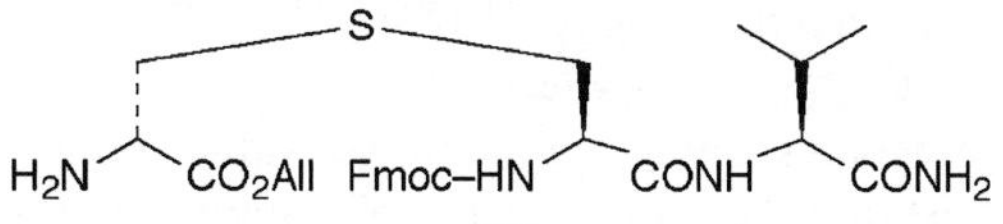

(74)

New pyrimidinyl peptidomimetics were synthesized and evaluated for their antimalarial activities against *Plasmodium falciparum*.[108] Compound (**75**), the most active compound of this class, showed comparable activity to that of chloroquine with an IC_{50} of 6–8 ng/ml. Conformationally restricted TRH (pGluHisProNH$_2$) analogues of varying steric size of constraint at the pyroglutamate region have been prepared.[109] Investigation of the analogue with a smaller ethylene bridge (**77**) and another one with a larger, more flexible propane bridge (**78**) supported the idea that the reduced activity of the ethane bridge - containing analogue (**76**) was not caused by the size of the bridge. A series of 4-amidinobenzylamine-based peptidomimetic inhibitors was designed, synthesized and tested as inhibitors of the plasminogen activator urokinase (uPA).[110] The most potent, benzylsulfonyl-*D*-Ser-Ala-4-amidinobenzylamide (**79**), inhibited uPA with a K_i of 7.7 nM but proved to be less selective than compound (**80**) with a Gly as P2 residue.

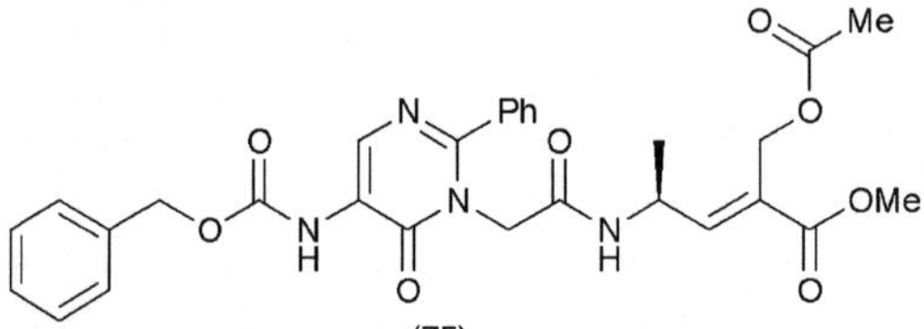

(75) 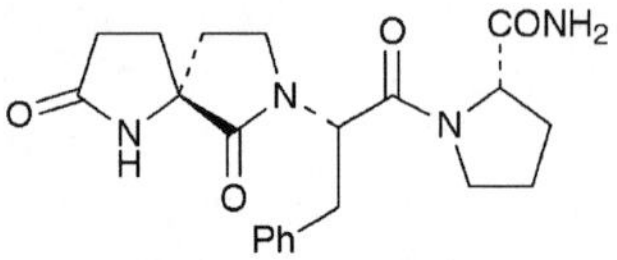 (76)

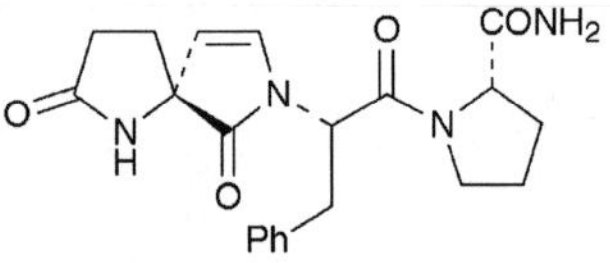

(77) 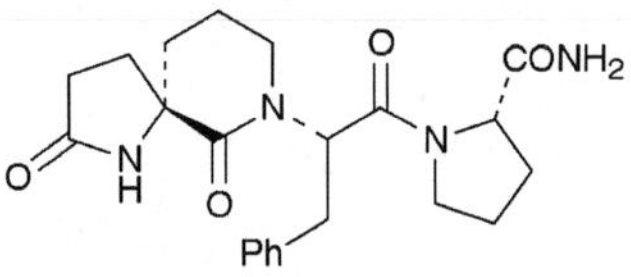 (78)

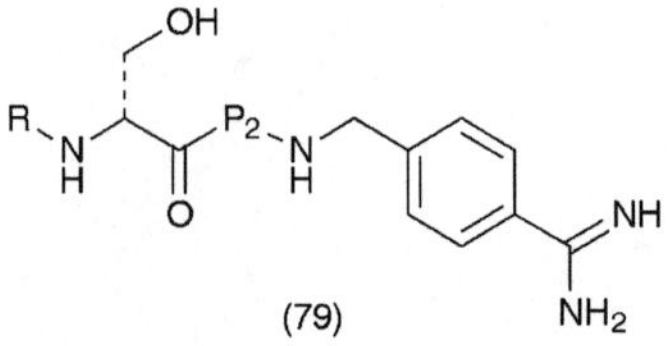

(79) 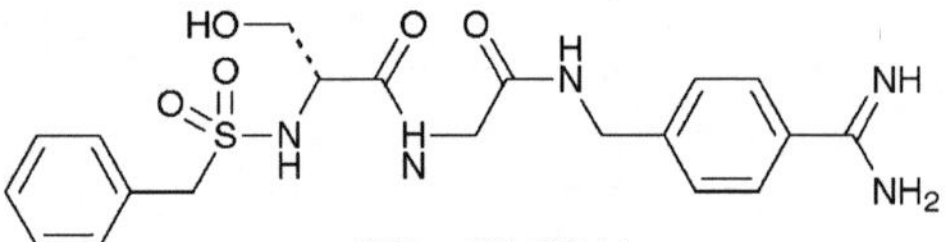 (80) x CF_3CO_2H

Starting from the tetrapeptide Ac-pYEEI-NHMe, the minimum peptide sequence of $p56^{lck}$ SH2 domain needed for optimal binding, conformationally restricted replacements for the two glutamate residues have been designed and synthesized.[111] Most of the analogues were of similar activity to the isoleucine-methylamide, but the 3(*R*)-aminoindan-1(*R*)-acetic acid replacement led to compound (**81**), the most potent isomer among these antagonists ($IC_{50} = 0.033$ μM). Structure-based design led to a bicyclic 2-pyridone-containing compound (**82**) which showed antiviral activity when tested against different human rhinovirus serotypes (EC_{50}: from 0.037 to 0.162 μM).[112] Calorimetric and X-ray studies were performed on constrained peptide mimetics of the pYEEI ligand for the Src Homology 2 domain of the Src kinase.[113] Structural and thermodynamic features of 1,2,3-trisubstituted cyclopropane-derived isosteres of pYEEI (**83, 84**) were compared to that of the flexible analogues.

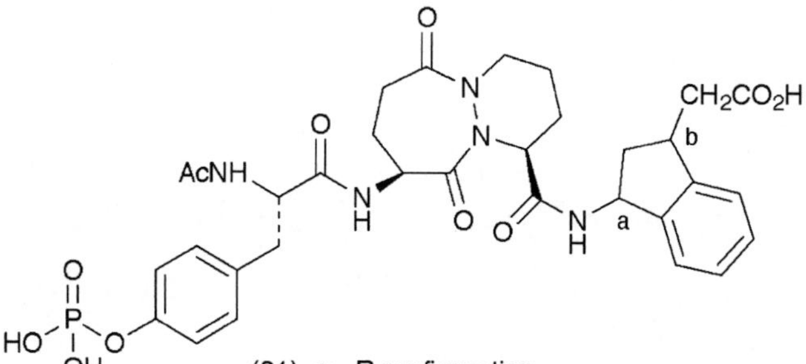

(81) a: R configuration
b: R configuration

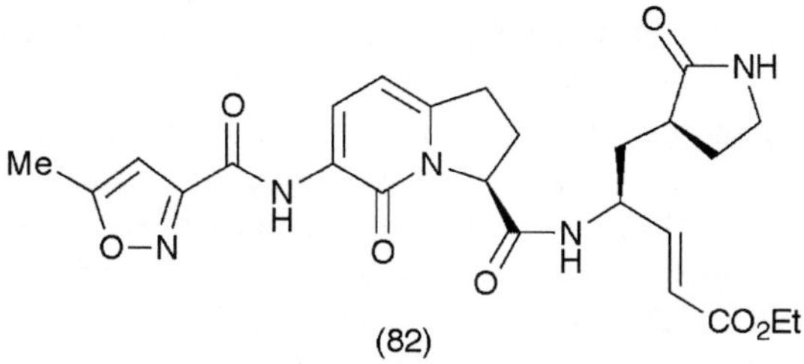

(82)

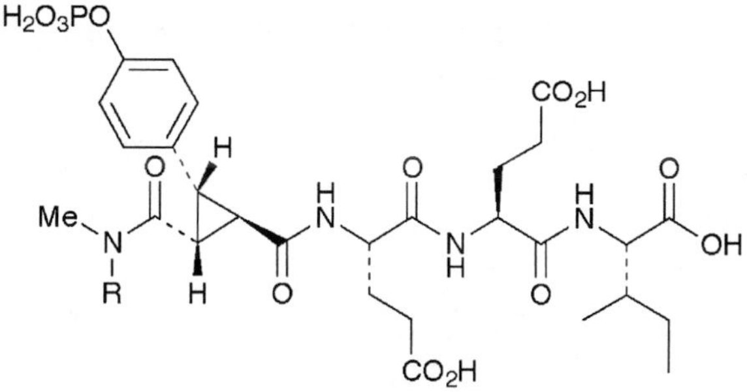

(83) R = H
(84) R = Me

The side-chain protected version of a negatively charged dibenzofuran-based β-turn mimetic (**85**) has been synthesized in 12 steps.[114] Incorporation of this mimetic into loop 1 of the PIN WW domain resulted in enhanced solubility compared to the uncharged analogue, without significantly perturbing the thermodynamic stability of the PIN WW domain. Compound (**86**), a conformationally constrained β-turn thiazolidine mimetic was prepared starting from *L*-phenylalanine and *L*-cysteine.[115] Cyclic ($\alpha_2\beta$)-tripeptides of general structure (**87**) have been designed and synthesized.[116]

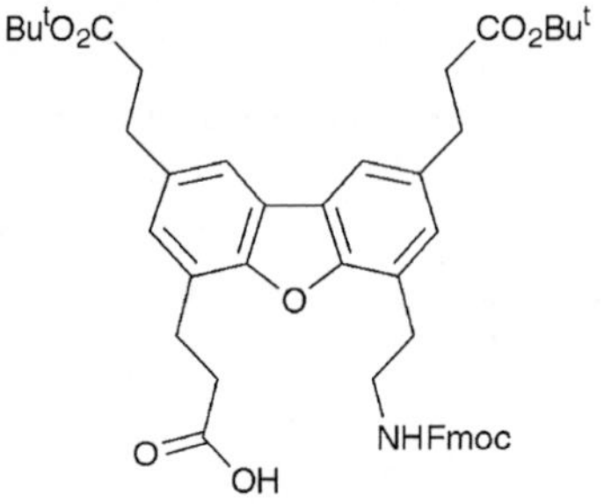

(85) 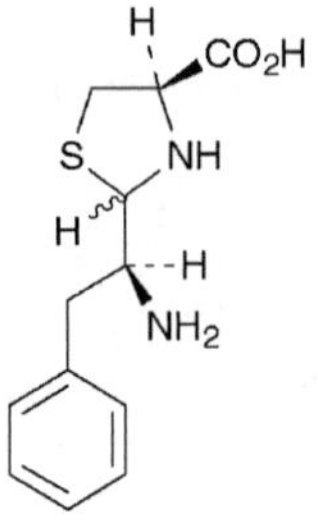 (86) 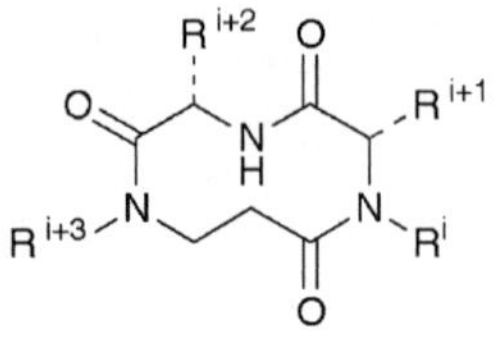 (87)

A stereoselective method has been developed for the synthesis of 7- and 8-substituted dipeptide β-turn mimetic azabicyclo[4.3.0]nonane amino acid esters.[117] A new turn mimic derived from prolyl-leucyl-glycine-amide containing a β-lactam in the turn area has been reported.[118] The synthesis started from a protected leucine-derived diazo ketone as seen in Scheme 4. Chemoselective Michael reactions on pyroglutamates led to spiro-bis-γ-lactams as β-turn mimetics.[119] The design and synthesis of novel, internal, putative β-turn mimetics were published.[120] Starting from *L*- and *D*-diaminopropionic acid, two epimeric series of mimetics (**88, 89**) were prepared. A novel strategy towards [6,5]-bicyclic dipeptides (**90, 91**) from δ,ε-unsaturated amino acid and cysteine has been developed.[121] Compounds (**92**) and (**93**), novel glycoamino acids as potent β-turn mimics, were prepared applying the glycoaminocyanation procedure using $Ti(OiPr)_4$ and TMSiCN.[122] β-Turn peptidomimetics, having the general formula of (**94**) with either *R*- or *S*-configuration at the sulfoxide function, were

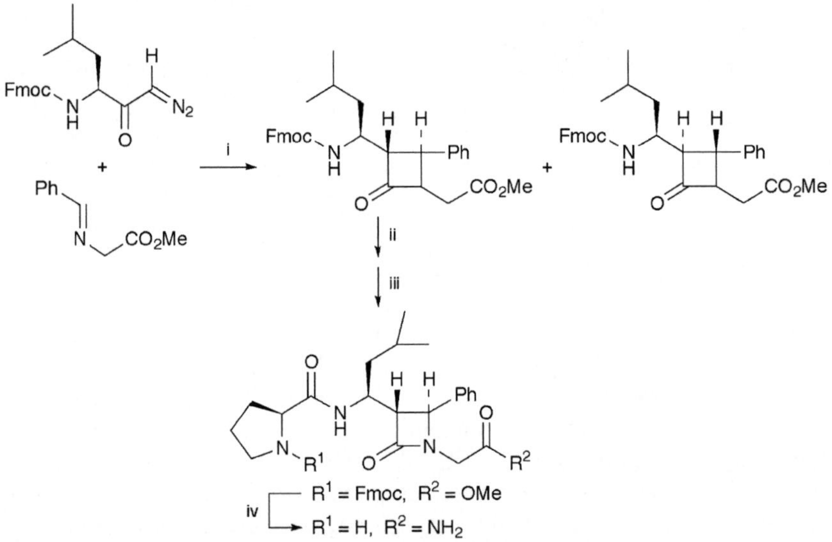

R^1 = Fmoc, R^2 = OMe
iv
R^1 = H, R^2 = NH_2

Reagents and conditions: i, Et_2O, *hν*, −30 °C, 1.5 h; ii, $HNEt_2$, THF, rt, 4.5 h; iii, R^1–Pro–OPfp (R^1 = Fmoc, Nsc), THF, rt, 2 h; iv, piperidine, $MeOH/NH_3$, 0 °C to rt, 16 h

Scheme 4

synthesized.[123] NMR and CD data for some of these compounds indicated that a transannular S*O* to *H*N hydrogen bond stabilizes β-turn conformations for the sulfones and one of the sulfoxide epimers.

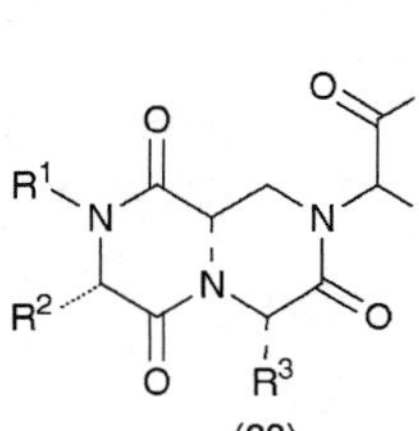

(88)

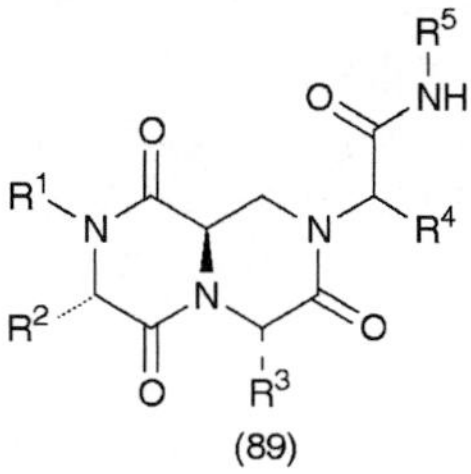

(89)

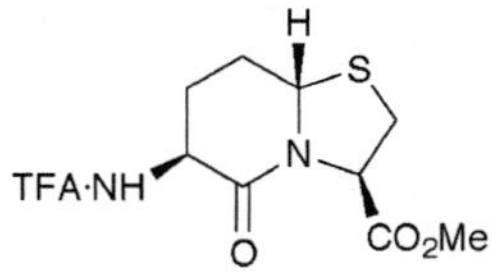

(90)

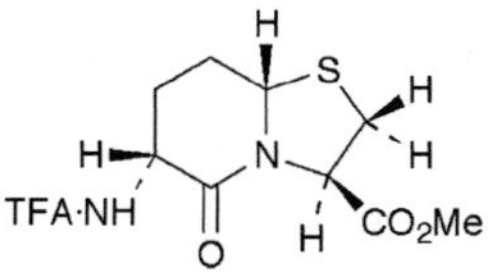

(91)

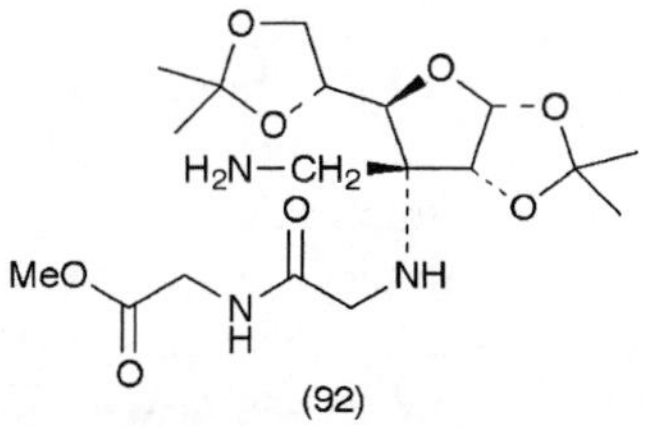

(92)

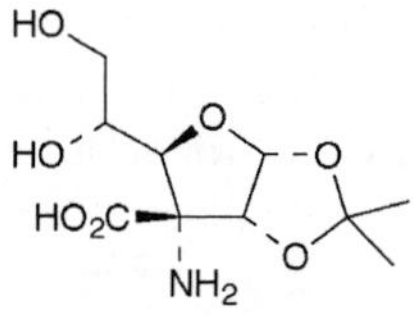

(93)

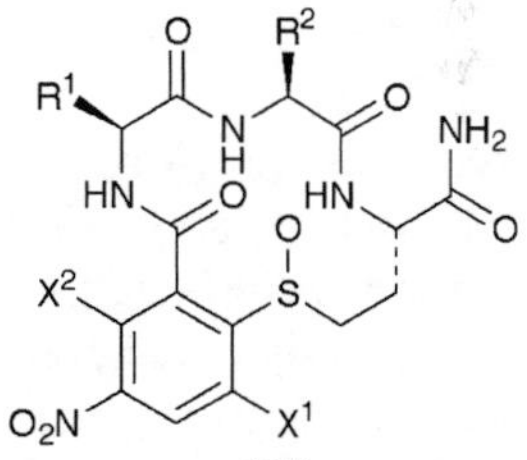

(94)

The peptide *N*-benzyloxycarbonyl-ΔLeu-*L*-Ala-*L*-Leu-OCH_3 was synthesized in solution phase and showed to adopt type II' β-turn conformation.[124] A study presented bicyclization of peptide acetals as a route to novel amino acid derived heterocycles and peptidomimetic scaffolds which show considerable promise as combinatorial motifs and β-turn mimetics.[125] A set of tripeptoids containing a residue of either *N*-methylglycine or *N*-isobutylglycine in position $i + 1/i + 2$ were synthesized and tested for intramolecular H-bonded β-turn formation.[126] Two new dipeptide isosteres, named 6-*endo*-BTL (**95**) and 6-*endo*-BtL (**96**), were inserted into a small peptide.[127] The presence of a reverse-turn conformation was observed in all the structures by NMR, IR, and molecular modeling techniques. A general methodology has been developed to prepare dipeptide reverse-turn mimetics based on both unsaturated (**97, 98**) and saturated (**99**) azabicyclo[4.3.0] alkane amino acid derivative.[128]

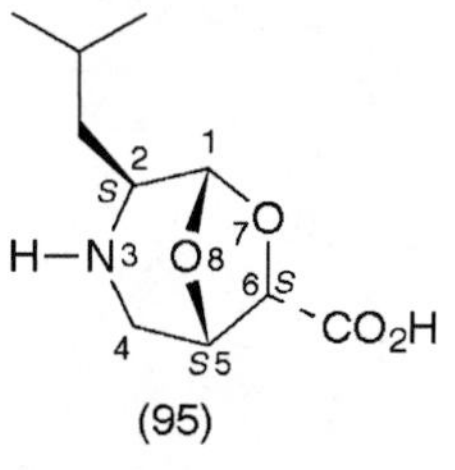

(95)

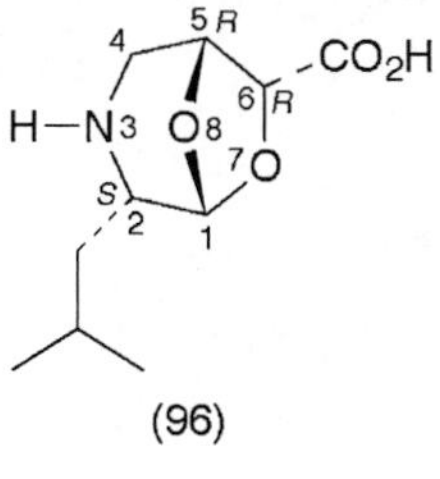

(96)

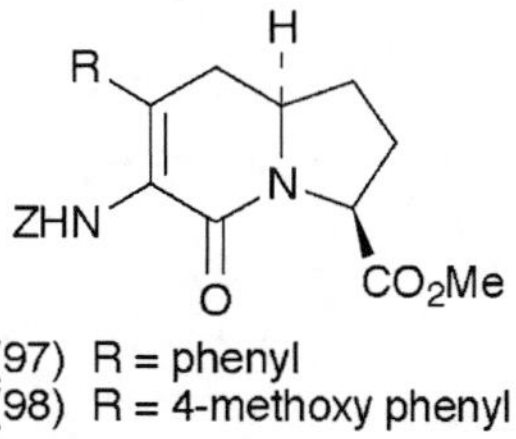

(97) R = phenyl
(98) R = 4-methoxy phenyl

The applications of β-amino acids as peptidomimetics in the design of bioactive peptide analogues.[129] Conformationally restricted β-amino acid analogues, homo-Freidinger lactams have been prepared by an olefin metathesis reaction.[130] The design and efficient synthesis of the first reported β-amino phosphotyrosyl mimetic (**100**) have been reported.[131] Asymmetric synthesis of $\beta^{2,3}$-amino acids of general structure (**101**) is seen as an entry to an molecules with interesting chemical and biological properties due to their similarities to serine and other unusual amino acids.[132]

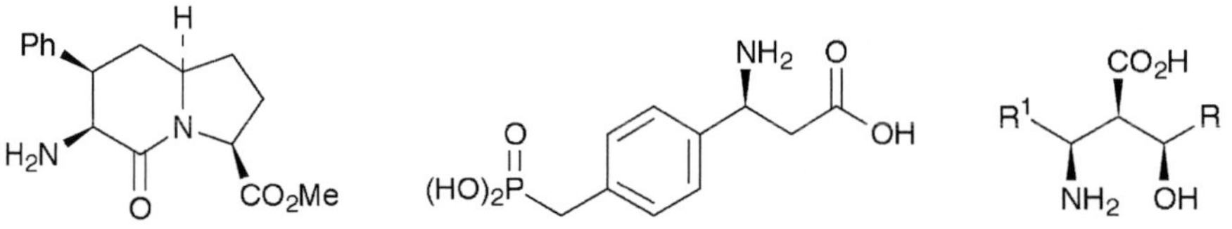

(99) (100) (101)

3 Cyclic Peptides

As in last previous volumes, the comprehensive coverage of this topic can be seen in Chapter 4. Apart from the discussion of new synthetic routes, this subsection is confined to cyclic pseudopeptides and peptidomimetics.

Synthesis and applications of monocyclic, dicyclic, and bicyclic side-chain lactam bridged peptides have been reviewed.[133] Chiral 3,4-aziridinolactams (**102**) were prepared and their reactivity with a range of nucleophiles was studied.[134] In all cases, nucleophilic attack occurred at the C(3) centre and led to the asymmetric synthesis of β-pseudopeptides (**103**). Aldol building blocks were successfully used for the synthesis of 12-membered lactams as mixtures of *E*/*Z* isomers with the general formula of (**104**).[135] Palladium(0)-catalyzed regioselective reaction of allyl acetates with amines afforded α-dehydro-β-amino esters which were cyclized via ring-closing metathesis, preorganized by an intramolecular hydrogen bonding, to a constrained β-turn mimic (**105**).[136] Intramolecular Mitsunobu protocol was used to prepare novel dipeptide macrocycles (Figure 19) from proline derivatives.[137] The solid-phase synthesis of spirobicyclic peptides, shown in (Figure 20), was performed using an orthogonally protected bis(aminomethyl)malonic acid building block (**106**) as the branching point.[138]

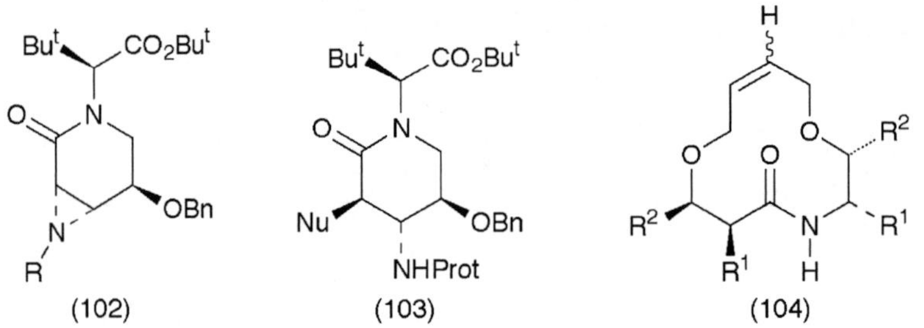

(102) (103) (104)

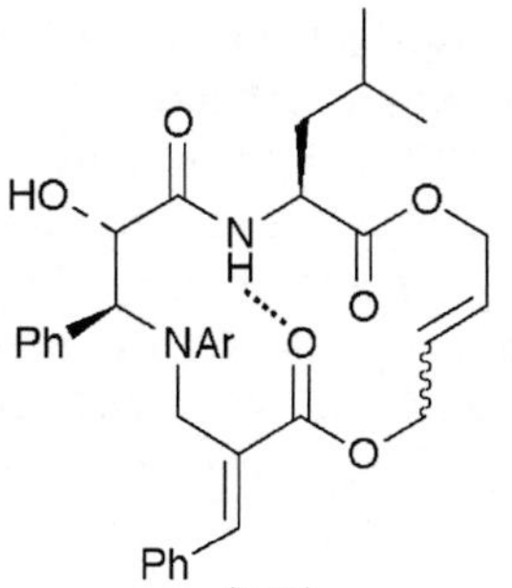

(105) 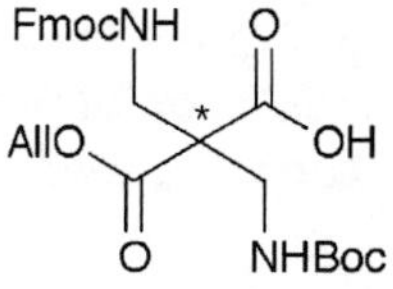 (106)

A novel method for peptide cyclization, in which the side chain of p-amino-phenylalanine was connected to that of either tyrosine or histidine by forming an azo bridge, has been presented.[139] Dimeric (2*R*)-2,5-dihydro-2-isopropyl-3,6-dimethoxypyrazine and 5-alkenyl derivatives were prepared by homocoupling.[140] A Ru^{II}-catalyzed ring-closing metathesis of the bisallyl derivative yielded the corresponding 1,2-bis(spiroannulated cyclic dipeptido)cyclohexane (**107**).

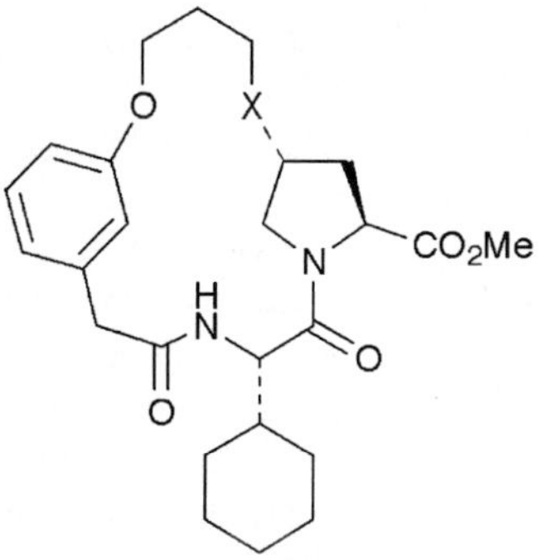

X = O, S, N–SO₂Ph

Figure 19

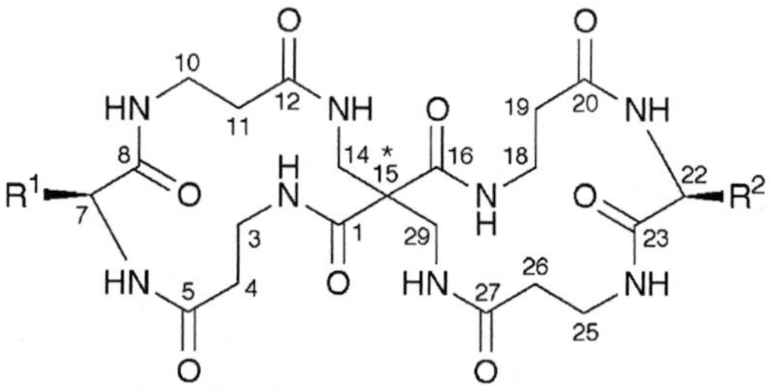

$R^1 = CH_2Ph$, $R^2 = H$
$R^1 = R^2 = CH_2Ph$
$R^1 = CH_2CH(CH_3)_2$, $R^2 = H$
$R^1 = R^2 = CH_2CH(CH_3)_2$
$R^1 = CH_2Ph$, $R^2 = CH_2CH(CH_3)_2$

Figure 20

Linear as well as cyclic forms of astressin, retro astressin, inverso astressin and retro inverso astressin were synthesized and tested for their binding at the corticotrophin releasing factor receptor, type I.[141] Among these analogues cyclic astressin, as an α-helical constraint, proved to be the most active ($pK_i = 9.20 \pm 0.12$). Triazinyl-amino acids, new heterocyclic building blocks were prepared and applied for the construction of macrocyclic pseudopeptides.[142] Design, synthesis and conformational analysis of N_i- to N_{i+3}-ethylene-bridged partially modified retro-inverso tetrapeptide β-turn mimetics (**108, 109**) were reported.[143]

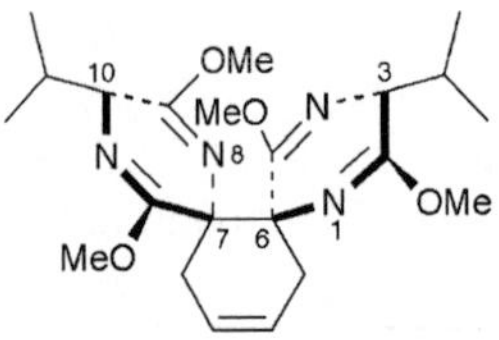

(107) 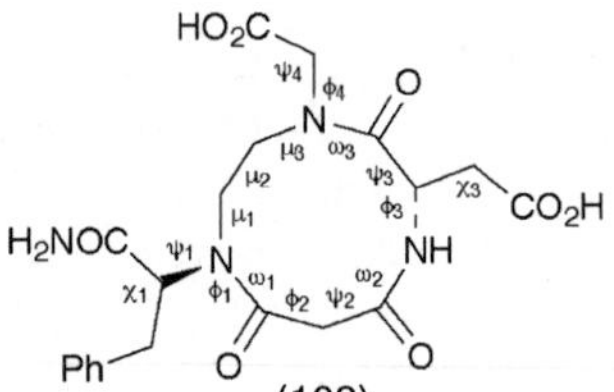 (108) 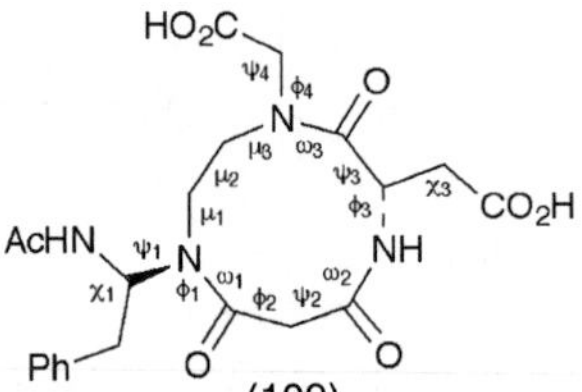 (109)

^{1}H-NMR studies showed that the 10-membered rings were conformationally constrained with well-defined structural features and that the three amide bonds in the ring were in *trans* orientation. Isolation, structure determination, and conformational analysis of two macrocyclic peptides (**110** and **111**) formed during lactamization of 1,1′-ferrocenylbis(alanine) have been presented.[144] NMR spectroscopy indicated C_2 symmetry for (**110**) and C_3 symmetry for (**111**). New linear and macrocyclic peptide-arene hybrids (**112**) were synthesized from N-tetrachlorophthaloyl protected amino acids in solid phase.[145] Ring-closing metathesis reactions were used for the synthesis of novel carbazole-linked[146] and binaphthyl-based (**113**)[147] cyclic peptoids with anti-bacterial activity. A novel 18-membered cyclic oligopeptide (**114**) based on 5-(amino-methyl)-2-furancarboxylic acid has been developed as an excellent receptor for carboxylate binding.[148] Cyclopeptides containing glucuronic acid methylamine (Figure 21) were synthesized and subjected to NMR studies.[149]

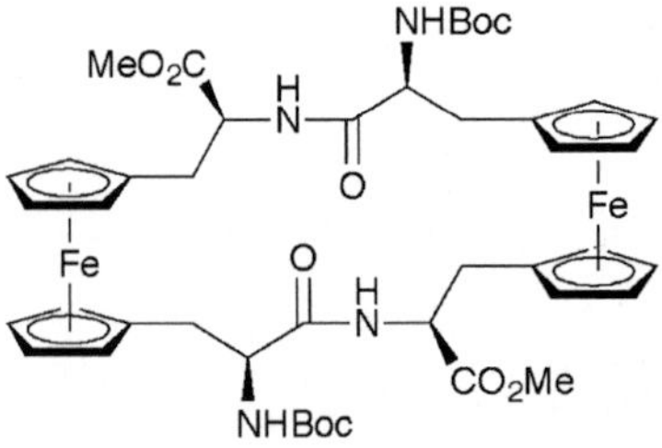

(110) 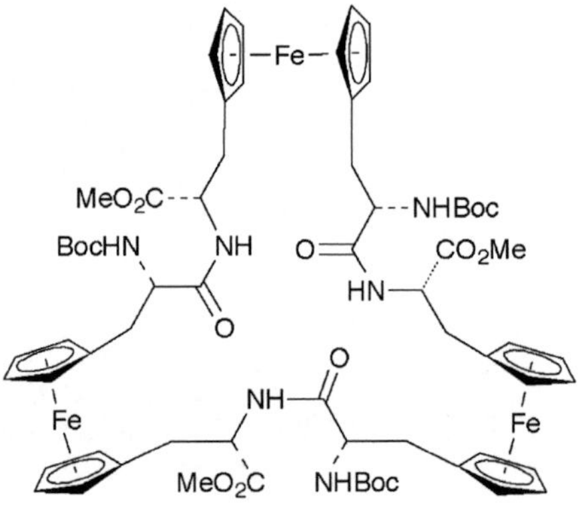 (111)

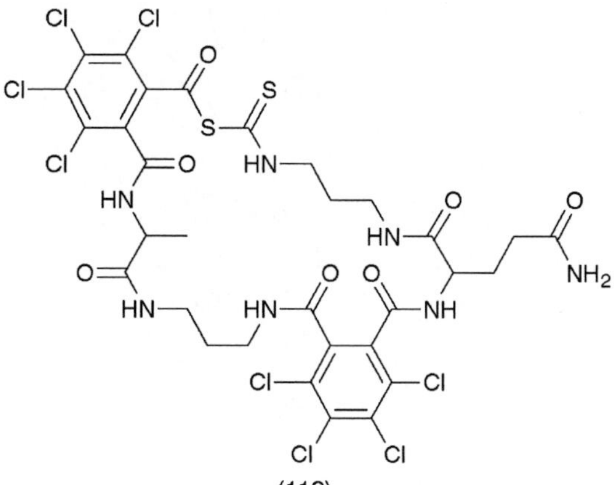

(112)

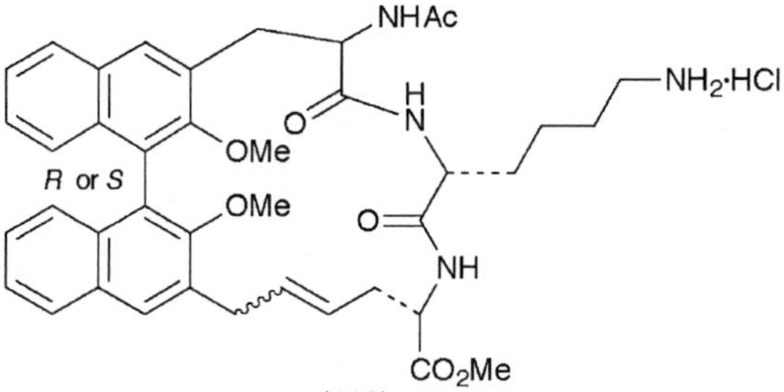

(113)

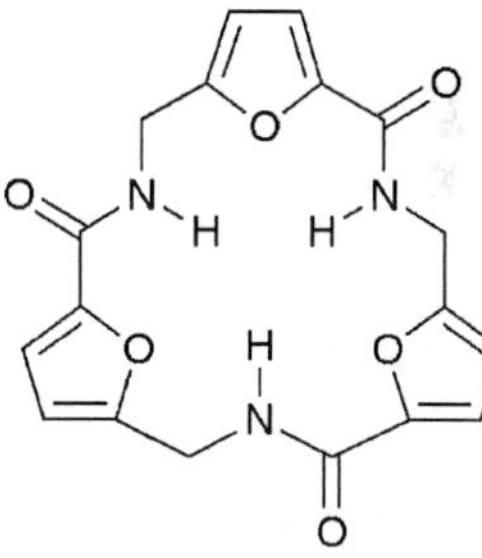

(114)

Ring-fused, C^{10} β-turn peptidomimetics led to compound (**115**), the first reported small-molecule mimic of neurotropin-3.[150] Three analogues of desmopressin, containing either γ-turn mimetics based on a morpholine-3-one framework (**116, 117**), or a methylene ether isostere in place of the amide bond between residues 3 and 4 (**118**), were designed and synthesized.[151]

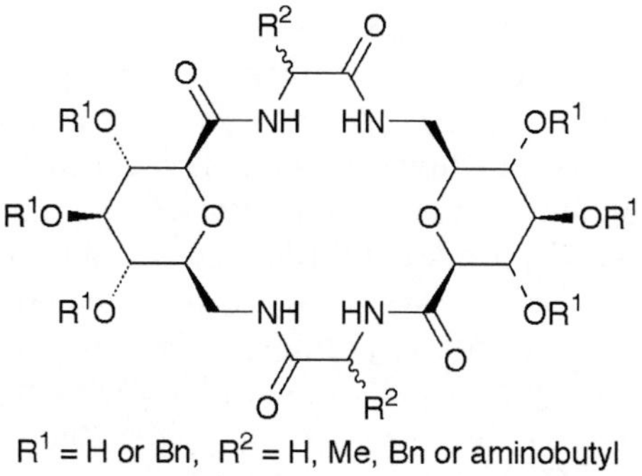

R^1 = H or Bn, R^2 = H, Me, Bn or aminobutyl

Figure 21

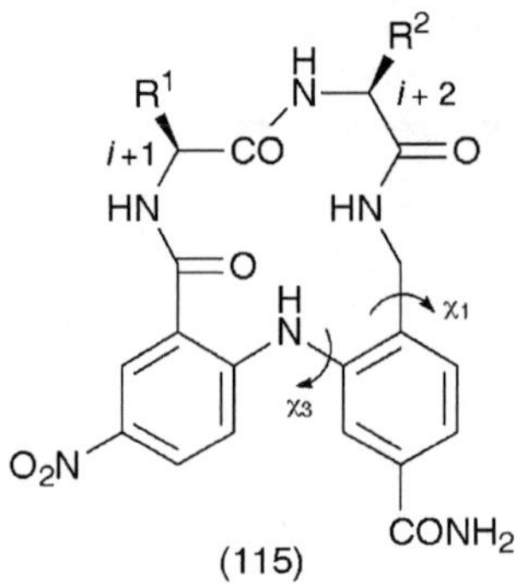

(115)

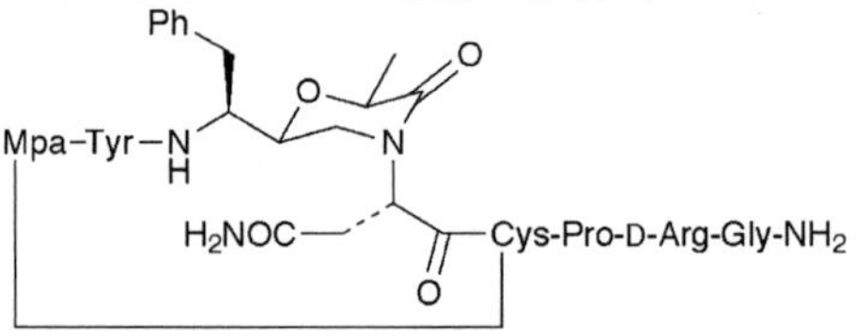

(116)

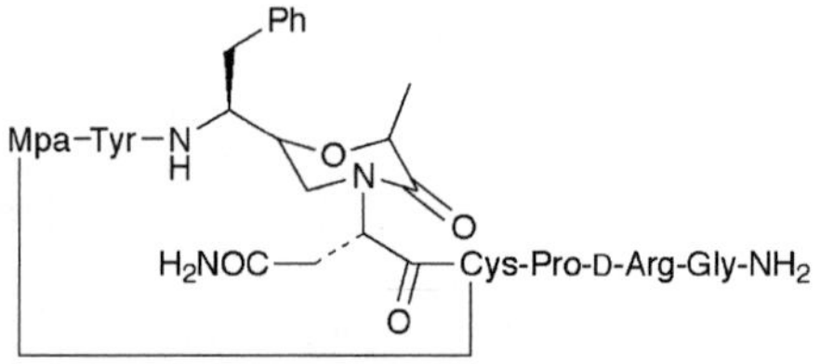

(117)

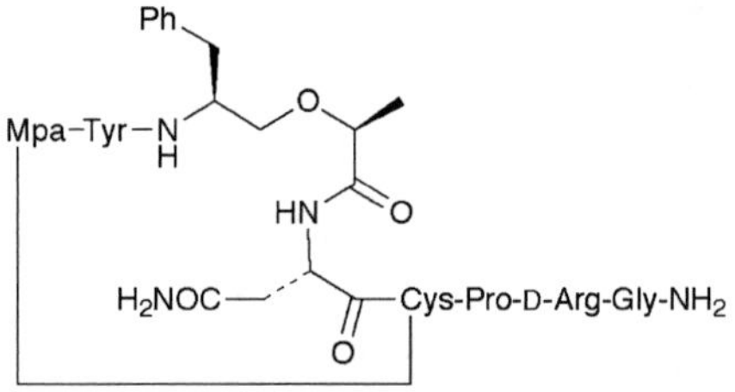

(118)

Comparison of structure and biological activity for analogues (**116**) and (**117**) did not support the presence of an inverse γ-turn involving residues 3 and 5 when desmopressin bound to the vasopressin V_2 receptor. An inverse γ-turn containing oxytocin mimetic (**119**), designed and prepared by the same research group, did not induce contractions of uterine tissue segments.[152] This might be due to the lack of amide bond between residues 3 and 4 in (**119**). A conformationally constrained cyclic peptide (**120**) designed as a novel amphi-receptor has been synthesized.[153] The stability of the cation and anion complexes was determined by circular dichroism. A novel cyclic prodrug (**122**) derived from the parent RGD eptidomimetic (**121**) was synthesized to improve its membrane permeation properties.[154] Chemical and enzymatic stabilities as well as antithrombotic activity of (**122**) were also determined.

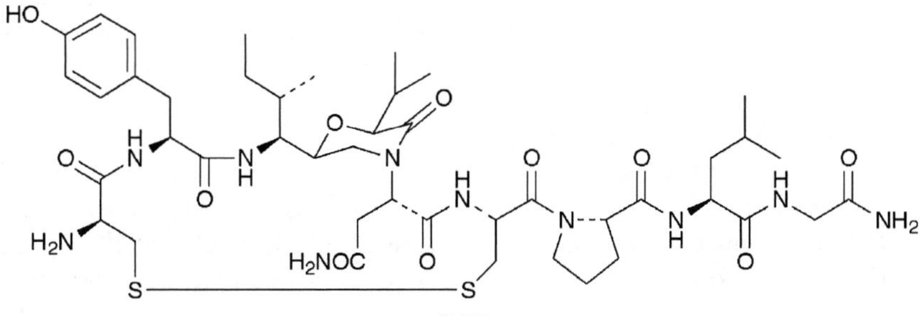

(119)

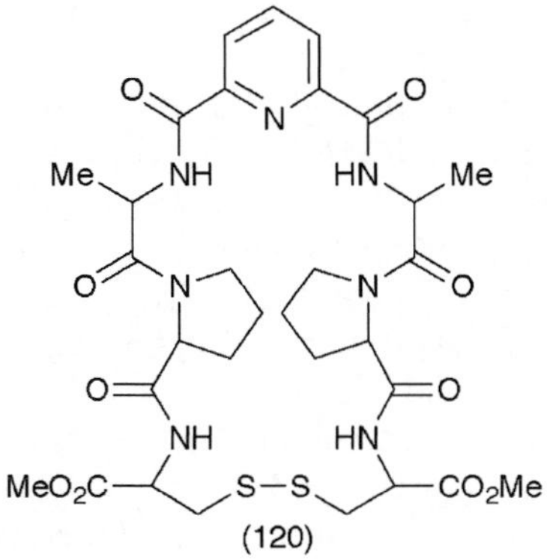

(120)

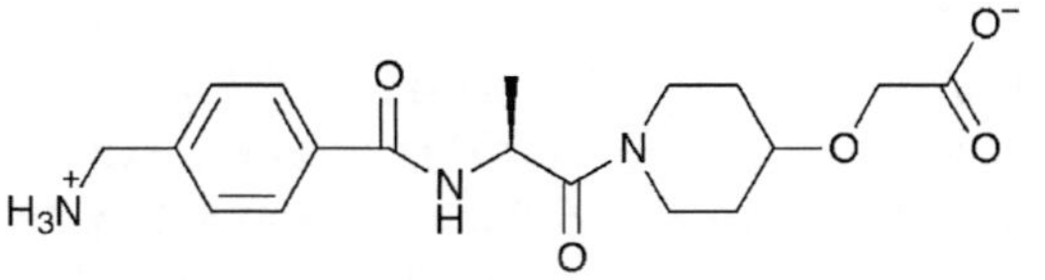

cis- and *trans*-isomers

(121)

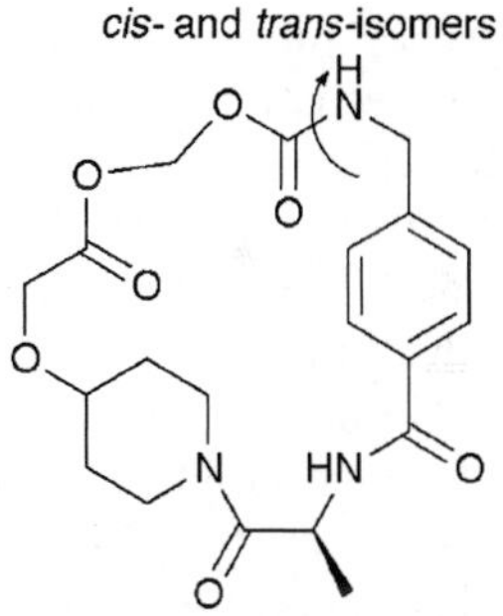

(122)

4 Biologically Active Peptides

4.1 Peptides Involved in Alzheimer's Disease. – Aggregation of polypeptide chains to fibrils having β-sheet structure seems to be a common mechanism of several neurodegenerative disorders. The β-amyloid (Aβ), a 39-43 residue peptide plays a central role by initiation of neuronal cell death during the very long period (15–20 years) of the Alzheimer's disease. The mechanism of neuronal death has not yet been fully discovered. Several studies support the hypothesis that Aβ fibrils activate microglial cells and the starting inflammation process and the neurotoxic microglial products cause neurodegeneration. This hypothesis would elucidate the very long starting phase of Alzheimer's disease. Other studies show that Aβ peptides have a direct toxic effect on neuronal cells by causing signalization disturbances.

Various Aβ 1–42 mutant peptides, found in different subtypes of familial Alzheimer's disease, were synthesized.[155] Amino acids of the native sequence of Aβ 1–42 were replaced, such as A21G (Flemish), E22Q(Dutch), E22K (Italian), E22G (Arctic) and D23N (Iowa) and the secondary structure, aggregation and neurotoxicity of the mutant peptides were studied. Mutant peptides in position 22 and 23 showed potent aggregative ability and neurotoxicity indicating that positions 22 and 23 play critical role in Dutch-, Italian-, Arctic-, and Iowa-subtype familial Alzheimer's disease.

A chemical, structural and biological study on the Aβ 12–28 peptide was performed using field-gradient NMR-spectroscopy, Fourier-transform infrared spectroscopy and transmission electron microscopy.[156] NMR studies showed that the soluble peptide fraction is composed of oligomeric intermediates, and Aβ 12–28 proved to be cytotoxic in PC12 cell cultures by MTT assay.

Protective effect of an APP 17-mer peptide against neuronal apoptosis induced by Aβ 25–35 was studied.[157]

Molecular simulation of the lipophilic and antigenic cores of the Aβ 1–42 showed that the Aβ 1–42 has two hydrophobic cores (Aβ 15–19 and Aβ 28–41).[158] The antigenic part seems to be at the N-terminus and at the middle region of Aβ 1–42. The middle area might be responsible for the aggregation of Aβ by forming noncovalent interactions between paired, antiparallel β-sheet conformations. Another computer simulation study using OM/MM and Amber calculations reveals that the self assembly of Aβ peptides to antiparallel β-sheets is supported by specific recognition of Gly29 and Gly33 by Met35.[159] The β-sheet conformation brings Met35 into the required close proximity to Gly29 or Gly33.

Homo-polyamino acids (poly-*L*-alanine, poly-*L*-glutamine) also can form fibrillary amyloid structures which are also neurotoxic. A novel artificial protein (poly-(γ-methyl-*L*-glutamate)-grafted polyallylamine) forms amyloid-like β-sheet assemblies.[160] Aggregation of various poly-*L*-glutamine sequences was studied.[161] CD spectra of poly-Gln aggregates exhibit a classical beta-sheet-rich structure, consistent with an amyloid-like structure. The fundamental unit of the *in vitro* aggregates is a filament about 3 nm in width, resembling the protofibrils of Aβ peptides.

Not only homogeneous but heterogeneous peptides may also undergo a self-initiated structural transition from α-helix to β-sheet and self-assemble into amyloid fibrils.[162,163] Complementary peptide pairs, triplets and quadruplets are able to form amyloid fibrils. Also a β-turn-forming tripeptide, Boc-Leu-Aib-βAla-OMe can self-assemble to a supramolecular β-sheet structure with amyloid-like fibrillar morphology via non-covalent interactions.[164] Solid-state NMR studies reveal that Aβ 1–40 has a specific supramolecular structure.[165] One-dimensional ^{13}C magic-angle spinning NMR spectra of ^{13}C

labeled Aβ 1–40 are presented, using the constant-time finite-pulse radiofrequency-driven recoupling (fpRFDR-CT) technique. The results indicate a parallel alignment of neighboring peptide chains in the predominantly β-sheet structure of the Aβ-fibrils. The fpRFDR-CT data and NMR spectra also indicate a structural disorder in the N-terminal sequence of Aβ 1–40, including the first 9 amino acids.

Congo red has long been used for staining amyloid fibrils and recently been found to inhibit amyloid formation. The binding mode of this pigment to Aβ 1–42 was examined by UV Raman spectroscopy.[166]

X-ray diffraction studies and theoretical consideration resulted in a new hypothesis, that amyloid fibrils are water-filled nanotubes.[167] The authors suggest that cylindrical β-sheets are the only structures consistent with the X-ray and electron microscope data. Fibers of Aβ peptides and variants are probably made of either two or three concentric cylindrical β-sheets. The Aβ peptide contains 42 amino acid residues, the best number for nucleating further growth. (Poly-*L*-glutamine structural studies show that over 37–40 glutamine residues are necessary for the formation of fibrillary structure).

Several organic compounds, peptides and non-peptides, can prevent Aβ-aggregation to fibrils or destroy the fibrillary structure (β-sheet breakers, BSB). A fast and efficient molecular docking computational method was used for mapping of possible binding sequences of two BSB-pentapeptides on Aβ 1–42 surface.[168] The aim of this study was to find proper binding sequences for the BSB peptides on Aβ 1–42 and characterize them. Novel Aβ-aggregation inhibitors composed of an Aβ-recognition element (the BSB-pentapeptide KLVFF) and a flexible hydrophilic disrupting element (aminoethoxy ethoxy acetate and aspartate) were designed and chemically synthesized.[169] The inhibitory effects were examined by a pigment binding assay using Congo red or thioflavin T. One of the compounds (**123**) is a very effective inhibitor.

Substitution of isoleucine-31 by helix-breaking proline abolishes oxidative stress and neurotoxic properties of Aβ 1–42.[170] Ile→Pro Substitution drastically changes the conformation of the important 30 to 35 hexapeptide region of Aβ 1–42.

Potent, small molecule Aβ inhibitors have been prepared that incorporate an alanine core bracketed by an N-terminal arylacetyl group and various C-terminal amino alcohols.[171] The compounds exhibit stereospecific inhibition in an *in vitro* assay. The structure of one of the best compounds is given as (**124**).

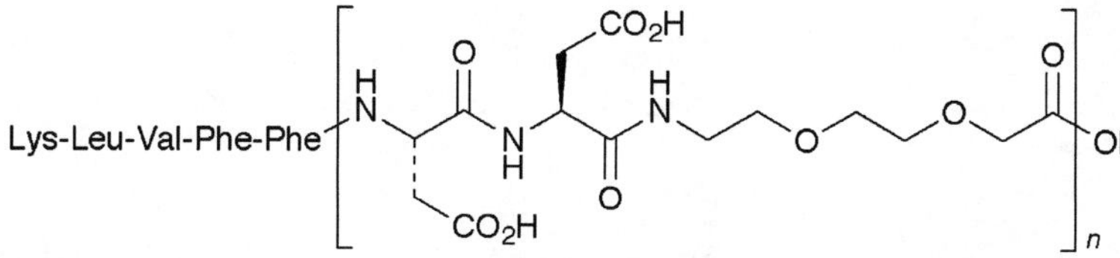

(123) Asp(D)-Asp(D)-AEEA linker(X), $n = 3$

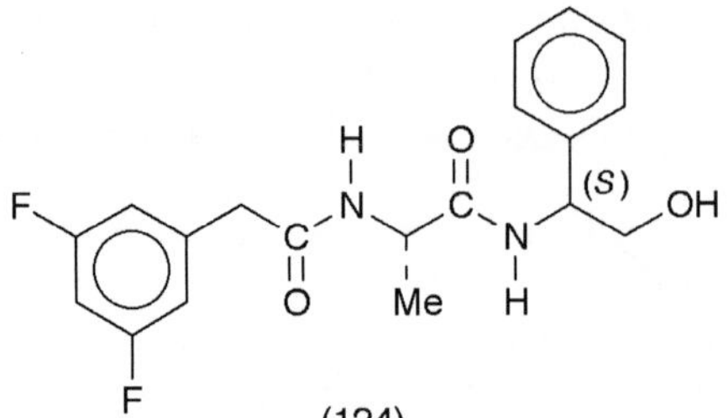

(124)

Low molecular weight (M_w of 700 to 1700) ligands for Aβ peptides were identified by using surface plasmon resonance.[172] The best of these ligands have a K_d of approximately 40 μM. The most potent ligands for immobilized Aβ-peptides are the most potent inhibitors of neurotoxicity of Aβ. Inhibition of amyloid fibril formation of human amylin by N-alkylated amino acid and α-hydroxy acid residue containing peptides are described.[173] Some of these peptides have ester moiety and depsipeptide structure. The authors give a general method for designing amyloid inhibitors by introducing molecular mutations.

α-Synuclein seems to play an important role in the etiology of both Alzheimer's and Parkinson's disease. β-Synuclein, a novel non-amyloidogenic homologue of α-synuclein was characterized as an inhibitor for α-synuclein aggregation.[174] A similar class of endogenous factors might regulate the aggregation state of other molecules (e.g. Aβ-peptides) involved in neurodegeneration.

Free radical formation induced by Aβ fibrils might play important role in neuronal cell damage in Alzheimer's disease. A novel concept of the free radicals states that Aβ 1–42 remaining as monomer exhibits antioxidant and neuroprotective effect.[175] In contrast, oligomeric and aggregated Aβ 1–40 and Aβ 1–42 lose their neuroprotective activity. Other authors suggest that the hydrophobic environment of Met-35 of Aβ 1–42 is very important for the neurotoxic and oxidative properties of the peptide.[176] Metal ion chelators and antioxidants suppress the Aβ 1–40/Fe^{2+}-ion-induced oxidative stress cascade and may be beneficial in reducing the severity of Alzheimer's disease.[177] Plasma membrane cholesterol controls the cytotoxicity of Aβ 1–40 and Aβ 1–42 peptides, by modifying the fluidity of the neuronal membranes.[178]

A novel concise synthesis of 5,8-dihydroxy-3*R*-methyl-2*R*-(dipropylamino)-1,2,3,4-tetrahydronaphthalene (**125**) has been developed. The hydroquinone is a potent inhibitor of the fibrillar aggregation of Aβ-peptides as determined in two different assay systems.[179]

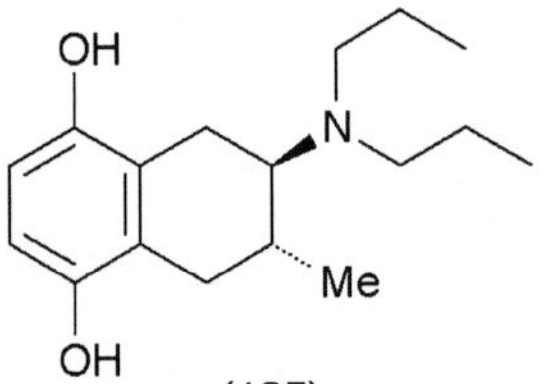

(125)

4.2 Antimicrobial Peptides. – Several reviews appeared in the field of antimicrobial peptides revealing that these compounds are still at the centre of interest. Different structural classes of antimicrobial peptides, emphasizing their molecular diversity, have been presented.[180] Intracellular targets of a new family of antibacterial peptides were reviewed in the course of developing selective antimicrobial agents.[181] A 126-reference review summarizes the physico-chemical and also biological properties of host-defense peptides and 'designer' peptides to explore their use as effective therapeutic agents.[182] Short proline-rich antimicrobial peptides, which enter the cells without destroying the cell membrane, were the subject of two reviews. While one of these articles reviewed the pyrrhocoricin derivatives which inhibit metabolic processes, such as protein synthesis and chaperone-assisted protein folding,[183] the other paper concentrated on PR-39, a member of the cathelicidin family, which appears to have an important role in wound repair and inflammation.[184] The classification, antimicrobial spectra, mechanism of action, and regulation of cathelicidins have been presented in another paper.[185] Most antimicrobial peptides use the bacterial phospholipids membrane as target of defense. Two aspects of this effect have also been reviewed. The role of cell membranes in the activities of cationic antimicrobial peptides[186] and a novel group of antimicrobial peptides derived from diastereomers of lytic peptides based on the "carpet" mechanism[187] were presented. Design, synthesis and characterization of cationic peptides as well as steroid antibiotics have been reviewed.[188] A 126-reference review highlights the development of short antimicrobial peptides from host defense peptides of natural origin or combinatorial libraries.[189] In addition to structure-activity relationships, methods for improving bioavailability and specificity of the peptides were also discussed.

4.2.1 Antibacterial peptides. Structure and biological activities of synthetic peptides corresponding to either human α-defensin HNP-1[190] or bovine neutrophil β-defensin BNBD-12[191] with disulphide connectivities of the natural peptides and their variants with one, two or three S–S bridges were investigated. Selective protection of cysteine thiols was necessary in both cases for the formation of disulphide linkages, as in the native peptides. Interestingly, antibacterial activities were observed for all peptides, irrespective of how the S–S bridges were linked. These results together with CD spectra suggested that a rigid β-sheet structure or the presence of three disulphide bridges was not essential for antibacterial activity. Isolation, synthesis, antimicrobial activities, and bacterial binding of homodimeric theta-defensins from Rhesus macaque leukocytes have been published.[192] Parallel and antiparallel dimers prepared from two monomeric magainin 2 precursors had a greater biological activity and a greater ability to interact with membranes than the monomers.[193] However, there was no significant difference between the activities of parallel and antiparallel dimers. Crucial residues for the antibacterial activity of pyrrhocoricin and a possible binding surface for this peptide on *Escherichia coli* DnaK have been identified.[194] A study demonstrated that the two

N-terminal residues of sarcotoxin IA, a 39-amino acid long cecropin-type antibacterial peptide, were necessary for its antibacterial activity against *Escherichia coli* and bacterial lipopolysaccharide (LPS) binding.[195] Using 18-membered analogues of the N-terminal fragment of sarcotoxin IA it was indicated that this amphiphilic region involved the residues responsible for the antimicrobial activity.[196]

P18, an α-helical antimicrobial peptide designed from a cecropin A-magainin 2 hybrid, showed salt resistance on antimicrobial activity under high (100–200 mM) NaCl concentrations and much greater salt resistance on antibacterial activity against Gram-negative bacteria at the physiological or elevated concentrations of $CaCl_2$ and $MgCl_2$ than magainin 2.[197] Two reversed peptides of P18 were found to have similar or higher antimicrobial activity against bacterial and fungal cells without hemolytic activity as compared with P18.[198] HP-MA hybrid peptide, consisting of 2-9 residues of *Helicobacter pyroli* ribosomal protein L1 and 1–12 residues of magainin 2, has more potent antifungal effects than the parent peptides and damages the plasma membranes of *Candida albicans*.[199] Antibiotic activities of HP-MA and a similar hybrid containing 1–12 residues of melittin (HP-ME) were evaluated using bacterial, tumor and human erythrocyte cells.[200] The hybrids had stronger or similar antibacterial activity against Gram-positive and Gram-negative bacteria than the parent peptides and, none of them had anti-tumor or hemolytic activity.

CRAMP, a cathelicidin-derived antimicrobial peptide consists of two amphipathic α-helices which are connected by a flexible region from Gly(11) to Gly(16) as determined by NMR spectroscopy.[201] Three analogues of an 18-amino acid region, corresponding to residues 16–33 of CRAMP have been synthesized and tested.[202] Leu-substitution in the hydrophobic helix face resulted in a dramatic increase in antibiotic activity without a significant increase in hemolytic activity. Structure-activity relationship of another cathelicidin-derived antimicrobial peptide, PMAP-23 was studied by NMR.[203] Trp(21) at the C-terminus of the peptide proved to be important for the primary binding to the membrane, and hence for the antibacterial activity. Tritrypticin and indolicidin are Trp- or Pro/Arg-rich members of the cathelicidin family. Structural and biological investigations of tritrypticin having a unique amino acid sequence (VRRFPWWWPFLRR) and its analogues revealed that the poly-proline II helical structure was crucial for bacterial cell selectivity.[204] Investigations on the interactions and structure of indolicidin and its retro analogue led to the conclusion that the key feature of the recognition between the cationic antibacterial peptides and endotoxin is the plasticity of molecular interactions.[205] A 13-residue indolicidin-based peptide, Rev4 possessed both antimicrobial and protease-inhibitory activities.[206] The lipopolysaccharide-neutralizing activity of the human cathelicidin CAP18/LL-37 could be augmented by modifying its hydrophobicity and cationicity, and the 18-mer LLKKK (KLFKRIVKRILKFLRKLV) was the most potent among them, with therapeutic potential for Gram-negative bacterial endotoxin shock.[207] The cathelicidin motif of protegrin-3, a common prosequence of

96–101 residues of many antibacterial peptides, has been overexpressed and subjected to structural studies.[208]

Lactoferrins, glycoproteins from many external secretions of mammals which are known to possess antimicrobial activity, were the subject of two minireviews. In one of them, structure and structure-function analysis of bovine lactoferrin have been discussed.[209] The other review discussed important structural features affecting the antimicrobial activity of 15-residue derivatives of lactoferrins.[210] The effects of charge and lipophylicity on the antibacterial activity of a bovine lactoferrin-derived undecapeptide (FKCRRWQWRMK) were investigated.[211] Size-related descriptors for eight non-proteinogenic amino acids were used to develop quantitative structure-activity relationships (QSARs) of bovine lactoferrin-(17–31)-pentadecapeptide.[212] In another article, tryptophan residues of this peptide were replaced by Trp(2-Pmc) (**126**) and antibacterial activity of the resulting compounds was investigated.[213]

Amphipathic α-helical peptides and their diastereomeric analogues were the subject of two papers. The mode of action of the diastereomers and their advantage over the all L-amino acid counterparts, the effects of sequence alteration on antibacterial and hemolytic activities, solubility, enzymatic degradation and structure have been discussed.[214] The other study analyzed in detail the structure of a model all *L*-peptide (KLLLKWLLKLLK) and its diastereomer and their interactions with membranes.[215] Oligomers of β-amino acids (**127–129**) were designed to mimic the biological activity of amphiphilic, cationic α-helical antimicrobial peptides.[216] The study revealed that an amphiphilic helix and the ratio of cationic to hydrophobic residues were the most important factors for activity of these β-peptides. The relationship between conformational stability and antimicrobial activity was investigated on 14-helical antibacterial β-peptides.[217]

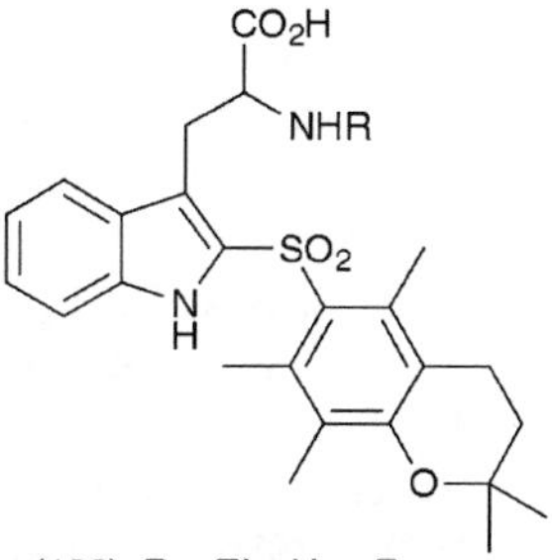

CO_2H NHR SO_2 N H O

(126) R = Tfa, H or Fmoc

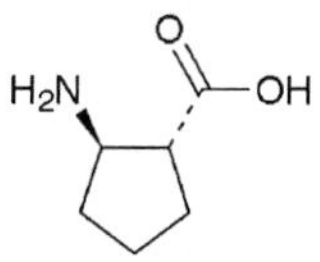

H_2N O OH ACPC (127)

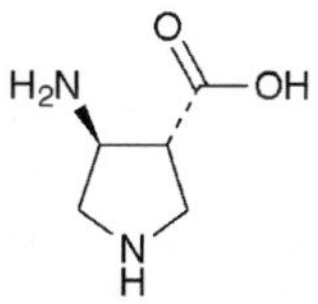

H_2N O OH N H APC (128)

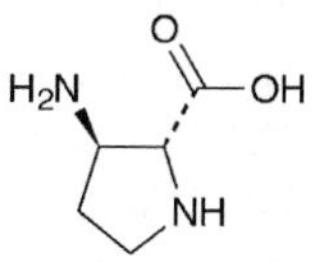

H_2N O OH NH AP (129)

Various experiments on intact cells and model membrane systems revealed that clavanin A, a cationic peptide antibiotic permeabilizes the antimicrobial membrane via two distinctly different pH-dependent mechanisms.[218] The histidine and glycine residues play an important role in these mechanisms. It was shown that the translocating regions of bactenecin 7 are capable of acting as a vector for delivering large cargo, and that there are multiple translocating regions on bactenecin 7 with most of them containing a novel repeating $(PX)_n$

motif.[219] A study discovered that the human antimicrobial peptide LL-37 induces chemokine production and surface expression of chemokine receptors, indirectly promoting the migration of immune cells.[220] A series of differently glycosylated and cyclic drosocin and apidaecin 1b analogues have been synthesized and tested against Gram-negative bacteria.[221] Indolizidin-2-one amino acids (**130, 131**) were introduced into the antimicrobial peptide gramicidin S [*cyclo*(Val-Orn-Leu-*D*-Phe-Pro)$_2$] to explore the relationship between configuration, conformation and biological activity.[222]

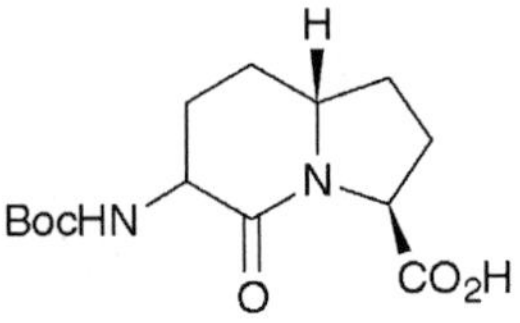

BocHN, H, N, O, CO_2H (130)

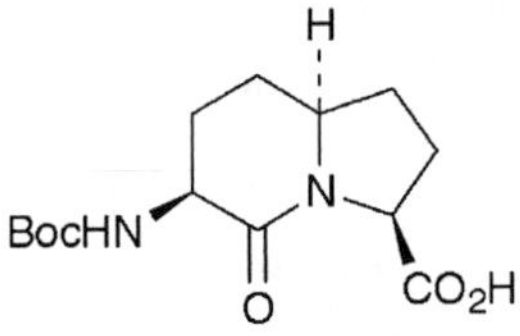

BocHN, H, N, O, CO_2H (131)

Three-dimensional solution structures of ovispirin-1, novispirin G-10 and novispirin T-7, determined by CD and 2D-NMR showed that simple modifications in the sequence can induce fine structural changes that evidently impacted their activities.[223] Three derivatives of the natural peptide dermaseptin S4 proved to be highly active against a series of multidrug-resistant Gram-negative clinical isolates as well as against Gram-positive strains.[224] In addition, these peptides caused less hemolysis than other tested peptides. Structure-function analysis on the frog skin antimicrobial peptide tigerinin 1 using cyclic and linear analogues of the parent peptide was the subject of a study.[225] The effect of amino acids within the disulfide loop of thanatin, a potent antibiotic peptide identified as the first antimicrobial peptide to show broad antibiotic activity against Gram-negative and Gram-positive bacteria, and fungi without cytotoxicity.[226] The role of peptide length and preassembly on the ability of diastereomeric cationic antimicrobial peptides to discriminate among different target cells has been investigated.[227] It was proved that preassembly of these cationic peptides increases their amphipathic structure, and activity toward zwitterionic membranes and therefore affects their target cell specificity.

Novel antimicrobial peptides, RP-1 (ALYKKFKKKLLKSLKRLG) and RP-11 (ALYKRLFKKLKKF) or gentamicin were tested in whole blood and blood-derived matrices as well as in conventional media at pH 5.5 and 7.2 in an *ex vivo* assay developed by the authors.[228] The solution structure of a peptide selected from phage-display libraries (PW2, HPLKQYWWRPSI) bound to SDS micelles determined by two-dimensional NMR methods was presented.[229] Libraries of cationic β-hairpin mimetics containing eight or twelve residues (Figure 22) have been synthesized.[230] Antimicrobial and hemolytic activity as well as conformational analysis of the mimetics were also published. In the example [(RLA)$_2$R]$_2$, a model synthetic α-helical all-D-peptide, the effect of net charge, helicity, and epimeric nature of the peptide on bacteriostatic activity has been examined.[231] L-Amino acids in a peptide consisting of five arginine residues and the 11-amino acid region (32–42) of human granulysin have been

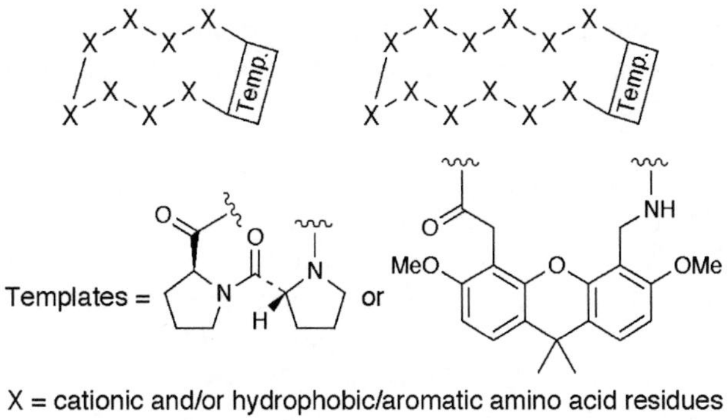
Temp.
Temp.
Templates =
or
MeO
OMe
NH
X = cationic and/or hydrophobic/aromatic amino acid residues

Figure 22

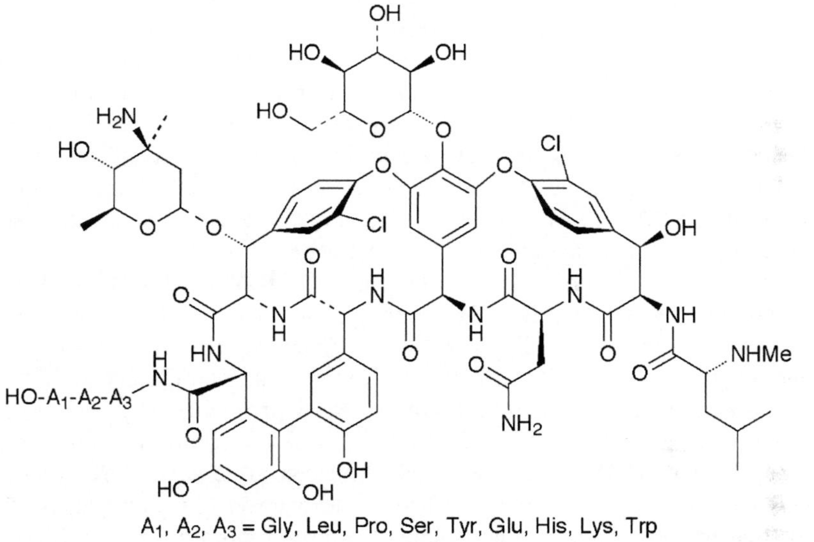
HO-A1-A2-A3
NHMe
NH2
A1, A2, A3 = Gly, Leu, Pro, Ser, Tyr, Glu, His, Lys, Trp

Figure 23

replaced partially or wholly with D-amino acids. Activity as well as stability of these peptides to proteolysis in serum were examined.[232] A simple and practical method for the preparation of a variety of peptide-bearing carboxamide derivatives of chloroorienticin B (Figure 23) has been presented.[233] Replacement of the peptide bond between the first and second amino acids of eremomycin, a glycopeptide antibiotic of the vancomycin family, by the aminoalkyl group resulted in a decrease in activity towards Gram-positive microorganisms.[234] SMAP-29, a mammalian antimicrobial peptide of the innate immunity, was successfully produced in transgenic tobacco plants using a modified VMA intein expression system.[235] Antibacterial, haemolytic and cytotoxic activities of temporin L (FVQWFSKFLGRIL-NH_2) isolated from the skin of the European red frog *Rana temporaria* were investigated.[236] The peptide showed the highest activity among temporins against human erythrocytes and different microbial strains, and caused perturbation of bilayer integrity of both neutral and negatively charged membranes. Phylogenic relationships and some biological properties of 34 peptides which are structurally similar to temporins and the derivation of the two new consensus sequences were published.[237] The solid phase synthesis of statherin SV2 and its dephosphorylated analogue, their purification and antibacterial activity have been presented.[238] The solution structure of the antibacterial and antifungal peptide hepcidin (LEAP-1) that acts as a signalling molecule in iron uptake was determined by standard two-dimensional ^{1}H-NMR spectroscopy.[239] The pore-forming microcin E492 was purified by solid-phase extraction and RP HPLC and its entire 84-amino-acid sequence was determined.[240] The sequence of this unmodified peptide showed homologies with the sequence of colicin V.

Cupiennin 1, a new family of highly basic antimicrobial peptides was isolated from the venom of the spider *Cupiennius salei.*[241] Biological effects and the structural properties indicated a membrane-destroying mode of action on prokaryotic as well as eukaryotic cells. Further investigation revealed that the hydrophobic N-terminus includes the major determinants of structure and activity of cupiennin 1 peptides.[242] Another spider, the wolf spider *Oxyopes kitabensis* was the source of five large amphipathic peptides, named oxyopinins, with antimicrobial, hemolytic, and insecticidal activity.[243] Oxyopinins showed disrupting activities towards both biological membranes and artificial vesicles, particularly to those rich in phosphatidylcholine.

Two peptides, referred to as alloferon 1 (HGVSGHGQHGVHG) and alloferon 2 (GVSGHGQHGVHG), were isolated from the blood of an experimentally infected insect, the blow fly *Calliphora vicina* (Diptera).[244] *In vitro* experiments revealed that the synthetic alloferon has stimulatory activities on natural killer lymphocytes. In addition, *in vivo* experiments in mice showed the antiviral and antitumor activities of alloferon. A novel pore-forming peptide, named bovicin HC5, which mediated antibacterial activity of *Streptococcus bovis* HC5 was identified and purified.[245] Two novel pore-forming peptides, opistoporin 1 and 2, have been isolated from the venom of the South-African scorpion *Opistophtalmus carinatus.*[246] Structure-activity studies were performed on the peptides in comparison with parabutoporin, a previously identified pore-forming peptide of another South-African scorpion *Parabuthus schlechteri.* Two groups of novel antimicrobial peptides named maximins have been isolated from the skin secretions of Chinese red belly toad *Bombina maxima.*[247] The peptides showed not only potent antimicrobial activity, cytotoxicity against tumor cells and spermicidal action but maximin 3 also possessed a significant anti-HIV activity. HE2, a gene expressed specifically in human epididymis, gave rise to multiple mRNAs that encode a group of small cationic secretory peptides. HE2α and HE2β1, the major peptide isoforms of this group, were recombinantly expressed and tested against *Escherichia coli.*[248]

A cationic antimicrobial peptide that is either identical or homologous to 40S Rp S30 eukaryotic ribosomal protein was purified and partially characterized.[249] The peptide displayed high activity against Gram-positive bacteria with an IC_{50} of 0.02 to 0.04 μM. Three-dimensional structure of moricin, an antibacterial peptide from the silkworm *Bombyx mori*, was determined by 2D ^{1}H-NMR spectroscopy.[250] The solution structure showed an elongated α-helix over nearly the whole length of the molecule. The results suggested that the N-terminal amphipatic α-helical fragment of moricin was responsible for the increased liposomal membrane permeability. The structure of the antimicrobial skin peptides of the New Guinea Tree Frog *Litoria genimaculata* and the Fringed Tree Frog *Litoria eucnemis* has been compared.[251]

4.2.2 Antifungal peptides. Classical antifungal agents as well as antimicrobial peptides that can have therapeutic potential for the treatment of *Candida albicans* infections have been discussed in a review.[252] Myristoyl-CoA:protein N-myristoyltransferase (Nmt) is the enzyme that is essential for the viability of *Candida albicans*. Thus, Nmt is a potential target for the development of an antifungal drug. The crystal structures of *C. albicans* Nmt with peptidic and nonpeptidic inhibitors have been presented.[253] Attachment of heptanoic, undecanoic, and palmitic acids to the N-terminus of magainin-2 resulted in lipopeptides with distinct oligomeric states and structures.[254] A direct correlation was found between the oligomerization of the lipopeptides and antifungal activity opening a new way for the design of potent antifungal agents. Conformation of histatin 3 was investigated by CD and NMR spectroscopy in the presence of *L*-α-dimyristoylphosphatidylcholine lipid vesicles.[255] The conformational information was then used to design histatin analogues possessing antifungal activity. Two-dimensional NMR study of tachystatin A, an antimicrobial peptide from the Japanese horseshoe crab *Tachypleus tridentatus*, showed a cysteine-stabilized triple-stranded β-sheet structure and an amphiphilic folding observed in many membrane-active peptide.[256] ^{1}H-NMR analysis of the antifungal peptide chromofungin revealed that it adopts a helical structure.[257] The mechanism of the interaction of chromofungin with fungi and yeast cells was also studied.

A research group has published its results on antimicrobial peptides such as HP(2-20), PMAP-23 and, SMAP-29. The synthetic peptide amide HP(2-20) corresponding to an N-terminal region of *Helicobacter pylori* ribosomal protein L1 exhibited potent antifungal activity against various fungi without hemolytic activity against human erythrocyte cells.[258,259] The fungicidal and anticancer activities of synthetic peptides obtained by replacing different amino acids by tryptophan at the hydrophilic face of PMAP-23, a peptide derived from porcine myeloid, were measured.[260] The results suggested that the increase of hydrophobicity of the peptides correlated with their fungicidal activity. SMAP-29, a cathelicidin-derived antimicrobial peptide deduced from the N-terminal sequence of sheep myeloid mRNA, damaged the plasma membranes of the pathogenic fungus *Trichosporon beigelii*.[261]

Two novel antifungal peptides, EAFP1 and EAFP2, were isolated from the bark of *Eucommia ulmoides* Oliv.[262] These are the first plant antifungal peptides with a five-disulphide motif. They exhibited relatively broad spectra of activities against eight pathogenic fungi with IC_{50} values in the micromolar range. Five disulphide bridges also stabilize a 45-amino acid peptide, named

E. europaeus chitin-binding protein (Ee-CBP), isolated from the bark of the spindle tree (*Euonymus europaeus* L.).[263] Ee-CBP was found to be more potent against fungi than Ac-AMP2 from *Amaranthus caudatus* seeds, which is considered one of the most potent antifungal hevein-type plant proteins. Angularin is an antifungal peptide with a novel N-terminal sequence (GEPGQKE) purified from adzuki beans.[264] Angularin exhibited strong antifungal activity against *Mycospharella arachidiocola* and *Rhizoctonia solani.* In addition to these fungi, a new 5 kDa antifungal peptide from rice beans inhibited mycelial growth of *Botrytis cinerea* and *Fusarium oxysporum.*[265] This antifungal peptide inhibited also the activity of HIV-1 reverse transcriptase. Two novel antifungal peptides, cicadin[266] and ascalin[267] were isolated from dried juvenile cicadas and the bulbs of the shallot *Allium ascalonicum*, respectively. Cicadin possesses an N-terminal sequence never found before among insect antifungal proteins. This peptide has potent antifungal activity with IC_{50} values at nanomolar concentrations against a wide variety of fungi. Ascalin inhibited mycelial growth of *Botrytis cinerea* but not of *Mycospharella arachidiocola* and *Fusarium oxysporum.* Therefore, ascalin expresses a unique species-specific antifungal activity. Both peptides suppressed the activity of HIV-1 reverse transcriptase.

Peptaibols are antibiotics known for their high α-aminoisobutyric acid content and as a mixture of isoforms ranging from 7 to 20 amino acids in length. A study demonstrated that a large peptide synthetase (2.3 MDa) is responsible for production of all classes of peptaibols in the fungus *Trichoderma virens.*[268] Several novel nonapeptides with potent antifungal activity against *Candida albicans* and *Cryptococcus neoformans* have been identified from combinatorial libraries derived from an antifungal hexapeptide pharmacophore Arg-*D*-Trp-*D*-Phe-Ile-*D*-Phe-His-NH_2.[269] Two of the nonapeptides exhibited approximately 17-fold increase in the activity compared to the lead molecule. The antifungal activity of a previously identified hexapeptide, Ac-RKTWFW-NH_2, was improved through screening of a combinatorial peptide library.[270] Distinct activity profiles were shown for peptides differentiated by just one or two residue substitutions. A study indicated that small peptides with broad antifungal activity could act synergistically with thiabendazole against thiabendazole-resistant *Fusarium sambusinum.*[271]

4.3 ACTH Peptides. – The use of a synthetic ACTH analogue (ACTH 1-24) in the treatment of acute aortic dissection were discussed in several correspondence.[272–274]

4.4 Angiotensin II Analogues and Non-peptide Angiotensin II Receptor Ligands. – The effect of angiotensin II and its 3-8 fragments (angiotensin IV) on tyrosine kinase activity in pituitary tumor homogenates were studied.[275] It was found that both peptides increase the kinase activity. Two nonpeptidic angiotensin type 1 receptor antagonists, irbesartan and losartan, were investigated for their antidipsogenic and antihypertensive efficacy.[276] Both compounds inhibited the drinking and pressor response to angiotensin II. Angiotensin II derivatives containing the 2,2,6,6-tetramethylpiperidine-*N*-oxyl-4-amino-4-carboxylic acid spin label were synthesized by solid-phase methodology.[277] The active paramagnetic analogue may be useful for conformational studies.

Nine *D*-amino acid containing analogues of the des-Asp[1]-angiotensin and ten alanine-substituted derivates were synthesized.[278] The above analogues were used for structure-activity and structure-binding studies on the rabbit pulmonary artery. Cyclic angiotensin II analogues with 3,5 side-chain bridges (**132**) and (**133**) have been designed, synthesized and investigated.[279] A non-peptide mimetic, 1-[2′-[(*N*-benzyl)tetrazol-5-yl]biphenyl-4-yl]methyl]-2-hydroxymethylbenzimidazole (**134**) was prepared for comparison. These molecules were used for conformational analysis and measuring their hypertensive response.

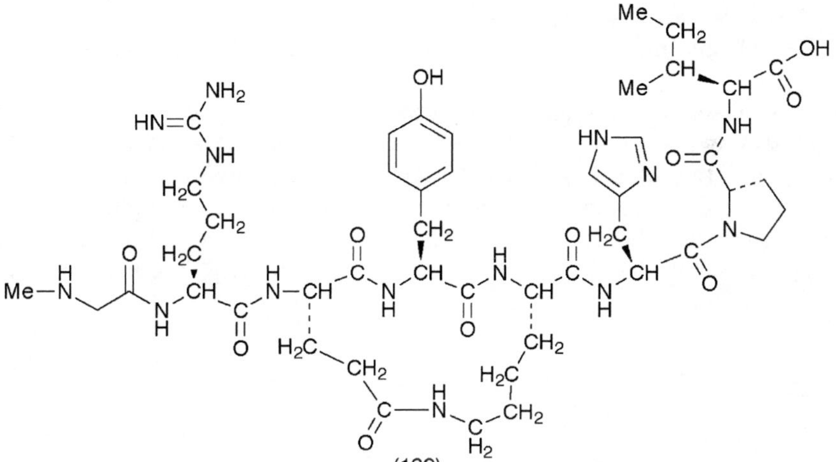

(132)

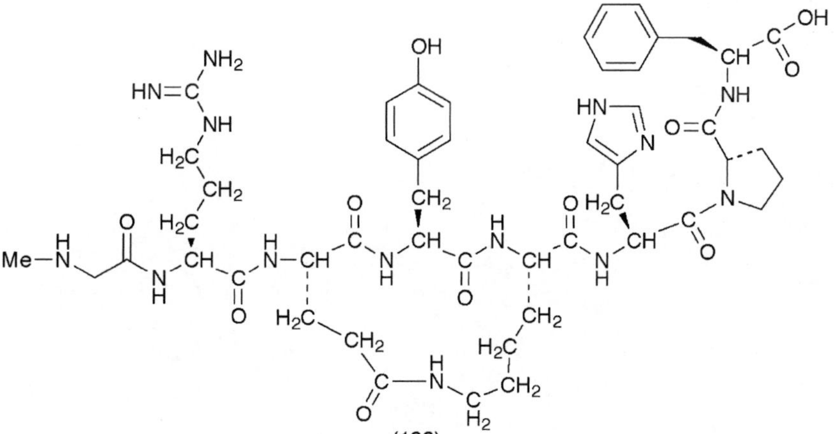

(133)

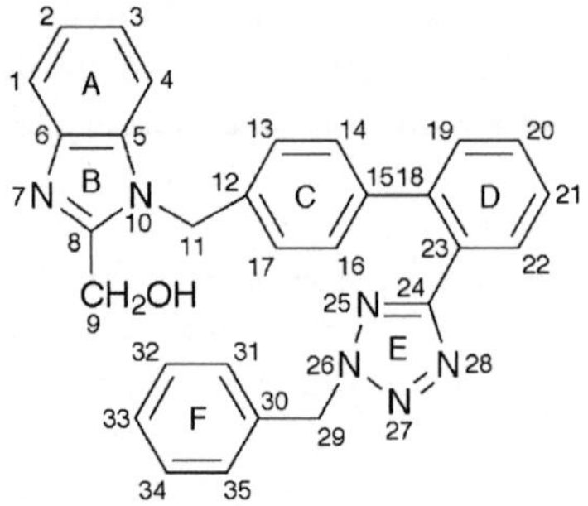

(134)

An angiotensin II degrading peptidase was isolated from the venom of *Conus geographus* marine snail. The main degradation product, angiotensin-(1-7), was analysed by mass spectrometry.[280]

The role and the generation of cardiac angiotensin II were reviewed.[281]

4.5 Bombesin / Neuromedin Analogues. – Macrocyclic chelater containing bombesin analogues, DO3A-amide-BBN (**135**) and DO3A-amide-βAla-BBN (**136**), and their Pm-149, Sm-153 and Lu-177 containing derivatives were prepared and evaluated for potential radiotherapy.[282] In an another attempt Tc-99 containing bombesin (7-14) analogue (**137**) was prepared and investigated for chemical and biological stability, binding assay, internalization and tissue distribution.[283]

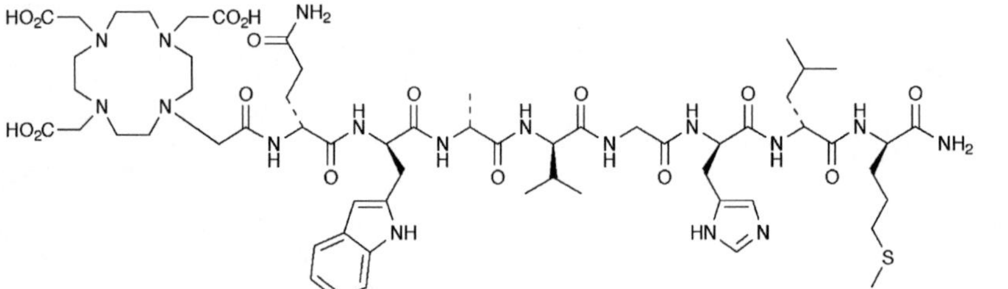

(135)

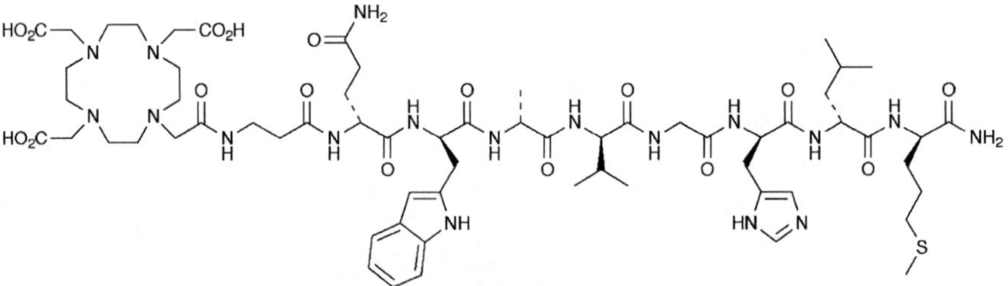

(136)

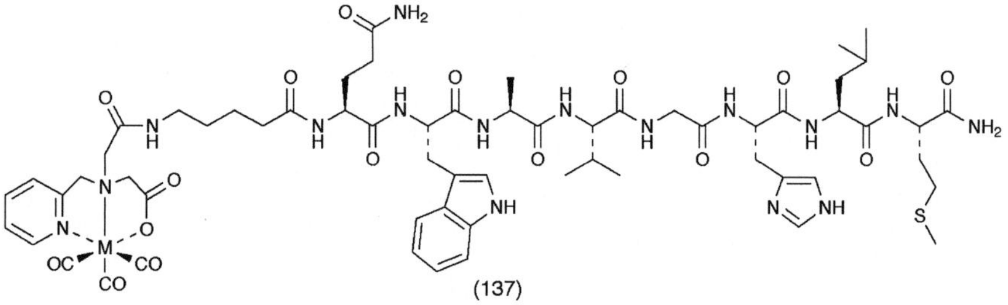

(137)

In a review article the potential use of different radiolabeled peptide including bombesin for the diagnosis and therapy of oncological diseases was discussed.[284]

The effects of bombesin and bombesin like peptides and antagonists on feeding behavior were reviewed.[285] Based on this investigation these peptides could serve as lead compound for the development of new therapeutics for the treatment of obesity.

Alanine-substituted derivatives of neuromedin U-8 peptide were synthesized and tested for induction of intracellular calcium flux.[286] Based on this measurement the arginine residue in position seven is essential for NmU-8 functional activity.

Recombinant large proneuromedin N was synthesized and characterized for its biological activity.[287] Based on these measurements the above molecule might represent an endogenous activator of the human subtype 1 neurotensin receptor.

4.6 Bradykinin Analogues. – The bradykinin receptors of osteoblastic cell line were characterized using ten different bradykinin analogues.[288] The results indicated that the receptor subtypes are linked to different signal transduction mechanisms.

The pharmacokinetics properties of a dimeric third generation bradykinin antagonist (**138**) were studied using LC-MS/MS experiments.[289] The highly sensitive method allows the quantization at nanogram / ml concentration. Nine linear and cyclic B_2 receptor antagonists involving the C-terminal moiety of Icatibant (**139**) and MEN 11270 (**140**) and their analogues were characterized for receptor binding.[290] Based on these measurements several suggestions were made for the mode of the peptide receptor interaction. Twelve new bradykinin antagonists were synthesized[291] and characterized by anti-cancer assay.[292] This class of compounds offers hopes for development of new drugs for prostate cancer. Bradykinin analogues containing 2,2,6,6-tetramethylpiperidine-N-oxyl-4-amino-4-carboxylic acid spin label were synthesized by solid-phase methodology.[277] The active paramagnetic analogue may be useful for conformational studies.

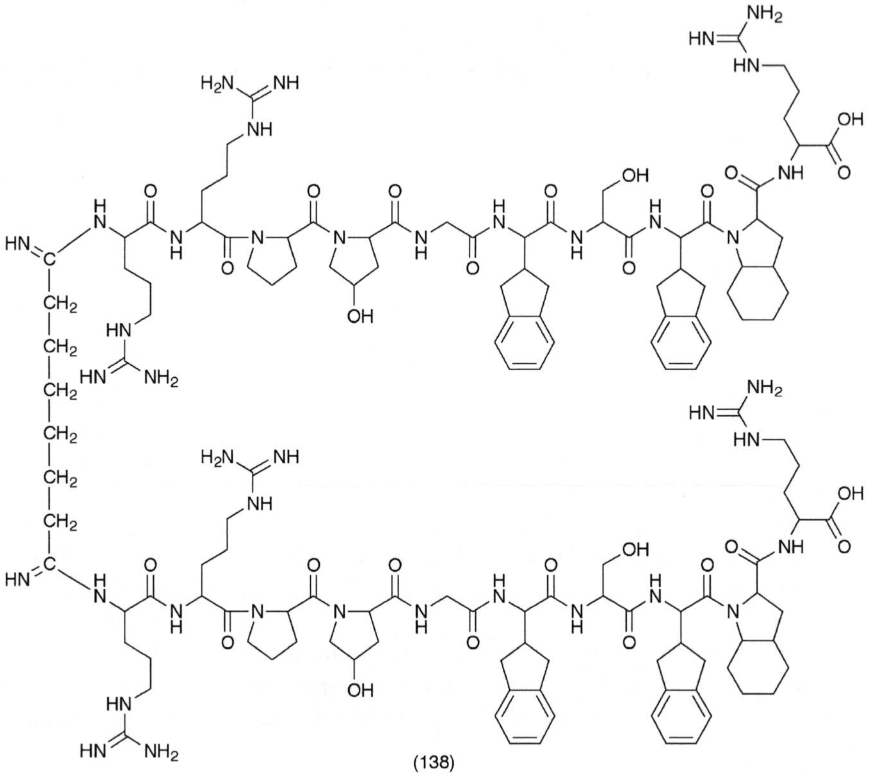

NH2 HN HN OH H2N NH NH O NH O OH O O O O H O H O O H N N N N N N N N N HN C H H H H O O O O CH2 OH HN CH2 HN NH2 CH2 CH2 CH2 CH2 HN C N N N N N N N N N HN H H H H O O O O OH HN HN NH2

(138)

H2N NH HN OH O N HO O 4 O H 5 H 10 H2N 1 N 2 3 N 6 N 7 N 8 9 N CO2H O O N H O H O O N H O S NH HN NH2

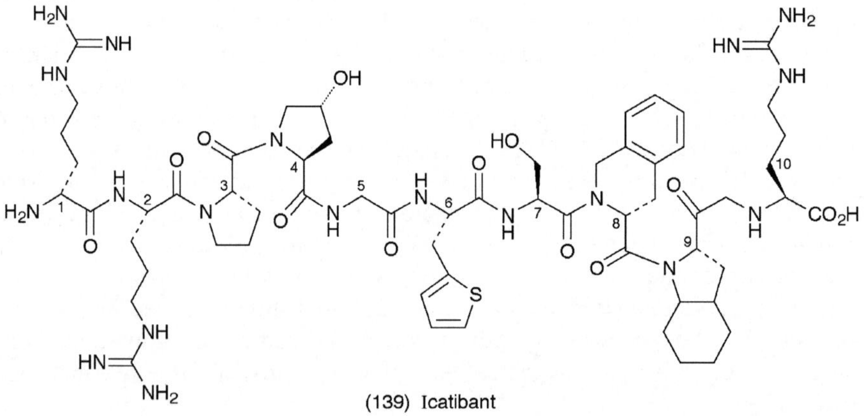

(139) Icatibant

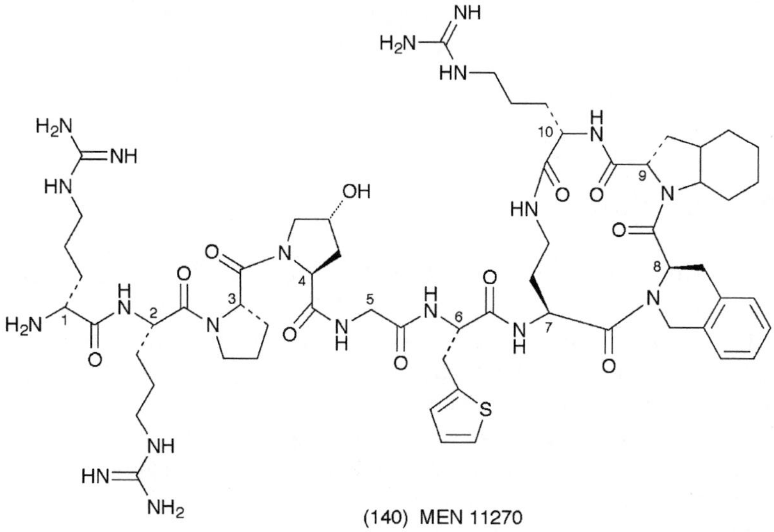

(140) MEN 11270

4.7 Cholecystokinin Analogues, Growth Hormone-Releasing Peptide and Analogues. – The degradation process of cholecystokinin-8 and tetragastrin were elucidated using jejunum-derived enzymes and various enzyme inhibitors.[293] According to this investigation at least three metabolic pathways occur independently at the peptide bonds G-W M-D and D-F.

In a review article the potential use of different radiolabeled peptides including cholecystokinin receptor subtype B ligands for the diagnosis and therapy of oncological diseases was discussed.[284]

Tripeptidyl peptidase-I proved to be essential for the degradation of sulphated cholecystokinin-8 in the brain by lysosomal metabolism.[294] The incubation of CCK8 with brain lysosomas results in the sequential removal of the tripeptides DYM and GWM form the N-terminus of CCK-8. Different enzyme inhibitors were used for increasing the life-time of CCK analogues in order to increase their bioavailability.[295]

The multimeric family of cholecystokinin peptides displays a high affinity for two pharmacologically distinct receptors (CCK_1-R and CCK_2-R). Molecular modelling and NMR conformational investigations were carried out in order to characterize the two different ligand/receptor complex.[296]

Nine analogues and shorter fragments of GHRH were tested for their binding to renal medulla.[297] This binding site exhibits biochemical characteristics different from those of anterior pituitary binding sites. The analogues were as follows: rGHRH(1-29)NH_2, N^{α}-Ac[*D*-Arg^2, Ala^{15}]rGHRH(1-29)NH_2, rGHRH(3-29)NH_2 hGHRH(1-29)NH_2, [Ala^{15}]rGHRH(1-29)NH_2, hGHRH(1-44)NH_2, [*D*-Val^{13}]hGHRH(1-29)NH_2, rGHRH(1-21)NH_2, [des^{13-15}]hGHRH(1-29)NH_2.

Numerous tripeptide based growth hormone secretagogues were designed and synthesized.[298] Their general structure is shown in Figure 24. Several of

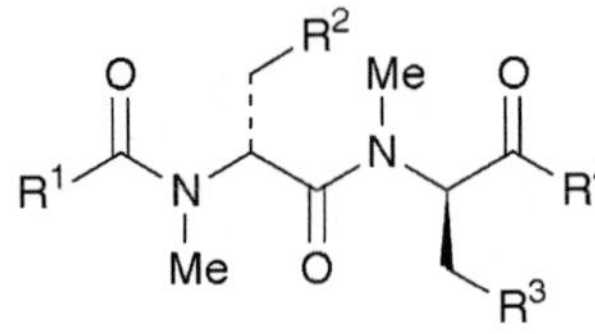

Figure 24

them proved to be more potent than the parent GHRP. Different ghrelins including chicken and bullfrog ghrelin, and GH-releasing peptide were tested as inhibitor of food intake.[299] In a short review the potential role of growth hormone antagonists in the treatment of acromegaly, cancer and chronic complications of diabetes mellitus was discussed.[300]

The action of GH secretagogues including ghrelin for different peripheral activities offers the possibility that GH secretagogue analogues acting as agonists or antagonists on appetite would represent new drug intervention for eating disorders.[301] Plasma ghrelin level and hunger has a strong correlation.[302] Fluorescent probes containing peptide agonist and antagonist molecules against cholecystokinin receptor were synthesized.[303] Investigation of these molecules supported the hypothesis of the molecular conformational change upon receptor activation.

4.8 Integrin-related Peptide and Non-Peptide Analogues. – Integrins are involved in a wide range of pathological processes. Potential therapeutic and diagnostic implications of β_1 ($\alpha_4\beta_1$ and $\alpha_5\beta_1$) and β_3 ($\alpha_{IIb}\beta_3$ and $\alpha_v\beta_3$) integrin antagonists have been reviewed.[304] Synthesis of ligands for the integrin glycoprotein receptors on platelets and difficulties of incorporation of a convenient nucleotide for imaging purpose were the subject of a 52-reference review.[305] Electron microscopy and physicochemical measurements showed a highly bent conformation of $\alpha_v\beta_3$ and $\alpha_{IIb}\beta_3$ integrins which tends to have low affinity for ligands.[306] However, addition of a high affinity ligand mimetic peptide resulted in a switch blade-like opening to an extended structure. The ability of the C-terminal thioesterase domain from a nonribosomal peptide synthetase, a catalyst for integrin binding peptides to catalyze cyclization of rationally designed peptide thioesters has been investigated.[307] The integrin inhibition ability of the products was also evaluated.

4.8.1 IIb/IIIa Antagonists. Chemical aspects of glycoprotein IIb/IIIa inhibitors as antiplatelet agents have been reviewed in 113 references.[308] A new synthetic route capable of producing several hundred kilograms of Lotrafiban (SB-214857) (**141**), an orally active GPIIb/IIIa receptor antagonist of high chiral purity has been developed.[309] A series of new GPIIb/IIIa antagonists were designed and synthesized modifying the side chains of sibrafiban and lamifiban.[310] Four of the synthesized peptides showed similar activity to the parent compounds. Two heterodimeric disintegrins, acostatin and piscivostatin were isolated and cloned from the venom of two *Agkistrodon* species.[311] Each subunit of these disintegrins is encoded on two different length cDNAs. A novel disintegrin, saxatilin, was purified from Korean snake (*Gloydius sayatilis*) venom.[312] This 73-amino acid polypeptide possessing the RGD sequence inhibited GPIIa/IIIb binding to immobilized fibrinogen with an IC_{50} of 2.0 nM. Locations of the six disulfide bonds of saxatilin and their structural and functional importance were determined in another study.[313] YM-57029 (**142**) has been found to be a potent GPIIb/IIIa antagonist which strongly inhibits various platelet functions.[314]

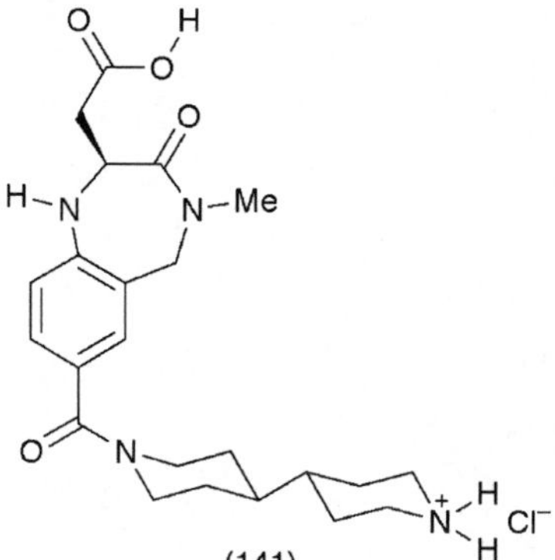

(141) 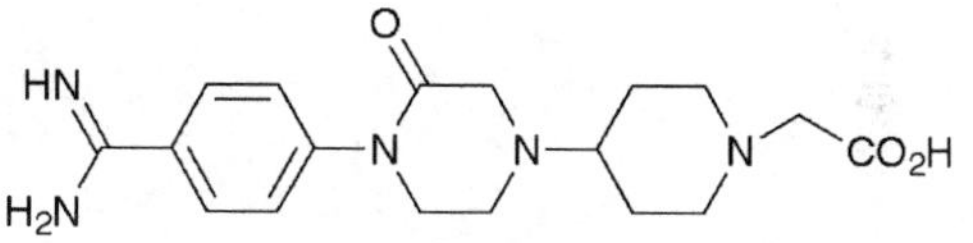 (142)

Biochemical properties of this novel GPIIb/IIIa antagonist as well as pharmacodynamics and pharmacokinetics of its prodrug, YM128 were also investigated.[315] Savignygrin, a platelet aggregation inhibitor that inhibits platelet by targeting the platelet integrin $\alpha_{IIb}\beta_3$ was purified from the soft tick *Ornithodoros savignyi*.[316] This is the first described platelet aggregation inhibitor with an RGD motif using the Kunitz-BPT1 protein fold.

4.8.2 $\alpha_v\beta_3$ Antagonists. Recent data on the role of integrins and the mechanism of action of $\alpha_v\beta_3$ integrin antagonists in order to help to develop an anti-integrin therapy in cancer have been reviewed.[317] A review of issued patents and published patent applications on novel $\alpha_v\beta_3$ ligands which have appeared since 1999 has been published.[318] An effective Heck reaction between a highly functionalized iodoquinazolinone and a series of unprotected allyl amidines and guanidines was applied to the synthesis of novel vitronectin receptor antagonist such as compound (**143**).[319] The synthetic procedure and the conformational analysis of sugar azido acids (**144** and **145**) has been reported[320]. Compound (**144**) was built into the cyclic RGD peptide providing a selective antagonist of the $\alpha_v\beta_3$ receptor. Synthesis and binding results of cyclic RGD analogues containing enantiomeric pure β-methylated (2*S*,3*S*) and (2*S*,3*R*) aspartic acid residues based on the $\alpha_v\beta_3$ selective Arg-Gly-Asp-Asp-Val, Arg-Gly-Asp-Asp-(tBuG), Arg-Gly-Asp-Tyr(Me)-Arg and the non-selective but effective Arg-Gly-Asp-Thr-Tic pharmacophores were published[321]. Novel tricyclic benzodiazepinedione-based RGD analogues (Figure 25) have been synthesized and tested in a solid-phase receptor assay.[322]

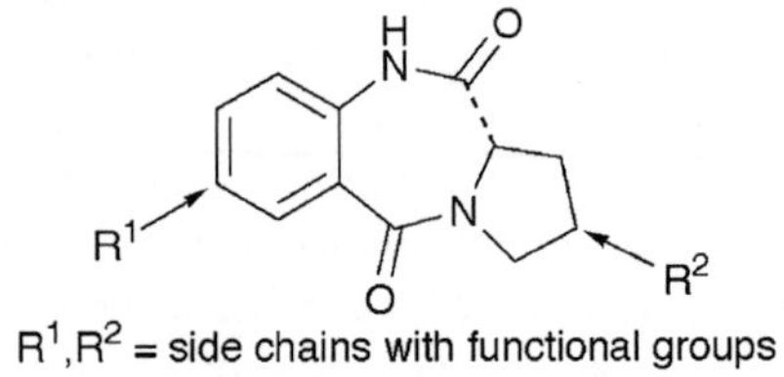

R^1,R^2 = side chains with functional groups

Figure 25

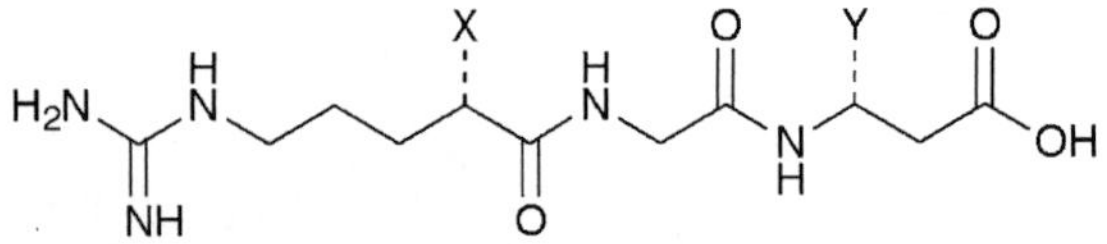

Figure 26

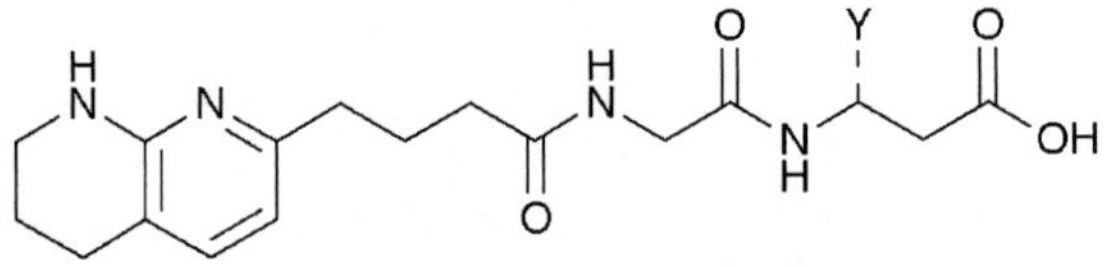

Figure 27

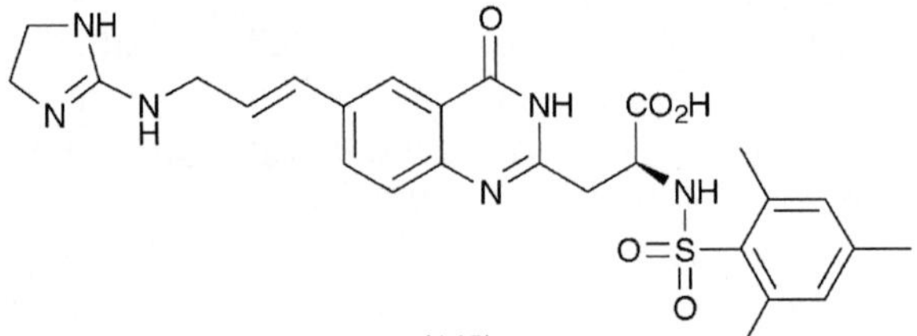

(143)

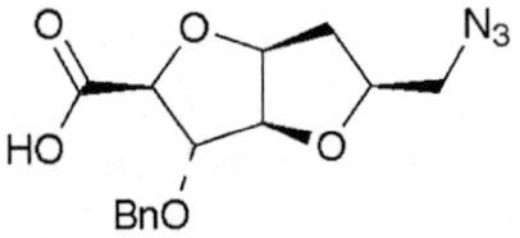

(144)

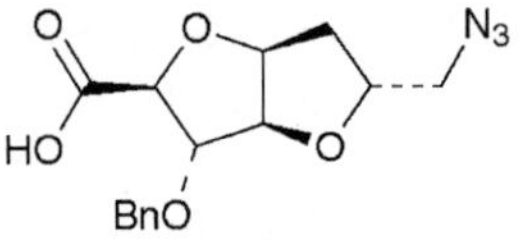

(145)

For the preparation of glycine-containing (Figure 26 and 27) as well as glycyl-constrained (Figure 28) non-peptide RGD antagonists, the use of 5,6,7,8-tetrahydro[1,8]naphthyridine group as a potency-enhancing N-terminus has been demonstrated.[323] In spite of the fact that most potent antagonist of this series shown in Figure 28a bound to human $\alpha_v\beta_3$ with an IC_{50} of 0.35 nM, its use is limited by poor oral pharmacokinetic parameters. To overcome this problem, more lipophilic, non-basic, novel 3-aryl-substituted β-alanyl residues were designed as aspartic acid replacements.[324] Among these $\alpha_v\beta_3$ receptor

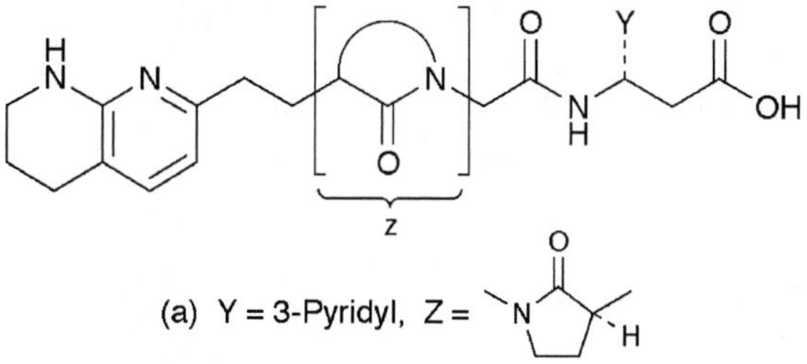

(a) Y = 3-Pyridyl, Z =

Figure 28

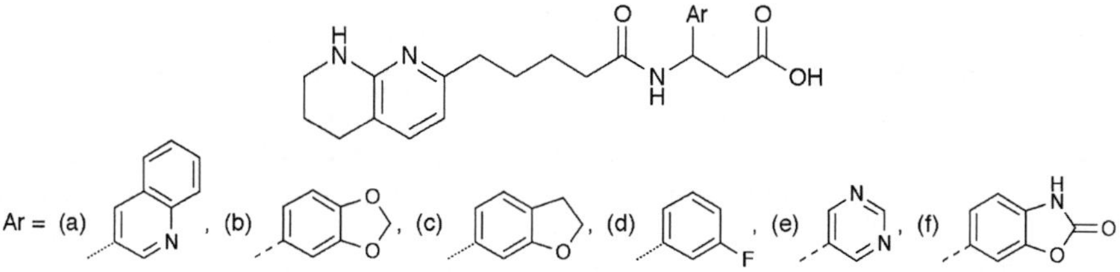

Ar = (a) , (b) , (c) , (d) , (e) , (f)

Figure 29

antagonists, the 3-fluorophenyl derivative (**146**), possessing an IC_{50} of 1.8 nM, displayed the best pharmacokinetic profile.

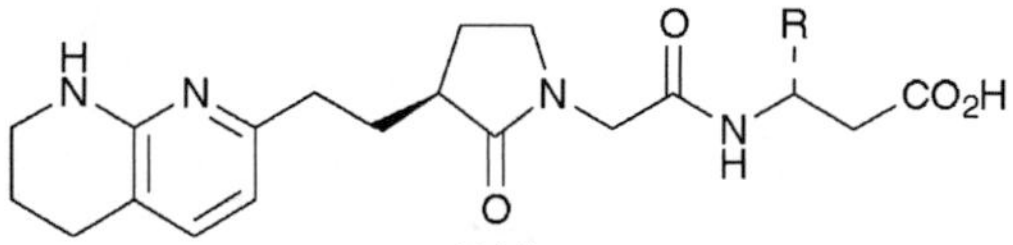

(146)

As continuation of this work, novel, chain-shortened vitronectin receptor antagonists (Figure 29) were prepared.[325] From this series, the dihydrobenzofuran derivative shown in Figure 29c represents an optimal compound in terms of antagonist activity, pharmacokinetics, and physicochemical properties. In another study by the same research group, the 3-aryl versus 2-aryl β-amino acid substitution as aspartic acid replacement in the chain-shortened series was examined.[326] In particular, the dihydrobenzofuran analogue (**147**) demonstrated excellent binding affinity (IC_{50}: 1.01 nM), while maintaining a good pharmacokinetic profile.

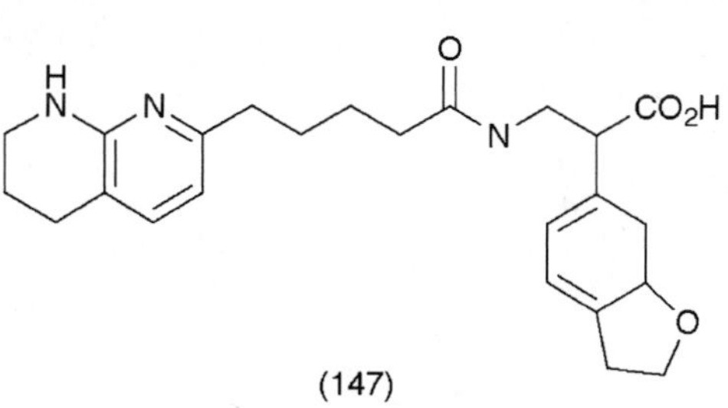

(147)

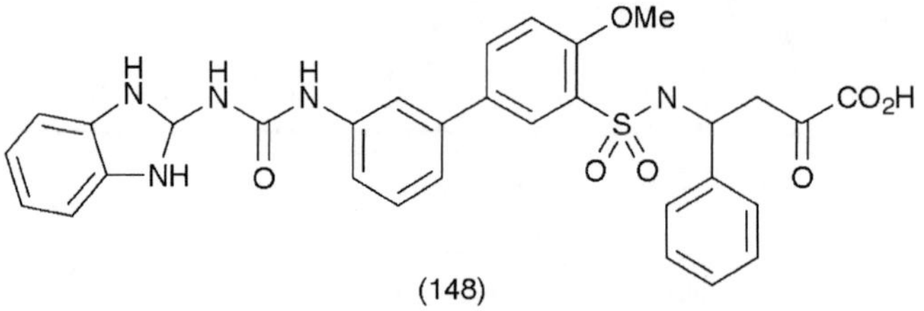

(148)

A series of potent vitronectin receptor antagonists based on a biphenyl motif has been identified from a combinatorial library.[327] Compound (**148**) was not only a potent antagonist of $\alpha_v\beta_3$, but exhibited more than 300-fold selectivity to the GPIIb/IIIa receptor. Kessler's c(RGDfK) $\alpha_v\beta_3$ integrin antagonist was synthesized on solid phase.[328] Lysine functionalization as well as cyclisation was conducted on-resin to avoid intermediate purification steps and solubility problems. The N-substituted dibenzazepinone scaffold was successfully applied for the synthesis of new vitronectin receptor antagonists.[329] Structure-activity relationship of the spacer and the guanidine pharmacophore led to novel, potent $\alpha_v\beta_3$ antagonists. The most potent compounds (**149**) and (**150**) showed high selectivity versus $\alpha_5\beta_1$ and $\alpha_4\beta_1$ integrins and displayed good absorption and metabolic stability.

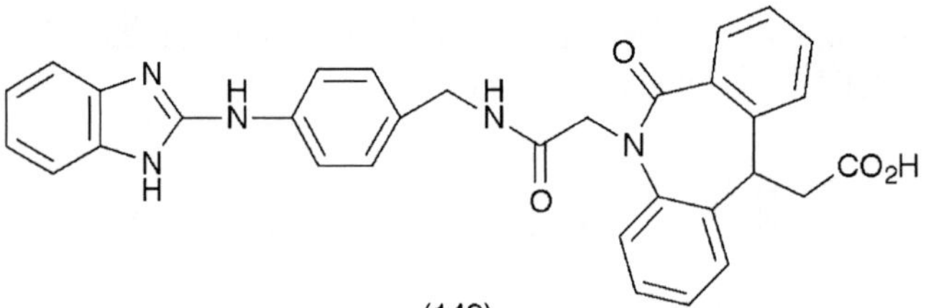

(149)

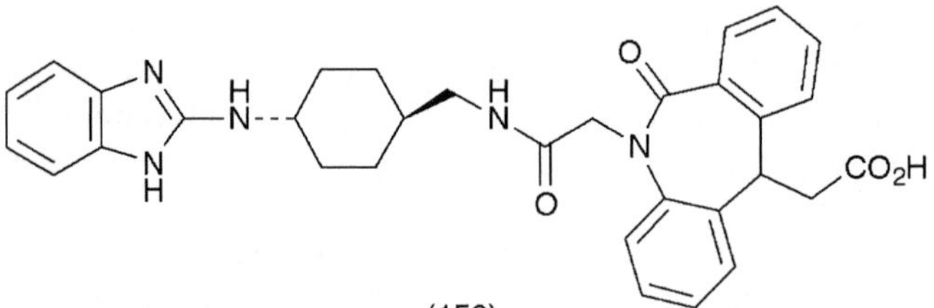

(150)

Solid-phase synthesis and structure-activity relationship of heteroaryl ureas as $\alpha_v\beta_3$ antagonists was described.[330] Two α-*Z*-diaminopropionic acid derivatives (**151**) and (**152**) (IC_{50}: 0.4 and 1.8 nM, respectively) were found to be selective versus related integrins like $\alpha_{IIb}\beta_3$, $\alpha_5\beta_1$ and $\alpha_4\beta_1$.

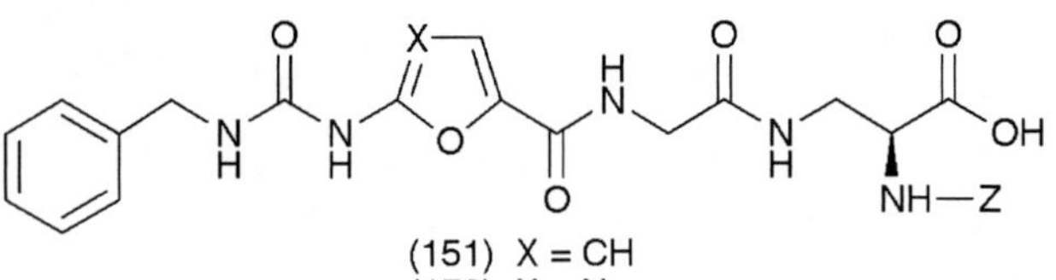

(151) X = CH
(152) X = N

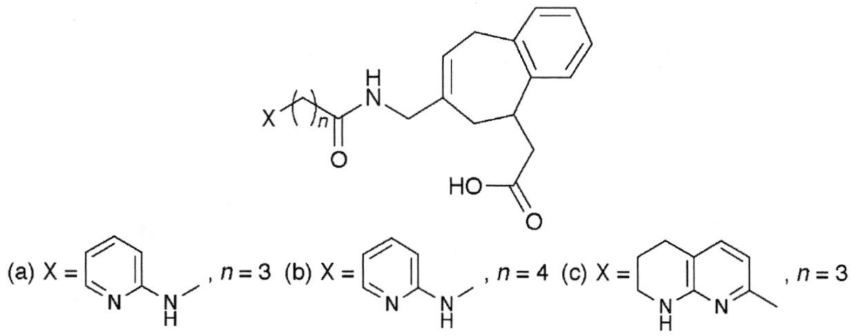

Figure 30

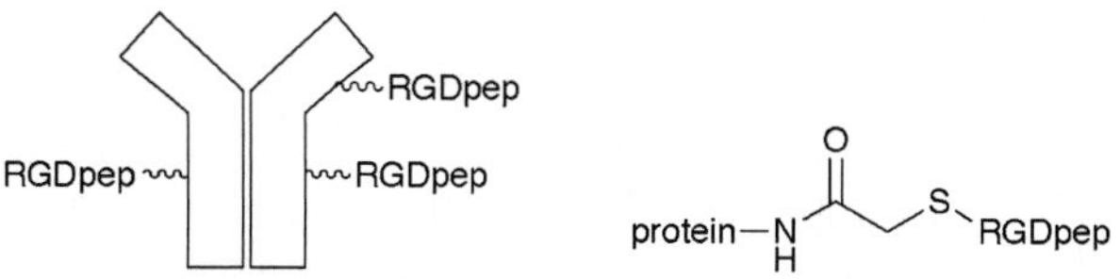

Figure 31

RGD peptidomimetic benzocycloheptene derivatives as potent vitronectin receptor antagonists have been designed and synthesized.[331] Three compounds of this novel series shown in Figure 30 displayed nano- to subnanomolar IC_{50} values on $\alpha_v\beta_3$ and $\alpha_v\beta_5$ integrins, with good selectivity over $\alpha_{IIb}\beta_3$ receptor and favorable pharmacokinetics. Multivalent derivatives of the cyclic RGD-peptide c(RGDfK) by covalent attachment of the peptide to a protein have been prepared and characterized.[332] These conjugates (Figure 31) showed increased affinity for $\alpha_v\beta_3/\alpha_v\beta_5$ integrins and inhibited the interaction between these integrins and vitronectin.

Starting from the crystal structure of the extracellular domain of integrin $\alpha_v\beta_3$, a three-state mechanism of integrin activation has been developed.[333] In addition, the ligand binding site in atomic detail has been written. The binding site for the RGD triad within the human $\alpha_v\beta_3$ integrin has been identified by synthesizing photoreactive p-benzoyl-phenylalanyl residue containing analogues of the disintegrin, echistatin.[334] Multivalent peptide-protein conjugates in which cyclic RGD-peptides were covalently attached to a protein backbone proved to be potent carriers for the intracellular delivery of pharmacologically active agents.[335] Inhibition of α_v-integrins by a methylated cyclic RGD-peptide resulted in retardation of tumor growth and metastasis *in vivo*.[336] The radiolabelled dimeric RGD peptides, ^{111}In-DOTA-E-[c(RGDfK)2] (**153**) and ^{99m}Tc-HYNIC- E-[c(RGDfK)2], demonstrated high and specific tumor uptake in a human tumor xenograft.[337] Another study showed that the tumor uptake of ^{99m}Tc-RGD-HYNIK in $\alpha_v\beta_3$ positive tumors growing in nude mice was marginal due to limited numbers of these integrins on the tumor receptors used.[338]

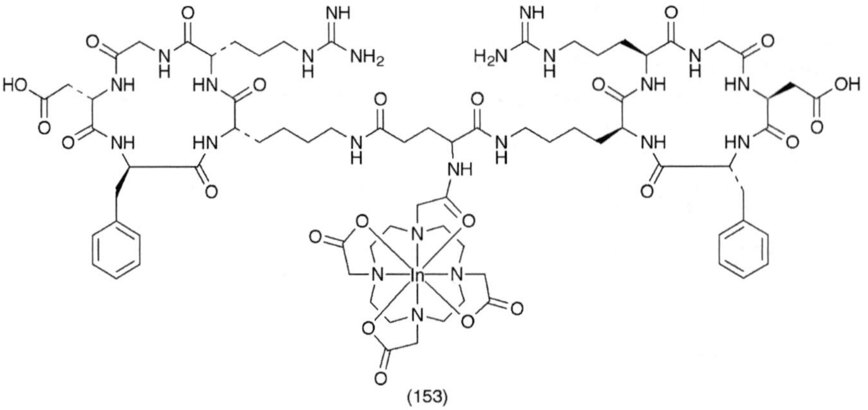

(153)

4.8.3 $\alpha_4\beta_1$, $\alpha_4\beta_7$ and $\alpha_5\beta_1$ Antagonists. Four reviews have been appeared on the field of VLA-4 antagonists showing the increasing interest toward α_4 integrins as drug targets in inflammation and autoimmune diseases. One of them reviewed VLA-4 antagonists in 56 references.[339] Among several types of molecules, peptide mimetics of the LDV sequence and 4-substituted N-acylphenylalanine derivatives have been summarized.[340] The rationale for targeting α_4 integrins for the treatment of autoimmune disorders was discussed.[341] A 162-reference review focused on the structural bases and clinical relevance of VLA-4 as well as LFA-1 inhibition.[342]

Three research groups published more than a dozen papers in Bioorganic and Medicinal Chemistry Letters about different molecules as VLA-4 antagonists. Most of these potent compounds were *N*-acylphenylalanine derivatives. Optimization of an acylated β-amino acid (**154**) identified from a combinatorial library led to potent and orally bioavailable VLA-4 antagonists (**155, 156** and **157**) having IC_{50} values of 0.49, 0.27 and 0.85 nM, respectively.[343] A series of *N*-(3,5-dichlorobenzenesulfonzl)-*L*-prolyl- (**158**) and α-methyl-*L*-prolyl-phenylalanine (**159**) derivatives proved to be excellent VLA-4 inhibitors having IC_{50} values in the nanomolar range and good oral bioavailability in several species.[344]

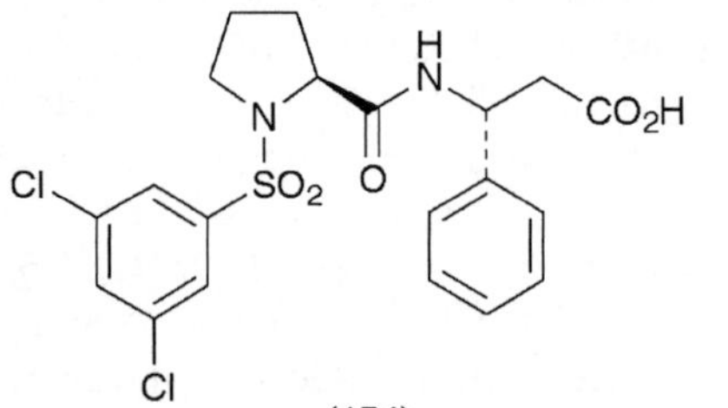

(154)

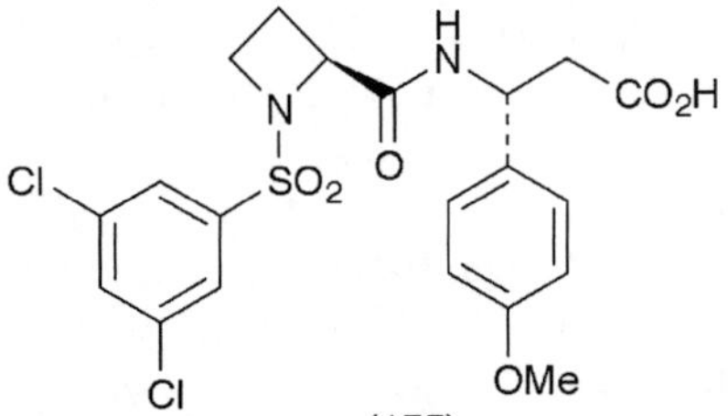

(155)

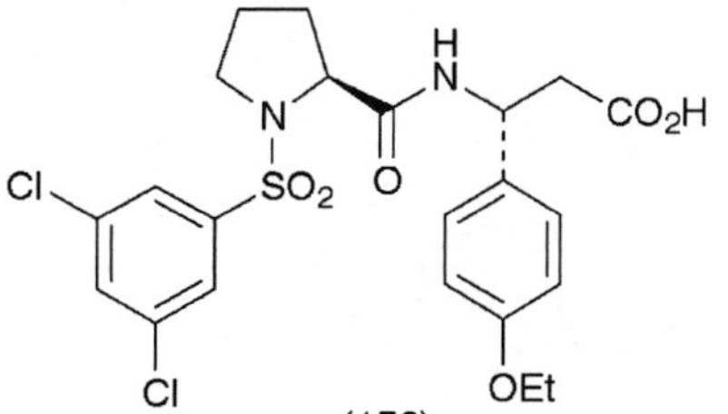

(156)

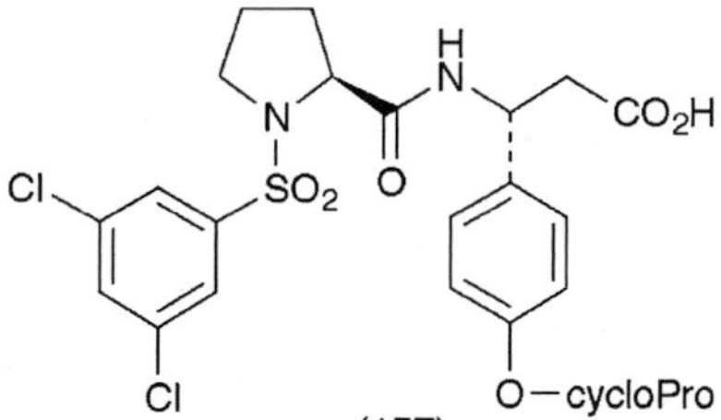

(157)

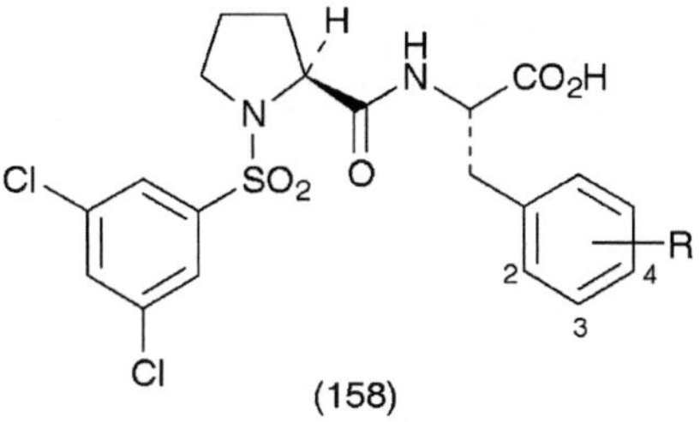

(158)

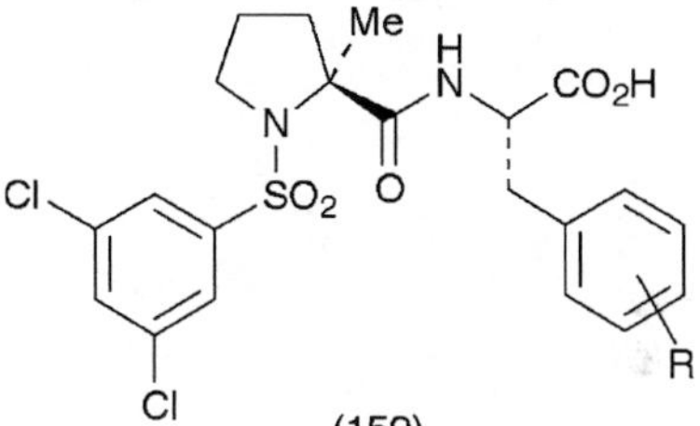

(159)

Three novel series of VLA-4 antagonists with low nanomolar activity were identified from a combinatorial library of 28 pools of 180 compounds.[345] Two of the most potent analogues (**160** and **161**) are urea-based compounds that had not been previously reported as α_4 integrin antagonists. *N*-Phenyl and *N*-heteroaryl derivatives of phenylalanine with hydrogen bond acceptors in the *meta* position having the general formula of (**162**) exhibited low nanomolar activity showing that molecules with less peptide character could also serve as VLA-4 inhibitors.[346] *N*-Tetrahydrofuroyl-L-phenylalanine derivatives (**147**) could also serve as potent VLA-4 antagonists.[347] Replacing several groups on (**163**) the inhibitory activity and selectivity among different α_4 integrins could be improved.[348]

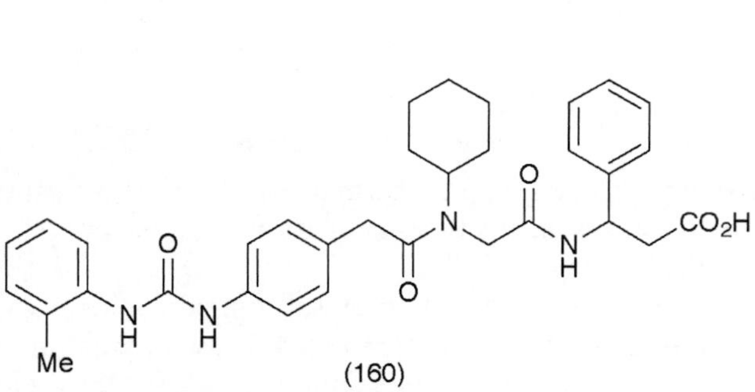

(160)

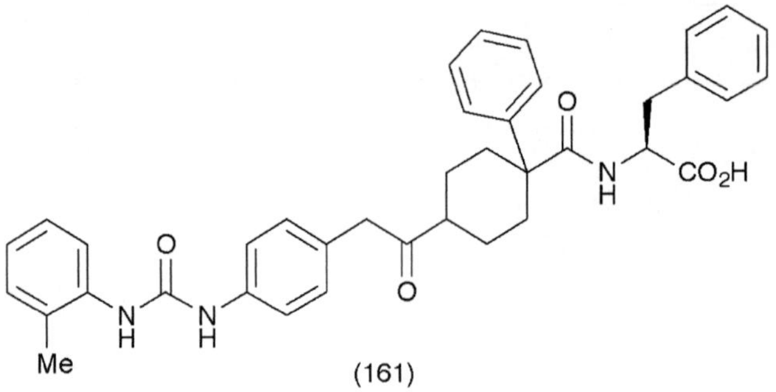

(161)

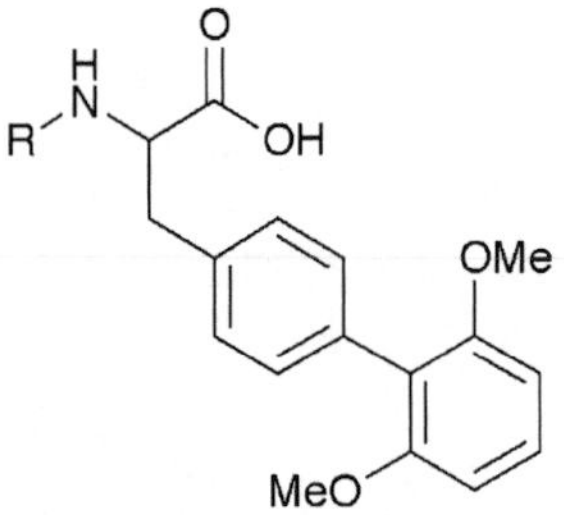

(162) 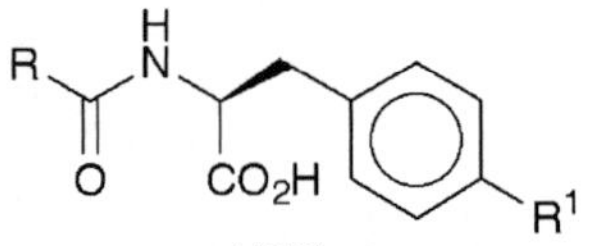 (163)

A series of potent *N*-(aralkyl-, aryl-cycloalkyl-, and heteroaryl-acyl)-4-biphenylalanine VLA-4 antagonists were prepared by rapid analogue methods on solid phase.[349] Several of these analogues showed high potency (IC_{50} < 1nM). *N*-Aryl-proline derivatives such as (**164**) and (**165**) represent small molecule VLA-4 antagonists having some advantages over the related *N*-(arylsulfonyl)-prolyl-dipeptide analogues.[350]

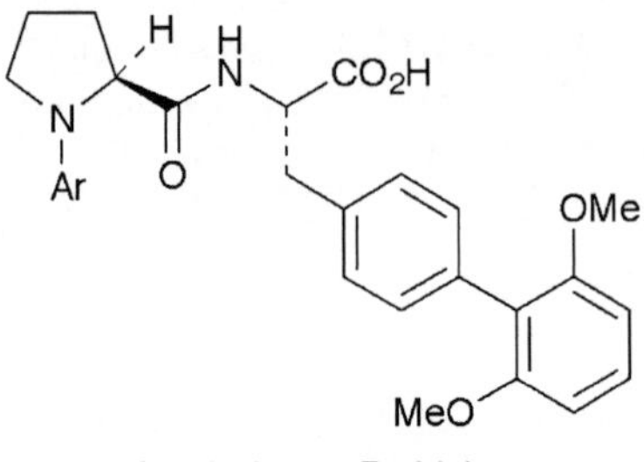

(164) Ar = 3-Pyridyl
(165) Ar = 2-Pyrazinyl

Finally, the same research group found that *N*-sulfonylated-dipeptide β-biaryl-β-amino acid derivatives having the general formulae (**166**) or (**167**) were potent and specific VLA-4 antagonists.[351]

Another research group worked on *N*-substituted-*L*-phenylalanine derivatives as VLA-4 antagonists. A structure-based focused library approach led to the discovery of two new classes of inhibitors: *N*-(α-phenylcyclopentanoyl)-(**168**) and *N*-(2,6-dimethylbenzoyl) (**169**) derivatives.[352] Using a series of *N*-(benzylpyroglutamyl)-derivatives having the general formula of (**170**) it was shown that a wide variety of sub-structures were tolerated on the aromatic ring.[353] A systematic structure-activity relationship investigation was performed on both *N*-cycloalkanoyl[354]- and *N*-aroyl-phenylalanine[355] derivatives having the general formula (**171**).

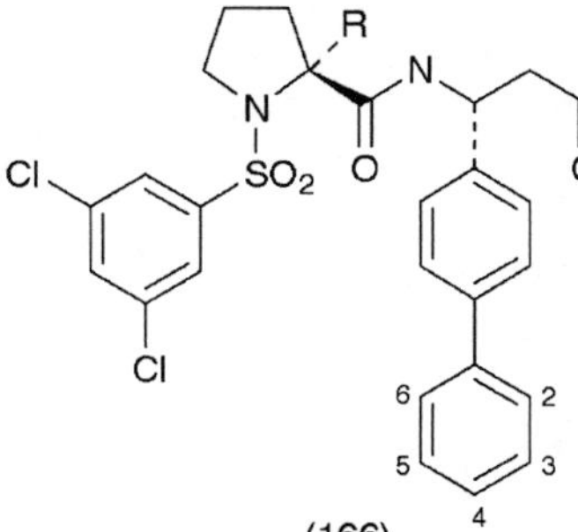

(166) 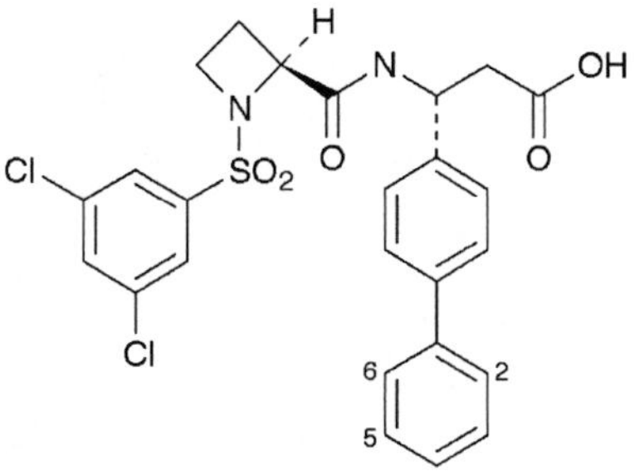 (167)

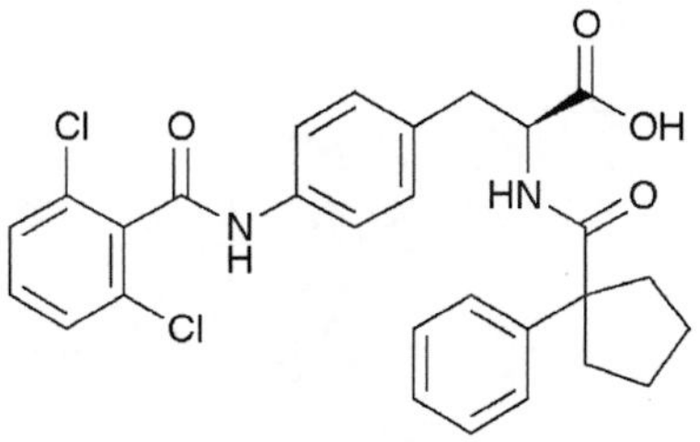

(168) 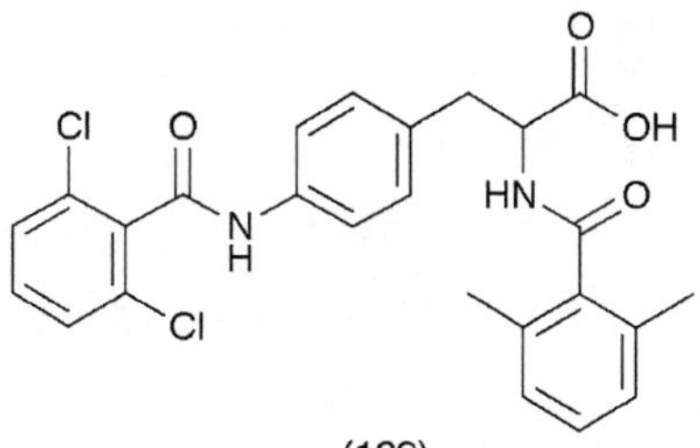 (169)

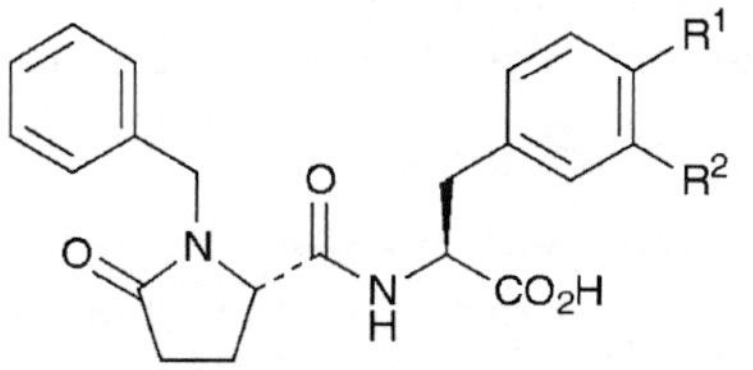

(170) 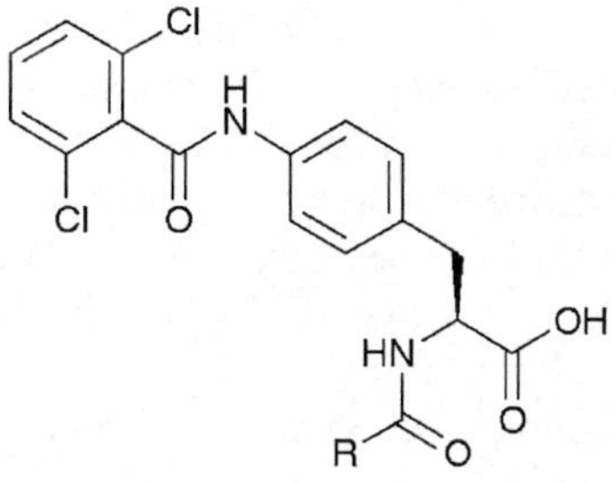 (171)

The replacement of the α-amido carbonyl moiety in VLA-4 antagonists with 3-cyclobutene-1,2-dione analogues provided potent antagonists with improved pharmacokinetic profile.[356] The *n*-propyl-analogue (**172**) proved to have the optimum combination of high potency and good rate of clearance. Compounds prepared by the introduction of a 1,3,5-triazine as an amide isostere to phenylalanine-derivatives displayed reduced clearance while showed only modest potency as VLA-4 antagonists.[357] Starting from this knowledge and some previously described *N*-(pyrimidin-4-yl) (**173**) and *N*-(pyridin-2-yl) (**174**) phenylalanine derivatives, SAR studies aimed at optimizing the potency and rate of clearance on these series of compounds.[358]

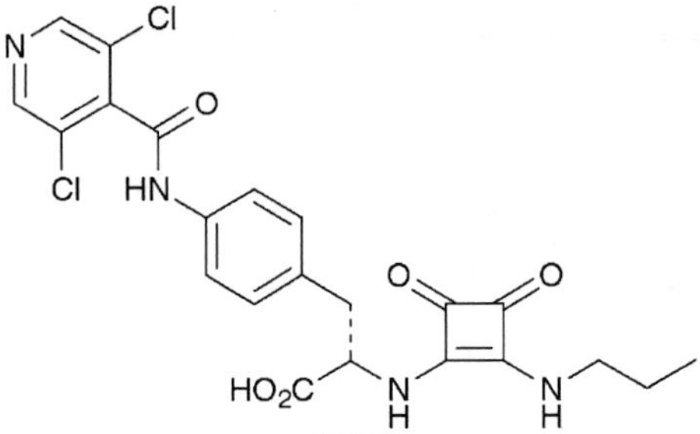

(172)

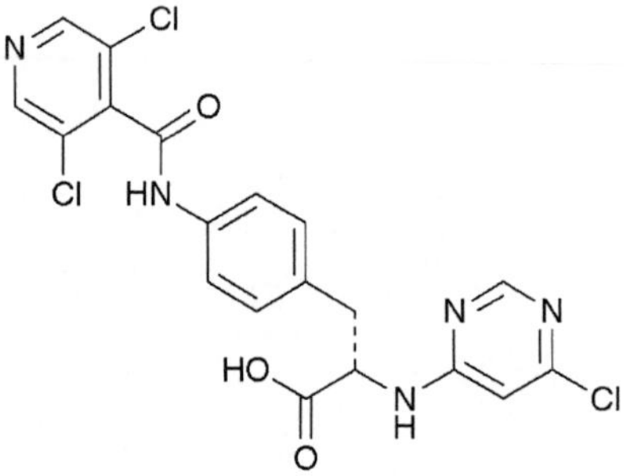

(173)

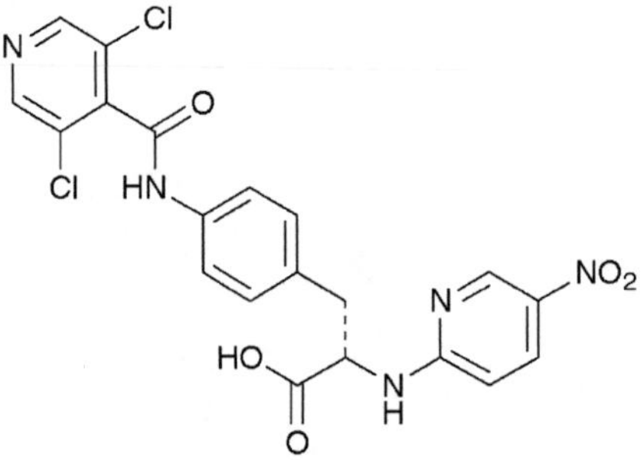

(174)

Aromatic β-amino acid esters were prepared and used as mimics of the Asp-Phg C-terminus in LDV derived VLA-4 antagonists.[359] *In silico* screening was used to identify potent and selective $\alpha_4\beta_1$ antagonists.[360] The most potent compound showed an IC_{50} of 1 nM. A possible structural basis of the receptor activity of a series of potent VLA-4 antagonists having the general formula of (**175**) has been described.[361] The Glaxo Group Ltd. has published an inhaled single peptidic VLA-4 antagonist.[362]

Highly potent and selective $\alpha_4\beta_7$ antagonists have been identified from a random cyclic peptide library.[363] Nonpeptidic $\alpha_4\beta_7$ antagonists such as compound (**176**) were designed from cyclic peptides using combinatorial and rational strategies.[364] Synthesis and NMR analysis of peptidomimetics act as $\alpha_4\beta_7$ antagonists provided information about the LDV recognition sequence for $\alpha_4\beta_7$ integrins.[365]

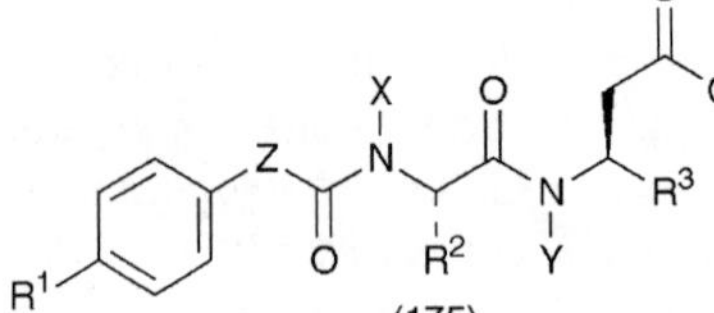

(175)

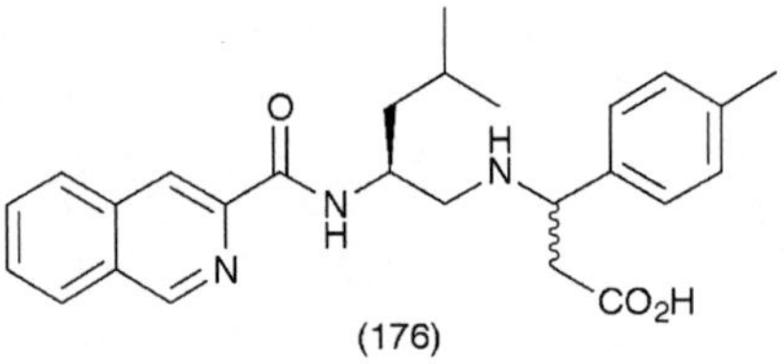

(176)

Dual $\alpha_4\beta_1/\alpha_4\beta_7$ integrin antagonists have also been described. Compound (**177**), having the *S* configuration at the thioproline moiety, inhibited $\alpha_1\beta_7$ binding with an IC_{50} of 0.2 nM and that of $\alpha_4\beta_7$ with an IC_{50} value of 2.1 nM.[366] Systematic SAR studies of a $\alpha_4\beta_1$ specific antagonist (**178**) led to small molecule carbamates as potent dual $\alpha_4\beta_1/\alpha_4\beta_7$ integrin antagonists, for example compound (**179**) [$\alpha_4\beta_1$ (VCAM-Ig) $IC_{50} = 0.14$ nM, and $\alpha_4\beta_7$ (VCAM-Ig) $IC_{50} = 2.83$ nM].[367] Two structural classes of dual $\alpha_4\beta_1/\alpha_4\beta_7$ integrin antagonists, namely biphenylalanine or tyrosine carbamate scaffolds having the general formulae (**180**) and (**181**), respectively, were optimized for inhibition of $\alpha_4\beta_1$/VCAM and $\alpha_4\beta_7$/MAdCAM.[368] A series of novel *N*-benzoyl-*L*-biphenylalanine derivatives as dual $\alpha_4\beta_1/\alpha_4\beta_7$ integrin antagonists has been discovered.[369] Compound **182** (TR-14035; IC_{50} $\alpha_4\beta_7/\alpha_4\beta_1 = 7/87$ nM) had completed Phase I studies in Europe.

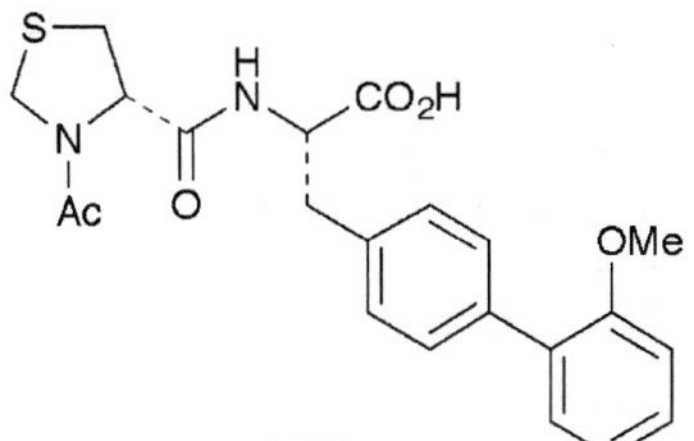

(177) 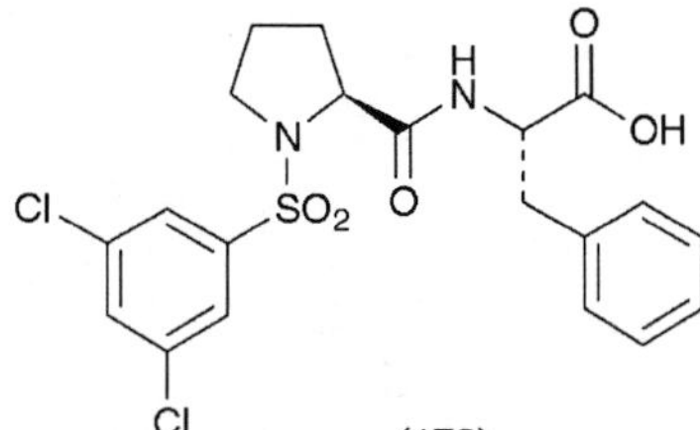 (178)

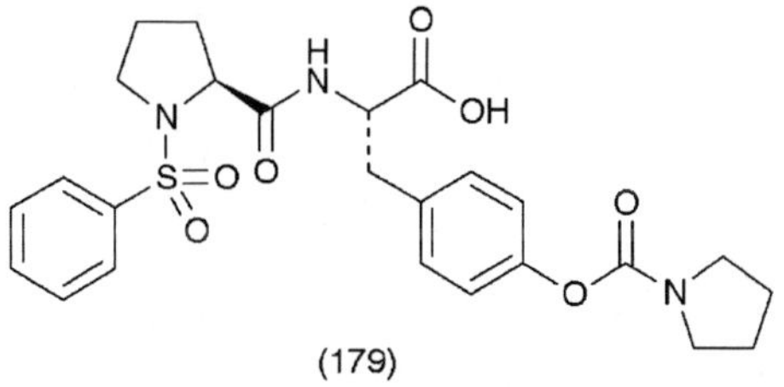

(179) 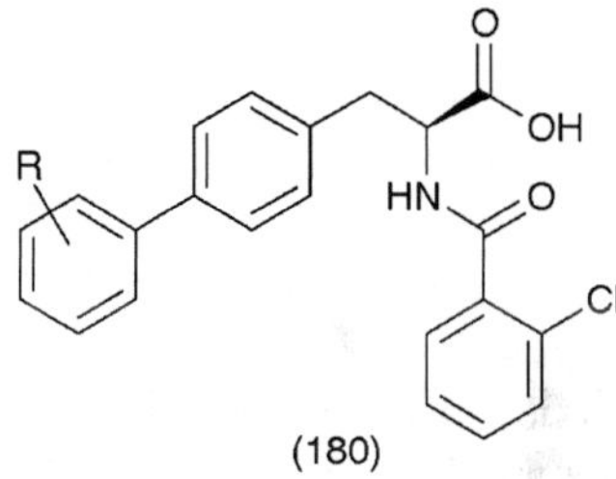 (180)

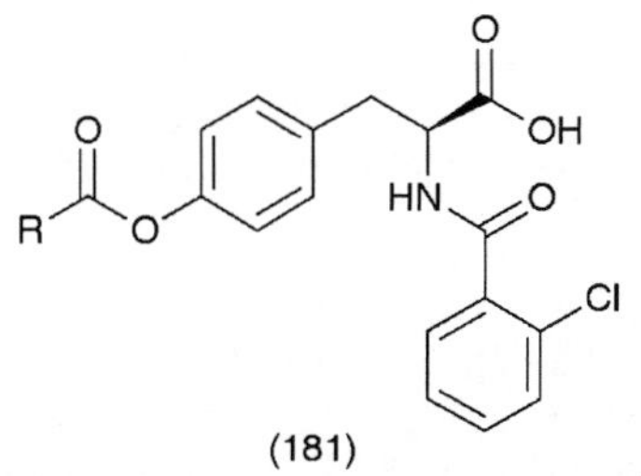

(181) 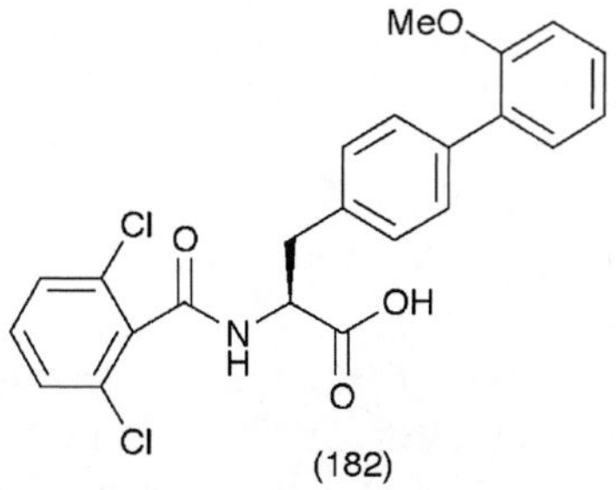 (182)

The discovery that certain RGD-containing peptides can suppress metastasis *in vitro* suggested that selective $\alpha_5\beta_1$ integrin antagonists could lead to new anticancer agents. Receptor bound conformation of an $\alpha_5\beta_1$ integrin antagonist was determined by ^{15}N-edited 2D transferred nuclear overhauser effects.[370] Ocellatusin, a new monomeric RGD-containing $\alpha_5\beta_1$ disintegrin, present in the venom of the Nigerian carpet viper, *Echis ocellatus*, has been characterized.[371] Highly active small molecule inhibitors of $\alpha_v\beta_6$ integrin were described for the first time.[372] Compound **(183)**, the most active member of this series possessing an IC_{50} of 0.04 nM on $\alpha_v\beta_6$ receptor can serve as starting point of structure-activity relationship investigations on this integrin.

4.9. LHRH and GnRH Analogues. – A cytotoxic analogue of LH-RH (*D*-Lys^{6}LH-RH coupled to a doxorubicin derivative) **(184)** was investigated and proved to be potent substance for the inhibition of *in vivo* proliferation of prostate cancer.[373] *D*-Lys^{6}LH-RH-doxorubicin derivative was also used for inhibition of different oral cancer types.[374]

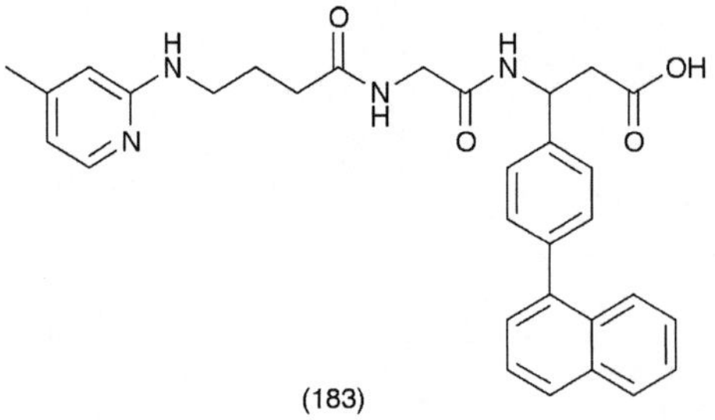

(183)

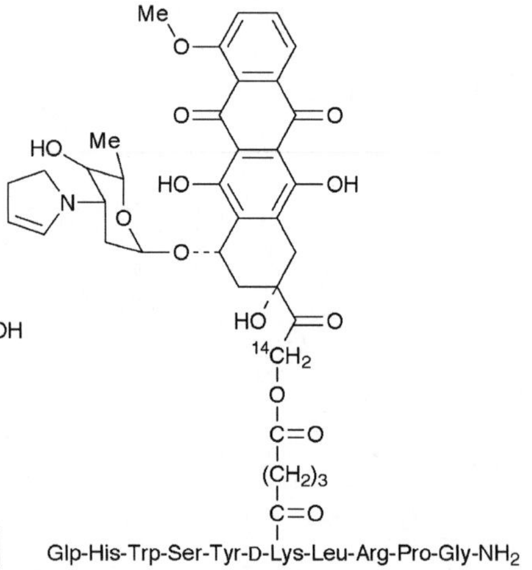

Glp-His-Trp-Ser-Tyr-D-Lys-Leu-Arg-Pro-Gly-NH_2

(184)

4.10 α-MSH Analogues. – Eight cyclic melanotropin analogues were prepared by solid-phase methodology and tested for binding to different melanocortin receptors.[375] The structures of the synthesized molecules are shown in (Figure 32). Two of them proved to be full agonist at the human melanocortin 5 receptor, while they were full antagonist at the human melanocortin 4 receptor. In a review article the role of the five different melanocortin receptors and the possibility for use of melanocortin for the body weight regulation and treatment of obesity was discussed.[376]

4.11 MHC Class I and II Analogues. – The first low-affinity HLA A2.1 restricted epitope (p572) of human telomerase reverse transcriptase (hTRT) has been identified.[377] It was predicted that the specific $CD8^+$ T cell repertoire against this epitope might be more intact than that of high affinity hTRT peptides and that cytotoxic T lymphocyte precursors can be expanded in cancer patients using the analogue epitope. In addition, a vaccine consisting of p572 together with a previously identified high-affinity epitope can have greater chance of success than vaccines containing only high-affinity epitopes.

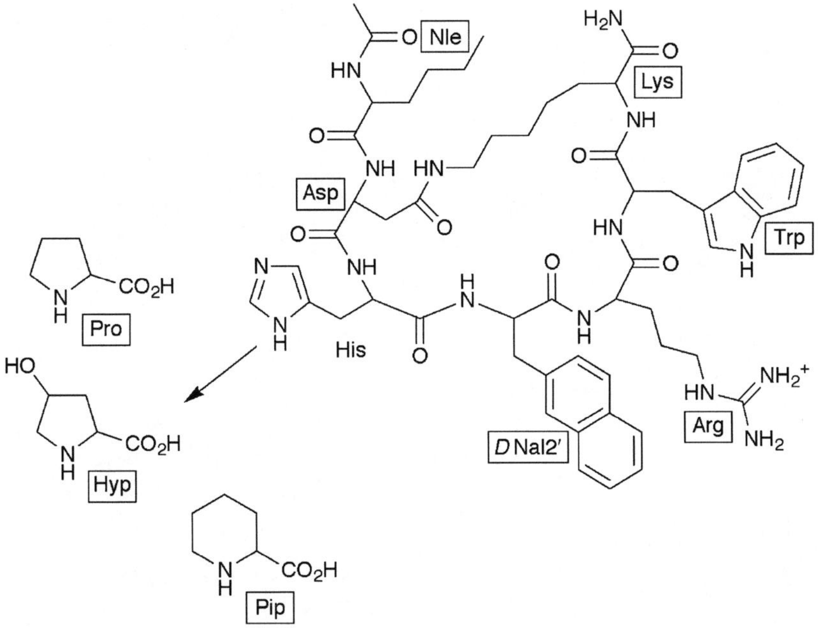

Figure 32

Class II MHC molecules undergo pH-dependent conformational changes. Therefore, low-affinity peptides tightly bound at pH 7.0 can be released at pH 5.0. The imidazole group of histidine is the only side-chain that is affected within this pH range. Conserved histidine residues of the soluble forms of HLA-DR have been substituted by tyrosine residues in order to identify crucial histidine residues by an increase in pH stability of the ligand complex.[378] Met and Cys residues that often reduce the stability of peptides in solution can be replaced by chemically related derivatives without affecting the binding affinity for the relevant MHC molecule.[379]

4.12 Neuropeptide Y (NPY) Analogues. – In a review article the connection between neuropeptide Y and depression was discussed.[380] The six different neuropeptide Y receptors and their involvement in the pathophysiology of several disorders were reviewed.[381]

The cardiovascular function of neuropeptide Y receptor antagonists was investigated.[382] The structure of the selective Y_1 and Y_2 receptor antagonists are shown in Figure 33. Another attempt has been made to use neuropeptide Y receptor antagonists for anti-obesity drug development.[383] Since the neuropeptide Y Y_1 and Y_5 receptors are involved in the regulation of food intake, their antagonists, shown in Figures 34 and 35, are potential pharmaceuticals. The affinities of neuropeptide Y and related peptides for the known neuropeptide receptors are summarized in the same paper. The structure and receptor binding properties of PYY analogues to different

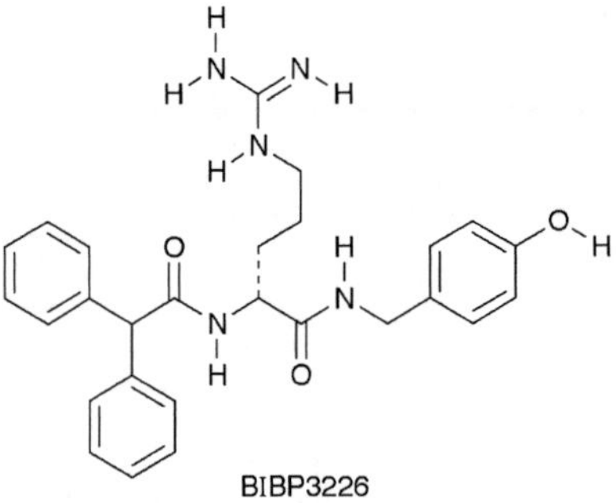

BIBP3226

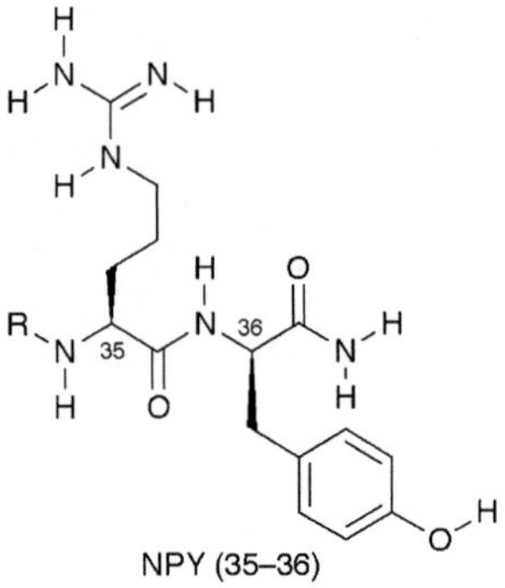

NPY (35–36)

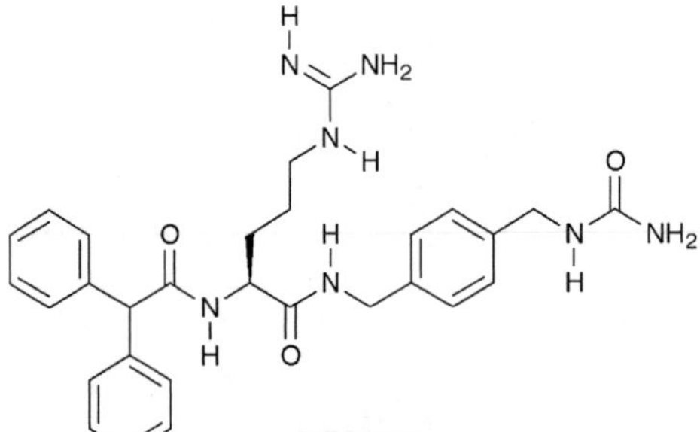

BIBO3304

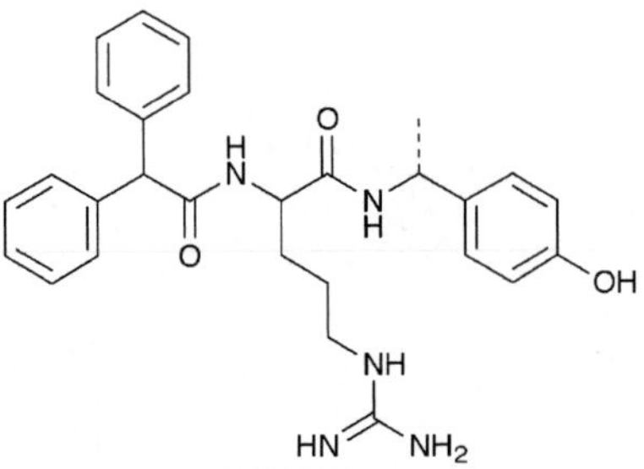

H 409/22

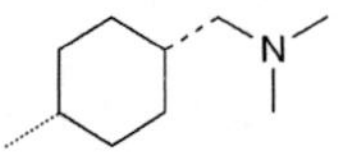

SR120819A

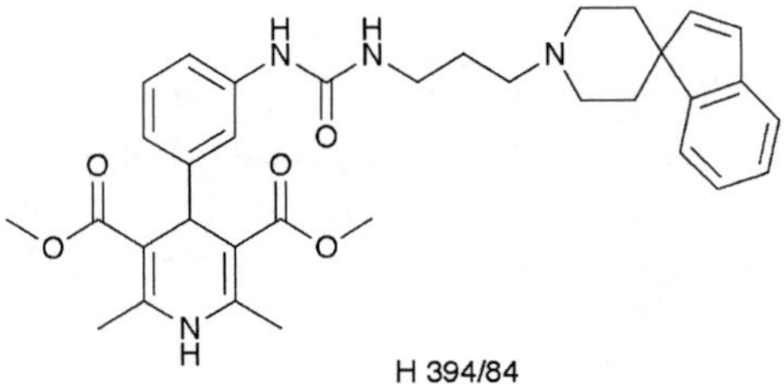

H 394/84

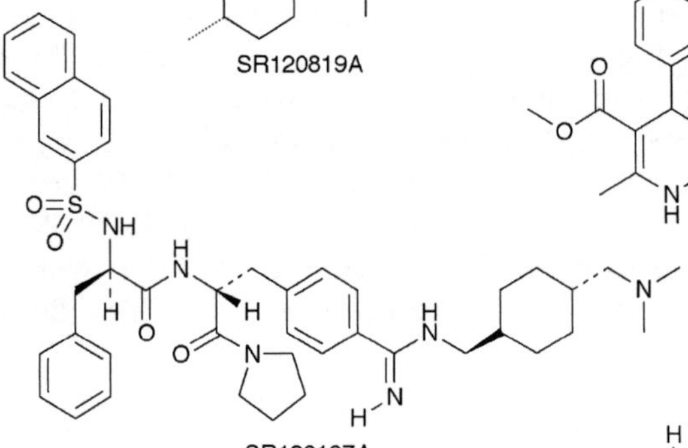

SR120107A

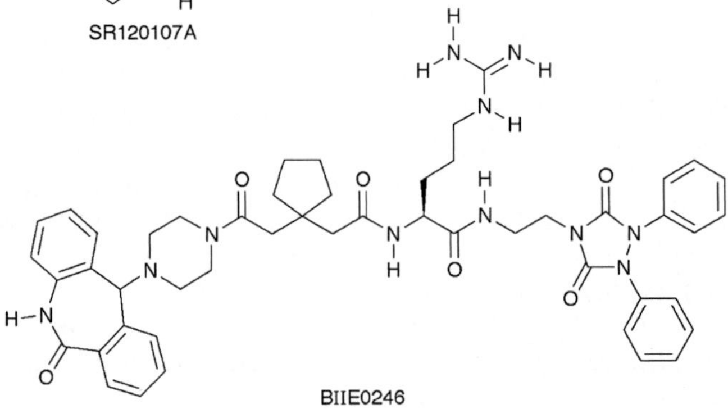

BIIE0246

Figure 33

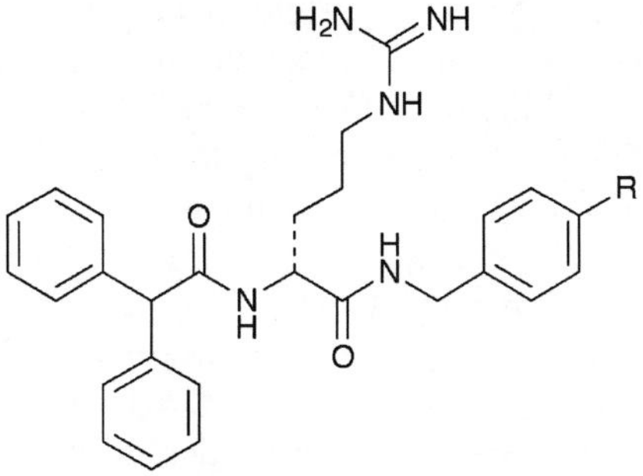

BIBP3226 R = –OH
BIBO3304 R = $-CH_2NHCONH_2$

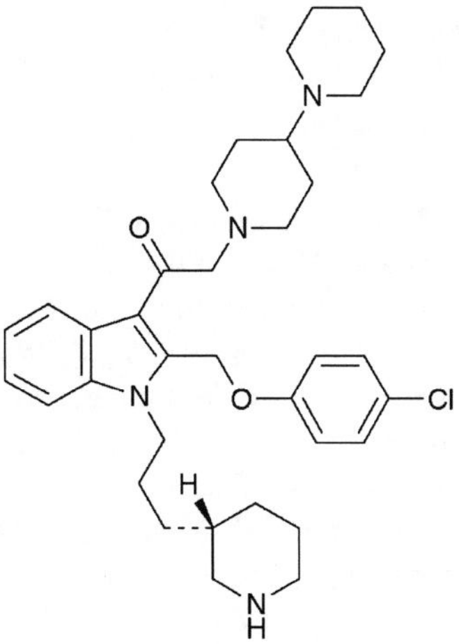

LY-357897

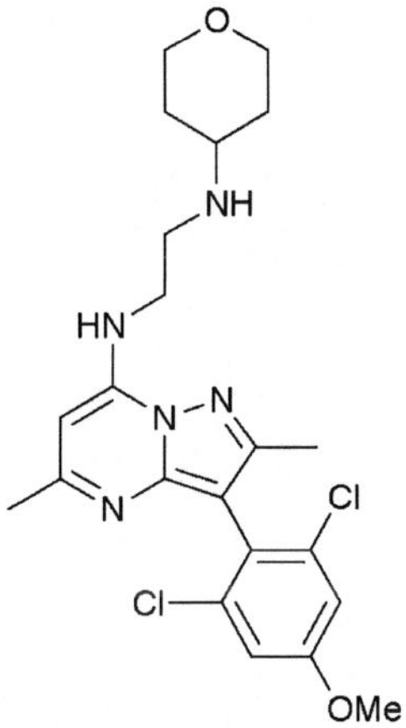

CP-671906

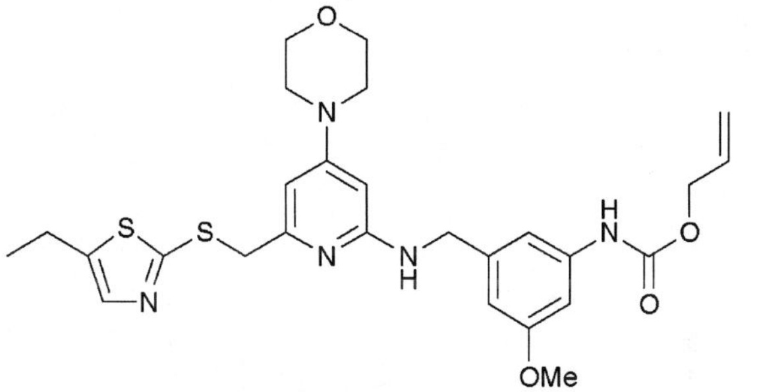

J-104870

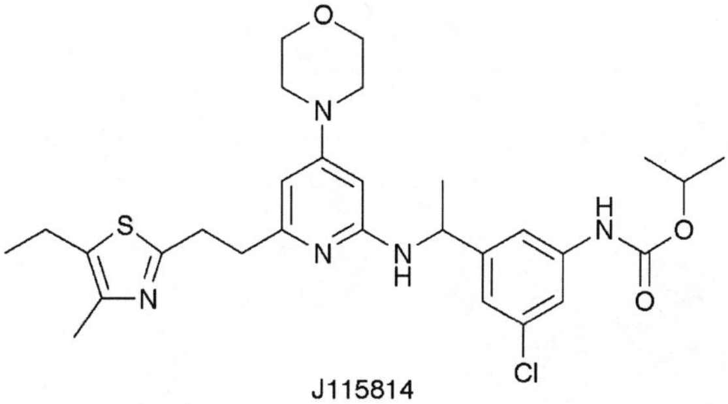

J115814

Figure 34

neuropeptide Y receptor subtypes were reviewed.[384] In addition to the natural NPY related peptides from different species, numerous unnatural analogues were characterized.

4.13 Opioid (Neuropeptide FF, Enkephalin, Nociceptin, Deltorphin and Dynorphin) Peptides. – 4-Imidazolidinone backbone constraints were incorporated into Leu-enkephalin analogues.[385] The resulting structures are shown in Figure 36. One of the analogues (Figure 36a) showed higher affinity to μ and kappa receptor and moderate receptor selectivity. Against the C-terminal tetrapeptide fragment of deltorphin I (H-Asp-Val-Val-Gly-NH_2) in mouse brain, immunoreactivity was observed.[386] Four 2,6-dimethylphenylalanine (Dmp)-containing dermorphin (DM) and deltorphin (DT) analogues were synthesized.[387] The structures of the analogues were as follows: [*L*-Dmp^3]DM, [*D*-Dmp^3]DM, [*L*-Dmp^3]DT, [*D*-Dmp^3]DT The conformationally constrained derivatives displayed increased receptor affinity and selectivity.

A *p*-fluorophenylalanine-containing nociceptin analogue was investigated for GTP γ^{35}S binding.[388] The degradation of nociceptin was studied and

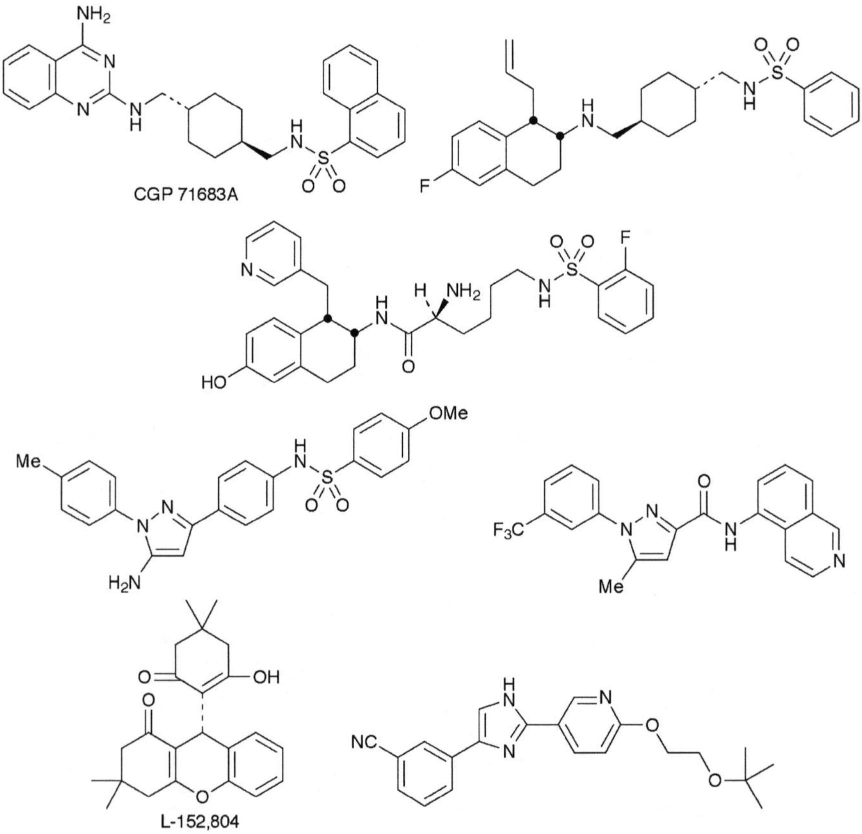

CGP 71683A

L-152,804

Figure 35

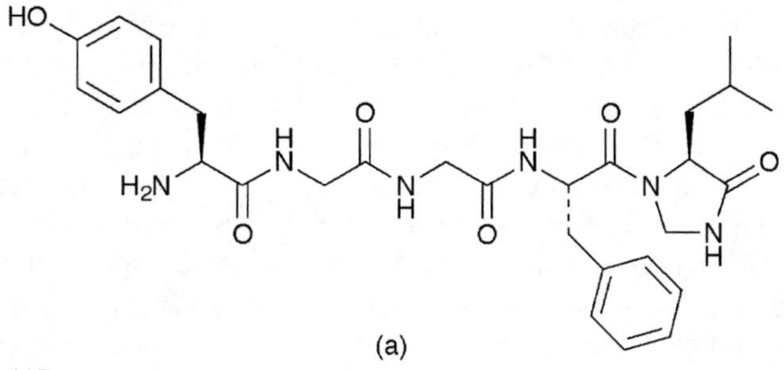

(a)

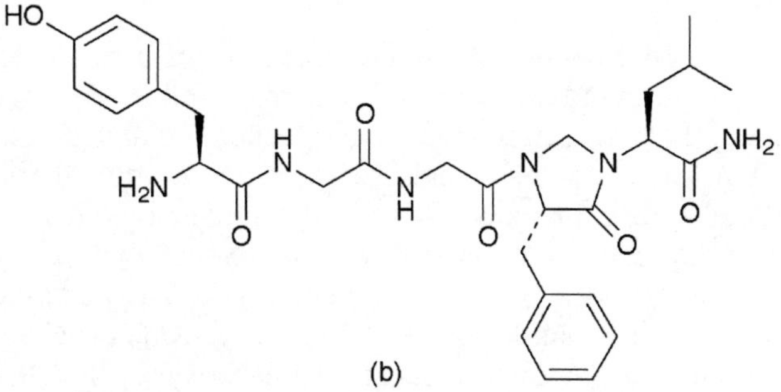

(b)

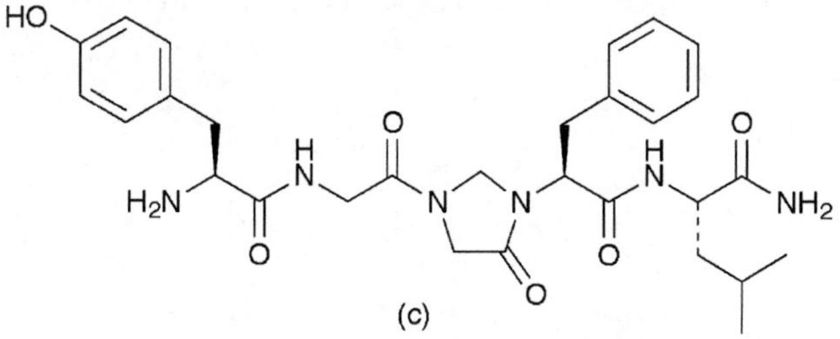

(c)

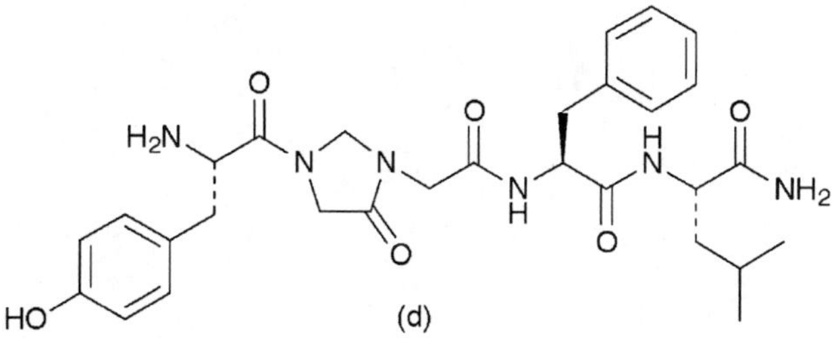

(d)

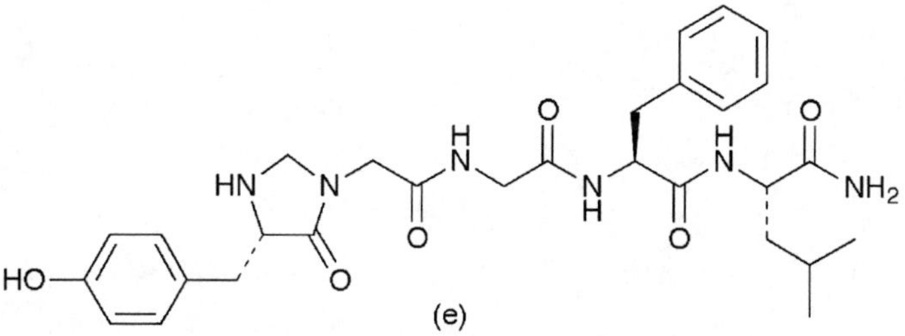

(e)

Figure 36

the cleavage between the Lys[13] and Leu[14] by endopeptidase-24.11 was observed.[389] N-Methyl phenylalanine and nicotinoyl proline containing neuropeptide FF analogues were prepared by manual solid-phase method and tested for opioid activity.[390] Dansyl-Pro-Glu-Arg-NH_2, a possible neuropeptide FF receptor antagonist was synthesized and tested.[391] Two neuropeptide FF related peptides, NPFA (AGEGLNSQFWSLAAPQRF-NH_2) and NPSF (SLAAPQRF-NH_2), were identified in human cerebrospinal fluid by mass spectrometry.[392] Two selective radioligand [^{125}I] YVP and [^{125}I] EYF were used for the quantitative autoradiographic distribution of $NPFF_1$ and $NPFF_2$ receptors.[393]

Lipidic and carbohydrate containing chimeras of Leu enkephalin were synthesized.[394] The lipophilicity and bioavailibility of the analogues summarized in Figure 37 were determined. A ferrocene containing Leu enkephalin analogue (**185**) was prepared[395] and used for conformational studies. Cyclic YKFA analogues were prepared and tested for their opioid activity.[396] The synthesized analogues are summarized were as follows: Tyr-c[*D*-Lys-Phe-Ala], Tyr-c[*D*-Lys-Phe-Val], Tyr-c[*D*-Lys-Phe-Leu], Tyr-c[*D*-Lys-Phe-Ile], Tyr-c[*D*-Lys-Phe-Asp], Tyr-c[*D*-Dab-Pheψ[CH_2NH_2]Ala]. 1-Aminocyclohexane-1-carboxylic acid (**186**) and 1-amino cyclopentane 1-carboxylic acid (**187**) containing Leu-enkephalin analogues were prepared and investigated for conformational properties.[397]

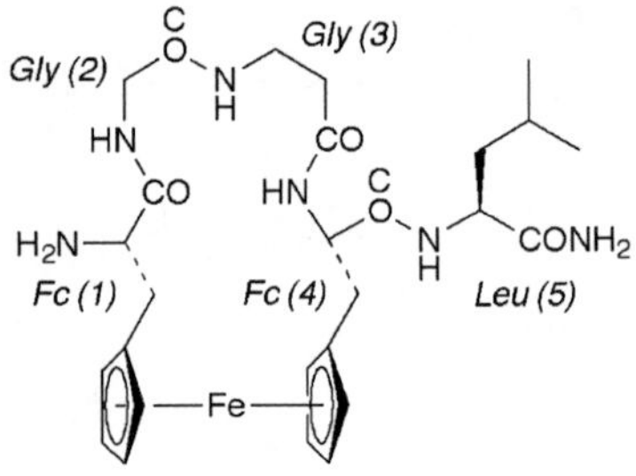

(185)

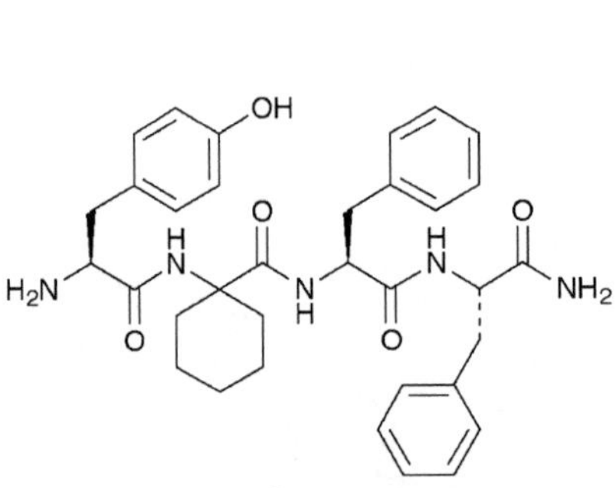

(186) [Chx^2]Endomorphin 2

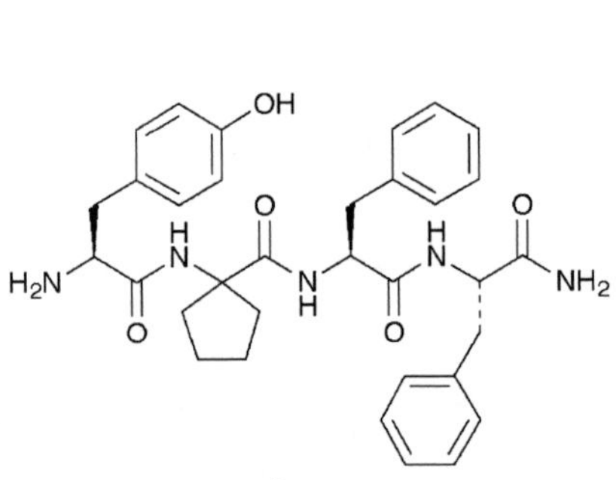

(187) [Cpn^2]Endomorphin 2

A set of endomorphin 2 analogues containing side chain modification (Figure 38) was prepared and tested for their receptor binding.[398]

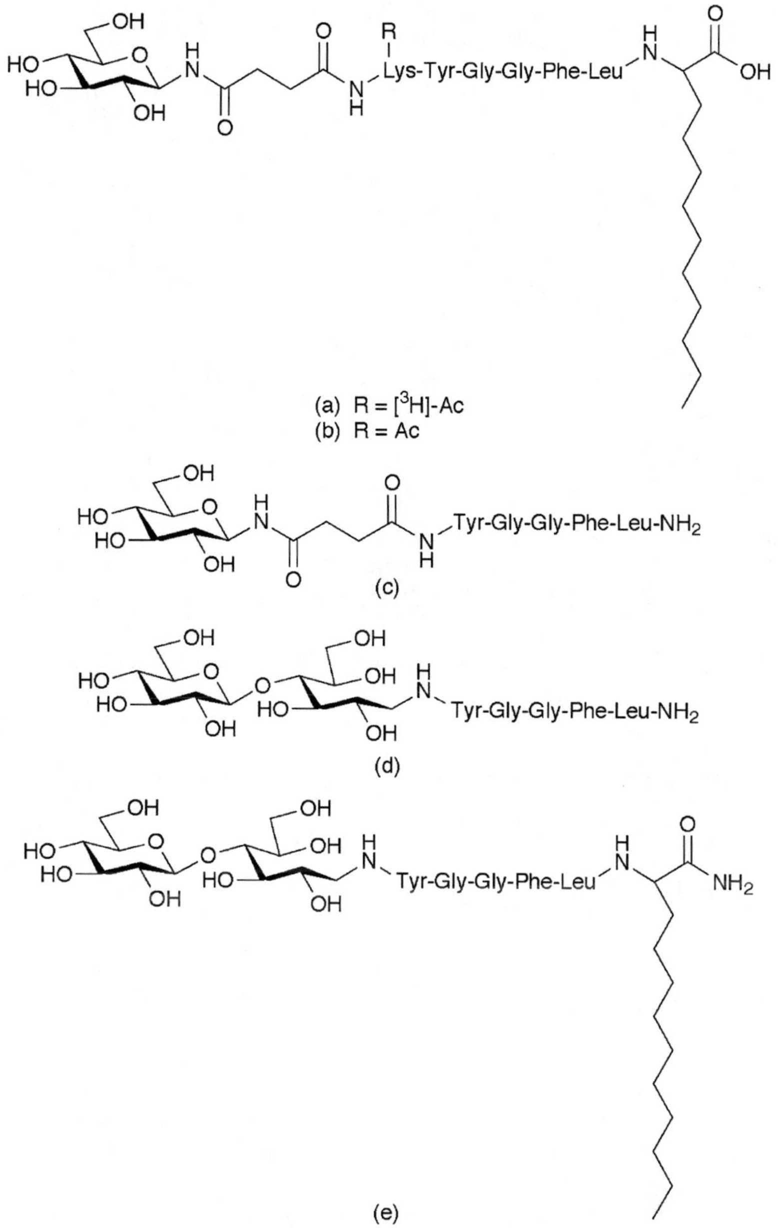

Lys-Tyr-Gly-Gly-Phe-Leu

(a) R = [^{3}H]-Ac
(b) R = Ac

Tyr-Gly-Gly-Phe-Leu-NH_2

(c)

Tyr-Gly-Gly-Phe-Leu-NH_2

(d)

Tyr-Gly-Gly-Phe-Leu

(e)

Figure 37

4.14 Somatostatin Analogues. – A nonapeptide somatostatin analogue with reduced size (**188**) was prepared and proved to be a high affinity agonist to all five somatostatin receptors.[399] The apoptotic effect of the somatostatin analogue, octreotide is described.[400]

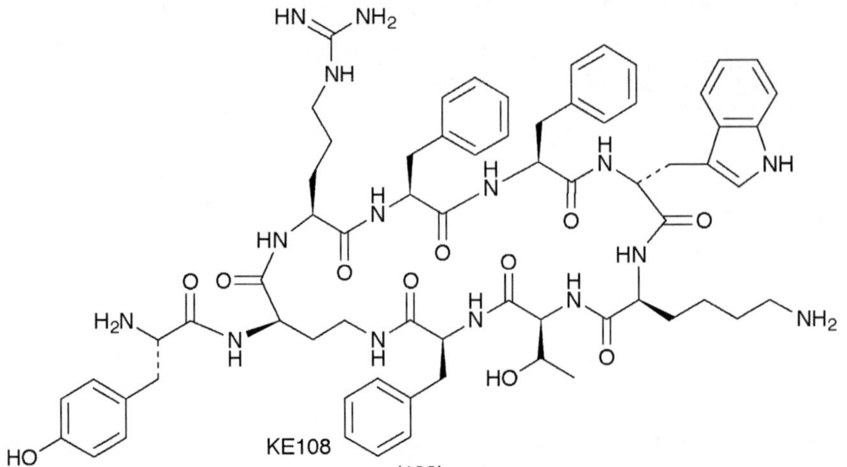

KE108

(188)

Twenty-three *D*-Trp-NMe-Lys motif-containing somatostatin analogues were synthesized during the search for potent antagonists.[401] The antiproliferative activity of TT232 (*D*-Phe-Cys-Tyr-*D*-Trp-Lys-Cys-Thr-NH_2) and six related analogues such as *D*-Phe-Cys-Tyr-*D*-Trp-Lys-Val-Cys-Thr-NH_2, *D*-Phe-Cys-Tyr-*D*-Trp-*D*-Trp-Lys-Cys-Thr-NH_2, *D*-Phe-Cys-Tyr-*D*-Trp-Ala-Cys-Thr-NH_2,

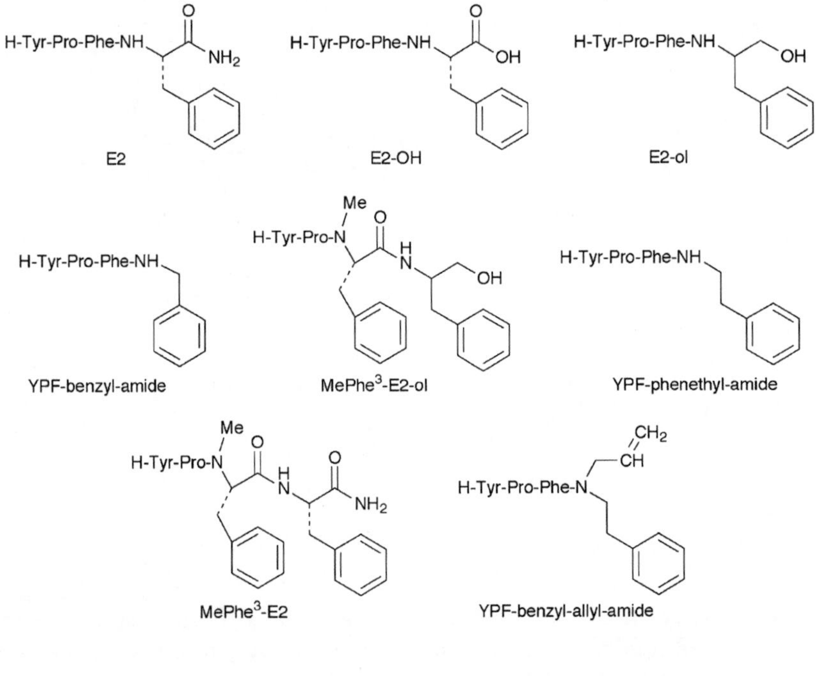

E2 E2-OH E2-ol

YPF-benzyl-amide MePhe[3]-E2-ol YPF-phenethyl-amide

MePhe[3]-E2 YPF-benzyl-allyl-amide

Figure 38

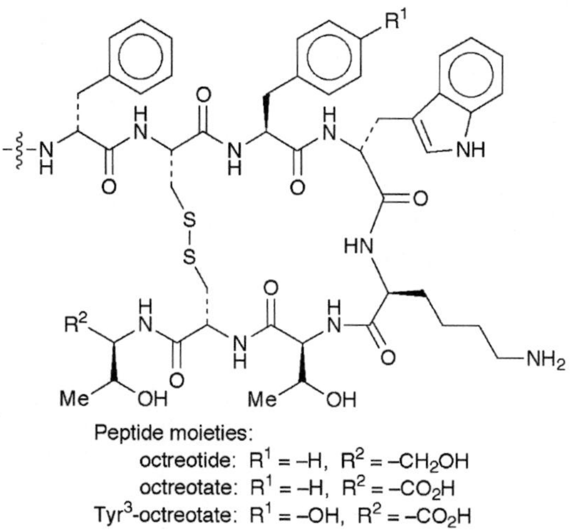

Peptide moieties:
octreotide: R^1 = –H, R^2 = –CH_2OH
octreotate: R^1 = –H, R^2 = –CO_2H
Tyr^3-octreotate: R^1 = –OH, R^2 = –CO_2H

Figure 39

D-Phe-Cys-Tyr-*D*-Trp-Cit-Cys-Thr-NH_2, *D*-Phe-Cys-Tyr-*D*-Trp-Arg-Cys-Thr-NH_2, *D*-Phe-Cys-Tyr-*D*-Trp-Orn-Cys-Thr-NH_2, were investigated.[402]

Fluorescent somatostatin receptor probes using the octapeptides octreotide, octreotate, Tyr^3-octreotate (Figure 39) and the four different fluorescent dyes (Figure 40) were prepared.[403] They might have value for intraoperative delineation of primary tumors and metastatic lesions. According to recent investigations, cyclo-[7-aminoheptanoyl-Phe-*D*-Trp-Lys-Thr(Bzl)], the classical somatostatin antagonist may express agonistic activity.[404]

Antisense peptide nucleic acids conjugated to somatostatin analogues can display enhanced cytotoxicity.[405] In a review the possible use of somatostatin analogues in the diagnosis and treatment of cancer was discussed.[406]

A chelator (1,4,7,10-tetraaza cyclododecane-N,N′,N″,N‴-tetraacetic acid) containing somatostatin analogue (DOTA-*D*-Phe^1-Tyr^3) was labeled with Ga-66 and used for PET-based diagnosis of somatostatin receptor positive tumors.[407] In addition to this, the upper analogue can be applied for radio-therapy of the above tumors. Similar attempt was made using In-111 and Tc-99.[284] The inhibitory properties of somatostatin in GLP-1 induced insulin secretion are described.[408] The therapeutic potential of somatostatin against schistosomiasis was discussed.[409]

4.15 Tachykinin (Substance P and Neurokinin) Analogues. – Tc-99 labeled substance P1 analogue was prepared and investigated.[410] For chelator molecule 1-imino-4-mercaptobutyl group was used. The resultant labeled molecule can be used for substance P receptor imaging.

Numerous, partially insect tachykinin related peptides were prepared and used for characterizing neurokinin receptor expressing cell lines.[411] The

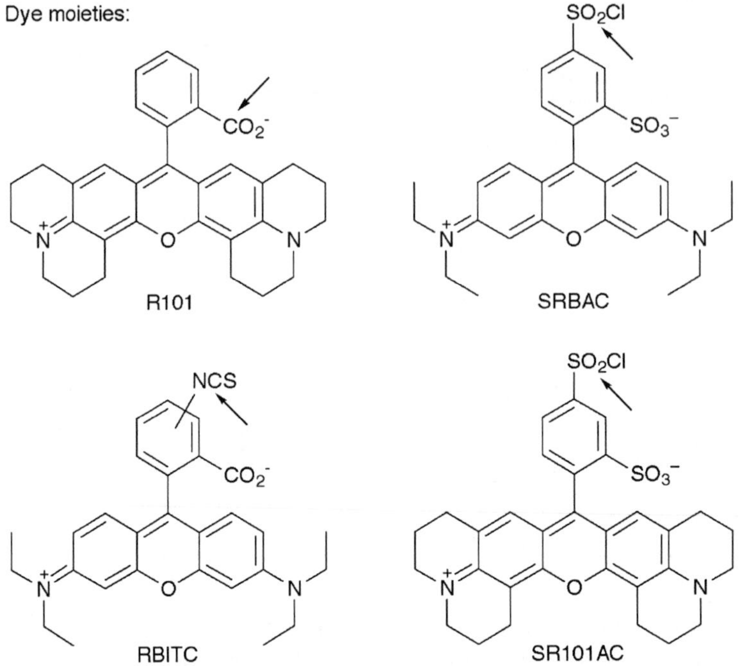

Dye moieties:

R101 SRBAC

RBITC SR101AC

Figure 40

N-terminal heptapeptide fragment of substance P (Arg-Pro-Lys-Pro-Glu-Gln-Phe) was investigated.[412] This active metabolite of substance P can enhance dopamine release in nucleus accumbens. Insect tachykinin-like peptides were prepared and tested.[413] According to these observations, the differential Arg-Met preference appears to be a major coevolutionary change between insect and human peptide-receptor couples. The *in vivo* metabolism of tachykinin and substance P was analysed using HPLC methods using tritiated peptides.[414]

Two antagonistic analogues of substance P [(Tyr^6, *D*-Phe^7, *D*-His^9)substance P-(6-11) and (*D*-Phe^7, His^9)substance P-(6-11)] were prepared and used for the investigation of the involvement of tachykinin NK_1 receptor in histamine induced hyperalgesia.[415] In a review article the G-protein dependent activation of mast cells by different peptides including substance P was discussed.[416]

4.16 Vasopressin and Oxytocin Analogues. – A novel arginine vasopressin derivative, NC-1900 (**189**) was reported as a candidate for relieving human amnesia and dementia.[417] A selective nonpeptide oxytocin receptor antagonist radioligand (1-(1-(2-(2,2,2-trifluoroethoxy)-4-(1-methylsulfonyl-4-piperidinyloxy) phenylacetyl)-4-piperidinyl)-3,4-dihydro-2(*1H*)-quinoline, (**190**) was described.[418] This compound appears suitable for studying the pharmacology of oxytocin receptors. The role of arginine-vasopressin in human labour was discussed.[419] According to this paper, there are ongoing debates about the involvement of V_{1a} receptor in term and preterm labour. A nonpeptide AVP receptor antagonist

(Z)-4′-(4,4-difluoro-5-[2- (4-dimethylaminopiperidine)- 2-oxoethylidene)-2,3,4,5-tetrahydro-1*H*-1-benzoazepine-1-carbonyl)-2-phenylbenzylanilide (**191**) was investigated for its action on human coronary artery smooth muscle cells.[420] In an article the decreasing effect of oxytocin for the TSH and TRH was described.[421] In a review article the preterm delivery, the possible treatment of it including the application of oxytocin antagonists was discussed.[422]

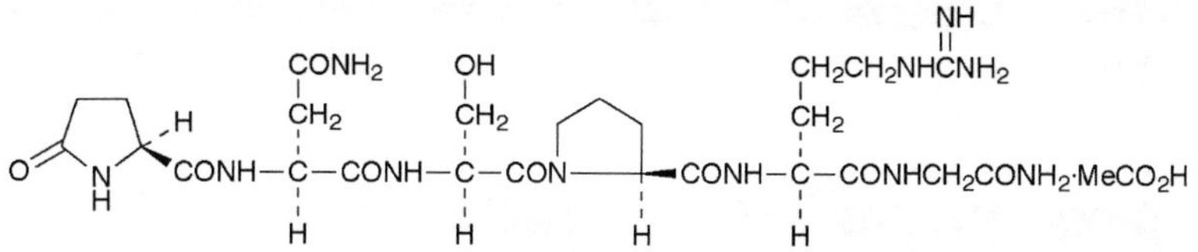

(189)

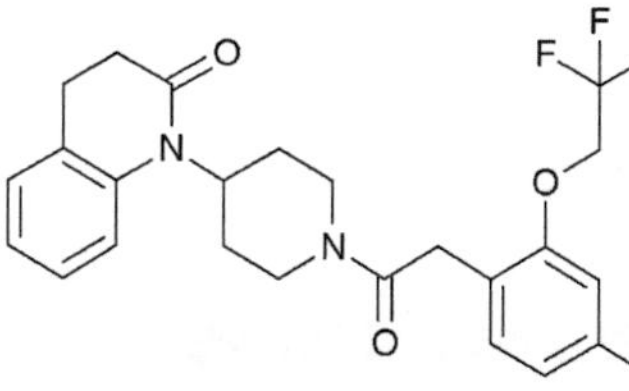

(190)

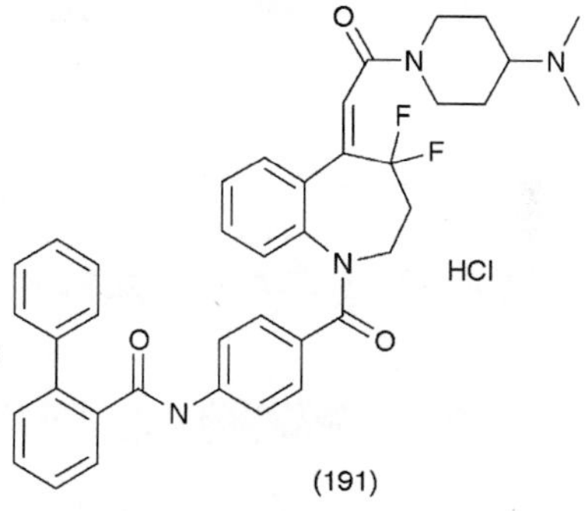

(191)

A set of indole and benzofuran structure-based potent and selective oxytocin antagonists was synthesized and evaluated.[423,424] The epidemiology, pathophysiology, and current therapeutic strategies utilized in the settings of preterm labour have been reviewed.[425] The incidence of preterm birth is increasing, and one of the promising management is the application of oxytocin/vasopressin antagonists.

Non competitive oxytocin antagonists with the general structure of $Mpa^1Xxy^2Sar^7Arg^8$ were synthesized and tested.[426]

4.17 Insulins and Chemokines. – *4.17.1 Insulins.* Chiral mutagenesis was used to elucidate the role of A2 isoleucine residue in the insulin-receptor interaction.[427] While the structure of the Ile^{A2} analogue retains the native conformation, the receptor affinity decreased by 50-fold. A detailed *in vitro* renauration process was described for recombinant human proinsulin, which can be converted to mature insulin.[428] The clinical applicability of three new synthetic analogues of insulin, insulin lispro (**192**), insulin aspart (**193**) and insulin glargine (**194**), was discussed.[429] Two new insulins were isolated and characterized from fish species (spotted dogfish and hammerhead shark).[430] The sequence of dogfish insulin was found to be as follows: A-chain GIVDHCCRNT CSLYDLEGYC NQ and B-chain LPSQHLCGSH LVETLYFVCG QKGFYYVPKV. The primary structure of hammerhead shark insulin differs from it in only two amino acids: $Arg^8 \rightarrow$ His in the A-chain and $Val^{30} \rightarrow$ Ile in the B-chain.

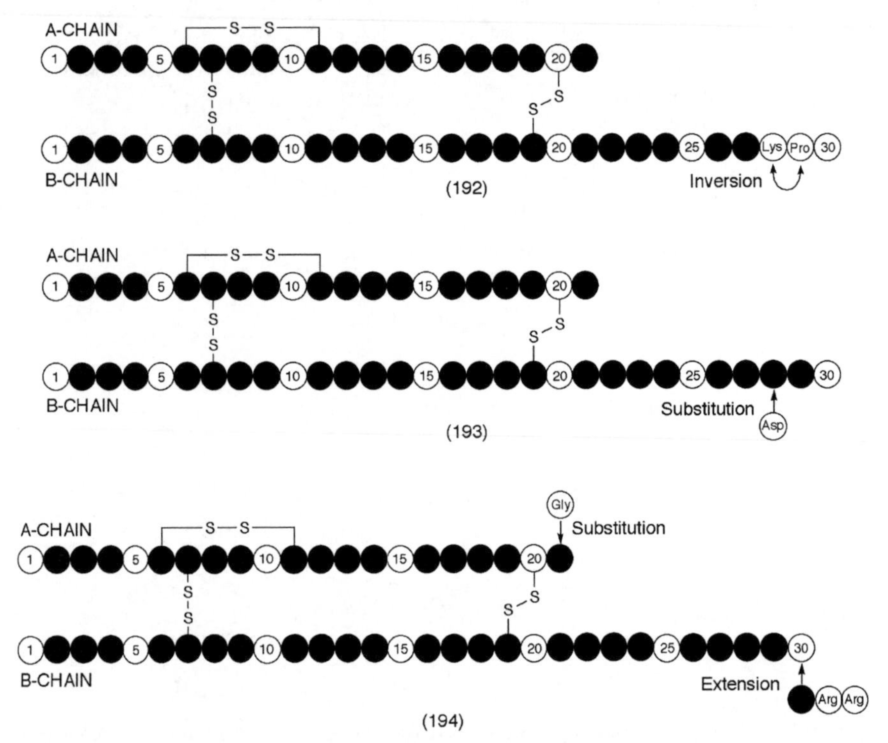

(192)

(193)

(194)

In addition to the previously mentioned new synthetic insulin analogues, several others considered as new drugs in the treatment of diabetes mellitus.[431] The role of histidine B10 in the activity of insulin was elucidated.[432] In a review the role of the immunogenicity of therapeutic proteins including different insulins was discussed.[433] Six human insulin recombinant analogues were used to compare their metabolic and mitogenic potencies.[434]

Low-molecular-weight polyethylene glycol was used as an insulin carrier molecule.[435] The resulting construct almost completely lost the original immunogenicity, allergenicity and antigenicity of insulin, while the steric structure and *in vivo* potency remained unaltered. Ser^{A7} and Ser^{A20} insulin A-chain was synthesized and used for *in vitro* assay.[436] Interestingly, the "mini insulin" structure lacking the native B-chain proved to be active.

Different acylated derivatives of the insulin analogue detemir were synthesized.[437] The incorporated albumin affinity tags increased the half life of the peptide. The applicability of B28 Asp human insulin (aspart) which is now in clinical practice was further investigated.[438]

4.17.2 Chemokines. The eighteen different chemokine receptors, their specific antagonists and recent developments in the associated diseases were reviewed.[439]

The CCR3 receptor agonist 2-[(6-amino-2-benzothiazolyl)thio]-N-[1-[(3,4-dichlorylphenyl)methyl-4-piperidinyl]acetamide (**195**) was characterized using [^{35}S] GTP γS binding test.[440] Pseudo peptide analogues of a CXCR4 inhibitor (T140) were synthesized and evaluated.[441] The general structure of the isostere containing derivatives is shown in Figure 41. In a short paper the implication of CCR5 and CXCR4 for therapy and vaccination against HIV infection is described.[442] A minireview discussed the involvement of different chemokine systems in human and veterinary health and disease.[443] The application of CCR3 antagonists as potent new therapeutic agents in the treatment of asthma was described.[444] The structures of the used substances can be divided into the following families: indolinopiperidinylalkylureas (Figure 42) and 4-benzylpiperidinealkylureas (Figure 43).

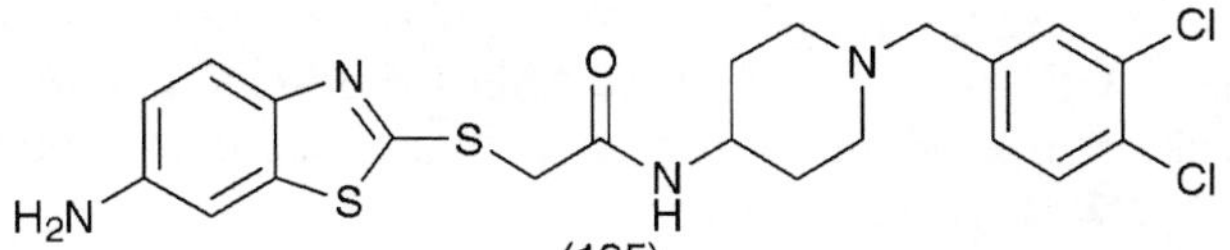

(195)

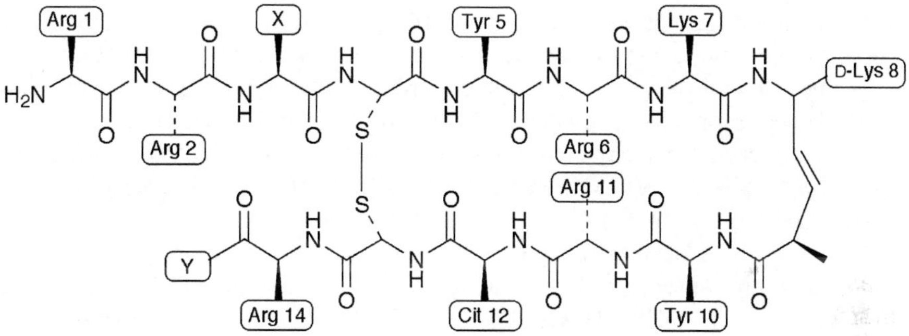

Figure 41

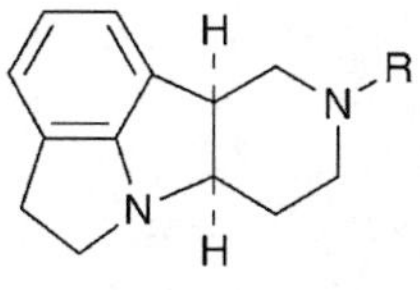

Figure 42

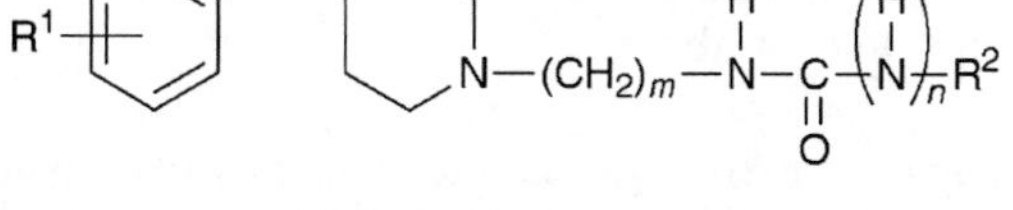

Figure 43

The advantages and disadvantages of chemokine receptor antagonists for anti-inflammatory therapy were discussed.[445] The published chemokine receptor antagonists are summarized in Figure 44. The treatment of experimental autoimmune encephalomyelitis with the chemokine receptor antagonist Met-RANTES is described.[446]

4.18 Peptide Toxins. – BmKAPi, a new type of scorpion venom peptides was isolated from *Buthus martensii* Karsch.[447] Although the peptide length of BmKAPi is similar to the long-chain scorpion toxins, its amino acid sequence showed no homology to these toxins. BmKAPi is the first venom peptide with five disulphide bridges identified from scorpion. In a parallel investigation three other unique cysteine rich peptides named BmTXKS3, BmTXLP2 and BmAP1 were isolated from the same scorpion.[448] A peptide toxin, PnTx2-6 was isolated and characterized from spider *Phonentria nigriventer* venom.[449] Its sequence is the following: MKVAILFLSI LVLAVASESI EESRDDFAVE ELGR. The solution structure of ω-grammotoxin, a 36 amino acid long peptide toxin was determined by NMR spectroscopy.[450]

In a review the structure of potassium channels, their structure and modeling was discussed.[451] Two new toxins named kurtoxin-like I and II were isolated from the scorpion *Parabuthus granulatus*.[452] The pharmacology and biochemistry of spider venoms has been reviewed.[453] The nature of the components is very complex, but most of them are polypeptides, proteins or enzymes.

A new 64 amino acid long, four disulphide-containing toxin was isolated from the scorpion *Tityus cambridgei*.[454] In addition to this, three other sodium channel affecting toxin can be found in its venom. The identified sequence is the following: KKEGYLVGND GCKYGCITRP HQYCVHECEL KKGTDGY-CAY WLACYCYNMP DWVKTWSSAT NKCK.

Conophysin-R, a new neurophysin-like toxin was isolated from *Conus radiatus* snail venom.[455] Its sequence is the following: HPTKPCMYCS FGQCVGPHIC CGPTGCEMGT AEANMCSEED EDPIPCQVFG SDCALNNPDN IHGHCVADGI CCVDDTCTTH LGCL. Conophysin-R is the first neurophysin family member isolated from an invertebrate source.

From *Centruroides sculpturatus* several short-chain peptide toxins inhibiting K^+ channels were isolated.[456] Atractotoxins isolated from the venom of Australian Funnel-web spiders (Figure 45) can serve as insect specific toxins.[457] This family contains a rare vicinal disulphide bridge between adjacent cysteines.

In an overview the toxins form the venom of the Asian scorpion *Buthus martensii* Karsch have been summarized.[458] A conus peptide, conotoxin MI inhibits the α-δ acetylcholine binding site of the purified nicotinic *Torpedo marmorate* receptor.[459] IsCT, a novel short linear peptide toxin was isolated and characterized from the venom of scorpion *Opisthacantus madagascariensis*.[460] Site-specifically conjugated μ-conotoxin GIIIA to BSA was used for generation of polyclonal antibody.[461]

4.19 Miscellaneous. – An analogue of the glucagon-like insulinotropic peptide (LY315902: 2-indolyl-propionyl-AEGTFTSDVSSYLEGQAAREF IAWLV-

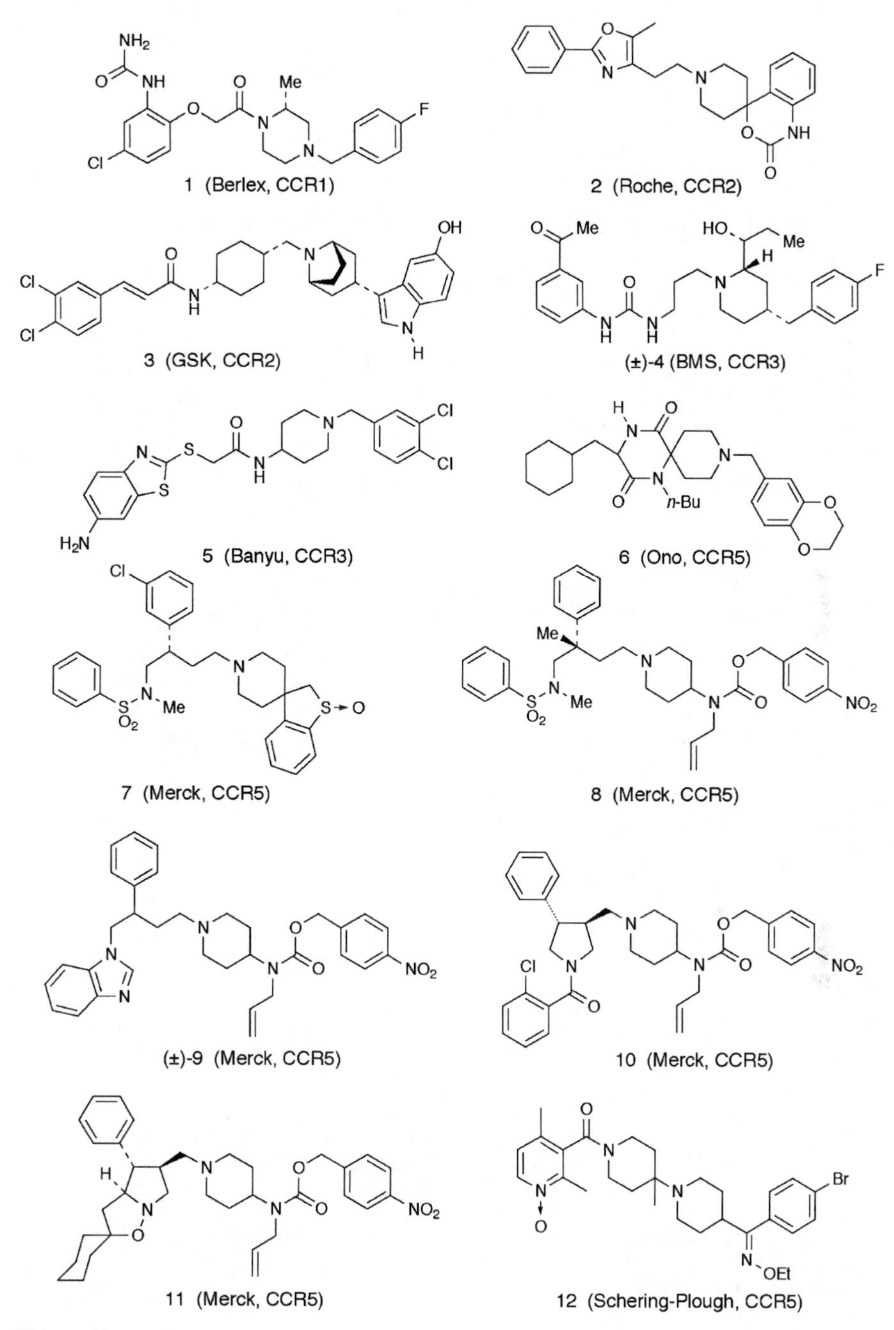
1 (Berlex, CCR1)
2 (Roche, CCR2)
3 (GSK, CCR2)
(±)-4 (BMS, CCR3)
5 (Banyu, CCR3)
6 (Ono, CCR5)
7 (Merck, CCR5)
8 (Merck, CCR5)
(±)-9 (Merck, CCR5)
10 (Merck, CCR5)
11 (Merck, CCR5)
12 (Schering-Plough, CCR5)

Figure 44

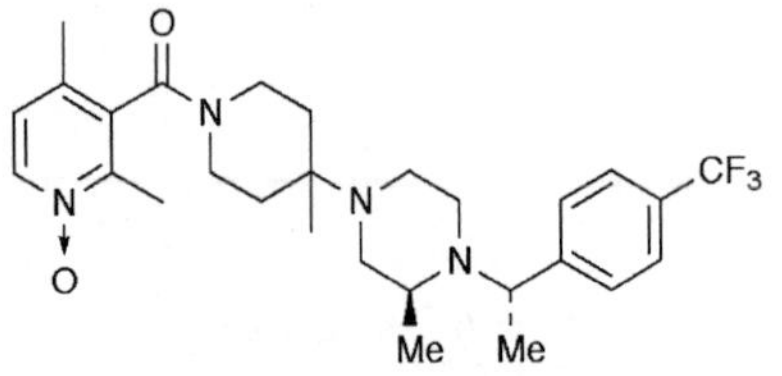

13 (Schering-Plough, CCR5)

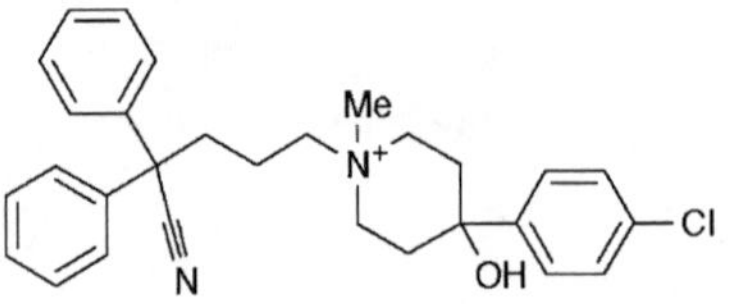

14 (Berlex, CCR1)

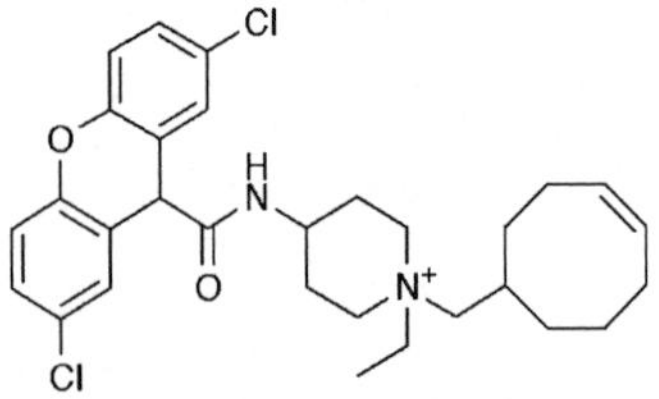

15 (Banyu, CCR1/3)

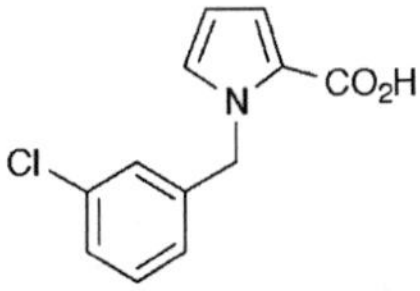

16 (Roche, CCR2)

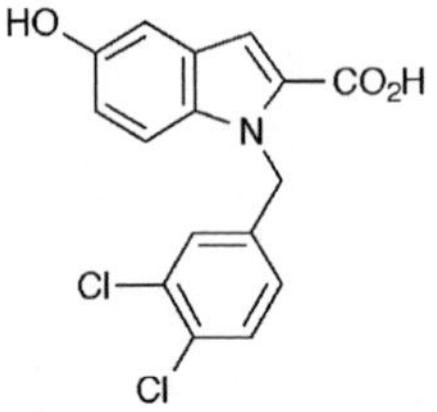

17 (Ast-Zen, CCR2/1)

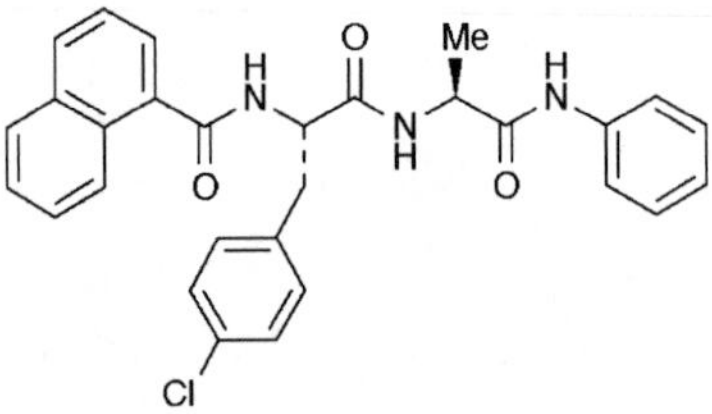

18 (GSK, CCR3)

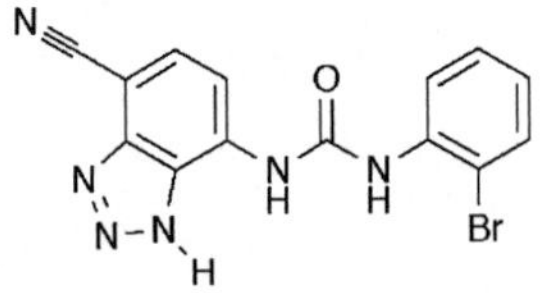

19 (GSK, CXCR2)

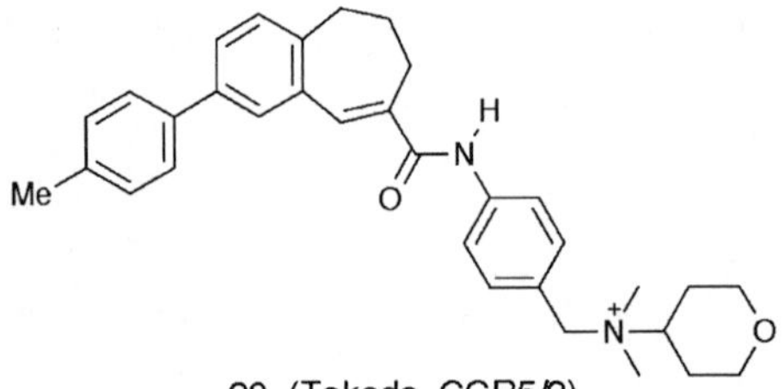

20 (Takeda, CCR5/2)

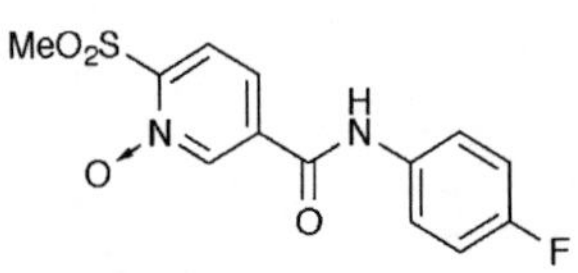

21 (Celltech, CXCR2)

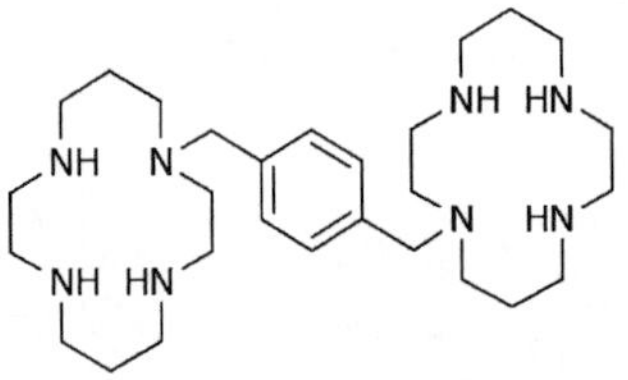

22 (AnorMED, CXCR4)

Figure 44 (continued)

K(Oct)-GRG-OH) was investigated for its insulin and somatostatin releasing activity.[462] The therapeutic perspectives of glucagons-like peptide as regulator of food intake and body weight have been discussed.[463] A review has appeared on the physiology and pharmacology of the orexins, two new 33 and 28 residue long neuropeptides.[464] While orexin-A (EPLPDCCRQK TCSCRLYELL

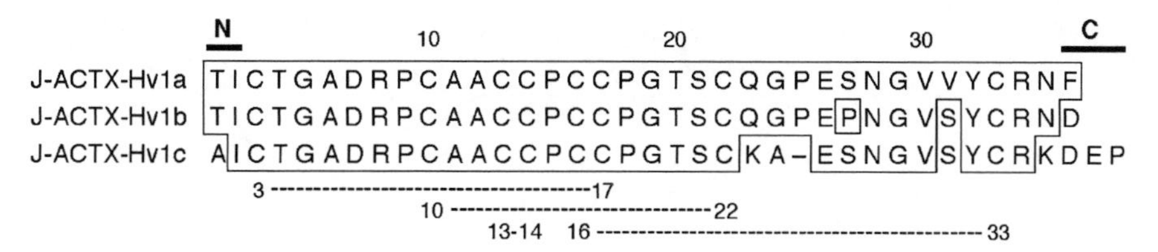

Figure 45

HGAGNHAAGI LTL-NH_2) is stabilized by two disulphide bridges, orexin-B(RSGPPGLQGR LQRLLQASGN HAAGILTM-NH_2) is a linear peptide.

A convenient new synthesis of human calcitonin, a calcium regulating peptide hormone **(196)** and its methionine sulfoxide derivative was described.[465] NMR and molecular dynamics simulation studies were applied for the elucidation of the conformation of immunostimulating tetrapeptide, rigin (H-Gly-Glu-Pro-Arg-OH).[466]

CGNLSTCMLGTYTQDFNKFHTFPQTAIGVGAP–NH_2

(196)

Five lactam analogues of urotensin-II (H-Glu-Thr-Pro-Asp-[Cys-Phe-Trp-Lys-Tyr-Cys]-Val-OH), a potent vasoconstrictor peptide, were designed and prepared.[467] The resulting analogues were used for NMR investigations, molecular modeling analysis and biological studies. Twelve acylated derivatives of the urotensin II (4-11) fragment were prepared.[468] The structures of the resulting analogues are shown in Figure 46. Novel analogues of antisauvagine-30, a specific antagonist for CRF_2 receptor, have been synthesized and characterized.[469] The ^{123}I-derivative of these analogues is a potential SPECT ligand.

Centrally active, selective, new TRH analogues (Figure 47) were prepared.[470] The most active compound of this series can serve as a new lead for the treatment of CNS disorders. Another set of TRH analogues were prepared using a substituted imidazole ring. (Figure 48).[471]

5 Enzyme Inhibitors

A great number of reviews were published in 2002 dealing with enzyme inhibitors. The use of synthetic oligonucleotides as protein inhibitors was reviewed[472] while specialized reviews on controlling apoptosis by inhibition of caspases[473,474] have been published. Very intensive work with induction of tumour cell apoptosis by matrix metalloproteinase inhibitors, has been reviewed in several papers.[475–478] More specialized reviews deal with inhibitors of protein phosphatase,[479] thrombin[480] and cyclooxygenase-2.[481]

5.1 Aminopeptidase and Deformylase Inhibitors. – The computer-aided design, synthesis and activity prediction of aminophosphonic acids as new leucine aminopeptidase (LAP) inhibitors has been described.[482] The same research group published the structure-based design, synthesis and activity prediction of

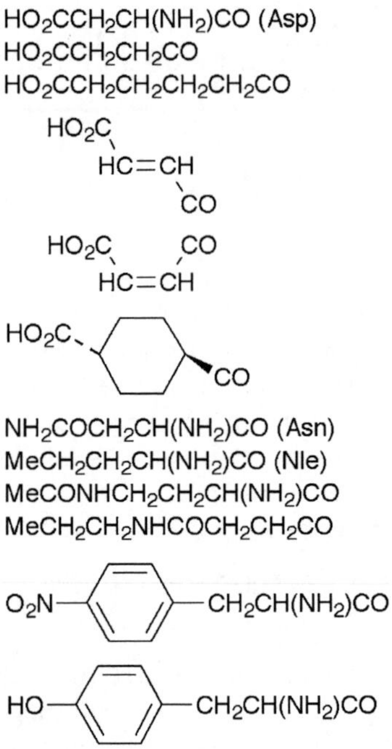

$HO_2CCH_2CH(NH_2)CO$ (Asp)
$HO_2CCH_2CH_2CO$
$HO_2CCH_2CH_2CH_2CH_2CO$
HO_2C–HC=CH–CO
HO_2C–HC=CH–CO
HO_2C–cyclohexyl–CO
$NH_2COCH_2CH(NH_2)CO$ (Asn)
$MeCH_2CH_2CH(NH_2)CO$ (Nle)
$MeCONHCH_2CH_2CH(NH_2)CO$
$MeCH_2CH_2NHCOCH_2CH_2CO$
O_2N–C_6H_4–$CH_2CH(NH_2)CO$
HO–C_6H_4–$CH_2CH(NH_2)CO$

Figure 46

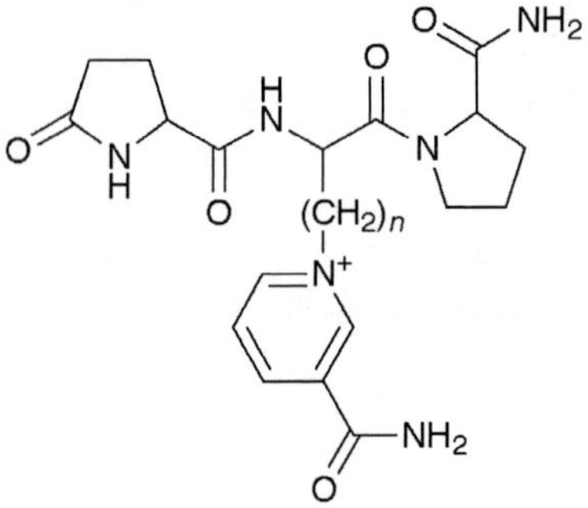

Figure 47

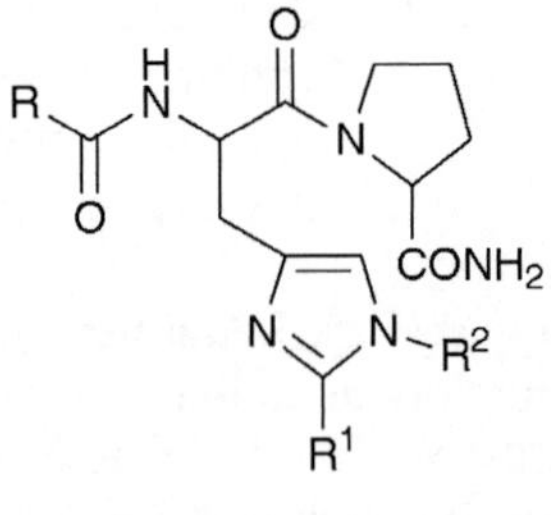

Figure 48

new LAP inhibitors such as the phosphonamidate and phosphinate dipeptide analogues.[483]

Various protease inhibitors block parasite development, suggesting that key proteases may be appropriate chemotherapeutic targets. Hemoglobin hydrolysis is mediated (among other enzymes) by neutral aminopeptidase, so inhibitors of this enzyme may serve as new weapons in the battle of malaria parasites.[484,485] Aqueous extract of the plant leaves of *Cistus incanus* dose-dependently inhibit the enzymatic activity of alanyl aminopeptidase (APN).[486] This inhibition is not reversible and very likely results from a covalent binding of reactive compounds to the enzymes. Bradykinin metabolism was studied in human and rat plasma by liquid chromatography with inductively-coupled plasma mass spectrometry and orthogonal acceleration time-of-flight mass spectrometry.[487] This method provides a powerful technique for monitoring the activity of many kinases involved in bradykinin metabolism such as aminopeptidase P. Another specific enzyme, the puromycin insensitive leucyl-specific aminopeptidase (PILSAP) was proven to be involved in the activation of endothelial integrins.[488] Puromycin-sensitive aminopeptidase can be inhibited by thalidomide (N-α-phthalimidoglutarimide) and its derivatives.[489]

A novel approach to antimicrobial therapy involves the application of peptide-deformylase inhibitors.[490] This enzyme represents one of the most exciting new targets for the development of novel antimicrobial chemotherapies, because it is an essential bacterial metalloenzyme that deformylates the N-formylmethionine of newly synthesized bacterial polypeptide chains.[490,491] One of the most encouraging compounds is BB-83698 that shows a very good *in vivo* efficacy against *Streptococcus pneumoniae*. A series of analogues of the potent peptide-deformylase inhibitor BB-3497 (**197**) containing alternative metal-binding groups (MBG) was synthesized.[492]

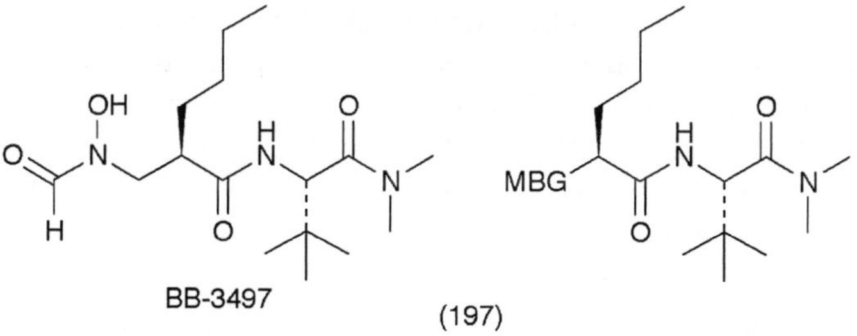

BB-3497

(197)

Enzyme inhibition and antibacterial data for BB-3497 and the MBG-analogues revealed that the bidentate hydroxamic acid and N-formyl hydroxylamine structural motifs represent the optimum chelating groups on the pseudo-peptidic BB-3497 backbone. The binding mode of BB-3497 to the active site of E. coli peptide-deformylase from X-ray data is also given. Another research group found N-alkyl urea hydroxamic acids as a new class of active PDF-enzyme inhibitors with antibacterial activity.[493] Patent highlights January to June 2002 for peptide-deformylase inhibitors are summarized in a review.[494]

5.2 Calpain Inhibitors. – Calpains, the calcium-dependent thiol proteases, are widely expressed with ubiquitous and tissue specific isoforms. Calpains have been implicated in basic cellular processes including cell proliferation, apoptosis and differentiation. One review summarizes recent findings implicating calpains in cytoskeletal rearrangements and cell migration.[495] Another review focuses on the recent patent literature of calpain inhibitors[496] showing very promising cytoprotective activity in a variety of disorders (cerebral stroke, cerebral and spinal cord trauma, and myocardial infarction).

The crystal structure of calpain is already known and it has given an insight into Ca^{2+}-dependent activation of the enzyme.[497] Calpains are involved in formation of long-term potentiation.[498] The analysis of the molecular mechanisms of the pathogenesis of traumatic brain injury shows that the activation of Cys-proteases, such as calpain and caspase-3, cleaves key cellular substrates and causes cell death.[499,500] A new pharmacological strategy in the therapy of traumatic brain injury might be the application of calpain inhibitors for preventing cell damage and death. It was found that calpain activates caspase-3 during UV-induced neuronal death.[501] UV neurotoxicity is mediated by a loss of Ca^{2+} homeostasis which leads to a calpain-dependent and caspase independent neuron death. Oncogenic transformation of cells is associated with deregulated growth control. Modulation of calpain-calpastatin proteolytic system plays an important role in focal adhesion disassembly, morphological transformation and cell cycle progression during cell transformation.[502]

Novel calpain inhibitors derived from phenylalanine aldehydes or ketoamides carrying a benzoyl residue were prepared and evaluated for their biological potency.[503] Structure-activity relationships elucidated the importance of *ortho* substituents in the benzoyl moiety. The most potent derivative (**198**) exhibited a K_i of 6 nM and represents a novel class of reversible, highly potent and non-peptidic calpain inhibitors. Peptidyl hydrazones of the general formula (**199**) were found to be water soluble calpain inhibitors.[504] One of the derivatives, *N*,*N*-dimethylglycylhydrazone possessed good water solubility and inhibited μ-calpain with IC_{50} of 0.37 μM. Novel peptidic compounds, having peptide chains linked to bi- and tricyclic heterocycles (peptide-heterocycle hybrids) were synthesized and studied for calpain inhibition.[505] Compound (**200**) has an IC_{50} value of 45 nM on calpain I.

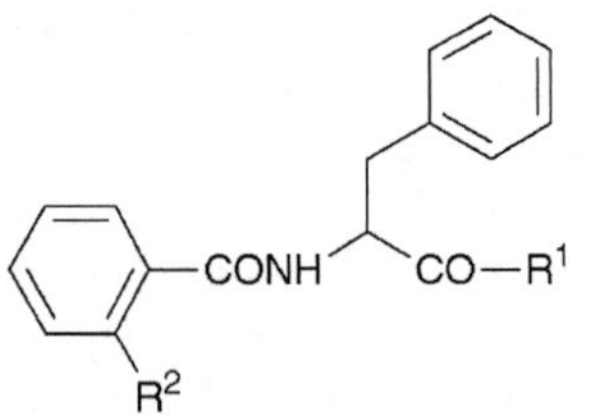

(198) $R^1 = CONH_2$, R^2 = E-2-Naphthyl–CH=CH

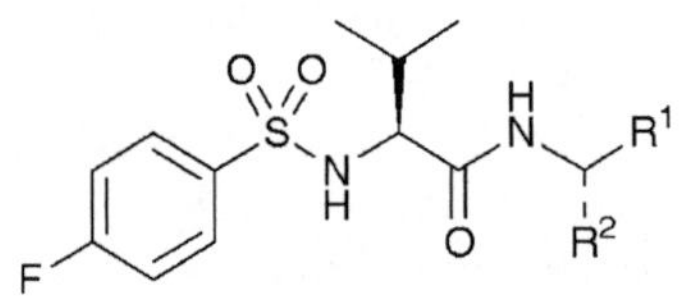

(199)

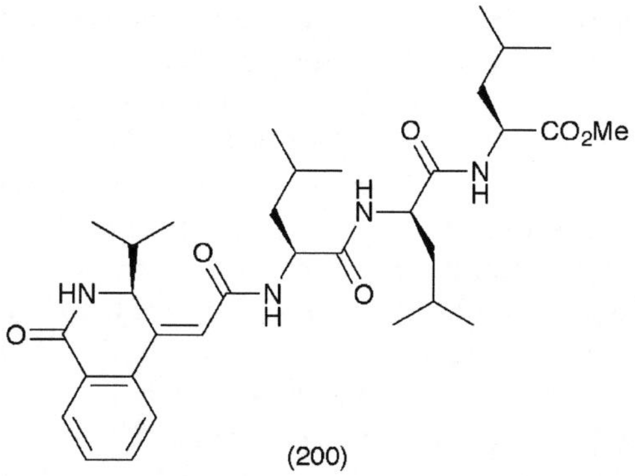

(200)

A recent review summarizes the advances in the synthesis, design and selection of cysteine protease inhibitors and includes calpain as target.[506]

5.3 Carboxypeptidase Inhibitors. – Carboxypeptidase A (CPA) cleaves the C-terminal amino acid residue having an aromatic side chain. It is a prototypic Zn-protease, the catalytically essential Zn^{2+} ion is coordinated to the side chain functional groups of His-69, Glu-72 and His-196. Phosphorus ester and amide, sulfoximine, sulfodiimine and sulfonimid amide have been previously incorporated into substrate analogues of CPA to obtain potent inhibitors. A novel type of transition state analogue inhibitors for CPA was published, in which a readily obtainable sulfamide moiety is incorporated into the basic structural frame of CPA substrate.[507] Compound (**201**) showed potent inhibitory activity with the K_i value of 0.64 μM.

The structure of CPA-**5** complex determined by single-crystal X-ray diffraction reveals that the sulfamoyl moiety interacts with the Zn^{2+} ion and functional groups at the active site of CPA. The same research group designed and synthesized a series of cysteine derivatives having an alkyl or arylalkyl moiety on the α-amino group of Cys as a novel type of inhibitor of CPA.[508] Structure-activity relationship investigations revealed that the inhibitors prepared from *D*-Cys are much more potent than those from *L*-Cys. The most potent inhibitor (**202**) with a K_i value of 55 ± 4 nM is obtained by introducing a phenetyl moiety on the amino group of *D*-Cys. Racemic and optically active 2-benzyl-2-methyl-3,4-epoxybutanoic acids have been synthesized and evaluated as CPA inhibitors.[509] Only the *threo*-form of the inactivator proved to be effective.

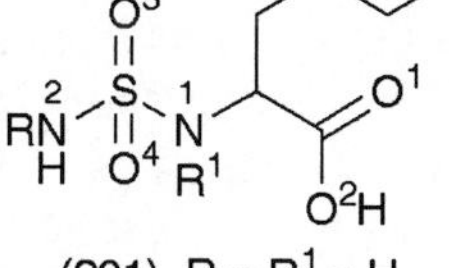

(201) $R = R^1 = H$,

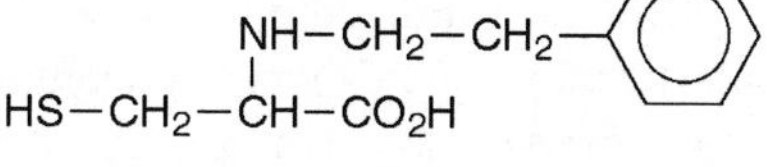

(202)

Assay for catalytic activity of metalloproteases is much simplified by employment of substrates incorporating the anisylformyl residue. Anisylazoformylarginine (**203**) has been used[510] as a substrate which is rapidly hydrolyzed at the acyl-arginine linkage by carboxypeptidase B. The catalytic reaction is readily monitored spectrophotometrically by disappearance of the intense absorption (349 nm, $\varepsilon = 19{,}100$) of the azochromophore.

Another enzyme, the angiotensin-converting enzyme-related carboxypeptidase (ACE 2) belongs also to this family. ACE 2 is an 805 amino acid protein with a transmembrane domain and has highest homology to ACE (60%). A new ACE 2 inhibitor (**204**) was synthesized and this compound inhibits ACE 2 in the picomolar range ($IC_{50} = 440$ pM).[511] All four stereoisomers of (**204**) were prepared and the greatest potency remained within the *S,S* isomer. (*R,R* $IC_{50} = 0.072$ μM; *R,S* $IC_{50} = 8.4$ μM; *S,R* $IC_{50} = 0.470$ μM).

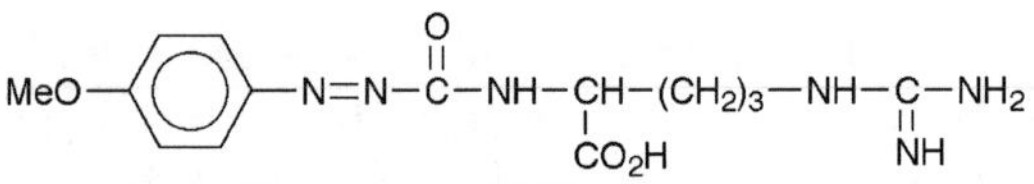

(203) 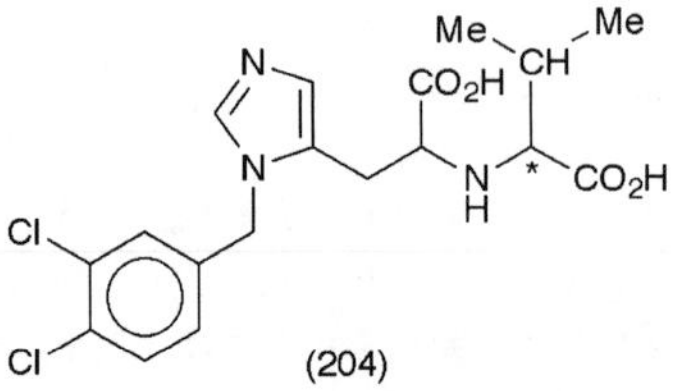 (204)

Carboxypeptidases participate also in the metabolism of bradykinin.[487]

5.4 Caspase Inhibitors. – The family of caspases includes highly conserved cysteine proteases and can be subdivided into 3 groups: (a) caspases involved in inflammation (caspase 1,4,5 and 13); (b) initiator caspases (6,8,9,10); and (c) effector caspases (2,3 and 7).

A review article focuses on the characteristics and the role of the proregions of peptidase zymogene structure including the members of the caspase family. It surveys novel zymogene structures determined in the past 5 years[512] and discusses nature's way of preventing undesired activation and proteolysis. Another review article summarizes the application of peptides in apoptosis research related to caspases and caspase regulatory proteins.[513] The use of peptidic substrates, caspase activator peptides, first generation caspase inhibitors and the next generation of caspase regulators (physiological inhibitors and activators) is discussed. Other authors focus on the role of caspase inhibitors as anti-inflammatory and antiapoptotic agents[514] and the importance of caspases in the neuroprotection.[515] A very detailed review article surveys natural and synthetic caspase-inhibitors that have been reported up to 2002 and discusses the control of apoptosis by inhibition of caspases.[516]

Interleukin-1β converting enzyme (ICE, caspase-1) activates interleukin-1β (IL-1β) and IL-18, important mediators of inflammation and infectious diseases. A novel dipeptide mimetic (**205**) (*S*) 3-tert-butyloxycarbonylamino-2-oxo-2,3,4,5-tetrahydro-1,5-benzodi-azepine-1-acetic acid methylester was prepared using a simple and versatile method for the synthesis.[517] A series of novel, potent, broad spectrum caspase inhibitors (**206**) was designed on the basis of the structure of the well known peptide inhibitor acetyl-DEVD-H.[518] Measurement of enzyme inhibitory activities shows that the tetrapeptide

Ac-DEVD-H inhibitor can be truncated to novel aryloxyacetyl dipeptides, which retain nanomolar, broad spectrum caspase inhibitory activity. The role of calpain- and caspase-mediated proteolysis of different initiation factors after brain ischemia and reperfusion has been reviewed[519] and implications for neuronal survival or death surveyed.

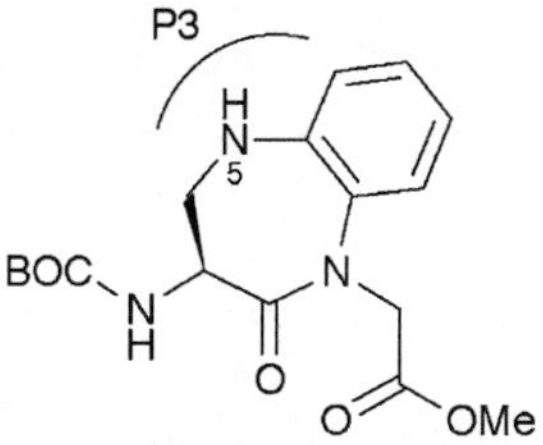

(205) 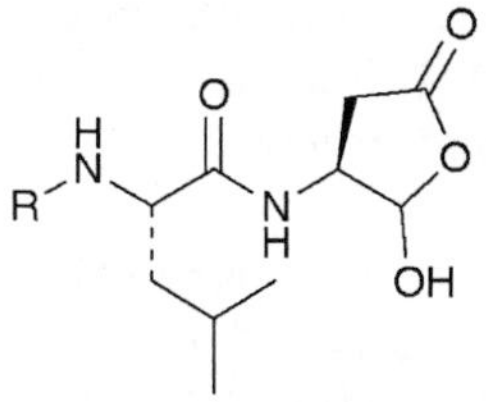 (206)

Phenylacetic acid and related compounds induce antiproliferation and differentiation of cancer cells. In order to find a new therapeutic strategy for treatment of lung cancer, phenylacetic acid derivatives have been synthesized.[520] 4-Fluoro-N-butylphenylacetamide activates caspase cascade (caspase-9 and -3) and induces apoptosis in humans. Mechanism of prion protein fragment PrP 106-126 and the participation of caspases were also analyzed. PrP 106-126 causes rapid depolarization of mitochondrial membranes, release of cytochrom C and caspase activation.[521] Parallel to this activation PrP 106-126 also cause a release of Ca^{2+} from the mitochondrial calcium stores and this results in activation of calpain. Experiments revealed that the mitochondrion was the primary site of action of PrP 106-126 and a combination of caspase and calpain inhibitors significantly inhibited apoptosis.

5.5 Cathepsin and Other Cysteine Protease Inhibitors. – Cysteine proteases have been grouped into families and clans. Most cysteine proteases are members of the papain clan (clan CA), which consists of enzymes such as calpains, cathepsins and papain. (Clan CD is smaller with several important enzymes, including caspases, legumain, etc.). Several reviews were published on the cathepsin enzyme family. Processing and activation of lysosomal proteinases[522] and the role of lysosomal enzymes (cathepsins) in brain tumor invasion[523] are summarized. Two other reviews focus on pathophysiological implications and recent advances in inhibitor design of thiol-dependent cathepsins.[524,525] A short review summarizes the new cysteine proteinase inhibitors that are homologous to the proregions of cysteine-proteinases.[526] The design, synthesis and selection of inhibitors of cysteine proteases (among them cathepsins) have been reviewed.[527]

Parasitic cysteine proteases are critical to the life cycle or pathogenicity of parasitic protozoans. A solid phase library synthesis (split-mix method), screening and selection of inhibitors of a *Leishmania mexicana* cysteine protease B have been described.[528] The inhibitor library was composed of octapeptides with a centrally located reduced bond introduced by reductive amination of the resin-bound amines with N^{α}-protected amino aldehydes.

Most of the inhibitors had micromolar K_i values and a few inhibited the enzyme in nanomolar concentrations.

A novel series of noncovalent inhibitors of cathepsin L have been designed to mimic the mode of autoinhibition of procathepsin-L.[529] The novel blocked tripeptide-size inhibitors show nanomolar potency. Moreover, these short peptides show higher selectivity (up to 310 fold) for inhibiting cathepsin L over K but only 2-fold selectivity over the 96-residue propeptide of cathepsin L. Compound (**207**) has a $K_i = 21$ nM for cathepsin L inhibition. A 1.9 Å X-ray crystallographic structure of the complex of cathepsin L with (**207**) confirmed the reverse binding mode of (**207**) as well as its noncovalent nature.

A series of N^{α}-benzyloxycarbonyl- and N^{α}-acyl-*L*-leucine (2-phenylamino-ethyl) amide derivatives was synthesized and evaluated for their inhibitory activity against rabbit and human cathepsins K, L and S.[530] Compound (**208**) shows a $K_i = 15$ nM against cathepsin K and has an excellent selectivity profile vs human cathepsins L and S. Cathepsin K is highly expressed in human osteoclasts, and is implicated in bone resorption. This makes it an attractive target for the treatment of osteoporosis. Peptides containing 2-amino-1′-hydroxymethyl ketones and 2-amino-1′-alkoxymethyl ketones are potent inhibitors of cathepsin K. A novel synthetic route was designed to facilitate the rapid elucidation of the structure-activity relationship (SAR) of these compounds[531] for easy identification of potent and selective cathepsin K inhibitors.

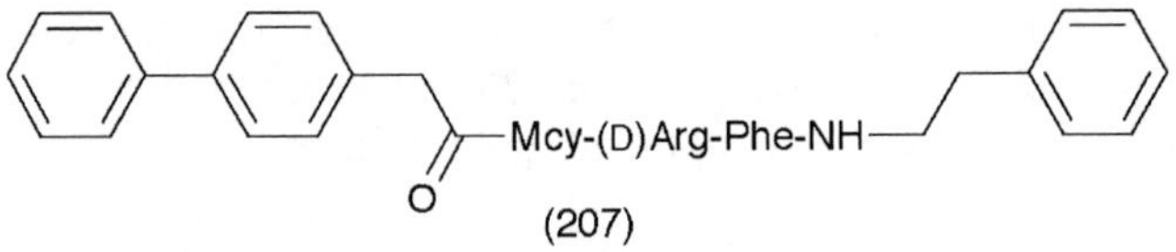

Mcy-(D)Arg-Phe-NH

(207)

OMe

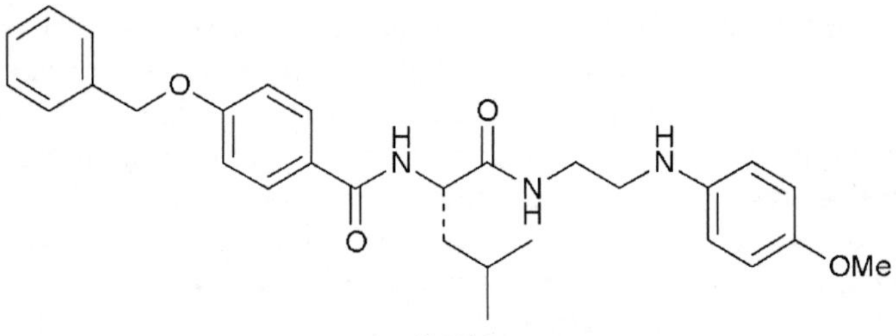

(208)

The specificity of immune response relies on processing of foreign proteins and presentation of antigenic peptides at the cell surface. Inhibition of antigen presentation modulates the immune response. Cathepsin S performs a fundamental step in antigen presentation and therefore represents an attractive target for inhibition. A novel cathepsin S inhibitor, paecilopeptin, isolated form the culture of the fungal strain *Paecilomyces carneus* has been synthesized in 6 steps. Paecilopeptin is a dipeptide aldehyde, acetyl-Leu-Val-CHO and shows an

IC_{50} value of 2 nM against cathepsin S.[532] A series of potent and reversible cathepsin S inhibitors, based on dipeptide nitriles, was synthesized.[533] Systematic investigation of the structure-activity relationship of the P1, P2 and P3 positions has provided picomolar inhibitors (**209**) of cathepsin S. The X-ray crystal structure of dipeptide nitriles cocrystallized with cathepsin S was presented.

Azaasparagine-containing halomethylketones were designed and synthesized for inhibition of legumain, a lysosomal cysteine peptidase.[534] The most potent inhibitor was the peptide derivative Cbz-L-Ala-L-Ala-Aza-Asn-chloromethylketone. Another research group also found that azapeptides were potent and selective inhibitors of cysteine proteases papain, cathepsin B and cathepsin K.[535] The structure of the novel azapeptides was based upon the enzyme-binding region of cystatins.

A series of 6-substituted amino-4-oxa-1-azabicyclo [3,2,0] heptan-7-one compounds was designed and synthesized as a new class of inhibitors for cathepsin B, L, K and S.[530] Compound (**210**) showed excellent cathepsin L and K inhibition activity with IC_{50} at a low nanomolar range.[536] The same research group published the synthesis of a new series of 6-acylamino penam derivatives for inhibition of cathepsin L, K and S.[537]

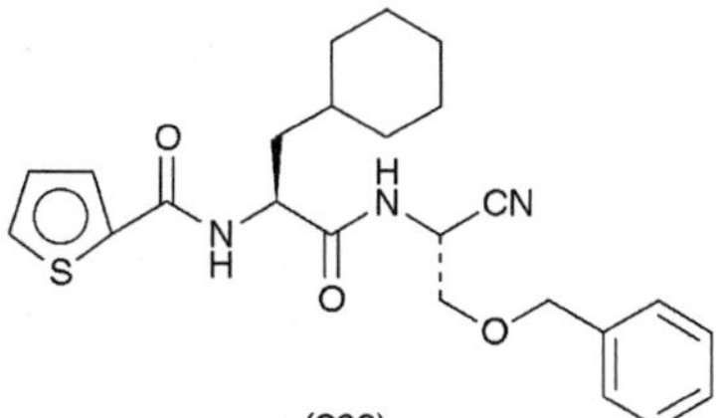

(209)

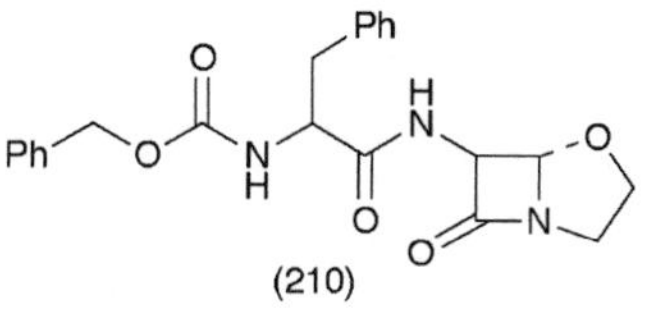

(210)

1-Cyanopyrrolidines have previously been reported to inhibit cathepsins. In a new publication the synthesis and structure-activity relationship studies of peptidic 1-cyanopyrrolidines have been summarized.[538] Two types of novel compounds were synthesized: either a small peptidic substituent was introduced to the 1-cyanopyrrolidine scaffold at the 2-position starting with proline (**211**), or at the 3-position of aminopyrrolidines (**212**). The resulting novel compounds proved to be good inhibitors of cathepsin B (in micromolar range) and very good inhibitors of cathepsins K, L and S (in nanomolar to picomolar range).

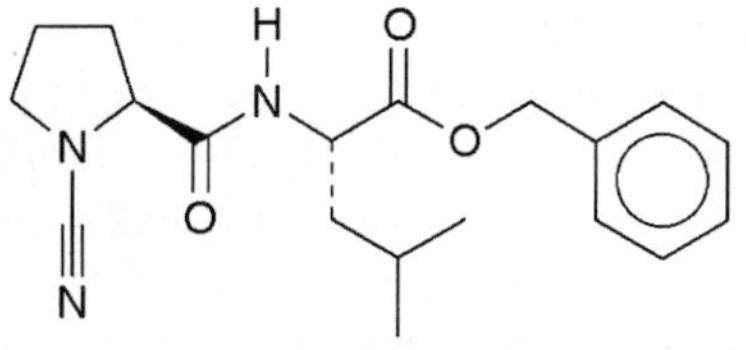

(211)

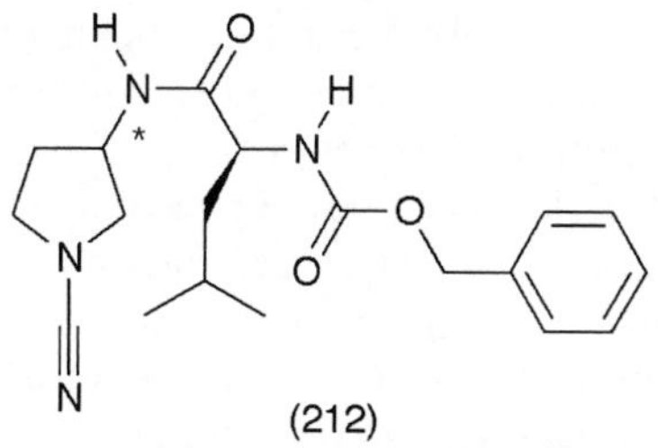

(212)

Extralysosomal cathepsin B has recently been implicated in apoptotic cell death. Highly specific irreversible cathepsin B inhibitors could be useful tools to elucidate the effects of cathepsin B in the cytosol. The epoxysuccinyl-based inhibitor MeO-Gly-Gly-Leu-(2*S*,3*S*)-tEps-Leu-Pro-OH was covalently coupled to the C-terminal heptapeptide of the cell permeable penetratin (Arg-Arg-Nle-Lys-Trp-Lys-Lys-NH_2) via an ε-aminocaproyl spacer. The conjugate was shown to efficiently penetrate into MCF-7 cells as an active enzyme inhibitor.[539] Synthetic DNA fragments specifically inhibited cathepsin D activity in a dose-dependent manner and the inhibition appeared to be electrostatic in nature.[540]

Inhibitor structure for the serine protease cathepsin G has been optimized. The lead was a novel β-ketophosphonic acid and optimization was performed by structure-based drug design.[541] The best compound (**213**) shows an IC_{50} value of 53 nM against cathepsin G. The first example of dual inhibitors for matrix metalloproteinase and cathepsin has been described.[542] An appropriate alignment of peptide parts and two different specific functional groups in one molecule led to the discovery of a potent dual inhibitor (**214**).

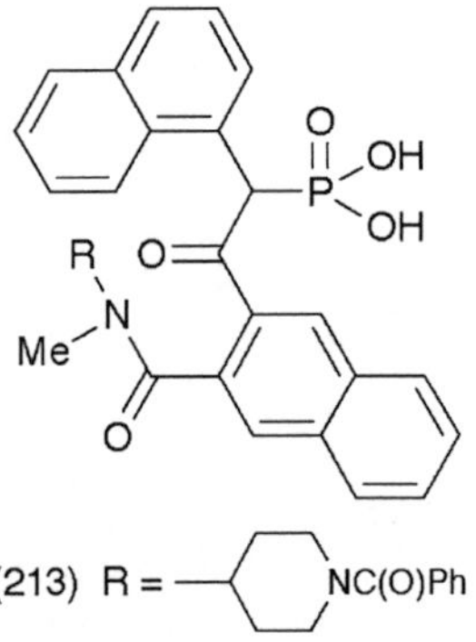

(213) R = NC(O)Ph

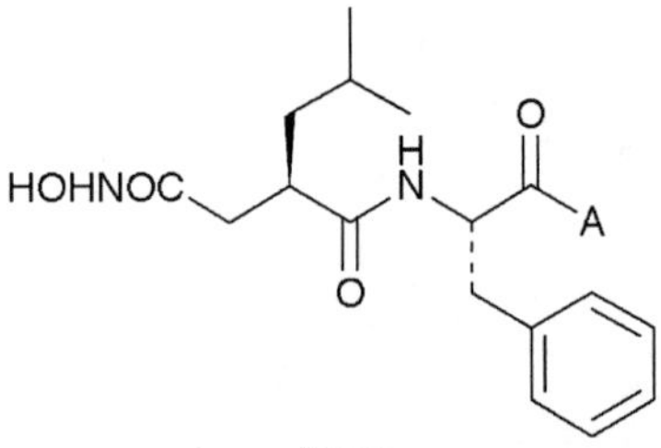

(214)

In the chemotherapy of Chagas disease (or American Trypanosomiasis) cathepsin inhibitors can be applied. A cathepsin-like Cys-protease, cruzain (cruzipain) is responsible for the major proteolytic activity in all stages of the life cycle of the parasite *Trypanosoma cruzi*. The therapeutic promise of cruzain inhibitors by blocking of proliferation of the parasite has been previously demonstrated. Two more recent publications [543,544]describe novel inhibitors against cruzain. First, a series of constrained ketone-based inhibitors has been developed that show low nanomolar K_i values. This inhibitor series includes constrained compounds, which can be assessed quickly and efficiently using combinatorial strategy.[543] Second, a series of potent mercaptomethyl ketones (**215**) as cruzain inhibitors was designed and synthesized showing K_i value of subnanomolar range.[544] The chemotherapeutic possibilities for curing chronic Chagas disease have been discussed.[545]

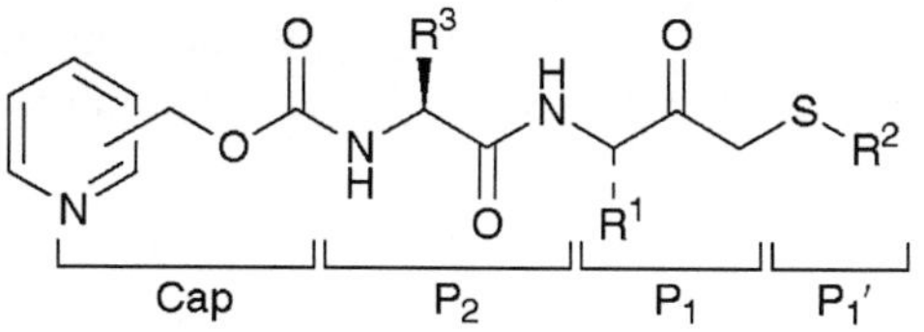

(215)

5.6 Cytomegalovirus and Rhinovirus 3C Protease Inhibitors. – Recent advances in the synthesis, design and selection of human cytomegalovirus inhibitors are reviewed in two publications dealing with thiol dependent enzyme and cysteine protease inhibitors.[506,525]

Human herpes viruses encode a serine protease, which is essential for viral replication. X-ray structures of the serine protease human cytomegalovirus protease (HCMV) revealed that this enzyme belongs to a novel class of Ser-proteases where the active site is composed of His, His and Ser triad. HCMV is an attractive target for potential anti-herpes virus agents with novel structures and new mechanisms. A new peptidomimetic skeleton and the synthesis of a chemical library containing 32 compounds with different substitutions on the skeleton has been published[546] Multicomponent condensation and liquid phase strategy was used for the preparation of the library from four types of building blocks: 4 carboxylic acid, 2 amines, 2 aldehydes and 2 isocyanides. The stereospecific synthesis of a series of α-methylpyrrolidine-5,5-*trans* lactam inhibitors of HCMV protease has been described and led to the discovery of a potent and highly selective inhibitor of HCMV **(216)**.[76] Compound **(216)** has low nanomolar activity against the viral enzyme. The crystal structure of HCMV protease was obtained and used to model the conformationally restricted novel HCMV-inhibitors into the active site groove of the enzyme.[547]

Human rhinoviruses are members of the picornavirus family and are the single most causative agents of the common cold. The virus 3C protease (HRV2-3CP) is a preferred target for drug design: inhibition of this enzyme by a small molecule agent should stop viral replication and thus control the extent of infection. Utilizing structure-based design, a new class of Michael acceptor-containing, irreversible inhibitors of HRV2-3CP was discovered.[548] The novel inhibitors exhibit antiviral activity when tested against HRV-14 infected H1-HeLa cells with EC_{50} values ranging from 1.94 to 0.15 μM. No cytotoxicity was observed at the limits of the assay concentration, and the most active compound was **(217)**.

The crystal structure of one of the more potent inhibitors covalently bound to HRV2-3CP has been detailed[548] and the evaluation of various 2-pyridone-containing HRV2-3CP protease inhibitors[549] carried out. These compounds comprised a peptidomimetic binding determinant and a Michael-acceptor moiety, which forms an irreversible covalent adduct with the active site cysteine-residue of the HRV2-3C enzyme. Optimization of the 2-pyridone containing compounds is shown to provide several highly active 3C enzyme inhibitors such as (**218**) that function as potent antirhinoviral agents (EC_{50} = <0.05 μM) against multiple virus serotypes in cell culture.

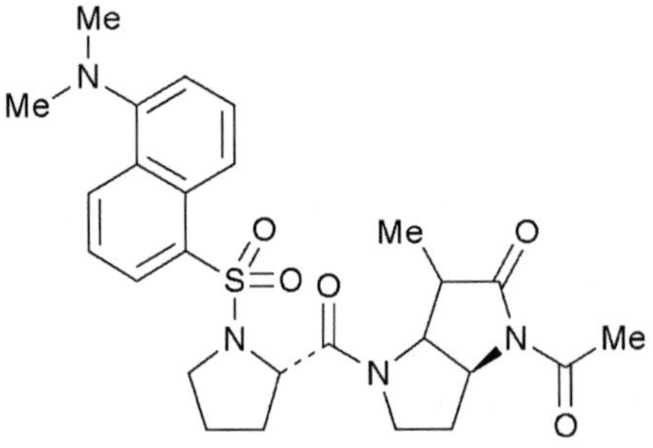

(216)

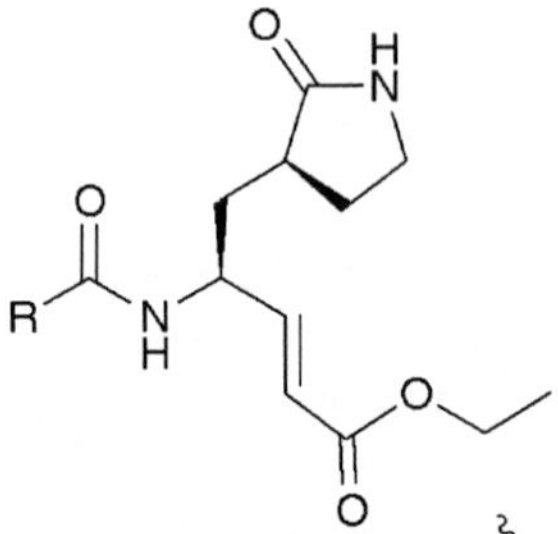

(217) R =

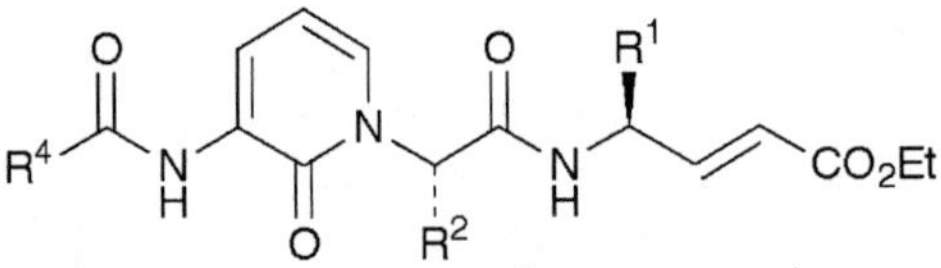

(218) $R^1 = CH_2CH_2CONH_2$, $R^2 = CH_2Ph$, $R^4 = PhCH_2O$

One of the 2-pyridone-containing 3CP inhibitors is shown to be bioavailable in the dog after oral dosing (F = 48%). The same research group published the structure-based design, chemical synthesis and biological evaluation of bicyclic 2-pyridone-containing HRV2-3CP inhibitors.[550] An optimized compound was shown to exhibit antiviral activity when tested against a variety of HRV serotypes (EC_{50} values are ranging from 0.037 to 0.162 μM).

5.7 Converting Enzyme Inhibitors. – *5.7.1 ACE and Related Enzyme Inhibitors.* The synthesis and enzyme inhibitory activities of novel peptide isosters acting on angiotensin converting enzyme (ACE) was reviewed.[551] Another publication describes the high prevalence of polymorphism of ACE in patients with systemic sclerosis.[552] A novel group of silicon-based ACE-inhibitors was published.[553] The silanediol tripeptide mimics such as (**219**) inhibited ACE with IC_{50} values as low as 14 nM. Methylsilanols, in contrast, were poor inhibitors. ACE, endothelin converting enzyme (ECE-1) and neprilysin are inhibited by *N*-[2-(indan-1-yl)-3-mercapto-propionyl] amino acids (**220**).[554] Optimization of the structure of the lead compound resulted in highly potent inhibitors of the above mentioned three enzymes with inhibitory potency in the nanomolar range.

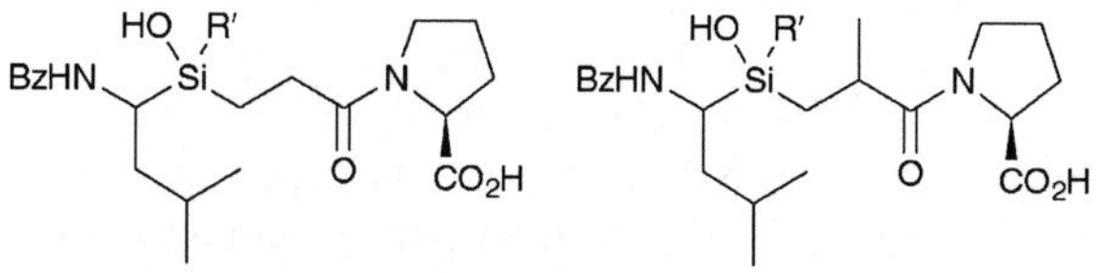

(219) R′ = Me or OH

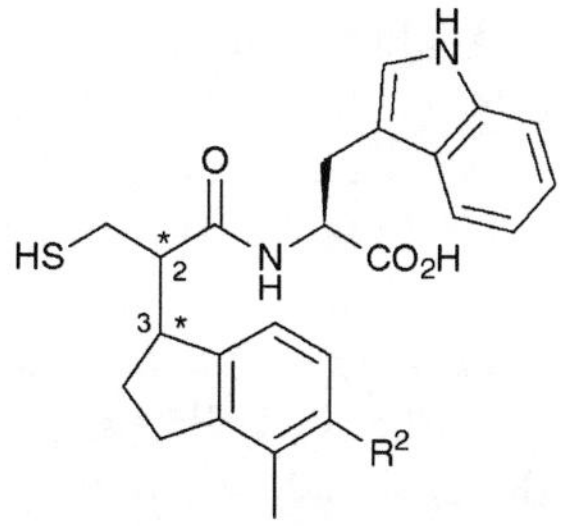

(220) R^2 = H, OMe or Br

The antithrombic effect of the ACE-inhibitor captopril is mediated by the heptapeptide angiotensin 1-7.[555] The preparation of enantiomerically pure converting enzyme inhibitors ecadotril, dexecadotril and fasidotril has been reviewed.[556] Angioedema and cough are known side effects of ACE-inhibitors, causing an increase of bradykinin level. Angiotensin II receptor blockers (ARBs) do not increase the level of bradykinin. However, angioedema can be associated also by taking an ARB in patients with a history of angioedema secondary to ACE-inhibitors.[557] Clinical studies showed that there was no negative interaction between ACE-inhibitors and aspirin: the latter did not reduce the beneficial effects of ACE-inhibitors in patients with heart failure.[558] Potential ACE-inhibitors of macrocyclic character were designed and synthesized; however, they were in fact enzyme substrates, rather than inhibitors.[559] Synthesis of novel hydroxamic acid-derived azepinones containing pendant mercaptoacyl groups of formyl hydroxamates (**221**) have been described.[560] These new analogues of therapeutically important ACE and NEP inhibitors act in the micromolar range on enzymes.

ACE-inhibitor-induced bronchial reactivity in patients with respiratory dysfunction was observed.[561] The effects of the dual ACE-NEP inhibitor MDL-100,240 and ramipril on hypertension and cardiovascular disease (CVD) in angiotensin II-dependent hypertension were studied in a transgenic rat model (TGRen2). Severe hypertension and CVD were regressed by MDL-100,240.[562] Tumor necrosis factor-α converting enzyme (TACE) is a membrane-anchored zinc metalloprotease. A series of novel, selective TACE-inhibitors based on 4-hydroxy and 5-hydroxy pipecolate hydroxamic acid scaffolds has been described.[563] Substituted 4-benzyloxybenzenesulfonamides exhibit excellent TACE potency with > 100 x selectivity over inhibition of matrix metalloprotease-1.

5.7.2 Endothelin Converting Enzyme. Endothelin converting enzyme-1 (ECE-1) is a zinc metalloprotease and catalyses the post-translational conversion of big endothelin-1 to endothelin-1 (ET-1), which is a very potent vasoconstrictor. A potent inhibitor of ET-1 biosynthesis is an attractive therapeutic tool for the treatment of diseases linked with elevated ET-1 levels. The therapeutic potential of ECE-1 inhibitors on the cardiovascular system has been reviewed.[564] Other reviews discussed nonpeptidic ECE-1 inhibitors and their potential therapeutic applications in cardiovascular diseases.[565,566] Vascular protection, current possibilities and future perspectives have also been summarized in a review article.[567]

A library of substituted 1,2,4-triazoles tethered to a 4-mercaptopyrrolidine core (**222**) has been produced[568] and are shown to be a novel class of non-peptidic compounds which inhibit the activity of ECE-1.

CGS 34,226 proved to be a potent dual inhibitor of ECE-1 and neprilysin both *in vitro* and *in vivo* and may represent a novel agent for the treatment of cardiovascular and renal dysfunction.[569] A novel non-peptidyl ECE-1 inhibitor was obtained through a pharmacophore analysis of known inhibitors and 3-dimensional structure database search. 1-Phenyl-tetrazole-formazan analogues showed activity after ip. administration in rats by suppressing the big ET-1 induced pressor responses.[570] Analogues of CGS 30084, a potent ECE-1 inhibitor were synthesized and evaluated for their biological activities in rats. One of the compounds (a prodrug) proved to be orally active.[571]

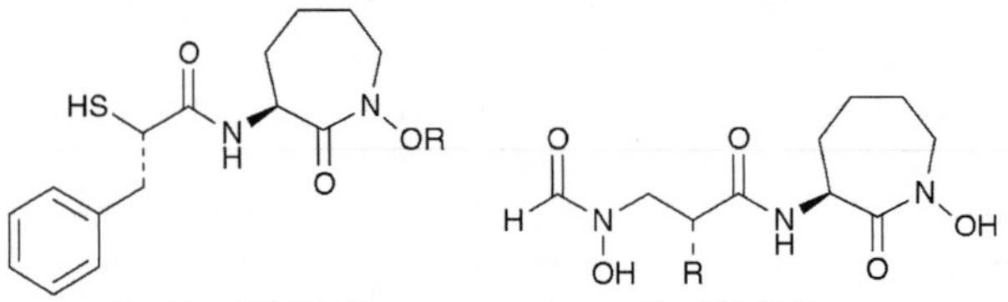

R = H or CH_2CO_2H (221) R = $CH_2C_6H_5$

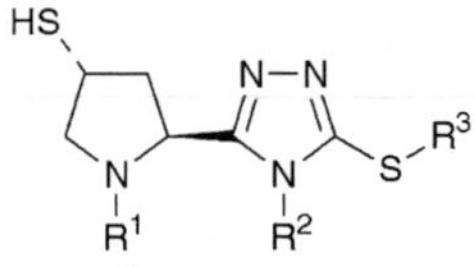

(222) R^1 = 2-Naphthyl–SO_2–
R^2 = 4-F-C_6H_4–CH_2–
R^3 = CH_3–CH_2–CH_2–

5.8 Elastase Inhibitors. – Human neutrophil elastase (HNE) is a neutral serine protease with blood substrate specificity. Although numerous peptidic and non-peptidic HNE-inhibitors have been investigated, no agent for clinical use has so far been developed. The current status and perspectives of HNE in acute lung injury have been reviewed.[572] The novel role of HNE inhibitors in infection and inflammation was also summarized in a review.[573]

For the treatment of acute respiratory failure, a series of peptide-based carboxylic acid-containing transition-state inhibitors was designed and synthesized.[574] The presence of a valyl moiety is found to be essential for potent *in vitro* inhibitory activity and also prevention of undesirable toxicity. The best compound (**223**) has the most potent *in vivo* effect, on HNE-induced lung hemorrhage. A 30 mg/kg (iv. or bolus administration) dose showed 99% inhibition in hamsters.

GW 311,616A, a derivative of pyrrolidine *trans*-lactams, is known as a potent, orally active inhibitor of HNE for the treatment of respiratory diseases. Further synthetic work has resulted in three back-up candidates for drug development, GW 447,631A, GW 469,002A and GW 475,151A (**224**).[575] All of these compounds are pyrrolidine *trans*-lactams. The new compounds have increased activity in inhibiting HNE in human whole blood and comparable pharmacokinetic properties, in particular clearance, in two species. The back-up candidate compounds are highly potent and selective HNE inhibitors, with a prolonged pharmacodynamic action. One of these compounds (**224c**) was cocrystallized with the enzyme and features of this structure were described.

The marine cycloheptapeptide hymenamide C (**225**) inhibits human neutrophil elastase degranulation release at micromolar levels. A systematic structure-activity relationship study of this cyclopeptide was performed using Ala-scan, in order to obtain useful information for the rational design of additional analogues.[576] The inhibitory effect of hymenamide C and its analogues could potentially be used in a therapeutic approach to pulmonary emphysema.

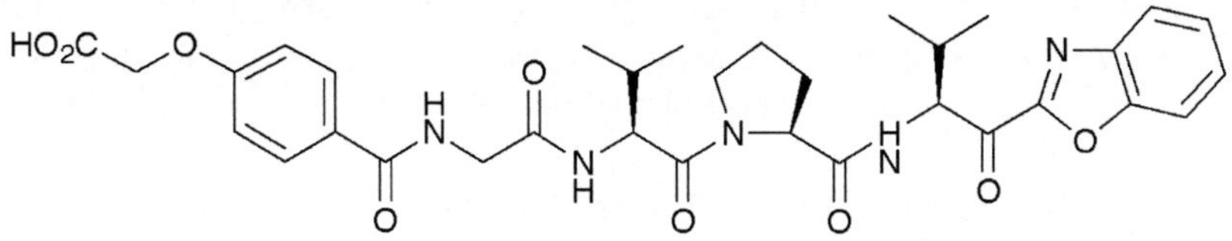

(223)

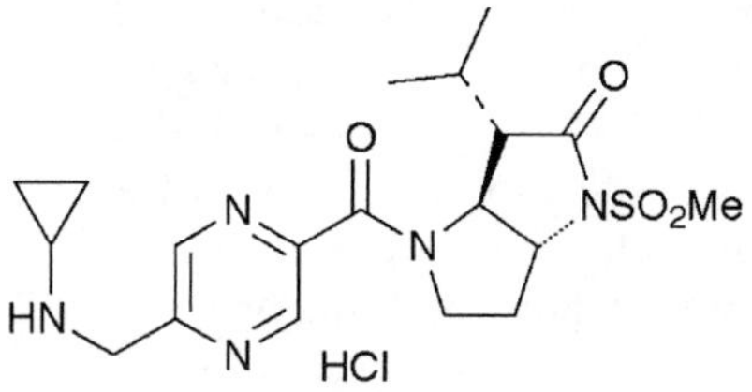

(224) a

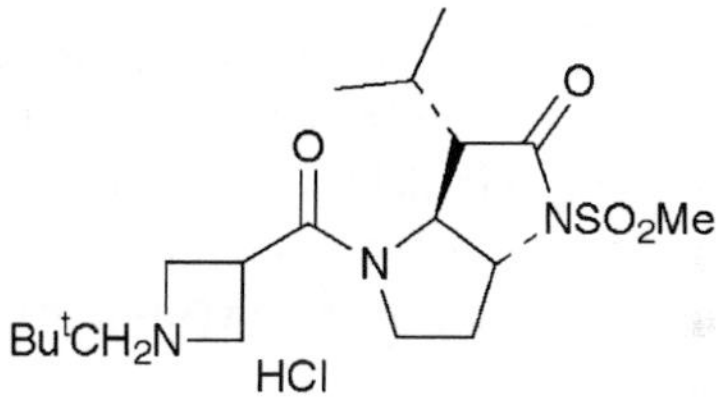

(224) b

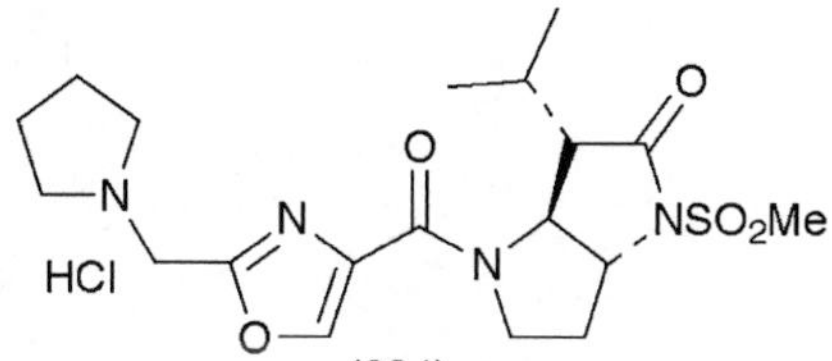

(224) c

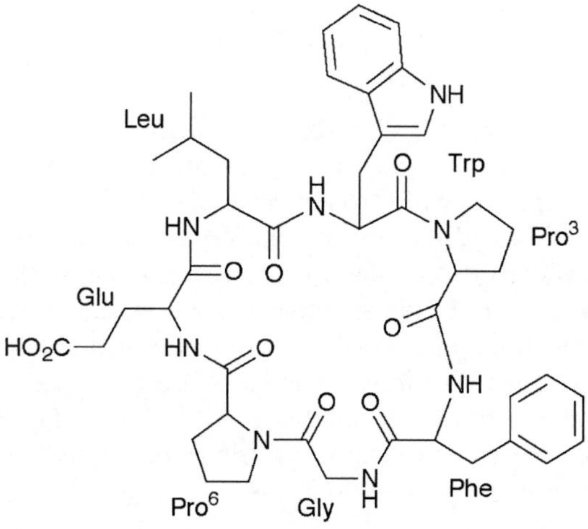

(225)

Crystallization and preliminary X-ray analysis of the complex of porcine pancreatic elastase and a hybrid squash inhibitor has been described.[577]

5.9 Farnesyltransferase Inhibitors. – Farnesylated Ras-proteins are involved in oncogenesis. The inhibition of the protein-modifying enzyme farnesyltransferase is a very attractive strategy in cancer therapy. A great number of review articles deal with farnesyltransferase inhibitors as anticancer agents,[578–583] and the promises and realities of application of inhibitor therapy.[584,585]

The aryl binding site of farnesyltransferase had already been described. So a 2-naphthylacryloyl residue was developed as an appropriate substituent for benzophenone-based peptidomimetics capable of occupying this aryl binding site.[586] The best new non-thiol farnesyltransferase-inhibitor (**226**) has biological activity in the nanomolar range (IC_{50} = 110 nM ± 40). The activity of this inhibitor is readily explained on the basis of docking studies which show the 2-naphthyl residue fitting into the aryl binding site. Non-thiol farnesyltransferase inhibitors and the results of the structure-activity relationships of benzophenone-based bisubstrate (peptidic and prenylic substrate) analogues have also been reported.[587]

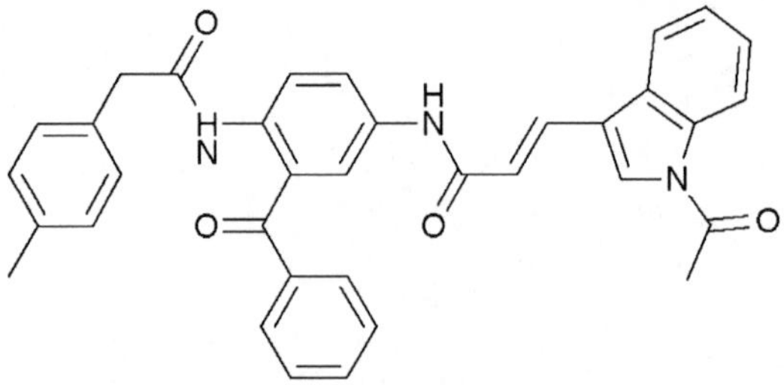

(226)

A stereocontrolled vinyl triflate-based synthetic route has been used to prepare four analogues of farnesyl diphosphate where the terminal isoprene units have been replaced with aromatic moieties.[588] Two of these analogues exhibit no interaction with the enzyme farnesyltransferase, but the 2-naphthyl derivative (**227a**) is a modest inhibitor of the enzyme, with a *para*-biphenyl derivative (**227b**), a surprisingly effective alternative substrate.

Highly potent dual inhibitors of farnesyltransferase and geranyltransferase-I were discovered. Molecular design was based on the structure of the known highly potent farnesyltransferase inhibitor (**228**)[589] with the *para*-cyano group being replaced by trifluoromethylphenyl, trifluoropentynyl and trifluoropentyl groups resulting in good enzyme inhibitors.

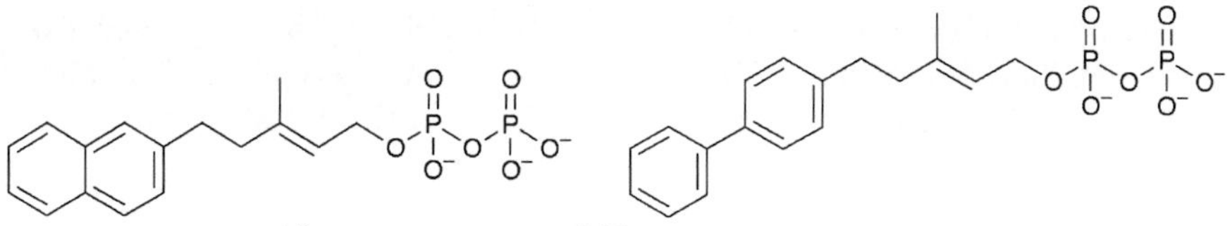

(a) (227) (b)

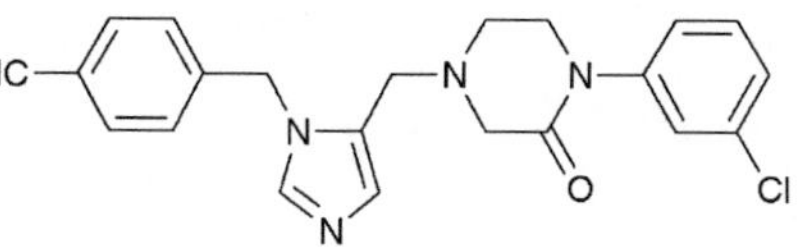

(228)

New results show that the farnesyltransferase inhibitor RPR-130,401 does not alter radiation susceptibility in human tumor cells with a K-Ras mutation.[590] The major outcome of drug action is the disruption of orderly progression through mitosis and cytokinesis as a result of lamin B farnesylation.

Potent suppression of proliferation of A10 vascular smooth muscle cells was achieved by combined treatment with lovastatin and 3-allylfarnesol, an inhibitor of farnesyltransferase.[591]

5.10 HIV-Protease Inhibitors. – Important viral enzymes for the chemotherapy of AIDS include HIV-1 reverse transcriptase and aspartic acid protease (HIV-1 PR). The latter enzyme has remained an attractive target for designing inhibitors for effective antiviral therapy. Dimerization of HIV-1 PR to enzymatically active C_2-symmetric homodimer is a necessary event to attain the active structure. A review[592] highlights the modulation of dimeric and oligomeric structures of HIV-1 PR by synthetic peptides and small molecules.

Pseudo-symmetric HIV-1 protease inhibitors containing a novel hydroxymethylcarbonyl isostere as the transition state mimic were designed and synthesized.[593] Most of the synthetic compounds, with varied structures at the P and P′ sites around this core unit, showed potent inhibitory activity against HIV-1 protease. One of the best compounds (**229**) has a K_i of 0.16 nM.

Z and Fmoc-*L*-Tetrahydrofuranyl glycines have been obtained from L-vinylglycine through dipolar cycloaddition reaction, and the Fmoc-derivative applied to the synthesis of modified S9 and S10 substrates of HIV-1 protease.[594] These compounds acted as strong inhibitors, rather than substrates of the enzyme, probably due to the favorable interactions of the tetrahydrofuranylglycine moiety at the S_2 site. Water soluble prodrugs (KNI-272 and KNI-279) were synthesized[595] consisting of allophenylnorstatine-thiazolidine-4-carboxylic acid core structure. The prodrugs could be converted to the parent compounds in aqueous solution, at slightly acidic to basic pH at 37°C, as a result of O → N intramolecular acyl migration. The synthetic prodrugs possess improved water solubility (> 300 mg/ml) more than 4000-fold better than the parent compounds.

The potent HIV-protease inhibitor JE-2147 (**230**) can be assembled from chiral building blocks Boc-AHPBA (**231** and Boc-DMTA **232**). The concise synthesis of the two chiral building blocks (**231**) and (**232**) is also described.[596]

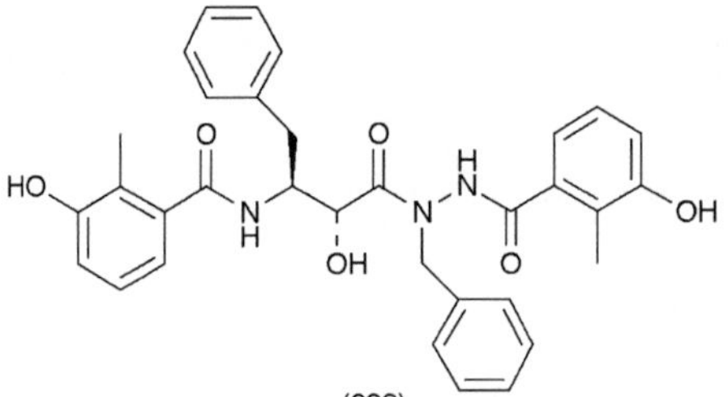

(229)

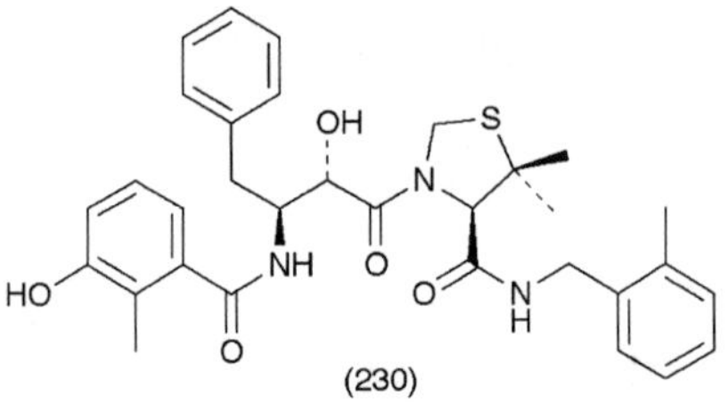

(230)

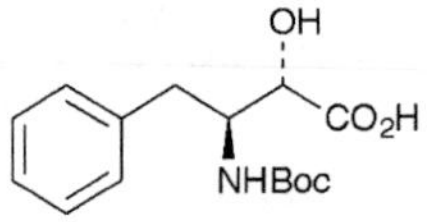

(231)

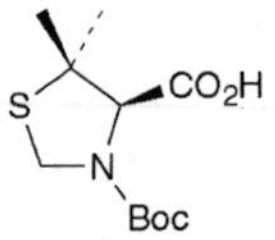

(232)

Structure-based design, synthesis and evaluation of peptidomimetic inhibitors of HIV-1 protease containing β-*D*-mannopyranoside scaffolds have been described.[597] The compounds prepared were moderate inhibitors ($IC_{50} = 4.48$–9.95 μM). The activity of compound (**233**) was encouraging since the synthetic route to this compound is relatively short and might facilitate preparation of a range of carbohydrate-based analogues with higher binding affinities. An α-methylene tetrazole based dipeptidomimetic has been prepared as a constrained and non-hydrolysable core for incorporation into peptides.[598] A single crystal X-ray structure determination revealed that its solid-state conformation closely resembles that of isosteric core of JG-365 bound to HIV-1 protease. The α-methylene tetrazole was incorporated into peptide sequences and in a HIV-1 protease assay, the results suggest that the longer the C-terminal substitution the greater the inhibitory potency.

N-Cinnamoyl-L-proline can be used as a template on which β-substituted phenylalanine and β-phenylisoserine residues can be stereoselectively synthesized leading to tripeptide derivatives as structural analogues of HIV-1 protease inhibitors.[599] Development of an efficient and scalable process for the HIV-1 protease inhibitor BMS-232632 (**234**) has been described.[600] Evaluation of a variety of salts and identification of bisulfate salt of BMS-232632 with enhanced bioavailability were also described.

A stereoselective synthesis of hydroxyethylene dipeptide isosteres based on the 1,4-diamino-2-hydroxybutane structure has been published.[601] Six hydroxyethylene dipeptide isosteres were synthesized, including the diamino alcohol core of the potent HIV-1 protease inhibitor Ritanovir (**235**) and its C-2 epimer. The new synthetic route provides a 35% overall yield for the preparation of the Boc-protected core structure of Ritanovir.

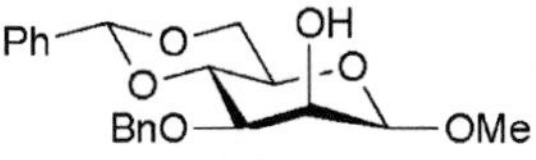

(233)

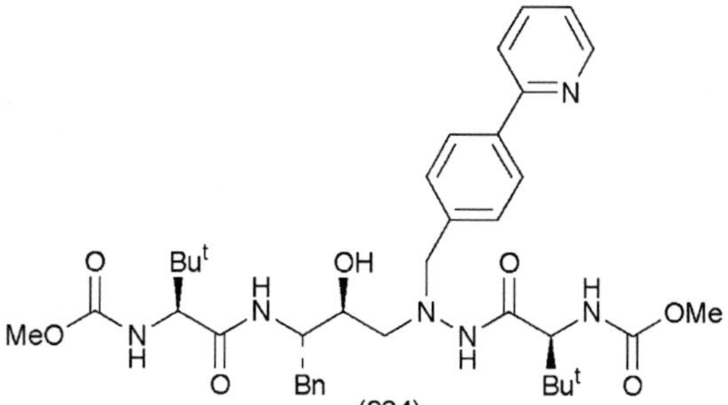

(234)

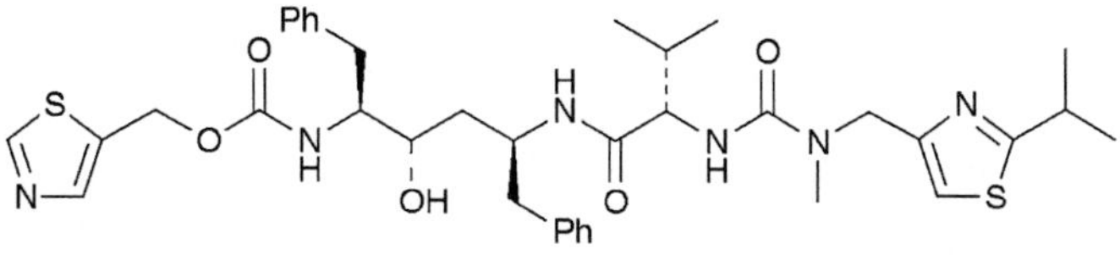

(235)

Novel analogues of the HIV-1 protease inhibitor Lopinavir (**236**) has a pseudosymmetric core unit incorporating benzyl groups at both P-1, P-1′ positions. A series of analogues incorporating non-aromatic side chains at the P-1 position were synthesized.[602] In the antiviral assay, all the compounds with alkyl side chains have enzymatic inhibitory potencies comparable to that of Lopinavir, but the latter was best overall. Lopinavir analogues formed by substitution of the 2,6-dimethylphenoxyacetyl group with various substituted phenyl or heteroaryl groups,[603] were not so potent as the parent molecule.

A new simple method for rapid preparation of novel putative inhibitors of HIV-1 protease has been described.[604] A library of 62 compounds were synthesized and screened in less than 1 hour. *Cis*-Aminoindanol, a key chiral precursor of the HIV-1 protease inhibitor Crixivan can be derived from indene oxidation products of (2*R*) stereochemistry. A number of different microorganisms (*Pseudomonas* and *Rhodococcus* strains) were isolated and used for oxygenation of indene to indandiol.[605] The bioconversion process provided indandiol with greater stereospecificity than is achievable through traditional chemical synthesis.

The structure of the known Merck HIV-1 protease inhibitor, Indinavir was changed by substitution of the tert-butylcarboxamide group with a trifluoroethylamide moiety.[606] This new analogue (**237**) had improved inhibitory potency against both the wild-type (NL4-3) virus and PI-resistant viral strains, and a slower clearance rate in dogs. Novel Indinavir analogues such as (**238**) with blocked metabolism sites showed highly improved pharmacokinetic profiles in animals,[607] and an excellent potency against both the wild type (NL4-3) virus and protease inhibitor-resistant HIV-1 strains.

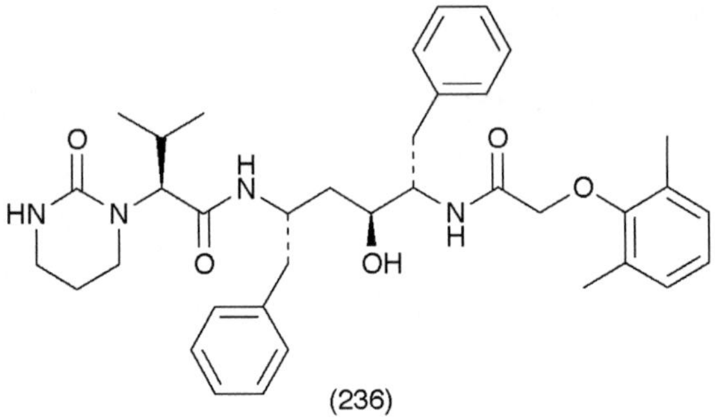

(236)

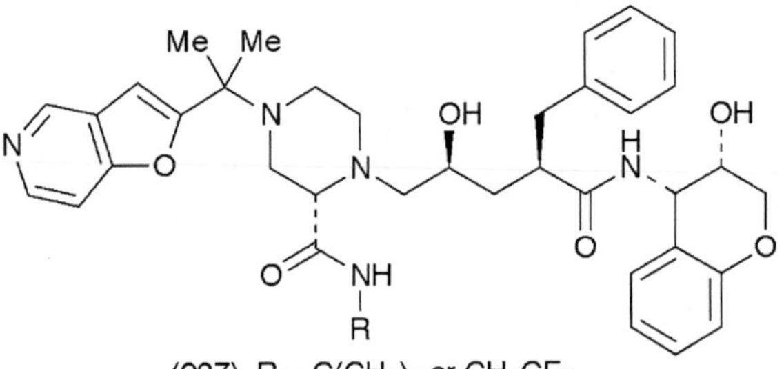

(237) R = $C(CH_3)_3$ or CH_2CF_3

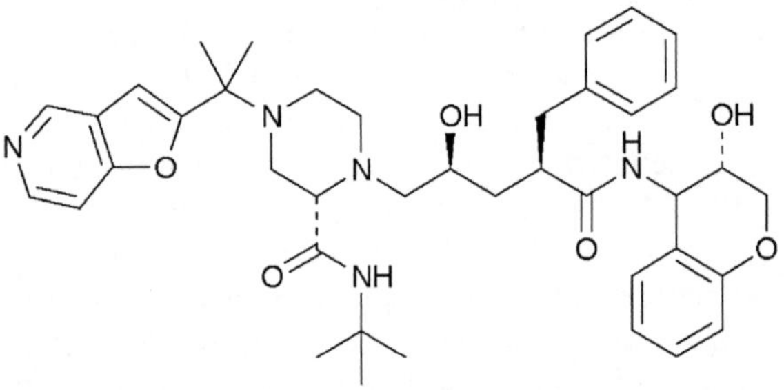

(238)

A new series of the HIV-1 protease inhibitor Amprenavir (Agenerase) was designed and synthesized by replacing the $-CH_2$-group of the benzyl moiety by a sulphur atom.[608] Introduction of S-atom in the molecule abolished or drastically decreased the inhibitory potency of the compound on HIV-1 protease. Fosamprenavir, a Vertex Pharmaceuticals phosphate ester prodrug of Amprenavir is under development by Glaxo Smith Kline.[609] Clinical pharmacology and pharmacokinetics of Amprenavir have been published.[610] Metal-organic compounds represent a new approach for design of inhibitors of HIV-1 protease.[611] N^1-(4-methyl-2-pyridyl)-2,3,6-trimethoxybenzamide copper (II) complex interacts with the active site of the enzyme leading to competitive inhibition.

The three-dimensional structures of Indinavir and three novel Indinavir analogues in complex with a multidrug-resistant variant of HIV-1 protease

(L63P, V82T and I84V) were determined to approximately 2.2 Å resolution.[612] The Lamarckian genetic algorithm of AutoDock 3.0 has been used to dock 3-(*S*)-amino-2-(*S*)-hydroxy-4-phenylbutanoic acids (AHPBA) into the active site of HIV-1 protease.[613]

5.11 Matrix Metalloproteinase Inhibitors. – The family of matrix metalloproteinases (MMPs) represents a group of tightly regulated metalloproteases. These enzymes are involved in neuroinflammation, some types of cancer, rheumatoid arthritis and other diseases. MMPs are preferred target proteins for inhibitor design. Metalloproteinase inhibitors, their actions and therapeutic opportunities have been reviewed.[614] The participation and role of MMPs in neuroinflammation were also summarized in a review article.[615] MMPs are multifunctional effectors of inflammation in multiple sclerosis and bacterial meningitis,[616] while two reviews exhibit the role and place of MMP inhibitors in anticancer therapy.[617,618]

In Section 5.5 we have already mentioned the synthesis of dual inhibitors for MMP and cathepsin,[542] for two proteinases belonging to different classes. This design could be useful for other new dual inhibitors. Several amines, amino acid derivatives and low molecular weight peptides containing an amide-bound oxal hydroxamic acid moiety have been synthesized[619] and their inhibitory effect tested towards native human gelatinase B (MMP-9) and the catalytic domains of the membrane-type MT1-MMP (MMP-14) and of neutrophil collagenase (MMP-8). A number of these compounds (e.g. **239, 240**) exhibited considerable inhibitory activity against the tested metalloproteinases.

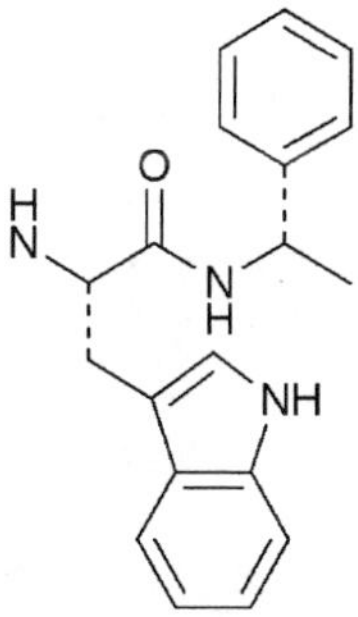

(239) 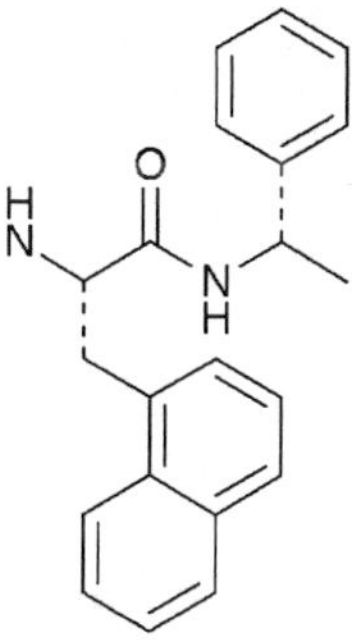 (240)

Fibroblast collagenase (MMP-1), a member of the MMP family, is believed to participate in the pathogenesis of arthritis, by cleaving triple-helical type II collagen in cartilage. On the basis of the similarity of the active site zinc binding mode with hydroxamate, new potential enzyme inhibitors were designed and synthesized containing α-mercaptocarbonyl group and various peptide sequences as enzyme recognition sites.[620] The best compounds had an inhibitor potency of $IC_{50} = 10^{-6}$ M. Cyclopropane-derived peptidomimetics as P1′ and P2′ replacements were designed and synthesized,[621] as analogues of the well-known and potent MMP inhibitors (**241**). The best compound (**242**) was designed explicitly to probe topological features of the S1′ or the S2′ binding pockets of the MMPs. However, this had only a moderate potency, suggesting, that there may be a loss of hydrogen bonding capability associated with introducing the P2′-P3′ retro amide group.

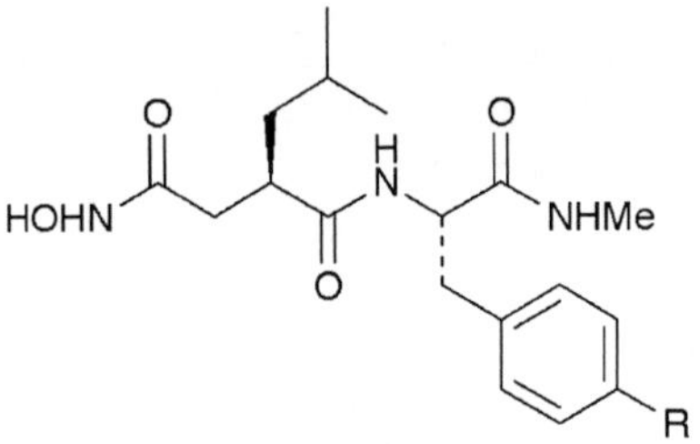

(241) R = OMe or H

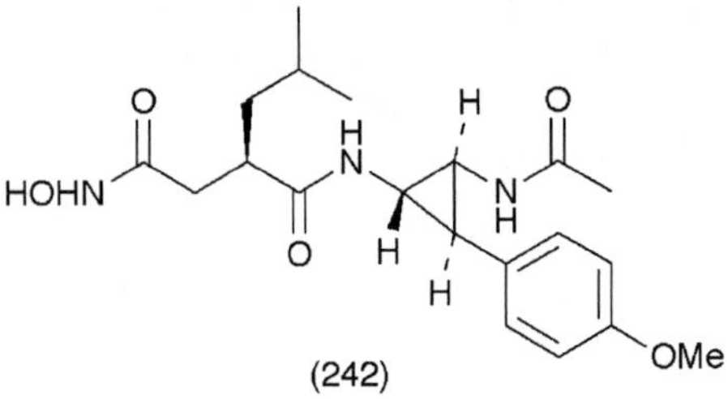

(242)

N-Arylation of electron deficient pyrroles has been achieved by Cu(II) mediated cross-coupling with arylboronic acid at room temperature.[622] This methodology has been used for the preparation of a key intermediate necessary for the synthesis of an MMP-inhibitor. Resveratrol (*trans*-3,4′,5-trihydroxystilbene) suppresses induced mammary carcinogenesis with down-regulation of nuclear factor κB and MMP-9 expression.[623] The activity of the novel anti-angiogenic compound AG3340 (Prinomastat), a selective inhibitor of matrix metalloproteases, was studied in an animal model (C57BL/6J mice) of retinal neovascularization.[624] AG3340 significantly inhibited oxygen induced retinal neovascularization in the mouse model and appears to be a promising candidate for the treatment of neovascular retinal diseases.

5.12 NO-synthase Inhibitors. – Nitric oxide (NO), an endogenous free radical, is produced by conversion of L-arginine to L-citrulline by a five-electron oxidation reaction catalyzed by the heme-containing enzyme, nitric oxide synthase (NOS). To date, three distinct human isoforms of NOS have been identified: the neuronal (hncNOS), the endothelial (hecNOS or eNOS), and inducible NOS (hiNOS). Selective inhibition of the isoforms of NOS could be beneficial in the treatment of certain diseases arising from the overproduction of NO by NOS.

Two new types of conformationally constrained S-[2-[(1-iminoethyl)-amino]ethyl] homocysteine derivatives were synthesized as potential NOS inhibitors[625](**243, 244**). These molecules represent the first attempt to probe the conformational constraint near the α-amino acid moiety of known homocysteine-based NOS inhibitors. A new prodrug, a selective inhibitor of hiNOS was synthesized. The 5-tetrazole amide of L-N^6-(1-iminoethyl) lysine (**245**) was prepared and obtained as a stable, nonhygroscopic, crystalline solid.[626] Compound (**245**) has minimal inhibitory activity *in vitro* on iNOS. However, it is rapidly converted *in vivo* to L-N^6-(1-iminoethyl) lysine (L-NIL) and produced dose-dependent inhibition of iNOS in acute and chronic models of inflammation in the rodent with an efficacy comparable to L-NIL. Both L-NIL and the 5-tetrazole amide derivative (**245**) exhibit significant and comparable *in vivo* selectivity for the inhibition of iNOS versus endothelial NOS.

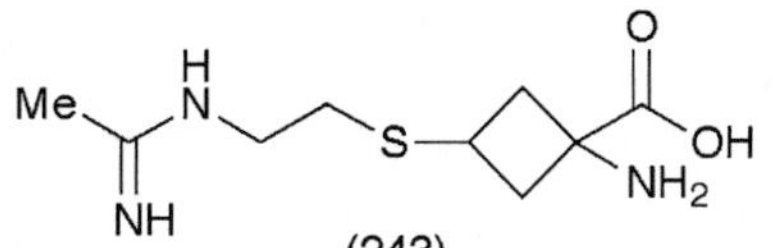

(243) 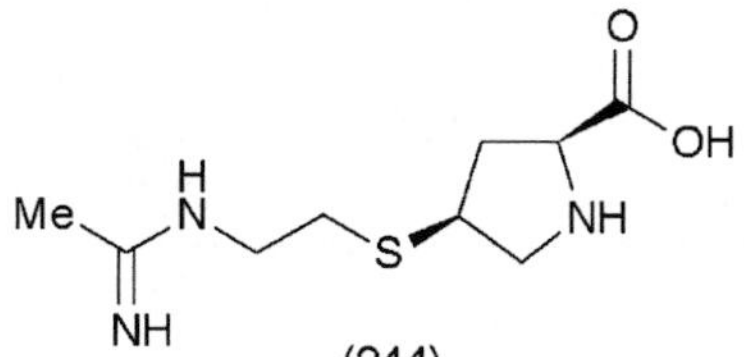 (244)

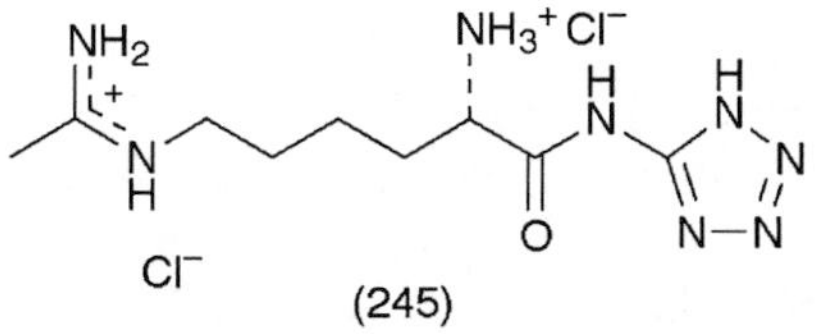

(245)

In macrophages, large amounts of NO are generated by iNOS; this is an important mechanism in macrophage-induced cytotoxicity. A synthetic carbazole compound, 9-(2-chlorobenzyl)-9H-carbazole-3-carbaldehyde (LCY-2-CHO) (**246**) was found to have an inhibitory effect on lipopolysaccharide (LPS)-stimulated NO-generation in RAW 264.7 macrophages (IC_{50} value of 1.3 ± 0.4 μM).[627] Compound (**246**) did not induce cytotoxicity and had a negligible effect on iNOS activity. The results of further studies indicate that (**246**) inhibits NO-generation via a decrease in the transcription of iNOS mRNA through a signaling pathway that does not involve NF-κ B activation.

Gabexalate mesylate (**247**), a non-antigenic synthetic inhibitor of trypsin-like serine-proteases, is a drug used efficiently in the treatment of pancreatitis. Compound (**247**) inhibited competitively L-arginine uptake and constituted NOS in a cell-free extract[628] so could be considered as an effective modulator of cellular NO synthesis.

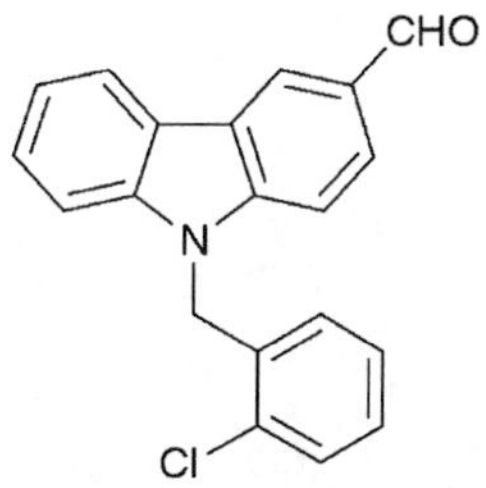

(246) 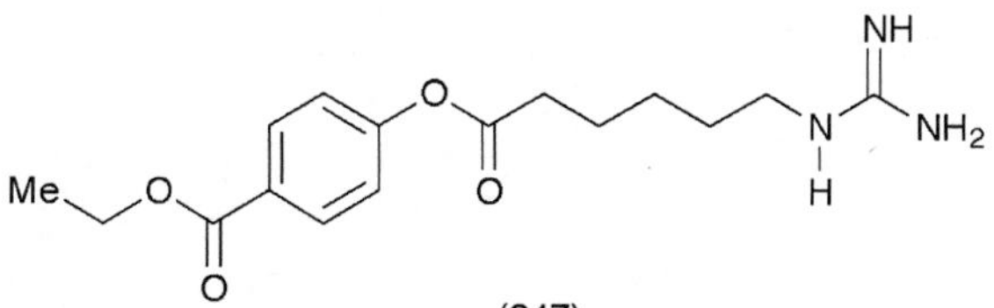 (247)

5.13 Proteasome Inhibitors. – Proteasomes are large cytosolic protease complexes, participating in many biological processes, including degradation of most of the cytosolic proteins in mammalian cells. Activation of proteasome may increase the proteolysis of cyclin and prevent retinoid cancer. Proteasome inhibitors are mostly short oligopeptide sequences or peptidomimetics. Mammalian proteasomes are composed of three distinct catalytically active species, β1 ("peptidyl-Glu-peptide hydrolyzing"), β2 ("tryptic-like") and β5 ("chymotryptic-like") individual subunits. An important goal of proteasome research is the development of inhibitors that are subunit-specific and cell permeable.

Natural ester-bond containing green-tea polyphenols, such as epigallocatechin-3-gallate (EGCG) and gallocatechin-3-gallate (GCG) have been shown to be potent and selective proteasome inhibitors both *in vitro* and *in vivo*.[629] Protection of the phenolic hydroxyl groups of EGCG renders the compound completely inactive. An enantioselective synthesis of the proteasome inhibitor (+)-lactacystin (**248**) has been achieved using an alkylidene carbene 1,5-CH insertion reaction as a key step.[630]

Based on the core structure of the natural proteasome inhibitor TMC-95A (**249**), a new lead compound (**250**) was synthesized containing a propylamide residue for interaction with the S1 pocket of the active sites, and the side chain of asparagine, which is present in the natural product, for occupancy of the S3 pocket.[631] To enhance cyclization, the linear precursor was oxidized to the keto form at the indole-2 position. Compound (**250**) retained almost full inhibition of proteosome β1 (peptidyl-glutamyl-peptide) and β2 (trypsin-like) hydrolase activities, but was significantly less potent against β5 (chymotrypsin-like) activity.

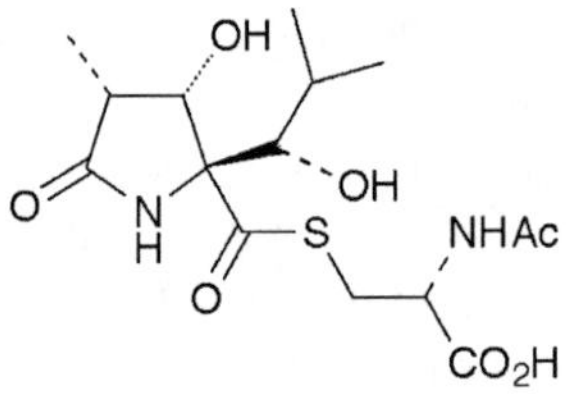

(248) (+)-lactacystin

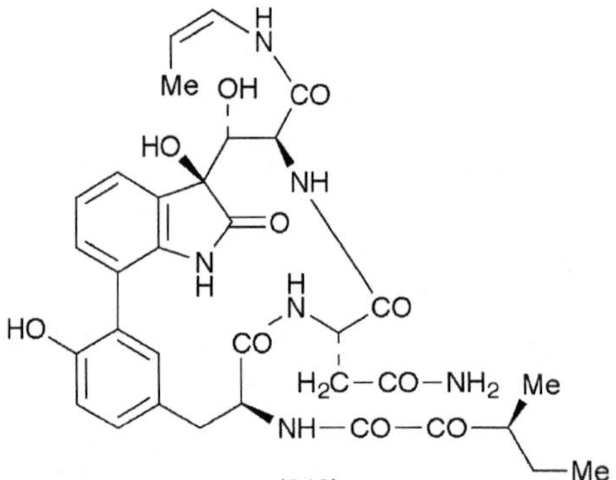

(249)

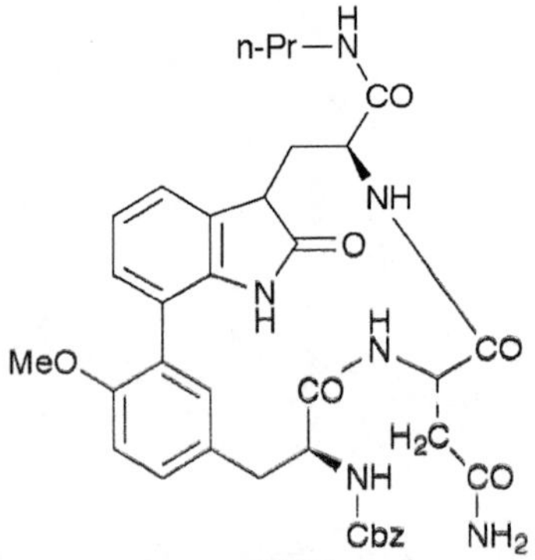

(250)

Another research group has published the total synthesis of TMC-95A (**249**) and TMC-95B.[632] [The side chain has *S* configuration in TMC-95A, and *R* in TMC-95B). TMC-95 compounds (from A to D) were isolated from the fermentation broth of *Apiospora montagnei*.]

5.14 Protein Phosphatase Inhibitors. – Modulation of the reversible phosphorylation of proteins, as catalyzed by protein kinases (PKs) and protein phosphatases (PPs), is the principal mechanism by which eukaryotic cells respond to external stimuli. The balance between phosphorylated and dephosphorylated proteins, which is controlled by protein kinase and phosphatase activities, is crucial for maintaining proper cellular function. Phosphatases exist as two major classes: protein phosphatase 1 (PP1) and phosphatase 2A (PP2A).

Motuporin (**251**) a cyclic pentapeptide isolated from the marine sponge *Theonella swinhoei* is one of the most potent PP1 inhibitors known, and shows strong *in vitro* cytotoxicity against a variety of human cancer cells. A highly convergent asymmetric synthesis of motuporin has been described.[633] Fostriecin (**252**) is a structurally unique phosphate ester produced by *Streptomyces pulveraceus*, displays potent *in vitro* activity against various cancerous cell lines, shows *in vivo* antitumor activity and is a potent phosphatase inhibitor. The chemistry and biology of fostriecin has been reviewed.[634] Due to concerns over stability and purification of the natural product, efficient and flexible methods for the synthesis of fostriecin have been studied. An enantioselective synthetic route to fostriecin secured all possible stereoisomeres.[635] Another research group also developed an efficient, multiconvergent asymmetric synthesis of fostriecin.[636]

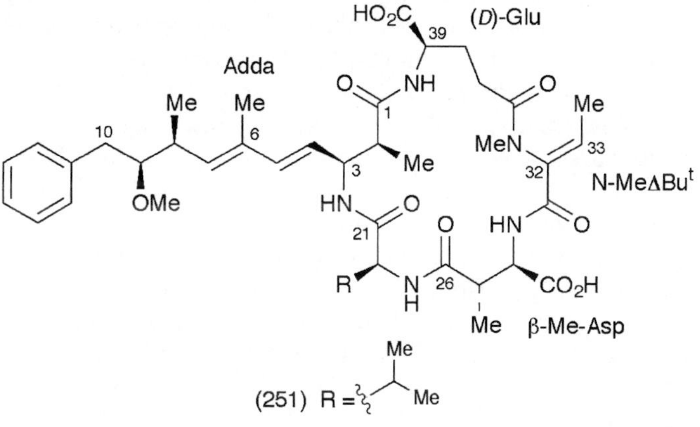

(251)

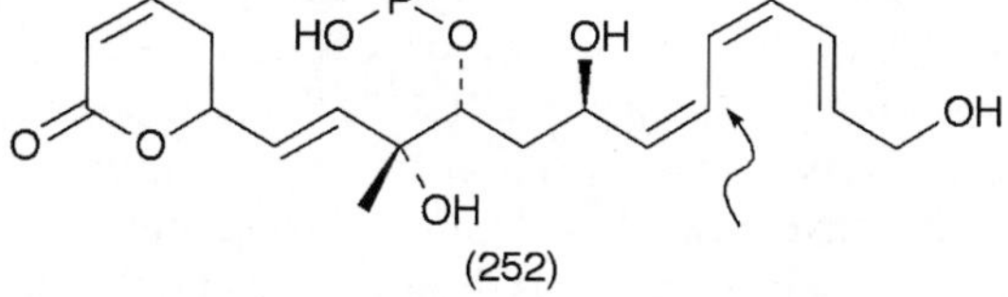

(252)

The marine natural product, Okadaic acid, a Ser/Thr protein phosphatase inhibitor laboratory has been synthesized and its analogues thoroughly reviewed.[637] Interpretation of the critical contacts observed between ocadaic acid and PP1 by X-ray crystallography was also discussed.

Cantharidin is a natural toxin of insect origin and an inhibitor of protein phosphatases (PP1 and PP2A), which plays a key role in cell cycle progression. The nephrotoxicity of the compound has prevented its use as an anticancer agent. However, the demethylated analogue, norcantharidin possesses anticancer activity and stimulates the bone marrow without nephrotoxicity. Two series of norcantharidine analogues have been synthesized, having either an open-ring lactone or a closed-ring lactone configuration.[638] The structures of two of these compounds are shown (**253, 254**). The open-ring series maintained the PP2A selectivity of cantharidine, although some were less potent. Inhibitors of the closed-ring series were considerably less potent, confirming the need of an open ring for inhibition. A number of analogues showed colon cancer cell selectivity, particularly compound (**254**), while remaining inactive in normal colon cells.

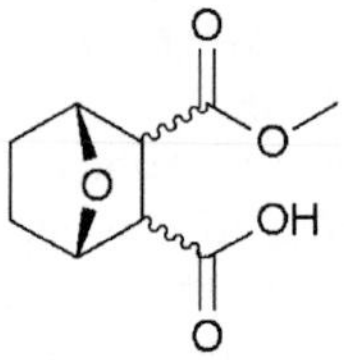

(253) 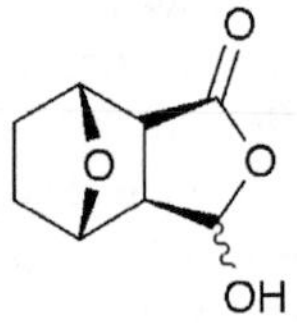 (254)

A small library containing 8 analogues of the protein phosphatase inhibitor dysidiolide was synthesized on solid phase.[639] Biological investigation of the analogues showed that they inhibit protein phosphatase $C_{dc}25$ in the low micromolar range and display considerable activities in cytotoxicity assays employing different cancer cell lines. Recently, there has been considerable interest in the development of inhibitors of protein tyrosine phosphatase 1B (PTP1B). Knock-out mice studies suggest that PPT1B is involved in the downregulation of insulin signaling and that PTP1B inhibitors may be useful for the treatment of type II diabetes and obesity. A series of [difluoro-(3-alkenylphenyl)-methyl]-phosphonates were prepared on non-crosslinked polystyrene.[640] The resulting phosphonic acids were studied for inhibition of PTP1B, with (**255**), bearing α-β unsaturated allyl ester moiety, being most potent of this series of compounds, being a reversible, competitive inhibitor with a K_i of 8.0 ± 1.4 μM.

In a series of α-chloro- and α-bromoacetophenone derivatives tested as neutral PPT1B inhibitors, the bromo-compounds proved much more potent. Derivatization of the phenyl ring with a tripeptide Gly-Glu-Glu resulted in a potent, selective inhibitor against PTP1B.[641] PTPs appear to be critically involved in major diseases such as cancer and autoimmunity. A broad range of compounds were investigated with high throughput screening given the diversity of PTPs and their potential as drug targets in different diseases. Using the previously identified 2-(oxalylamino) benzoic acid structure, a number of new chemical scaffolds was synthesized for the development of inhibitors of different members of the PTP-family.[642] Good oral bioavailability has been observed in rat for some compounds showing an inhibitory effect of K_i in micromolar range against PTP1B. Targetting protein tyrosine phosphatase to promote therapeutic neovascularization may be a potential strategy since α,α-difluoro(phenyl)methylphosphonic acids as PTP inhibitors increase neo-vascularization in mice.[643] Inhibitors of PTP1B would prolong the phosphorylated (activated) state of the insulin receptor and are anticipated to be a novel treatment of the insulin resistance characteristic of type 2 diabetes. A series of small molecular weight peptidomimetics as competitive inhibitors of PTP1B have been described.[644] A tetrazole-containing compound (**256**; $K_i = 2.0$ μM) had good cell permeability in Caco-2 cells as compared to all previous compounds.

Previous reports showed that the tripeptide Ac-Asp-Tyr(SO_3H)-Nle-NH_2 is a surprisingly effective inhibitor of PTP1B ($K_i = 5$ μM). Analogues of this peptide were synthesized for improving the stability and potency of the lead compound, with the most potent analogue[645] being the triacid (**257**), which inhibits PTP1B competitively with a $K_i = 0.22$ μM. An X-ray cocrystal structure of a nonpeptide analogue bound to PTP1B confirms a mode of binding similar to that of peptidic substrates.

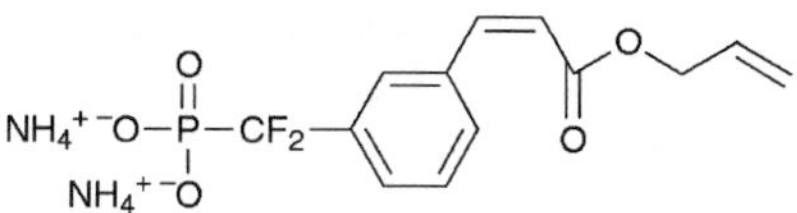

(255)

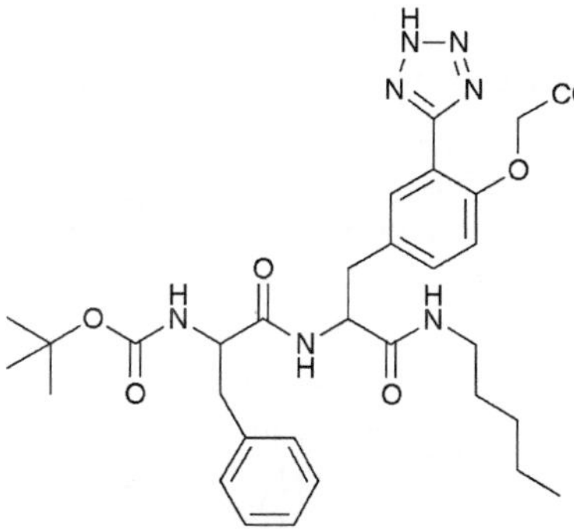

(256)

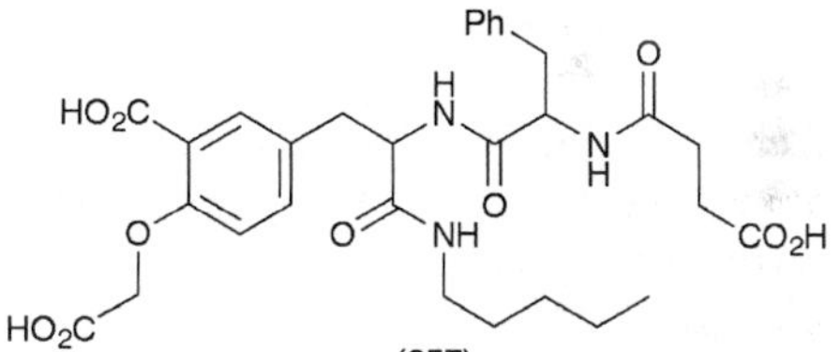

(257)

The efficient total synthesis of a CD45 PTP-inhibitor, pulchellalactam has been described.[646] The phosphomonoesterase activity of bacteriophage lambda protein phosphatase was inhibited by orthophosphate and the oxoanion analogues orthovanadate, tungstate, molybdate, arsenate and sulfate. Small organic anions, e.g. phosphonoacetohydroxamic acid, proved to be strong competitive inhibitors ($K_i = 5.1 \pm 1.6$ μM) of lambda protein phosphatase.[647]

5.15 Renin and Other Aspartyl Proteinase Inhibitors. – Almost all the development programs of renin inhibitors have been closed. Several cyclic peptides, as renin inhibitors containing the Glu-DPhe-Lys structural motif, have been investigated by NMR spectroscopy and molecular dynamic calculations

(MD).[648] The structure of the cyclic moiety is similar in all the peptides, which takes at least two conformations around C_α-C_β in the Glu side-chain. The restrained molecular dynamic calculations further support such observations and show that the macrocycle is fairly rigid, with two conformations around C_α-C_β in the Glu side chain. Two strategies were developed for the stereocontrolled synthesis of 8-aryl-3-hydroxy-4-amino-2,7-diisopropyloctanoic acids. A potent renin inhibitor, CGP-60536B, was synthesized using L-pyroglutamic acid as a chiron.[649] Another publication described the synthesis of four new potential renin inhibitors with pseudodipeptidic units in P(1)-P(1′) and P(2′)-P(3′) positions.[650] All positions of the 8–13 fragment of human angiotensinogen were occupied by unnatural units including unnatural amino acids for enzymatic resistance and metabolic stability.

Hydroxyethylene dipeptide isosteres L-685,434, L-682,679 and L685,458 have been synthesized as aspartyl protease inhibitors,[651] and an efficient method for the preparation of hydroxyethylamine-based aspartyl protease inhibitors with diverse P1 side chains[652] from commercially available amino acids has been described. Immobilized amino acids and their derivatives (Tyr, 3-iodo-Tyr, 3,5-diiodo-Tyr, Phe, 4-iodo-Phe, N-acetyl-Phe) were used as ligands for chromatographic isolation of aspartyl proteases.[653]

Proteolysis of the amyloid precursor protein (APP) with β- and γ-secretases generates β-amyloid (Aβ) peptides in an alternative pathway. β- and γ-Secretases are not only aspartyl proteases but also preferred target enzymes of the inhibitor design for the treatment of Alzheimer's disease at the early stage. Human β-secretase-1 (BACE-1) was produced in insect cells to generate pure enzyme for protein crystallography and subsequent structure-based drug design.[654] The purified protein was successfully used to generate BACE-1/inhibitor co-crystals and to determine the crystal structure of the complex by X-ray analysis. Another publication described the crystal structure of memapsin 2 (β-secretase) bound to an inhibitor OM00-3 (Glu-Leu-Asp-Leu*-Ala-Val-Glu-Phe; $K_i = 0.3$ nM; the asterisk denotes the hydroxyethylene transition-state isostere) determined at 2.1 Å resolution.[655] Emerging Alzheimer's disease therapies using inhibition of β-secretase have been reviewed.[656–660]

A closely homologous enzyme to BACE, memapsin-1 can also hydrolyze the β-secretase site of APP. It is not significantly present in the brain and there is no direct evidence links for its involvement in Alzheimer's disease.[661] The residue specificity of eight memapsin 1 subsites has been described. By use of an effectively cleaved peptide substrate of β-secretase, novel BACE inhibitors were designed and synthesized.[662] Incorporation of statine in P_1 resulted in a weak inhibitor of the enzyme. Further substitution of P_1′-Asp by P_1′-Val led to a potent inhibitor of BACE.. The octapeptide inhibitor, H-Glu-Val-Asn-Sta-Val-Ala-Glu-Phe-OH, showed an $IC_{50} = 40$ nM (*in vitro* bioassay).

Substrate and inhibitor profile of β-secretase has been compared to those of other mammalian Asp-proteases such as cathepsin D, pepsin and renin.[663] BACE accepts polar or acidic residues at positions P2′0 and P1, but prefers bulky hydrophobic residues at position P3. Inhibitor profiling, site-directed mutagenesis and affinity labeling together have suggested that the multi-pass membrane protein presenilins are γ-secretases, activated through autoproteolysis. The current knowledge of biochemistry and cell-biology of γ-secretase have been highlighted and the development of inhibitors of this important therapeutic target reviewed.[664] Presenilin/γ-secretase as a potential target for Alzheimer's disease and its role in regulated intramembrane proteolysis have been discussed.[665] Combinations of surface enhanced laser desorption/ ionization time-of-flight mass spectrometry (SELDI-TOFMS) and a specific inhibitor compound of γ-secretase were used for proving the important role of γ-secretase in production of β-amyloid peptides.[666] Pharmaceutical application of γ-secretase inhibitors might be dangerous: such inhibitors block Notch-signaling *in vivo* and affect embryonic development in zebra fish causing a severe neurogenic phenotype.[667] Cleavage of APP by γ-secretase may contribute to Alzheimer's disease in two ways: by release of Aβ-peptides and by liberation of a bioactive carboxyl terminal domain from membrane-bound APP.[668] Presenilin-1 (PS1)/γ-secretase mutations significantly increased Aβ 1–40 and Aβ 1–42 levels in familial Alzheimer's disease mutants. Direct generation of Aβ-peptides by PS1/γ-secretase may play an important role in the pathogenesis of Alzheimer's disease.[669]

5.16 Thrombin and Factor-Xa Inhibitors. – The control of blood coagulation has become a major target for new therapeutic agents preventing thrombosis, a major cause of mortality in the industrialized world. Historical perspective of the development of thrombin inhibitors[670] and new targets for antithrombotic drugs[671] have been reviewed. The traditional versus modern anticoagulant strategies have been compared,[672] and the synthetic direct and indirect factor-Xa inhibitors have been reviewed.[673] Results and experiences with argatobran, a direct thrombin inhibitor[674,675] have been highlighted. Fondaparinux, the synthetic pentasaccharide, the first in a new class of antithrombin agents selectively inhibiting coagulation factor Xa, has been highlighted in several publications.[676–679] DPC423, a highly potent and orally bioavailable pyrazole antithrombotic agent[680] and bivalirudin, a direct thrombin inhibitor[681] have been discussed.

Novel non-peptidic thrombin-inhibitors, bioisosteres of the inhibitory tripeptide D-Phe-Pro-Arg have been examined.[682] *p*-Amidinobenzylamine, an elongated homologue, and 2,5-dichloro benzylamine were used to replace the P1 (Arg). The P2-P3 (D-Phe-Pro) was replaced with a novel tartaric acid template coupled to a series of readily available, mainly lipophilic, amines. Some of these compounds exhibit promising thrombin inhibition activity *in vitro* with an IC_{50} between 5.9-14 μM. One of the most active thrombin inhibitors was compound (**258**). Dysinosin A1, a marine natural product is a new member of thrombin and Factor VIIa[683] inhibition family. The total synthesis of dysinosin A utilizing a carbon construct strategy that generates subunits originating from L-glutamic acid, butyrolactone, D-leucine and D-mannitol (**259**) was described.[684]

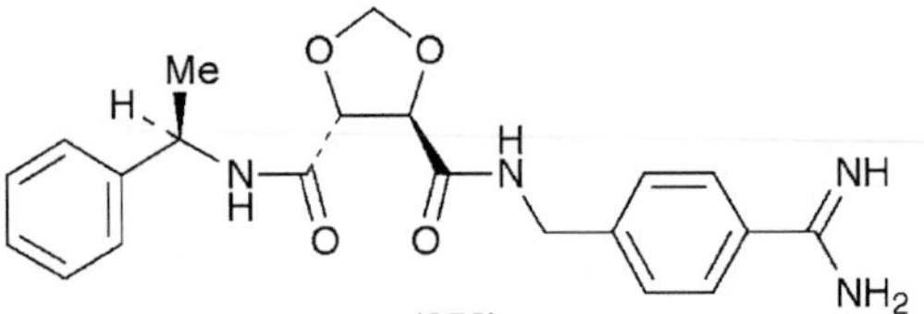

(258)

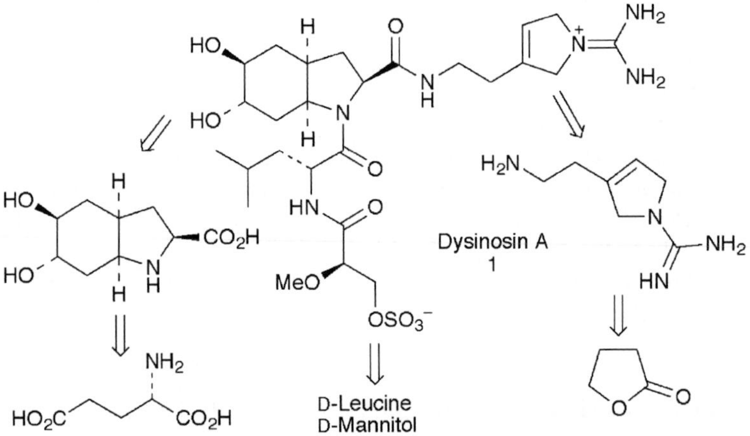

(259) Disconnection of dysinosin A to subunits and chirons

In the design of targeting antithrombotic agents, 3S-1,2,3,4-tetrahydro-β-carboline-3-carboxylic acid, an anti-aggregation compound, was coupled to the tetrapeptides RGDS, RGDV and RGDF.[685] Another research group has reviewed RGD-analogues as antithrombin and anticancer agents.[686] Novel, potent non-covalent thrombin inhibitors incorporating P_3-lactam scaffolds have been designed and synthesized.[687]

Despite intense research over the last 10 years only a very few orally active thrombin inhibitors have been found. Compound (**260**), a tripeptide inhibitor of thrombin, which shows an oral bioavailability of 32 % in rats and 71 % in dogs, the ED_{50} was 5.4 nmol/kg. x min has been synthesized.[688]

The inhibition of the serine protease component of the prothrombinase complex, factor Xa(fa), is generally recognized as a viable strategy for the development of novel antithrombotic agents. The discovery of a novel series of β-aminoester derived fXa inhibitors was published previously. Now the same research group report the systematic modification of the C_3 side-chain of the β-aminoester class of inhibitors and examined different P_4 variations.[689] These changes have resulted in the identification of subnanomolar inhibitors such as (**261**) with improved selectivity versus related proteases. Synthesis and evaluation of 1-arylsulfonyl-3-piperazinone derivatives as fXa inhibitors have been described.[690] Compound M55551, (R)-4-[(6-chloro-2-naphthalenyl)sulfonyl]-6-oxo-1[(4-pyridinyl)-4-piperidinyl]methyl)-2-piperazinecarboxylic acid is ten times more powerful on inhibitor than the lead compound with high selectivity for fXa over trypsin and thrombin. A novel series of triaryloxypyridines have been designed to inhibit fXa.[691] Inhibitor (**262**) has a K_i against fXa of 0.12 nM and is greater than 8000- and 2000-fold selective over thrombin and trypsin, respectively. The 4-position of the central pyridine ring is an ideal site for chemical modifications without deleterious effects on potency and selectivity, as seen in (**263**) which is an inhibitor of fXa with an oral bioavailability of 6% in dogs.

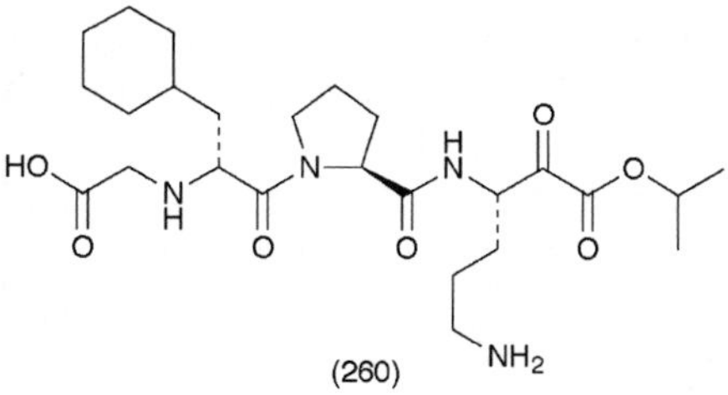

(260)

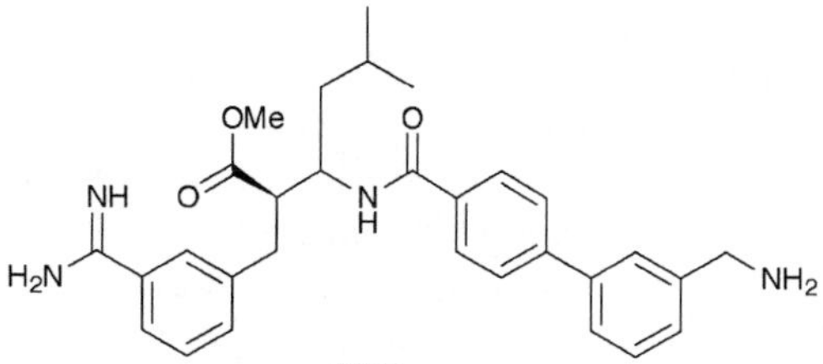

(261)

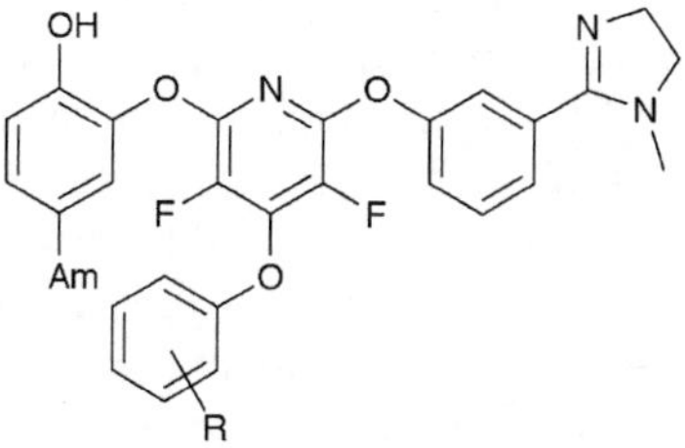

(262) R = 2-OH-4-CO_2H
(263) R = 2-OMe-4-CO_2H

5.17 Trypsin and Other Serine Protease Inhibitors. – Structural and functional properties of a novel serine-protease inhibiting peptide family in arthropods have been reviewed.[692]

Radiosumin (**264**), a strong trypsin inhibitory dipeptide (IC_{50} of 0.14μg/mL) isolated from the freshwater blue-green alga *Plectonema radiosum* was synthesized for the first time by use of the hetero Diels-Alder reaction, the Horner-Wadsworth-Emmons reaction, the Corey-Winter reaction, regioselective hydrogenation, and reduction with zinc and formic acid as key steps.[693] The absolute configuration of (**264**) was unambiguously determined.

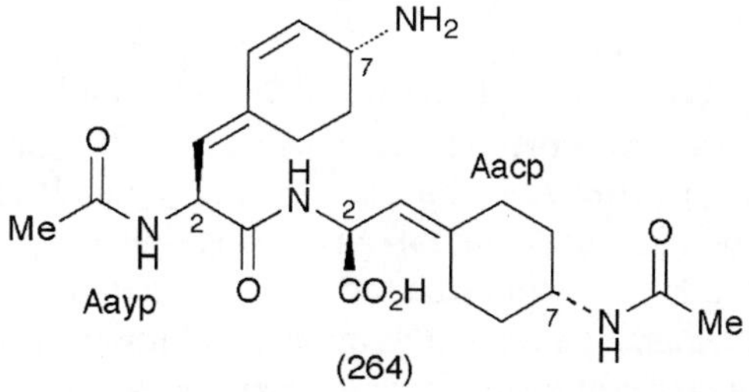

(264)

Interactions in the contact region of the trypsin-pancreatic trypsin inhibitor complex have been evaluated using free energy simulation methods and appropriate thermodynamic cycles.[694] The results obtained reveal that the strength of a specific interaction increases with the charge separation between the amino acid residues involved, and with the hydrophilicity of the surrounding environment.

Two analogues, Pen[19,21,27]-EETI-II and Homocys[19,21,27]-EETI-II, of a peptidic trypsin inhibitor isolated from the seeds of *Ecballium elatherium* (EETI-II) were synthesized: using the Fmoc/Bu^t strategy.[695] Both analogues showed reduced trypsin inhibitory activity. The observed differences might reflect the role of disulfide bridges in the interaction of inhibitors with trypsin. The same research group described the synthesis of two new analogues of *Cucurbita maxima* trypsin inhibitor III (CMTI-III).[696] (D-Arg^5)-CMTI-III displayed a K_a three orders of magnitude, and (D-Lys^5)-CMTI-III four orders of magnitude lower than CMTI-III. The configuration of basic amino acid residues (Arg or Lys) played an important role for the stabilization of the active structure of the inhibitor.

O-Phenyl phosphonamidates have been designed to bind covalently by nucleophilic substitution to the serine residue in the active site of Ser-proteases.[697] The synthesis of these new classes of potential inhibitors of Ser-proteases as well as their role in investigating the mechanisms of inhibition, has been described. N-Terminal truncation of a hexapeptide gave rise to potent tripeptide inhibitors of the hepatitis C virus NS3 protease/NS4A cofactor complex.[698] Optimisation of the tripeptide structure led to ketoacid (**265**) with an IC_{50} of 0.38 μM. Extension of the previously reported modification of Passerini multicomponent reaction (involving condensation with Boc-α-aminoaldehydes followed by a deprotection-transacylation step) to α-amino acid derived carboxylic or isocyanide components was reported.[699] This reaction allowed the highly convergent and short synthesis of complex peptidomimetic structures, including known potent inhibitors of Ser-proteases.

P_3 *c*-Pentylgly

P_2 Leu

(265)

Mutilin (**266**), derived from the natural product pleuromutilin, was used as a lead for designing conformationally restricted eight membered ring diketones as potential Ser-protease inhibitors.[700] Compound (**267**) shows significant inhibition of plasmin and urokinase in an enzyme rate assay. The X-ray crystal structure of this diketone confirmed the conformational predictions made by molecular modeling. The screening of artificial libraries, created by mixing different resin-bound peptidosteroid derivatives, was performed in two stages for evaluation of Ser-protease-like activity.[701] The use of a photocleavable linker allowed the unambiguous structural characterization of the selected members via application of single-bead electrospray tandem mass spectrometry.

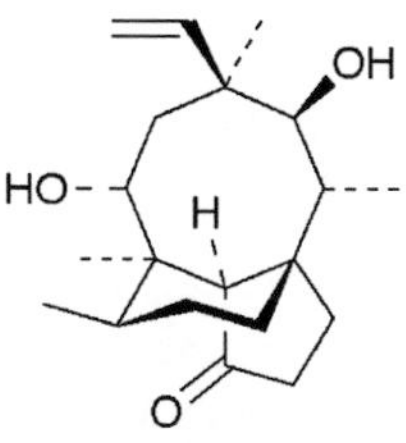

(266) 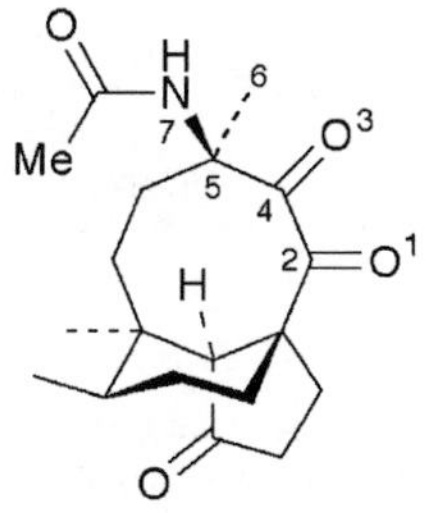 (267)

A new subtilisin-like serine protease (named SEP-1) was isolated from germinated seeds of barley (*Hordeum vulgare*) and characterized.[702] The enzyme has a molecular weight of 70 kDa and its amino acid sequence is similar to that of other plant subtilisin-like Ser-proteases.

5.18 tRNA Synthetase Inhibitors. – In the antiviral and antibacterial area, increasing drug resistance means that there is an ever growing need for novel approaches towards structures and mechanisms which avoid the current problems. The huge increase in high resolution structural data is set to make a dramatic impact on targeting RNA as a drug target. Examples of the RNA-binding antibiotics have been reviewed, particularly, the totally synthetic oxazolidinones, that are clinically useful, selective drugs.[703] The manipulation of tRNA properties by structure-based and combinatorial compounds has been highlighted.[704]

The antimicrobial natural product chuangxinmycin (**268**) has been found to be a potent and selective inhibitor of bacterial tryptophanyl tRNA synthase (WRS). A number of analogues have been synthesized. The interaction of the chuangxinmycin and its derivatives with WRS is highly constrained and only sterically smaller analogues afforded significant inhibition. The only analogue showing interaction is the lactone (**269**) which also had antibacterial activity.[705] Indolmycin, a secondary metabolite of *Streptomyces griceus* and a selective inhibitor of prokaryotic tryptophanyl-tRNA synthetase TrpRS was used to explore the mechanism of inhibition and to explain the resistance of a naturally occurring strain.[706] The lysine (position 9) in the Trp-binding site of the enzyme is essential for making TrpRS indolmycin-resistance. Replacement of Lys^9 by Gln which is conserved in most bacterial TrpRS proteins at this position abolished the ability of the mutant trpS gene to confer indolmycin resistance *in vivo*.

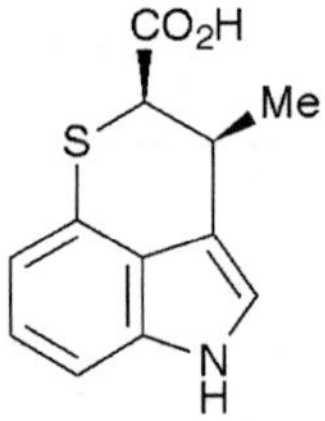

(268) 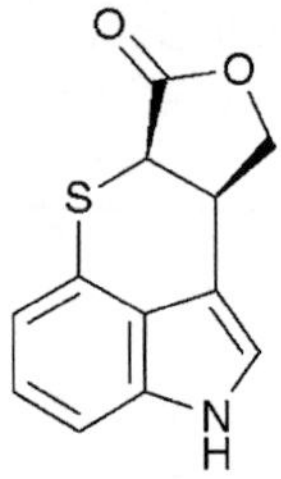 (269)

5.19 Miscellaneous. – *Topoisomerases* are nuclear enzymes stabilizing DNA during replication, transcription and recombination. (5Z, 9Z)-5,9-Hexadecadienoic acid, an inhibitor of human topoisomerase I was synthesized in stereochemically pure form in six steps.[707] A potential route to pentacyclic hypoxyxylerone, a topoisomerase I inhibitor, was demonstrated by a highly convergent synthesis of penta (O-methyl)-hypoxyxylerone.[708] 5-Fluorouracil and Topotecan, a topoisomerase I inhibitor show therapeutic synergy.[709] Camptotechin, as a topoisomerase I inhibitor, induces DNA-strand break but does not provoke apoptosis in quiescent leukocytes.[710] ICRG-193, a catalytic inhibitor of topoisomerase II, inhibits re-entry into the cell division cycle from the quiescent state in mammalian cells.[711] A new chemosensitization strategy for melanoma therapy might be an adenovirus – mediated E2F-1 gene transfer which sensitizes melanoma cells to topoisomerase II inhibitors.[712] Etoposide, a topoisomerase II inhibitor triggers a sequence of biochemical events that mediates the apoptosis induced by etoposide (activation of PS3→ Bax synthesis→Bax translocation to mitochondria→induction of mitochondrial permeability transition→release of Cytochrom-C).[713] Two 4′-propylcarbonoxy derivatives of etoposide were synthesized and evaluated as potential prodrugs for anticancer therapy.[714] XR 11576 is a dual inhibitor of topoisomerase I and II. Novel 3,4-benzofused phenazine compounds were synthesized based on the structure of XR 11576 and the structure-activity relationship was studied.[715] F 11782, a novel fluorinated epipodophylloid and dual catalytic inhibitor of topoisomerases I and II was identified as a potent inhibitor of nucleotide excision repair which inhibits the incision rather than the repair synthesis step.[716]

Human *chymase* has exoproteolytic activity and possesses a wide variety of actions (histamine release from mast cells, angiotensin II production, etc.). Recent efforts to discover novel chymase inhibitors have produced orally bioavailable compounds. Chymase inhibitors have prevented atherosclerosis and are promising for the treatment of cardiovascular as well as inflammatory diseases.[717] Two papers have dealt with the role of chymase in the development of vascular proliferation in injured vessels and effect of chymase inhibitors.[718,719] The chymase inhibitor NK 3201 was demonstrated to have oral activity against neointimal hyperplasia in dog models. Experimental data suggest that chymase has a vital role in tissue remodeling through promotion of the inflammatory response and tissue fibrosis. Therefore, the orally active chymase inhibitor NK 3201 may have protective effects on tissue remodeling in several diseases.[720] Y40613, the orally active specific inhibitor may have therapeutic application in atopic dermatitis.[721] BCEAB, 4-[1-[bis-(4-methyl-phenyl)-methyl]-carbamoyl]-3-(2-ethoxybenzyl)-4-oxo azetidine-2-yloxy]-benzoic acid, a new orally active chymase inhibitor was developed and pharmacologically characterized in inflamed tissue remodeling and fibrosis.[722] A pharmacophore-based database search resulted in the identification of a benzo[b]tiophen-2-sulfonamide derivative which is a stable chymase inhibitor.[723] The previously mentioned two novel chymase inhibitors, NK3201 and BCEAB, prevent postoperative peritoneal adhesion formation in a hamster model.[724–726] BCEAB has a beneficial effect during the acute phase of myocardial infarction[727] and strongly suppresses the angiogenesis induced by basic fibroblast growth factor.

Inhibitor design against *viral proteases* must take into account the peculiar characteristic of viral biology and the plasticity of the replicative mechanism of the viruses. Combinatorial ligand ensembles may be a powerful tool against viruses for contrasting the adaptive potential of virus strains.[728] *L*-Chicoric acid proved to be a potent inhibitor of *HIV-integrase*. Seventeen analogues of chicoric acid were studied and the best compounds inhibit HIV-1 integrase in non-toxic micromolar concentration.[729] *Hepatitis C virus NS3 protease (HCV-NS3)* is essential for viral replication and infectivity and represents a target for the development of antiviral drugs. Serine and threonine β-lactones form a new class of HCV-NS3 protease inhibitors. N-Cbz-*D*-Serine β-lactone displays competitive reversible inhibition with a K_i value of 1.5 μM.[730] Trifluoromethylphenetyl peptide boronic acids inhibit HCV NS3 protease.[731] Homochiral α-ketoamide amino acid containing peptide analogues were synthesized both in solution and on solid phase. These compounds are potent inhibitors of HCV NS3 protease.[732] NMR spectroscopy was used to characterize the HCV protease in a complex with a 24-residue peptide cofactor from NS4A and the boronic acid inhibitor, Ac-Asp-Glu-Val-Val-Pro-boroAlg-OH. The HCV NS3 requires its NS4A cofactor peptide for optimal binding of the inhibitor.[733] A difluromethyl-cysteine derivative was designed by computational chemistry as a mimetic of the canonical P1 Cys-thiol for inhibiting HSV NS3 protease.[734] Screening of a diverse set of bisbenzimidazoles for inhibition HSV NS3 led to the identification of highly-potent Zn^{2+}-dependent inhibitors.[735]

Hepatitis A virus (HAV) 3C enzyme is a picornaviral Cys-proteinase and plays very important role in viral maturation and infectivity. The enzyme recognizes peptide substrates with a glutamine residue at the P_1 site. A great number of ketoglutamine analogues show inhibitory effect against HAV 3C enzyme.[736] An azetidinone scaffold was incorporated into the glutamine fragment (**270**) resulting in a modest inhibitory effect. However, the introduction of phthalhydrazide (**271**) group to the ketone moiety produced significantly better inhibition.

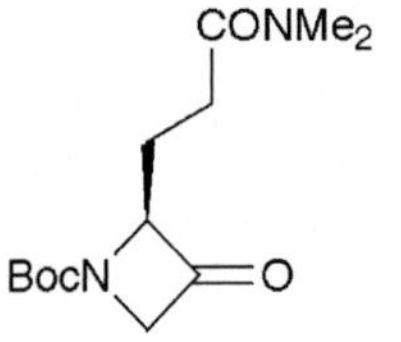

(270) 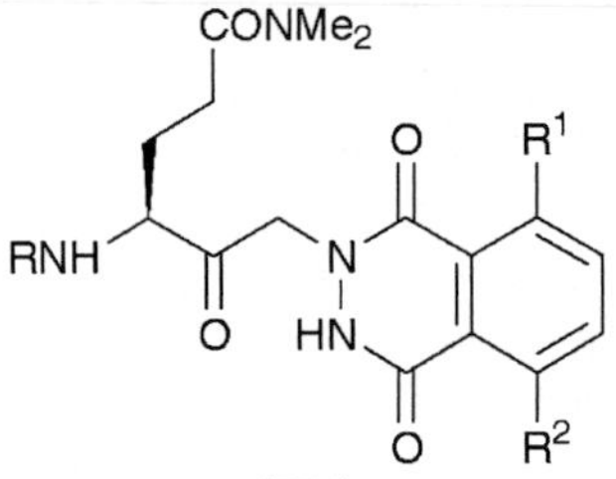 (271)

Neuraminidase inhibitors are candidates for development of antiinfluenza drugs. Structure-based design has led to the synthesis of a novel analogue of GS 4071, an influenza neuraminidase inhibitor, (K_i = 45nM), in which the basic amino group has been replaced by a hydrophobic vinyl group.[737] The same research group performed a concise, stereocontrolled synthesis of the neuraminidase inhibitor A-315675 consisting of a highly functionalized *D*-proline scaffold.[738] Another group also published an enantioselective synthesis of A-315[739] Peramivir (BCX-1812, RW7-270201), a cyclopentane derivative is a highly selective inhibitor of influenza A and B virus neuraminidases and a potent inhibitor of influenza A and B virus replication in cell culture. In clinical trials with patients experimentally infected with influenza A or B viruses, oral treatment with Peramivir significantly reduced nasal wash virus titers without adverse effects.[740]

Carbonic anhydrase inhibitors and activators were intensively studied. A series of aromatic and heterocyclic sulfonamides incorporating valproyl moieties were prepared to design carbonic anhydrase inhibitors as antiepileptic compounds.[741] The same research group described the synthesis and evaluation of carbonic anhydrase activators[742,743] as enhancers of synaptic efficacy, spatial learning and memory. New sulfonamide inhibitors (e.g. 1,2-bis[(4-sulfonamidobenzamide)ethoxy] ethane, SBAM) were synthesized and evaluated.[744] SBAM proved to be a potent inhibitor, approximately 13 times more active against carbonic anhydrase isoform II than isoform I.

Bile acid amides (cholan-24-amides) of 5-substituted 1,3,4-thiadiazole-2-sulfonamide have been prepared. The novel compounds showed significant inhibition of carbonic anhydrase.[745] A new method for identifying enzyme inhibitors is to conduct their synthesis in the presence of the targeted enzyme. Good inhibitors will be formed in larger amounts than poorer ones, because the binding speeds up the synthesis (target accelerated synthesis) and shifts the synthesis equilibrium. The method was applied for a static library of several sulfonamide inhibitors of carbonic anhydrase.[746]

Histone deacetylase (HDAC) inhibitors inhibit cancer cell proliferation, induce apoptosis and regulate gene expression involved in the cell cycle.

Phenylalanine-containing HDAC inhibitors were synthesized and evaluated in Friend leukemic cells. The best compound, a biphenylalanine derivative blocked cell proliferation in the submicromolar range and proved also to be a potent inducer of terminal cell differentiation.[747] HDAC inhibitor trichostatin-A (TSA) induced DNA-damage gene 45 (GADD45), this induction may play important role in TSA-induced cellular effects.[748] FR 901228, an inhibitor of histone deacetylases, increased the cellular responsiveness to IL-6 type cytokines by enhancing the expression of receptor proteins.[749] A new synthetic histone deacetylase inhibitor, 3-(4-benzoyl-1-methyl-1*H*-2-pyrrolyl)-*N*-hydroxy-2-propenamide induced histone hyperacetylation, growth inhibition and cellular differentiation.[750] Suberoylanilide hydroxamic acid (SAMA), a histone deacetylase inhibitor exhibits anti-inflammatory properties via suppression of cytokines.[751] Another HDAC-inhibitor, apicidin induced apoptosis and Fas/Fas ligand expression in human acute promyelotic leukemia cells.[752]

α-*Glucosidase inhibitors* have been designed and synthesized, e.g. acarbose (dTDP-4-amino-4,6-dideoxy-α-*D*-glucose) was prepared to study the biosynthesis of this pseudotetrasaccharide.[753] A library of 72 compounds related to *N*[4-(benzyloxy) benzoyl]alanine was synthesized and screened for α-glucosidase inhibitory activity.[754] One compound, *N*-[4-(benzyloxy)benzoyl]-serine was found to be a potent inhibitor (100 % inhibition at 1 μM). 2,5-Dideoxy-2,5-imino-glycero-*D*-mannoheptitol, a common alkaloid in *Hyacinthaceae* family, is a potent inhibitor of β-glucosidase and β-galactosidase.[755]

A detailed review has described some of the structure- and mechanism-based approaches to design of new β-lactamase inhibitors and the study of probable mechanisms of inhibition using X-ray and electrospray ionization mass spectrometry, as well as molecular modeling techniques.[756] Dipeptide-bound epoxides and α,β-unsaturated amides were synthesized and evaluated as potential irreversible transglutaminase inhibitors.[757] New dicarboxylic acid bis (*L*-prolyl-pyrrolidine) amides were synthesized and evaluated as prolyl-oligopeptidase inhibitors.[758]

Novel 4,1-benzoxazepine derivatives were synthesized as squalene synthase inhibitors.[759] The glycine (**272**) and β-alanine derivatives (**273**) exhibited the most potent inhibition of squalene synthase. A small GTP-binding protein, Ras is an important switch in signal transduction pathways that mediate cell proliferation. Blocking prenylation by enzyme inhibitors severely impairs the biological functioning of Ras. Geranylgeranyltransferase I (PGGT I) inhibitors were designed and synthesized containing CaX motif (C = Cys, aa = any aliphatic, hydrophobic dipeptide).[760] A new class of PGGT I inhibitor with high selectivity was synthesized from a natural and highly selective enzyme inhibitor (**274**) of *Candida albicans* by replacement of the two thiol groups in (**274**) with an imidazole ring.[761] 4,5-Disubstituted thiophene-2-amidines were designed and synthesized as potent urokinase inhibitors.[762]

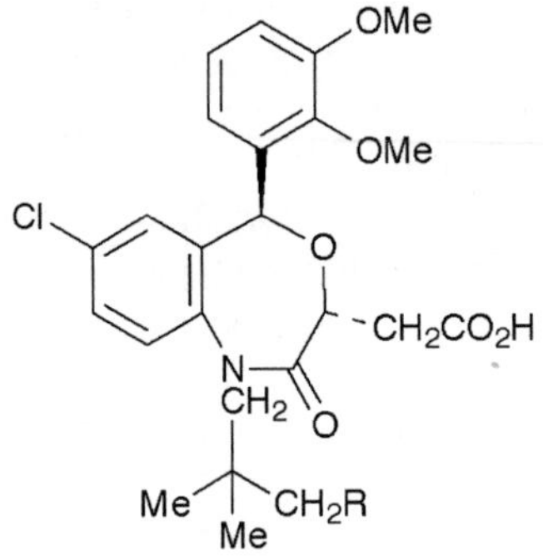

(272) 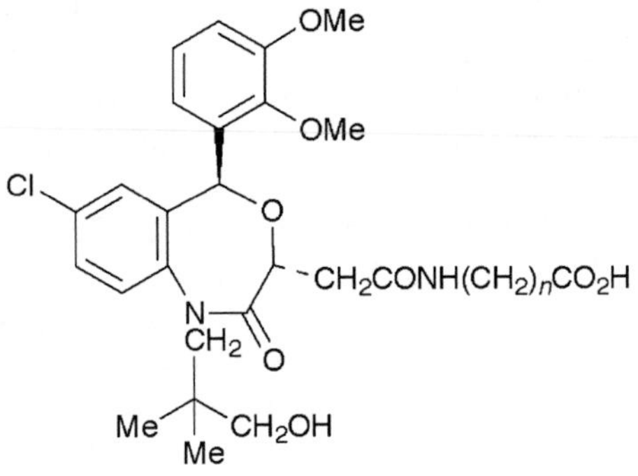 (273)

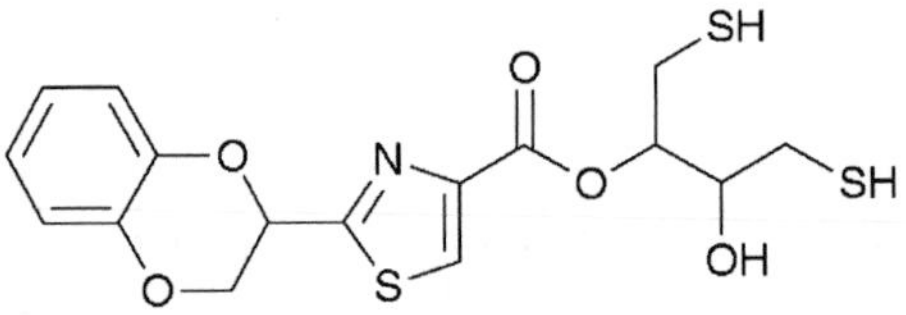

(274)

6 Phage Library Leads

Functional structured peptides could be made by combining the power of selection methods with combinatorial peptide libraries and the *de novo* design of small peptide scaffolds for the selectable sequences. A construction of a library of *de novo* designed helix-loop-helix peptides on minor coat protein of M13 filamentous phage has been described.[763] From this library, cytokine receptor binding proteins were successfully isolated. A polyvalent, lytic phage display system (T7 Select415-1b) displaying a random peptide library has been investigated for its ability to discover novel mimotopes reactive with the therapeutic monoclonal antibody C595.[764] Sequence analysis of enriched phage leads to the identification of a predominant sequence RNREAPRGKICS. This dodecapeptide was synthesized, coupled to beaded Agarose and used for mimotope affinity chromatography.

The catalytic domain of TNF-α converting enzyme (TACE) was expressed in a phage display system to determine whether stable and active enzyme could be made for high throughput screening.[765] Successful application of the phage display system proved that the method might be a useful tool for expressing proteins that have proper stability.

7 Protein-Protein Interaction Inhibitors

7.1 SH2 and SH3 Domain Ligands. – Protein domains mediate protein-protein interactions through binding to similar peptide motifs in their corresponding ligands. As a consequence, protein-domain microarray technique is suitable for identifying novel protein-protein interactions. SH3 domain has been used as protein-interacting module in the microarray technique.[766] The protein domain chip can "fish" proteins out of total cell lysate; the domain bound proteins can be detected on the chip with a specific antibody. The protein-domain chip not only identifies potential binding partners for proteins, but also promises to recognize qualitative differences in protein ligands, caused by post-translational modification.

The NMR solution structure of a regulatory complex involving Abelson kinase (Abl) SH3 domain, the Crk SH2 domain and a Crk-derived phosphopeptide was determined.[767] This terner complex is a good example of the extremely modular nature of regulatory proteins that provides a rich repertoire of mechanisms for control of biological function. Another NMR-study[768] of the orientation of SH3-SH2 domains in Scr kinases (FynSH32) showed that cross communication between SH3 and SH2 domains is small. One of the first steps of intracellular signal transduction is the recognition of a Tyr-phosphorylated target by the SH2 domain of Grb2. The crystal structure of complexes between Grb2-SH2 domain and Shc-derived peptides was determined.[769] The autophosphorylation of protein tyrosine kinase Src significantly reduced the ability of its SH2 domain to bind phosphotyrosine.[770] Interleukin-2 tyrosine kinase (Itk) contains the conserved SH3, SH2 catalytic domains common to many kinase families.

Itk catalytic activity is inhibited by the peptidyl-prolyl isomerase cylophilin A. NMR structural studies showed that a proline-dependent conformational switch within the Itk SH2 domain regulates substrate recognition.[771] Starting form known Src SH2 inhibitors incorporating five-member heterocycles or benzamide scaffolds, a series of tetrasubstituted imidazole compounds were synthesized. The best compound (**275**) has an IC_{50} value of 5.3 to 8.6 μM.[772] Starting form the tetrapeptide Ac-pXEEI-$NHCH_3$ and using a structure-based approach, a peptidomimetic ligand for $p56^{lck}$ SH2 domain was designed containing conformationally restricted replacements for the two Glu-residues.[773] The X-ray crystal structures of two of the novel antagonists were also reported.

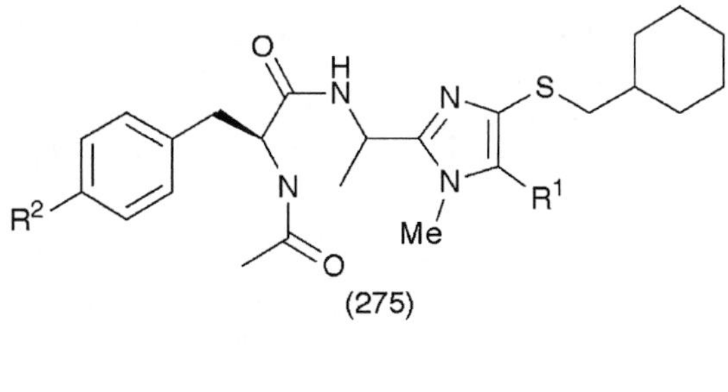

(275)

8 Advances in Formulation/Delivery Technology

Synthesis of novel polycationic lipophilic peptide cores was accomplished for complexation and transport of oligonucleotides.[774] Tandem ligation of multipartite peptides were designed and synthesized for transporting several functional cargoes. The multipartite peptides contain a cell-permeable sequence. The strategy was successful in generating cell permeable peptides with one-, two-and three-chain architectures.[775] Synthetic fragments of membrane translocating proline-rich peptide fragments of the antimicrobial peptide bactanecin 7 were synthesized. A 10 residue proline-rich peptide with two arginine residues was capable of delivering a noncovalently linked protein into cells.[776] Guanidine-rich oligocarbamate molecular transporters were designed and synthesized. Biological evaluation of the peptides proved that these compounds represent a new class of transporters for drug delivery.[777] Sulfhydryl cross-linking poly (ethylene-glycol)-peptides and glycopeptides were prepared and tested for spontaneous polymerization by –S–S- bond formation when bound to plasmid DNA, resulting in stable PEG-peptide and glycopeptide DNA-condensates.[778] Decamer peptides containing seven Arg and three nonarginine residues were synthesized and their membrane translocation effect was compared to that of heptaarginine. The decamers with seven arginines performed almost without exception better then heptaarginine itself, suggesting that spacing between residues is also important for transport and membrane translocation.[779] A promising development of p53 protein transduction therapy using membrane-permeable peptides was applied to oral cancer cells; the transporter was a poly-Arg peptide.[780]

Membrane permeable basic peptides and the potential of these peptides for the intracellular delivery of proteins and drugs have been reviewed.[781] Branched-chain Arg-peptides were synthesized and their membrane translocation was studied. The results strongly suggested that a linear structure of the peptide was not indispensable for the translocation and there could be a considerable flexibility in the location of the Arg residues in the molecules.[782] Antisense oligonucleotides were complexed to an SV40 nuclear localization signal (NLS) coupled to protamin fragment peptides. The fused Arg-rich peptide was a useful vehicle for the cellular delivery of antisense oligonucleotides.[783] The hydrophobicity threshold for peptide insertion into membranes has been thoroughly studied.[784] A new type of cell-permeable "karyophilic" peptides has been derived from dermaseptins, a family of antimicrobial peptides that lyse target bacterial cells by destabilization of their membranes.[785] Two reviews have concentrated on the application of recombinant polypeptide carriers[786] and synthetic block copolypeptides[787] for targeted drug delivery.

References

1. B. Penke, G. Tóth and G. Váradi, *in Specialist Periodical Reports, Amino Acids, Peptides and Proteins*, 2003, **34**, 55–128.
2. C. A. Selects on *Amino Acids, Peptides and Proteins*, published by the American Chermical Society and Chermical Abstract Service, Colombus, Ohio.
3. *Web of Science Service* on http://www.ncbi.nml.nih.gov/entrez/query.fcgi.
4. *Web of Science Service* on http://isinet.com/isi/products/citation/wos.

5. 'Peptides 2002', Proceedings of the 27th European Peptide Symposium, Sorrento. eds. E. Benedetti and C. Pedone, Editioni Ziino, Castellamare di Stabia,(Na) Italy, 2002,1057pp.
6. '*Peptides: The Wave of the Future*', Proceedings of the 17th American Peptide Symposium and 2nd International Peptide Symposium, San Diego, Ca. eds. M. Lebl and R. A. Houghten, American Peptide Society, 2001, 1111pp.
7. '*Peptide Science 2002*', Proceedings of the 39th Japanese Peptide Symposium, ed. T. Yamada, Japanese Peptide Society, 2003, 444pp.
8. Synthesis of Peptides and Peptidomimetics, in '*Houben-Weyl Methods of Organic Chemistry*' 4th Edition, eds. M. Goodman, A. Felix, L. Moroder and C. Toniolo, Georg Thieme Verlag, Stuttgart and New York, 2001-2003, Vols. E22 a-e.
9. J.A. Patch and A.E. Barron, *Curr. Op. in Chem. Biol.*, 2002, **6**, 872.
10. N. Venkatesan and B.H. Kim, *Curr. Med. Chem.*, 2002, **9**, 2243.
11. C. Adessi and C. Soto, *Curr. Med. Chem.*, 2002, **9**, 963.
12. J.J. Perez, F. Corcho and O. Florens, *Curr. Med. Chem.*, 2002, **9**, 2209.
13. J.W. Payne, N.J. Marshall, B.M. Grail and S. Gupta, *Curr. Med. Chem.*, 2002, **6**, 1221.
14. V. Apostolopoulos, J. Matsoukas, A. Plebanski and T. Mavromoustakos, *Curr. Med. Chem.*, 2002, **9**, 411.
15. R. Zhang, J.P. Durkin and W.T. Windsor, *Bioorg. Med. Chem. Lett.*, 2002, **12**, 1005.
16. J.L. Asgian, K.E. James, Z.Z. Li, W. Carter, A.J. Barrett, J. Mikolajczyk, G.S. Salvesen and J.C. Powers, *J. Med. Chem.*, 2002, **45**, 4958.
17. H.J. Lee, J.W. Song, Y.S. Choi, H.M. Park and K.B. Lee, *J. Am. Chem. Soc.*, 2002, **124**, 11881.
18. T.K. Chakraborty, A. Ghosh, A.R. Sankar and A.C. Kunwar, *Tetrahedron Lett.*, 2002, **43**, 5551.
19. T. Morwick, M. Hrapchak, M. DeTuri and S. Campbell, *Organic Letters*, 2002, **4**, 2665.
20. S. Rajaram and M.S. Sigman, *Organic Letters*, 2002, **4**, 3399.
21. Y. Hitotsuyanagi, S. Motegi, H. Fukaya and K. Takeya, *J. Org. Chem.*, 2002 **67**, 3266.
22. K. Dabak and A. Akar, *Heterocyclic Commun.*, 2002, **8**, 385.
23. R.J. Nachman, J. Zabrocki, J. Olczak, H.J. Williams, G. Moyna, A.I. Scott and G.M. Coast, *Peptides*, 2002, **23**, 709.
24. B.C.H. May and A.D. Abell, *J. Chem. Soc.–Perkin Transactions 1*, 2002, 172.
25. P. Zubrzak, K. Kociolek, M. Smoluch, J. Silberring, M.L. Kowalski, B. Szkudlinska and J. Zabrocki, *Acta Polonica Chimica*, 2002, **48**, 1151.
26. S. Oishi, A. Niida, T. Kamano, Y. Miwa, T. Taga, Y. Odagaki, N. Hamanaka, M. Yamamoto, K. Ajito, H. Tamamura, A. Otaka and N. Fujii, *J. Chem. Soc.–Perkin Transactions 1*, 2002, 786.
27. S. Oishi, A. Niida, T. Kamano, Y. Odagaki, H. Tamamura, A. Otaka, N. Hamanaka and N. Fujii, *Org. Lett.*, 2002, **4**, 1055.
28. S. Oishi, T. Kamano, A. Niida, Y. Odagaki, H. Tamamura, A. Otaka, N. Hamanaka and N. Fujii, *Org. Lett.*, 2002, **4**, 1051.
29. S. Oishi, T. Kamano, A. Niida, Y. Odagaki, N. Hamanaka, M. Yamamoto, K. Ajito, H. Tamamura, A. Otaka and N. Fujii, *J. Org. Chem.*, 2002, **67**, 6162.
30. A. Otaka, F. Katagiri, T. Kinoshita, Y. Odagaki, S. Oishi, H. Tamamura, N. Hamanaka and N. Fujii, *J. Org. Chem.*, 2002, **67**, 6152.

31. H. Tamamura, K. Hiramatsu, K. Miyamoto, A. Omagari, S. Oishi, H. Nakashima, N. Yamamoto, Y. Kuroda, T. Nakagawa, A. Otaka and N. Fujii, *Bioorg. Med. Chem. Lett.*, 2002, **12**, 923.
32. V. Boucard, H. Sauriat-Dorizon and F. Guibé, *Tetrahedron*, 2002, **58**, 7275.
33. M. Okada, Y. Nakamura, A. Saito, A. Sato, H. Horikawa and T. Taguchi, *Tetrahedron Letters*, 2002, **43**, 5845.
34. F. Benedetti, F. Berti and S. Norbedo, *J. Org. Chem.*, 2002, **67**, 8635.
35. H. Tamamura, T. Hori, A. Otaka and N. Fujii, *J. Chem. Soc.–Perkin Trans. 1*, 2002, 577.
36. M. Sani, P. Bravo, A. Volonterio and M. Zanda, *Coll. of Czech. Chem. Commun.*, 2002, **67**, 1305.
37. A. Volonterio, S. Bellosta, P. Bravo, M. Canavesi, E. Corradi, S.V. Meille, M. Monetti, N. Mousseir and M. Zanda, *Eur. J. Org. Chem.*, 2002, **3**, 428.
38. J. Marik, M. Budesinsky, J. Slaninova and J. Hlavacek, *Coll. of Czech. Chem. Commun.*, 2002, **67**, 373.
39. M. Hedenstrom, L. Holm, Z.Q. Yuan, H. Emnetas and J. Kihlberg, *Bioorg. Med. Chem. Lett.*, 2002, **12**, 841.
40. V. Kesavan, N. Tamilarasu, H. Cao and T.M. Rana, *Bioconjugate Chem.*, 2002, **13**, 1171.
41. N. Brosse, A. Grandeury and B. Jamart-Grégoire, *Tetrahedron Lett.*, 2002, **43**, 2009.
42. D. Burg, D.V. Filippov, R. Hermanns, G.A. van der Marel, J.H. van Boom and G.J. Mulder, *Bioorg. Med. Chem.*, 2002, **10**, 195.
43. D. Burg, P. Wielinga, N. Zelcer, T. Saeki, G.J. Mulder and P. Borst, *Mol. Pharmacol.*, 2002, **62**, 1160.
44. I. Shin and K. Park, *Org. Lett.*, 2002, **4**, 869.
45. N.M. Khan, M. Cano and S. Balasubramanian, *Tetrahedron Letters*, 2002, **43**, 2439.
46. G.E. Feurle, J.W. Metzger, A. Grudinski and G. Hamscher, *Peptides*, 2002, **23**, 523.
47. A. Reichelt, C. Gaul, R.R. Frey, A. Kennedy and S.F. Martin, *J. Org. Chem.*, 2002, **67**, 4062.
48. L. Demange, M. Moutiez and C. Dugave, *J. Med. Chem.*, 2002, **45**, 3928.
49. H.-J. Cristau, A. Coulombeau, A. Genevois-Borella, F. Sanchez and J.-L. Pirat, *J. Organometallic Chem.*, 2002, **643–644**, 381.
50. T. Ben-Yedidia, A.S. Beignon, C.D. Partidos, S. Muller and R. Arnon, *Mol. Immunol.*, 2002, **39**, 323.
51. Y.C. Chen, A. Muhlrad, A. Shteyer, M. Vidson, I. Bab and M. Chorev, *J. Med. Chem.*, 2002, **45**, 1624.
52. M.G. Bursavich and D.H. Rich, *J. Med. Chem.*, 2002, **45**, 541.
53. V.A. Soloshonok, *Current Org. Chem.*, 2002, **6**, 341.
54. W. Maison and G. Adiwidjaja, *Tetrahedron Lett.*, 2002, **43**, 5957.
55. S.T. Phillips, M. Rezac, U. Abel, M. Kossenjans and P.A. Bartlett, *J. Am. Chem. Soc.*, 2002, **124**, 58.
56. L. Peterlin-Masic, A. Jurca, P. Marinko, A. Jancar and D. Kikelj, *Tetrahedron*, 2002, **58**, 1557.
57. C. Leung, J. Grzyb, J. Lee, N. Meyer, G. Hum, C. Jia, S. Liu and S.D. Taylor, *Bioorg. Med. Chem.*, 2002, **10**, 2309.
58. M. Ruiz, V. Ojea, J.M. Quintela and J.J. Guillín, *Chem. Comm.*, 2002, **15**, 1600.

59. L. Yaouancq, L. René, M.-E. Tran Huu dau and B. Badet, *J. Org. Chem.*, 2002, **67**, 5408.
60. R.A. Stalker, T.E. Munsch, J.D. Tran, X. Nie, R. Warmuth, A. Beatty and C.B. Aakeröy, *Tetrahedron*, 2002, **58**, 4837.
61. S.R. Chhabra, A. Mahajan and W.C. Chan, *J. Org. Chem.*, 2002, **67**, 4017.
62. W. Wang, J. Zhang, C. Xiong and V.J. Hruby, *Tetrahedron Lett.*, 2002, **43**, 2137.
63. W. Wang, M. Cai, C. Xiong, J. Zhang, D. Trivedi and V.J. Hruby, *Tetrahedron*, 2002, **58**, 7365.
64. S. Kotha, S. Halder and E. Brahmachary, *Tetrahedron*, 2002, **58**, 9203.
65. S. Kotha and E. Brahmachary, *Bioorg. Med. Chem.*, 2002, **10**, 2291.
66. D.-G. Liu, Y. Gao, X. Wang, J.A. Kelly and T.R. Burke Jr., *J. Org. Chem.*, 2002, **67**, 1448.
67. S. Royo, P. Lopez, A.I. Jiménez, L. Oliveros and C. Cativiela, *Chirality*, 2002, **14**, 39.
68. A. Cordomi, J. Gomez-Catalan, A.I. Jimenez, C. Cativiela and J.J. Perez, *J. Pept. Sci.*, 2002, **8**, 253.
69. Q. Xia and B. Ganem, *Tetrahedron Lett.*, 2002, **43**, 1597.
70. J.-F. Guichou, L. Patiny and M. Mutter, *Tetrahedron Lett.*, 2002, **43**, 4389.
71. V.J. Huber, T.W. Arroll, C. Lum, B.A. Goodman and H. Nakanishi, *Tetrahedron Lett.*, 2002, **43**, 6729.
72. I. Ujváry and R.J. Nachman, *Peptides*, 2002, **23**, 795.
73. A. Donella-Deana, P. Ruzza, L. Cesaro, A.M. Brunati, A. Calderan, G. Borin and L.A. Pinna, *FEBS Letters*, 2002, **523**, 48.
74. K.E. Jenkins, A.P. Higson, P.H. Seeberger and M.H. Caruthers, *J. Am. Chem. Soc.*, 2002, **124**, 6584.
75. T.K. Chakraborty, S. Ghosh and S. Jayaprakash, *Curr. Med. Chem.*, 2002, **9**, 421.
76. T.K. Chakraborty, S. Jayaprakash and S. Ghosh, *Comb. Chem. & High Throughput Screening*, 2002, **5**, 373.
77. T.K. Chakraborty, P. Srinivasu, S. Kiran Kumar and A.C. Kunwar, *J. Org. Chem.*, 2002, **67**, 2093.
78. T.K. Chakraborty, S. Jayaprakash, P. Srinivasu, S.S. Madhavendra, A. Ravi Sankar and A.C. Kunwar, *Tetrahedron*, 2002, **58**, 2853.
79. A.D. Abell, *Letters in Peptide Science*, 2002, **8**, 267.
80. Y.K. Kang, *J. Phys. Chem.*, 2002, **106**, 2074.
81. M. Rinnová, A. Nefzi and R.A. Houghten, *Tetrahedron Lett.*, 2002, **43**, 2343.
82. Y. Han, D.F. Mierke and M. Chorev, *Biopolymers*, 2002, **64**, 1.
83. P. Gloanec, Y. Hervé, N. Brémand, J.-P. Lecouvé, F. Bréard and G. De Nanteuil, *Tetrahedron Lett.*, 2002, **43**, 3499.
84. P.R. Guzzo, M.P. Trova, T. Inghardt and M. Linschoten, *Tetrahedron Lett.*, 2002, **43**, 41.
85. M.M. Vasbinder and S.J. Miller, *J. Org. Chem.*, 2002, **67**, 6240.
86. P. Ruzza, A. Calderan, A. Osler, S. Elardo and G. Borin, *Tetrahedron Lett.*, 2002, **43**, 3769.
87. T.K. Chakraborty, B.K. Mohan, S. Kiran Kumar and A.C. Kunwar, *Tetrahedron Lett.*, 2002, **43**, 2589.
88. R. Millet, J. Domarkas, P. Rombaux, B. Rigo, R. Houssin and J.-P. Hénichart, *Tetrahedron Lett.*, 2002, **43**, 5087.
89. C.E. Elliott, D.O. Miller and D. Jean Burnell, *J. Chem. Soc. Perkin Trans. 1*, 2002, 217.
90. W. Maison, D. Kuntzer and D. Grohs, *SYNLETT*, 2002, **11**, 1795.

91. M.M. Fernández, A. Diez, M. Rubiralta, E. Montenegro and N. Casamitjana, *J. Org. Chem.*, 2002, **67**, 7587.
92. P.S. Dragovich, R. Zhou and T.J. Prins, *J. Org. Chem.*, 2002, **67**, 741.
93. C. Alvarez-Ibarra, A.G. Csákÿ and C.G. de la Oliva, *J. Org. Chem.*, 2002, **67**, 2789.
94. A. Guarna, I. Bucelli, F. Machetti, G. Menchi, E.G. Occhiato, D. Scarpi and A. Trabocchi, *Tetrahedron*, 2002, **58**, 9865.
95. N. Cini, F. Machetti, G. Menchi, E.G. Occhiato and A. Guarna, *Eur. J. Org. Chem.*, 2002, 873.
96. H. Tye and C.L. Skinner, *Helvetica Chim. Acta*, 2002, **85**, 3272.
97. J. Grembecka, A. Mucha, T. Cierpicki and P. Kafarski, *Phosphorus, Sulfur and Silicon*, 2002, **177**, 1739.
98. M. Weigl and B. Wünsch, *Tetrahedron*, 2002, **58**, 1173.
99. S. Gupta, B.M. Grail and J.W. Payne, *Prot. Pept. Lett.*, 2002, **9**, 133.
100. O. Michielin, V. Zoete, T.M. Gierasch, J. Eckstein, A. Napper, G. Verdine and M. Karplus, *J. Am. Chem. Soc.*, 2002, **124**, 11131.
101. S. Herrero, M.L. Suárez-Gea, M.T. García-López and R. Herranz, *Tetrahedron Lett.*, 2002, **43**, 1421.
102. S. Herrero, A. Salgado, M.T. García-López and R. Herranz, *Tetrahedron Lett.*, 2002, **43**, 4899.
103. W.A. Murray, P.K. Mishra, S. Sun and A. Maden, *Tetrahedron Lett.*, 2002, **43**, 7389.
104. T.E. Renau, R. Léger, R. Yen, M.W. She, E.M. Flamme, J. Sangalang, C.L. Gannon, S. Chamberland, O. Lomovskaya and V.J. Lee, *Bioorg. Med. Chem. Lett.*, 2002, **12**, 763.
105. B.M. Eisenhauer and S.M. Hecht, *Biochemistry*, 2002, **41**, 11472.
106. H. Sun and K.D. Moeller, *Org. Lett.*, 2002, **4**, 1547.
107. V. Swali, M. Matteucci, R. Elliot and M. Bradley, *Tetrahedron*, 2002, **58**, 9101.
108. S. Zhu, T.H. Hudson, D.E. Kyle and A.J. Lin, *J. Med. Chem.*, 2002, **45**, 3491.
109. J.C. Simpson, C. Ho, E.F. Berkley Shands, M.C. Gershengorn, G.R. Marshall and K.D. Moeller, *Bioorg. Med. Chem.*, 2002, **10**, 291.
110. S. Künzel, A. Schweinitz, S. Reißmann, J. Stürzebecher and T. Steinmetzer, *Bioorg. Med. Chem. Lett.*, 2002, **12**, 645.
111. C.J. Hobbs, R.A. Bit, A.D. Cansfield, B. Harris, C.H. Hill, K.L. Hilyard, I.R. Kilford, E. Kitas, A. Kroehn, P. Lovell, D. Pole, P. Rugman, B.S. Sherborne, I.E.D. Smith, D.R. Vesey, D.L. Walmsley, D. Whittaker, G. Williams, F. Wilson, D. Banner, A. Surgenor and N. Borkakoti, *Bioorg. Med. Chem. Lett.*, 2002, **12**, 1365.
112. P.S. Dragovich, T.J. Prins, R. Zhou, T.O. Johnson, E.L. Brown, F.C. Maldonado, S.A. Fuhrman, L.S. Zalman, A.K. Patick, D.A. Matthews, X. Hou, J.W. Meador III, R.A. Ferre and S.T. Worland, *Bioorg. Med. Chem. Lett.*, 2002, **12**, 733.
113. J.P. Davidson, O. Lubman, T. Rose, G. Waksman and S.F. Martin, *J. Am. Chem. Soc.*, 2002, **124**, 205.
114. R. Kaul, S. Deechongkit and J.W. Kelly, *J. Am. Chem. Soc.*, 2002, **124**, 11900.
115. P. Grieco, P. Campiglia, I. Gomez-Monterey and E. Novellino, *Tetrahedron Lett.*, 2002, **43**, 1197.
116. B. Wels, J.A.W. Kruijtzer and R.M.J. Liskamp, *Org. Lett.*, 2002, **4**, 2173.
117. W. Wang, J. Yang, J. Ying, C. Xiong, J. Zhang, C. Cai and V.J. Hruby, *J. Org. Chem.*, 2002, **67**, 6353.
118. T.C. Maier, W.U. Frey and J. Podlech, *Eur. J. Org. Chem.*, 2002, 2686.

119. M.F. Brana, M. Garranzo, B. de Pascual-Teresa, J. Pérez-Castells and M.R. Torres, *Tetrahedron*, 2002, **58**, 4825.
120. A. Golebiowski, J. Jozwik, S.R. Klopfenstein, A.-O. Colson, A.L. Grieb, A.F. Russell, V.L. Rastogi, C.F. Diven, D.E. Portlock and J.J. Chen, *J. Comb. Chem.*, 2002, **4**, 584.
121. X. Gu, X. Tang, S. Cowell, J. Ying and V.J. Hruby, *Tetrahedron Lett.*, 2002, **43**, 6669.
122. A.N. Van Nhien, H. Ducatel, C. Len and D. Postel, *Tetrahedron Lett.*, 2002, **43**, 3805.
123. L. Jiang and K. Burgess, *Tetrahedron*, 2002, **58**, 8743.
124. J. Makker, S. Dey, P. Kumar and T.P. Singh, *Acta Crystallographica Section C*, 2002, **58**, o212.
125. M.H. Todd, C. Ndubaku and P.A. Bartlett, *J. Org. Chem.*, 2002, **67**, 3985.
126. M. Rainaldi, V. Moretto, M. Crisma, E. Peggion, S. Mammi, C. Toniolo and G. Cavicchioni, *J. Pept. Sci.*, 2002, **8**, 241.
127. A. Trabocchi, E.G. Occhiato, D. Potenza and A. Guarna, *J. Org. Chem.*, 2002, **67**, 7483.
128. J. Zhang, C. Xiong, W. Wang, J. Ying and V.J. Hruby, *Org. Lett.*, 2002, **4**, 4029.
129. D.L. Steer, R.A. Lew, P. Perlmutter, A.I. Smith and M.I. Aguilar, *Curr. Med. Chem.*, 2002, **9**, 811.
130. T. Hoffmann and P. Gmeiner, *SYNLETT*, 2002, **6**, 1014.
131. K. Lee, M. Zhang, D. Yang and T.R. Burke, Jr., *Bioorg. Med. Chem. Lett.*, 2002, **12**, 3399.
132. M. Anzai, R. Yanada, N. Fujii, H. Ohno, T. Ibuka and Y. Takemoto, *Tetrahedron*, 2002, **58**, 5231.
133. J.W. Taylor, *Biopolymers*, 2002, **66**, 49.
134. J. Piró, P. Forns, J. Blanchet, M. Bonin, L. Micouin and A. Diez, *Tetrahedron: Asymmetry*, 2002, **13**, 995.
135. S. Sasmal, A. Geyer and M.E. Maier, *J. Org. Chem.*, 2002, **67**, 6260.
136. S. Rajesh, B. Banerji and J. Iqbal, *J. Org. Chem.*, 2002, **67**, 7852.
137. A. Arasappan, K.X. Chen, F.G. Njoroge, T.N. Parekh and V. Girijavallabhan, *J. Org. Chem.*, 2002, **67**, 3923.
138. P. Virta, J. Sinkkonen and H. Lönnberg, *Eur. J. Org. Chem.*, 2002, 3616.
139. G. Fridkin and C. Gilon, *J. Pept. Res.*, 2002, **60**, 104.
140. J. Efskind, H. Hope and K. Undheim, *Eur. J. Org. Chem.*, 2002, 464.
141. D.T.S. Rijkers, J.A.J. den Hartog and R.M.J. Liskamp, *Biopolymers*, 2002, **63**, 141.
142. J.A. Zerkowski, L.M. Hensley and D. Abramowitz, *SYNLETT*, 2002, **4**, 557.
143. Y. Han, C. Giragossian, D.F. Mierke and M. Chorev, *J. Org. Chem.*, 2002, **67**, 5085.
144. S. Maricic and T. Frejd, *J. Org. Chem.*, 2002, **67**, 7600.
145. M. Planas, E. Cros, R.-A. Rodríguez, R. Ferre and E. Bardají, *Tetrahedron Lett.*, 2002, **43**, 4431.
146. J.B. Bremner, J.A. Coates, P.A. Keller, S.G. Pyne and H.M. Witchard, *SYNLETT*, 2002, **2**, 219.
147. J.B. Bremner, J.A. Coates, D.R. Coghlan, D.M. David, P.A. Keller and S.G. Pyne, *New J. Chem.*, 2002, **26**, 1549.
148. T.S. Chakraborty, S. Tapadar and S.K. Kumar, *Tetrahedron Lett.*, 2002, **43**, 1317.
149. M. Stöckle, G. Voll, R. Günther, E. Lohof, E. Locardi, S. Gruner and H. Kessler, *Org. Lett.*, 2002, **4**, 2501.

150. M. Pattarawarapan, M.C. Zaccaro, U.H. Saragovi and K. Burgess, *J. Med. Chem.*, 2002, **45**, 4387.
151. M. Hedenström, Z.Q. Yuan, K. Brickmann, J. Carlsson, K. Ekholm, B. Johansson, E. Kreutz, A. Nilsson, I. Sethson and J. Kihlberg, *J. Med. Chem.*, 2002, **45**, 2501.
152. Z.Q. Yuan, D. Blomberg, I. Sethson, K. Brickmann, K. Ekholm, B. Johansson, A. Nilsson and J. Kihlberg, *J. Med. Chem.*, 2002, **45**, 2512.
153. H. Huang, L. Mu, J. He and J.-P. Cheng, *Tetrahedron Lett.*, 2002, **43**, 2255.
154. X.P. Song, C.R. Xu, H.T. He and T.J. Siahaan, *Bioorg. Chem.*, 2002, **30**, 285.
155. K. Murakami, K. Irie, A. Morimoto, H. Ohigashi, M. Shindo, M. Nagao, T. Shimizu and T. Shirasawa, *Biochem. Biophys. Res. Commun.*, 2002, **294**, 5.
156. F. Rabanal, J.M. Tusell, L. Sastre, M.R. Quintero, M. Cruz, D. Grillo, M. Pons, F. Albericio, J. Serratosa and E. Giralt, *J. Pept. Sci.*, 2002, **8**, 578.
157. Z. Cai, Z. Ji, Y. Xu and S. Sheng, *Zhongguo Yaolixue Tongbao*, 2002, **18**, 31.
158. P.P. Mager and K. Fischer, *Mol. Simul.*, 2001, **27**, 237.
159. P. Brunelle and A. Rauk, *J. Alzheimer's. Dis.*, 2002, **4**, 283.
160. T. Koga, K. Taguchi, T. Kinoshita and M. Higuchi, *Chem. Commun. (Cambridge, UK)*, 2002, **3**, 242.
161. S. Chen, V. Berthelier, J.B. Hamilton, B. O'Nuallain and R. Wetzel, *Biochemistry*, 2002, **41**, 7391.
162. Y. Takahashi, T. Yamashita, A. Ueno and H. Mihara, *Pept. Sci.*, 2002, Volum Date 2001, **38th**, 353.
163. Y. Takahashi, A. Ueno and H. Mihara, *Chem. Biol. Chem.*, 2002, **3**, 637.
164. A. Banerjee, S.K. Maji, M.G.B. Drew, D. Haldar and A. Banerjee, *Tetrahedron Lett.*, 2002, **44**, 335.
165. J.J. Balbach, A.T. Petkova, N.A. Oyler, O.N. Antzutkin, D.J. Gordon, S.C. Meredith and R. Tycko, *Biophys. J.*, 2002, **83**, 1205.
166. T. Miura, C. Yamamiya, M. Sasaki, K. Suzuki and H. Takeuchi, *J. Raman Spectr.*, 2002, **33**, 530.
167. M.F. Perutz, J.T. Finch, J. Berriman and A. Lesk, *Proc. Nat. Acad. Sci. USA*, 2002, **99**, 5591.
168. C. Hetenyi, T. Kortvelyesi and B. Penke, *Bioorg. Med. Chem.*, 2002, **10**, 1587.
169. K. Watanabe, K. Nakamura, S. Akikusa, T. Okada, M. Kodaka, T. Konakahara and H. Okuno, *Biochem. Biophys. Res. Commun.*, 2002, **290**, 121.
170. J. Kanski, M. Aksenova, C. Schoneich and D.A. Butterfield, *Free Radic. Biol. Med.*, 2002, **32**, 1205.
171. A.W. Garofalo, D.W.G. Wone, A. Phuc, J.E. Audia, C.A. Bales, H.F. Dovey, D.B. Dressen, B. Folmer, E.G. Goldbach, A.C. Guinn, L.H. Latimer, T.E. Mabry, J.S. Nissen, M.A. Pleiss, S. Sohn, E.D. Thorsett, J.S. Tung and J. Wu, *Bioorg. Med. Chem. Lett.*, 2002, **12**, 3051.
172. C.W. Cairo, A. Strzelec, R.M. Murphy and L.L. Kiessling, *Biochemistry*, 2002, **41**, 8620.
173. D.T.S. Rijkers, J.W.M. Hoppener, G. Posthuma, C.J.M. Lips and R.M.J. Liskamp, *Chemistry - A European Journal*, 2002, **8**, 4285.
174. E. Masliah and M. Hashimoto, *Neurotoxicology*, 2002, **23**, 461.
175. K. Zou, J.S. Gong, K. Yanagisawa and M. Michikawa, *J. Neurosci.*, 2002, **22**, 4833.
176. J. Kanski, M. Aksenova and D.A. Butterfield, *Neurotoxicity Res.*, 2002, **4**, 219.
177. F. Kuperstein and E. Yavin, *Eur. J. Neurosci.*, 2002, **16**, 44.
178. N. Arispe and M. Doh, *FASEB J.*, 2002, **16**, 1526.

179. M.H. Parker, R. Chen, K.A. Conway, D.H. Lee, C. Luo, R.E. Boyd, S.O. Nortey, T.M. Ross, M.K. Scott and A.B. Reitz, *Bioorg. Med. Chem.*, 2002, **10**, 3565.
180. A. Tossi and L. Sandri, *Current Pharm. Design*, 2002, **8**, 743.
181. M. Cudic and L. Otvos, *Current Drug Targets*, 2002, **3**, 101.
182. N. Sitaram and R. Nagaraj, *Current Drug Targets*, 2002, **3**, 259.
183. L. Otvos, *Cell. Mol. Life Sci.*, 2002, **59**, 1138.
184. R. Gennaro, M. Zanetti, M. Benincasa, E. Podda and M. Miani, *Current Pharm. Design*, 2002, **8**, 763.
185. B. Ramanathan, E.G. Davis, C.R. Ross and F. Blecha, *Microbes & Infection*, 2002, **4**, 361.
186. R.E.W. Hancock and A. Rozek, *FEBS Microbiol. Lett.*, 2002, **206**, 143.
187. Y. Shai, *Current Pharm. Design*, 2002, **8**, 715.
188. P.B. Savage, *Eur. J. Org. Chem.*, 2002, 759.
189. K.H. Lee, *Current Pharm. Design*, 2002, **8**, 795.
190. M. Mandal and R. Nagaraj, *J. Pept. Res.*, 2002, **59**, 95.
191. M. Mandal, M.V. Jagannadham and R. Nagaraj, *Peptides*, 2002, **23**, 413.
192. D. Tran, P.A. Tran, Y.Q. Tang, J. Yuan, T. Cole and M.E. Selsted, *J. Biol. Chem.*, 2002, **277**, 3079.
193. Y. Mukai, Y. Matsushita, T. Niidome, T. Hatekeyama and H. Aoyagi, *J. Pept. Sci.*, 2002, **8**, 570.
194. G. Kragol, R. Hoffmann, M.A. Chattergoon, S. Lovas, M. Cudic, P. Bulet, B.A. Condie, K.J. Rosengren, L.J. Montaner and L. Otvos, *Eur. J. Biochem.*, 2002, **269**, 4226.
195. K. Okemoto, Y. Nakajima, T. Fujioka and S. Natori, *J. Biochem.*, 2002, **131**, 277.
196. S.A. Taran, T.Z. Esikova, L.G. Mustaeva, M.B. Baru and Y.B. Alakhov, *Russian J. Bioorg. Chem.*, 2002, **28**, 357.
197. S.Y. Shin, S.T. Yang, E.J. Park, S.H. Eom, W.K. Song, Y. Kim, K.S. Hahm and J.I. Kim, *Biochem. Biophys. Res. Commun.*, 2002, **290**, 558.
198. S.H. Lee, D.G. Lee, S.T. Yang, V. Kim, J.I. Kim, K.S. Hahm and S.Y. Shin, *Prot. Pept. Lett.*, 2002, **9**, 395.
199. D.G. Lee, Y. Park, P.I. Kim, H.G. Jeong, E.R. Woo and K.S. Hahm, *Biochem. Biophys, Res. Commun.*, 2002, **297**, 885.
200. H.K. Kim, D.G. Lee, Y. Park, H.N. Kim, B.H. Choi, C.H. Choi and K.S. Hahm, *Biotechnology Lett.*, 2002, **24**, 347.
201. K. Yu, K. Park, S.W. Kang, S.Y. Shin, K.S. Hahm and Y. Kim, *J. Pept. Res.*, 2002, **60**, 1.
202. S.W. Kang, D.G. Lee, S.F.T. Yang, Y. Kim, J.I. Kim, K.S. Hahm and S.Y. Shin, *Prot. Pept. Lett.*, 2002, **9**, 275.
203. K. Park, D. Oh, S.Y. Shin, K.S. Hahm and Y. Kim, *Biochem. Biophys. Res. Commun.*, 2002, **290**, 204.
204. S.T. Yang, S.Y. Shin, Y.C. Kim, Y.M. Kim, K.S. Hahm and J.I. Kim, *Biochem. Biophys. Res. Commun.*, 2002, **296**, 1044.
205. S. Nagpal, K.J. Kaur, D. Jain and D.M. Salunke, *Protein Science*, 2002, **11**, 2158.
206. Q.S. Li, C.B. Lawrence, H.M. Davies and N.P. Everett, *Peptides*, 2002, **23**, 1.
207. I. Nagaoka, S. Hirota, F. Niyonsaba, M. Hirata, Y. Adachi, H. Tamura, S. Tanaka and D. Heumann, *Clin. Diagn. Lab. Immun.*, 2002, **9**, 972.
208. J.F. Sanchez, F. Wojcik, Y.S. Yang, M.P. Strub, J.M. Strub, A. Van Dorsselaer, M. Martin, R. Lehrer, T. Ganz, A. Chavanieu, B. Calas and A. Aumelas, *Biochemistry*, 2002, **41**, 21.

209. H.J. Vogel, D.J. Schibli, W.G. Jing, E.M. Lohmeier-Vogel, R.F. Epand and R.M. Epand, *Biochem. Cell Biol.*, 2002, **80**, 49.
210. M.B. Strom, B.E. Haug, O. Rekdal, M.L. Skar, W. Stensen and J.S. Svedsen, *Biochem. Cell Biol.*, 2002, **80**, 65.
211. M.B. Strom, O. Rekdal and J.S. Svedsen, *J. Pept. Sci.*, 2002, **8**, 36.
212. T. Lejon, J.S. Svedsen and B.E. Haug, *J. Pept. Sci.*, 2002, **8**, 302.
213. B.E. Haug, J. Andersen, O. Rekdal and J.S. Svedsen, *J. Pept. Sci.*, 2002, **8**, 307.
214. N. Papo, Z. Oren, U. Pag, H.G. Sahl and Y. Shai, *J. Biol. Chem.*, 2002, **277**, 33913.
215. Z. Oren, J. Ramesh, D. Avrahami, N. Suryaprakash, Y. Shai and R. Jelinek, *Eur. J. Biochem.*, 2002, **269**, 3869.
216. E.A. Porter, B. Weisblum and S.H. Gellman, *J. Am. Chem. Soc.*, 2002, **124**, 7324.
217. T.L. Raguse, E.A. Porter, B. Weisblum and S.H. Gellman, *J. Am. Chem. Soc.*, 2002, **124**, 12774.
218. E.J.M. van Kan, R.A. Demel, E. Breukink, A. van der Bent and B. de Kruijff, *Biochemistry*, 2002, **41**, 7529.
219. K. Sadler, K.D. Eom, J.L. Yang, Y. Dimitrova and J.P. Tam, *Biochemistry*, 2002, **41**, 14150.
220. M.G. Scott, D.J. Davidson, M.R. Gold, D. Bowdish and R.E.W. Hancock, *J. Immunol.*, 2002, **169**, 3883.
221. M. Gobbo, L. Biondi, F. Filira, R. Gennaro, M. Benincasa, B. Scolaro and R. Rocchi, *J. Med. Chem.*, 2002, **45**, 4494.
222. S. Roy, H.G. Lombart, W.D. Lubell, R.E.W. Hancock and S.W. Farmer, *J. Pept. Res.*, 2002, **60**, 198.
223. M.V. Sawai, A.J. Waring, W.R. Kearney, P.B. McCray, W.R. Forsyth, R.I. Lehrer and B.F. Tack, *Prot. Eng.*, 2002, **15**, 225.
224. S. Navon-Venezia, R. Feder, L. Gaidukov, Y. Carmeli and A. Mor, *Antimicrob. Agents Chemother.*, 2002, **46**, 689.
225. N. Sitaram, K.P. Sai, S. Singh, K. Sankaran and R. Nagaraj, *Antimicrob. Agents Chemother.*, 2002, **46**, 2279.
226. M.K. Lee, L. Cha, S.H. Lee and K.S. Hahm, *J. Biochem. Mol. Biol.*, 2002, **35**, 291.
227. N. Sal-Man, Z. Oren and Y. Shai, *Biochemistry*, 2002, **41**, 11921.
228. M.R. Yeaman, K.D. Gank, A.S. Bayer and E.P. Brass, *Antimicrob. Agents Chemother.*, 2002, **46**, 3883.
229. L.W. Tinoco, A. de Silva, A. Leite, A.P. Valente and F.C.L. Almeida, *J. Biol. Chem.*, 2002, **277**, 36351.
230. S.C. Shankaramma, Z. Athanassiou, O. Zerbe, K. Moehle, C. Mouton, F. Bernardini, J.W. Wrijbloed, D. Obrecht and J.A. Robinson, *Chembiochem*, 2002, **3**, 1126.
231. M.G. Ryadnov, O.V. Degtyareva, I.A. Kashparov and Y.V. Mitin, *Peptides*, 2002, **23**, 1869.
232. K. Hamamoto, Y. Kida, Y. Zhang, T. Shimizu and K. Kuwano, *Microbiol. and Immunol.*, 2002, **46**, 741.
233. T. Yasukata, H. Shindo, O. Yoshida, Y. Sumino, T. Munekage, Y. Narukawa and Y. Nishitani, *Bioorg. Med. Chem. Lett.*, 2002, **12**, 3033.
234. S.S. Printsevskaya, E.N. Olsuf'eva, E.I. Lazhko and M.N. Preobrazhenskaya, *Russian J. Bioorg. Chem.*, 2002, **28**, 65.
235. C. Morassutti, F. De Amicis, B. Skerlavaj, M. Zanetti and S. Marchetti, *FEBS Lett.*, 2002, **519**, 141.
236. A.C. Rinaldi, M.L. Mangoni, A. Rufo, C. Luzi, D. Barra, H.X. Zhao, P.K.J. Kinnunen, A. Bozzi, A. Di Giulio and M. Simmaco, *Biochem. J.*, 2002, **368**, 91.

237. D. Wade, *Internet J. Chem.*, 2002, **5**, Art. No. 5.
238. W. Kamysz, B. Kochanska, A. Kedzia, J. Ochocinska, Z. Mackiewicz and G. Kupryszewski, *Polish J. Chem.*, 2002, **76**, 801.
239. H.N. Hunter, D.B. Fulton, T. Ganz and H.J. Vogel, *J. Biol. Chem.*, 2002, **277**, 37597.
240. A.M. Pons, N. Zorn, D. Vignon, F. Delalande, A. Van Dorsselaer and G. Cottenceau, *Antimicrob. Agents Chemother.*, 2002, **46**, 229.
241. L. Kuhn-Nentwig, J. Muller, J. Schaller, A. Walz, M. Dathe and W. Nentwig, *J. Biol. Chem.*, 2002, **277**, 11208.
242. L. Kuhn-Nentwig, M. Dathe, A. Walz, J. Schaller and W. Nentwig, *FEBS Letters*, 2002, **527**, 193.
243. G. Corzo, E. Villegas, F. Gomez-Lagunas, L.D. Possani, O.S. Belokoneva and T. Nakajima, *J. Biol. Chem.*, 2002, **277**, 23627.
244. S. Chernysh, S.I. Kim, G. Bekker, V.A. Pleskach, N.A. Filatova, V.B. Anikin, V.G. Platonov and P. Bulet, *Proc. Nat. Acad. Sci. USA*, 2002, **99**, 12628.
245. H.C. Mantovani, H.J. Hu, R.W. Worobo and J.B. Russell, *Microbiology-SGM*, 2002, **148**, 3347.
246. L. Moerman, S. Bosteels, W. Noppe, J. Willems, E. Clynen, L. Schoofs, K. Thevissen, J. Tytgat, J. Van Eldere, J. van der Walt and F. Verdonck, *Eur. J. Biochem.*, 2002, **269**, 4799.
247. R. Lai, Y.T. Zheng, J.H. Shen, G.J. Liu, H. Liu, W.H. Lee, S.Z. Tang and Y. Zhang, *Peptides*, 2002, **23**, 427.
248. H.H. von Horsten, P. Derr and C. Kirchhoff, *Biol. Repr.*, 2002, **67**, 804.
249. J.M.O. Fernandes and V.J. Smith, *Biochem. Biophys. Res. Commun.*, 2002, **296**, 167.
250. H. Hemmi, J. Ishibashi, S. Hara and M. Yamakawa, *FEBS Letters*, 2002, **518**, 33.
251. C.S. Brinkworth, J.H. Bowie, M.J. Tyler and J.C. Wallace, *Australian J. Chem.*, 2002, **55**, 605.
252. A. Lupetti, R. Danesi, J.W. van't Wout, J.T. van Dissell, S. Senesi and P.H. Nibbering, *Exp. Opin. Invest. Drugs*, 2002, **11**, 309.
253. S. Sogabe, M. Masubuchi, K. Sakata, T.A. Fukami, K. Morikami, Y. Shiratori, H. Ebiike, K. Kawasaki, Y. Aoki, N. Shimma, A. D'Arcy, F.K. Winkler, D.W. Banner and T. Ohtsuka, *Chem. Biol.*, 2002, **9**, 1119.
254. D. Avrahami and Y. Shai, *Biochemistry*, 2002, **41**, 2254.
255. D. Brewer and G. Lajoie, *Biochemistry*, 2002, **41**, 5526.
256. N. Fujitani, S. Kawabata, T. Osaki, Y. Kumaki, M. Demura, K. Nitta and K. Kawano, *J. Biol. Chem.*, 2002, **277**, 23651.
257. K. Lugardon, S. Chasserot-Golaz, A.E. Kieffer, R. Maget-Dana, G. Nullans, B. Kieffer, D. Aunis and M.H. Metz-Boutigue, *Ann. N. Y. Acad. Sci.*, 2002 **971**, 359.
258. D.G. Lee, Y. Park, H.N. Kim, H.K. Kim, P.I. Kim, B.H. Choi and K.S. Hahm, *Biochem. Biophys. Res. Commun.*, 2002, **291**, 1006.
259. D.G. Lee, P.I. Kim, Y. Park, S.H. Jang, S.C. Park, E.R. Woo and K.S. Hahm, *J. Pept. Sci.*, 2002, **8**, 453.
260. D.G. Lee, P.I. Kim, Y.K. Park, E.R. Woo, J.S. Choi, C.H. Choi and K.S. Hahm, *Biochem. Biophys. Res. Commun.*, 2002, **293**, 231.
261. D.G. Lee, P.I. Kim, Y. Park, S.C. Park, E.R. Woo and K.S. Hahm, *Biochem. Biophys. Res. Commun.*, 2002, **295**, 591.
262. R.H. Huang, Y. Xiang, X.Z. Liu, Y. Zhang, Z. Hu and D.C. Wang, *FEBS Letters*, 2002, **521**, 87.

263. K.P.B. Van der Bergh, P. Proost, J. Van Damme, J. Coosemans, E.J.M. Van Damme and W.J. Peumans, *FEBS Letters*, 2002, **530**, 181.
264. X.Y. Ye and T.B. Ng, *J. Pept. Sci.*, 2002, **8**, 101.
265. X.Y. Ye and T.B. Ng, *J. Pept. Res.*, 2002, **60**, 81.
266. H.X. Wang and T.B. Ng, *Peptides*, 2002, **23**, 7.
267. H.X. Wang and T.B. Ng, *Peptides*, 2002, **23**, 1025.
268. A. Wiest, D. Grzegorski, B.W. Xu, C. Goulard, S. Rebuffat, D.J. Ebbole, B. Bodo and C. Kenerley, *J. Biol. Chem.*, 2002, **277**, 20862.
269. B. Kundu, T. Srinavasan, A.P. Kesarwani, A. Kavishwar, S.K. Raghuwanshi, S. Batra and P.K. Shukla, *Bioorg. Med. Chem. Lett.*, 2002, **12**, 1473.
270. B. Lopez-Garcia, E. Perez-Paya and J.F. Marcos, *Appl. Environ. Microbiol.*, 2002, **68**, 2453.
271. C.F. Gonzalez, E.M. Provin, L. Zhu and D.J. Ebbole, *Phytopathology*, 2002, **92**, 917.
272. C. Olsson, *The Lancet*, 2002, **359**, 168.
273. G. Noera, M. Lamarra, S. Guarini and A. Bertolini, *The Lancet*, 2002, **359**, 168.
274. E.L. Altschuler, *The Lancet*, 2002, **359**, 169.
275. L. Ochedalska, E. Rebas, J. Kunert-Radek, M.C. Fournie-Zaluski and M.B. Pawlikowski, *Biochem. Biophys. Res. Commun.*, 2002, **297**, 931.
276. J. Grippo, R.F. Kirby, T.G. Beltz and A.K. Johnson, *Pharmacology, Biochemistry and Behavior*, 2002, **71**, 139.
277. R. Nakaie, E.G. Silva, E.M. Cilli, R. Marchetto, S. Schreier, T.B. Paiva and A.C.M. Paiva, *Peptides*, 2002, **23**, 65.
278. W.S. Chen, M.K. Sim and M.L. Go, *Regulatory Peptides*, 2002, **106**, 39.
279. P. Roumelioti, L. Polevaya, P. Zoumpoulakis, N. Giatas, I. Mutule, T. Keivish, A. Zoga, D. Vlahakos, E. Iliodromatis and D. Kremastinos, *Bioorg. Med. Chem. Lett.*, 2002, **12**, 2627.
280. M.T. Le, P.M.L. Vanderheyden, G. Baggerman, J.V. Broeck and G. Vauquelin, *Regulatory Peptides*, 2002, **105**, 101.
281. M.P. Schuijt and A.H.J. Danser, *Am. J. Hypertens.*, 2002, **15**, 1109.
282. F. Hu, C.S. Cutler, T. Hoffman, G. Sieckman, W.A. Volkert and S.S. Jurisson, *Nuclear Medicine and Biology*, 2002, **29**, 423.
283. R. La Bella, E.G. -Garayoa, M. Langer, P. Bläuenstein, A.G. Beck-Sickinger and P.A. Schubiger, *Nuclear Medicine and Biology*, 2002, **29**, 553.
284. R.E. Weiner and M.L. Thakur, *Applied Radiation and Isotopes*, 2002, **57**, 749.
285. K. Yamada, E. Wada, Y. Santo-Yamada and K. Wada, *Eur. J. Pharm.*, 2002, **440**, 281.
286. S. Funes, J.A. Hedrick, S.J. Yang, L.X. Shan, M. Bayne, F.J. Monsma and E.L. Gustafson, *Peptides*, 2002, **23**, 1607.
287. C. Friry, S. Feliciangeli, F. Richard, P. Kitabgi and C. Rovere, *Biochem. Biophys. Res. Commun.*, 2002, **290**, 1161.
288. B. Brechter and U.H. Lerner, *Regulatory Peptides*, 2002, **103**, 39.
289. W.Y. Feng, K.K. Chan and J.M. Covey, *J. Pharmaceut. Biomed. Anal.*, 2002, **28**, 601.
290. P. Cucchi, S. Meini, L. Quartara, A. Giolitti, S. Zappitelli, L. Rotondaro and C.A. Maggi, *Peptides*, 2002, **23**, 1457.
291. J.M. Stewart, L. Gera, D.C. Chan, P.A. Bunn Jr. and E.J. York, *et al.*, *Can. J. Physiol. Pharm.*, 2002, **80**, 275.
292. J.M. Stewart, D.C. Chan, V. Simkeviciene, P.A. Bunn, Jr., B. Helfrich, E.J. York, L. Taraseviciene-Stewart, D. Bironaite and L. Gera, *Int. Immunopharmacol.*, 2002, **2**, 1781.
293. S.-F. Su, G.L. Amidon and H.J. Lee, *Life Sciences*, 2002, **72**, 35.

294. M.J. Warburton and F. Bernardini, *Neuroscience Letters*, 2002, **331**, 99.
295. S.F. Su, G.L. Amidon and H.J. Lee, *Biochem. Biophys. Res. Commun.*, 2002, **292**, 632.
296. S.F. Su, G.L. Amidon and H.J. Lee, *Life Sciences*, 2002, **72**, 35.
297. L. Boulanger, N. Girard, J. Strecko and P. Gaudreau, *Peptides*, 2002, **23**, 1187.
298. B. Peschke, M. Ankersen, M. Bauer, T.K. Hansen, B.S. Hansen, K.K. Nielsen, K. Raun, L. Richter and L. Westergaard, *Eur. J. Med. Chem.*, 2002, **37**, 487.
299. E. Saito, H. Kaiya, T. Takagi, I. Yamasaki, D.M. Denbow, K. Kangawa and M. Furuse, *Eur. J. Pharmacol.*, 2002, **453**, 75.
300. J. van der Lely, *Curr. Opin. Pharmacol.*, 2002, **2**, 730.
301. G. Muccioli, M. Tschöp, M. Papotti, R. Deghenghi, M. Heiman and E. Ghigo, *Eur. J. Pharmacol.*, 2002, **440**, 235.
302. J. Pinkney and G. Williams, *The Lancet*, 2002, **359**, 1360.
303. K.G. Harikumar, D.I. Pinon, W.S. Wessels, F.G. Prendergast and L.J. Miller, *J. Biol. Chem.*, 2002, **277**, 18552.
304. S.A. Mousa, *Curr. Opin. Chem. Biol.*, 2002, **6**, 534.
305. J.E. Blum and H. Handmaker, *Curr. Pharm. Design*, 2002, **8**, 1815.
306. J. Takagi, B.M. Petre, T. Walz and T.A. Springer, *Cell*, 2002, **110**, 599.
307. R.M. Kohli, J. Takagi and C.T. Walsh, *Proc. Nat. Acad. Sci. USA*, 2002, **99**, 1247.
308. J.M. Dogne, X. de Leval, P. Benoit, J. Delarge, B. Masereel and J.L. David, *Curr. Med. Chem.*, 2002, **9**, 577.
309. T.C. Walsgrove, L. Powell and A. Wells, *Org. Proc. Res. Dev.*, 2002, **6**, 488.
310. H.W. Zhang and Z.R. Guo, *Chinese J. Org. Chem.*, 2002, **22**, 754.
311. D. Okuda, H. Koike and T. Morita, *Biochemistry*, 2002, **41**, 14248.
312. S.Y. Hong, Y.S. Koh, K.H. Chung and D.S. Kim, *Thrombosis Res.*, 2002, **105**, 79.
313. S.Y. Hong, Y.D. Sohn, K.H. Chung and D.S. Kim, *Biochem. Biophys. Res. Commun.*, 2002, **293**, 530.
314. Y. Moritani, K. Sato, T. Shigenaga, N. Hisamichi, M. Ichihara, S. Akamatsu, K. Suzuki, T. Nii, S. Kaku, T. Kawasaki, Y. Matsumoto, O. Inagaki, K. Tomioka and I. Yanagisawa, *Eur. J. Pharmacol.*, 2002, **439**, 43.
315. K. Suzuki, Y. Moritani, N. Hisamichi, M. Ichihara, S. Akamatsu, H. Arai, H. Matsushima, T. Nii, K. Sato, Y. Taniuchi, T. Shigenaga, S. Kaku, T. Kawasaki, Y. Matsumoto, O. Inagaki, K. Tomioka and I. Yanagisawa, *Drug Dev. Res.*, 2002, **55**, 149.
316. B.J. Mans, A.I. Louw and A.W. H. Neitz, *J. Biol. Chem.*, 2002, **277**, 21371.
317. G.C. Tucker, *Curr. Opin. Pharm.*, 2002, **2**, 394.
318. P.J. Coleman and L.T. Duong, *Expert Opin. Ther. Pat.*, 2002, **12**, 1009.
319. E.C. Lawson, W.A. Kinney, D.K. Luci, S.C. Yabut, D. Wisnoski and B.E. Maryanoff, *Tetrahedron Lett.*, 2002, **43**, 1951.
320. F. Peri, R. Bassetti, E. Caneva, L. de Gioia, B. La Ferla, M. Presta, E. Tanghetti and F. Nicotra, *J. Chem. Soc. Perkin Trans. 1*, 2002, 638.
321. S. Schabbert, M.D. Pierschbacher, R.H. Mattern and M. Goodman, *Bioorg. Med. Chem.*, 2002, **10**, 3331.
322. E. Addicks, R. Mazitschek and A. Giannis, *Chembiochem*, 2002, **3**, 1078.
323. R.S. Meissner, J.J. Perkins, L.T. Duong, G.D. Hartman, W.F. Hoffman, J.R. Huff, N.C. Ihle, C.T. Leu, R.M. Nagy, A. Naylor-Olsen, G.A. Rodan, S.B. Rodan, D.B. Whitman, G.A. Wesolowski and M.E. Duggan, *Bioorg. Med. Chem. Lett.*, 2002, **12**, 25.
324. P.J. Coleman, K.M. Brashear, C.A. Hunt, W.F. Hoffman, J.H. Hutchinson, M.J. Breslin, C.A. McVean, B.C. Askew, G.D. Hartman, S.B. Rodan, G.A. Rodan,

C.T. Leu, T. Prueksaritanont, C. Fernandez-Metzler, B. Ma, L.A. Libby, K.M. Merkle, G.L. Stump, A.A. Wallace, J.J. Lynch and M.E. Duggan, *Bioorg. Med. Chem. Lett.*, 2002, **12**, 31.
325. P.J. Coleman, B.C. Askew, J.H. Hutchinson, D.B. Whitman, J.J. Perkins, G.D. Hartman, G.A. Rodan, C.T. Leu, T. Prueksaritanont, C. Fernandez-Metzler, K.M. Merkle, R. Lynch, J.J. Lynch, S.B. Rodan and M.E. Duggan, *Bioorg. Med. Chem. Lett.*, 2002, **12**, 2463.
326. K.M. Brashear, C.A. Hunt, B.T. Kucer, M.E. Duggan, G.D. Hartman, G.A. Rodan, S.B. Rodan, C.T. Leu, T. Prueksaritanont, C. Fernandez-Metzler, A. Barrish, C.F. Homnick, J.H. Hutchinson and P.J. Coleman, *Bioorg. Med. Chem. Lett.*, 2002, **12**, 3483.
327. K. Urbahns, M. Harter, M. Albers, D. Schmidt, B. Stelte-Ludwig, U. Bruggemeier, A. Vaupel and C. Gerdes, *Bioorg. Med. Chem. Lett.*, 2002, **12**, 205.
328. C.F. McCusker, P.J. Kocienski, F.T. Boyle and A.G. Schatzlein, *Bioorg. Med. Chem. Lett.*, 2002, **12**, 547.
329. A. Kling, G. Backfisch, R. Delzer, H. Geneste, C. Graef, U. Holzenkamp, W. Hornberger, U.E.W. Lange, A. Lauterbach, H. Mack, W. Seitz and T. Subkowski, *Bioorg. Med. Chem. Lett.*, 2002, **12**, 441.
330. U.E.W. Lange, G. Backfisch, J. Delzer, H. Geneste, C. Graef, W. Hornberger, A. Kling, A. Lauterbach, T. Subkowski and C. Zechel, *Bioorg. Med. Chem. Lett.*, 2002, **12**, 1379.
331. F. Perron-Sierra, D. Saint Dizier, M. Bertrand, A. Genton, G.C. Tucker and P. Casara, *Bioorg. Med. Chem. Lett.*, 2002, **12**, 3291.
332. R.J. Kok, A.J. Schraa, E.J. Bos, H.E. Moorlag, S.A. Asgeirsdottir, M. Everts, D.K.F. Meijer and G. Molema, *Bioconj. Chem.*, 2002, **13**, 128.
333. K.E. Gottschalk, R. Gunther and H. Kessler, *Chembiochem*, 2002, **3**, 470.
334. D. Yahalom, A. Wittelsberger, D.F. Mierke, M. Rosenblatt, J.M. Alexander and M. Chorev, *Biochemistry*, 2002, **41**, 8321.
335. A.J. Schraa, R.J. Kok, A.D. Berendsen, H.E. Moorlag, E.J. Bos, D.K.F. Meijer, L.F.M.H. de Leij and G. Molema, *J. Control. Release*, 2002, **83**, 241.
336. M.A. Buerkle, S.A. Pahernik, A. Sutter, A. Jonczyk, K. Messmer and M. Dellian, *Brit. J. Cancer*, 2002, **86**, 788.
337. M.L. Janssen, W.J. Oyen, I. Dijkgraaf, L.F. Massuger, C. Frielink, D.S. Edwards, M. Rajopadhye, H. Boonstra, F.H. Corstens and O.C. Boerman, *Cancer Res.*, 2002, **62**, 6146.
338. Z. Su, G.Z. Liu, S. Gupta, Z.H. Zhu, M. Rusckowski and D.J. Hnatowich, *Bioconj. Chem.*, 2002, **13**, 561.
339. G.W. Holland, R.J. Biediger and P. Vanderslice, *Ann. Rep. Med. Chem.*, 2002, **37**, 65.
340. J.W. Tilley, *Exp. Opin. Ther. Patients*, 2002, **12**, 991.
341. D.Y. Jackson, *Curr. Pharm. Design*, 2002, **8**, 1229.
342. H. Yusuf-Makagiansar, M.E. Anderson, T.V. Yakovleva, J.S. Murray and T.J. Siahaan, *Med. Res. Rev.*, 2002, **22**, 146.
343. L.S. Lin, I.E. Kopka, R.A. Mumford, P.A. Magriotis, T. Lanza, P.L. Durette, T. Kamenecka, D.N. Young, S.E. de Laszlo, E. McCauley, G. Van Riper, U. Kidambi, L.A. Egger, X.C. Tong, K. Lyons, S. Vincent, R. Stearns, A. Colletti, Y. Teffera, J. Fenyk-Melody, J.A. Schmidt, M. MacCoss and W.K. Hagmann, *Bioorg. Med. Chem. Lett.*, 2002, **12**, 611.
344. I.E. Kopka, D.N. Young, L.S. Lin, R.A. Mumford, P.A. Magriotis, M. MacCoss, S.G. Mills, G. Van Riper, E. McCauley, L.A. Egger, U. Kidambi, J.A. Schmidt, K.

Lyons, R. Stearns, S. Vincent, A. Colletti, Z. Wang, X.C. Tong, J.Y. Wang, S. Zheng, K. Owens, D. Levorse and W.K. Hagmann, *Bioorg. Med. Chem. Lett.*, 2002, **12**, 637.
345. S.E. de Laszlo, B. Li, E. McCauley, G. Van Riper and W.K. Hagmann, *Bioorg. Med. Chem. Lett.*, 2002, **12**, 685.
346. G.A. Doherty, T. Kamenecka, E. McCauley, G. Van Riper, R.A. Mumford and W.K. Hagmann, *Bioorg. Med. Chem. Lett.*, 2002, **12**, 729.
347. G.X. Yang, L.L. Chang, Q. Truong, G.A. Doherty, P.A. Magriotis, S.E. de Laszlo, B. Li, M. MacCoss, U. Kidambi, E. McCauley, G. Van Riper, R.A. Mumford, J.A. Schmidt and W.K. Hagmann, *Bioorg. Med. Chem. Lett.*, 2002, **12**, 1497.
348. G.A. Doherty, G.X. Yang, E. Borges, L.L. Chang, M. MacCoss, S. Tong, U. Kidambi, L.A. Egger, E. McCauley, G. Van Riper, R.A. Mumford, J.A. Schmidt and W.K. Hagmann, *Bioorg. Med. Chem. Lett.*, 2002, **12**, 1501.
349. B. Li, S.E. de Laszlo, T. Kamenecka, I.E. Kopka, P.L. Durette, T. Lanza, M. MacCoss, S. Tong, R.A. Mumford, E. McCauley, G. Van Riper, J.A. Schmidt and W.K. Hagmann, *Bioorg. Med. Chem. Lett.*, 2002, **12**, 2141.
350. T. Kamenecka, T. Lanza, S.E. de Laszlo, B. Li, E. McCauley, G. Van Riper, L.A. Egger, U. Kidambi, R.A. Mumford, S. Tong, M. MacCoss, J.A. Schmidt and W.K. Hagmann, *Bioorg. Med. Chem. Lett.*, 2002, **12**, 2205.
351. I.E. Kopka, L.S. Lin, R.A. Mumford, T. Lanza, P.A. Magriotis, D.N. Young, S.E. de Laszlo, M. MacCoss, S.G. Mills, Van Riper, E. McCauley, K. Lyons, S. Vincent, L.A. Egger, U. Kidambi, R. Stearns, A. Colletti, Y. Teffera, S. Tong, K. Owens, D. Levorse, J.A. Schmidt and W.K. Hagmann, *Bioorg. Med. Chem. Lett.*, 2002, **12**, 2415.
352. L. Chen, R. Trilles, D. Miklowski, T.N. Huang, D. Fry, R. Campbell, K. Rowan, V. Schwinge and J.W. Tilley, *Bioorg. Med. Chem. Lett.*, 2002, **12**, 1679.
353. L. Chen, J.W. Tilley, R. Trilles, W.Y. Yun, D. Fry, C. Cook, K. Rowan, V. Schwinge and R. Campbell, *Bioorg. Med. Chem. Lett.*, 2002, **12**, 137.
354. A. Sidduri, J.W. Tilley, K. Hull, J.P. Lou, G. Kaplan, A.A. Sheffron, L. Chen, R. Campbell, R. Guthrie, T.N. Huang, N. Huby, K. Rowan, V. Schwinge and L.M. Renzetti, *Bioorg. Med. Chem. Lett.*, 2002, **12**, 2475.
355. A. Sidduri, J.W. Tilley, J.P. Lou, L. Chen, G. Kaplan, F. Mennona, R. Campbell, R. Guthrie, T.N. Huang, K. Rowan, V. Schwinge and L.M. Renzetti, *Bioorg. Med. Chem. Lett.*, 2002, **12**, 2479.
356. J.R. Porter, S.C. Archibald, K. Childs, D. Critchley, J.C. Head, J.M. Linsley, T.A. H. Parton, M.K. Robinson, A. Shock, R.J. Taylor, G.J. Warrellow, R.P. Alexander and B. Langham, *Bioorg. Med. Chem. Lett.*, 2002, **12**, 1051.
357. J.R. Porter, S.C. Archibald, J.A. Brown, K. Childs, D. Critchley, J.C. Head, B. Hutchinson, T.A.H. Parton, M.K. Robinson, A. Shock, G.J. Warrellow and A. Zomaya, *Bioorg. Med. Chem. Lett.*, 2002, **12**, 1591.
358. J.R. Porter, S.C. Archibald, J.A. Brown, K. Childs, D. Critchley, J.C. Head, B. Hutchinson, T.A.H. Parton, M.K. Robinson, A. Shock, G.J. Warrellow and A. Zomaya, *Bioorg. Med. Chem. Lett.*, 2002, **12**, 1595.
359. V. Wehner, H. Blum, M. Kurz, H.U. Stilz, *Synthesis-Stuttgart*, 2002, Sp. Iss. SI 2002, 2023.
360. J. Singh, H. van Vlijmen, Y.S. Liao, W.C. Lee, M. Cornebise, M. Harris, I.H. Shu, A. Gill, J.H. Cuervo, W.M. Abraham and S.P. Adams, *J. Med. Chem.*, 2002, **45**, 2988.
361. J. Singh, H. van Vlijmen, W.C. Lee, Y.S. Liao, K.C. Lin, H. Ateeq, J. Cuervo, C. Zimmerman, C. Hammnod, M. Karpusas, R. Palmer, T. Chattopadhyay and S.P. Adams, *J. Computer-Aided Mol. Des.*, 2002, **16**, 201.

362. Glaxo Group Ltd., *Exp. Opin. Ther. Pat.*, 2002, 12, 755.
363. N.J.P. Dubree, D.R. Artis, G. Castanedo, J. Marsters, D. Sutherlin, L. Caris, K. Clark, S.M. Keating, M.H. Beresini, H. Chiu, S. Fong, H.B. Lowman, N.J. Skelton and D.Y. Jackson, *J. Med. Chem.*, 2002, **45**, 3451.
364. D.J. Gottschling, J. Boer, A. Schuster, B. Holzmann and H. Kessler, *Angew. Chem. – Int. Ed.*, 2002, **41**, 3007.
365. D.J. Gottschling, J. Boer, L. Marinelli, G. Voll, M. Haupt, A. Schuster, B. Holzmann and H. Kessler, *Chembiochem*, 2002, **3**, 575.
366. L.S. Lin, T. Lanza, E. McCauley, G. Van Riper, U. Kidambi, J. Cao, L.A. Egger, R.A. Mumford, J.A. Schmidt, M. MacCoss and W.K. Hagmann, *Bioorg. Med. Chem. Lett.*, 2002, **12**, 133.
367. L.L. Chang, Q. Truong, R.A. Mumford, L.A. Egger, U. Kidambi, K. Lyons, E. McCauley, G. Van Riper, S. Vincent, J.A. Schmidt, M. MacCoss and W.K. Hagmann, *Bioorg. Med. Chem. Lett.*, 2002, **12**, 159.
368. G.M. Castanedo, F.C. Sailes, N.J.P. Dubree, J.B. Nicholas, L. Caris, K. Clark, S.M. Keating, M.H. Beresini, H. Chiu, S. Fong, J.C. Marsters, D.Y. Jackson and D.P. Sutherlin, *Bioorg. Med. Chem. Lett.*, 2002, **12**, 2913.
369. I. Sircar, K.S. Gudmundsson, R. Martin, J. Liang, S. Nomura, H. Jayakumar, B.R. Teegarden, D.M. Nowlin, P.M. Cardarelli, J.R. Mah, S. Connell, R.C. Griffith and E. Lazarides, *Bioorg. Med. Chem.*, 2002, **10**, 2051.
370. L. Zhang, R.H. Mattern, T.I. Malaney, M.D. Pierschbacher and M. Goodman, *J. Am. Chem. Soc.*, 2002, **124**, 2862.
371. J.B. Smith, R.D.G. Theakston, A.L.J. Coelho, C. Barja-Fidalgo, J.J. Calvete and C. Marcinkiewicz, *FEBS Letters*, 2002, **512**, 111.
372. S.L. Goodman, G. Holzemann, G.A.G. Sulyok and H. Kessler, *J. Med. Chem.*, 2002, **45**, 1045.
373. Q. Plonowski, A.V. Schally, A. Nagy, K. Groot, M. Krupa, N.M. Navone and C. Logothetis, *Cancer Lett.*, 2002, **176**, 57.
374. L.J. Krebs, X. Wang, A. Nagy, A.V. Schally, P.N. Prasad and C. Liebow, *Oral Oncology*, 2002, **38**, 657.
375. P. Grieco, G. Han, D. Weinberg, T. MacNeil, L.H.T. Van der Ploeg and V.J. Hruby, *Biochem. Biophys. Res. Commun.*, 2002, **292**, 1075.
376. D.J. MacNeil, A.D. Howard, X. Guan, T.M. Fong, R.P. Nargund, M.A. Bednarek, M.T. Goulet, D.H. Weinberg, A.M. Strack, D.J. Marsh, H.Y. Chen, Chun-Pyn Shen, A.S. Chen, C.I. Rosenblum, T. MacNeil, M. Tota, E.D. MacIntyre and L.H.T. Van der Ploeg, *Eur. J. Pharmacol.*, 2002, **440**, 141.
377. J. Hernandez, F. Garcia-Pons, Y.C. Lone, H. Firat, J.D. Schmidt, P. Langlade-Demoyen and M. Zanetti, *Proc. Natl. Acad. Sci. USA*, 2002, **99**, 12275.
378. O. Rotzschke, J.M. Lau, M. Hofstatter, K. Falk and J.L. Strominger, *Proc. Natl. Acad. Sci. USA*, 2002, **99**, 16946.
379. M. Baratin, M. Kayibanda, M. Ziol, R. Romieu, J.P. Briand, J.G. Guillet and M. Viguier, *J. Pept. Sci.*, 2002, **8**, 327.
380. J.P. Redrobe, Y. Dumont and R. Quirion, *Life Sciences*, 2002, **71**, 2921.
381. P. Silva, C. Cavadas and E. Grouzmann, *Clinica Chimica Acta*, 2002, **326**, 3.
382. R.E. Malmström, *Eur. J. Pharmacol.*, 2002, **447**, 11.
383. E. Parker, M. Van Heek and A. Stamford, *Eur. J. Pharmacol.*, 2002, **440**, 173.
384. D.A. Keire, C.W. Bowers, T.E. Solomon and J.R. Reeve, *Peptides*, 2002, **23**, 305.
385. M. Rinnová, A. Nefzi and R.A. Houghten, *Bioorg. Med. Chem. Lett.*, 2002, **12**, 3175.

386. L. D'Este, A. Casini, S. Puglisi-Allegra, S. Cabib, I. Tooyama, H. Kimura and T.G. Renda, *J. Chem. Neuroanatomy*, 2002, **24**, 189.
387. A. Ambo, H. Murase, H. Niizuma, H. Ouchi, Y. Yamamoto and Y. Sasaki, *Bioorg. Med. Chem. Lett.*, 2002, **12**, 879.
388. J. McDonald, T.A. Barnes, G. Calo', R. Guerrini, D.J. Rowbotham and D.G. Lambert, *Eur. J. Pharm.*, 2002, **443**, 7.
389. Sakurada, S. Sakurada, T. Orito, K. Tan-No and T. Sakurada, *Biochem. Pharm.*, 2002, **64**, 1293.
390. I. Quelven, A. Roussin, O. Burlet-Schiltz, C. Gouardères, J.A.M. Tafani, H. Mazarguil and J.M. Zajac, *Eur. J. Pharmacol.*, 2002, **449**, 91.
391. E.Y.K. Huang, J.Y. Li, C.H. Wong, P.P.C. Tan and J.C. Chen, *Peptides*, 2002, **23**, 489.
392. O. Burlet-Schiltz, H. Mazarguil, J.C. Sol, P. Chaynes, B. Monsarrat, J.M. Zajac and A. Roussin, *FEBS Letters*, 2002, **532**, 313.
393. C. Gouardères, I. Quelven, C. Mollereau, H. Mazarguil, S.Q.J. Rice and J.M. Zajac, *Neuroscience*, 2002, **115**, 349.
394. K. Wong, B.P. Ross, Y.N. Chan, P. Artursson, L. Lazorova, A. Jones and I. Toth, *Eur. J. Pharmaceut. Sci.*, 2002, **16**, 113.
395. S. Maricic, U. Berg and T. Frejd, *Tetrahedron*, 2002, **58**, 3085.
396. J.E. Burden, P. Davis, F. Porreca and A.F. Spatola, *Bioorg. Med. Chem. Lett.*, 2002, **12**, 213.
397. M. Doi, A. Asano, E. Komura and Y. Ueda, *Biochem. Biophys. Res. Commun.*, 2002, **297**, 138.
398. A. Lengyel, G. Orosz, D. Biyashev, L. Kocsis, M. Al-Khrasani, A. Rónai, Cs. Tömböly, Zs. Fürst, G. Tóth and A. Borsodi, *Biochem. Biophys. Res. Commun.*, 2002, **290**, 153.
399. J.C. Reubi, K.P. Eisenwiener, H. Rink, B. Waser and H.R. Mäcke, *Eur. J. Pharmacol.*, 2002, **456**, 45.
400. D. Lattuada, C. Casnici, A. Venuto and O. Marelli, *J. Neuroimmunol.*, 2002, **133**, 211.
401. W.G. Rajeswaran, W.A. Murphy, J.E. Taylor and D.H. Coy, *Bioorg. Med. Chem.*, 2002, **10**, 2023.
402. J.-U. Lee, R. Hosotani, M. Wada, R. Doi, T. Koshiba, K. Fujimoto, Y. Miyamoto, S. Tsuji, S. Nakajima, M. Hirohashi, T. Uehara, Y. Arano, N. Fujii and M. Imamura, *Eur. J. Cancer*, 2002, **38**, 1526.
403. W. Mier, B. Beijer, K. Graham and W.E. Hull, *Bioorg, Med, Chem.*, 2002, **10**, 2543.
404. J. Stirnweis, F.D. Boehmer and C. Liebmann, *Peptides*, 2002, **23**, 1503.
405. L. Sun, J.A. Fuselier, W.A. Murphy and D.H. Coy, *Peptides*, 2002, **23**, 1557.
406. S.W.J. Lamberts, W.W. de Herder and L.J. Hofland, *Trends in Endocrinology and Metabolism*, 2002, **13**, 451.
407. Ö. Ugur, P.J. Kothari, R.D. Finn, P. Zanzonico, S. Ruan, I. Guenther, H.R. Maecke and S.M. Larson, *Nuclear Medicine and Biology*, 2002, **29**, 147.
408. Stark and R. Mentlein, *Regulatory Peptides*, 2002, **108**, 97.
409. S. Chatterjee, A. Mbaye, J.G. deMan and E.A.E. Van Marck, *TRENDS in Parasitology*, 2002, **18**, 295.
410. S.K. Ozker, R.S. Hellman and A.Z. Krasnow, *Applied Radiation And Isotopes*, 2002, **57**, 29.
411. H. Torfs, K.E. Åkerman, R.J. Nachman, H.B. Oonk, M. Detheux, J. Poels, T. Van Loy, A. De Loof, R.H. Meloen, G. Vassart, M. Parmentier and J. Vanden Broeck, *Peptides*, 2002, **23**, 1999.

412. Q. Zhou and F. Nyberg, *Neuroscience Letters*, 2002, **320**, 117.
413. H. Torfs, M. Detheux, H.B. Oonk, K.E. Akerman, J. Poels, T. Van Loy, A. De Loof, G. Vassart, M. Parmentier and J. Vanden Broeck, *Biochem. Pharmacol.*, 2002, **63**, 1675.
414. T. Michael-Titus, K. Fernandes, H. Setty and R. Whelpton, *Neuroscience*, 2002, **110**, 277.
415. S. Sakurada, T. Orito, C. Sakurada, T. Sato, T. Hayashi, J.I. Mobarakeh, K. Yanai, K. Onodera, T. Watanabe and T. Sakurada, *Eur. J. Pharmacol.*, 2002, **434**, 292.
416. X. Ferry, S. Brehin, R. Kamel and Y. Landry, *Peptides*, 2002, **23**, 1507.
417. E. Hori, T. Uwano, R. Tamura, N. Miyake, H. Nishijo and T. Ono, *Cognitive Brain Research*, 2002, **13**, 1.
418. W. Lemaire, J.A. O'Brien, M. Burno, A.G. Chaudhary, D.C. Dean, P.D. Williams, R.M. Freidinger, D.J. Pettibone and D.L. Williams, *Eur. J. Pharmacol.*, 2002, **450**, 19.
419. S. Thornton, P.J. Baldwin, P.A. Harris, F. Harding, J.M. Davison, P.H. Baylis, P.M. Timmons and D.C. Wathes, *BJOG-An International Journal of Obstetrics And Gynaecology*, 2002, **109**, 57.
420. B. Tahara, J. Tsukada, Y. Tomura, K. Wada, T. Kusayama, N. Ishii, T. Yatsu, W. Uchida, N. Taniguchi and A. Tanaka, *Peptides*, 2002, **23**, 1809.
421. M. Petersson, *Regulatory Peptides*, 2002, **108**, 83.
422. M.M. Slattery and J.J. Morrison, *Lancet*, 2002, **360**, 1489.
423. P.G. Wyatt, M.J. Allen and J. Chilcott, et al., *Bioorg. Med. Chem. Lett.*, 2002, **12**, 1399.
424. P.G. Wyatt, M.J. Allen and J. Chilcott, et al., *Bioorg. Med. Chem. Lett.*, 2002, **12**, 1405.
425. R.L. Goldenberg, *Obstetrics and Gynecology*, 2002, **100**, 1020.
426. J. Havass, K. Bakos, Á. Márki, R. Gáspár, L. Gera, J.M. Stewart, F. Fülöp, G.K. Tóth, I. Zupkó and G. Falkay, *Peptides*, 2002, **23**, 1419.
427. X. Bin, Q. Hua, S.H. Nakagawa, W. Jia, Y.C. Chu, P.G. Katsoyannis and A.M. Weiss, *J. Mol. Biol.*, 2002, **316**, 435.
428. J. Winter, L. Hauke and R. Rudolph, *Anal. Biochem.*, 2002, **310**, 148.
429. J.E. Gerich, *Am. J. Med.*, 2002, **113**, 308.
430. W.G. Anderson, M.F. Ali, I.E. Einarsdóttir, L. Schäffer, N. Hazon and J.M. Conlon, *General and Comparative Endocrinology*, 2002, **126**, 113.
431. F. Lindholm, *Best Practice & Research Clinical Gastroenterology*, 2002, **16**, 475.
432. M. Wilchek and T. Miron, *Biochem. Biophys. Res. Commun.*, 2002, **290**, 775.
433. H. Schellekens, *Clinical Therapeutics*, 2002, **24**, 1720.
434. R.B. Weinstein, N. Eleid, C. LeCesne, B. Durando, J.T. Crawford, M. Heffner, C. Layton, M. O'Keefe, J. Robinson and S. Rudinsky *et al.*, *Metabolism*, 2002, **51**, 1065.
435. K.D. Hinds and S.W. Kim, *Advanced Drug Delivery Reviews*, 2002, **54**, 505.
436. G. Le Flem, F.Y. Dupradeau, J.P. Pujol, J.P. Monti and P. Bogdanowicz, *Bioorg. Med. Chem.*, 2002, **10**, 2111.
437. M.F.T. Koehler, K. Zobel, M.H. Beresini, L.D. Caris, D. Combs, B.D. Paasch and R.A. Lazarus, *Bioorg. Med. Chem. Lett.*, 2002, **12**, 2883.
438. I. Russo, P. Massucco, L. Mattiello, F. Cavalot, G. Anfossi and M. Trovati, *Thromb. Res.*, 2002, **107**, 31.
439. J. Onuffer and R. Horuk, *Trends in Pharmacol. Sci.*, 2002, **23**, 459.

440. Y. Wan, J.P. Jakway, H. Qiu, H. Shah, C.G. Garlisi, F. Tian, P. Ting, D. Hesk, R.W. Egan, M.M. Billah and S.P. Umland, *Eur. J. Pharmacol.*, 2002, **456**, 1.
441. H. Tamamura, K. Hiramatsu, K. Miyamoto, A. Omagari, S. Oishi, H. Nakashima, N. Yamamoto, Y. Kuroda, T. Nakagawa, A. Otaka and N. Fujii, *Bioorg. Med. Chem. Lett.*, 2002, **12**, 923.
442. P. Lusso, *Vaccine*, 2002, **20**, 1964.
443. V. Gangur, N.P. Birmingham and S. Thanesvorakul, *Veterinary Immunology and Immunopathology*, 2002, **86**, 127.
444. D.A. Wacker, J.B. Santella III, D.S. Gardner, J.G. Varnes, M. Estrella, G.V. DeLucca, S.S. Ko, K. Tanabe, P.S. Watson, P.K. Welch, M. Covington, N.C. Stowell, E.A. Wadman, P. Davies, K.A. Solomon, R.C. Newton, L. Trainor, S.M. friedman, C.P. Decicco and J.V. Duncia, *Bioorg. Med. Chem. Lett.*, 2002, **12**, 1785.
445. P.H. Carter, *Curr. Opin. Chem. Biol.*, 2002, **6**, 510.
446. M. Matsui, J. Weaver, A.E.I. Proudfoot, J.R. Wujek, T. Wei, E. Richer, B.D. Trapp, A. Rao and R.M. Ransohoff, *J. Neuroimmunol.*, 2002, **128**, 16.
447. Z. Xian-Chun, W. San-Xia and L. Wen-Xin, *Toxicon*, 2002, **40**, 1719.
448. Z. Shunyi and L. Wenxin, *Comp. Biochem. Physiol. Part B: Biochem. Mol. Biol.*, 2002, **131**, 749.
449. Matavel, J.S. Cruz, C.L. Penaforte, D.A.M. Araújo, E. Kalapothakis, V.F. Prado, C.R. Diniz, M.N. Cordeiro and P.S.L. Beirão, *FEBS Letters*, 2002, **523**, 219.
450. K. Takeuchi, E.J. Park, C.W. Lee, J.I. Kim, H. Takahashi, K.J. Swartz and I. Shimada, *J. Mol. Biol.*, 2002, **321**, 517.
451. M.S.P. Sansom, I.H. Shrivastava, J.N. Bright, J. Tate, C.E. Capener and P.C. Biggin, *Biochim. Biophys. Acta (BBA) - Biomembranes*, 2002, **1565**, 294.
452. T. Olamendi-Portugal, B.I. García, I. López-González, J. Van Der Walt, K. Dyason, C. Ulens, J. Tytgat, R. Felix, A. Darszon and L.D. Possani, *Biochem. Biophys. Res. Commun.*, 2002, **299**, 562.
453. L.D. Rash and W.C. Hodgson, *Toxicon*, 2002, **40**, 225.
454. C.V.F. Batista, F.Z. Zamudio, S. Lucas, J. W. Fox, A. Frau, G. Prestipino and L.D. Possani, *Toxicon*, 2002, **40**, 557.
455. M. Lirazan, E.C. Jimenez, A. Grey Craig, B.M. Olivera and L.J. Cruz, *Toxicon*, 2002, **40**, 901.
456. W. Nastainczyk, H. Meves and D.D. Watt, *Toxicon*, 2002, **40**, 1053.
457. F. Maggio and G.F. King, *Toxicon*, 2002, **40**, 1355.
458. C. Goudet, C.W. Chi and J. Tytgat, *Toxicon*, 2002, **40**, 1239.
459. L.M. Cortez, S.G. del Canto, F.D. Testai and M.J. Biscoglio de Jiménez Bonino, *Biochem. Biophys. Res. Commun.*, 2002, **295**, 791.
460. L. Dai, G. Corzo, H. Naoki, M. Andriantsiferana and T. Nakajima, *Biochem. Biophys. Res. Commun.*, 2002, **293**, 1514.
461. M. Nakamura, Y. Oba, T. Mori, K. Sato, Y. Ishida, T. Matsuda and H. Nakamura, *Biochem. Biophys. Res. Commun.*, 2002, **290**, 1037.
462. E. Naslund, S. Skogar, S. Efendic and P.M. Hellstrom, *Regulatory Peptides*, 2002, **106**, 89.
463. J.J. Meier, B. Gallwitz, W.E. Schmidt and M.A. Nauck, *Eur. J. Pharmacol.*, 2002, **440**, 269.
464. D. Smart and J.C. Jerman, *Pharmacology & Therapeutics*, 2002, **94**, 51.
465. T. Shi and D.L. Rabenstein, *Bioorg. Med. Chem. Lett.*, 2002, **12**, 2237.
466. Ashish and R. Kishore, *Bioorg. Med. Chem.*, 2002, **10**, 4083.

467. P. Grieco, A. Carotenuto, R. Patacchini, C.A. Maggi, E. Novellino and P. Rovero, *Bioorg. Med. Chem.*, 2002, **10**, 3731.
468. D.H. Coy, W.J. Rossowski and B.L. Cheng, et al., *Peptides*, 2002, **23**, 2259.
469. A. Ruhmann, J. Chapman, J. Higelin, B. Butscha and F.M. Dautzenberg, *Peptides*, 2002, **23**, 453.
470. K. Prokai-Tatrai, P. Perjesi, A.D. Zharikova, X.X. Li and L. Prokai, *Bioorg. Med. Chem. Lett.*, 2002, **12**, 2171.
471. R. Jain, J. Singh, J.H. Perlman and M.C. Gershengorn, *Bioorg. Med. Chem.*, 2002, **10**, 189.
472. M. Faria and H. Ulrich, *Curr Cancer Drug Targets*, 2002, **2**, 355.
473. N.O. Concha and S.S. Abdel-Meguid, *Curr. Med. Chem.*, 2002, **9**, 713.
474. J. Salgado, A.J. Garcia-Saez, G. Malet, I. Mingarro and E. Perez-Paya, *J. Pept. Sci.*, 2002, **8**, 543.
475. M. Bloomston, E.E. Zervos and A.S. Rosemurgy, *Ann. Surg. Oncol.*, 2002, **9**, 668.
476. F. Tosetti, N. Ferrari, S. De Flora and A. Albini, *FASEB J.*, 2002, **16**, 2.
477. L.M. Coussens, B. Fingleton and L.M. Matrisian, *Science.*, 2002, **295**, 2387.
478. M.A. Rudek, J. Venitz and W.D. Figg, *Pharmacotherapy*, 2002, **22**, 705.
479. H. Oikawa, *Curr. Med. Chem.*, 2002, **9**, 2033.
480. F. Markwardt, *Pathophysiol. Haemost. Thromb.*, 2002, **32**, 15.
481. B.S. Reddy and C.V. Rao, *J. Environ. Pathol. Toxicol. Oncol.*, 2002, **21**, 155.
482. M. Drag, J. Grembecka and P. Kafarski, *Phosphorus Sulfur and Silicon and the Related Elements*, 2002, **177**, 1591.
483. J. Grembecka, A. Mucha, T. Cierpicki and P. Kafarski, *Phosphorus, Sulfur and Silicon and the Related Elements*, 2002, **177**, 1739.
484. P.J. Rosenthal, *Curr. Opin. Hematol.*, 2002, **9**, 140.
485. D. Ferker, *Science*, 2002, **295**, 433.
486. U. Lendecke, M. Arndt, C. Wolke, D. Reinhold, T. Kahne and S. Ansorge, *J. Ethnopharmacol.*, 2002, **79**, 221.
487. P. Marshall, O. Heudi, S. Mckeown, A. Amour and F. Abou- Shakra, *Rapid Commun. Mass Spectrom.*, 2002, **16**, 220.
488. T. Akada, T. Yamazaki, H. Miyashita, O. Niizeki, M. Abe, A. Sato, S. Satomi and Y. Sato, *J. Cell Physiol.*, 2002, **193**, 253.
489. Y. Hashimoto, *Bioorg. Med. Chem. Lett.*, 2002, **10**, 461.
490. A.S. Waller and J.M. Clements, *Curr. Opin. Drug. Discov. Rev.*, 2002, **5**, 785.
491. L. Dubreuil, *Presse Med.*, 2002, **31**, 1810.
492. H.K. Smith, R.P. Beckett, J.M. Clements, S. Doel, S.P. East, S.B. Launchbury, L.M. Pratt, Z.M. Spavold, W. Thomas, R.S. Todd and M. Whittaker, *Bioorg. Med. Chem. Lett.*, 2002, **12**, 3595.
493. C.J. Hackbarth, D.Z. Chen, J.D. Lewis, K. Clark, J.B. Mangold, J.A. Cramer, P.S. Margolis, W. Wang, J. Koehn, C. Wu, S. Lopez, G. Withers, H. Gu, E. Dunn, R. Kulathila, S.H. Pan, W.L. Porter, J. Jacobs, J. Trias, D.V. Patel, B. Weidmann, R.J. White and Z. Yuan, *Antimicrob. Agents Chemother.*, 2002, **46**, 2752.
494. O.A. Philips and W.C. Matowe, *Curr. Opin. Investig. Drugs*, 2002, **3**, 1701.
495. B.J. Perrin and A. Huttenlocher, *Int. J. Biochem. Cell. Biol.*, 2002, **34**, 722.
496. P.B. DePetrillo, *Drugs*, 2002, **5**, 568.
497. Z. Jia, C.M. Hosfield, P.L. Davies and J.S. Elce, *Methods Mol. Biol.*, 2002, **172**, 51.
498. Y. Tomimatsu, S. Idemoto, S. Moriguchi, S. Watanabe and H. Nakanishi, *Life Sci.*, 2002, **72**, 355.
499. S.K. Ray, C.E. Dixon and N.L. Banik, *Histol. Histopathol.*, 2002, **17**, 1137.

500. R.S. Morrison, Y. Kinoshita, M.D. Johnson, S. Ghatan, J.T. Ho and G. Garden, *Adv. Exp. Med. Biol.*, 2002, **513**, 41.
501. A.T. McCollum, P. Nasr and S. Estus, *J. Neurochem.*, 2002, **82**, 1208.
502. N.O. Carragher, M.A. Westhoff, D. Riley, D.A. Potter, J.S. Elce, P.A. Greer and M.C. Frame, *Mol. Cell Biol.*, 2002, **22**, 257.
503. W. Lubisch and A. Moller, *Bioorg. Med. Chem. Lett.*, 2002, **12**, 1335.
504. M. Nakamura and J. Inoue, *Bioorg. Med. Chem. Lett*, 2002, **12**, 1603.
505. E. Mann, A. Chana, F. Sanchez-Sancho and O. Puerta, *Adv. Synt. Catal.*, 2002, **344**, 855.
506. A.A. Hernandez and W.R. Roush, *Curr Opin. Chem. Biol.*, 2002, **6**, 459.
507. J.D. Park, D.H. Kim, S.J. Kim, J.R. Woo and S.E. Ryu, *J. Med. Chem.*, 2002, **45**, 5295.
508. J.D. Park and D.H. Kim, *J. Med. Chem.*, 2002, **45**, 911.
509. M. Lee and D.H. Kim, *Bioorg. Med. Chem.*, 2002, **10**, 913.
510. W. Mock, L. Stanford and J. Daniel, *Bioorg. Med. Chem. Lett.*, 2002, **12**, 1193.
511. N.A. Dales, A.E. Gould, J.A. Brown, E.F. Calderwood, B. Guan, C.A. Minor, J.M. Gavin, P. Hales, V.K. Kaushik, M. Stewart, P.J. Tummino, C.S. Vickers, T.D. Ocain and M.A. Patane, *J. Am. Chem. Soc.*, 2002, **124**, 11852.
512. C. Lazure, *Curr. Pharm. Des.*, 2002, **8**, 511.
513. J. Salgado, A.J. García-Sáez, G. Malet, I. Mingarro and E. Pérez-Payá, *J. Pept. Sci.*, 2002, **8**, 543.
514. P.P. Graczyk, *Prog. Med. Chem.*, 2002, **39**, 1.
515. J. Bilsland and S. Harper, *Curr. Opin. Investig. Drugs*, 2002, **3**, 1745.
516. N.O. Concha and S.S. Abdel-Meguid, *Curr. Med. Chem.*, 2002, **9**, 713.
517. D.J. Lauffer and M.D. Mullican, *Bioorg. Med. Chem. Lett.*, 2002, **12**, 1225.
518. S.D. Linton, D.S. Karanewsky, R.J. Ternansky, J.C. Wu, B. Pham, L. Kodandapani, R. Smidt, J.-L. Diaz, L. C Fritz and K.J. Tomaselli, *Bioorg. Med. Chem. Lett.*, 2002, **12**, 2969.
519. D.J. DeGracia, R. Kumar, C.R. Owen, G.S. Krause and B.C. White, *J. Cereb. Blood Flow Metab.*, 2002, **22**, 127.
520. H.C. Chan, S.C. Kuo, S.C. Liu, C.H. Liu and S.L. Hsu, *Cancer Lett.*, 2002, **186**, 211.
521. C.N. O'Donovan, D. Tobin and T.G. Cotter, *J. Biol. Chem.*, 2001, **276**, 43516.
522. K. Ishidoh and E. Kominami, *Biol. Chem.*, 2002, **383**, 1827.
523. N. Levicar, T. Strojnik, J. Kos, R.A. Dewey, G.J. Pilkington and T.T. Lah, *J. Neurooncol.*, 2002, **58**, 21.
524. D. Bromme and J. Kaleta, *Curr. Pharm. Des.*, 2002, **8**, 1639.
525. R. Leung-Tound, W. Li, T.F. Tam and K. Karimian, *Curr. Med. Chem.*, 2002, **9**, 979.
526. Y. Yamamoto, M. Kurata, S. Watabe, R. Murakami and S.Y. Takahashi, *Curr. Prot. Pept. Sci.*, 2002, **3**, 231.
527. A.A. Hernandez and W.R. Roush, *Curr. Opin. Chem. Biol.*, 2002, **6**, 459.
528. P.M.S. Hilaire, L.C. Alves, F. Herrera, M. Renil, S.J. Sanderson, J.C. Mottram, G.H. Coombs, A. Maria, L. Juliano, J. Arevalo and M. Medal, *J. Med. Chem.*, 2002, **45**, 1971.
529. S.F. Chowdhury, J. Sivaraman, J. Wang, G. Devanathan, P. Lachance, H. Qi, R. Menard, J. Lefebvre, Y. Konishi, M. Cygler, T. Sulea and E.O. Purisima, *J. Med. Chem.*, 2002, **45**, 5321.
530. E. Altman, J. Renaud, J. Green, D. Farley, B. Cutting and W. Jahnke, *J. Med. Chem.*, 2002, **45**, 2352.

531. R.V. Mendonca, S. Venkatraman and J.T. Palmer, *Bioorg. Med. Chem. Lett.*, 2002, **12**, 2887.
532. K. Shindo, H. Suzuki and T. Okuda, *Biosci. Biotechnol. Biochem.*, 2002, **66**, 2444.
533. Y.D. Ward, D.S. Thomson, L.L. Frye, C.L. Cywin, T. Morwick, M.J. Emmanuel, R. Zindell, D. McNeil, Y. Bekkali, M. Girardot, M. Hrapchak, M. DeTuri, K. Crane, D. White, S. Pav, Y. Wang, M.H. Hao, C.A. Grygon, M.E. Labadia, D.M. Freeman, W. Davidson, J.L. Hopkins, M.L. Brown, D.M. Spero and M. Giradot, *J. Med. Chem.*, 2002, **45**, 5471.
534. A.J. Niestroj, K. Feussner, U. Heiser, P.M. Dando, A. Barrett, B. Gerhartz and H.U. Demuth, *Biol. Chem.*, 2002, **383**, 1205.
535. M. Wieczerzak, P. Drabik, L. Lankiewicz, S. Oldziej, Z. Grzonka, M. Abrahamson, A. Grubb and D. Bromme, *J. Med.. Chem.*, 2002, **45**, 4202.
536. N.E. Zhou, D. Gou, J. Kaleta, E. Purisima, R. Menard, R.G. Micetich and R. Singh, *Bioorg. Med. Chem. Lett.*, 2002, **12**, 3413.
537. N.E. Zhou, J. Kaleta, E. Purisima, R. Menard, R.G. Micetich and R. Singh, *Bioorg. Med. Cherm. Lett.*, 2002, **12**, 3417.
538. R.M. Rydzewski, C. Bryant, R. Oballa, G. Wesolowski, S.B. Rodan, K.E. Bass and D.H. Wong, *Bioorg. Med. Chem. Lett.*, 2002, **6**, 459.
539. N. Schaschke, D. Deluca, I. Assfalg-Machleidt, C. Hohneke, C.P. Sommerhoff and W. Machleidt, *Biol. Chem.*, 2002, **383**, 849.
540. M. Shibata, M. Koike, S. Waguri, G. Zhang, T. Koga and Y. Uchiyama, *FEBS Lett.*, 2002, **517**, 281.
541. M.N. Greco, M.J. Hawkins, E.T. Powell, H.R. Almond, Jr., T.W. Corcoran, L. de Garavilla, J.A. Kauffman, R. Recacha, D. Chattopadhyay, P. Andrade-Gordon and B.E. Maryanoff, *J. Am. Chem. Soc.*, 2002, **124**, 3810.
542. M. Yamamoto, S. Ikeda, H. Kondo and S. Inoue, *Bioorg. Med. Chem. Lett.*, 2002, **12**, 375.
543. L. Huang and J.A. Ellman, *Bioorg. Med. Chem. Lett.*, 2002, **12**, 2993.
544. L. Huang, A. Lee and J.A. Ellman, *J. Med. Chem.*, 2002, **45**, 676.
545. J.A. Urbina, *Curr. Pharm. Des.*, 2002, **8**, 287.
546. P. Xu, W. Lin and X. Zou, *Synthesis*, 2002, 1017.
547. A.D. Borthwick, A.J. Crame, P.F. Ertl, A.M. Exall, T.M. Haley, G.J. Hart, A.M. Mason, A.M. Pennell, O.M. Singh, G.G. Weingarten and J.M. Woolven, *J. Med. Chem.*, 2002, **45**, 1.
548. T.O. Johnson, Y. Hua, H.T. Luu, E.L. Brown, F. Chan, S.S. Chu, P.S. Dragovich, B.W. Eastman, R.A. Ferre, S.A. Fuhrman, T.F. Hendrickson, F.C. Maldonado, D.A. Matthews, J.W. Meador, A.K. Patick, S. H. Reich, D.J. Skalitzky, S.T. Worland, M. Yang and L.S. Zalman, *J. Med. Chem.*, 2002, **45**, 2016.
549. P.S. Dragovich, T.J. Prins, R. Zhou, E.L. Brown, F.C. Maldonado, S.A. Fuhrman, L.S. Zalman, T. Tuntland, C.A. Lee, A.K. Patick, D.A. Matthews, T.F. Hendrickson, M.B. Kosa, B. Liu, M.R. Batugo, J.P. Glesson, S.K. Sakata, L. Chen, M.C. Guzman, J.W. Meador, R.A. Ferre and S.T. Worland, *J. Med. Chem.*, 2002, **45**, 1607.
550. P.S. Dragovich, T.J. Prins, R. Zhou, T.O. Johnson, E.L. Brown, F.C. Maldonado, S.A. Fuhrman, L.S. Zalman, A.K. Patick, D.A. Matthews, X. Hou, J.W. Meador, R.A. Ferre and S.T. Worland, *Bioorg. Med. Chem. Lett.*, 2002, **12**, 733.
551. N. Venkatesan and B.H. Kim, *Curr. Med. Chem.*, 2002, **9**, 2243.
552. C. Fatini, F. Gensini, E. Sticchi, B. Battaglini, C. Angotti, M.L. Conforti, S. Generim, A. Pignone, R. Abbate and M. Matucci-Cerinic, *Am. J. Med.*, 2002, **112**, 540.

553. M.W. Mutahi, T. Nittoli, L. Guo and S.M. Sieburth, *J. Am. Chem. Soc.*, 2002, **124**, 7363.
554. N. Inguimbert, H. Poras, F. Teffo, F. Beslot, M. Selkti, A. Tomas, E. Scalbert, C. Bennejean, P. Renard, M.C. Fournié-Zaluski and B.P. Roques, *Bioorg. Med. Chem. Lett.*, 2002, **12**, 2001.
555. J. Kucharewicz, R. Pawlak, T. Matys, D. Pawlak and W. Buczko, *Hypertension*, 2002, **40**, 774.
556. T. Monteil, D. Danvy, M. Sihel, R. Leroux and J.C. Plaquevent, *Mini Rev. Med. Chem.*, 2002, **2**, 209.
557. R. Abdi, V.M. Dong, C.J. Lee and K.A. Ntoso, *Pharmacotherapy*, 2002, **22**, 1173.
558. A. Ahmed, *J. Am. Geriatr. Soc.*, 2002, **50**, 1293.
559. E. Dumez, J.S. Snaith, R.F. Jackson, A.B. McElroy, J. Overington, M.J. Wythes, J.M. Withka and T.J. McLellan, *J. Org. Chem.*, 2002, **67**, 4882.
560. A.J. Walz and M.J. Miller, *Org. Lett.*, 2002, **4**, 2047.
561. K.A. Packard, R.L. Wurdeman and A.J. Arouni, *Ann. Pharmacother.*, 2002, **36**, 1058.
562. G. P Rossi, S. Bova, A. Sacchetto, D. Rizzoni, E. Agabiti-Rosei, G. Neri, G.G. Nussdorfer and A.C. Pessina, *Am. J. Hypertens.*, 2002, **15**, 181.
563. M.A. Letavic, M.Z. Matt, J.T. Barberia, J. Thomas, D.E. Danley, K.F. Geoghegan, N.S. Halim, L.R. Hoth, V.A. Kamath, R. Ellen, L.L. Lopesti-Morrow, K.F. McClure, P.G. Mitchell, V. Natarajan, M.C. Noe, J. Pandit, L. Reeves, G.K. Schulte, S.L. Snow, F.J. Sweeney, D.H. Tan and C.H. Yu, *Bioorg. Med. Chem. Lett.*, 2002, **12**, 1387.
564. S.A. Doggrell, *Expert Opin. Investig. Drugs*, 2002, **11**, 1537.
565. A.Y. Jeng, P. Mulder, A.L. Kwan and B. Battistini, *Can. J. Physiol. Pharmacol.*, 2002, **80**, 440.
566. R. Veelken and R.E. Schmieder, *J. Hypertens.*, 2002, **20**, 599.
567. T.F. Luscher, *Int. J. Clin. Pract. Suppl.*, 2001, **3**.
568. E.A. Kitas, B.-M. Loffler, S. Daetwyler, H. Dehmlow and J.D. Aebi, *Bioorg. Med. Chem. Lett.*, 2002, **12**, 1727.
569. A.Y. Jeng, P. Savage, M.E. Beil, C.W. Bruseo, D. Hoyer, C. A Fink and A.J. Trapani, *Clin. Sci. (London)*, 2002, **103 Suppl 48**, 98S.
570. K. Yamazaki, H. Hasegawa, K. Umekawa, Y. Ueki, N. Ohashi and M. Kanaoka, *Bioorg. Med. Chem. Lett.*, 2002, **12**, 1275.
571. F. Firooznia, C. Gude, K. Chan, J. Tan, C.A. Fink, P. Svage, M. E. Beil, C.W. Bruseo, A.J. Trapani and A.Y. Jeng, *Bioorg. Med. Chem. Lett.*, 2002, **12**, 3059.
572. K. Kawabata, T. Hagio and S. Matsuoka, *Eur. J. Pharmacol.*, 2002, **451**, 1.
573. P.S. Hiemstra, *Biochem. Soc. Trans.*, 2002, **30**, 116.
574. F. Sato, Y. Inoue, T. Omodani, K. Imano, H. Okazaki, T. Takemura and M. Komiya, *Bioorg. Med. Chem. Lett.*, 2002, **12**, 551.
575. S.J. MacDonals, M.D. Dowle, L.A. Harrison, G.D. Clarke, G.G. Inglis, M.R. Johnson, P. Shah, R.A. Smith, A. Amour, G. Fleetwood, D.C. Humphreys, C.R. Molloy, M. Dixon, R.E. Godward, A.J. Wonacott, O.M. Singh, S.T. Wonacott, O.M. Singh, S.T. Hodgson and G.W. Hardy, *J. Med. Chem.*, 2002, **45**, 3878.
576. A. Napolitano, I. Bruno, P. Rovero, R. Lucas, M.P. Peris, L. Gomez-Paloma and R. Riccio, *J. Pept. Sci.*, 2002, **8**, 407.
577. K. Hilpert, J. Schneider-Mergener and J. Ay, *Acta Crystallogr. D. Biol. Crystallogr.*, 2002, **58**, 672.
578. S. Ayral-Kaloustian and E.J. Salaski, *Curr. Med. Chem.*, 2002, **9**, 1003.
579. A.A. Adjei, *Cancer Chemother. Biol. Response Modif.*, 2002, **20**, 151.

580. P. Haluska, G.K. Dy and A.A. Adjei, *Eur. J. Cancer*, 2002, **38**, 1685.
581. J. Ohkanda, D.B. Knowles, M.A. Blaskovich, S. M Sebti and A.D. Hamilton, *Curr. Top. Med. Chem.*, 2002, **2**, 303.
582. A.D. Cox and C.J. Der, *Curr. Opin. Pharmacol.*, 2002, **2**, 388.
583. W.T. Purcell and R.C. Donehower, *Curr. Oncol. Rep.*, 2002, **4**, 29.
584. K.N. Cho and K.I. Lee, *Arch. Pharm. Res.*, 2002, **25**, 759.
585. J.A. Gietema and E.G. de Vries, *Eur. Resist. Updat.*, 2002, **5**, 192.
586. A. Mitsch, M. Bohm, P. Wissner, I. Sattler and M. Schlitzer, *Bioorg. Med. Chem.*, 2002, **10**, 2657.
587. M. Schlitzer, M. Bohm and I. Sattler, *Bioorg. Med. Chem.*, 2002, **10**, 615.
588. C. Zhou, Y. Shao and R.A. Gibbs, *Bioorg. Med. Chem. Lett.*, 2002, **12**, 1417.
589. D.N. Nguyen, C.A. Stump, E.S. Walsh, C. Fernandes, J.P. Davide, M. Ellis-Hutchings, R.G. Robinson, T.M. Williams, R.B. Lobell, H.E. Huber and C.A. Buser, *Bioorg. Med. Chem. Lett.*, 2002, **12**, 1269.
590. F. Megnin-Chanet, F. Lavelle and V. Favaudon, *BMC Pharmacol.*, 2002, **2**, 2.
591. R.R. Mattingly, R.A. Gibbs, R.E. Menard and J.J. Reiners, Jr., *J. Pharmacol. Exp. Ther.*, 2002, **303**, 74.
592. N. Bogetto and M. Reboud-Ravaux, *Biol. Chem.*, 2002, **383**, 1321.
593. K. Hidaka, T. Kimura, Y. Hayashi, K.F. McDaniel, T. Dekhtyar, L. Colletti and Y. Kiso, *Bioorg. Med. Chem. Lett.*, 2002, **13**, 93.
594. S. Rajesh, E. Ami, T. Kotake, T. Kimura, Y. Hayashi and Y. Kiso, *Bioorg. Med. Chem. Lett.*, 2002, **12**, 3615.
595. Y. Hamada, J. Ohtake, Y. Sohma, T. Kimura, Y. Hayashi and Y. Kiso, *Bioorg. Med. Chem. Lett.*, 2002, **10**, 4155.
596. M. Ikunaka, J. Matsumoto and Y. Nishimoto, *Tetrahedron: Asymmetry*, 2002, **13**, 1201.
597. B.V. Murphy, J.L. O'Brien, L.J. Gorey-Feret and A.B. Smith, *Bioorg. Med. Chem. Lett.*, 2002, **12**, 1763.
598. B.C. H. May, A.D. Abell and D. Andrew, *J. Chem. Soc., Perkin Trans. 1*, 2002, 172.
599. B. Saha, J.P. Nandy, S. Shukla, I. Siddiqui and J. Iqbal, *J. Org. Chem*, 2002, **67**, 7858.
600. Z. Xu, J. Singh, M.D. Schwinden, B. Zheng, T.P. Kissick, B. Patel, M.J. Humora, F. Quiroz, L. Dong, D.-M. Hsieh, J.E. Heikes, M. Pudipeddi, M.D. Lindrud, S.K. Srivastava, D.R. Kronenthal and R.H. Mueller, *Org. Process Res. Dev.*, 2002, **6**, 323.
601. F. Benedetti, F. Berti and S. Norbedo, *J. Org. Chem.*, 2002, **67**, 8635.
602. H.L. Sham, C. Zhao, D.A. Betebenner, A. Saldivar, S. Vasavanonda, D.J. Kempf, J.J. Plattner and D.W. Norbeck, *Bioorg. Med. Chem. Lett.*, 2002, **12**, 3101.
603. H.L. Sham, D.A. Betebenner, X. Chen, A. Saldivar, S. Vasavanonda, D.J. Kempf, J.J. Plattner and D.W. Norbeck, *Bioorg. Med. Chem. Lett.*, 2002, **12**, 1185.
604. A. Brik, Y.C. Lin and J. Elder. C.H. Wong, *Chem. Biol.*, 2002, **9**, 891.
605. X.M. O'Brien, J.A. Parker, P.A. Lessard and A.J. Sinskey, *Appl. Microbiol. Biotechnol.*, 2002, **59**, 389.
606. J.L. Duffy, N.J. Kevin, B.A. Kirk, K.T. Chapman, W.A. Schleif, D.B. Olsen, M. Stahlhut, C.A. Rutkowski, L.C. Kou, L. Jin, J.H. Lin, E.A. Emini and J.R. Tata, *Bioorg. Med. Chem. Lett.*, 2002, **12**, 2423.
607. Y. Cheng, F. Zhang, T.A. Rano, Z. Lu, W.A. Schleif, L. Gabryelski, D.B. Olsen, M. Stahlhut, C.A. Rutkowski, J.H. Lin, L. Jin, E.A. Emini, K.T. Chapman and J.R. Tata, *Bioorg. Med. Chem. Lett.*, 2002, **12**, 2419.

608. L. Rocheblave, F. Bihel, C. De Michelis, G. Priem, J. Courcambeck, B. Bonnet, J.C. Chermann and J.L. Kraus, *J. Med. Chem.*, 2002, **45**, 3321.
609. A.H. Corbett and A.D. Kashuba, *Curr. Opin. Investig. Drugs*, 2002, **3**, 384.
610. F. Lebon, N. Boggetto, M. Ledecq, F. Durant, Z. Benatallah, S. Sicsis, R. Lapouyade, O. Kahn, A. Mouithys-Mickalad, G. Deby-Dupont and M. Reboud-Ravaux, *Biochem. Pharmacol.*, 2002, **63**, 1863.
611. B.M. Sadler and D.S. Stein, *Ann. Pharmacother.*, 2002, **36**, 102.
612. N.M. King, L. Melnick, M. Prabu-Jeyabalan, E.A. Nalivaika, S.S. Yang, Y. Gao, X. Nie, C. Zepp, D.L. Heefner and C.A. Schiffer, *Protein Sci.*, 2002, **11**, 418.
613. X. Huang, L. Xu, X. Luo, K. Fan, R. Ji, G. Pei, K. Chen and H. Jiang, *J. Med. Chem.*, 2002, **45**, 333.
614. A.H. Baker, D.R. Edwards and G. Murphy, *J. Cell Sci.*, 2002, **115**, 3719.
615. G.A. Rosenberg, *Glia*, 2002, **39**, 279.
616. S. Das, T. Chakraborti, M. Mandal, A. Mandal and S. Chakraborti, *Mol. Cell. Biochem.*, 2002, **237**, 85.
617. M.A. Rudek, J. Venitz and W.D. Figg, *Pharmacotheraphy*, 2002, **22**, 705.
618. L.M. Coussens, B. Fingleton and L.M. Matrisian, *Science*, 2002, **29**, 2387.
619. D. Krumme and H. Tschesche, *Bioorg. Med. Chem. Lett.*, 2002, **12**, 933.
620. T. Fujisawa, S. Odake, Y. Ogawa, J. Yasuda and T. Morikawa, *Pept. Sci.* (pub. 2002) 2001, **38**, 171.
621. A. Reichelt, C. Gaul, R.R. Frey, A. Kennedy and S.F. Martin, *J. Org. Chem.*, 2002, **67**, 4062.
622. S. Yu, J. Saenz and J.K. Srirangam, *J. Org. Chem.*, 2002, **67**, 1699.
623. S. Banerjee, C. Bueso-Ramos and B.B. Aggarwal, *Cancer Res.*, 2002, **62**, 4945.
624. C. Garcia, D.U. Bartsch, M.E. Rivero, M. Hagedorn, C.D. McDermott, G. Bergeron-Lynn, L. Cheng, K. Appelt and W.R. Freeman, *Curr. Eye Res.*, 2002, **24**, 33.
625. L.J. Wang, M.L. Grapperhaus, B.S. Pitzele, T.J. Hagen, K.F. Fok, A.J. Scholten, D.P. Spangler, M.V. Toth, G.M. Jerome, W.M. Moore, P.T. Manning and J.A. Sikorski, *Heteroatom Chem.*, 2002, **13**, 77.
626. E.A. Hallinan, S. Tsymbalov, C.R. Dorn, B.S. Pitzele, D.W. Hansen, Jr., W.M. Moore, G.M. Jerome, J.R. Connor, L.F. Branson, D.L. Widomski, Y. Zhang, M.G. Currie and P.E. Manning, *J. Med. Chem.*, 2002, **45**, 1688.
627. L.T. Tsao, C.Y. Lee, L.J. Huang, S.C. Kuo and J.P. Wang, *Biochem. Pharmacol.*, 2002, **63**, 1961.
628. G. Leoncini, R. Pascale and M.G. Signorello, *Biochem. Pharmacol.*, 2002, **64**, 277.
629. D.M. Smith, Z. Wang, A. Kazi, L.H. Li, T.H. Chan and Q.P. Dou, *Mol. Med.*, 2002, **8**, 382.
630. M.P. Green, A.C. Prodger and C.J. Hayes, *Tetrahedron Lett.*, 2002, **43**, 6609.
631. M. Kaiser, M. Groll, C. Renner, R. Huber and L. Moroder, *Angewandte Chemie International Edition*, 2002, **41**, 780.
632. S. Lin and S.J. Danishefsky, *Angewandte Chemie International Edition*, 2002, **41**, 512.
633. T. Hu and J.S. Panek, *J. Am. Chem. Soc.*, 2002, **124**, 11368.
634. D.S. Lewy, C.M. Gauss, D.R. Soenen and D.L. Boger, *Curr. Med. Chem.*, 2002, **9**, 2005.
635. T. Esumi, N. Okamoto and S. Hatakeyama, *Chem. Commun. (Camb.)*, 2002, **3042**.
636. Y.K. Reddy and J.R. Flack, *Org. Lett.*, 2002, **4**, 969.
637. A.B. Dounay and C.J. Forsyth, *Curr. Med. Chem.*, 2002, **9**, 1939.
638. J.A. Sakoff, S.P. Ackland, M.L. Baldwin, M.A. Keane and A. McCluskey, *Invest. New Drugs*, 2002, **20**, 1.

639. D. Brohm, N. Philippe, S. Metzger, A. Bhargava, O. Muller, F. Lie and H. Waldmann, *J. Am. Chem. Soc.*, 2002, **124**, 13171.
640. G. Hum, J. Lee and S.D. Taylor, *Bioorg. Med. Chem. Lett.*, 2002, **12**, 3471.
641. G. Arabaci, T. Yi, H. Fu, M.E. Porter, K.D. Beebe and D. Pei, *Bioorg. Med. Chem. Lett.*, 2002, **12**, 3047.
642. H.S. Andersen, O.H. Olsen, L.F. Iversen, A.L. Sorensen, S.B. Mortensen, M.S. Christensen, S. Branner, T.K. Hansen, J.F. Lau, L. Jeppesen, E.J. Moran, J. Su, F. Bakir, L. Judge, M. Shahbaz, T. Collins, T. Vo, M.J. Newman, W.C. Ripka and N.P. Moller, *J. Med. Chem.*, 2002, **45**, 4443.
643. S. Soeda, T. Shimada, S. Koyanagi, T. Yokomatsu, T. Murano, S. Shibuya and H. Shimeno, *FEBS Lett.*, 2002, **524**, 54.
644. C. Liljebris, S.D. Larsen, D. Ogg, B. Palazuk and J.E. Bleasdale, *J. Med. Chem.*, 2002, **45**, 1785.
645. S.D. Larsen, T. Barf, C. Liljebris, P.D. May, D. Ogg, T.J. O'Sullivan, B.J. Palazuk, H.J. Schostarez, F.C. Stevens and J.E. Bleasdale, *J. Med. Chem.*, 2002, **45**, 598.
646. W.R. Li, T.S. Lin, N.M. Hsu and M.S. Chern, *J. Org. Chem.*, 2002, **67**, 4702.
647. N.J. Reiter, D.J. White and F. Rusnak, *Biochemistry*, 2002, **41**, 1051.
648. A.V.S. Sarma, M.H.V.R. Rao, J.A.R.P. Sarma, R. Nagaraj, A.S. Dutta and A.C. Kunwar, *J. Biochem. Biophys. Methods*, 2002, **51**, 27.
649. S. Hanessian, S. Claridge and S. Johnstone, *J. Org. Chem.*, 2002, **67**, 426.
650. R. Paruszewski, P. Jaworski, I. Winiecka, J. Tautt and J. Dudkiewicz, *Chem. Pharm. Bull.*, 2002, **50**, 850.
651. L.C. Dias, A.A. Ferreira and G. Diaz, *Synlett*, 2002, 1845.
652. M. Chino, M. Wakao and J. Ellman, *Tetrahedron*, 2002, **58**, 6305.
653. Z. Kucerova and M. Ticha, *J. Chromatogr. B. Anal, Technologies in the Biomed. and Life Sciences*, 2002, **770**, 121.
654. W. Bruinzeel, J. Yon, S. Giovannelli and S. Masure, *Prot. Expres. Purif.*, 2002, **26**, 139.
655. L. Hong, R.T. Turner, G. Koelsch, A.K. Ghosh and J. Tang, *Biochemistry*, 2002, **41**, 10963.
656. M. Citron, *Neurobiol. Aging*, 2002, **23**, 1017.
657. M. Citron, *J. Neurosci. Res.*, 2002, **70**, 373.
658. L. Hong, R.T. Turner, G. Koelsch, A.K. Ghosh and J. Tang, *Biochem. Soc. Trans.*, 2002, **30**, 530.
659. A.K. Ghosh, L. Hong and J. Tang, *Curr. Med. Chem.*, 2002, **9**, 1135.
660. S. Roggo, *Curr. Top. Med. Chem.*, 2002, **2**, 359.
661. R.T. Turner, J.A. Loy, C. Nguyen, T. Devasamudram, A.K. Ghost, G. Koelsch and J. Tang, *Biochemistry*, 2002, **41**, 8742.
662. J.S. Tung, D.L. Davis, J.P. Anderson, E.D. Walker, S. Mamo, N. Jewett, R.K. Hom, S. Sinha, E.D. Thorsett and V. John, *J. Med. Chem. Soc.*, 2002, **45**, 259.
663. F. Gruninger-Leitch, D. Schlatter, E. Kung, P. Nelbock and H. Dobelo, *J. Biol. Chem.*, 2002, **277**, 4687.
664. J.Y. Tsai, M.S. Wolfe and W. Xia, *Curr. Med. Chem.*, 2002, **9**, 1087.
665. M. Xu, M.T. Lai, Q. Huang, J. DiMuzio-Mower, J.L. Castro, T. Harrison, A. Nadin, J.G. Neduvelil, M.S. Shearman, J.A. Shafer, S.J. Gardell and Y.M. Li, *Neurobiol. Aging*, 2002, **23**, 1023.
666. D. Beher, J.D. Wrigley, A.P. Owens and M.S. Shearman, *J. Neurochem.*, 2002, **82**, 563.
667. A. Geling, H. Steiner, M. Willem, L. Bally-Cuif and C. Haass, *EMBO Rep.*, 2002, **3**, 688.

668. A. Kinoshita, C.M. Whelan, O. Berezovska and B.T. Hyman, *J. Biol. Chem.*, 2002, **277**, 28530.
669. M. Shizuka-Ikeda, E. Matsubara, M. Ikeda, M. Kanai, Y. Tomidokoro, Y. Ikeda, M. Watanabe, T. Kawarabayashi, Y. Harigaya, K. Okamoto, K. Maruyama, E.M. Castano, P.S. George-Hyslop and M. Shoji, *Biochem. Biophys. Res. Commun.*, 2002, **292**, 571.
670. F. Markwardt, *Pathophysiol. Haemost. Thromb.*, 2002, **32**, *Suppl. 3*, 23.
671. P. Thiagarajan, *Am. J. Cardiovasc. Drugs*, 2002, **2**, 227.
672. E. Nutescu and E. Racine, *Am. J. Health-Syst. Pharm.*, 2002, **15**(20 *Suppl.* 6), S7.
673. M.M. Samama, *Thromb. Res.*, 2002, **106**, V267.
674. J.M. Walenga, *Pathophysiol. Haemost. Thromb.*, 2002, **32**, Suppl. 3, 9.
675. S. Kathiresan, J. Shiomura and I.K. Jang, *J. Thrombolysis*, 2002, **13**, 41.
676. M. Petitou, P. Duchaussoy, J.M. Herbert, G. Duc, M. El Hajji, G.F. Branellec, F. Donac, J. Necciari, R. Cariou, J. Boutier and E. Garrigou, *Semin. Thromb. Hemost.*, 2002, **28**, 393.
677. K.A. Bauer, D.W. Hawkins, P.C. Peters, M. Petitou, J.M. Herbert, C.A. van Boeckel and D.G. Meuleman, *Cardiovasc. Drug Rev.*, 2002, **20**, 37.
678. J.M. Walenga, W.P. Jeske, M.M. Samama, F.X. Frapaise, R.L. Bick and J. Fareed, *Exp. Opin. Investig. Drugs*, 2002, **11**, 397.
679. K.A. Bauer, *Am. J. Orthop.*, 2002, **31**(11 Suppl.), 4.
680. P.C. Wong, D.J. Pinto and R.M. Knabb, *Cardiovasc. Drug Rev.*, 2002, **20**, 137.
681. T.D. Gladwell, *Clin. Ther.*, 2002, **24**, 38.
682. A. Dahlgren, J. Branalt, I. Kvarnstrom, I. Nilsson, D. Musil and B. Samuelsson, *Bioorg. Med. Chem. Lett.*, 2002, **10**, 1567.
683. S. Hanessian, R. Margarita, A. Hall, S. Johnstone, M. Tremblay and L. Parlanti, *J. Am. Chem. Soc.*, 2002, **124**, 13342.
684. A.R. Carroll, G.K. Pierens, G. Fechner, P. de Leone, A. Ngo, M. Simpson, E. Hyde, J.N.A. Hooper, S.-T. Bostroem, D. Musil and R.J. Quinn, *J. Am. Chem. Soc.*, 2002, **124**, 13340.
685. N. Lin, M. Zhao, C. Wang and S. Peng, *Bioorg. Med. Chem. Lett.*, 2002, **12**, 585.
686. K. Belekoukias, J. Sarigiannis, K.G. Stavropoulos and T. Mavromoustakos, *Pharmakeutike*, 2002, **15**, 2.
687. J.Z. Ho, T.S. Gibson and E.J. Semple, *Bioorg. Med. Chem. Lett.*, 2002, **12**, 743.
688. A.E.P. Adang, A.P.A. de Man, G.M.T. Vogel, P.D.J. Grootenhuis, M.J. Smit, C.A.M. Peters, A. Visser, J.B.M. Rewinkel, T. Van Dinther, H. Lucas, J. Kelder, S. Van Aelst, D.G. Meuleman, C.A.A. van Boeckel and A.A. Constant, *J. Med. Chem.*, 2002, **45**, 4419.
689. M. Czekaj, S.I. Klein, K.R. Guertin, C.I. Gardner, A.L. Zulli, H.W. Pauls, A.P. Spada, D.L. Cheney, K.D. Brown, D.J. Colusi, V. Chu, R.J. Leadley and C. Dunwiddie, *Bioorg. Med. Chem. Lett.*, 2002, **12**, 1667.
690. H. Nishida, Y. Miyazaki, T. Mukaihira, F. Saitoh, M. Fukui, K. Harada, M. Itoh, A. Muraoka, T. Matsusue, A. Okamoto, Y. Hosaka, M. Matsumoto, S. Ohnishi and H. Mochizuki, *Chem. Pharm. Bull.*, 2002, **50**, 1187.
691. H.P. Ng, B.O. Buckman, K.A. Eagen, W.J. Guilford, M.J. Kochanny, R. Mohan, K.J. Shaw, S.C. Wu, D. Lentz, A. Liang, L. Trinh, E. Ho, D. Smith, B. Subramanyam, R. Vergona, J. Walters, K.A. White, M.E. Sullivan, M.M. Morrissey and G.B. Phillips, *Bioorg. Med. Chem.*, 2002, **10**, 657.
692. G. Simonet, I. Claeys and J.V. Broeck, *Comp. Biochem. Physiol. B - Biochem. Mol. Biol.*, 2002, **132**, 247.
693. H. Noguchi, T. Aoyama and T. Shioiri, *Heterocycles*, 2002, **58**, 471.

694. A. Melo and M.J. Ramos, *THEOCHEM*, 2002, **580**, 251.
695. P. Fraszczak, K. Kazmierczak, M. Stawikowski, A. Jaskiewicz, G. Kupryszewski and K. Rolka, *Polish. J. Chem.*, 2002, **76**, 1441.
696. P. Fraszczak, K. Rolka and G. Kupryszewski, *Polish J. Chem.*, 2002, **76**, 519.
697. A. Mucha and P. Kafarski, *Tetrahedron*, 2002, **58**, 5855.
698. S. Colarusso, B. Gerlach, U. Koch, E. Muraglia, I. Conte, I. Stansfield, V.G. Matassa and F. Narjes, *Bioorg. Med. Chem. Lett.*, 2002, **12**, 705.
699. L. Banfi, G. Guanti, R. Riva, A. Basso and E. Calcagno, *Tetrahedron Lett.*, 2002, **43**, 4067.
700. N.D. Pearson, D.S. Eggleston, R.C. Haltiwanger, M. Hibbs, A.J. Laver and A.C. Kaura, *Bioorg. Med. Chem. Lett.*, 2002, **12**, 2359.
701. A. Madder, L. Li, H. De Muynck, N. Farcy, D. Van Haver, F. Fant, G. Vanhonacker, P. Sandra, A.P. Davis and P.J. De Clercq, *J. Comb. Chem.*, 2002, **4**, 552.
702. D. Fontanini and B.L. Jones, *Planta*, 2002, **215**, 885.
703. M.J. Drysdale, G. Lentzen, N. Matassova, A.I. Muechie, F. Aboul-Ela and D.M. Afshar, *Prog. Med. Chem.*, 2002, **39**, 73.
704. T. Shimizu, T. Usui, K. Machida, K. Furuya, H. Osada and T. Nakata, *Bioorg. Med. Chem. Lett.*, 2002, **12**, 3363.
705. M.J. Brown, P.S. Carter, A.S. Fenwick, A.P. Fosberry, D.W. Hamprecht, M.J. Hibbs, R.L. Jarvest, L. Mensah, P.H. Milner, P.J. O'Hanlon, A.J. Pope, C.M. Richardson, A. West and D.R. Wiity, *Bioorg. Med. Chem. Lett.*, 2002, **12**, 317.
706. M. Kitabatake, K. Ali, A. Demain, K. Sakamoto, S. Yokoyama and D. Soll, *J. Biol. Chem.*, 2002, **277**, 23882.
707. N.M. Carballeira, J.E. Betancourt, E.A. Orellano and F.A. Gonzalez, *J. Nat. Prod.*, 2002, **65**, 1715.
708. A. Piettre, E. Chevenier, C. Massardier, Y. Gimbert and A.E. Greene, *Org. Lett.*, 2002, **4**, 3139.
709. E.I. Sbar, J. Khatri, W.D. Rodman, L. Tritschler, J. Goldberg, G. Grana, L. Devereux and A. Hageboutros, *Cancer Invest.*, 2002, **20**, 644.
710. P. Daza, J. Torreblanca, G. Garcia-Herdugo and F.J. Moreno, *Cell. Biol. Int.*, 2002, **26**, 707.
711. M.S. Hossain, N. Akimitsu, T. Takaki, H. Hirai and K. Sekimizu, *Genes Cells*, 2002, **7**, 285.
712. Y.B. Dong, H.L. Yang, M.J. Elliott and K.M. McMasters, *Cancer Res.*, 2002, **62**, 1776.
713. N.O. Karpinich, M. Tafani, R.J. Rothman, M.A. Russo and J.L. Farber, *J. Biol. Chem.*, 2002, **277**, 16547.
714. W. Wrasidlo, U. Schroder, K. Bernt, N. Hubener, D. Shabat, G. Gaedicke and H. Lode, *Bioorg. Med. Chem. Lett.*, 2002, **12**, 557.
715. S. Wang, W. Miller, J. Milton, N. Vicker, A. Stewart, P. Charlton, P. Mistry, D. Hardick and W.A. Denny, *Bioorg. Med. Chem. Lett.*, 2002, **12**, 415.
716. J.M. Barret, M. Cadou and B.T. Hill, *Biochem. Pharmacol.*, 2002, **63**, 251.
717. T. Muto and H. Fukami, *J. Drugs*, 2002, **5**, 1141.
718. S. Takai and M. Miyazaki, *Jpn. J. Pharmacol.*, 2002, **90**, 223.
719. S. Takai and M. Miyazaki, *Drug News Perspect.*, 2002, **15**, 278.
720. Y. Sukenga, K. Kamoshita, S. Takai and M. Miyazaki, *Jpn. J. Pharmacol.*, 2002, **90**, 218.
721. T. Imada, N. Komorita, F. Kobayashi, K. Naito, T. Yoshikawa, M. Miyazaki, N. Nakamura and T. Kondo, *Jpn. J. Pharmacol.*, 2002, **90**, 214.

722. M. Nakajima and N. Naya, *Jpn. J. Pharmacol.*, 2002, **90**, 206.
723. Y. Koide, A. Tatsui, T. Hasegawa, A. Murakami, S. Satoh, H. Yamada, S. Kazayama and A. Takahashi, *Bioorg. Med. Chem. Lett.*, 2002, **13**, 25.
724. Y. Okamoto, S. Takai and M. Miyazaki, *J. Surg. Res.*, 2002, **107**, 219.
725. Y. Okamoto, S. Takai and M. Miyazaki, *Jpn. J. Pharmacol.*, 2002, **90**, 94.
726. Y. Okamoto, S. Takai, M. Yamada and M. Miyazaki, *Fertil. Steril.*, 2002, **77**, 1044.
727. D. Jin, S. Takai, M. Yamada, M. Miyazaki and M. Sakaguchi, *Life Sci.*, 2002, **71**, 437.
728. E. Bianchi and A. Pessi, *Biopolymers*, 2002, **66**, 101.
729. R.A. Reinke, P.J. King, J.G. Victoria, B.R. McDougall, G. Ma, Y. Mao, M.G. Reinecke and W.E. Robinson, Jr., *J. Med. Chem.*, 2002, **45**, 3669.
730. M.S. Lall, Y.K. Ramtohul, M.N.G. James and J.C. Vederas, *J. Org. Chem.*, 2002, **67**, 1536.
731. E.S. Priestley, I. De Lucca, B. Ghavimi, S. Erickson-Viitanem and C.P. Decicco, *Bioorg. Med. Chem. Lett.*, 2002, **12**, 3199.
732. R. Beevers, M.G. Carr, P.S. Jones, S. Jordan, P.B. Kay, R.C. Lazell and T.M. Raynham, *Bioorg. Med. Chem. Lett.*, 2002, **12**, 641.
733. S.J. Archer, D.M. Camac, Z.J. Wu, N.A. Farrow, P.J. Domaille, Z.R. Wasserman, M. Bukhtiyarova, C. Rizzo, S. Jagannathan, I.J. Mersinger and C.A. Kettner, *Chem. Biol.*, 2002, **9**, 79.
734. F. Narjes, K.F. Koehler, U. Koch, B. Gerlach, S. Colarusso, C. Steinkuhler, M. Brunetti, S. Altamura, R. De Francesco and V.G. Matassa, *Bioorg. Med. Chem. Lett.*, 2002, **12**, 701.
735. D. Sperandio, A.R. Gangloff, J. Litvak, R. Goldsmith, J.M. Hataye, V.R. Wang, E.J. Shelton, K. Elrod, J.W. Janc, J.M. Clark, K. Rice, S. Weinheimer, K.S. Yeung, N.A. Meanwell, D. Hernandez, A.J. Staab, B.L. Venables and J.R. Spencer, *Bioorg. Med. Chem. Lett.*, 2002, **12**, 3129.
736. Y.K. Ramatohul, M.N.G. James and J.C. Vederas, *J. Org. Chem.*, 2002, **67**, 3169.
737. S. Hanessian, J. Wang, D. Montgomery, V. Stoll, K.D. Stewart, W. Kati, C. Maring, D. Kempf, C. Hutchins and W.G. Laver, *Bioorg. Med. Chem. Lett.*, 2002, **12**, 3425.
738. S. Hanessian, M. Bayrakdarian and X. Luo, *J. Am. Chem. Soc.*, 2002, **124**, 4716.
739. D.A. DeGoey, H.J. Chen, W.J. Flosi, D.J. Grampovnik, C.M. Yeung, L.L. Klein and D.J. Kempf, *J. Org. Chem.*, 2002, **67**, 5445.
740. R.W. Sidwel and D.F. Smee, *Exp. Opin. Invest. Drugs*, 2002, **11**, 859.
741. B. Masereel, S. Rolin, F. Abbate, A. Scozzafava and C.T. Supuran, *J. Med. Chem.*, 2002, **45**, 312.
742. A. Scozzafava and C.T. Supuran, *J. Med. Chem.*, **45**, 284.
743. A. Scozzafava and C.T. Supuran, *Bioorg. Med. Chem. Lett.*, 2002, **12**, 1177.
744. O. Arslan, U. Cakir and H.I. Ugras, *Biochemistry*, 2002, **67**, 1055.
745. M. Bulbul, N. Saracoglu, O.L. Kufrevioglu and M. Ciftci, *Bioorg. Med. Chem.*, 2002, **10**, 2561.
746. J.D. Cheeseman, A.D. Corbett, R. Shu, J. Croteau, J.L. Gleason and R.J. Kazlauskas, *J. Am. Chem. Soc.*, 2002, **124**, 5692.
747. S. Wittich, H. Scherf, C. Xie, G. Brosch, P. Loidl, C. Gerhauser and M. Jung, *J. Med. Chem.*, 2002, **45**, 3296.
748. Z. Chen, S. Clark, M. Birkeland, C.M. Sung, A. Lago, R. Liu, R. Kirkpatrick, K. Johanson, J.D. Winkler and E. Hu, *Cancer Lett.*, 2002, **188**, 127.
749. F. Blanchard, E. Kinzie, Y. Wang, L. Duplomb, A. Godard, W.A. Held, B.B. Asch and H. Baumann, *Oncogene*, 2002, **21**, 6264.

750. A. Mai, S. Massa, R. Ragno, M. Esposito, G. Sbardella, G. Nocca, R. Scatena, F. Jesacher, P. Loidl and G. Brosch, *J. Med. Chem.*, 2002, **45**, 1778.
751. F. Leoni, A. Zaliani, G. Bertolini, G. Porro, P. Pagani, P. Pozzi, G. Dona, G. Fossan, S. Sozzani, T. Azam, P. Bufler, G. Fantuzzi, I. Goncharov, S.H. Kim, B.J. Pomerantz, L.L. Reznikov, B. Siegmund, C.A. Dinarello and P. Mascagni, *Proc. Nat. Acad. Sci. USA*, 2002, **99**, 2995.
752. S.H. Kwon, S.H. Ahn, Y.K. Kim, G.U. Bae, J.W. Yoon, S. Hong, H.Y. Lee, Y.W. Lee, H.W. Lee and J.W. Han, *J. Biol. Chem.*, 2002, **277**, 2073.
753. S.G. Bowers, T. Mahmud and H.G. Floss, *Carbohydr. Res.*, 2002, **337**, 297.
754. B. Kundu, S.K. Rastogi, R. Ahmad and A.K. Srivastava, *Comb. Chem. High Throughput Screen.*, 2002, **5**, 545.
755. T. Yamashita, K. Yasuda, H. Kizu, Y. Kameda, A.A. Watson, R.J. Nash, G.W. Fleet and N. Asano, *J. Nat. Prod.*, 2002, **65**, 1875.
756. V.P. Sandanayaka and A.S. Prashad, *Curr. Med. Chem.*, 2002, **9**, 1145.
757. P. De Macedo, C. Marrano and J.W. Keillor, *Bioorg. Med. Chem.*, 2002, **10**, 355.
758. E.A. Wallen, J.A. Christiaans, M.M. Forsberg, J.I. Venalainen, P.T. Mannisto and J. Gynther, *J. Med. Chem.*, 2002, **45**, 4581.
759. T. Miki, M. Kori, H. Mabuchi, R.-I. Tozawa, T. Nishimoto, Y. Sugiyama, K. Teshima and H. Yukimasa, *J. Med. Chem.*, 2002, **45**, 4571.
760. F.E. Oualid, L. Bruining, I.M. Leroy, L.H. Cohen, J.H. van Boom, G.A. van der. Marel, H.S. Overkleeft and M. Overhand, *Helvetica Chimica Acta*, 2002, **85**, 3455.
761. S. Sunami, M. Ohkubo, T. Sagara, J. Ono, S. Asahi, S. Koito and H. Morishima, *Bioorg. Med. Chem. Lett.*, 2002, **12**, 629.
762. M.J. Rudolph, C.R. Illig, N.L. Subasinghe, K.J. Wilson, J.B. Hoffman, T. Randle, D. Green, C.J. Molloy, R.M. Soll, F. Lewandowski, M. Zhang, R. Bone, J.C. Spurlino, I.C. Deckman, C. Manthey, C. Sharp, D. Maguire, B.L. Grasberger, R.L. DesJarlais and Z. Zhou, *Bioorg. Med. Chem. Lett.*, 2002, **12**, 491.
763. Y. Zenitani, Y. Takaoka, M. Arai and I. Fujii, *Pept. Sci.*, 2002, **38**, 351.
764. R.G. Smith, S. Missailidis and M.R. Price, *J. Chromatogr. B - Analyt. Technol. Biomed. Life Sci.*, 2002, **766**, 13.
765. Y. Chen, K. Diener, I.R. Patel, J.K. Kawooya, G.A. Martin, P. Yamdagni, X. Zhang, A. Sandrasalphaa, S. Sahasrabudhe and S.J. Busch, *J. Biomol. Screen.*, 2002, **7**, 433.
766. A. Espejo, J. Cote, A. Bednarek, S. Richard and M.T. Bedford, *Biochem. J.*, 2002, **367**, 697.
767. L.W. Donaldson, G. Gish, T. Pawson, L.E. Kay and J.D. Forman-Kay, *Proc. Natl. Acad. Sci. USA*, 2002, **99**, 14053.
768. T.S. Ulmer, J.M. Werner and I.D. Campell, *Structure (Camb.)*, 2002, **10**, 901.
769. P. Nioche, W.Q. Liu, I. Broutin, F. Charbonnier, M.T. Latreille, M. Vidal, B. Roques, C. Garbay and A. Ducruix, *J. Mol. Biol.*, 2002, **315**, 1167.
770. G. Sun, L. Ramdas, W. Wang, J. Vinci, J. McMurray and R.J. Budde, *Arch. Biochem. Biophys.*, 2002, **397**, 11.
771. K.N. Brazin, R.J. Mallis, D.B. Fulton and A.H. Andreotti, *Proc. Natl. Acad. Sci. USA*, 2002, **99**, 1899.
772. P. Deprez, E. Mandine, A. Vermond and D. Lesuisse, *Bioorg. Med. Chem. Lett.*, 2002, **12**, 1287.
773. C. Hobbs, R.A. Bit, A.D. Cansfield, B. Harris, C.H. Hill, K.L. Hilyard, I.R. Kilford, E. Kitas, A. Kroehn, P. Lovell, D. Pole, P. Rugman, B.S. Sherborne, I.E.D. Smith, D.R. Vesey, D.L. Walmsley, D. Whittaker, G. Williams, F. Wilson, D. Banner, A. Surgenor and N. Borkakoti, *Bioorg. Med. Chem. Lett.*, 2002, **12**, 1365.

774. N. Wimmer, R.J. Marano, P.S. Kearns, E.P. Rakoczy and I. Toth, *Bioorg. Med. Chem. Lett.*, 2002, **12**, 2635.
775. A. Yamamoto, K. Setoh, M. Murakami, M. Shironoshita and T. Kobayashi, *Int. J. Pharm.*, 2002, **250**, 119.
776. K. Sadler, K.D. Eom, J.–L. Yang, Y. Dimitrova and J.P. Tam, *Biochemistry*, 2002, **41**, 14150.
777. P.A. Wender, J.B. Rothbard, T.C. Jessop, E.L. Kreider and B.L. Wylie, *J. Am. Chem. Soc.*, 2002, **124**, 13382.
778. Y. Park, K.Y. Kwok, C. Boukarim and K.G. Rice, *Bioconjugate Chemistry*, 2002, **13**, 232.
779. J.B. Rothbard, E. Kreider, C.L. VanDeusen, L. Wright, B.L. Wylie and P.A. Wender, *J. Am. Chem. Soc.*, 2002, **45**, 3612.
780. T. Takenobu, K. Tomizawa, M. Matsushita, S.T. Li, A. Moriwaki, Y.F. Lu and H. Matsui, *Mol. Cancer Ther.*, 2002, **1**, 1043.
781. S. Futaki, *Tanpakushitsu Kakusan Koso*, 2002, **47**, 1415.
782. S. Futakai, I. Nakase, T. Suzuki, Z. Youjun and Y. Sugiura, *Biochemistry*, 2002, **41**, 7925.
783. L. Benimetskaya, N. Guzzo-Pernell, S.-T. Liu, C.H.J. Ai, P. Miller and C.A. Stein, *Bioconjugate Chemistry*, 2002, **13**, 177.
784. C.M. Deber, L.-P. Liu, C. Wang, N.K. Goto and R.A.F. Reithmeier, *Curr. Top. Membr.*, 2002, **52**, 465.
785. E. Hariton-Gazal, R. Feder, A. Mor, A. Graessmann, R. Brack-Werner, D. Jans, C. Gilon and A. Loyter, *Biochemistry*, 2002, **41**, 9208.
786. A. Chilkoti, M.R. Dreher and D.E. Meyer, *Adv. Drug Deliv. Rev.*, 2002, **54**, 1093.
787. T.J. Deming, *Adv. Drug Deliv. Rev.*, 2002, **54**, 1145.

Cyclic, Modified and Conjugated Peptides

BY JOHN S. DAVIES
Department of Chemistry, University of Wales Swansea, Singleton Park, Swansea SA2, 8PP, UK

1 Introduction

As an aid to researchers who wish to review work year by year, it has again been possible to sub-divide this Chapter into traditional sub-divisions. This format has been retained even though the productivity under the various categories has often been cyclical. However overall the year under review (2002) has again seen good productivity (10% increase over last year), reflecting the central role the subject matter holds in the peptidomimetic sector. The reduced flexibility of the cyclic structures often allows further insight into the development of optimised conformations for receptor targeting.

Collecting the core references has again benefited from scanning CA Selects on Amino Acids, Peptides and Proteins (up to issue 14, 2003),[1] while the development of computer scanning of the Web of Knowledge data bases[2] has contributed greatly to the aim of securing comprehensive coverage. Proceedings from various symposia have not been rigorously reviewed until material arrives in refereed Journals. Nevertheless International gatherings do provide feedback on contemporary trends in the field, as reflected in the books of the Proceedings of Symposia organised by the European,[3] American[4] and Japanese[5] Peptide Societies.

Most of the subject matter of this Chapter is also featured in authoritative reviews within the five-volume set (Vol. E22 a-e) on 'Synthesis of Peptides and Peptidomimetics' of the 4th Edition of *Houben-Weyl Methods of Organic Chemistry*.[6] These volumes also concentrate on the practical aspects, which are so often underestimated in the mastering of synthetic skills in this demanding research endeavour. The present and the future of peptide science has also been the subject of comment from senior workers in the area. Thus the future role of peptides in drug discovery has been highlighted[7] while the question of the market for peptides as drugs has been discussed.[8,9] A status report on the large scale manufacturing of peptides has appeared[10] while the commercial feasibility of chemical synthesis of natural cyclic peptides has been explored.[11] In this context, optimising the yields of cyclisation, the use of cheaper reagents,

Amino Acids, Peptides and Proteins, Volume 35

advantageous chromatographic separations may well push the commercial synthesis of cyclosporin A into the chemical, and away from the microbiological domain. Designing non-peptide peptidomimetics in the 21st Century, by targeting conformational ensembles, is the subject of a review,[12] while development of potential therapeutic strategies using serine-threonine protein phosphatase[13] refers to the possible application of cyclic peptides and depsipeptides. Amongst the 218 references compiled by Lewis[14] in a review on alkaloids from the plant families *Amaryallidacae* and *Scletium*, are references to oxazole, thiazole and peptide alkaloids of relevance to this Chapter.

2 Cyclic Peptides

2.1 General Considerations. – Cyclic peptides have recently acquired the status of being referred to as ‘privileged’ (they bind to multiple, unrelated classes of receptor with high affinity), and ‘smart’ (they exhibit a variety of properties which are unique compared to commonly produced peptides). These definitions were picked up from reviews on the strategies by which cyclic peptides have been synthesised in combinatorial libraries,[15] and on the use of natural and synthetic peptide libraries in combination with enzymatic *des*- and *trans* amidation, oxidative decarboxylation and phenol coupling, in making peptide collections[16] for screening for new drugs. Synthetic approaches to peptide and peptidomimetic cyclisation have also been reviewed.[17] Progress in cyclopeptide research,[18] and in research on cyclic RGD peptides[19] has been reviewed in Chinese Journals not readily available for content appraisal. The principle of minimal side chain protection has been exemplified[20] by the synthesis of the cyclohexapeptide (1), using DPPA/$NaHCO_3$/DMF for cyclisation at position (a) in (1), when a yield of 70% was recorded for the cyclisation. As part of a study to find ligand peptides for Human Leukocyte Elastase, various methods were tried[21] to incorporate Met-Ile-Phe-Gly-Ile into a cyclic structure. Solution phase head to tail cyclisation to a cyclopentapeptide could only be achieved in 7% yield, but cyclisation on solid phase using Met-Ile-Phe-Gly-Ile-Asp-Tentagel gave more than 35% yield. A tri-orthogonal strategy has also been applied[22] to solid support cyclisation utilising an azide group, which can then be reduced and used to extend the chain. This strategy is summarised in Scheme 1. It has been shown[23] that triphosgene is a highly efficient and racemisation free coupling agent, for sterically hindered N-alkylated amino acids, and was used successfully in the solid phase synthesis of cyclosporin O. Backbone cyclisation of unprotected polypeptides can be carried out[24] provided the C-terminal residue bears a thioester while the N-terminal residue is selenocysteine. The latter allows further processing to give alanine or dehydroalanine containing cyclic peptides, as summarised in Scheme 2. A subset-orientated algorithm for minimising the number of steps required for the synthesis of cyclic peptide libraries has been devised,[25] while the valuable information gained by designing biologically active cyclic RGDS peptides has led to successful combinatorial and rational strategies[26] for making low MW non-peptidic $\alpha_4\beta_7$ integrin antagonists having the general structure (2).

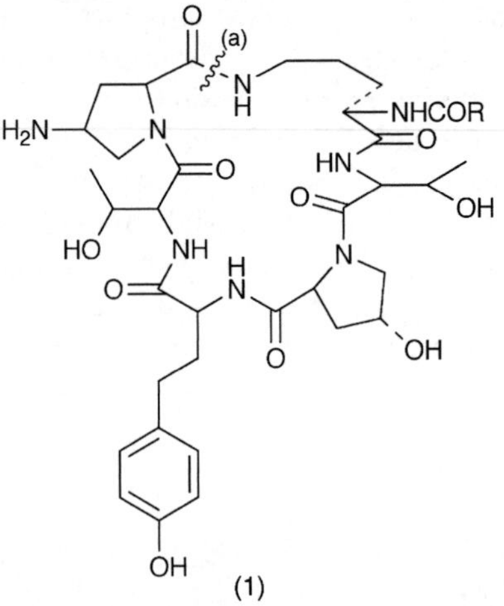

(1)

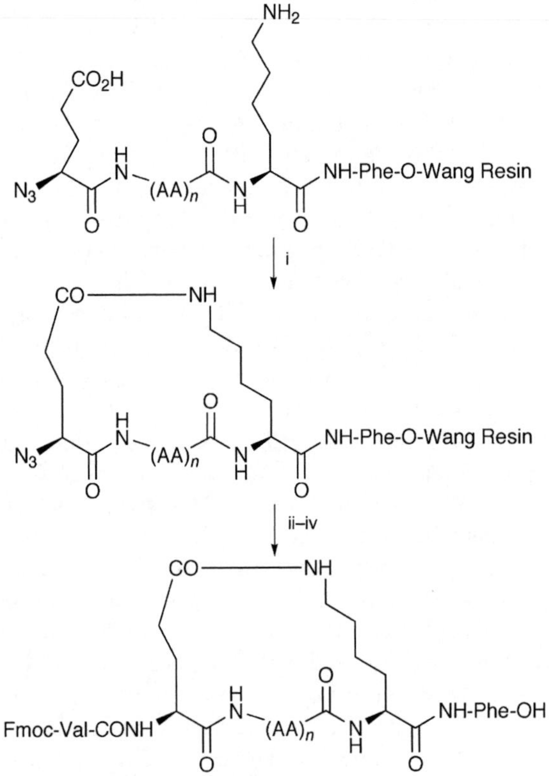

Reagents: i, DCC/HOBt; ii, Me_3P/dioxan; iii, Fmoc-Val-ONSu; iv, TFA

Scheme 1

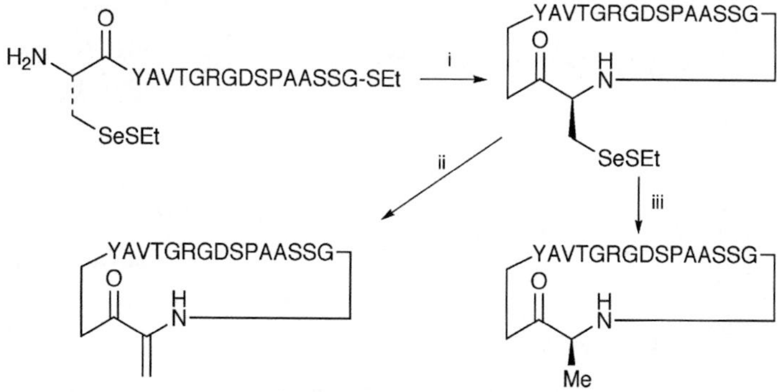

Reagents: i, 0.1M phosphate pH 7.5/thiophenol 3% v/v; ii, $H_2O_2/H_2O/CH_3CN$;
iii, Raney Ni, H_2/tricarboxyphosphine/AcOH

Scheme 2

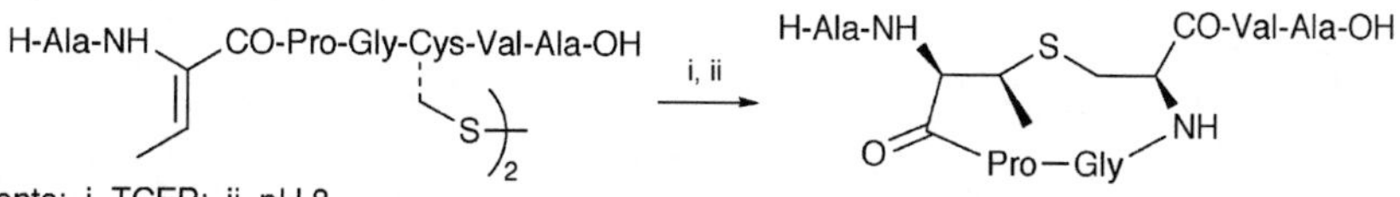

Reagents: i, TCEP; ii, pH 8

Scheme 3

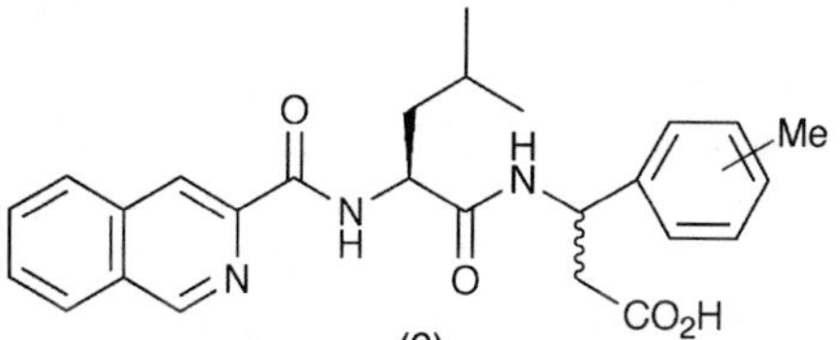

(2)

Lanthionine with its single sulfur bridge is proving to be a more robust, and conformationally more rigid analogue of cystine in the design of cyclic analogues. Hence there has been increased interest in ways and means of incorporating the unit into peptides. The more biomimetic approach installs[27] a *Z*-dehydrobutyrine residue into the peptide starting from Fmoc-(2*R*,3*S*)-3-methyl-Se-phenylselenocysteine, and this dehydropeptide can then undergo a Michael addition of sulfur as exemplified in the final stage of the synthesis of the B-ring of lantibiotic subtilin given in Scheme 3. Michael addition has also featured[28] as the last step conversion from disulfide bridges, as shown in the formation of lanthionine peptides as potential inhibitors of protein-protein interactions (Scheme 4). Since earlier studies had found deficiencies in the biomimetic approach (e.g. lack of stereospecificity), the synthesis[29] of four monosulfide bridged analogues of enkephalins, H-Tyr-cyclo (D-Ala_L-Gly-Phe-D-Ala_L)-OH, (3), H-Tyr-cyclo (D-Val_L-Gly-Phe-D-Ala_L)-OH, (4), H-Tyr-cyclo (D-Ala_L-Gly-Phe-Ala_L)-OH, (5), H-Tyr-cyclo (D-Val_L-Gly-Phe-Ala_L)-OH, (6), has been based on the method of Ménez et.al. Analogues (4) and (6) showed

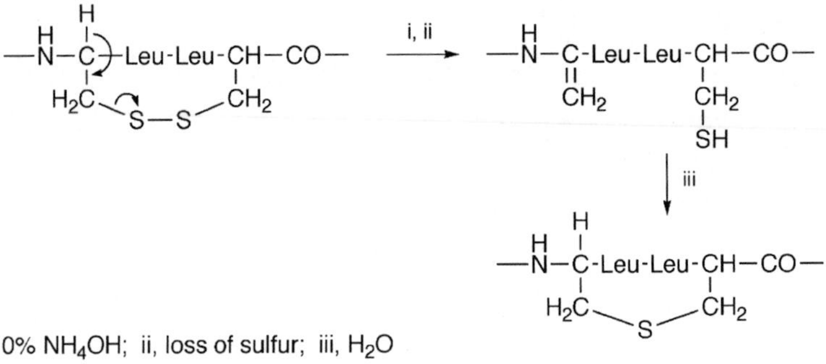

Reagents: i, 10% NH_4OH; ii, loss of sulfur; iii, H_2O

Scheme 4

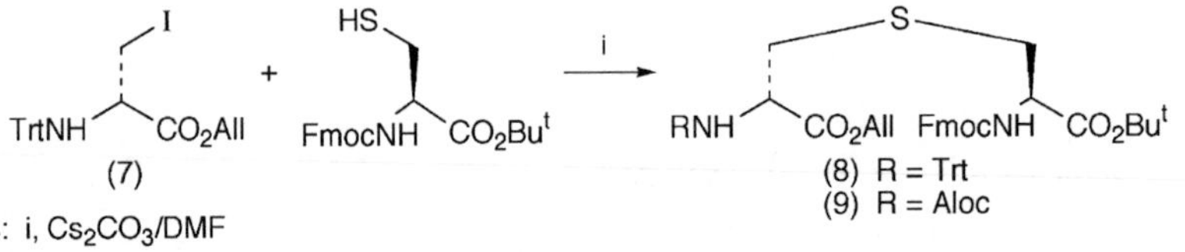

(7)

(8) R = Trt
(9) R = Aloc

Reagents: i, Cs_2CO_3/DMF

Scheme 5

substantial reactivity towards the δ-opioid receptor, while the unsubstituted analogues (3) and (5) bound to both μ- and δ-receptors. The stereospecific synthesis of orthogonally-protected lanthionines for incorporation into peptides has not been easy. One successful synthetic scheme[30] has given (8) in 74% yield (Scheme 5), while others[31] report that in the synthesis of (9) that the iodoalanine derivative (7) is prone to rearrangement to its regio-isomer. The latter has enabled[32] nor-lanthionine bridges to be incorporated as shown by the on-resin synthesis of (10), using PyAOP for the cyclisation step.

A variety of cyclic di- to penta-peptides has been obtained[33] using the photo-induced electron transfer initiated decarboxylation of N-phthaloylpeptides as illustrated in Scheme 6. Split inteins have been used[34] to create random cyclic peptides in mammalian cells using retro-viral technology. Cyclisation was effected by *Ssp* Dna E intein in BJAB cells.

In carrying out their annual assessment of the synthetic capabilities of a number of participating peptide laboratories, the Association of Biomolecular Resource Facilities' Peptide Science Research Group, requested each laboratory to produce cyclo (Tyr-Glu-Ala-Ala-Arg-D-Phe-Pro-Glu-Asp-Asn) via side-chain anchoring of Fmoc-Asp(OH)-ODmab to a Rink amide resin, linear assembly, Dmab and Fmoc removal, on-resin cyclisation with an uronium-based coupling reagent, followed finally by cleavage and deprotection using trifluoroacetic acid. On the basis of mass spectrometric analysis and other methods to determine purity,[35] unexpected side reactions uncovered were, (i) N-terminal guanidine formation; (ii) C-terminal piperidyl formation; (iii) elimination of the 4,4-dimethyl-2,6-dioxocyclohexylidene moiety to form a

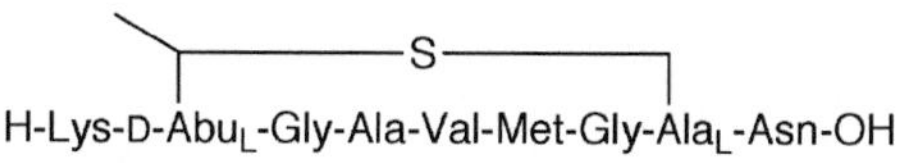

H-Lys-D-Abu$_L$-Gly-Ala-Val-Met-Gly-Ala$_L$-Asn-OH

(10)

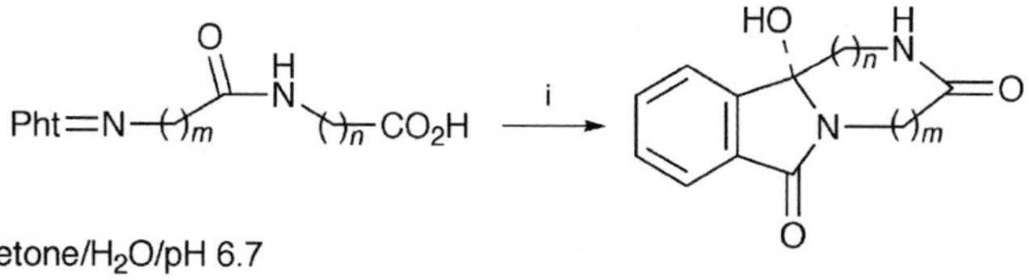

Reagents: i, *hν*/acetone/H_2O/pH 6.7

Scheme 6

4-aminobenzyl ester of the C-terminal amino acid. A report has also appeared[36] on the use of tandem mass spectrometry for identifying cyclic peptides.

2.2 Cyclic Dipeptides (Dioxo- or Diketo-piperazines). – Although it just squeezes the coverage into 2003, it is opportune to include the comprehensive coverage of the dioxopiperazines in a review[37] which critically summarises the revived interest in these compounds as scaffolds in combinatorial work. Another 107-reference review[38] concentrates on their synthesis, including strategies for 2,3-, 2,5- and 2,6-dioxo-piperazines.

E-Dehydrophenylahistin (11) has been prepared[39] from its hydrogenated precursor (-)-phenylahistin by enzymic conversion using the cell-free extract of *Streptomyces albulus* KO-23. The dehydro form (11) is 10,000 times better than albonoursin as a cell-cycle inhibitor, and in its enzymic synthesis is formed via the Z-isomer. A marine actinomycete has yielded[40] the previously known (6S, 9S)-cyclo (Pro-Val), which was purified by countercurrent techniques, while *Pseudomonas putida* WCS358 has been shown to produce,[41] a mixture of dioxopiperazines, cyclo (Pro-Tyr), cyclo (Pro-Leu), cyclo (Phe-Pro) and cyclo (Val-Leu), detected using bacterial biosensors. Thaxtomins, 4-nitroindole-containing dioxopiperazines, have been shown to alter the deposition of monocot and dicot plant cell walls.[42] The Murchison and Yamato-791198 carbonaceous chondrites have been shown[43] to contain 23 and 18 pmol g^{-1} respectively of cyclo (Gly-Gly), which is believed to be native to the chondrites and not from terrestial contaminants.

The thrust generated by combinatorial work has resulted in an increasing number of solid phase syntheses of dioxopiperazines. Thus biologically active unsaturated 3-substituted piperazine-2,5-diones have been obtained via Scheme 7, with alkylation of intermediate carbamates and aldehyde condensations being carried out on-resin.[44] Nucleophilic attack on a phenacyl ester (Scheme 8) represents the key step in the synthesis[45] of eighteen diketo-piperazines, while cyclisation as in Scheme 9, produced[46] a series of moderate inhibitors of acetyl cholinesterase, which could be lead compounds for further development as drug candidates to treat Alzheimer's disease. The scope and limitations of adapting the Ugi 4CC multicomponent reaction to solid-phase

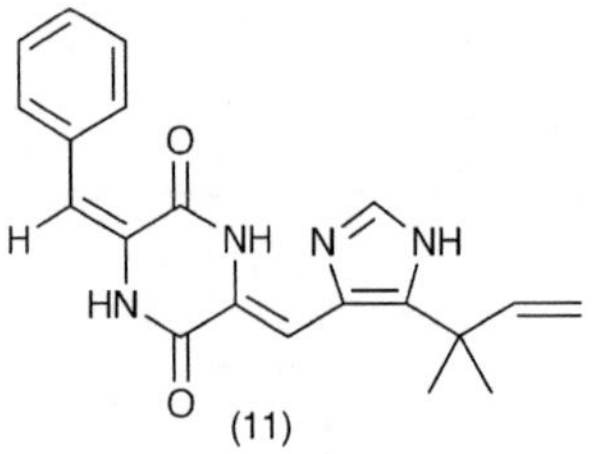

(11)

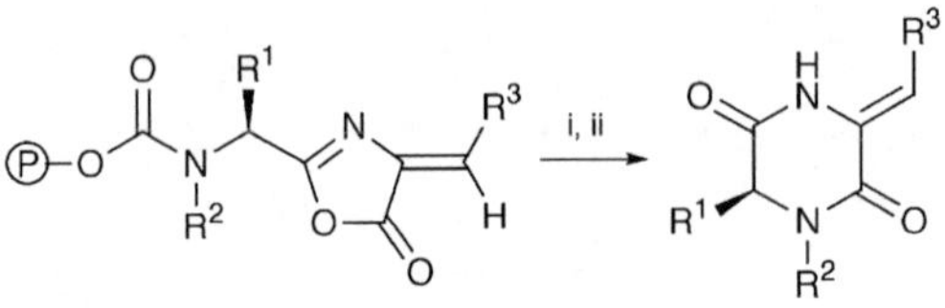

Reagents: i, $TFA.CH_2Cl_2$; ii, Et_3N/toluene/80° C

Scheme 7

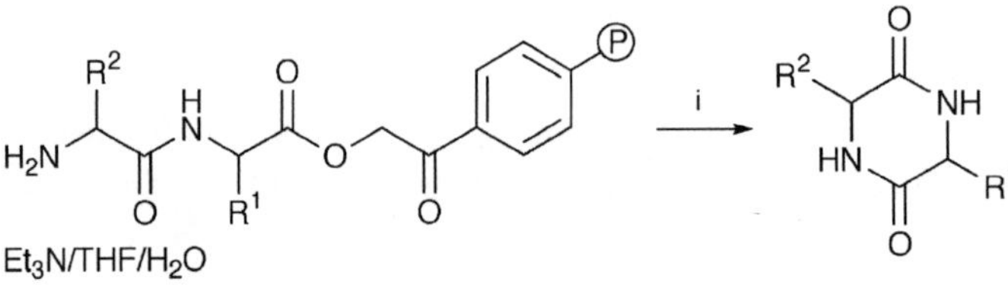

Reagents: i, 5% $Et_3N/THF/H_2O$

Scheme 8

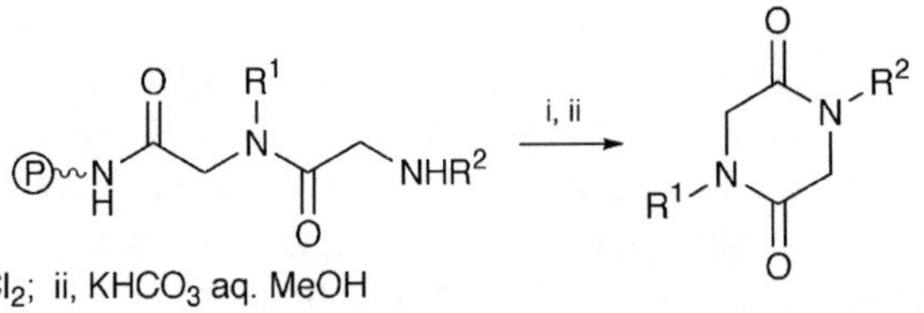

Reagents: i, $TFA.CH_2Cl_2$; ii, $KHCO_3$ aq. MeOH

Scheme 9

techniques have been presented[47] as a result of its use in the synthesis of the β-turn mimetics (12) and (13). The isonitrile component of the Ugi reaction has also been used[48] resin-bound as a Rink-isonitrile resin in the synthesis of dioxopiperazines as summarised in Scheme 10. A novel version (14) of the resin-bound isonitrile has also been tested out[49] in an Ugi reaction to make 2,5-diketopiperazines.

Procedures exemplified by Scheme 11 have enabled[50] a number of 4-substituted piperazine-2,5-diones to be synthesised. $LiAlH_4$ reduction of one of the analogues, gave (15) which was shown to have promising affinity for σ_1-receptors (K_i=66.1 nM). Twenty four bifunctional dioxopiperazines, based on Glu/Asp and Lys/Orn/Dab combinations, of which (16–21) are representative, have been synthesized[51] using the method of Suzuki et. al.(weak acid catalysed cyclisation of dipeptide esters under reflux conditions). Excessive racemisation experienced in some cases, could be controlled by cooling the reaction after 3–4 hr, and filtering off the dioxopiperazine. Two diastereoisomeric forms (22) of an analogue of cyclo(His-Pro) have been synthesized[52] for testing as neuroprotective agents. They were active at picomolar concentrations in preventing neuronal death in an *in vitro* model of traumatic injury.

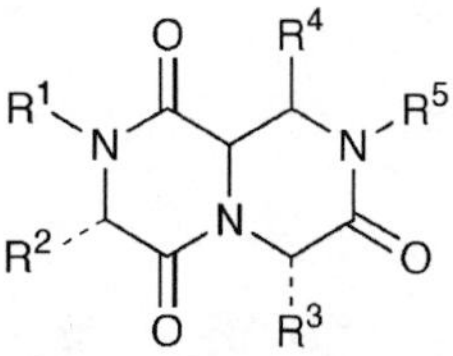

O R^4 R^1 N R^5 N R^2 N O O R^3

(12) $R^4 = H$, $R^5 = CH(R^6)CONHR^7$
(13) $R^4 = CH(R^6)CONHR^7$, $R^5 = H$

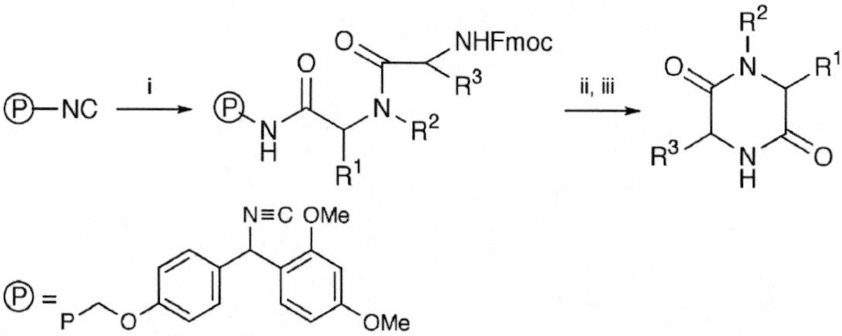

Ⓟ—NC —i→ Ⓟ N H O O NHFmoc R^3 N R^2 R^1 —ii, iii→ R^2 O N R^1 R^3 N H O

Ⓟ = P O N≡C OMe OMe

Reagents: i, Fmoc-AA-OH, RCHO, R^2NH_2, MeOH/THF; ii, 20% piperidine/DMF; iii, 10% HOAc/DCE/60° C, 16 h

Scheme 10

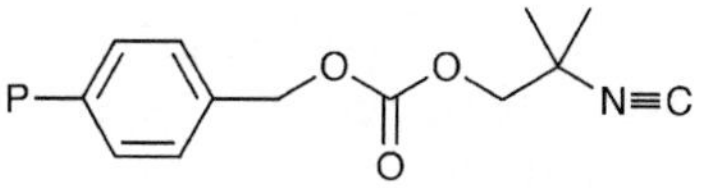

P O O N≡C O

(14)

CO_2Me Bzl N H H CO_2Me —i→ Cl O N Bzl CO_2Me CO_2Me —ii→ Bzl N O O N Bzl CO_2Me

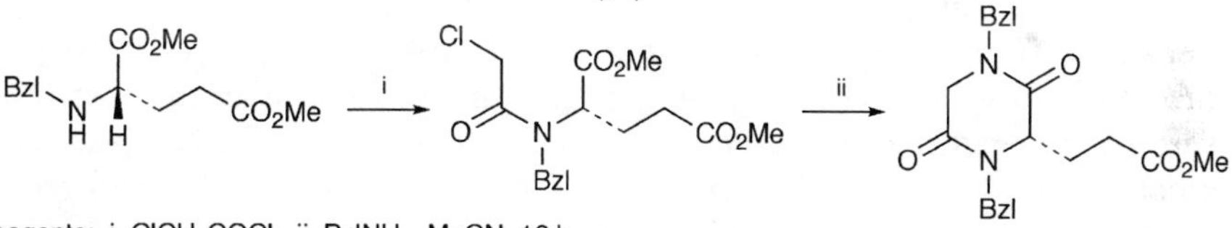

Reagents: i, $ClCH_2COCl$; ii, $BzlNH_2$, MeCN, 16 h

Scheme 11

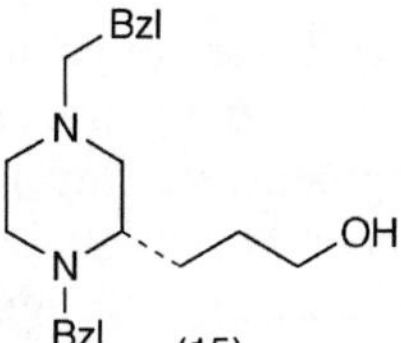

Bzl N N Bzl OH (15)

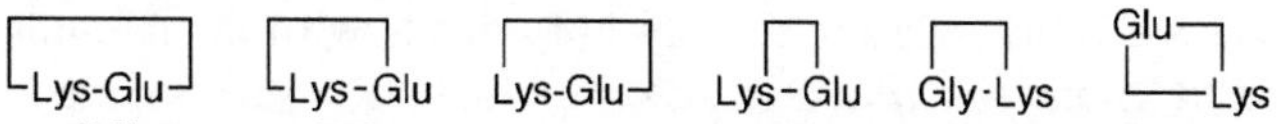

Lys-Glu (16)

Lys-Glu (17)

Lys-Glu (18)

Lys-Glu (19)

Gly-Lys (20)

Glu Lys (21)

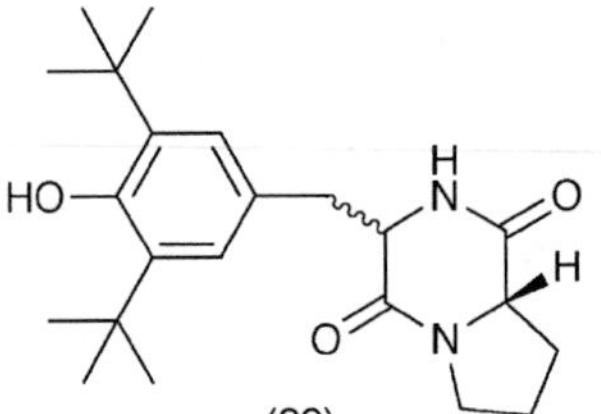

(22) 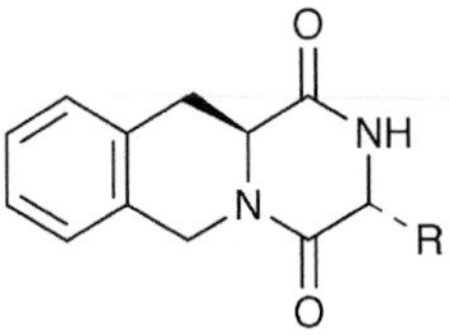 (23) 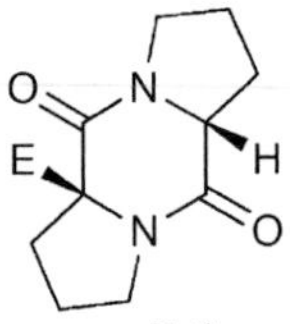 (24)

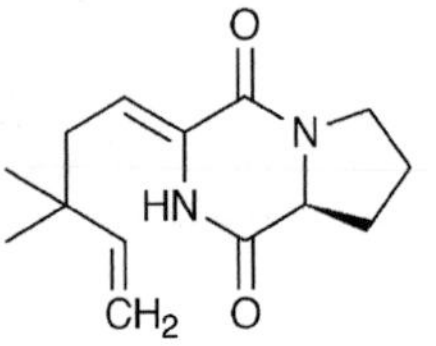

(25) 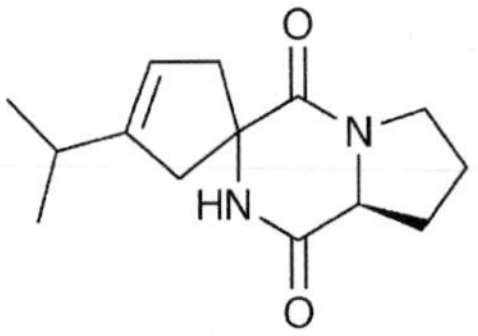 (26) 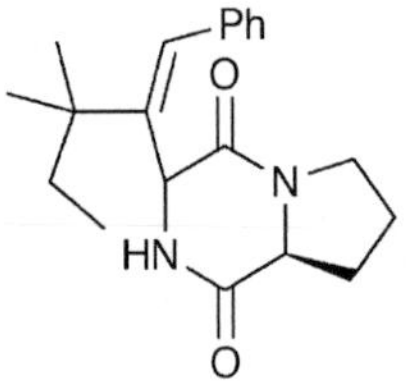 (27)

Alkylation studies on 1-methoxy-3, 6, 11, 11a-tetrahydro-4H-pyrazino[1,2] isoquinolin-4-one have produced[53] alkylated lactim ethers, which in the presence of acid, form a series of dioxopiperazines represented by (23). Low temperature base–catalysed deprotonation of cyclic (D-Pro-D-Pro), gave enolates,[54] which could be trapped by a variety of electrophiles (E), to yield precursors (24) of interest to the synthesis of marine alkaloids phakellstatin and phakellin. Cope rearrangement products of dioxopiperazine (25) have been revealed[55] to have structures (26) and (27) which can on hydrolysis be transformed to new quaternary α-amino acids. The intermediacy of N-acyliminium species derived from (28) has been used[56] as an explanation for the stereoselective conversion to form (29) in 60% diastereomeric excess, using allyltrimethylsilane under Lewis acid conditions. Piperazine-2, 6-diones (30) have been prepared[57] from amino acid methyl ester hydrochlorides via [3+3] annulation with 2-bromoacetamide, while preliminary studies[58] have shown that 2-oxopiperazines such as (31) can be obtained by on-bead intramolecular cyclisation. A new class of β-tryptase inhibitors (32) have used[59] *m*-aminomethyl-Phe as an arginine mimetic, and the K_i values were optimal when the terminal amino groups had 29–31 bond between them (when m=n=1, and X and Y were Gly and β-Ala respectively. The 5- or 6-membered rings in pyroglutamic acid dioxopiperazine (33) can be selectively opened up under different conditions.[60] In mild basic conditions, the dioxopiperazine ring is opened with retention of chirality, while with nucleophiles under slightly acidic conditions bis glutamate dioxopiperazines were formed. The influence of high pressure[61] on dioxopiperazine formation has been assessed.

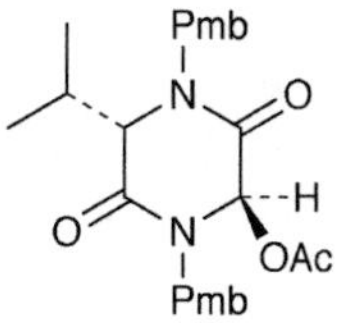

(28) 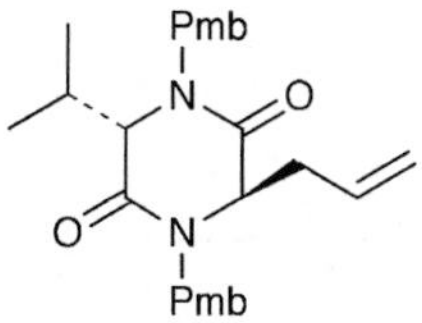 (29)

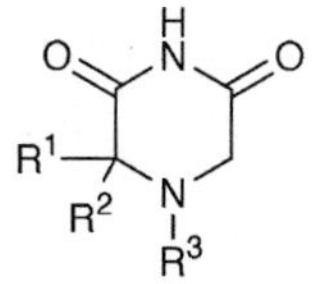

(30) R^1 = H, CH_2CO_2Me
R^2 = H, Me, Et, hexyl, Me_2CHCH_2
R^3 = H, Me

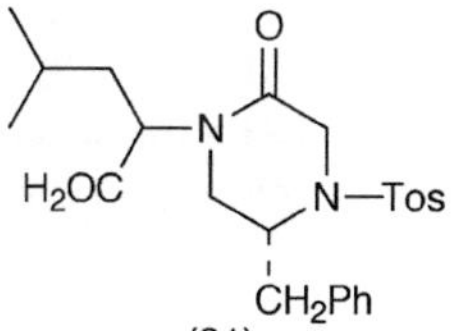

(31) 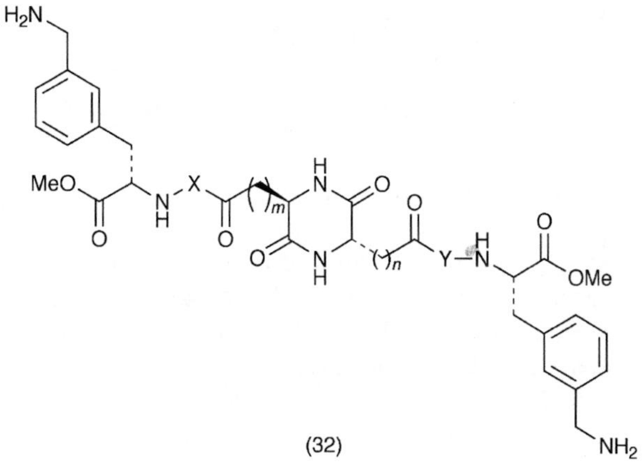 (32)

X-ray crystallography has been used to study[62] the formation of homochiral supramolecular interactions of *p*-xylylene-linked dioxopiperazines such as (34), as well as the complexation characteristics of the crown ether macro-ring dioxopiperazines (35).[63] The central dioxopiperazine rings in the latter remain planar. Cyclo (Ala-Ser) units have been attached to calix[4]arenes to study their potential as gas-sensors.[64] These compounds were found to bind more favourably with (*R*)-methyl lactate than with the (*S*) form. A library of 'two-armed' dioxopiperazines has been constructed[65] on polystyrene resin, and the binding properties of individual members elucidated, using an attached dye marker molecule. Results showed that receptor (36) bound strongly to D-Asn(Trt)-D-Ala or D-Asn(Trt)-D-Val, while (37) bound predominantly to tripeptides.

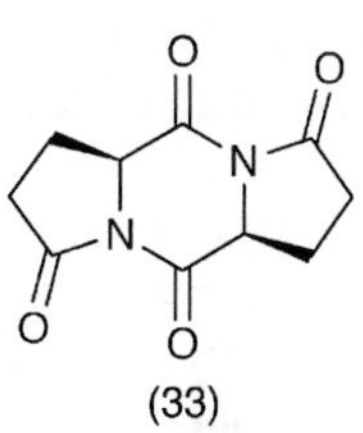

(33) 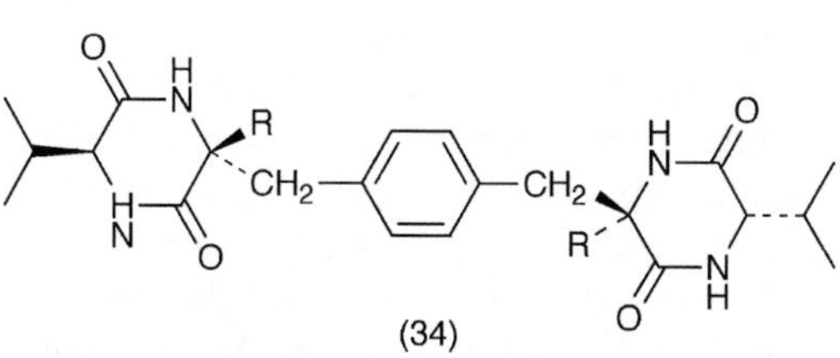 (34)

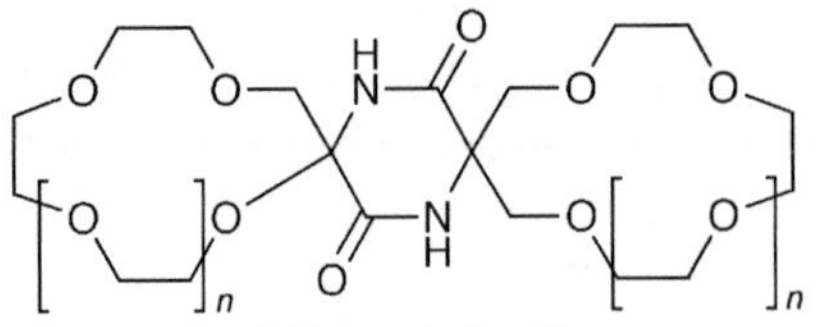

(35) $n = 1, 2$ or 3

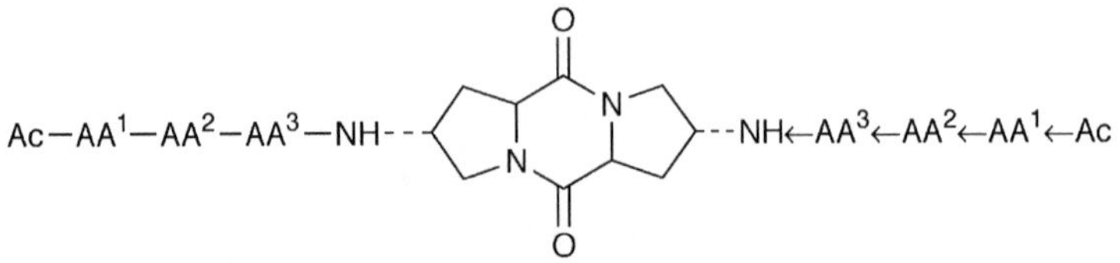

(36) AA^1 = Phe, AA^2 = Asn(Trt), AA^3 = Tyr(Dye)
(37) AA^1 = Phe, AA^2 = Asn(Trt), AA^3 = D-Tyr(Dye)

A crystal structure has been obtained[66] for cyclo (Gly-Gln), a mimic of Gly-Gln, (β-endorphin 30-31), which has been identified as an inhibitor of β-endorphin-induced cardiorespiratory depression. In an NMR and Raman Optical Activity (ROA) study[67] of cyclo (Pro-Pro), the spin-spin NMR coupling constants were found to be the most suitable for establishing conformational ratios. Secondary donor-acceptor interactions in co-crystals of cyclo (Gly-Gly) have been shown to dictate[68] the location of guest molecules in the crystal lattice.

2.3 Cyclotripeptides. – The conformational tightness of a homodetic cyclotripeptide ring mitigates against many examples being known. Nature overcomes this conformational restriction by forming heterodetic rings of which renieramide (38) from the Vanatu sponge *Reniera n.* sp is a typical example.[69] This is typical of the class that includes OF4949 I-IV, K13 and the eurypamides.

The total synthesis of the proteasome inhibitors TMC-95A (39) and −95B (40) has been the focus of much attention. Key to one total synthesis[70] is the macrocyclisation step summarised in Scheme 12, brought about by EDC/HOAt/DIEA in greater than 50% yield. Further elaboration of the macrocycle thus formed, involved generation of the *Z*-propenamide side chain using a silatropic re-organisation of 1-silylated-2-propen-1-amides, and the acylation of the free amine generated from removal of the *Z*-group shown in Scheme 12 with racemic 3-methyl-2-oxo-pentanoic acid. Hplc separated (39) from (40) formed in the reaction mixture. Based on the core structure of TMC-95A (39), other authors[71] have synthesised analogue (41), containing a propylamide residue for interaction with the S1 pocket of the active sites, and the side chain of asparagine for occupancy of the S3 pocket. Analogue (41) was formed via a macrocyclisation route similar to Scheme 12, but using PyBOP/HOBt, and retained the full inhibition of proteasome trypsin-like and peptide hydrolase activities, but was less potent against the chymotrypsin-like activity. It was only when the C_7-oxindole/phenol junction was present that the monomeric cyclic

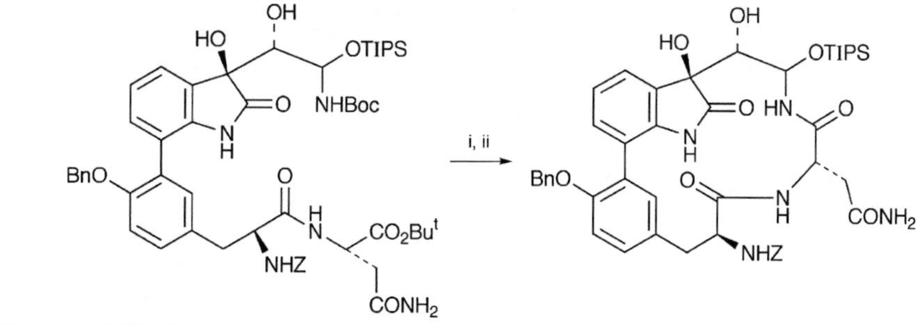

Reagents: i, TFA/CH_2Cl_2; ii, EDC/HOAt/DIEA

Scheme 12

species was obtained.[72] A C6-indole/phenol junction gave rise exclusively to the related dimer.

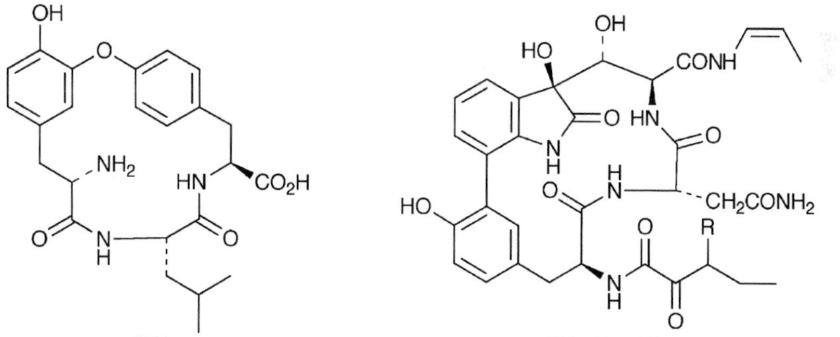

(38)

(39) R = (*S*)-Me
(40) R = (*R*)-Me

A novel route has been developed[73] for the cyclisation of tripeptides on solid support (Rink amide resin), which uses a nucleophilic aromatic substitution strategy as summarised in Scheme 13. The need for pre-organisation of linear precursors such as *N*-allyl β-phenylisoserine-Leu-Pro-allyl (42) into γ- or β-turns, prior to metathesis with Grubbs' catalyst has been recognized[74] from the results of carrying out Scheme 14 when X=NH (35% yield) and when X=O (0% yield). Use has been made[75] of the increased *cis-trans* flexibility of *N*-alkylated amides to enable macrocyclisation of tripeptide β-turn mimetics to occur as summarised in Scheme 15. A series of motilin antagonists bearing structures based on (43) have been synthesized[76] off-resin using BOP as the cyclisation reagent. Analogues with t-butyl substituents on the tyrosine ring (43, R^1=t-Bu or H, R^2=t-Bu or H.) gave enhanced binding to the motilin receptor.

Renin inhibitors, based on the Glu-D-Phe-Lys motif (44), have been investigated[77] by NMR spectroscopy and molecular dynamics calculations. At least two conformational forms around the C_α–C_β Glu side chain were observed, with a H-bond between LysNH and Glu side chain CO. Condensation of each of the tripeptide esters, Gly-Gly-β-Ala-OMe, Val-Gly-β-Ala-OMe and

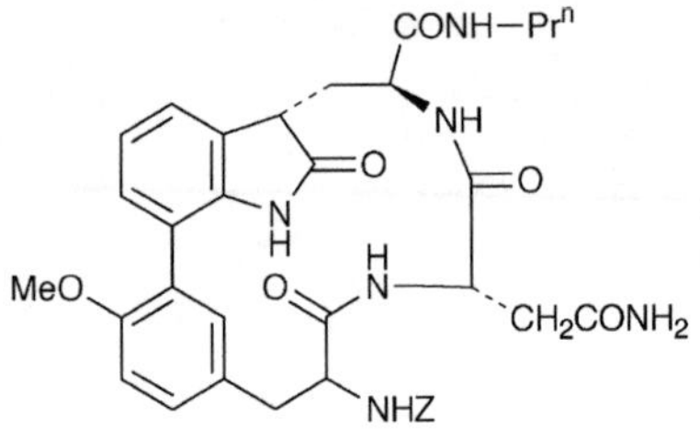

(41)

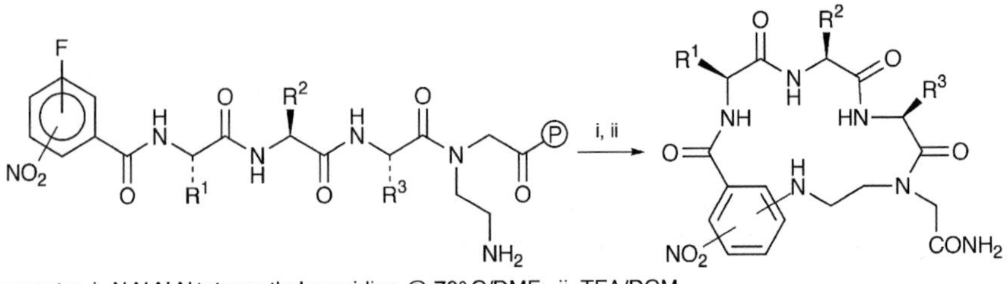

Reagents: i, *N,N,N,N*-tetramethylguanidine @ 70° C/DMF; ii, TFA/DCM

Scheme 13

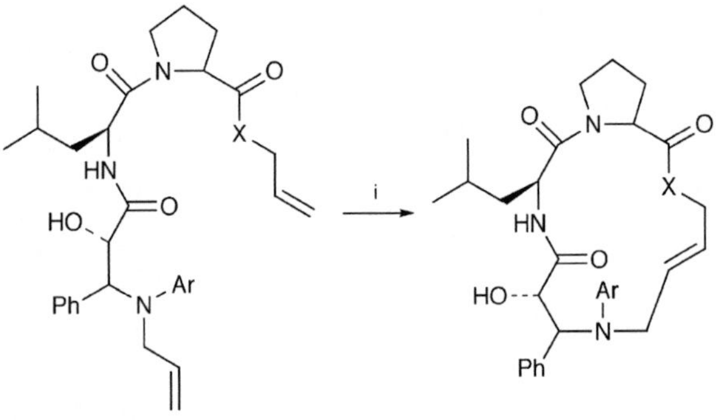

(42)

Reagents: i, $Cl_2(Cy_3P)_2Ru{=}CHPh$; when X = NH (35% yield), X = O (0% yield)

Scheme 14

Phe-Gly-β-Ala-OMe with (C6Me6)Ru complex using sodium methanolate has given rise to triple deprotonated cyclic tripeptides.[78]

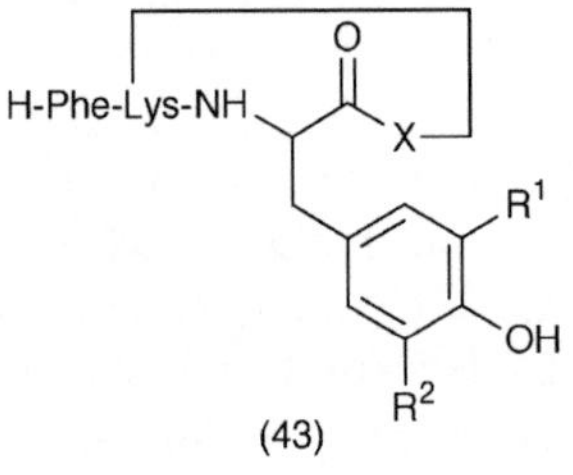

(43)

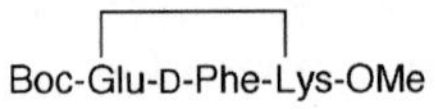

Boc-Glu-D-Phe-Lys-OMe

(44)

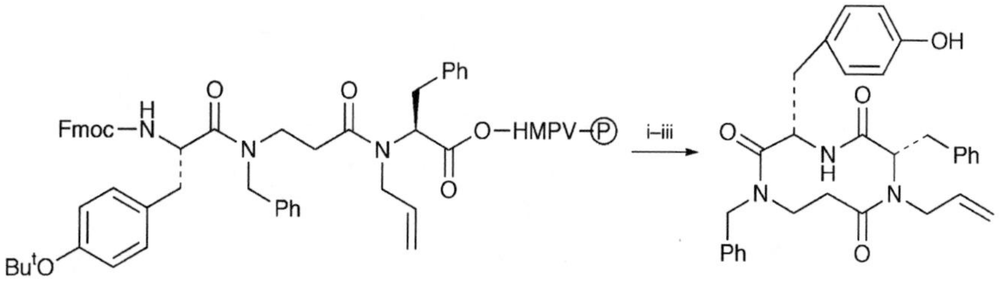

Reagents: i, 20% piperidine/DMF; ii, TFA/DCM; iii, BOP/DIEA 16 h

Scheme 15

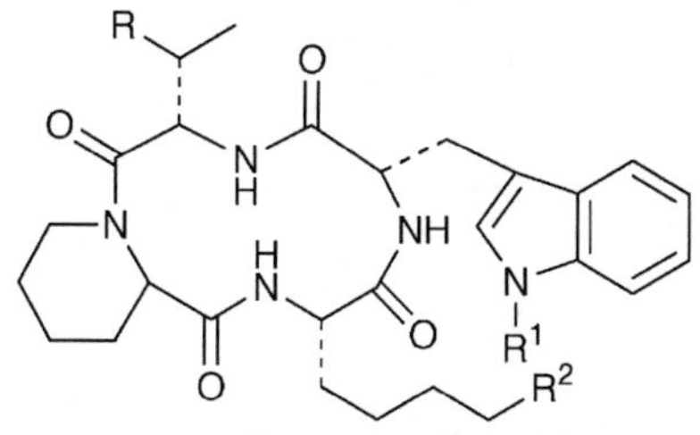

apicidin R = Et, R^1 = OMe, R^2 = CH_2COEt
apicidin A R = Et, R^1 = H, R^2 = CH_2COEt
apicidin C R = Me, R^1 = OMe, R^2 = CH_2COEt
apicidin D_1 R = Et, R^1 = OMe, R^2 = $CH_2COCH(OH)Me$
apicidin D_2 R = Et, R^1 = OMe, R^2 = (S)$CH_2CH(OH)Et$
apicidin D_3 R = Et, R^1 = OMe, R^2 = $CH_2CH_2CH(OH)Me$

Figure 1

2.4 Cyclotetrapeptides. – Two cyclotetrapeptides, halolitoralins B [cyclo (Leu-Ile-Leu-Ile)] and C [cyclo (Val-Ile-Leu-Ile)] have been found[79] together with known cyclodipeptides in the fermentation broth of *Halobacillus litoralis* YS3106. The cyclotetrapeptides showed moderate antifungal activity. Further members of the apicidin family of cyclotetrapeptides have been identified,[80] which now include the congeners listed in Figure 1. Apicidin and apicidins A and D_1 are the most active inhibitors of histone deacetylase in the series, although they do not contain the α-keto epoxide group usually associated with such activity. A solid phase protocol[81] has been added to methods for the synthesis of apicidin A, and is summarised in Scheme 16, with solution phase macrocyclisation under high dilution conditions with PyBOP/DMF the best way of reducing the competitive formation of the cyclic octamer. Introduction of a reduced amidic linkage in conjunction with an *o*-nitrophenylsulfonamide group, gave cyclotetrapeptoid analogues without the competing cyclo-octamer being formed. Apicidin's phytotoxic and cytotoxic properties have been compared[82] with other mycotoxins (sambutoxin, wortmannin and HC-toxin). The relative order of phytotoxicity towards duckweed was wortmannin > HC-

Boc-NH-Aoda-Trp-Ile-D-Pip-(2-Cl-Trt)-Ⓟ

↓ i

NH_2-Aoda-Trp-Ile-D-Pip-OH —ii→ apicidin A (Fig. 1)

Aoda = *S*-2-amino-8-oxodecanoic acid

Reagents: i, TFA; ii, activator/HOBt/$Pr^i{}_2EtN$

Scheme 16

toxin > apicidin ≫ sambutoxin, and all four proved cytotoxic to four mammalian cell cultures. Cyclotetrapeptide JM47 from a marine Fusarium fungus, has been shown[83] to have structure (45), and is therefore an analogue of HC-toxin.

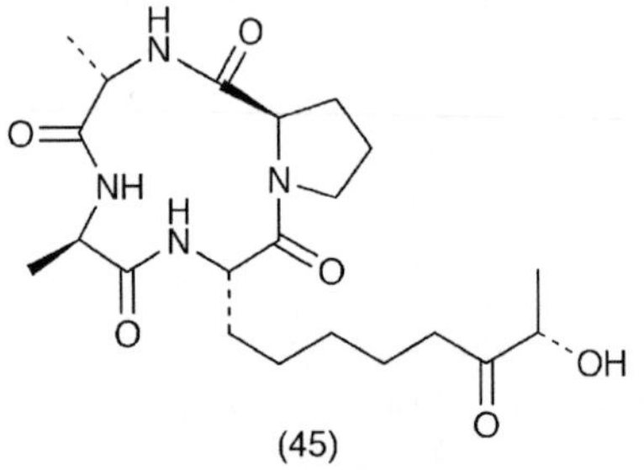

(45) 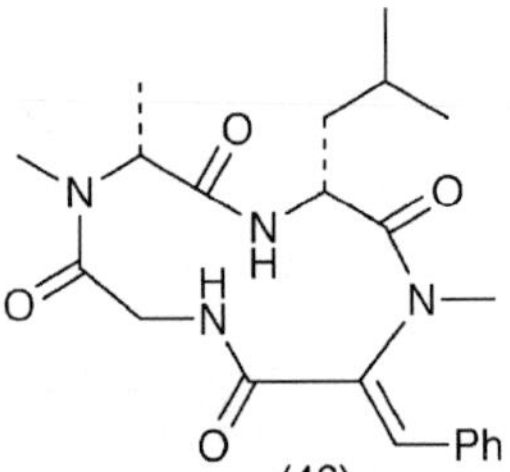 (46)

The cyclisation step and the introduction of the dehydro amino acid unit have proved to be the 'Achilles' Heel' of past syntheses of tentoxin (46) which have given poor yields. An improved approach,[84] using a modified Erlenmeyer aldolisation for the introduction of Z-Δ–Phe, gave a good yield of the linear precursor, Boc-MeAla-Leu-MeΔPhe-Gly-OMe. After deprotection, this tetrapeptide was cyclised using four different reagents with the best yield (81%), coming from cyclisation with HATU. The other cyclisation yields were DPPA (29%), DCC/PfpOH (60%) and HBTU (73%). The interaction of three membrane proteins with tentoxin derivatives has been discussed.[85] A four-dimensional orthogonal solid phase synthesis has been worked out[86] for cyclotetra-β-peptides such as (47), which allow the scaffolds to be further elaborated, using the side-chain amino groups. A summary of the protocol appears in Scheme 17.

2.5 Cyclopentapeptides. – The synthesis of two cyclic pentapeptides, cyclo (Ala-Tyr-Leu-Ala-Gly) and cyclo (Pro-Tyr-Leu-Ala-Gly) has been used as targets[87] for fine-tuning the thioester method for macrocyclisation. The new conditions use the phenacyl derivative of 3-mercaptopropionic acid for activation, with the cyclisation carried out in sodium acetate buffer from pH 4.6–5.8. The efficiency of the reagent DEPBT (48) in cyclising linear precursors to give the naturally-occurring cyclopentapeptides, cyclo (Gly-Pro-Tyr-Leu-Ala) and cyclo (Gly-Ala-Tyr-Leu-Ala), together with a cycloheptapeptide cyclo (Gly-Tyr-Gly-Gly-Pro-Phe-Pro) has been explored.[88] All cyclisations were carried

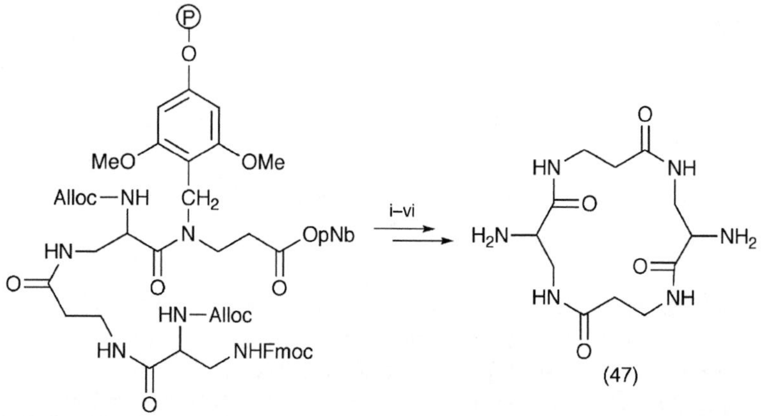

Reagents: i, $SnCl_2$/HOAc-phenol/DMF; ii, piperidine/DMF; iii, PyAOP/DIEA-DMF; iv, Pd(O); v, solid phase organic synthesis; vi, TFA

Scheme 17

cyclo[Trp(Boc)-Phg-Arg(Tos)-Aph(Boc)-D-Ala]
cyclo[Trp(Boc)-Phg-Arg(Tos)-D-Aph(Boc)-D-Ala]
cyclo[D-Nal-Phg-Arg(Tos)-L-Aph(Boc)-D-Ala]
cyclo[D-Nal-Phg-Arg(Tos)-D-Aph(Boc)-D-Ala]
cyclo[D-Pal-Phg-Arg(Tos)-L-Aph(Boc)-D-Ala]
cyclo[Cit-Arg(Tos)-Aph(Boc)-D-Ala-Trp-(Boc)]

where Nal = β-(2-naphthyl)-alanine, Aph = β-(4-aminophenyl)-alanine, Pal = β-(3-pyridyl)-alanine, Phg = phenylglycine

Figure 2

out in the solution phase(DMF), under high dilution, and under these conditions, DEPBT proved superior to DPPA, BOP, HBTU and TBTU. The structural influences on the cyclisation yields could be summarised as: (i) precursors with a small amino acid at one terminus and a larger amino acid at the other, give satisfactory yields, (ii) Pro could be chosen for either terminus, and (iii) any turn-inducing properties within the sequence are beneficial. In addition it was found that the addition of alkali metal ions increased the rate of cyclisation, Na^+ being better for the cyclopentapeptides, while Cs^+ was better for the cycloheptapeptide. Factors affecting cyclisation have also been deduced[89] from the synthesis of cyclic pentapeptides that exhibit LHRH antagonism. The six cyclopentapeptides synthesised, are listed in Figure 2, with the third in the list being the subject of a detailed survey of its linear precursor sequence. All reactions were carried out in solution at high dilution, and the main deductions for the sequence of the linear precursors, were, (i) at the C-terminus, steric bulk plays an important role, and (ii) at the N-terminus, residue configuration was a dominant factor. For the cyclisation, the reagent EDCI/

DMAP had a wider range of applicability than others such as HBTU or DEPBT.

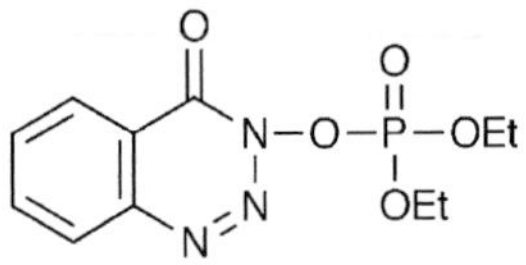

(48)

The success of cyclo (Arg-Gly-Asp-D-Phe-Lys) (49) as a selective RGD ligand for $\alpha_v\beta_3$ integrin continues to prompt great interest in this field. A third synthesis[90] of this cyclopentapeptide has been reported and was conducted on solid phase. Polymer-mounted (50) reflects the strategy and protecting groups used, and its conversion to (49) involved successive treatment with, $Pd(PPh_3)_4$/N-Me morpholine, piperidine/DMF and on-resin cyclisation with HATU (2 eq)/DIPEA for 16 hr, to give after deprotection (49) in 46% yield. The synthesis of the alkene isosteres [(51) and (52)] of cyclo (Arg-Gly-Asp-D-Phe-MeVal) has been carried out[91] using the diastereoselective alkylation of β-methylated-γ-mesyloxy-α,β-enoates by an organo-copper reagent. When tested against integrins $\alpha_v\beta_3$ and $\alpha_{IIb}\beta_3$, *E*-forms of both (51) and (52) showed good antagonistic properties but showed less selectivity. In contrast, the *Z*-form of (51) showed low antagonistic activity, suggesting a *cis*-form of peptide bond between MeVal and Phe might be inappropriate. Macrocyclisations at position (a), were achieved using nitrosation of a hydrazide of the linear precursors. A selective antagonist (53) of $\alpha_v\beta_3$ integrin has been obtained by incorporating[92] arabinose-derived bicyclic amino acids into the macrocycle, by firstly incorporating the bicyclic unit into a linear precursor on solid phase. A βII'– turn mimetic based on (3*R*, 5*S*, 6*E*, 8*S*, 10*R*)-11-amino-3, 5, 8, 10-tetramethylundec-6-enoic acid has been incorporated[93] into a cyclic RGD peptide as shown in (54), but this has lower activity towards integrins, than the cyclo (Arg-Gly-Asp-D-Phe-Val) it was meant to mimic. Another β-turn mimetic group has been incorporated[94] into a cyclopentapeptide as shown in (55), together with a cyclohexapeptide analogue. Although (55) was less active towards integrins than lead structures, it did show a 200-fold higher inhibition of echistatin binding to $\alpha_v\beta_5$ than to $\alpha_v\beta_3$. The linear precursor of (55) was assembled on a Sasrin resin starting with the Gly residue, with cyclisation off-resin using HATU/HOAt which gave a 35% yield at the cyclisation stage. The group of Kessler and co-workers have successfully developed active cyclo RGD peptides over many years and this experience have led them[95] to non-peptidic small molecule inhibitors for $\alpha_v\beta_6$, $\alpha_v\beta_3$ and $\alpha_v\beta_5$ integrins. For example, subnanomolar activity was shown by (56), selective towards $\alpha_v\beta_6$. A first synthetic RGD receptor (57) has been synthesised and tested,[96] and shown to form 1:1 stoichiometric complex with cyclo (Arg-Gly-Asp-D-Phe-Val) with an association constant of 1000 M^{-1}. A modified lysine side-chain in (58) has been used[97] to study the expression patterns of integrins on the cell surface of chondrocytes, as models for tissue engineering of articular cartilage.

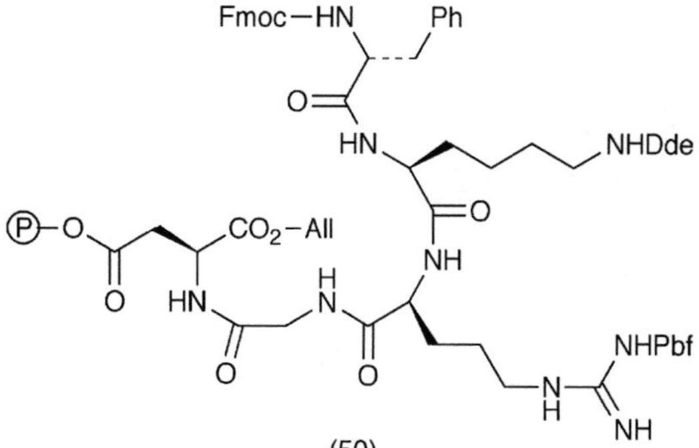

(50)

Pbf = pentamethyldihydrobenzofuran-5-sulfonyl

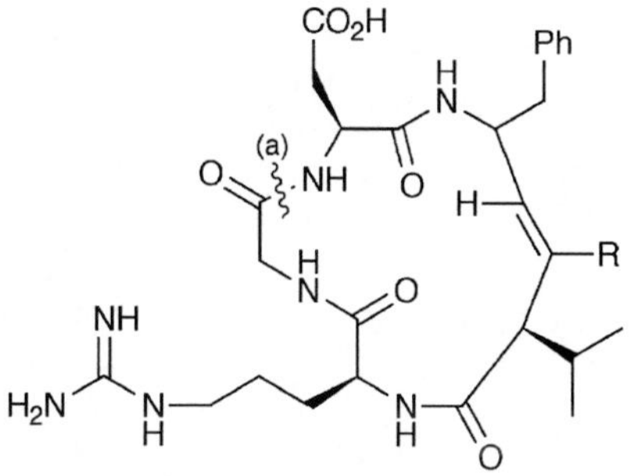

(51) *E* and *Z*-forms with R = Me

(52) *E*-form with R = H

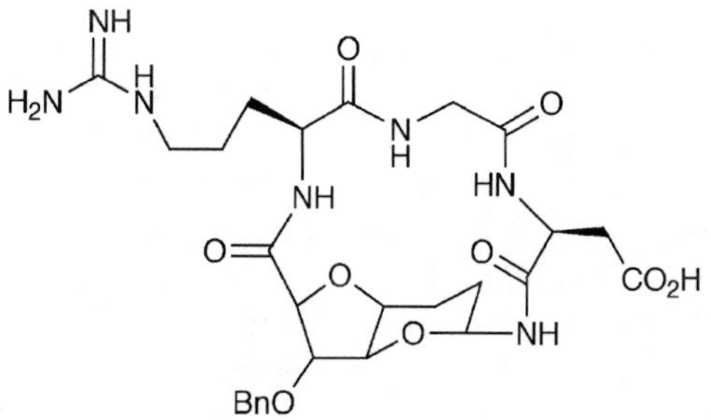

(53)

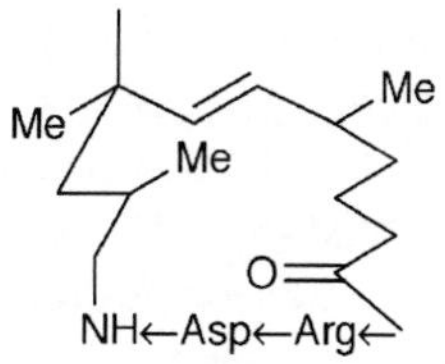

(54)

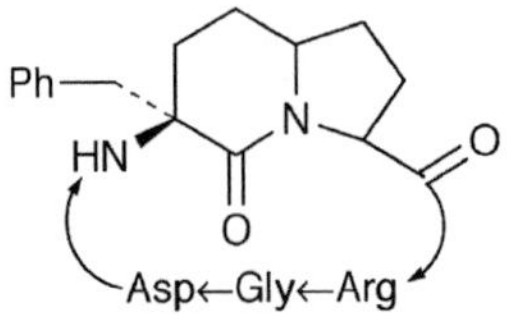

(55)

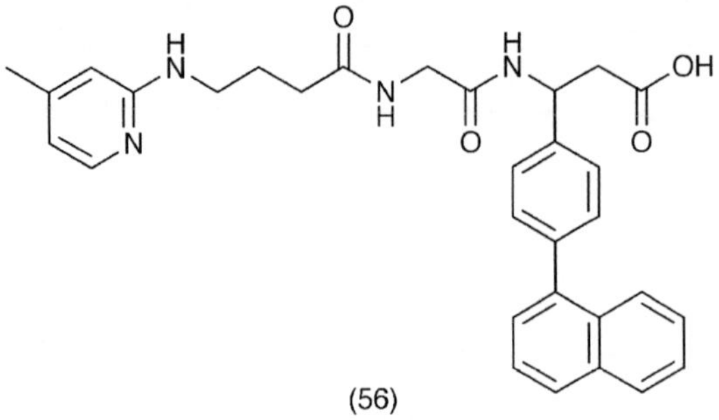

(56)

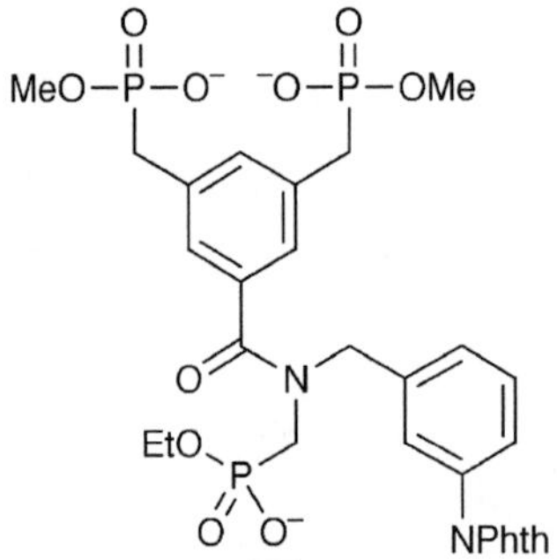

(57)

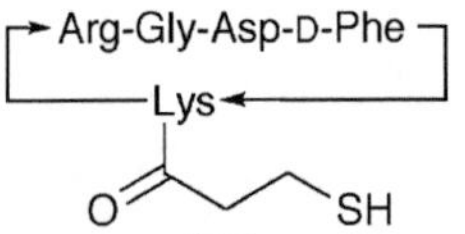

(58)

Full details[98] for the synthesis of two congeners E_2 and E_3 (59 and 60) within the cyclotheonamide family have appeared. Other congeners have already been synthesised, but in the current approach, the formation of the α-ketoamide linkage at point (a) in the molecules has been carried out via the cyano-ylide activation of a carboxyl group. Cyclic templates[99] of α-MSH carrying an enlarged cyclopentapeptide ring have been developed as antagonists of hMC3 and hMC4 receptors. Thus (61) had a K_i=57 nM and a pA_2 of 9.5 as antagonist of hMC4 while (62) had values of 5.9 and 10.6 respectively against hMC3. High-resolution crystal structures have been obtained[100] of complexes formed between naturally-occurring argifin (63) and argadin (64) and the carbohydrate-processing enzyme chitinase. Both cyclopentapeptides interact with Asp-142, Glu-144 and Tyr-214 in the chitinase active site, which are conserved residues in all 18 chitinases.

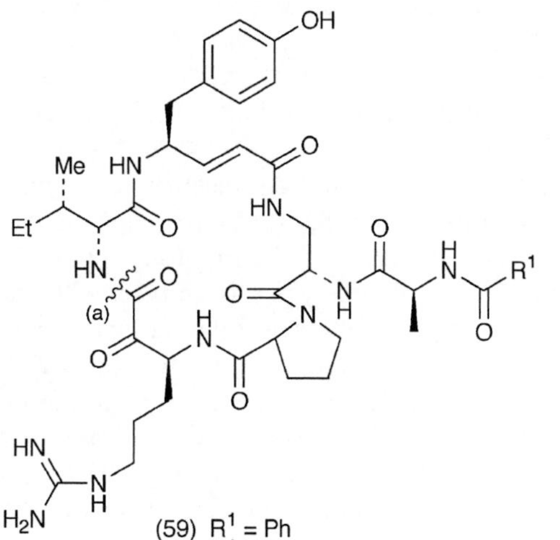

(59) R^1 = Ph
(60) R^1 = CH_2CHMe_2

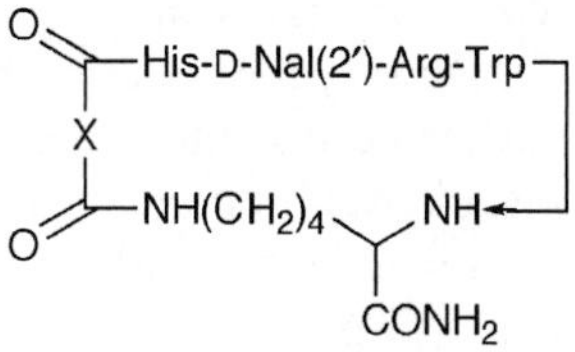

(61) X = [o-phenylene]
(62) X = $—(CH_2)_3—$

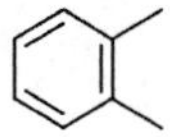

(63)

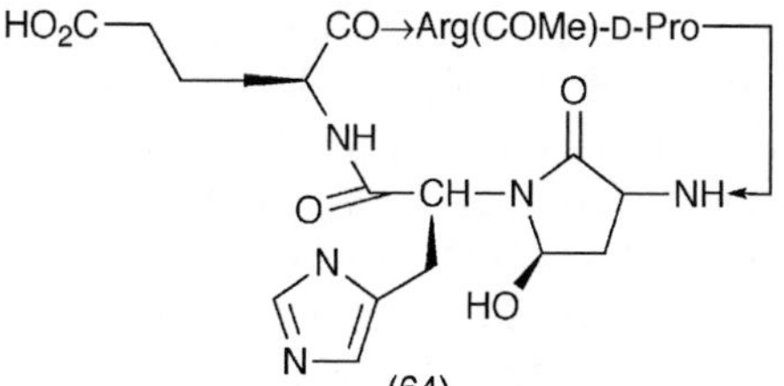

(64)

2.6 Cyclohexapeptides. – Spectroscopic analysis and chemical conversions have enabled[101] structures to be deduced for the mannopeptimycins [α, β, γ, δ and ε (65–69)], a series of novel antibacterial glycopeptides from *Streptomyces hygroscopicus* LL-AC98. The fruit peels of the higher plant *Citrus aurantum* have yielded[102] cyclo (Gly-Leu-Val-Leu-Pro-Ser) and the cyclo-octapeptide cyclo (Gly-Gly-Leu-Leu-Leu-Pro-Pro-Phe).

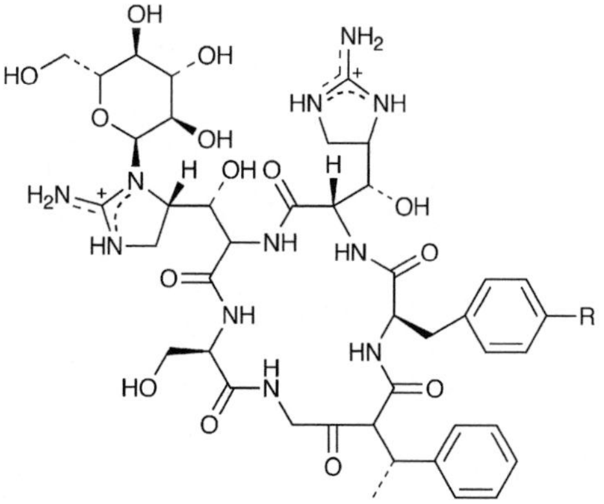

(65) R 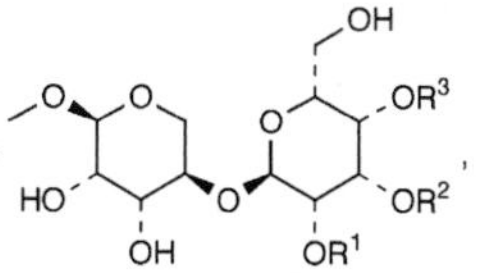 = ; $R^1 = R^2 = R^3 = H$
(66) R = OH
(67) R = (65) but R^1 = 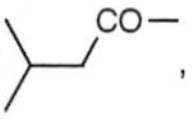 , $R^2 = R^3 = H$
(68) R = (65) but $R^1 = R^3 = H$, R^2 = 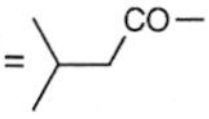
(69) R = (65) but $R^1 = R^2 = H$, R^3 =

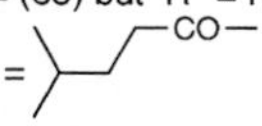

Several novel analogues of SHU9119, [AcNle-cyclo (Asp-His^{6}-D-Nal(2')-Arg-Trp-Lys)NH_2], a high affinity nonselective antagonist at hMC3R and hMC4R, and with potent agonist activity at hMC1R and hMC5R, modified at position 6 have been synthesized.[103] Substitution of a cyclohexyl or cyclopentyl ring at 6, gave highly selective hMC4R antagonists (IC_{50}=0.48 nM and 0.51 nM respectively), while the cyclopropyl analogue showed selective antagonism at hMC3R (IC_{50}=2.5 nM). Echinocandins are of interest as anti-fungal agents, and a new series of analogues have been discovered[104] by initiating a pinacol-type rearrangement of echinocandins [with diol moiety at position A in (70) and (71)] to give the resulting ketones which are themselves useful intermediates for further derivatisation. A *cis* amide bond surrogate, in the form of a 1, 2, 4-triazole has been incorporated[105] into the RA VII analogue (72) by treating the native molecule with Lawesson's reagent followed by formic hydrazide. A synthetic protocol to make the strained 14-membered L, L-cyclodityrosine ring in (72) and in other natural products such as piperazinomycin, bouvardin and RP66453, has been published.[106] Continuing with the role α-aminoxy acids might have in the design of new foldamers, the cyclic hexamer (73) has been synthesized.[107] Previously published methodologies were used to construct a linear precursor, which after end-deprotection was cyclised with PyAOP in 35% yield. A conformational study on (73) showed that it adopted a bracelet-like structure similar to valinomycin, but in contrast to the latter, it complexed halide ions more strongly than cations. Efforts to design and synthesise artificial siderophores in place of desferrioxamine B as treatment of iron overload in the liver and pancreas, have struck on (74) which forms a 1:1 iron-III complex,[108] with useful characteristics. The cyclic hexapeptide was synthesised according to Scheme 18. Enantioselective synthesis[109] of (75) and (76) (macrocyclisation with HATU/HOAt) has enabled uv/vis and emission spectroscopy to show evidence of 2:1 sandwich complexation with alkaline earth metal cations. The disulfide bridge in human urotensin II (hU-II), a

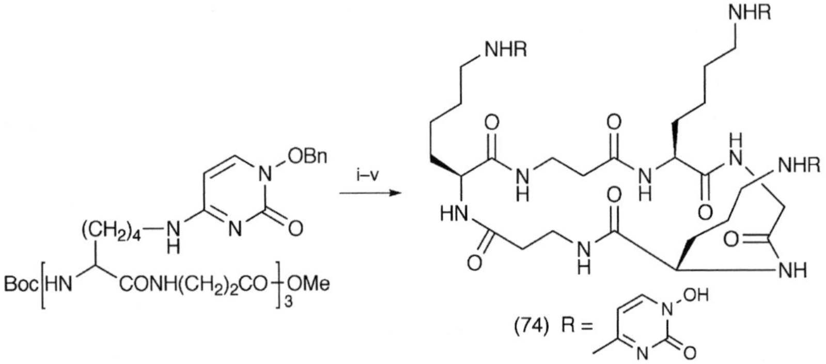

Reagents: i, 1 MNaOH/MeOH; ii, WSC·HCl/HOSu; iii, TFA/DCM; iv, pyridine under high dilution; v, H_2/10% Pd/C in DMF/MeOH

Scheme 18

ligand of an orphan G-protein-coupled receptor, has been replaced with lactam bridges[110] as analogues of the active fragment hU-II (4-11). The most active analogue [H-Asp-cyclo (Orn-Phe-Trp-Lys-Tyr-Asp)-Val-OH (with lactam bridge joining Orn-Asp) was still two orders of magnitude less potent than the natural material, which was rationalised as the inability of the Trp residue to maintain an active conformational role.

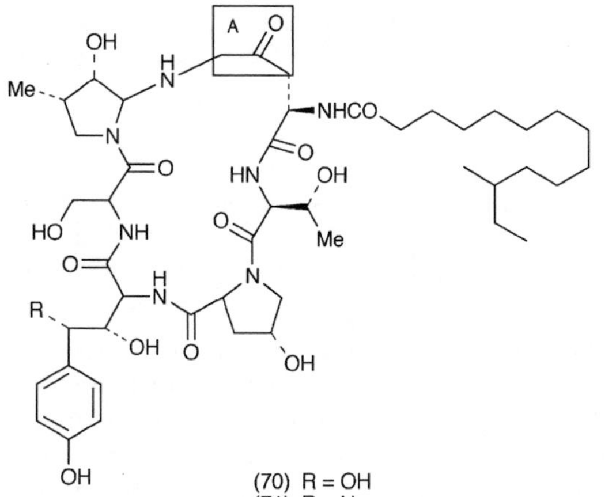

(70) R = OH
(71) R = H

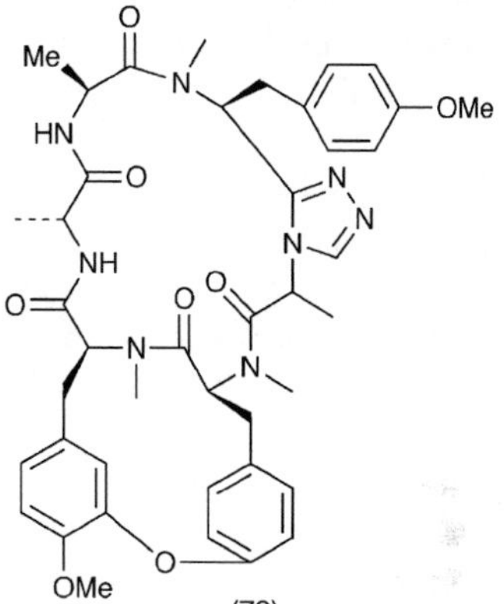

(72)

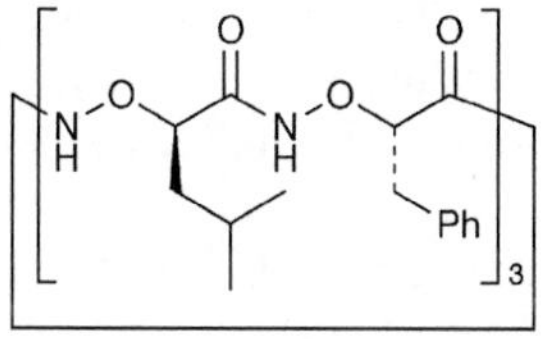

(73)

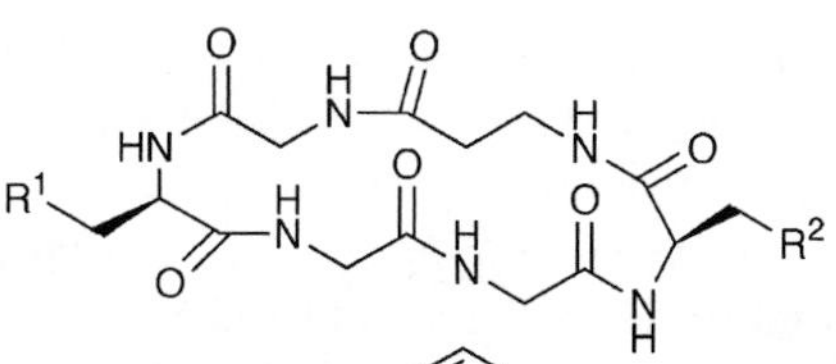

(75) $R^1 = R^2 =$ 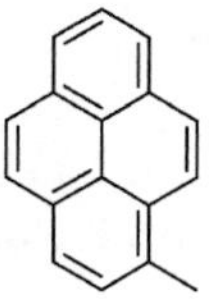 (pyrenyl)

(76) R^1 = pyrenyl, R^2 = H

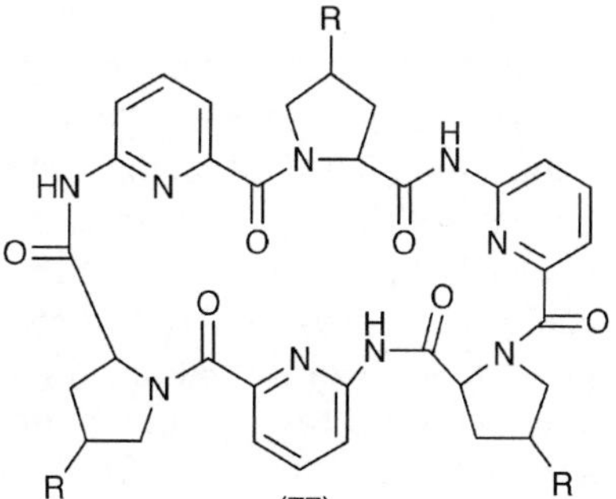

(77)

The conformational equilibrium of cyclo (Gln-Trp-Phe-Gly-Leu-Met), a NK-2 tachykinin antagonist has been studied[111] using 2D-NMR and theoretical methods. The conformations agreed well with published data, but this approach also provided values for the statistical weights of the most populated conformations, enabling the conformational equilibrium to be worked out. MO AM1 and density functional B3LYP computational methods have been applied[112] to cyclo $(Phe\text{-}D\text{-}Ala)_n$ and cyclo $(Phe\text{-}D\text{-}MeAla)_n$ with n varied between 3 and 6 to study their potential for self-assembly to nanotubes. The conclusions deduced from the calculations were: (i) a perfectly planar ring with outwardly expanded side groups [occurs with cyclo (Phe D-Ala)4] is required for self-assembly;(ii) the intrinsic driving force is ability to form H-bonds between monomers stacked along C4 axes and (iii) N-Me groups block 'congregation'. *Ab initio* calculations, molecular mechanics and molecular dynamics simulations have been used[113] on cyclohexaalanyl, simulating both aqueous and gas phase conditions. Amphi-ionophore properties were found in both phases. In the presence of cations (Li^+, Na^+) the CO groups tend to fold inward to capture a cation, whereas in the presence of anions the N-H groups tend to fold inward to capture an anion (F^-, or Cl^-). The conformations of anion-binding cyclohexapeptides bearing 4-hydroxyproline and 6-aminopicolinic acid residues as in (77) (R=*R* or *S*- OH) have been investigated[114] and their complexing properties assessed. While the analogue with R=(*R*-)OH forms 1:1 sandwich-type with halide and sulfate anions, the non-natural configuration, R=(*S*-)OH is much less pre-organised for anion binding. Two cyclohexapeptides related to (77) have been linked[115] using an adipic acid spacer to yield a 'molecular oyster' with remarkable sulfate anion affinity. The compound formed a 1:1 complex with halides, sulfate and nitrate, with stability constants in the range 10^5–10^2 M^{-1}, decreasing in the order $SO_4^{2-} > I^- > Br^- > Cl^- > NO_3^-$. Cyclic analogues (78–81) of the active site hexapeptides of glutaredoxin (Grx), thioredoxin (Trx), protein disulfide isomerase (PDI) and thioredoxin reductase have been synthesized[116] and their redox-potentials were found to increase in the order (81) < (78) < (79) < (80) with E'_o values ranging between −204 mV and −130 mV. The yield and rate of folding of reduced RNase A in the presence of these cyclic peptides increased with the oxidising character of the peptides.

Intramolecular quenching of Trp fluorescence by the peptide bond in mono-Trp-substituted cyclohexapeptides such as cyclo (D-Pro-pTyr-Thr-Phe-Trp-Phe) has been studied[117] which included determining the aqueous solution conformation by NMR techniques. As the cyclohexapeptide has a single predominant conformation, the three fluorescence lifetimes have been correlated with the rotamer populations of the side-chains. Crystallographic data accumulated[118] for cyclo $[(D, L\text{-}Pro)_2\text{-}(Ala)_4]$ monohydrate confirm that the topology of the experimental charge density shows good agreement with the topology of the constituent amino acids. The oxytocin receptor antagonist cyclohexapeptide L-366,948 has been computer-docked[119] to reliable oxytocin receptor and arginine/vasopressin receptor models via a genetic algorithm.

Identification of key amino acid residues regularly identified with the interactions, could help with site-directed mutagenesis analysis.

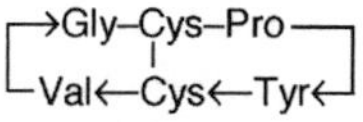

→Gly–Cys–Pro
Val←Cys←Tyr←
(78) (Grx peptide 1)

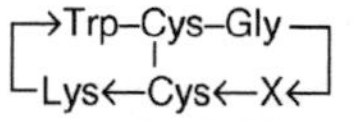

→Trp–Cys–Gly
Lys←Cys←X←
(79) X = Pro (Trx peptide 2)
(80) X = His (PDI peptide 3)

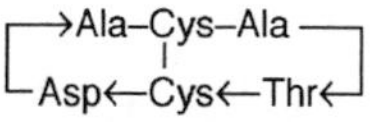

→Ala–Cys–Ala
Asp←Cys←Thr←
(81) (Trr peptide)

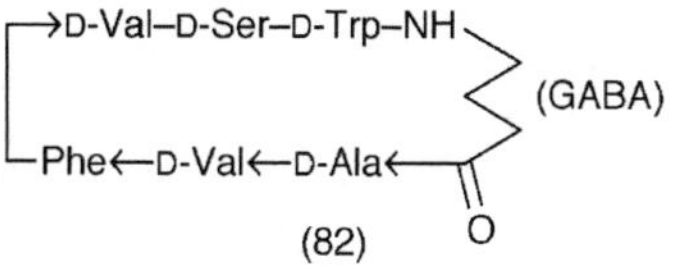

→D-Val–D-Ser–D-Trp–NH (GABA)
Phe←D-Val←D-Ala←C(=O)
(82)

2.7 Cycloheptapeptides. – In addition to the known unguisins A and B, the fungus *Emericella unguis* also produces[120] unguisin C (82) which contains a GABA-derived moiety, and the related unguisin D, which has one of the D-Val residues in (82) replaced by D-Leu. Fruit peels of *Citrus medica var. sarcodactylis* Swingle have yielded[121] cyclo (Gly-Asp-Leu-Thr-Val-Tyr-Phe) as well as the cyclooctapeptide, cyclo (Gly-Leu-Pro-Trp-Leu-Ile-Ala-Ala). Cyclosenegalin A [cyclo (Pro-Gly-Leu-Ser-Ala-Val-Thr)] and the cyclooctapeptide, cyclosenegalin B [cyclo (Pro-Gly-Tyr-Val-Tyr-Pro-Pro-Val)] have been isolated[122] from the methanol extract of the seeds of *Annona senegalensis Pers.* along with the known glabrin A.

Systematic replacement[123] of residues in hymenamide C [cyclo (Leu-Trp-Pro-Phe-Gly-Pro-Glu)] by alanine residues, showed no dramatic changes in biological activity, although the *cis-trans* conformations around Pro residues were changed by some of the substitutions. The linear precursors of the analogues were assembled on chlorotrityl resin, followed by macrocyclisation with HATU. Cyclo (Gly-Ile-Pro-Tyr-Ile-Ala-Ala) from *Stellaria yunnanensis* Franch (M) has been synthesised[124] in the solution phase from its H-Gly-Ile-Pro-Tyr-Ile-Ala-Ala-OH precursor using DEPBT in the presence of several metal ions. Cs^+ ions enhanced both the yield and cyclisation rate, while bivalent metal ions such as Mg^{2+}, Ca^{2+} and Ni^{2+} decreased the yield drastically, with Cr^{3+} completely preventing cyclisation. Progress has been made[125] with synthesising suitably protected units corresponding to the regions A-D in (83), representing cyclomarin A, which has been licensed for pre-clinical trials. A series of cyclohepta RGD-containing peptides, have been modified[126] by incorporation of (2*S*, 3*S*) and (2*S*, 3*R*) forms of β-MeAsp instead of the aspartyl residue, to give Ac-cyclo[Cys-Arg-βMeAsp-Tyr(Me)-Arg-Cys]-NH2. The (2*S*, 3*S*) analogues had a higher binding affinity to $\alpha_{IIb}\beta_3$ and $\alpha_v\beta_3$ integrins than their (2*S*, 3*R*) counterparts.

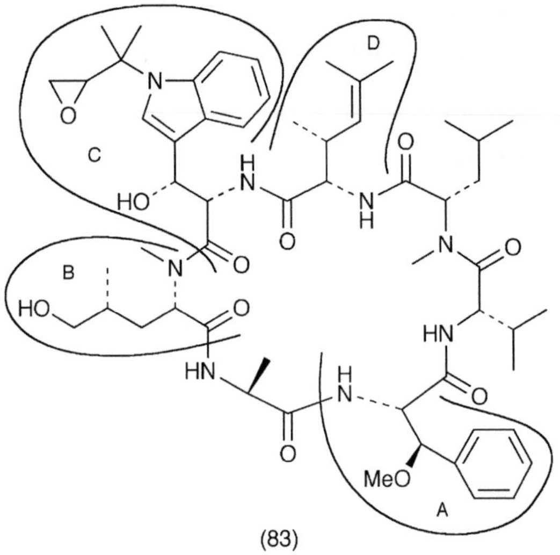

(83)

Two different conformational forms of phakellistatin 2 [cyclo (Phe-Pro-Ile-Ile-Pro-Tyr-Pro)] have actually been separated by hplc,[127] and are stable in methanol solution. They have different folds, H-bonding patterns and solvent accessible surfaces in their conformations. Changes in solvent polarity from $CDCl_3$ to CD_3OD, causes a large change in the conformation, and might explain the disparity between the biological activity of synthetic and extracted samples of phakellistatin 2. The core sequence ((His-Phe-Arg-Trp) of α-melanotropin (α-MSH) has been conserved in a cycloheptapeptide analogue,[128] cyclo (Nle-Gly-His-D-Phe-Arg-Trp-Gly), and shown to have three orders of magnitude higher biological activity than a cyclohexapeptide equivalent. The conformation of the cycloheptapeptide ring was revealed to contain two β-turns, running across the residues D-Phe-Gly and Gly-His. Another microcystin ([D-Leu]^1-microcystin LR) has been isolated[129] from a *Microcystis aeruginosa* strain, and its conformation completely assigned by NMR techniques. With only one residue change (D-Leu instead of D-Ala), the data were very similar to those for microcystin LR studied previously. The pros and cons of applying ion trap and quadrupole/orthogonal time of flight mass spectrometry to determining the sequences of polymyxin and colistin antibiotics have been reported.[130]

2.8 Cyclooctapeptides. – A new proline-rich cyclic octapeptide, axinellin C, cyclo (Thr-Val-Pro-Trp-Pro-Phe-Pro-Leu) has been isolated[131] from the Fijian sponge, *Stylotella aurantium* and showed weak activity against cancer tumours. Its crystal structure was surprisingly similar to a cyclic decapeptide, phakellistatin 8. Electrospray–MS/MS techniques[132] have detected two new cyclic octapeptides, CLF cyclo (Pro-Phe-Phe-Trp-Val-Met-Leu-Met) and CLG cyclo (Pro-Phe-Phe-Trp-Ile-Met-Leu-Met) in extracts of linseed seeds of *Linum usitatissimum.* Also present were oxidised forms (methionine oxide), CLD and CLE, of cyclo (Pro-Phe-Phe-Trp-Ile-Met-Leu-Leu) and cyclo (Pro-Leu-Phe-Ile-Met-Leu-Val-Phe). An analogue (84) of triostin A has been

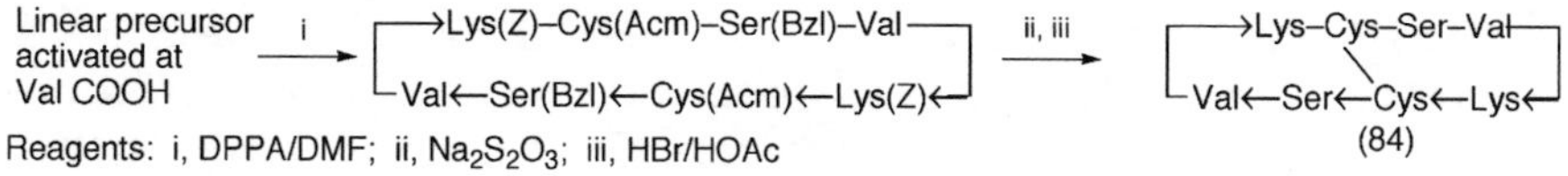

Linear precursor activated at Val COOH —i→ cyclo[→Lys(Z)–Cys(Acm)–Ser(Bzl)–Val–Val←Ser(Bzl)←Cys(Acm)←Lys(Z)←] —ii, iii→ cyclo[→Lys–Cys–Ser–Val–Val←Ser←Cys←Lys←] (84)

Reagents: i, DPPA/DMF; ii, $Na_2S_2O_3$; iii, HBr/HOAc

Scheme 19

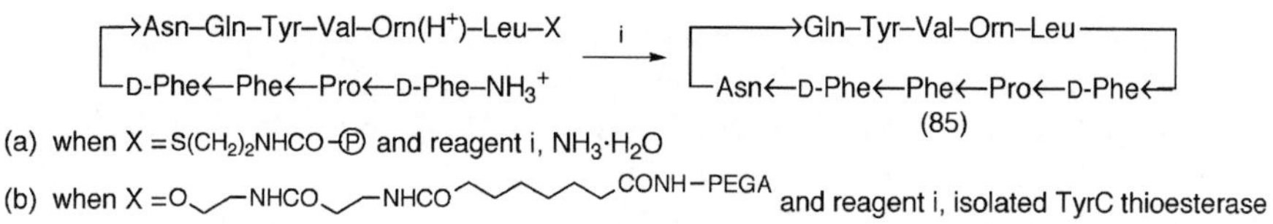

cyclo[→Asn–Gln–Tyr–Val–Orn(H^+)–Leu–X / D-Phe←Phe←Pro←D-Phe–NH_3^+] —i→ cyclo[→Gln–Tyr–Val–Orn–Leu–Asn←D-Phe←Phe←Pro←D-Phe←] (85)

(a) when X = $S(CH_2)_2NHCO$-Ⓟ and reagent i, $NH_3 \cdot H_2O$

(b) when X = O–CH₂CH₂–NHCO–CH₂CH₂–NHCO–(CH₂)₅–CONH–PEGA and reagent i, isolated TyrC thioesterase

Scheme 20

synthesised[133] via a solution phase cyclisation with DPPA as summarised in Scheme 19. Because of the presence of the disulfide bond, the final set of deprotections had to be carried out by HBr/HOAc.

2.9 Cyclononapeptides. – Extracts from *Linum usitatissimum* seeds have again yielded[132] cyclolinopeptides A [CLA, cyclo (Pro-Pro-Phe-Phe-Leu-Ile-Ile-Leu-Val)] and B [CLB, cyclo (Pro-Pro-Phe-Phe-Val-Ile-Met-Ile-Leu). Conformational analysis[134] using X-ray crystallography and nOe constraints from an NMR study has been carried out on CLA and CLB, and confirmed the presence of similar structures in both solid and solution, made up of two β-turns (type III and type I) and five intramolecular H-bonds. Tetrazole surrogates[135] have been inserted at positions Val^5-Pro^6 and Pro^6-Pro^7 in CLA with linear surrogate precursors being cyclised using TBTU. NMR studies confirmed that the analogue, cyclo (Leu-Ile-Ile-Leu-Val-Pro-Ψ[CN_4]-Ala-Phe-Phe) resembled the conformation of native CLA, and showed similar immunosuppressive activity. A heterodetic (disulfide link between two cysteines) cyclic nonapeptide, cyclo (Cys-Pro-Gly-Pro-Glu-Gly-Ala-Gly-Cys) has been shown[136] by *in vivo* phage display to bind specifically to aminopeptidase P in breast vasculature. It could have an influence on the design of drugs for the prevention and treatment of breast cancer.

2.10 Cyclodecapeptides. – Synthesis of tyrocidine A (85) and its analogues has been achieved[137] on-resin by spontaneous cyclisation of a thioester linear precursor using aqueous ammonia (Scheme 20a). Residue replacement confirmed that the method was conformation dependent, and always required a D-amino acid at position 4 in the sequence. A chemo-enzymatic approach[138] has also revealed a requirement for the D-configuration at residue-4, and was carried out under the conditions outlined in Scheme 20b, and shows that an isolated thioesterase can catalyse the cyclisation of linear peptides immobilised on a solid support modified with a biomimetic linker. The thioesterase Tyc C responsible for the natural production of tyrocidine A has also been shown[139] to be a versatile catalyst for other linear precursors, and could accommodate an

Arg-Gly-Asp sequence as shown for the synthesis of cyclo (D-Phe-Ala-D-Ala-MeArg-Gly-Asp-Gly-Trp-Orn-Leu) which has a nanomolar potency in preventing fibrinogen binding to $\alpha_{II}\beta_3$ integrin. The timing of epimerisation and condensation reactions in the non-ribosomal peptide assembly lines have been explored[140] for tyrocidine synthetase B, drawing the conclusion that epimerisation of the Phe residue occurs at the dipeptidyl-S-enzyme rather than at the aminoacyl-S-enzyme stage. The challenges and opportunities provided by these 'combinatorial biosynthetic' approaches have been reviewed.[141]

Loloatins A-C, cyclo (Val-Orn-Leu-D-Tyr-Pro-Phe-D-Phe-Asn-Asp-Tyr), cyclo (Val-Orn-Leu-D-Tyr-Pro-Phe-D-Phe-Asn-Asp-Trp) and cyclo (Val-Orn-Leu-D-Tyr-Pro-Trp-D-Phe-Asn-Asp-Trp) repectively, isolated from cultures of tropical marine bacteria have been synthesised.[142] On-resin cyclisation of a linear precursor attached to the resin via the side chain of aspartic acid gave the cyclic decapeptides in overall yields of 31-37%. Loloatin A has also been isolated from the IGM52 strain of *Brevibacillus laterosporus* and its structure in solution assessed.[143] An X-ray crystallographic determination[144] on crystals (formed after standing for many years!) of a tetra N-methyl derivative of gramicidin S, [cyclo (Val-MeOrn(Boc)-Leu-D-MePhe-Pro)$_2$] has shown an unexpected mode of H-bonding between the substituted Orn side chain (urethane NH's) and the main chain. The main chain adopts an antiparallel β-pleated sheet, but is slightly twisted. Conformational restrictions based on indolizidine-2-one amino acids (86) (Boc-6R I^{2}aa) and (87) (Boc-6S I^{2}aa) have been incorporated[145] into gramicidin S (GS) as shown in the general structure (88). Analogues, [6*S*-I^{2}aa$^{4-5,4'-5'}$] GS, [Lys$^{2,2'}$, 6*S*-I^{2}aa$^{4-5,4'-5'}$] GS and [6*R*-I^{2}aa$^{4-5,4'-5'}$] GS were synthesised via a convergent approach with final cyclisation carried out at point (a) in (88) using TBTU/HOBt in yields of 60%. All analogues showed a β-pleated sheet structure but had 2-fold less antimicrobial and antifungal activity than gramicidin S, with the first two listed analogues being more active than the third. The biological activities of 'ornithine-rich' gramicidin S and analogues with enlarged rings have been studied.[146] The Gram-negative activity of the cyclic tetradecapeptide analogue was greater than that of gramicidin S. The effect of ring size analogues of gramicidin S on phospholipid model membranes and on the growth of *Acholeplasma laidlawii* has also been explored.[147] Differential scanning calorimetric techniques showed that the three analogues of gramicidin S with ring sizes of 10, 12 and 14 amino acid residues, all perturbed the gel/liquid-crystalline phase transition of zwitterionic phosphatidyl ethanolamine or of anionic phosphatidylglycerol vesicles in contrast to GS. The order of growth inhibition on *A. laidlawii* was GS14 > GS10 > GS12 and is qualitatively correlated with abilities to affect the permeability of phospholipid membranes. This gives further support for the hypothesis that the primary target of these antimicrobial peptides is the lipid bilayer of the bacterial membrane. The structural determinants of type II' β-turns has been probed[148] using CD, NMR and molecular dynamics analysis of ten analogues e.g. cyclo (Val-D-Tyr-Pro-Leu-Lys-Val-Lys-Leu-D-Tyr-Pro-Val-Lys-Leu-Lys) of a 14-residue gramicidin S analogue. This was done by changing i+1 and i+2 residues of one β-turn and assessing

%β–sheet character of the resulting compound. The main conclusion drawn were: (i) heterochirality is essential – Gly or Sar don't work; (ii) equatorial/axial rule of Rose *et al.* accounts for about 60% of type II' β-turn stabilisation, and (iii) static rotational barriers are a contribution to β-turn stabilisation with Pro being better than N-methylation. Various ring–sized gramicidin S analogues have also featured[149] in determining the role of peptide-membrane interactions in the biological activity of cyclic cationic peptides. The chosen GS10, GS12 and GS14 cyclic analogues this time were, cyclo (Val-Lys-Leu-D-Tyr-Pro)2, cyclo (Val-Lys-Leu-Lys-D-Tyr-Pro-Lys-Val-Lys-Leu-D-Tyr-Pro) and cyclo (Val-Lys-Leu-Lys-Val-D-Tyr-Pro-Leu-Lys-Val-Lys-Leu-D-Tyr-Pro) respectively, as well as a [D-Lys^4]GS14 analogue. A combination of electrostatic and hydrophobic interactions induced aggregation of PE/PG vesicles in the order, [D-Lys^4]GS14 > GS14 > GS12 > GS10. The structural framework of [D-Lys^4]GS14 has been systematically altered[150] by residue replacement in the hydrophobic sites of the molecule to give a range of hydrophobicity values. The hydrophobicity of [D-Lys^4]GS14 represented the mid-point for hemolytic activity, and it was deduced that the balance between activity and specificity can be optimised for each microorganism by systematic modulation of hydrophobicity.

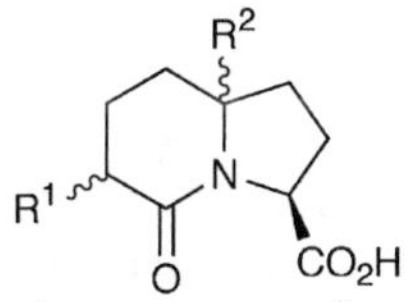

(86) R^1 = BocHN---, R^2 = —H
(87) R^1 = BocHN—, R^2 = ---H

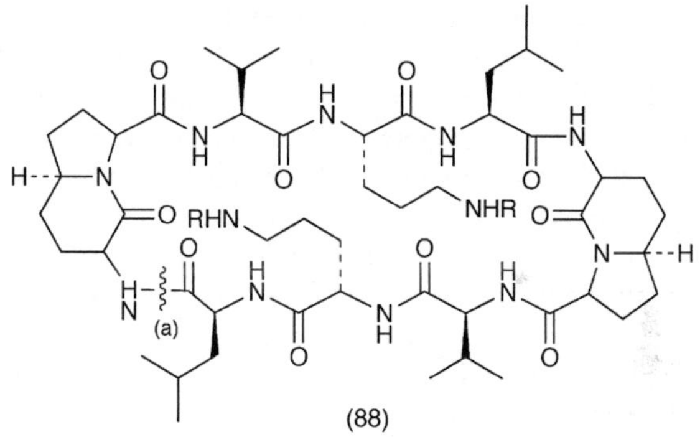

(88)

Pairs of Zn and Co(III) porphyrins [analogues(91–93)] linked to gramicidin S (89) have been shown[151] to form stable host-guest complexes with 4.4′-dipyridyl derivatives in dichloromethane. Tannins and related polyphenols associate[152] with gramicidin S *via* regioselective interactions with the β-turn structure as perceived by changes in the proton signals of Pro and Phe moieties. Carbohydrate binding, such as that between the aromatic side chains of [Gln$^{1,1'}$,Trp$^{3,3'}$]gramicidin S and mannose has been detected in nOe NMR studies.[153] Gramicidin S has also been used[154] as a model system to study the glycation of proteins using electrospray ionisation mass spectrometry. Formation of advanced glycation and diglycation of the ornithine residues was observed.

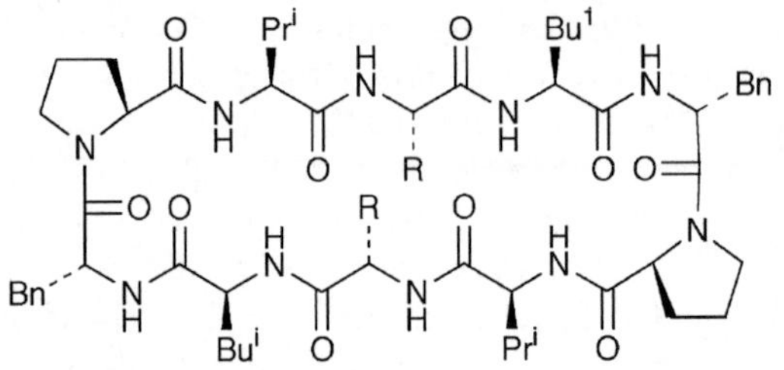

(89) R = $(CH_2)_3NH_3^+$
(90) R = $(CH_2)_3NHCO$-*p*-C_6H_4-tritolylporphoryl
(91) (90) Zn_2
(92) (90) [CoIII$(Py)_2Br]_2$
(93) (90) [CoIII$Cl_2]_2$

2.11 Higher Cyclic Peptides. – Work on the immunosuppressant drug cyclosporin A (CsA), (known commercially as sandimmune) and its analogues, once again dominates this sub-section. The immunochemical and crystallographic studies of the interaction between CsA [cyclo (MeBmt^1-Abu^2-MeGly^3-MeLeu^4-Val^5-MeLeu^6-Ala^7-D-Ala^8-MeLeu^9-MeLeu^{10}-MeVal^{11})] and the Fab of monoclonal antibody R45-45-11 have been reviewed.[155] A pseudo-proline residue has been incorporated 'post-synthetically' into cysteine-containing peptides as exemplified[156] by the formation of the CsA analogue (94) featured in Scheme 21. Using the Grubbs ruthenium catalyst, olefin metathesis has provided[157] a link between CsA and a polymer to give an affinity reagent (95), which was used to identify new members of the cyclophilin superfamily from cellular extracts. Stucture-activity studies[158] for the inhibition of the human MDR1 P-glycoprotein ABC reporter by the cyclosporins have shown that inhibition is favoured by larger hydrophobic residues in positions 1, 4, 6 and 8 and a smaller one on residue 7. N-Desmethylation of any of the seven N-methylated amides regularly led to decreased inhibitory activity. Of a series of CsA analogues tested[159] for the inhibition of the human FPR1 formyl receptor, [D-MeVal^{11}]-CsA (cyclosporin H) proved to be the most active inhibitor, although it is devoid of immunosuppresive and antifungal activities.

Two reports have appeared recording the crystal structure of calcineurin complexed with cyclosporin and human cyclophilin. Calcineurin (CN), a protein phosphatase, is the common target for two immunophilin-immunosuppressant complexes, the cyclophilinA(CyPA)-cyclosporin A and FKBP-FK506 complexes. The crystal structure of the latter complex with CN has already been reported, and it is now the turn of the CN-CyPA-CsA complex to be studied. In one report[160] the CyPA-CsA complex was shown to bind to a composite surface formed by the catalytic and regulatory units of CN, similar to where the FKBP-FK506 also binds, but CyPA-CsA also interacts with Arg-122 at the active site of CN. In the other report[161] it is also revealed that residues 3–9 of CsA, particularly MeLeu^4 and MeLeu^6 form a composite surface with Trp-121 of cyclophilin. The results do suggest that the best position to target improvement to CsA would be by alteration of the butenyl moiety of MeBmt^1 or by increasing the H-bonding to the carbonyls at positions

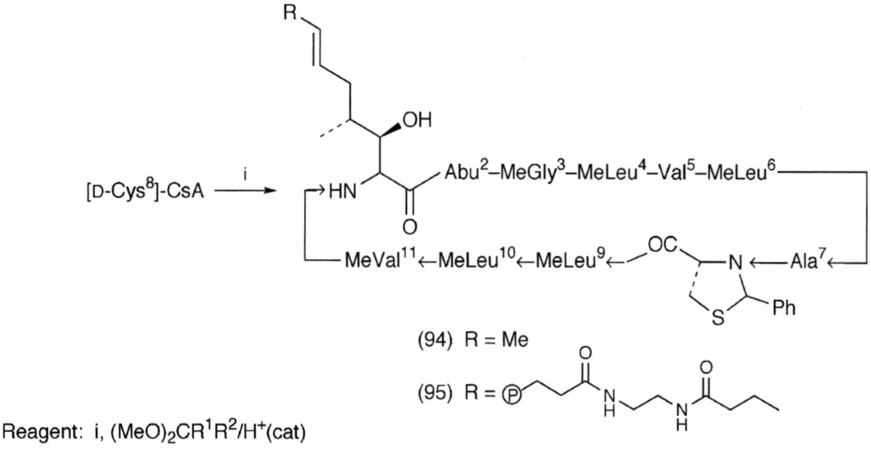

[D-Cys[8]]-CsA $\xrightarrow{i}$

Abu[2]–MeGly[3]–MeLeu[4]–Val[5]–MeLeu[6]

MeVal[11]←MeLeu[10]←MeLeu[9]← OC N ← Ala[7]←

(94) R = Me

(95) R = Ⓟ

Reagent: i, $(MeO)_2CR^1R^2/H^+$(cat)

Scheme 21

342 and 312 of CN. 11-Demethyl-cyclosporins A (CsE) and B exist[162] as single conformations in polar and non-polar solvents. NMR studies showed that the new NH present, forms a bond to the D-Ala[8] carbonyl group. The role of N-methylation on ring opening and fragmentation mechanisms in collision induced dissociation mass spectrometry has been recorded,[163] and the solubility of CsA in water has been enhanced[164] using a matrix consisting of stearic acid, phosphatidylcholine and taurocholate LIP. A conjugate of CsA with 4-benzoylbenzoic acid has been produced[165] under uv radiation, and then linked to poly-L-lysine using water soluble carbodiimide to generate a CsA immunogen.

A major breakthrough in obtaining good yields during the coupling of sterically hindered N-methylated amino acids has resulted from the careful tuning[166] of the triphosgene activation method. A successful solid phase synthesis of the nematicidal cyclododecapeptide omphalotin A, cyclo (Trp-MeVal-Ile-MeVal-MeVal-Sar-MeVal-MeIle-Sar-Val-MeIle-Sar) was made possible using this strategy, which is a major achievement, considering that nine out of the 12 amino acid residues are N-methylated. Assembly was commenced with a Sar residue attached to a trityl linker, followed by addition of Fmoc-protected residues using the triphosgene reagent. After cleavage of the resin under mild conditions, macrocyclisation was carried out in solution using HOAt/EDCI/DIEA in 31% yield. The triphosgene-based *in situ* formation of Fmoc-acid chlorides in this way has been the subject of a short review[167] of its success in the synthesis of omphalotin A and cyclosporin O.

Over the last decade a large family of circular disulfide-rich polypeptides have been discovered in plants and explored from the point of view of their potential in uterotonic action, anti-HIV and neurotensin antagonism. Now collectively named, cyclotides, their discovery, chemical syntheses , folding and biosynthetic origins have been reviewed.[168]

2.12 Peptides containing Thiazole/Oxazole Rings. – New structures based on these ring systems continue to be discovered in various environments. Thus the marine ascidian *Didemnum molle* has sourced[169] cyclodidemnamide B (96), and in order to confirm the stereochemical details, two forms were synthesised, using D-aIle and D-Ile by cyclisation (DPPA) at point (a) in (96). The natural compound was identical to the D-aIle analogue. *Streptomyces* sp. RSP9 has been shown[170] to be a source of radamycin (97) which is structurally related to the berminamycins, sulfomycins, promothiocins and A10255 complex. Leucamide (98), a moderately cytotoxic compound (GI50 = 5.1 μg/mL) has been isolated and characterised[171] from the Australian marine sponge *Leucetta microraphis*, while swinhoeiamide A (99) from sponge *Theonella swinhoei*, just about qualifies for this sub-section as its structure[172] only contains one oxazole ring. Cyanobacteria have been associated with the presence of bioactive cyclic peptides, but in a localisation study[173] patellamides A-C were found not to be present in cyanobacetrium *Prochloron* sp. but were distributed throughout the ascidian tunic *Lissoclinum patella*. The structure of micrococcin P (100) has been re-investigated[174] and although the Bycroft-Gowland hypothesis has been confirmed experimentally, some fine differences between the structure and the data for the natural product still persist. An 'advanced' Marfey's method has been developed[175] for the determination of the absolute configuration of thiazole amino acids and has been used to determine the absolute configuration of microcyclamide (101) and for the confirmation of the configuration of waikeamide and goadsporin.

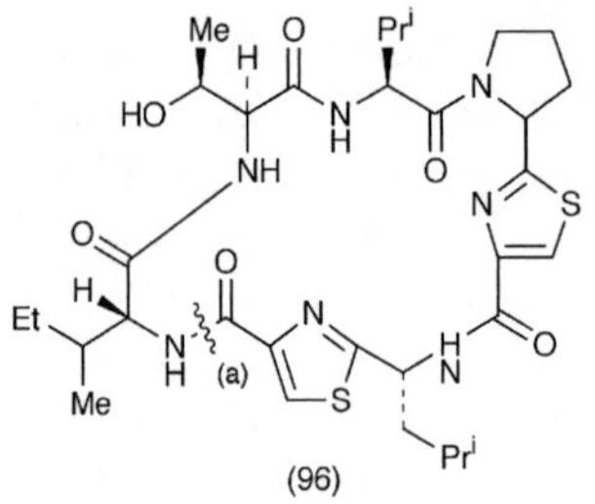

(96)

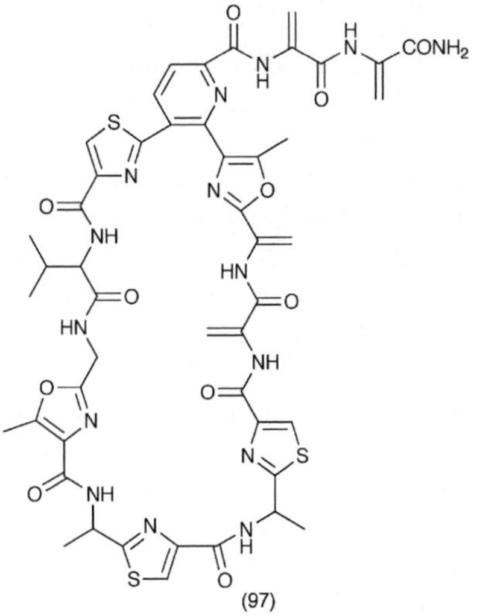

(97)

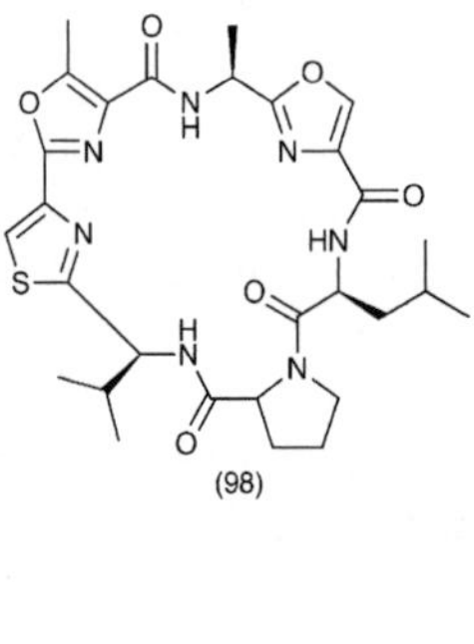

(98)

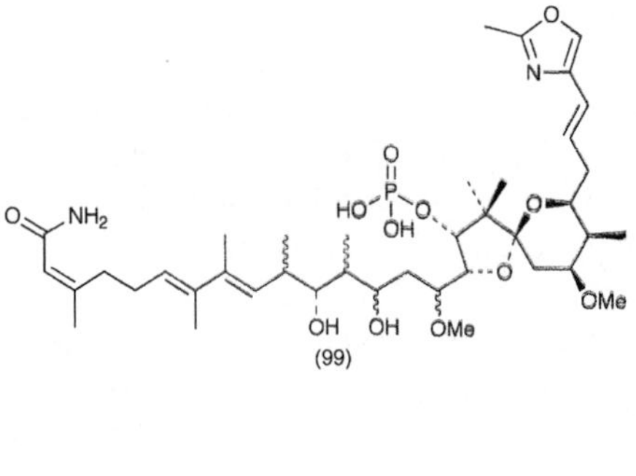

(99)

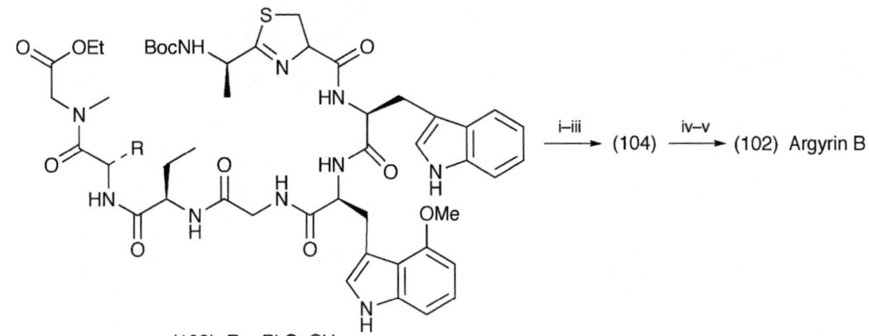

(103) R = $PhSeCH_2$
(104) R = $=CH_2$

i–iii → (104) iv–v → (102) Argyrin B

Reagents: i, $LiOH/THF/MeOH/H_2O$, r.t.; ii, anisole/TFA, r.t.; iii, TBTU/HOBt/DIPEA; iv, $NaIO_4$/dioxan/H_2O; v, $NaHCO_3/CH_3CN/H_2O$

Scheme 22

Details[176] of the synthesis of dendroamide A, nostocyclamide and other cyclic trimers, which have previously been presented in preliminary form have now appeared as a full paper with experimental details. A range of physico-chemical techniques has been used[177] to deduce the structure of the argyrin family obtained from the myxobacterium, *Archnagium gephyra*, and can now be represented as structures A-H in structure (102). Argyrins A and B are identical to antibiotics A21459A and B. The total synthesis[178] of argyrin B (102) has been successfully achieved, with the strategy used in the later stages, [from (103) via (104)] as summarised in Scheme 22. The group of Harran and co-workers in 2001, after a demanding multistage synthesis, recommended that the original structure of diazonamide A needed modification to (105). The background leading to the re-assignation of structure has been reviewed briefly,[179] and although the new structure (105) posed greater challenges than the original, Nicolaou *et al.*[180] have successfully made diazonamide A (105). Scheme 23 summarises highlights from this demanding synthetic task. The synthesis[181] of a bis-indole and an indole salicylate with the required axial chirality for diazonamide A has also been reported.

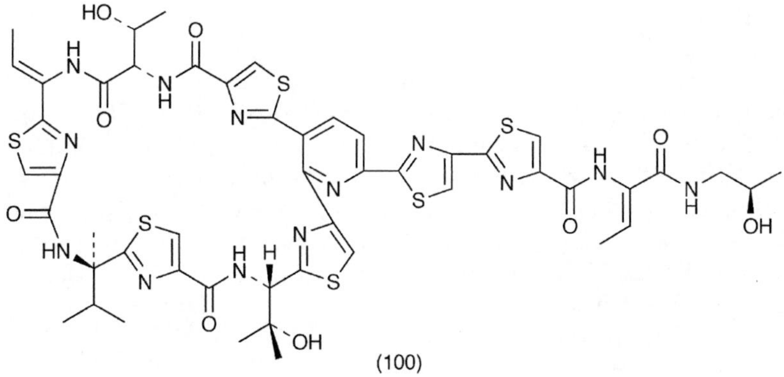

(100)

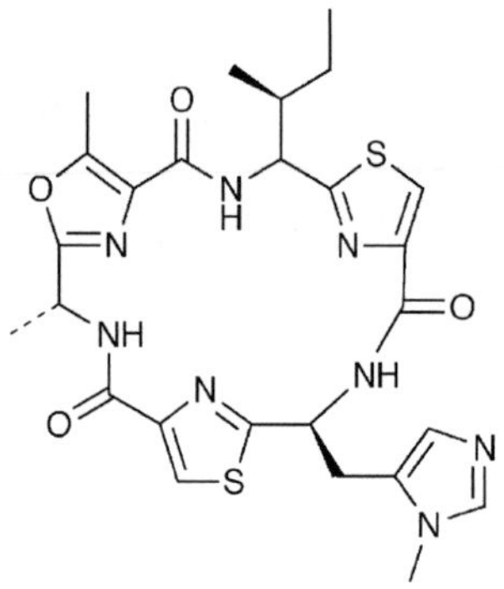

(101)

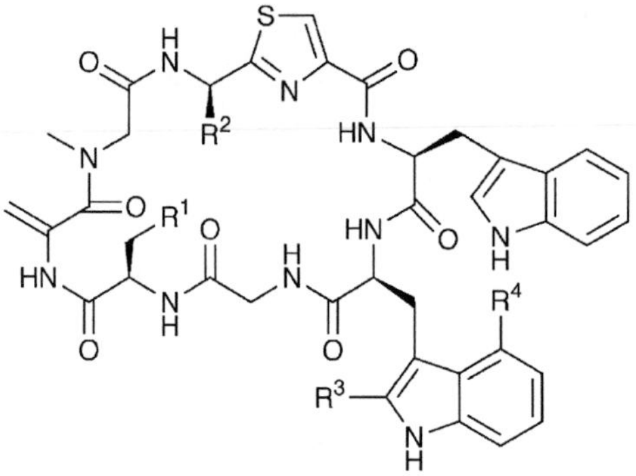

(102)

Argyrin A $R^1 = R^3 = H$, $R^2 = Me$, $R^4 = OMe$
Argyrin B $R^1 = R^2 = Me$, $R^3 = H$, $R^4 = OMe$
Argyrin C $R^1 = H$, $R^2 = R^3 = Me$, $R^4 = OMe$
Argyrin D $R^1 = R^2 = R^3 = Me$, $R^4 = OMe$
Argyrin E $R^1 = R^3 = R^4 = H$, $R^2 = Me$
Argyrin F $R^1 = R^3 = H$, $R^2 = CH_2OH$, $R^4 = OMe$
Argyrin G $R^1 = Me$, $R^2 = CH_2OH$, $R^3 = H$, $R^4 = OMe$
Argyrin H $R^1 = R^2 = R^3 = H$, $R^4 = OMe$

Ceratospongamide possesses the interesting feature that rotational isomers around the Phe-Pro bond can be isolated separately as *cis*, *cis*-and *trans*, *trans*-ceratospongamide. Crystal structures for the *cis*, *cis* form (106) have been obtained[182] A full paper[183] has appeared on the synthesis of (106), which involved surveying two possible macrocyclisation positions. The best yields came from cyclisation at position (a) in (106), and after attempts to cyclise with HATU, FDDP (pentafluorophenyl diphenylphosphinate), and DPPA, it was FDDP/iPr$_2$Net/DMF for 50 hr that gave the best yield (60%). The oxazoline ring was constructed post-cyclisation. The Shiori group[183] have also reported problems in converting the *cis*, *cis*- to the *trans*, *trans*-form, and have eventually concluded that thermodynamic isomerisation is initiated by C_α epimerisation of Ile and the product is *trans*, *trans*-[D-aIle] ceratospongamide and not the direct *trans*, *trans*-isomer. However in a detailed study[184] of rotamer interconversion in ceratospongamide it has been found that *cis*, *cis* could be transformed to *trans-trans* at 175°C in degassed D^6 DMSO, giving an equilibrium mixture 5:1 in favour of *trans*, *trans*. It was also shown in this work that using Fmoc protection in a linear precursor [ready to cyclise at point (a) in (106) using BOP/DMAP] it allowed the formation of the oxazoline ring prior to cyclisation, and on cyclisation this yielded a 1:10 ratio in favour of the trans, trans form. Another total synthesis of (106) has also been reported[185] using a strategy of cyclisation of a threonine-containing linear precursor at point (b) in (106) using DPPA/DIPEA/DMF. Dehydration of the cyclic threonine analogue to (106) used Deoxo-Fluor [bis (2-methoxyethylaminosulfur trifluoride)], giving a mixture of two diastereoisomers.

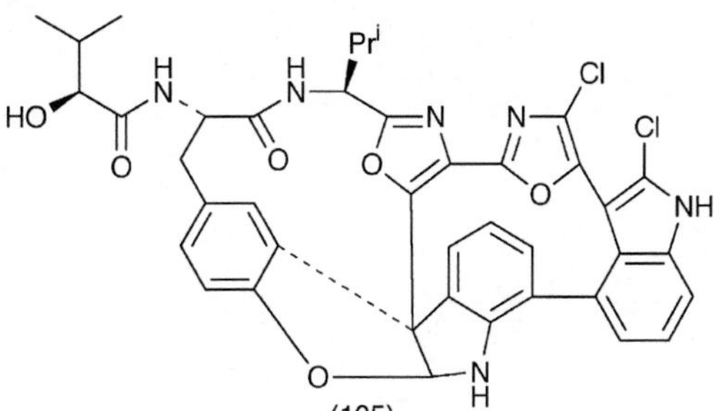

(105)

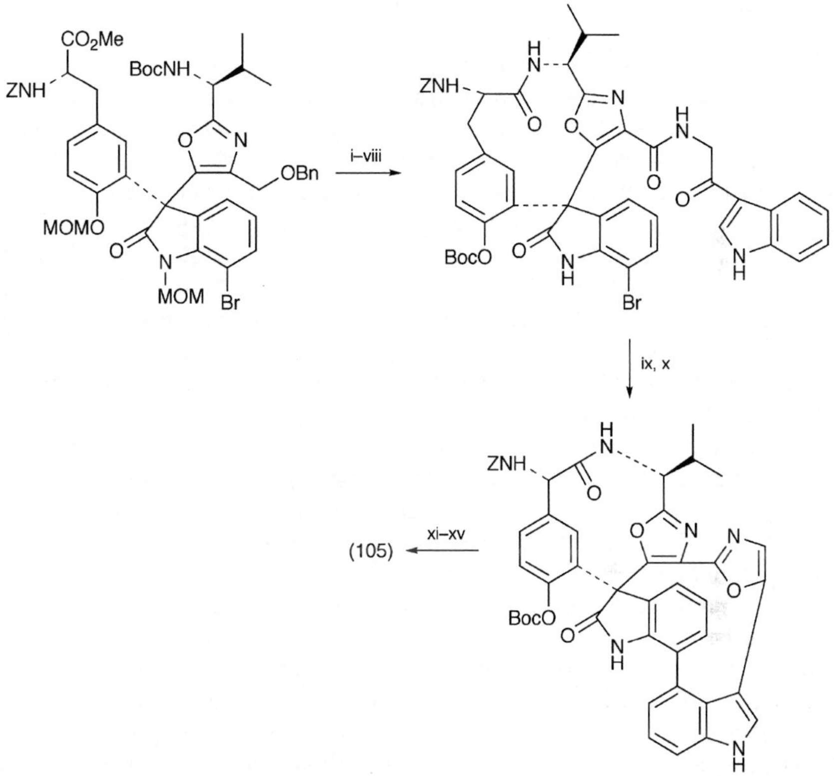

i–viii

ix, x

xi–xv

Reagents: i, LiOH; ii, TFA; iii, HATU/collidine; iv, BCl_3 (–78 °C); v, NaOH; vi, Boc_2O; vii, IBX

viii, $NaClO_2$/ (3-($COCH_2N_2$)indole) /EDC/HOBt; ix, $POCl_3$/Py; x, *h*ν, HOAc; xi, NCS; xii, TFA;

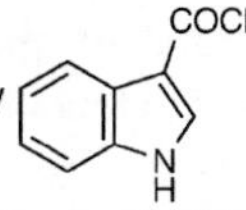

xiii, DIBAL–H; xiv, H_2/Pd/C; xv, $Me_2CHCH(OH)CO_2H$ /EDC/HOBt

Scheme 23

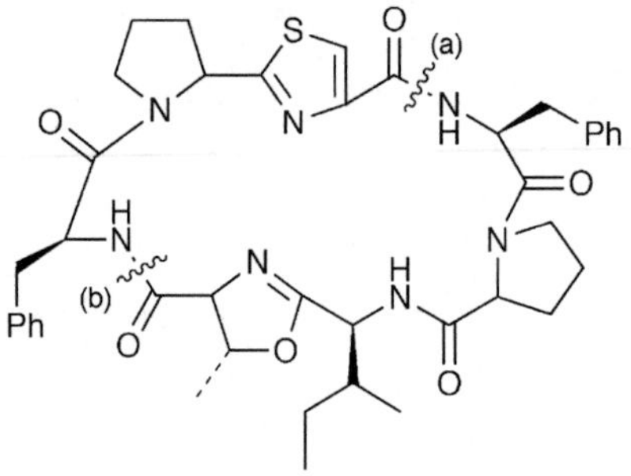

(106)

Based on a previously published solid phase protocol for synthesising deglycobleomycins, a new report covers the total synthesis[186] of bleomycin A5 (107) and three monosaccharide analogues with R=D-mannose, L-gulose or L-rhamnose. A search[187] for the major catabolite of bleomycin A_2 (108), has involved the synthesis of demethyl deamidobleomycin A_2 and deamidobleomycin A_2, as well as their aglycones. The most significant alteration of function noted in this study was a reduction in the ability of deamidobleomycin A_2 to mediate double-strand DNA cleavage relative to the original bleomycin A_2. A convenient synthesis[188] of a protected P-2 fragment (109) of macrocyclic antibiotic berninamycin A has been accomplished. The dehydroalanine residues were constructed during a last step dehydration (MsCl/Et_3N/DBU) of seryl residues. Several fragments of the more complicated structures associated with this sub-section have become available. Thus, methyl (2S)-2-[1-(N-Boc)amino-2-hydroxy-2-methyl]propylthiazole-4-carboxylate, a fragment C section of thiocilline has been synthesised,[189] and tetrasubstituted dehydropiperidine and piperidine cores (110)[190] as well as dihydroquinoline portions[191] of the thiostrepton family of antibiotics have also been prepared[190] ready for further elaboration. Modified cyclopeptide cages can be obtained[192] via scaffolded cyclooligomerisations of heterocyclic amino acids as summarised in Scheme 24. Tris-macrocyclic rings such as (111) have been synthesised[193] and NMR solution structures determined, to assess their potential as prototypes for mimicking multiple loops of proteins. A new class of imidazole mimics of marine cyclopeptides such as westiellamide and ascidiacyclamide, have been presented[194] while smart peptide libraries made available via enzyme modifications can be produced for the micrococcin B17 amongst many other applications.[195] Modifications in the ascidiacyclamide [cyclo (Ile-Oxz-D-Val-Thz)$_2$] skeleton, have been extended[196] to include replacements at Ile^1 and Val^3 residues. Extensive X-ray, NMR and CD studies carried out on the analogues, showed most of the replacements at position 1 were folded, except for the D-Val^1 analogue which had a square form. All analogues except Val^3 showed positive Cotton curves, and in cytotoxicity tests using P388 cells Val^1 analogue excelled, whereas epimers at positions 1 and 3 only showed fairly moderate activities. Chiral modification[197] of an oxazoline residue in ascidiacyclamide has yielded cyclo (Ile-aThr-D-Val-Thz)$_2$ which on

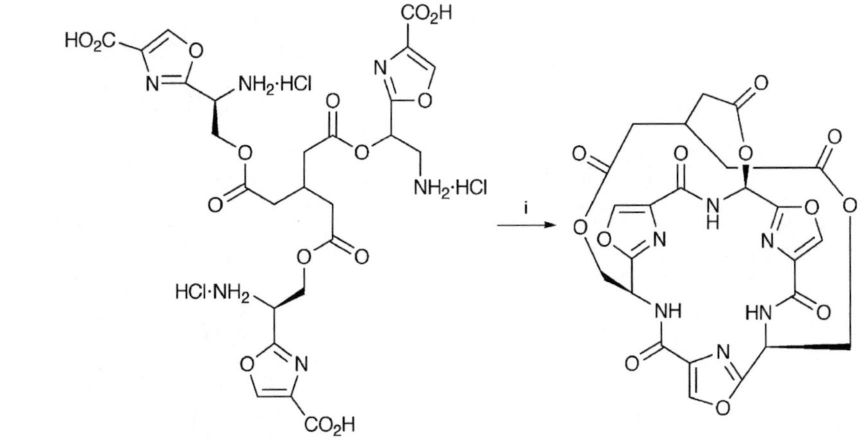

Reagents: i, FDPP/DMAP/DMF 5 days

Scheme 24

dehydration to the *SS*, *RR* form of ascidiacyclamide is a novel flat-squared conformation, retaining 60% of the cytotoxicity of the parent.

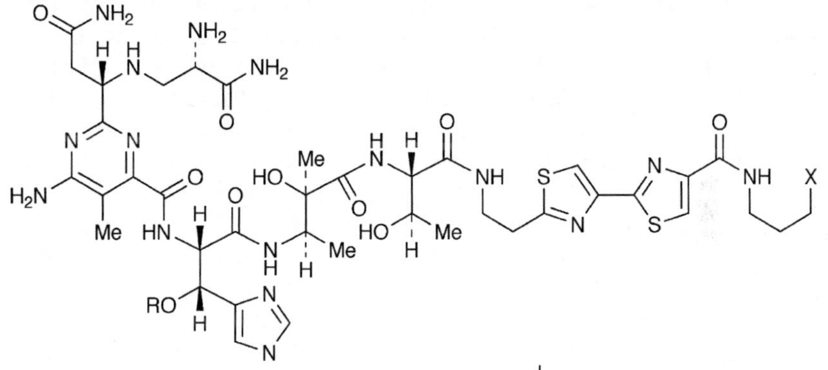

(107) X = NH–$(CH_2)_4NH_2$, R =

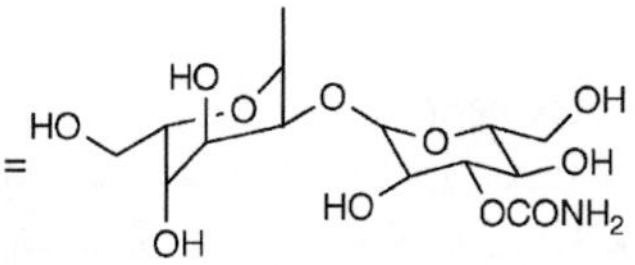

(108) as (107) except X = $S^+Me_2X^-$

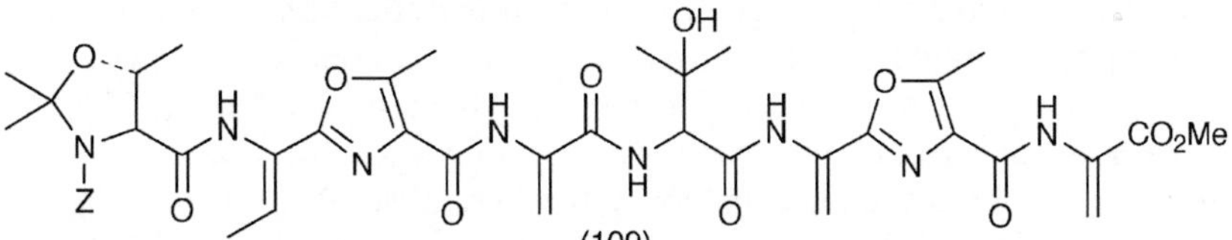

(109)

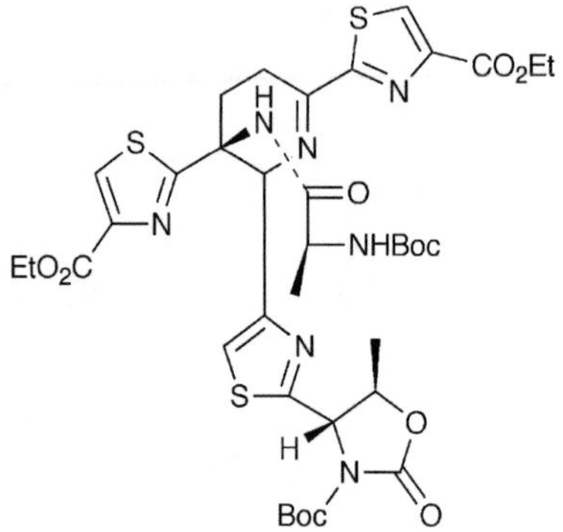

It has been confirmed,[198] by CD and nOe–restrained molecular dynamics studies that the patellamides (112) change conformation to complex with one equivalent of Cu^{2+}, in a form which pre-organises the molecule to complex with a second Cu^{2+} ion. Zn^{2+} does not give a conformational change, and it has been shown[199] that solvent polarity has a significant effect on conformational changes in the patellamides. Symmetrical patellamides in polar solvents gave an open type I conformation, whereas in non-polar solvents, symmetrical and asymmetrical forms gave a folded type II conformation. Three new cyclic peptide ligands based on patellamide A but without its oxazoline rings have been synthesised[200] by solid phase techniques, prior to cyclisation. The three, Pat J^1 [cyclo (Ile-Thr-Gly(Thz)-Ile-Thr-Gly(Thz)], Pat J^2 [cyclo (Ile-Thr-Gly(Thz)-D-Ile-Thr-Gly(Thz)], and Pat L cyclo (Ile-Ser-Gly(Thz)-Ile-Ser-Gly(Thz)], showed a similar saddle conformation when compared with the X-ray crystallographic structure of Pat J^1, and when coordinated with Cu^{2+} showed various mono and dinuclear Cu^{2+} complexes. Solution and calcium-binding properties have also been investigated[201] for cyclo [D-Thr-D-Val(Thz)-Ile]$_2$ and cyclo [Thr-Gly(Thz)-Ile-Ser-Gly(Thz)-Ile], which were shown to have a 'twisted figure of eight' conformation, and the strength of their 1:1 calcium binding depended on the degree of conformational flexibility in the system.

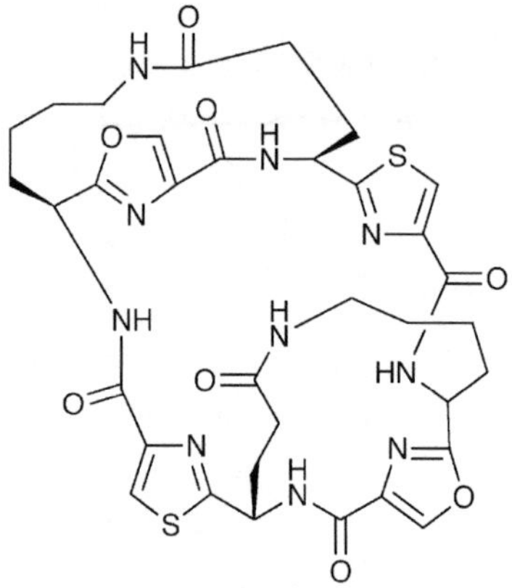

2.13 Cyclodepsipeptides. – The period under review has once again seen a significant increase in the diverse structures found in nature, and some of the known structural families have matured to be the subject of extensive reviews. Thus the synthesis, biosynthesis, biotransforamation and biological activity of the destruxins have been reviewed,[202] and an investigation[203] carried out into the ionophoric characteristics of destruxin A. Novel approaches to the synthesis of cryptophycin and related depsipeptides have been reviewed,[204] and the biosynthesis of mycotoxins, such as the enniatins, in *Fusarium* has been discussed.[205] The mechanism of the latter resembles fatty acid biosynthesis utilising stepwise condensation of dipeptidyl building blocks.

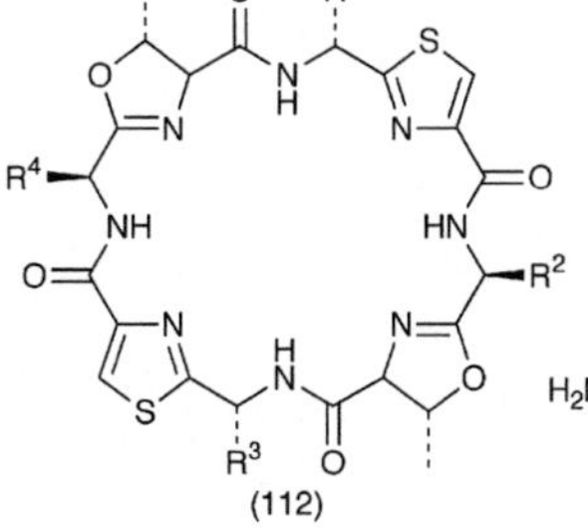

(112)

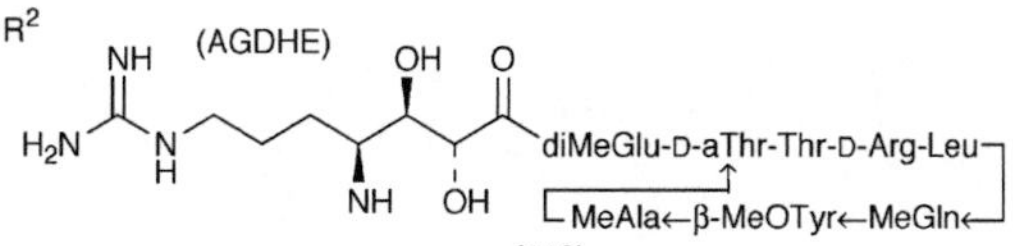

(AGDHE) diMeGlu-D-aThr-Thr-D-Arg-Leu-MeGln→β-MeOTyr→MeAla

(113)

RCHCH$_2$CO-Glu-Leu-Leu-Leu-Asp-Leu-X (O-linked ester)

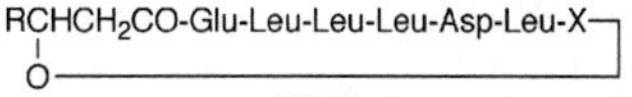

(114)

PA R = C$_{12}$-iso and anteiso, X = Ile
PB R = C$_{12}$-iso and anteiso, X = Val
PC R = C$_{14}$-iso and anteiso, X = Ile
PD R = C$_{14}$-iso, X = Val
PE R = C$_{13}$-iso, X = Ile

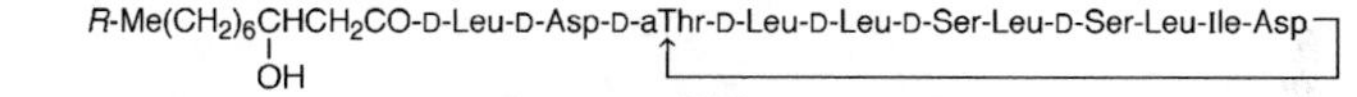

R-Me(CH$_2$)$_6$CH(OH)CH$_2$CO-D-Leu-D-Asp-D-aThr-D-Leu-D-Leu-D-Ser-Leu-D-Ser-Leu-Ile-Asp

(115)

Application[206] of Marfey's approach to determine the configuration of callipeltin A, from the marine sponge *Latrunculia* Sp, has brought about revision of the configuration of two amino acid residues, giving the revised structure (113). On the road to total synthesis of callipeltin A, progress has been made with the synthesis of (2*R*, 3*R*, 4*S*)-3-hydroxy-2,4,6-trimethylheptanoic acid[207] and the AGDHE unit in (113).[208] The marine bacterium *Bacillus pumilis* produces[209] a series of cyclic depsipeptides, that differ from surfactin in having a Leu residue at position 4 with Val or Ile at position 7. Their general structures are summarised in (114). Lokisin (115) has been isolated[210] from *Pseudomonas* sp. strain DSS41, and *Beauveria felina* has been identified[211] as a source of [Phe^3, MeVal^5]-destruxin B. Cyano bacteria *Lyngya* sp. from Palau, and from the Mariana Islands have been shown to be a source of ulongamides A-F (116-121)[212] and of obyanamide (122)[213] respectively. Another family of compounds, antanapeptins A-D (123-126) have been isolated[214] from *Lyngbya majuscula* collections from Madagascar. A new enniatin designated G, (127), has been isolated[215] from a mangrove fungus *Halosarpheia* sp (strain 732), while a *Fusarium* sp has yielded[216] HA23 (128). ESI-LC/MS has been instrumental in identifying[217] a family of cyclic peptides (anabaenopeptins), microcystins and the cyclodepsipeptides (anabaenopeptolides) (129-132) in *Anabaena* sp (Strains 90 and 202A2). Kulokekahilide (133) has shown cytoxicity against P388 murine leukaemia cells (IC50 = 2.1 μg/mL) and was isolated and characterised[218] from the mollusk, *Philinopsis speciosa.* Two southern Australian sponges, *Phoriospongia* sp. and *Callyspongia bilamellata* have yielded[219] phoriospongins A (134) and B (135), while the sponge *Theonella swinhoei* has yielded[220] the antibacterial nagahamide A (136), related to YM-47522. Isolated[221] from the root bark of *Tripterygium wilfordii* Hook. F. for the first time was triptotine L (137).

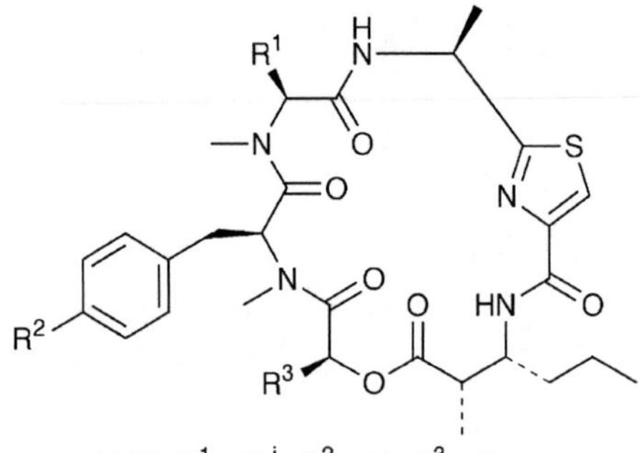

(116) $R^1 = Pr^i$, $R^2 = H$, $R^3 = Me$
(117) $R^1 = Pr^i$, $R^2 = OH$, $R^3 = Me$
(118) $R^1 = Bn$, $R^2 = OH$, $R^3 = Me$
(119) $R^1 = R^3 = Pr^i$, $R^2 = OH$
(120) $R^1 = Pr^i$, $R^2 = OH$, R^3 = s-But
(121) (120) with Tyr replaced by Val

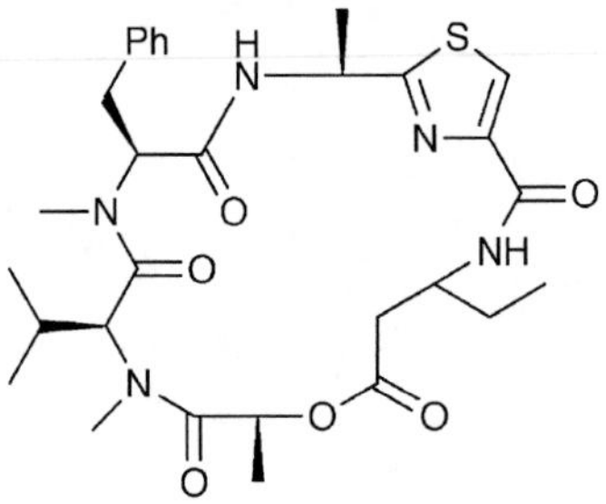

(122)

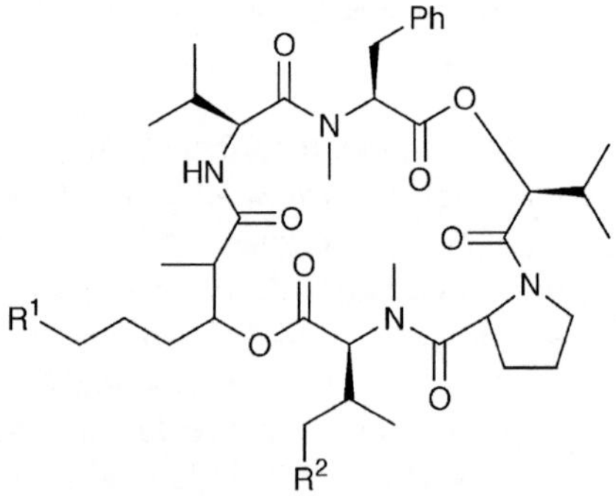

(123) R^1 = ≡—, $R^2 = Me$
(124) R^1 = =—, $R^2 = Me$
(125) R^1 = ⁄⁀, $R^2 = Me$
(126) R^1 = ≡—, $R^2 = H$

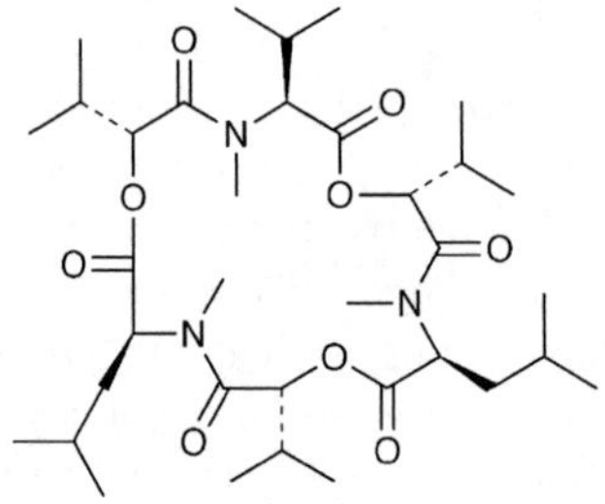

(127)

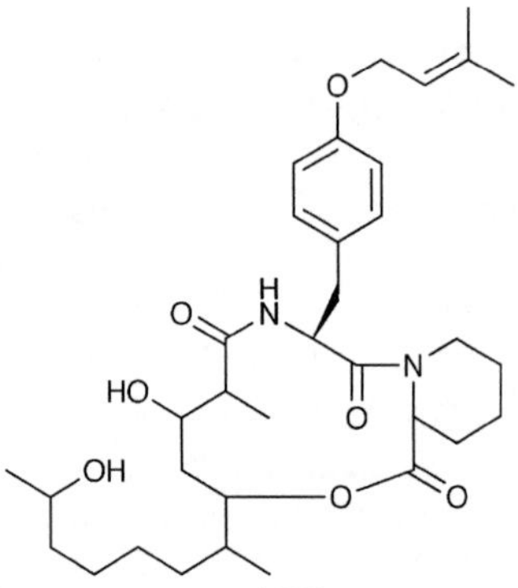

(128)

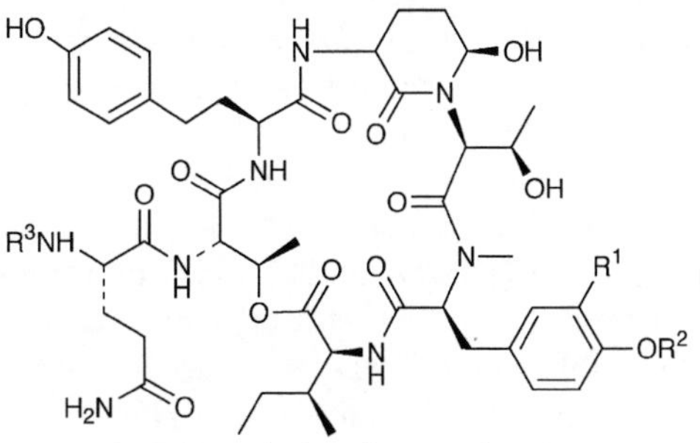

(129) 90A, $R^1 = H$, $R^2 = Me$, $R^3 = HCO$
(130) 90B, $R^1 = Cl$, $R^2 = H$, $R^3 = HCO$
(131) 202A, $R^1 = Cl$, $R^2 = H$, $R^3 = Me$
(132) 202B, $R^1 = Cl$, $R^2 = H$, $R^3 = HCO\text{-}Pro$

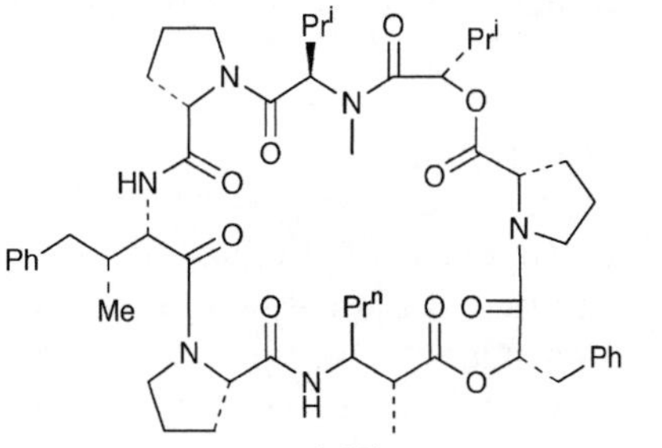

(133)

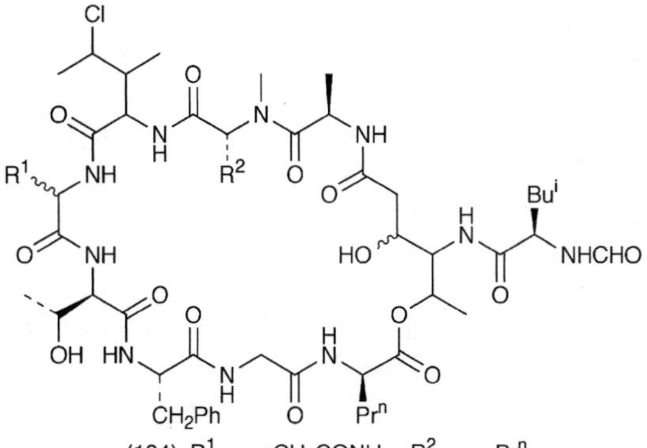

(134) $R^1 = CH_2CONH_2$, $R^2 = Pr^n$
(135) $R^1 = CH_2CONH_2$, $R^2 = Bu^i$

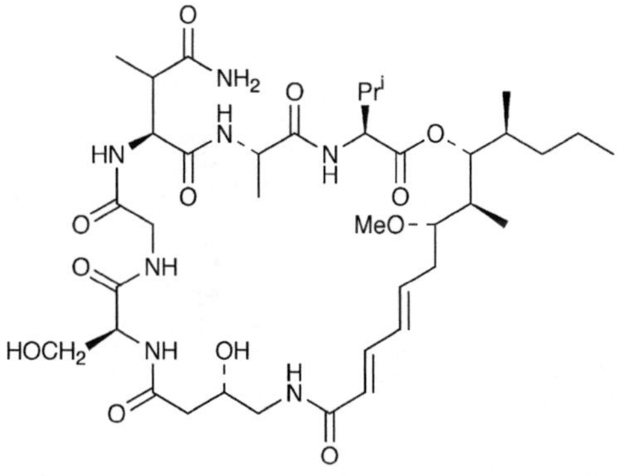

(136)

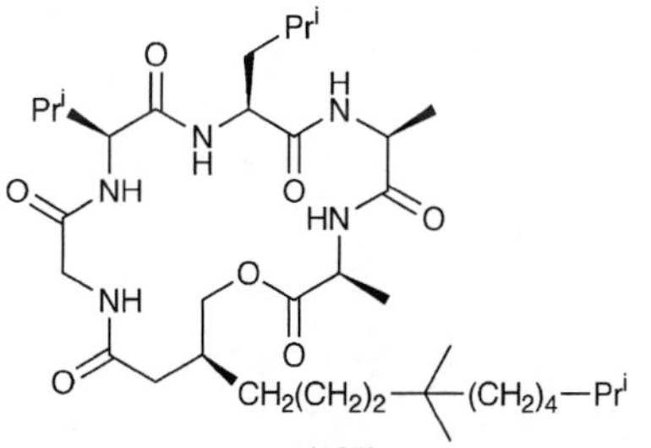

(137)

An alternative synthesis to that published in 2001 has been reported[222] for stevastelin B (138). Macrolactamisation was achieved at point (a) in structure (138) from a precursor activated by the Shioiri method (diethylphosphorocyanidate DEPC). Macrolactonisation (using 2,4,6-trichlorobenzoyl chloride/DMAP, 90% yield) for final ring closure was adopted[223] for the synthesis of (139), which was then transformed to stevastelin C3 (140). Another route has been worked out[224] for the synthesis of cryptophycin 52 (141), which is in advanced clinical evaluation. While previous syntheses have incorporated epoxide formation as a final step, the new approach uses Shi's method of incorporating the epoxide, masked as a fructose-derived chiral dioxirane, earlier in the synthesis. When cryptophycin 52 was reacted with thiols,[225] acidic conditions caused attack at the epoxide ring (1), and under basic conditions addition of thiol at point (2) in (141) occurred. In the first total synthesis[226] of somamide A (142) from the marine cyanobacteria *Lyngbya majuscula*, the 3-amino-6-hydroxy-2-piperidone residue was constructed from 5-HO-norVal post macrocyclisation, which had been carried out at point (a) in (142) using FDPP [$Ph_2P(O)OC_6F_5$] in 64% yield. Another cyclodepsipeptide, ramoplanin A2 (143), in phase III clinical trials for oral treatment of vancomycin–resistant *Enterococcus faecium* has been totally synthesised[227] for the first time. To aid convergency of the synthesis, three sub-units were constructed to form a linear precursor which was cyclised at point (x) in (143) using EDCI/HOAt. The high yield of 89% for this step was attributed to the probable pre-organisation of the linear precursor into a β-pleated sheet prior to cyclisation. With the total synthesis of HUN-7293 (144) already accomplished research has moved on to solution phase parallel synthesis[228] of a library of its analogues. Amongst the analogues synthesised was HUN 7239B (145), which proved to be a natural product. Each macrocyclisation was carried out at position (y) using DPPA/DIEA. An amide analogue of the depside bond in (144) has also been synthesised[229] by chemoselective cleavage of the 2-hydroxyacid amide, followed by 'zipping up' the ring using an amino group attack on a BOP/DIEA activated carboxyl group.

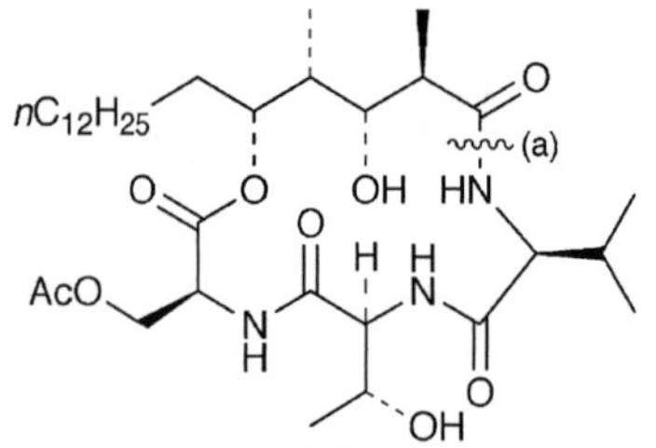

(138)

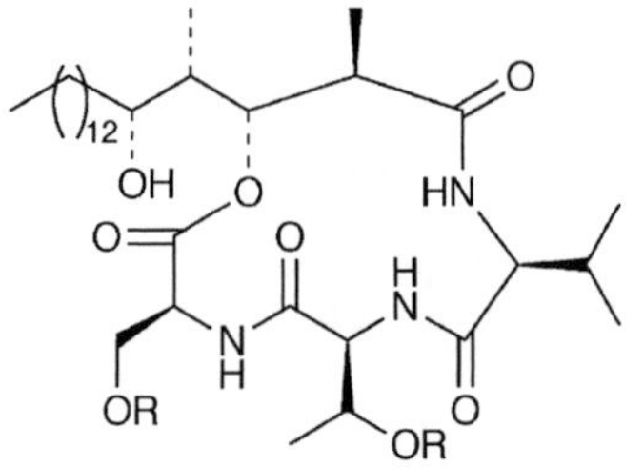

(139) R = Bn
(140) R = H

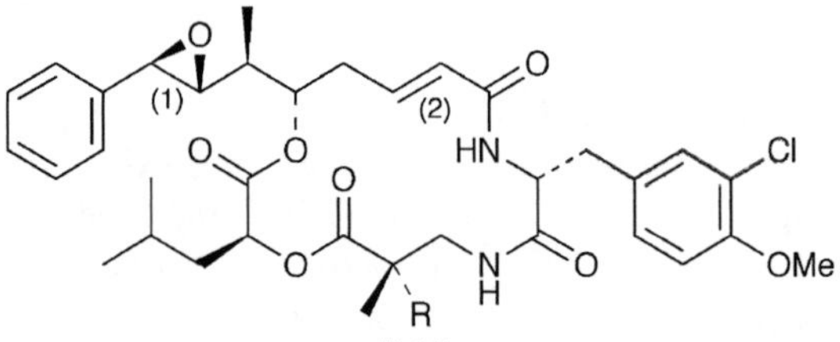

(141)

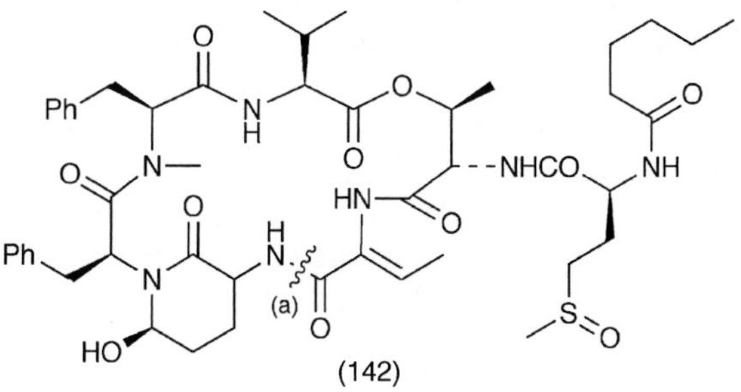

(142)

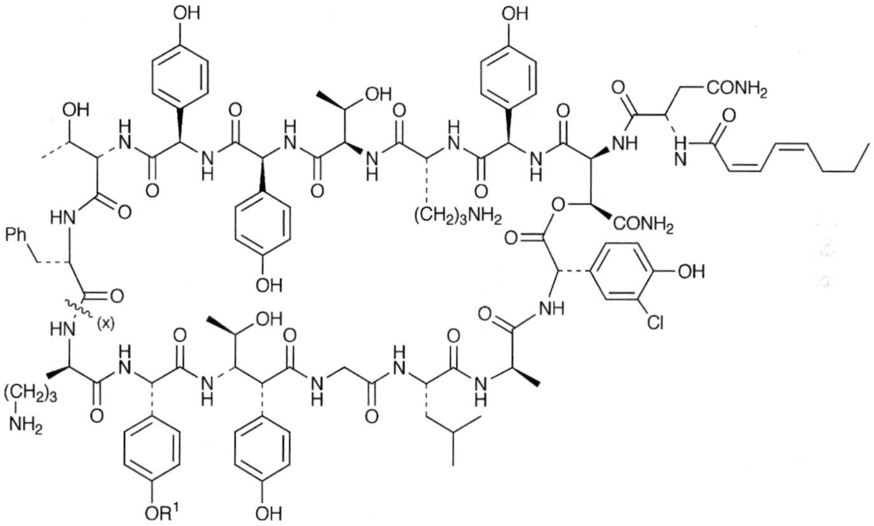

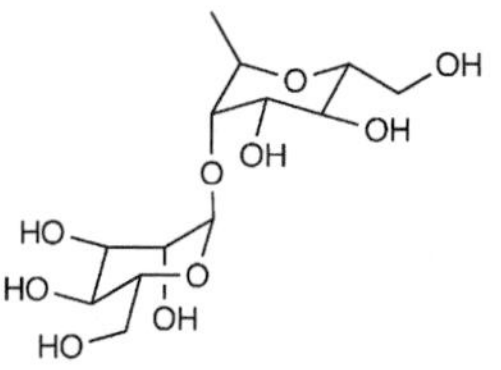

(143) R^1 =

Piperazic acid-containing cyclodepsipeptides have been regularly found in nature of late, and many syntheses have now been reported. A further synthesis[230] of the immunosuppressant (-) sanglifehrin A (146) has been achieved via a convergent three-component synthetic plan. Part of this plan was to bring segments through that could be joined up at point (z) via an iodovinyl group. Prior to that, the macro ring had been formed by lactonisation at the depside link using EDC/DIEA. The C_{13}–C_{19} segment of (146) has also been stereoselectively synthesised.[231] The cyclodepsipeptide ring analogue of A83586C, with a Pro residue replacing the piperazic acid residue B in (147) has been synthesised[232] via a HATU-catalysed macrocyclisation step at the Pro C-terminus of the linear precursor. The ring section of GE3 (148) has also been synthesised[233] efficiently by macrolactamisation at the C-terminus of piperazic acid residue B in (148), and in a separate report[234] stereo selective construction of the acyl side chain R in (148) has been carried out. Kaiser's oxime resin has been the polymer support of choice for constructing[235] a library of PF1022A analogues such as (149). Lactonisation was carried out on-resin at the lactic acid carboxyl group. Amidoxamine analogues (151) of PF1022A were derived[236] from previously synthesised mono thionated derivatives based on (150). PF 1022A, known commercially as emodepside, belongs to a new class of anthelmintically active compounds, whose role has been placed into context by a mini-review.[237]

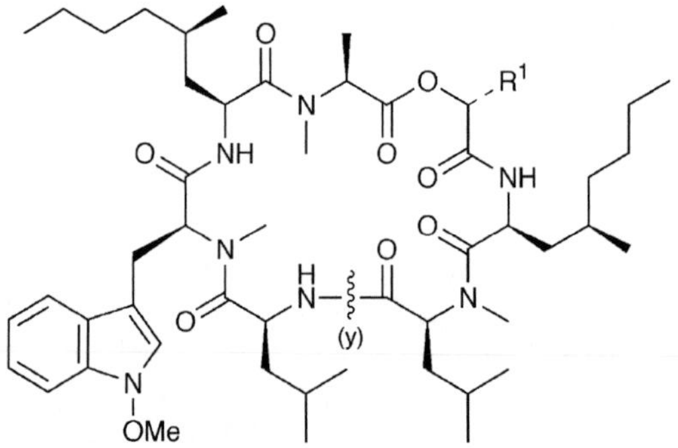

(144) $R^1 = (CH_2)_2CN$
(145) $R^1 = Me$

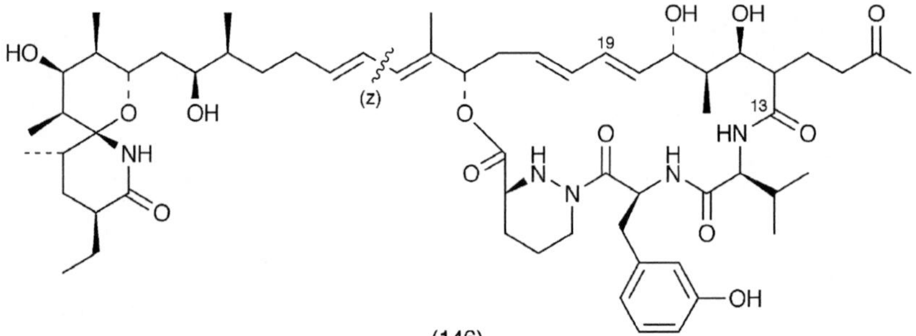

(146)

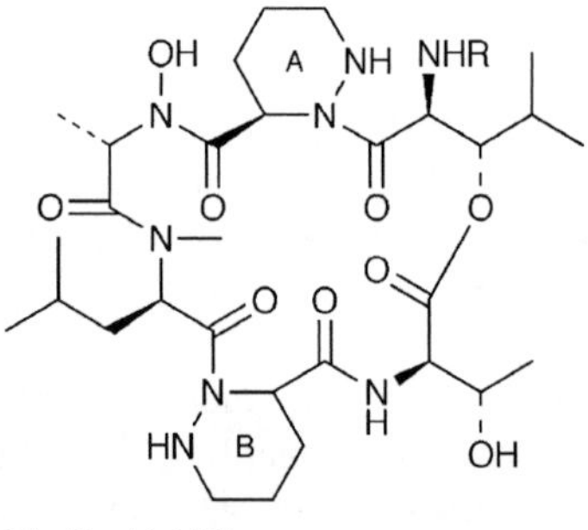

(147) R = H, HCl

(148) R =

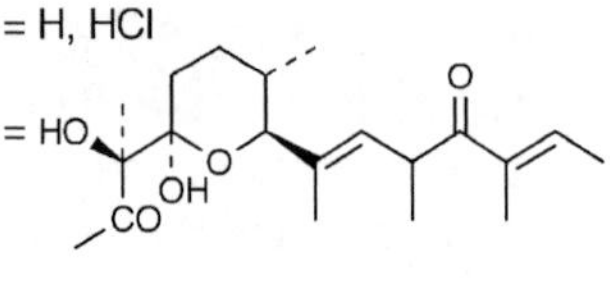

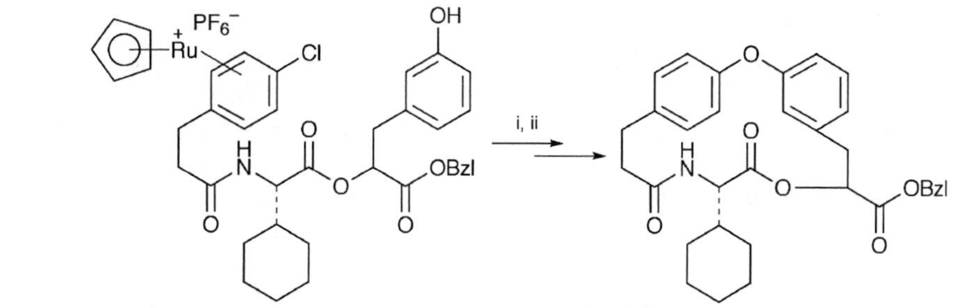

Reagents: i, Cs_2CO_3/DMF/12 h; ii, λ = 350 nm/DMF

Scheme 25

Detailed biochemical studies[238] on non-ribosomal peptide synthetases, capable of catalysing intramolecular cyclisation to form macrolactone and macrolactam cyclic peptides have been carried out[239] for surfactin. Structures represented in (152) summarise the current mechanistic thoughts. It has been shown,[240] as summarised in Scheme 25, that the depside link can survive the conditions for cyclisation via a biaryl ether using a ruthenium–based strategy, although the yield (6%) was low.

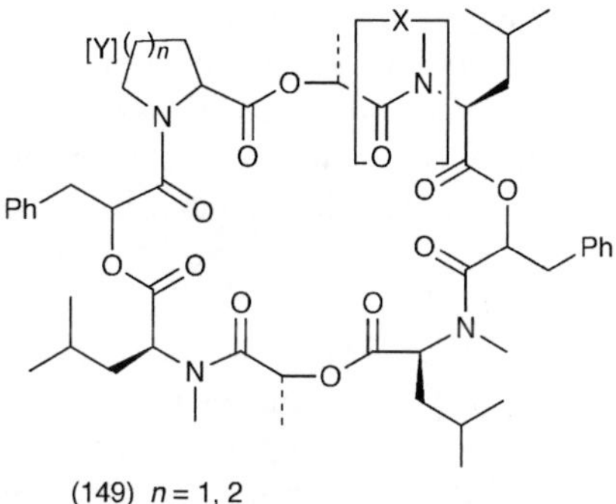

(149) $n = 1, 2$

(150) [X] = —CSN—, [Y] = iso-butyl

(151) [X] = —C(=N–OR)–N—, [Y] = iso-butyl

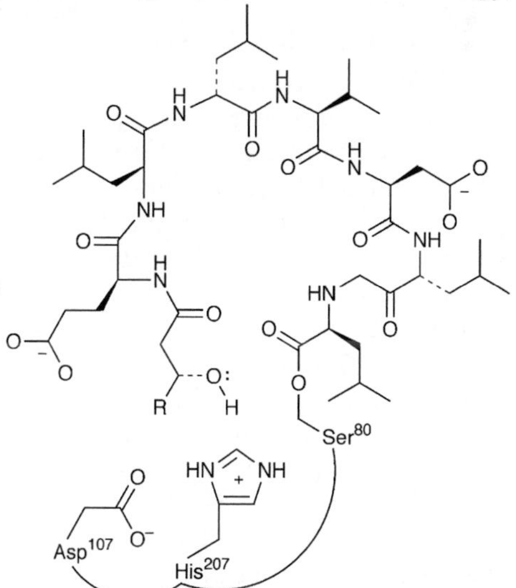

(152)

Crystals of montanastatin, cyclo [(Val-D-Hyv-D-Val-Lac)$_2$] have been shown[241] to have a structure similar to valinomycin, and LC/ESI-MS/MS has proved a useful technique[242] for quantitative analysis of the anti-cancer drug kahalalide F in human blood plasma. The apparent binding of the diverse group of structures, (-)-doliculide, jasplakinolide, phalloidin and chondramide C at the same site on F-actin, has been rationalised[243] using computer-driven shape descriptor analysis. The portion of the macrocycle incorporating the phenyl and isopropyl side chains in (-)-doliculide gave the best overlap with the other molecules. The side chain segment of polyoxypeptins A and B has been prepared from a chiral 2,3 epoxy alcohol,[244] and the pathogenic fungus

Paecilomyces tenuipes BCC 1614 can take up either stereoisomer of Ile or aIle to give diastereoisomeric beauvericin analogues.[245] Biodegradable amphiphilic AB-type diblock co-polymers can be obtained[246] from the syntheis of poly (Glc-Lys)-block-poly(L-lactide) and poly (Glc-Asp)-block-poly(L-lactide).

3 Modified and Conjugated Peptides

In this sub-section it has now been traditional for a number of Volumes of these Reports, to cover groups of molecules that possess additional conjugated groups. These are often fundamental to their biological activity, yet are only added post-translationally, and sometimes can be very labile and rather difficult to detect in their native intact forms.

3.1 Phosphopeptides. – Several tissues from mamushi, a venomous snake in Japan, have yielded[247] the O-phosphorserylethanolamine (153). A O-phosphohydroxylysine (HylP)-containing peptide, H-Phe-DL-HylP-Gly-Gln-Pro-Ala-Ile-Gly-Phe-OH, has been synthesised as a model for collagen, to test out some analytical techniques, which could serve to locate phosphate residues in the protein.[248] Solid phase synthesis using racemic Fmoc-δ-hydroxy-Lys (Boc)-OH was used for the assembly, and global phosphitylation/oxidation was used to introduce the phosphate group. The phosphorylated peptide was resistant to hydrolysis by lysine-specific endopeptidases. There has been increasing interest in assemblying peptides containing phosphorylated amino acid mimetics. The fluorinated building block (154) has been synthesised[249] as a mimetic of phospho-serine and – threonine by fluorination of a lithiated bis-lactim ether, while the conformationally-constrained phosphonomethylphenylalanyl building block (155) has also been made available synthetically.[250] A first synthesis[251] of a β-amino phosphotyrosyl mimetic (156) has been reported, and the building block has been incorporated into a tripeptide involved in Grb2SH2 domain binding. Probing the phosphopeptide specificities of protein tyrosine phosphatases, SH2 and PTB domains with combinatorial library methods have been reviewed.[252] To overcome the problems associated with the difficulty to genetically encode phosphoamino acids, phosphorylated peptide sections have been incorporated[253] into a receptor by chemical ligation of a tetraphosphorylated analogue, as summarised in Scheme 26.

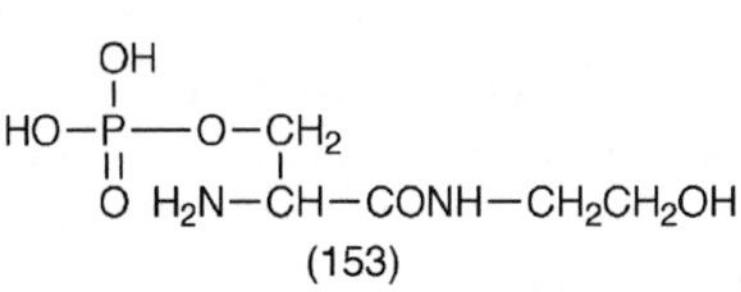

$HO-P(O)(OH)-O-CH_2-CH(NH_2)-CONH-CH_2CH_2OH$

(153)

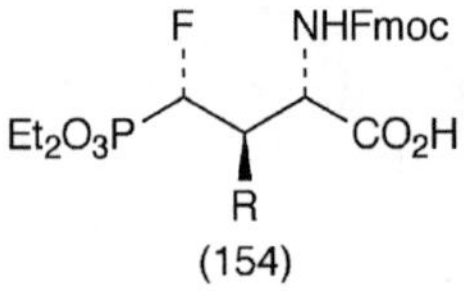

F NHFmoc Et_2O_3P CO_2H R

(154)

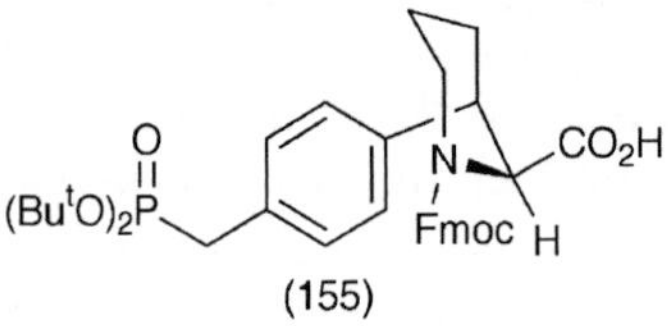

$(Bu^tO)_2P(O)$ N Fmoc CO_2H H

(155)

$(BnO)_2P(O)$ $NHSO-Bu^t$ CO_2H

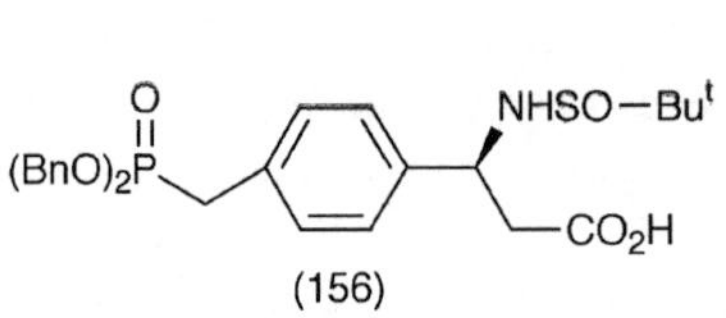

(156)

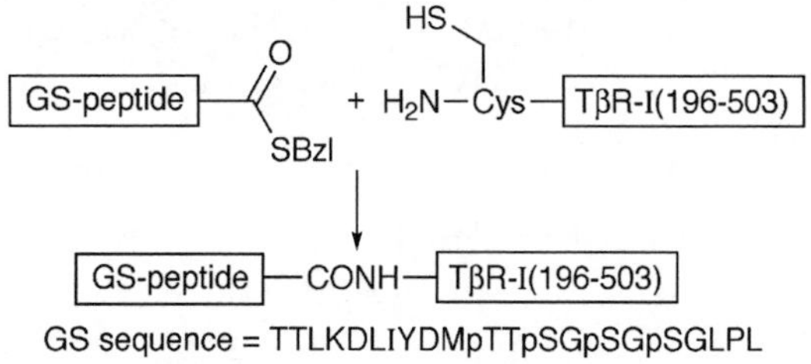

GS-peptide —C(O)SBzl + H_2N—Cys(SH)— TβR-I(196-503)

↓

GS-peptide —CONH— TβR-I(196-503)

GS sequence = TTLKDLIYDMpTTpSGpSGpSGLPL

Scheme 26

Mass spectrometry techniques are making significant inroads into identifying the presence and location of phosphate groups in peptides and proteins. While the phosphorylated peptides are still considered rather labile for mass spectrometric analysis, it is possible to use this lability to generate β-eliminated dehydro peptides which by sulfite addition, can be analysed as cysteic and β-methylcysteic acid residues[254] using tandem mass spectrometry. Differentiation between sulphated and phosphorylated sites has been possible[255] from a study of fragmentation signatures generated from model sulphated and phosphorylated peptides using in-source dissociation and tandem nanoelectrospray MS/MS. Phosphopeptides isolated from mixtures using immobilised metal ion affinity (IMAC), have been analysed[256] directly as IMAC beads using MALDI MS/MS, and it has been shown[257] that MALDI/ionisation quadrupole time-of-flight can successfully analyse pure phosphoserine and phosphothreonine peptides, but mixtures were more challenging. Fragmentation in all cases involved conversion to dehydropeptides. A detailed study[258] of MALDI and electrospray/ion trap fragmentations of phosphopetides revealed that while phosphoserine and phosphothreonine peptides experience a loss of 98 Da regardless of peptide composition, phosphate loss from phosphotyrosine seemed dependent on the presence of arginine and lysine in the sequence. Enough evidence has been accumulated[259] to prove that ion-mobility/mass spectrometry can distinguish between phosphorylated and non-phosphorylated peptides, and characterisation of phosphorylated peptoid and peptide hybrids has been made possible through the use[260] of nano-electrospray tandem mass spectrometry.

3.2 *O*-Sulfated Peptides. – As for the phosphopeptides in the above subsection, mass spectrometry is a powerful technique for detecting post-translational conjugates. As previously referred to,[255] sulfated and phosphorylated peptides can be distinguished using positive electrospray ionisation tandem mass spectrometry, since in all cases analysed 100% loss of sulfate occurs and the sequence information can be obtained from the de-sulfated peptides. Electrospray ionisation broadband FTICR mass spectrometry[261] has been found to give a similar mass spectral resolution of phosphorylated and sulphated peptides.

The acid sensitivity of *O*-sulfated peptides (e.g. to TFA), has hampered the routine synthesis of these peptides, so novel strategies have to be developed.

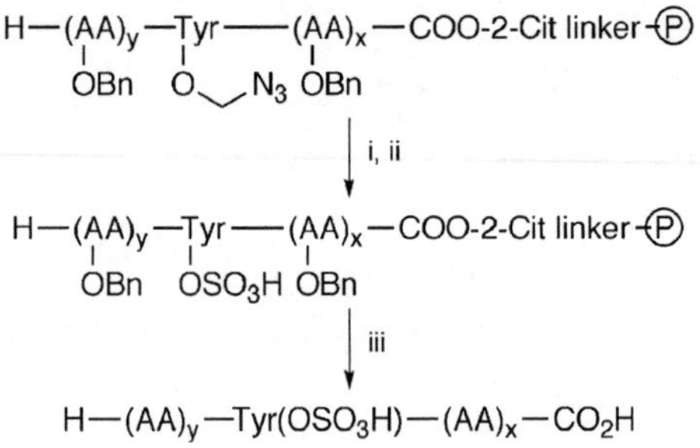

H—(AA)$_y$—Tyr——(AA)$_x$—COO-2-Cit linker-Ⓟ
OBn O$\smile$N$_3$ OBn

i, ii

H—(AA)$_y$—Tyr——(AA)$_x$—COO-2-Cit linker-Ⓟ
OBn OSO_3H OBn

iii

H—(AA)$_y$—Tyr(OSO_3H)—(AA)$_x$—CO_2H

Reagents: i, $SnCl_2$/PhSH/Et_3N; ii, SO_3·DMF; iii, TFE/CH_2Cl_2/AcOH; iv, $Pd(OH)_2$/C

Scheme 27

One strategy[262] used the azidomethyl group for initial protection of tyrosyl residues and post-assembly sulfation was carried out as summarised in Scheme 27. Stabilised *O*-sulfated building blocks have been prepared[263] for Fmoc protocols using 5 equivalents of SO_3.DMF complex as optimal sulfation conditions. The stability of the building blocks, Fmoc-Ser($SO_3^-N^+Bu_4$)-OH, Fmoc-Thr($SO_3^-N^+Bu_4$)-OH and Fmoc-Hyp($SO_3^-N^+Bu_4$)-OH is derived from the use of the tetrabutyl ammonium counter ions, which protect against TFA cleavage. It has been reported[264] that tyrosine-sulfated peptides but not their unsulfated analogues can restore the HIV-1 coreceptor activity of a CCR5 variant lacking critical residues 2–17 of its amino terminus.

3.3 Glycopeptide Antibiotics. – The considerable literature now available on vancomycin has been authoritatively reviewed[265] and includes the development of synthetic inhibitors of vancomycin resistant enzymes. Another review[266] has concentrated on the synthetic protocols, which have been developed during the total synthesis of vancomycin. The 30-year reign of vancomycin as an ‘antibiotic of last resort’ has been placed under strain with the increasing number of vancomycin-resistant strains, so alternative strategies to overcome this resistance are welcomed, such as investigations on antibiotics related to vancomycin in structure. Balhimycin (157) and degluco-balhimycin (158) have been crystallised[267] in complexes with di-, tri- and penta-peptides, which emulate bacterial cell wall precursors. In addition to previously found binding modes, two dipeptide complexes and face to face oligomerisation were detected. In the pentapeptide complex, the positions are close to those required for D-Ala elimination, so it could provide a realistic model for the prevention of the enzyme-catalysed cell wall cross-linking by antibiotic binding. A null mutant gene OP696 (from *Amycolatopsis mediterranei*) has accepted[268] several analogues of β-hydroxytyrosine and has been used to generate the fluoro bahlhimycin (159). New derivatives of eremomycin (160) have also been prepared. Thus Edman degradation of eremomycin, removed[269] the N-terminal *N*-Me-D-Leu which was replaced by more hydrophobic derivatives, e.g *N*-decyl or *p*-(*p*-chlorophenyl) benzylamides. When binding studies were carried out interaction with cell wall intermediates was seriously decreased yet the analogues were equally active against glycopeptide resistant enterococci, and more active against staphylococci. It is suggested that the analogues inhibit the transglycosylase stage even in the absence of dipeptide or depsipeptide binding. N-Benzyl substitution of the two amino groups in the sugar units and (*R*) or (*S*)–2-amino-4-methylpentyl replacement of the N-terminal *N*-D-MeLeu residue[270] have given eremomycin derivatives which have decreased antibacterial activity towards vancomycin-resistant strains. A number of doubly modified water-soluble derivatives of chloroorienticin B have been semi-synthesised[271] and the most potent against MRSA and vancomycin-resistant enterococci (VRE) was analogue (161). Combinatorial parallel synthesis[272] of over 80 compounds bearing carboxamide derivatives of chloroorienticin B has shown that derivatives having both Tyr and Trp residues in positions 1-3 possessed the most potent antibacterial activity against VRE. An evaluation[273] has been carried out on a set of vancomycin aglycones having modifications at the Asn residue in position 3. Replacement of the amide in the Asn side chain with CN had already shown promising biological activity, but further modifications on this theme proved detrimental to antimicrobial activity. More potency against VanB in particular, was obtained in aglycone analogues, which lacked a lipid anchor.

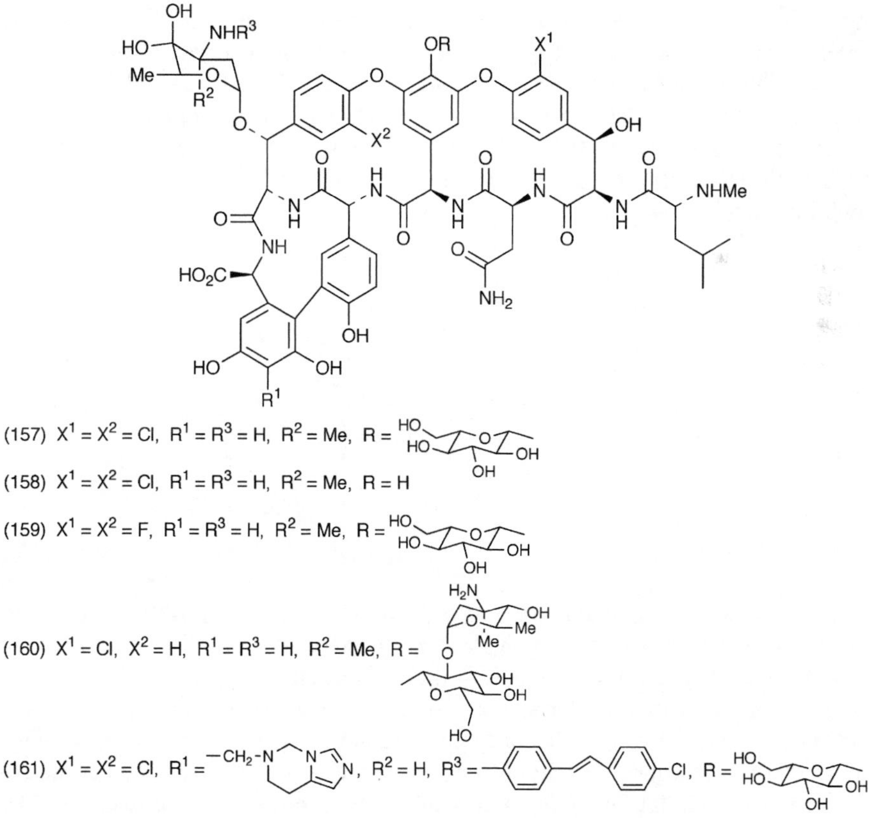

(157) $X^1 = X^2 = Cl$, $R^1 = R^3 = H$, $R^2 = Me$, R =

(158) $X^1 = X^2 = Cl$, $R^1 = R^3 = H$, $R^2 = Me$, R = H

(159) $X^1 = X^2 = F$, $R^1 = R^3 = H$, $R^2 = Me$, R =

(160) $X^1 = Cl$, $X^2 = H$, $R^1 = R^3 = H$, $R^2 = Me$, R =

(161) $X^1 = X^2 = Cl$, $R^1 = -CH_2-N$, $R^2 = H$, $R^3 =$ –Cl, R =

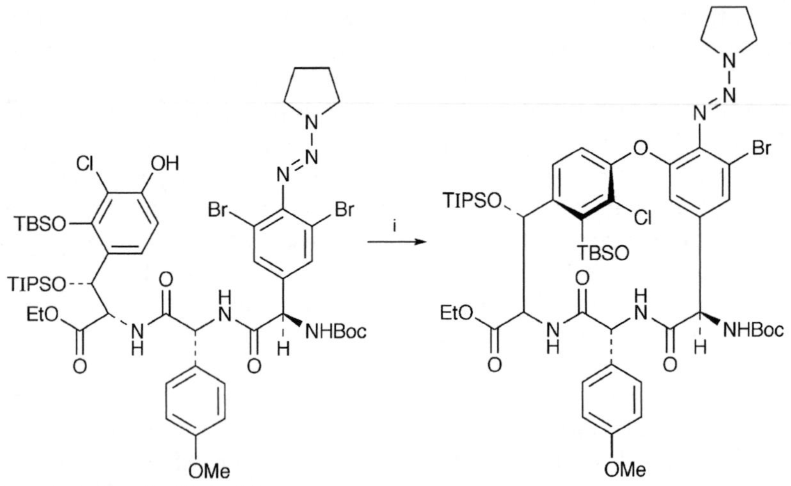

Reagents: i, $CuBr{\cdot}SMe_2/K_2CO_3$/pyridine

Scheme 28

The strategic placement of a bulky substituent on the aromatic ring of the tyrosine residue has led to a single atropisomer being produced[274] in the diaryl ether formation of the C-O-D macrocycle vancomycin model exemplified in Scheme 28. A highly stereoselective construction of the AB biaryl fragment of vancomycin [e.g. (162)] can be achieved[275] via the dynamic kinetic diastereomeric resolution of the lactone (163). A ruthenium-promoted intermolecular S_NAr reaction[276] has secured the formation of the diaryl ether link in building block (164) corresponding to the F-O-G fragment of ristocetin A.

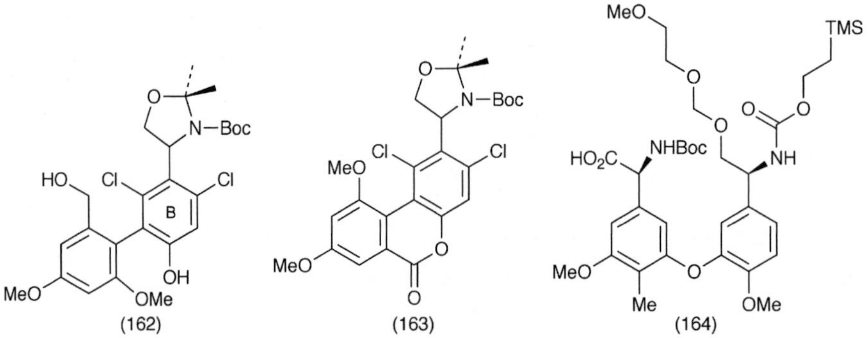

(162) (163) (164)

A wide variety of analogues corresponding to functionalisation of the 6-position of the glucose residue have been prepared[277] but all proved inactive against van A phenotype VREF strains, although they retained substantial activity against other bacteria. Des-N-methylleucyl derivatives of LY264826 have been used[278] to study the mechanism of action of glycopeptide antibiotics. Loss of the N-terminal residue lowers the activity, but the activity can be re-enhanced by alkylation of the 4-epivancosamine moiety of the disaccharide.

This alkylation enhanced dimerisation 7000-fold and enhanced binding to bacterial membranes.

Chlorobiphenyl derivatives of the sugar moiety in vancomycin have generated a great deal of interest[279] as they have higher activity than the parent antibiotic, and have been defined as hybrid, in the context that they might function as two biologically active components. This hypothesis has been developed further[280] by the synthesis of other chlorobiphenyl analogues, and the results provide compelling evidence that there is an interaction between the disaccharide portion of vancomycin and bacterial transglycosylases. A mass spectrometric study has focussed[281] on the gas phase stability of vancomycin complexes with bacterial cell wall precursor peptides such as Ac2-Lys-D-Ala-D-Ala. Small changes in the peptide binding pocket of vancomycin affect gas phase stability in a manner which parallels known solution binding affinity.

LY 154989, closely related to teicoplanin, has been the subject of an NMR-based study.[282] It forms concentration-dependent aggregates in aqueous solution, similar to teicoplanin, even though it does not possess a C-11 acyl chain, but these can be disrupted by the addition of bacterial cell wall analogues such as, Ac-2-KDADA, Ac-DADA or Ac-DA. Binding constants between teicoplanin and D-Ala-D-Ala terminus peptides have also been determined by affinity capillary electrophoresis (ACE).[283] Following up on the idea that glycopeptide antibiotics target enzymes necessary for cell wall biosynthesis as well as immediate cell wall precursors, solid state NMR experiments[284] with stable isotope-labelled *Staphylococcus aureus* have provided a model of the peptidoglycan binding site of the potent 4-fluorobiphenyl derivative of chloroeremomycin (LY 329332). The model positions the vancomycin cleft around an un-crosslinked D-Ala-D-Ala peptide stem, with the fluorobiphenyl moiety near the base of a second proximate stem in a locally ordered peptidoglycan matrix. The interplay between two sets of non-covalent bonds formed at separate interfaces between indole-2-carboxylic acid (L) and chloroeremomycin (CE) have been studied[285] as ligand/CE dimer interactions (L/CE/CE/L). They have been found to occur with cooperativity and structural tightening at the dimer surface. Details of the biosynthesis of glycopeptide antibiotics have been emerging slowly. Four proteins required for the biosynthesis by actinomycetes of (*S*)-3,5-dihydroxyphenylglycine, the cross-link in the maturation of vancomycin and teicoplanin scaffolds, have been expressed[286] in *E. coli*, and their detailed functions explored. Biosynthetic gene clusters, already reported for chloroeremomycin, bahlymicin and GPA-like complestatin have also been sequenced and annotated[287] for the glycopeptide antibiotic A47934 from *Streptomyces toyocaensis* NRRL 15009.

A series of novel antibiotics with activity against methicillin-resistant staphylococci and vancomycin-resistant enterococci have been identified[288] in *Streptomyces hygroscopicus* LL-AC98. They have been designated mannopeptimycins α–ε (165–169). The presence and position of an isovaleryl group in the terminal mannose unit seems critical for retaining antibacterial potency.

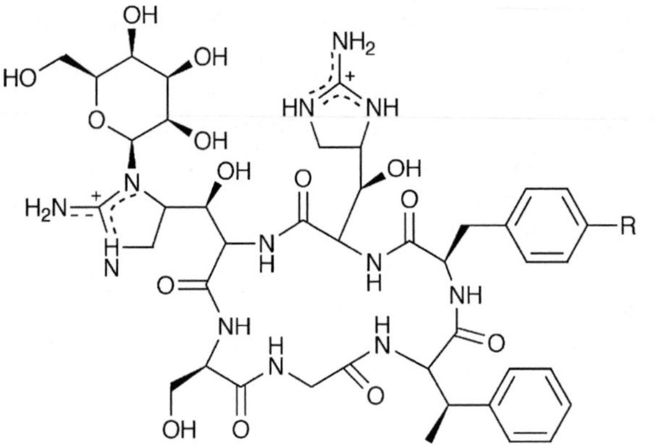

(165) R = ManA-ManB where $R^1 = R^2 = R^3 = H$ in

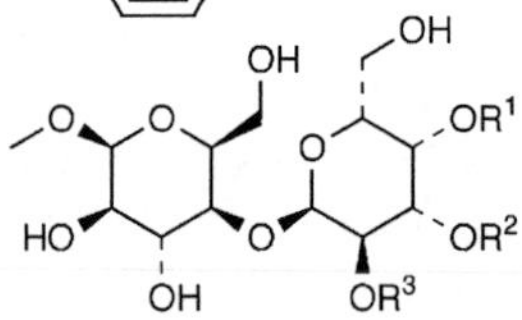

(166) R = H
(167) R = ManA-ManB with $R^1 = R^2 = H$, $R^3 = Me_2CHCH_2CO$–
(168) R = ManA-ManB with $R^1 = R^3 = H$, $R^2 = Me_2CHCH_2CO$–
(169) R = ManA-ManB with $R^2 = R^3 = H$, $R^1 = Me_2CHCH_2CO$–

The glycopeptide antibiotics, especially teicoplanin, have been useful as chiral stationary phases for a number of HPLC based chiral resolutions. Details of the parameters required have been reviewed[289] and it has been shown[290] that teicoplanin columns can achieve chiral resolution of cromakalim, a typical K^+ channel opener. Enantioselective separation of two 2-arylpropionic acids was carried out semi-preparatively[291] using teicoplanin and A-40,926. HPLC columns, and teicoplanin-bonded stationary phases have also been used[292] for the resolution of native or derivatised amino acids. Vancomycin silica stationary phases have provided[293] good separation of basic compounds in capillary electrochromatography.

3.4 Glycopeptides. – The review period has seen a great deal of activity in this area, with some of the newer innovations proving difficult to categorise under the usual headings. So in addition to the traditional *O*-, *N*- and *C*-glycoside sub-divisions, this year a few papers under *S*-glycosides and *glycoso*–amino acids deserve separate categories. The latter should not strictly be defined as conjugates, as they represent hybrid structures of amino acids and carbohydrates that are inserted into peptide backbones generating peptidomimetics. However it seemed appropriate to include them in this section. Even after the sub-division of topics, some papers still cover aspects that overlap many themes, so these deserve reporting up-front. Recent developments in glycopeptide synthesis have been reviewed[294] by means of selected examples focussing on glycopeptides carrying tumour-associated carbohydrate motifs. Coupling reactions of glycosyl amino acids and glycans and enzymic ligation have been covered by a syntheis and applications review,[295] while the analytical aspects of glycoconjugates have been well covered in a 478-reference review.[296] Contributions made to conformational studies of glycopeptides by NMR, X-ray crystallography and molecular modelling have been reviewed,[297] and the effectiveness of glycosylated opioid peptides (enkephalins, deltorphins and dermorphins) in blood-brain barrier penetration has been assessed.[298] The novel linker (170), cleaved by the fluoride,TBAF.3H_2O, has been shown[299] to offer advantages in solid phase synthesis of glycopeptides, and *O*- and *N*-glycosylated α-aminooxy units have been successfully incorporated into pentamers such as (171) on solid support.[300]

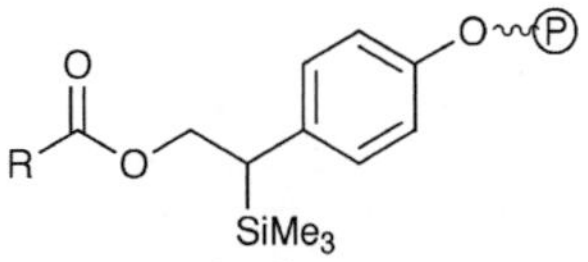

(170)

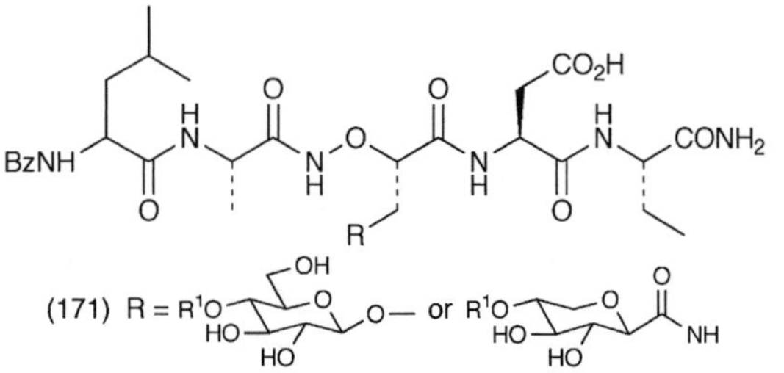

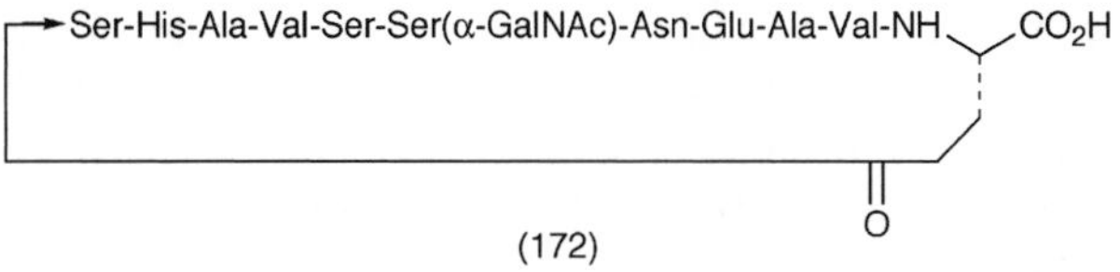

(171)

(172)

3.4.1 O-Glycopeptides. The homophilic recognition domain (Ser^{78}–Glu^{89}) of E-cadherin has been converted[301] to a cyclic analogue (172), from a linear precursor assembled on Tentagel solid support using the allylic HYCRON linker. Cyclisation via the side chain of the Glu residue was carried out using PfPyU, and the glycosylated (172) adopted a conformation in D_2O, which showed contacts between the 6-H of the N-acetylgalactosamine, and the methyl group of Ala and the β-methylene group of Glu. D-Galactopyranosides of trans 4-hydroxyproline have been synthesized[302] using the sulfoxide glycosylation method, while fluorine was used as a leaving group in the tetrasaccharide glycosylation of Fmoc-Ser-allyl esters,[303] to make core-class 2 *O*-linked glycopeptides. Fmoc, allyl and Dde protecting groups have enabled[304] a cyclic analogue (173) of muramyl dipeptide, to be made using the 'Meshed-Bag-gathered-Bunch' method. Side-chain cyclisation between D-Glu and Lys was carried out on-resin using HBTU/HOBt. As part of the demand for more information on heparin biosynthesis, a plausible precursor, H-Ser[β-D-GlcA (1 → 3)β-D-Gal(1 → 3)-β-D-Gal(1 → 4)-β-D-Xyl(1 →.O)]-Gly-Trp-Pro-Asp-Gly-OH has been synthesized[305] from a glycosylated Ser-Gly coupled with the tetrapeptide Trp-Pro-Asp-Gly-OH. Extended carbohydrate side chains have been built on to Fmoc-Ser/Thr-Opfp in either α or β form, for the synthesis of tri and heptasaccharide analogues of peptide 31D, (174) and (175) respectively.[306] When the T-helper cell activity of the glycopeptides was determined, the α-linked forms, which mimic the natural glycoproteins, showed a reduction in stimulator activity, while the β-form retained activity, and as the trisaccharide it was as active as the unglycosylated peptide. So glycosylation seems to have abrogated the antigenicity of the protein fragment. For the solid phase synthesis[307] of mucin-type glycopeptide, β hCG 130-145, the β-subunit of human chorionic gonatropin, it was possible to use the pentafluorophenyl ester of (176) as building block without any protection on the carbohydrate OH groups.

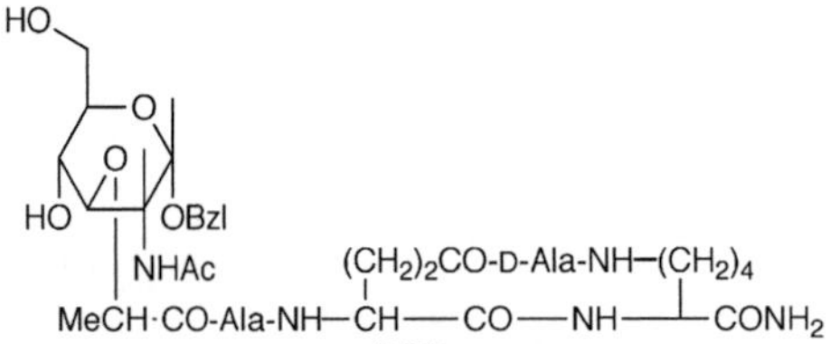

(173)

AVYT(X)RIMMNGGRLKR

(174) X = Glc(α-1→4)Glc(α-1→4)Glc(β-1→0)

(175) X = [Glc(α-1→4)Glc]$_3$-Glc(β-1→0)

Of the two insect antimicrobial peptides, drosocin and apidaecin Ib, only drosocin (177) contains a carbohydrate conjugate at position 11. Through synthesis[308] of a series of different glycosylated drosocin analogues it has been shown that antimicrobial activity against Gram-negative bacteria was modulated by the choice of carbohydrate and the type of glycosidic linkage. Head to tail cyclic analogues, *i.e.* (177) joined between Val and Gly but without the sugar unit were also synthesised, but deglycosylation reduced the antimicrobial activity. Glycopeptides having a recurring sequence of [Glu(OMe)-Ser (β-D-GlcNAc)-Aib] have been made[309] by polymerisation of a glycosylated tripeptide using DPPA or active ester methods. The monomers used were, H-Glu(OMe)-Ser[β-D-GlcNAc(Ac)3]-Aib-OH and H-Glu(OMe)-Ser[β-D-GlcNAc(Ac)$_3$]-Aib-ONp respectively, and the average molecular weights of the polymers formed were higher using the active ester method. After deacetylation with hydrazine/methanol, CD spectra indicated an α-helical conformation. The protocol for the syntheses of glycoclusters will benefit from the availability[310] of the orthogonally protected α,α-bis(aminomethyl)-β-alanine building block (178).

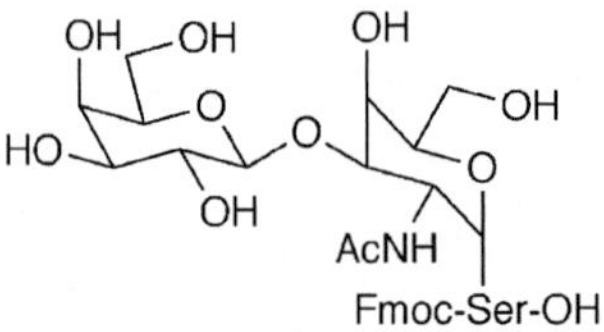

(176)

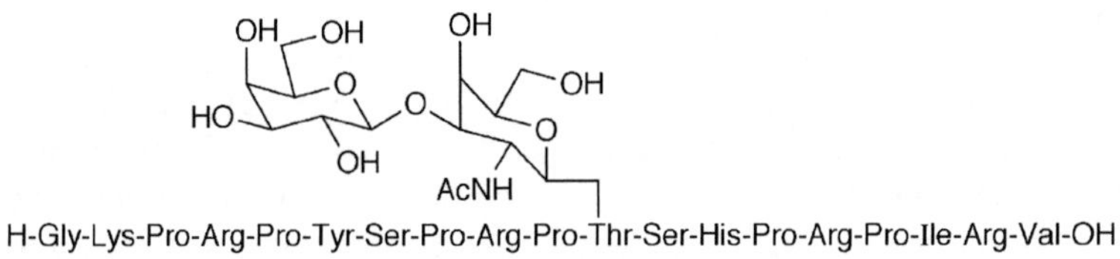

H-Gly-Lys-Pro-Arg-Pro-Tyr-Ser-Pro-Arg-Pro-Thr-Ser-His-Pro-Arg-Pro-Ile-Arg-Val-OH

(177)

Chemoenzymatic synthesis has made a useful contribution to glycopeptide synthesis, and in this period of review, two reports of its application have been found. In the synthesis[311] of N-protected MUC II oligosaccharide serine (179), the enzymic reaction of GalNAcα1-O-(Z)-Ser-OAll with pNP-β-Gal in the presence of recombinant β1,3-galactosidase from *Bacillus circulans* gave Galβ-(1 → 3)-GalNAcα1O-(Z)-Ser-OAll. The branched Galβ-(1 → 3)-[NeuAcα-(2 → 6]-GalNAcα1-O-(Z)-Ser-OAll was constructed chemically, with the terminal Gal residue being added using α2,3-(O)-sialyltransferase. Using recombinant sialyl transferases,[312] chicken GalNAcα2,6- and porcine Galβ(1-3)GalNAcα-2,3-sialyl transferase, the common *O*-linked sialosides, Neu5Acα(2-6)GalNAcα(1-1)Thr, Galβ(1-3)Neu5Acα(2-6)GalNAcα(1-1)Thr, Neu5Acα(2-3)Galβ(1-3)GalNAcα(1-1)Thr, and Neu5Acα(2-3)Galβ(1-3)[Neu5Acα(2-6)GalNAcα(1-1)Thr have been made on a preparative scale. *In vitro* protein synthesis has been found useful in incorporating glycosylated amino acid residues in a site-specific manner[313] and for this purpose glycosylated amino acyl nucleotides had to be prepared. Cyanomethyl esters such as (180) have been found to be key intermediates for this methodology.

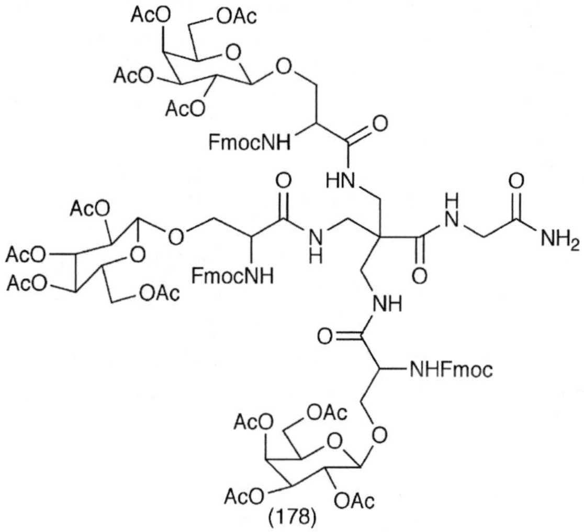

AcO OAc O AcO O AcO FmocNH O O HN O AcO H N NH2 O N AcO O H AcO FmocNH O OAc NH O NHFmoc AcO OAc O O AcO OAc

(178)

OH OH HO CO2H AcNH O chemical construction OH O OH OH CO2H OH OH OH HO O O O O O OH enzymic construction HO AcNH O CO2H NH2

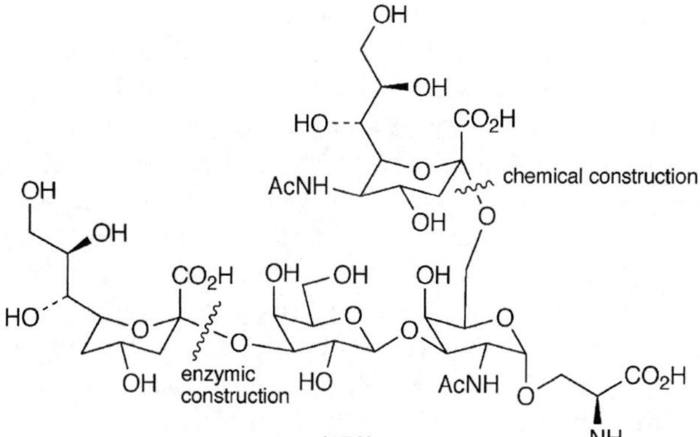

(179)

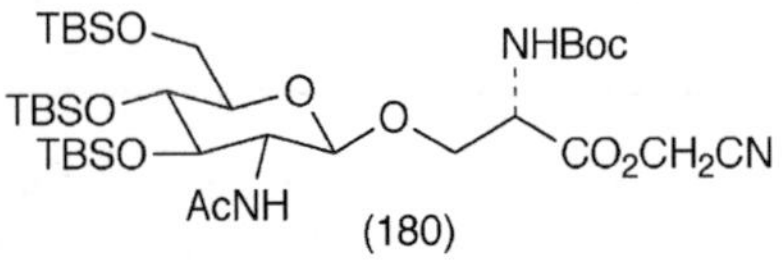

The relative stabilities of rotamers of N-acetyl-O-(2-acetamido-2-deoxy-α-D-galactopyranosyl)-L-seryl-N'Me amide and eleven analogous molecules containing β-galactose, α- and β-mannose, α- and β-glucose and L-threonine have been calculated[314] using *ab initio* methods and molecular mechanics. The most stable interactions with serine were with the glucose moiety, while with threonine the largest stabilisation effect was with the α-anomer of ManNAc. NMR methods[315] have been used to study the binding mode and binding epitope of the glycopentapeptide antigen, Pro-Asp-Thr(O-α-D-GalNAc)-Arg-Pro and its non-glycosylated analogue, to the SM3 antibody. TrNOESY build up rates showed that the glycopeptide was interacting with SM3 utilising all the amino acid residues, and in the bound state the glycopeptide has an extended conformation. Two fluorophores, dansyl (D) (di-Me amino-naphthalenesulfonyl) and naphthyl (N) have been linked[316] to the glucose ring and side chain of lysine respectively in the model glycopeptides, Boc-Ala-Thr[Gluc-(D)]-A1-A2-Leu-Leu-Lys(N)-Ala-OMe. Using fluorescence resonance energy transfer studies it has been shown that the peptides adopt an α-helical conformation in solution and in the presence of liposomes.

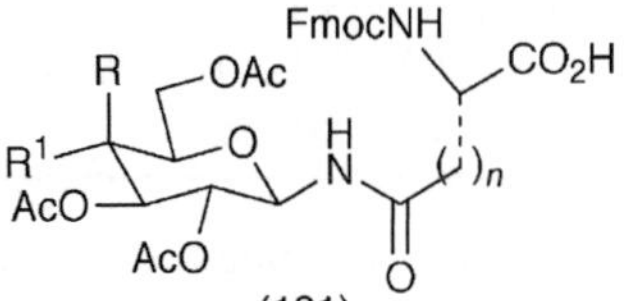

3.4.2 N-Glycopeptides. Several N-linked glycoamino acid building blocks typified by (181) have been prepared[317] by coupling a sugar unit having an amino group at the anomeric centre to a Boc-protected amino acid using BOP/DIEA, followed by further coupling to protected aspartyl or glutamyl residues. These building blocks were used to make the N-linked dodecaglycopeptide Ac-(Gly-Pro-Asn-[Gal])$_4$-NH$_2$. Starting from β-GlcNAcN$_3$, the building block (182) has been synthesised[318] and used for the solid phase synthesis of Ac-Ser-Gln-Thr-(GlcNAc-Asp)-Glu-Thr-His-NH$_2$. The effect of glycosylation on yeast *Saccharomyces cerevisiae* α-mating factor has been studied[319] after successfully extending the glycoside moiety using enzymic catalyses. After assembling H-Trp-His-Trp-Leu-Gln-(GlcNAc)-Leu-Lys-Pro-Gly-Gln-Pro-Met-Tyr-OH on solid phase, a sialo complex was added using *Mucor hiemalis* endo-β-N-acetylglucosaminidase to give (183), but this had less activity than that of the native α-mating factor. A biocatalytic method[320] has allowed Gln or Asn to be modified with structurally diverse carbohydrates. The methodology uses the protease clostripain from *Clostridium histolyticum* to activate a substrate mimetic such as (184) to acylate monosaccharides such as D-glucosamine, D-galactosamine and muramic acid. Stepwise synthesis in the solution phase[321] using EDC/HOBt couplings has produced two series of homo-oligomers, Boc-[(GlcNAc(Ac)$_3$β)Asn]$_n$ –NHMe (n=1–8) , and their deacetylated series Boc-[(GlcNAcβ)Asn]$_n$ –NHMe (n=1–8). NMR and IR measurements indicated the presnce of β-turn conformations in both glycosylated series.

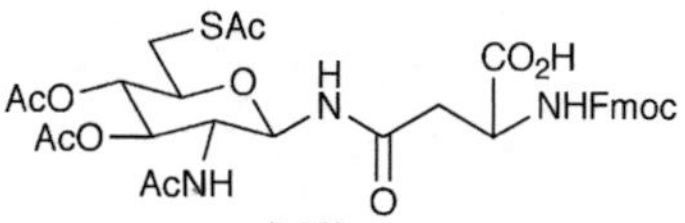
SAc
AcO
AcO
O
H
N
CO2H
NHFmoc
AcNH
O

(182)

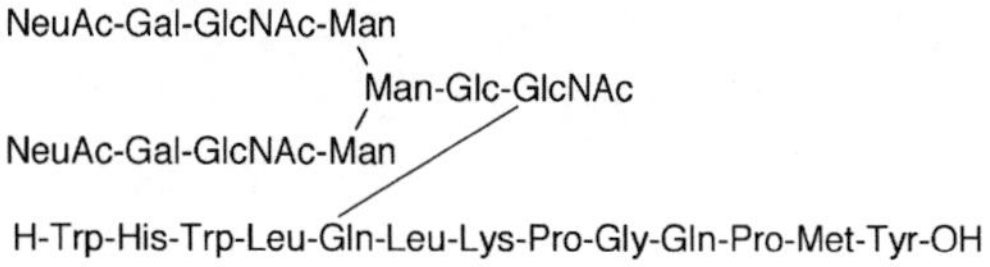
NeuAc-Gal-GlcNAc-Man
Man-Glc-GlcNAc
NeuAc-Gal-GlcNAc-Man
H-Trp-His-Trp-Leu-Gln-Leu-Lys-Pro-Gly-Gln-Pro-Met-Tyr-OH

(183)

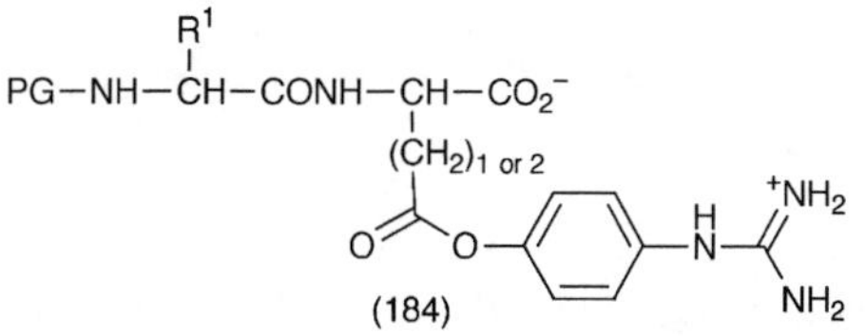
R1
PG—NH—CH—CONH—CH—CO2−
(CH2)1 or 2
+NH2
NH2

(184)

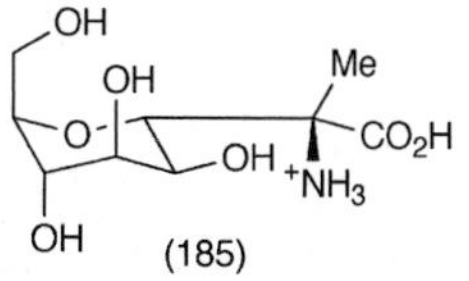
OH
OH
Me
CO2H
+NH3
OH
OH

(185)

3.4.3 C-Linked Glycopeptides. The sensitivity of *O*-, *N*- and *S*-glycosyl links to basic and acidic conditions has generated significant synthetic activity in the synthesis of *C*-glycosyl analogues with their more stable carbon-carbon anomeric links. Many of the synthetic schemes for enabling these links to be made have been reviewed.[322]

Many new syntheses of *C*-linked building blocks have been put forward during 2002. α-D-C-Mannosyl (*R*)-alanine (185) has been synthesised[323] in four steps from protected D-ribo-hex-1-enitol and N-benzoylalanine. A Heck-coupling reaction[324] between 1-[4,6-di-O-t-butyldimethylsilyl)-2,3-dideoxy-α-(+)-erythro-hex-2-enopyranosyl]-4-bromobenzene and methyl 2-acetamidoacrylate in the presence of $Pd(OAc)_2$, $P(PhOMe)_3$ and $AgNO_3$ gave an enamido ester, which could be reduced in the presence of chiral catalysts to give (*R*) and (*S*) forms of (186). Stereoisomerically pure forms of C-glycosyl tyrosine derivatives (188) were made[325] by building an aromatic ring on to the allylic compounds (187). Allylic intermediates have also featured[326] in an olefine cross metathesis reaction using an improved Grubb's catalyst to give (189). Ethylene isosteres of N-glycosyl asparagines have evolved[327] from the coupling of metalated sugar acetylenes with N-Boc-D-serinal acetonide (Garner aldehyde) giving (190). Stereoselective *C*-Linked D-gluco- and D-galactosyl-serines have been obtained[328,329] from *C*-allyl glycopyranosides after conversion into iodoethyl derivatives that are then substituted with the Williams' chiral glycine enolate equivalent to give (191) after deprotection. Reductive ring-opening of an epoxide ring using titanocene followed by reaction with a trapping agent gives rise to *C*-glycosides as summarised in Scheme 29. Another addition to the few known examples of a C–C link being made in a non-anomeric position has appeared[330] with the synthesis of the fully protected α-*C*-galactosyl dipeptide (192), but when α-*C*-glucosylpropargyl glycine underwent[331] Pd-catalysed heteroannulation with iodoaniline, α-*C*-glucosyl isotryptophan was obtained, and not the expected tryptophan analogue. A fully automated method[332] for the synthesis of artificial glycopeptides having two similar or different carbon-linked (e.g. 193) has been developed, and a fully unprotected α-*C*-glycosyl analogue of N-acetylgalactosamine has been conjugated[333] by a non-natural

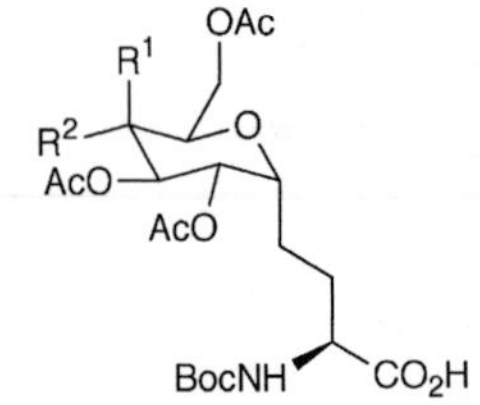

(191) R^1 = H, R^2 = OBn or R^1 = OBn, R^2 = H

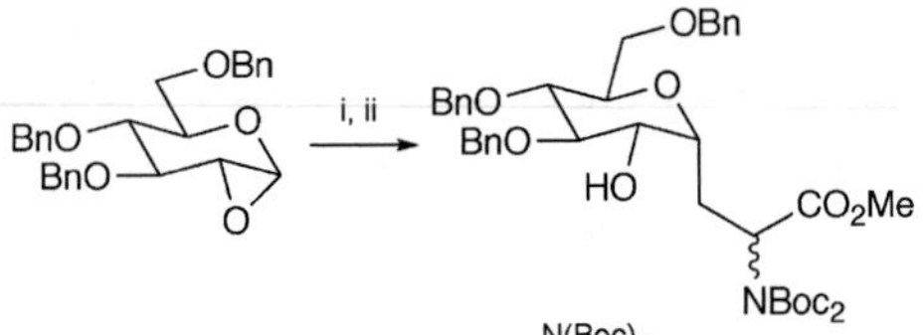

Reagents: i, Cp_2TiCl_2Mn; ii, $=C(N(Boc)_2)CO_2Me$

Scheme 29

oxime bond to the segment peptides OVA 328-340 and OVA 327-339, as given in (194). Attachment of the sugar unit does not interfere with the capacity of dendritic cells to present the anti-genic peptide to OVA-specific T cells.

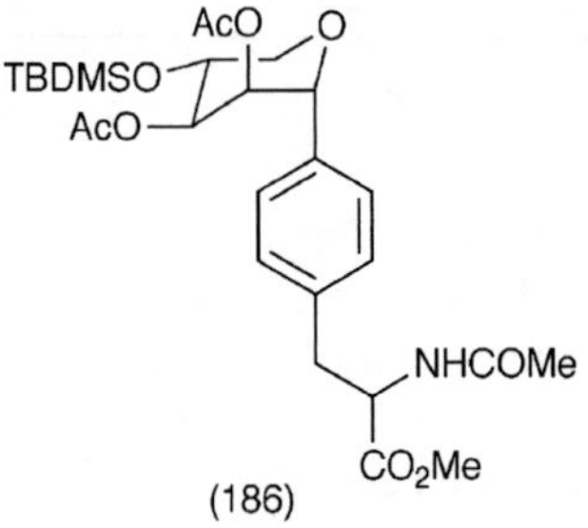

(186)

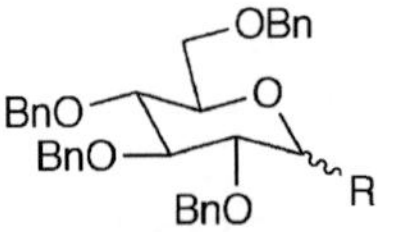

(187) R = α- or β-allyl

(188) R = α- or β- —CH_2—C_6H_4—CH_2–CH(NHCOMe)CO_2H

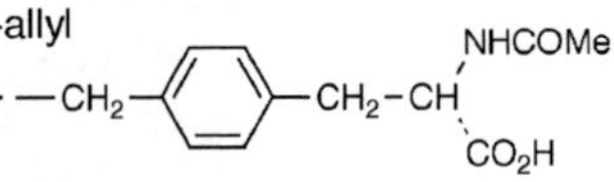

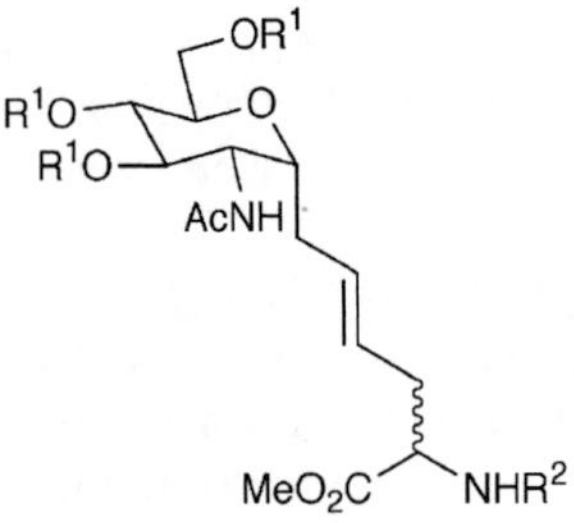

(189)

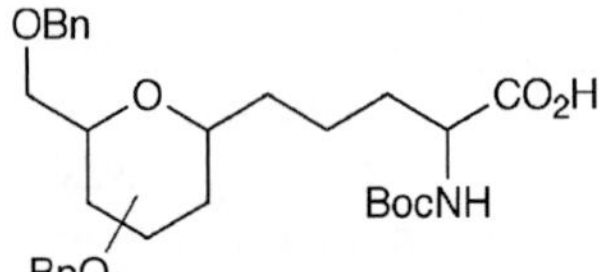

(190)

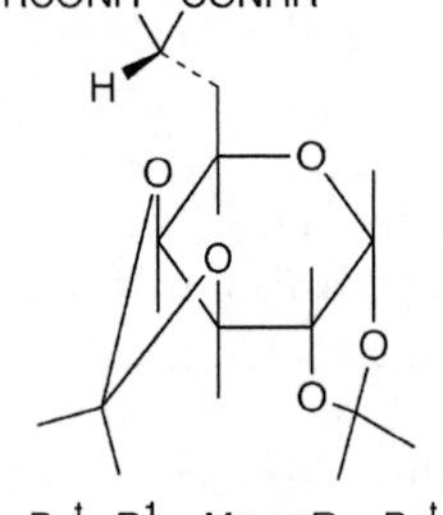

(192) R = Bu^t, R^1 = Me or R = Bu^t, R^1 = Pr^i

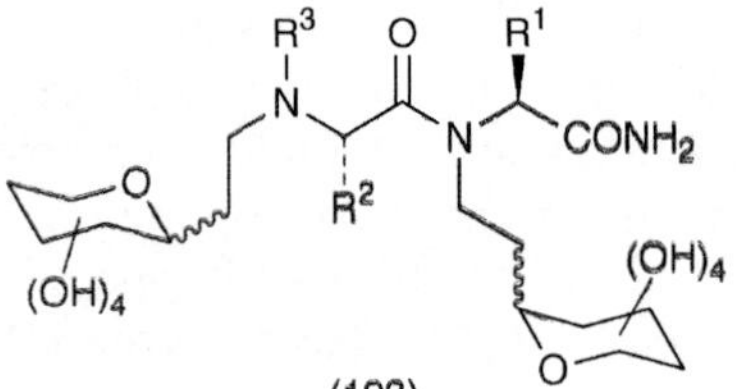

(193)

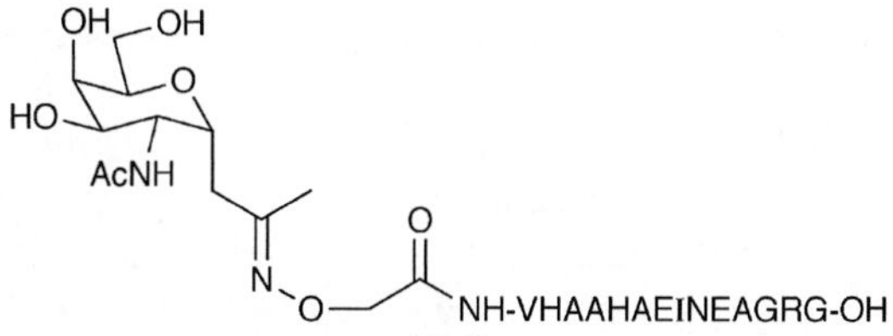

(194)

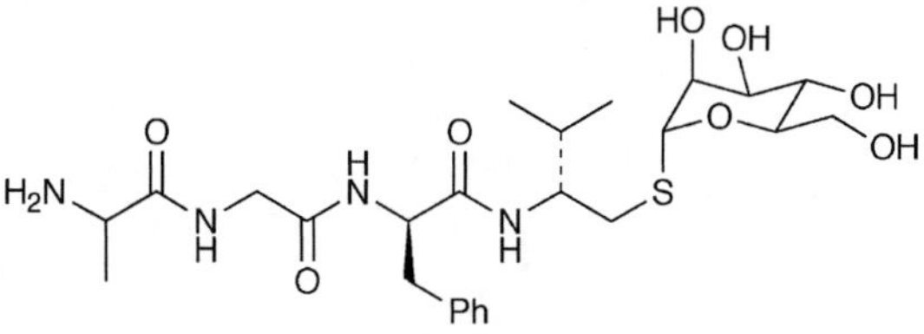

(195)

3.4.4 Thio-linked Peptides. Replacement of O-linked glycosyl amino acids with their S-linked isosteres results in glycopeptides that are more chemically stable and are more resistant to the action of glycosidases. Such a linkage has been tried out[334] as a C- or N-terminal blocking group to inhibit recognition by exopeptidases. The thiosugar was attached via a solid-phase Mitsunobu reaction between a resin-bound 1-thiosugar and a Fmoc-protected amino alcohol, to give (195). Glycosyl iodoacetamides, normally used for the modification of bacterially derived proteins, have been applied[335] to solid phase synthesis of glycopeptides, by providing an access to glycopeptide α-thioesters which can be processed as summarised in Scheme 30. Cyclic sulfamidates derived from serine and threonine have been reacted[336] with 1-thiosugars in aqueous buffer (pH 8) to afford the corresponding S-linked serine and threonine-glycosyl amino acids,

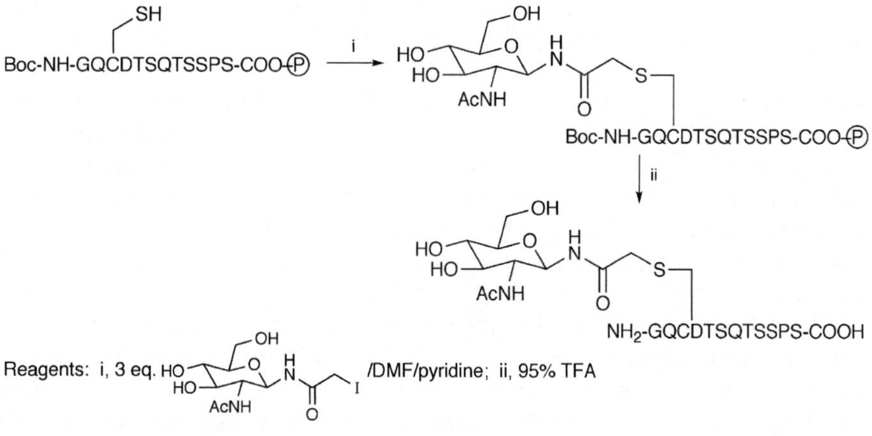

Reagents: i, 3 eq. [glycosyl iodoacetamide] /DMF/pyridine; ii, 95% TFA

Scheme 30

with good diastereoselectivity. The reaction could be adapted also for solid phase work, and the support-bound S-linked glycopeptide extended using standard solid phase techniques. An attractive building block (196) for the synthesis of α-GalNAc thioconjugates[337] has emerged from S-substitution of the sulfide anion of 2-acetamido-2-deoxy-1-thio-α-D-galactopyranose. Glyco-clusters –peptide conjugates such as (197) have been prepared[338] by hydrazone chemoselective ligation to N-chloroacetylated-Lys trees and 2-thioethyl-α-D-mannopyranoside, using the epitopic peptide $TT^{830-846}$ as a model. The S-xanthenyl group could well be a good protecting group for the synthesis of thioglycosides.[339]

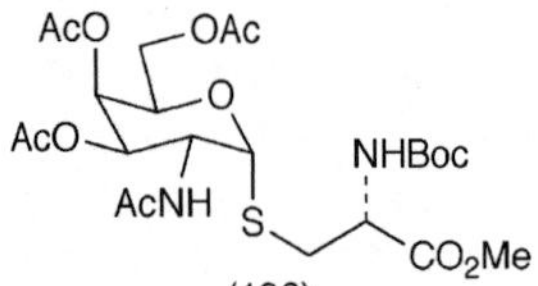

(196)

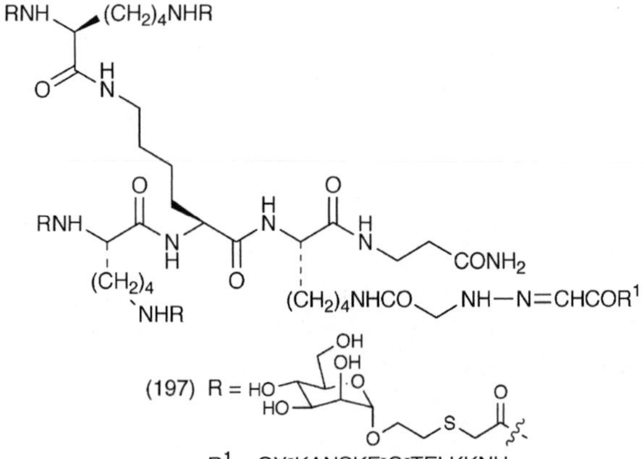

(197) R =

R^1 = QYIKANSKFIGITELKKNH$_2$

3.4.5 Glycosoamino acid residues. These residues have become increasingly popular as peptidomimetic units, and their role as building blocks for combinatorial synthesis has been reviewed.[340] The dipeptide isostere, glucuronic acid methylamine (Gum) has been included[341] in a new structural scaffold (198) by using the building block Fmoc-Gum$(Bn)_3$ –OH. The macrocyclisation to (198) was carried out either using DIC/HOAt/NMM or HATU/HOAt/collidine, but somewhat surprisingly the NMR data on (198) showed no evidence that Gum functions as a β-turn mimetic. Novel furanoid sugar amino acids have been incorporated[342,343] into a series of cyclic oligomers e.g. (199), which have shown a great deal more solubility in organic solvents than is typical of β-peptides. Glyco-α-aminonitriles have given rise to two glyco amino acids (200) and (201) which can be considered to be mimetics of α,α,-disubstituted amino acids. With oligomers bearing β(1 → 2) and β(1 → 6) linkages already synthesised, the β(1 → 3) (202) and β(1 → 4) (203)versions have now been prepared.[344] On *O*-sulfation these oligomers effectively inhibited HIV infection sialyl Lewis x-mediated cell adhesion and heparanose activity. The list of glycosyl amino acid building blocks has been further augmented by the availability of (204) and (205), for use in combinatorial synthesis of neoglycoconjugates.[345]

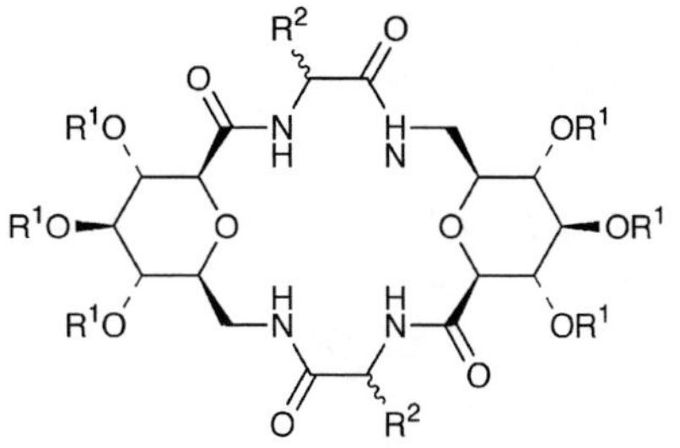

(198)

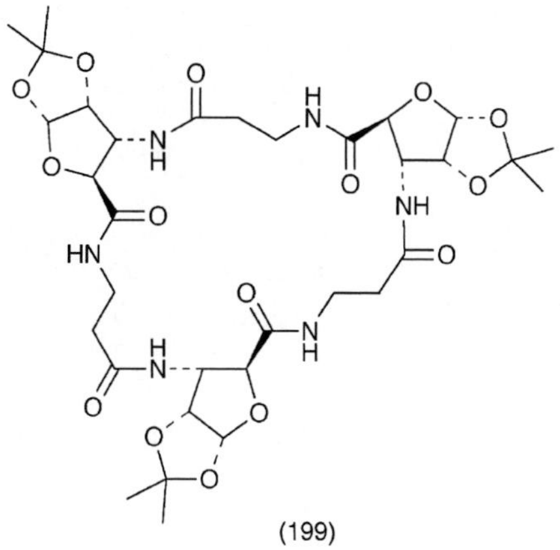

(199)

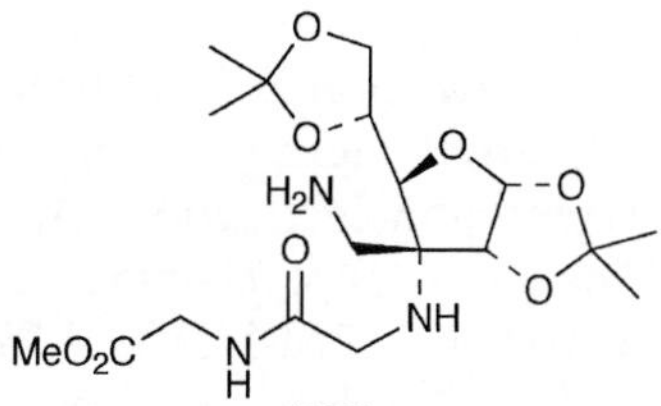

(200)

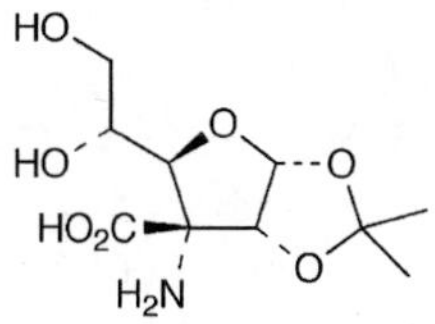

(201)

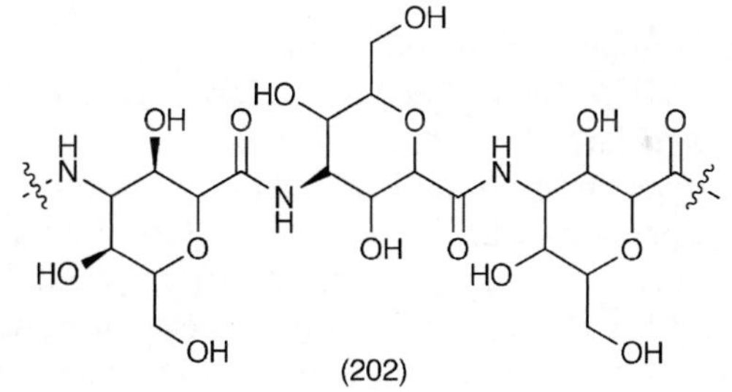

(202)

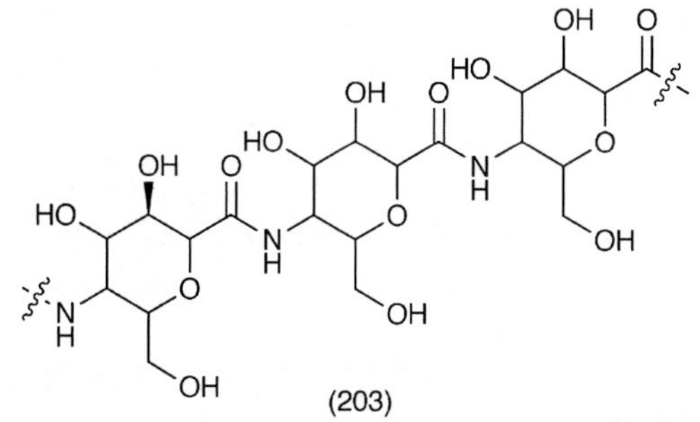

(203)

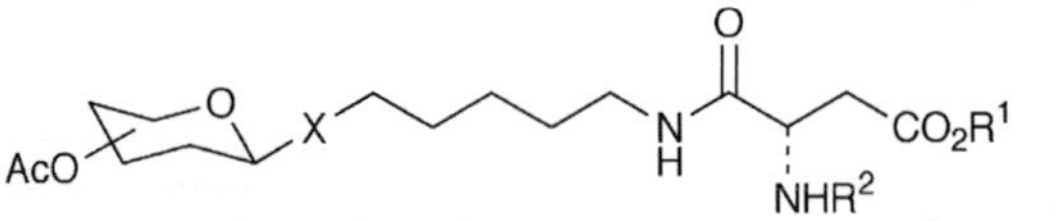

(204) X = α- or β-O or β-NHCO, R^1 = Bu^t, H, pentafluorophenyl, R^2 = Z, H, Boc, Fmoc

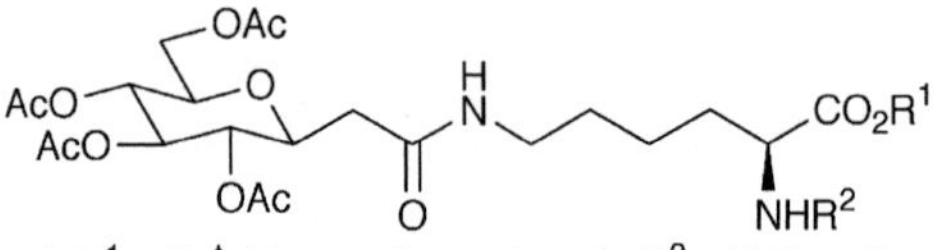

(205) R^1 = Bu^t, H, pentafluorophenyl, R^2 = Z, Boc, Fmoc

3.5 Lipopeptides. – The use of lipidated peptide conjugates as molecular probes for investigation of biological phenomena, and the development of methods for their synthesis have been reviewed.[346] Lobocyclamides A (206), B (207) and C (208) have been identified[347] in the cryptic cyanobacterial mat of *Lyngya confervoides*. Congeners B and C are thought to be the first peptides reported to have 4-hydroxy threonine as part of their structure. Their structures are related to laxaphycins A and B, but the lobocyclamides only showed moderate anti-fungal activity. The combination of ESI-FTICR-MS and NMR spectroscopy has been key to the elucidation[348] of the structures of new biaryl-bridged arylomycins A (209) and B (210). *Cis-trans* isomerism was detected around the fatty acid-MeSer bond. Seven known iturins and one new congener (A-8, 211) have been isolated[349] from *Bacillus amyloliquefaciens* RC-2, and have been found to inhibit the development of mulberry anthracnose caused by the fungus *Colletotrichum dematium*. Lipopeptide biomarkers isolated from *Bacillus globigii* have been found[350] to be cyclic peptides with similar sequences to fengycins and plipastatins previously identified in other Bacillus species. Cyclic lipopeptides with antibiotic and biosurfactant properties are produced[351] by fluorescent *Pseudomonas* spp. from sugar beet rhizosphere, while the fluorescent lipophilic moiety (212) has been introduced[352] to position 75 of the immunodominant epitope GpMBP (74–75). The fluorescent lipopeptide interacted with more affinity with the membrane than the lipid-free analogue.

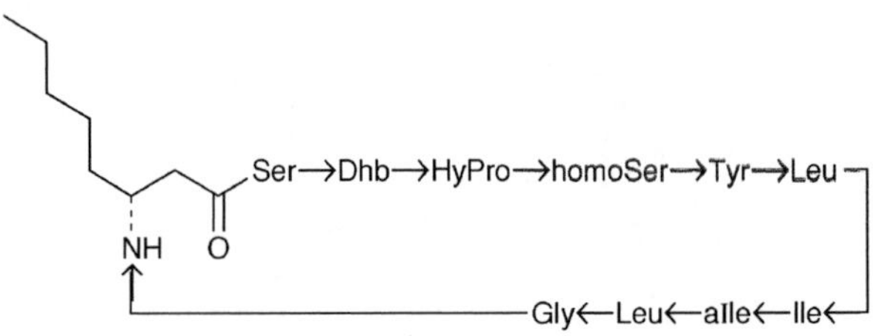

Ser→Dhb→HyPro→homoSer→Tyr→Leu
Gly←Leu←aIle←Ile←

(206)

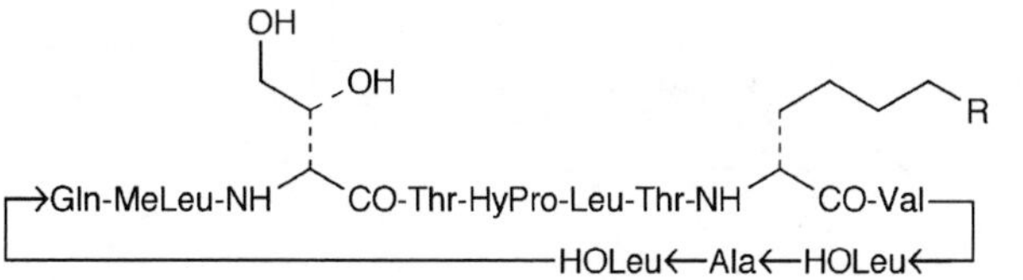

(207) R = Et
(208) R = H

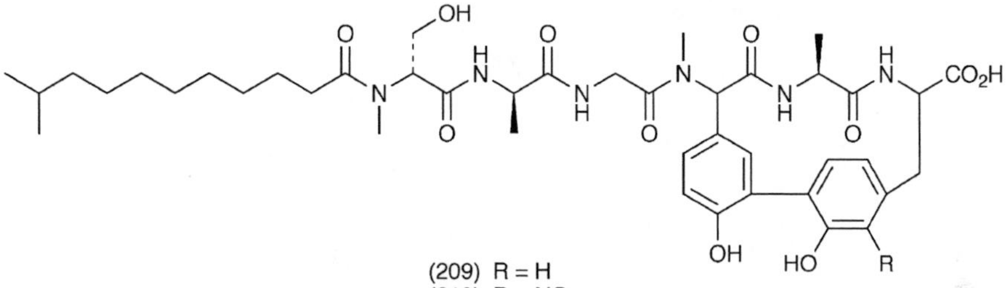

(209) R = H
(210) R = NO_2

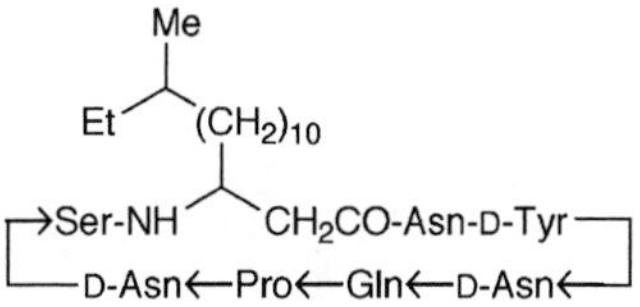

(211)

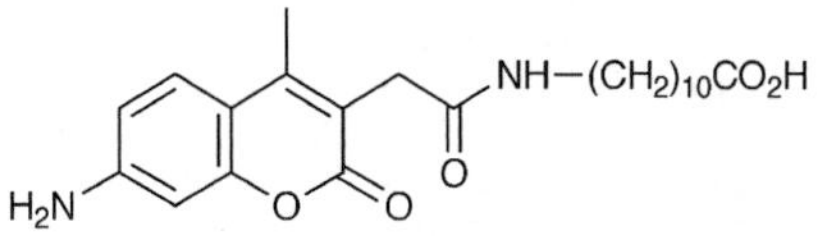

(212)

Carefully chosen orthogonally-stable protecting groups, together with a linker (hydrazide) for the solid support, that is cleavable under the mildest conditions, have been successfully applied[353] to the solid phase synthesis of (213). Initial protection of Cys used the Trt group followed after its removal by farnesylation and palmitoylation on-resin. The cytotoxic lyngbyabellins A (214) and B (215) from the marine cyanobacteria, *Lyngbya majuscula* have undergone total synthesis,[354] with macrocyclisation in (214) being carried out at point (a) using DPPA. In (215) the thiazoline ring was constructed post-macrocyclisation, the latter being carried out at (a) using FDPP. Solid phase synthesis of cell-membrane lipopeptide (216) has provided[355] a mimic of the Myr-Gly-Cys N-terminus of Src family protein, which can be used for assay of the efficiency of protein palmitoylation. Compound (216) is palmitoylated intracellularly to yield radio-labelled (217). Several glycero-conjugates of Arg, Asp, Glu, Asn, Gln and Tyr have been produced[356] by a lipase-catalysed esterification of one or two hydroxy groups of amino acid glyceryl ester derivatives. Scheme 31 summarises the lauroylation process.

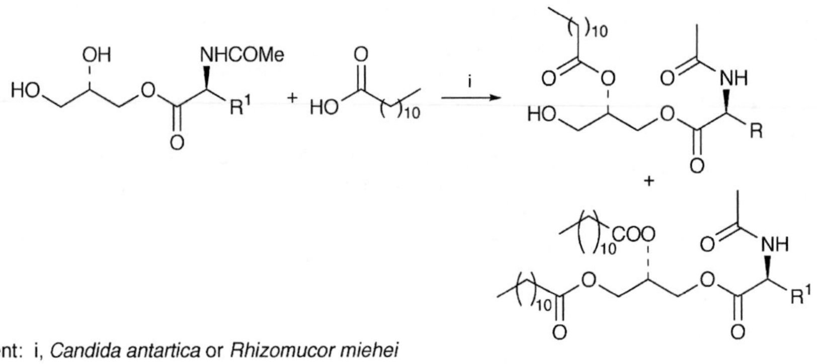

Reagent: i, *Candida antartica* or *Rhizomucor miehei*

Scheme 31

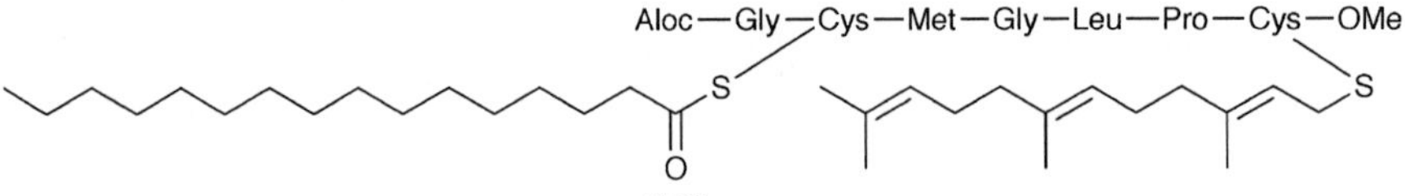

(213)

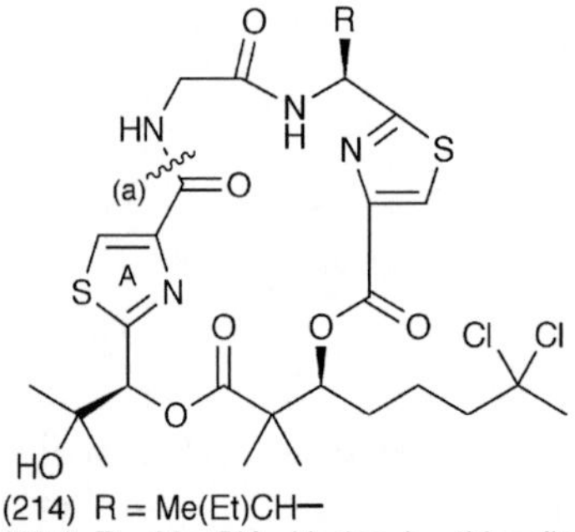

(214) R = Me(Et)CH–
(215) R = Me_2CH with ring A = thiazoline

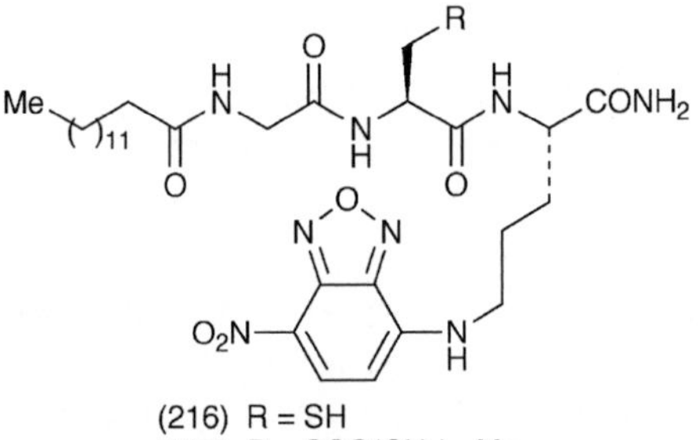

(216) R = SH
(217) R = $SCO(CH_2)_{14}Me$

4 Miscellaneous Structures

A number of interesting papers appear under this section, only because they do not follow themes of the previous sections. However in this review period, quite significant themes have emerged even in this category. For example, quite a few papers report the developing trend towards applying ring-closing metathesis using Grubbs' catalyst [benzylidene(bis (tricyclohexyl)dichlororuthenium] to restrict the conformation of peptides by macrocyclisation. Examples of this approach are, the novel binaphthyl-based cyclic peptide (218),[357] and cyclic RGD pseudopeptide (219).[358] Deprotected versions of the latter inhibit integrin $\alpha_v\beta_3$-dependent cell adhesion with an IC 100% at 0.25 mM. Carbo-carbon bond cyclisation in the context of RGD peptides has also been reviewed.[359] Further examples include cyclic peptidomimetics such as (220) from aldol building blocks,[360] and the cyclic peptoid (221),[361] which incorporates some of the binding features of vancomycin towards the Lys-D-Ala-D-Ala and the Lys-D-Ala-D-Lac terminii. (*Z*)-Alkene isosteres of Phe-Phe, Phe-D-Phe, Phe-Val and Phe-D-Val have been produced[362] using the Grubbs' catalyst, and the cyclohexenyl ring in (222), a dimeric cyclic peptide,[363] was also the product of the same reaction. NMR Studies have revealed[364] that the products of carrying out ring-closure metathesis on N- and C- terminal ω-alkenylamino acid linked on to D-Pro-L-Pro, have a number of conformationally stable motifs. Conformational energy computations[365] have been carried out on dipeptides containing C^{α}-Me, C^{α}-allylglycine to ascertain the parameters needed to involve the allyl groups in ring-closing metathesis.

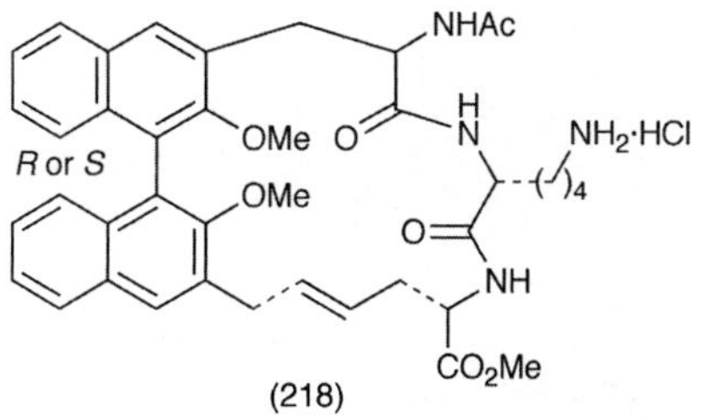

NHAc OMe R or S OMe NH₂·HCl NH CO₂Me

(218)

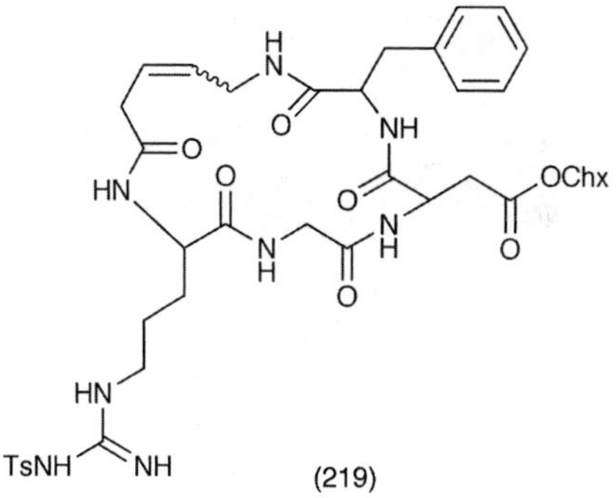

NH OChx HN TsNH NH

(219)

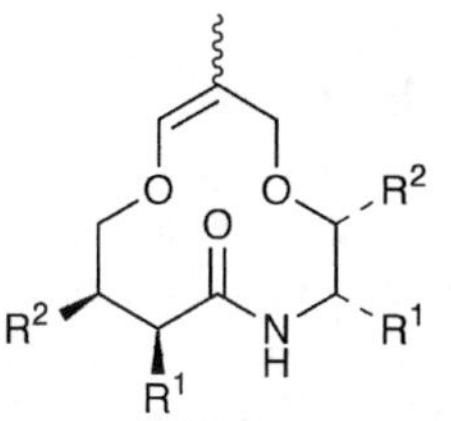

(220)

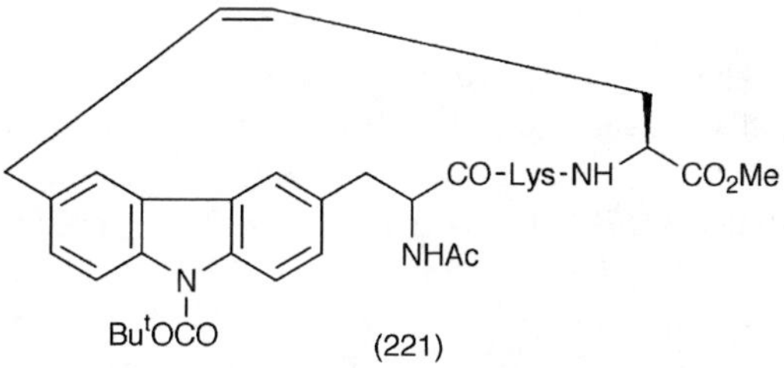

CO-Lys-NH CO₂Me NHAc Bu^tOCO

(221)

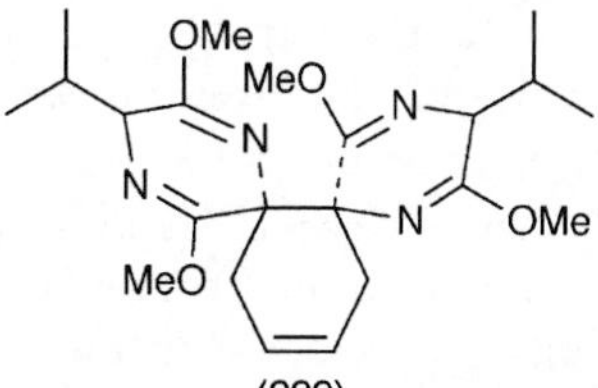

OMe MeO N OMe MeO

(222)

(a) OH Me₂N Ph

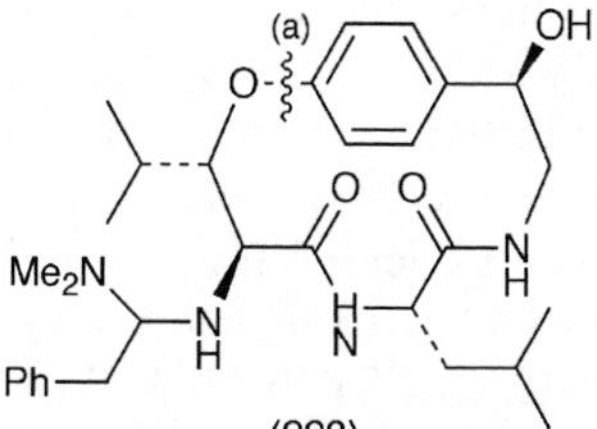

(223)

Past sub-sections of this Chapter usually contain reference to peptide alkaloids, and this year is no exception. It has been reported in a review[366] that up to the year 2000, that the genus Zizyphus has been the source of 81 peptide alkaloids. The total synthesis[367] of sanjoinine G1 (223) has been achieved in 7 steps, using an intramolecular S_NAr reaction [at position (a)] involving a displacement of fluorine in a nitro/fluorinated aromatic ring to achieve macrocyclisation. In the synthesis[368] of ustiloxin D (224), isolated from the fungus *Ustilaginoidea virens*, the ether link was installed at an early stage via a nitrile activated aryl fluoride. Macrocyclisation was via amide bond formation at position (a) in (224) using EDCI/HOBt in 78% yield.

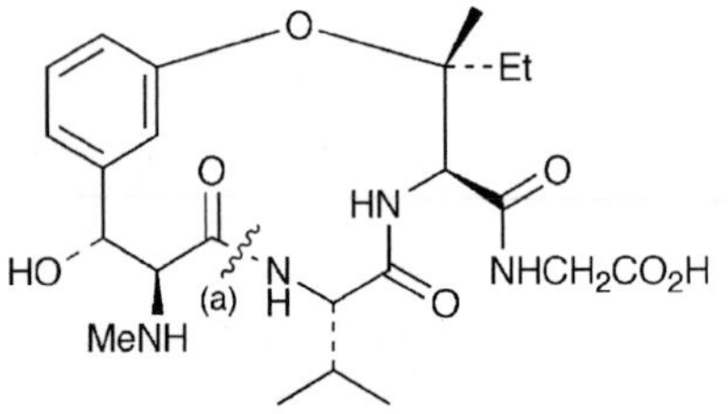

(224)

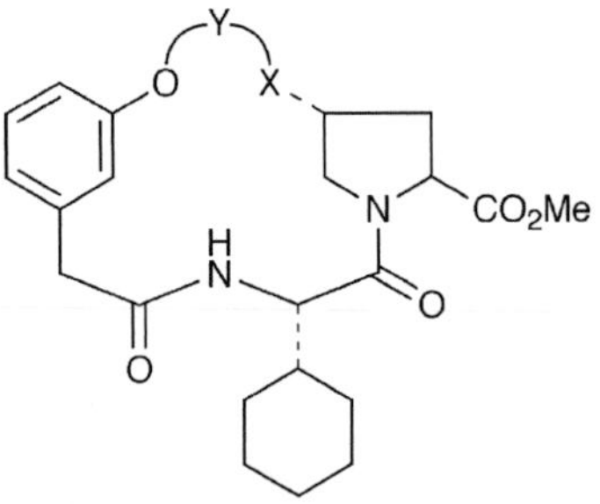

(225) X = O, S or NSO_2Ph, Y = $(CH_2)_2$
(226) X = O, Y = $(CH_2)_3CMe_2$

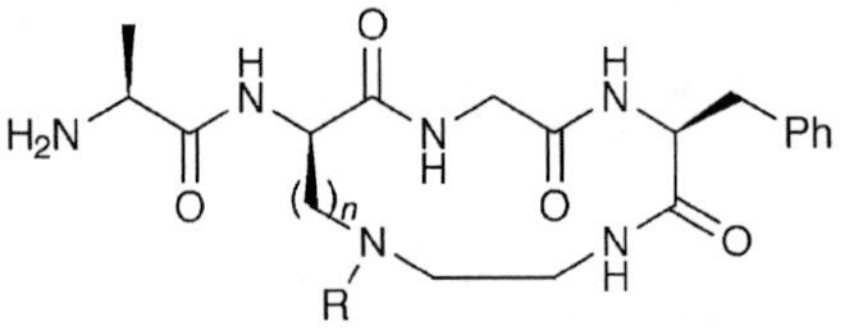

(227) *n* = 1 or 2, R = Me, allyl or 1-naphthylmethyl

The Mitsunobu strategy of ether formation between phenols and alcohols, has achieved success in the synthesis of the novel macrocycles (225)[369] and (226).[370] The solid phase synthesis[371] of amine-bridged cyclic enkephalin analogues (227), also relied on a Fukuyama-Mitsunobu reaction, to prepare a resin-bound sulfonamide-protected secondary amine. The constrained somatostatin analogues (228) and (229) have been assessed[372] spectroscopically, and the conclusion made that an upfield shift of Lys γ-protons was indicative of a bioactive turn conformation. Replacement[373] of the CH_2-S-S-element of the AT_1 receptor agonist cyclo[$Hcy^{3,5}$] angiotensin II by the S-CH=CH unit as in (230), did not influence the agonistic activity, with K_i values of less than 2 nM. The unit was built up from the reaction of the thiol group of cysteine reacting with an aldehyde group on the side chain of lysine. The well known coupling reaction between aryl diazonium salts and phenols (or imidazoles) have been used to great effect[374] for azo-bridge formation, such as (231) in a number of situations such as the constraining of the sequences derived from RGD, GnRH, tuftsin, VIP and SV40 NLS. An orthogonally-protected spiro building block has been used to achieve[375] what is believed to be the first description of the synthesis of spirocyclic peptides represented by (232). A cyclic prodrug (233) based on an RGD peptidomimetic has been developed[376] to improve cell membrane permeation, and can be converted back to the parent molecule using an esterase. The prodrug had an IC_{50}=4μM compared with 1.9 μM for the parent. Fertilin β binds directly to $\alpha_6\beta_1$ integrin, and it has been proposed that fertilin-$\alpha_6\beta_1$ binding is a precursor to sperm-egg fusion. It has been shown that mimetics inhibit this binding *in vitro*, which has initiated research[377] into the conformation of cyclic mimetics such as (234)[cyclo(EC^2DC^1)YNH_2] based on Glu-Cys-Asp being the minimum inhibitory sequence. After solid phase synthesis[378] of the cyclopeptidomimetics (235–238), conformational analysis revealed that it was only sulfone (236) and sulfoxide (238) existed as β-turn conformations in solution.

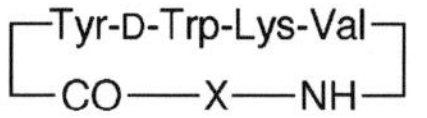

Tyr-D-Trp-Lys-Val–CO–X–NH (cyclic)

(228) X =

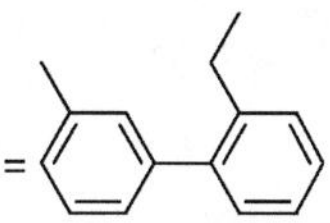

(229) X =

H-Asp-Arg-NH CO-Tyr-NH CO-His-Pro-Phe-OH

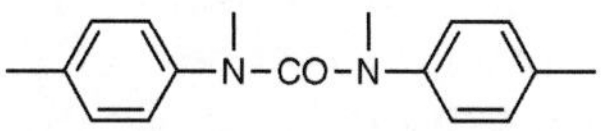

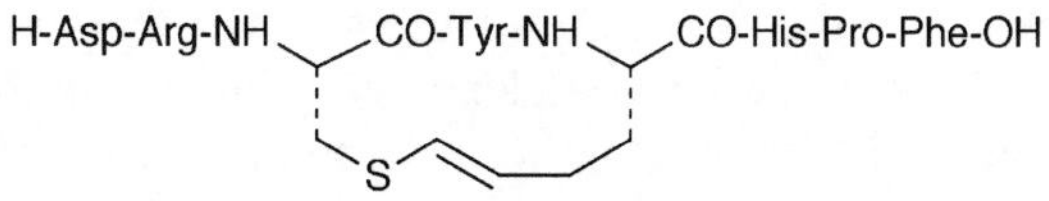

(230)

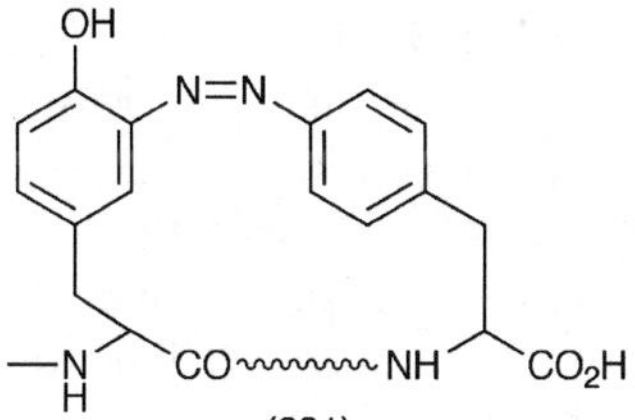

(231)

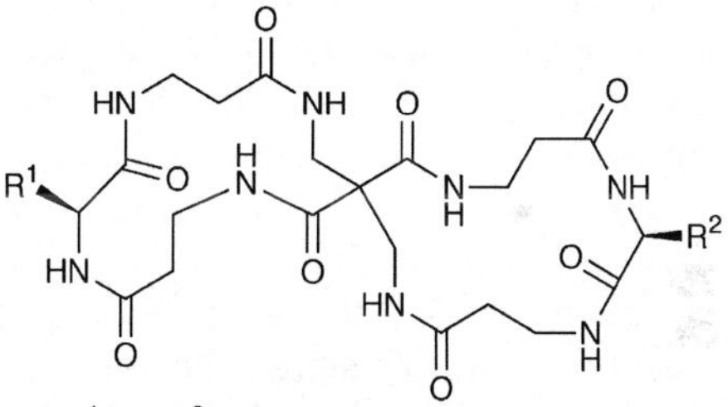

(232) R^1 and R^2 ranging from H to CH_2CMe_2 and CH_2Ph

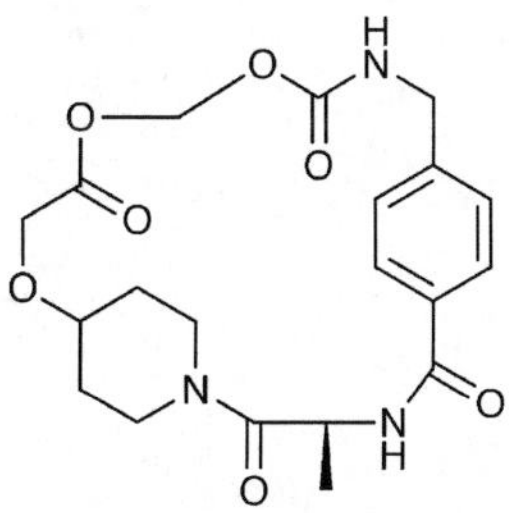

(233)

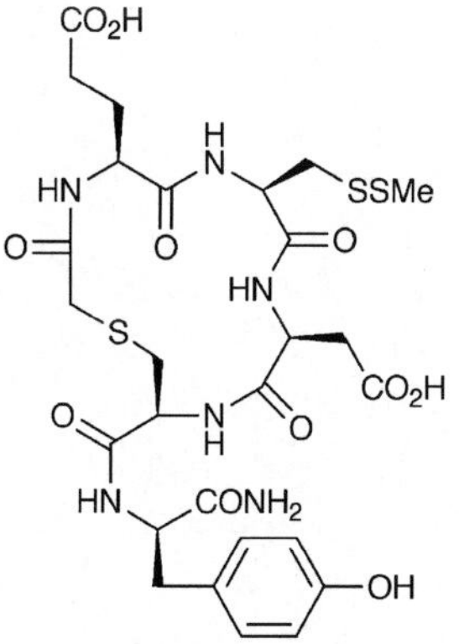

(234)

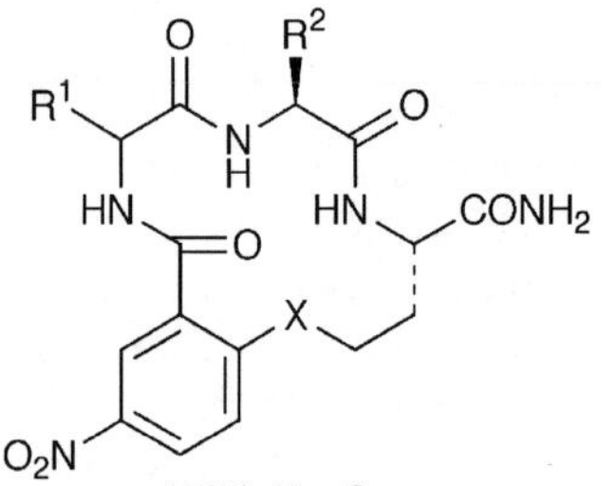

(235) X = S
(236) X = SO_2
(237)
(238)

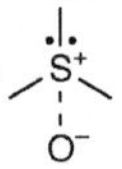

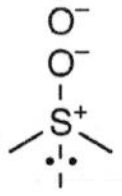

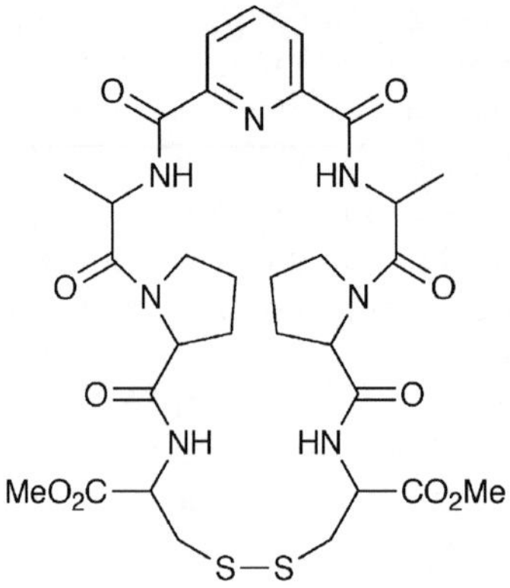

(239)

Another recurring theme in this sub-section is the design and application of cyclopeptides as receptor molecules, be they mimics of natural receptors, or molecules able to complex to small ions. One such example (239) was designed as a novel amphi-receptor,[379] which has been shown to bind to cations and anions. The stability of the cation complexes relied largely on the polarisiability of the cation, whereas the stability of the anion complexes relied on the strength of H-bonding as well as polarisability. An excellent receptor for carboxylate binding having an association constant of 8.64×10^3 M^{-1} for tetrabutylammonium acetate in CD_3CN has been achieved with the synthesis[380] of the furan-based cyclopeptide (240). Binding measurements on (241) showed[381] it to be an enantioselective receptor for N-protected L-glutamate on a 1:1 basis in MeCN/DMSO, but the binding was sensitive to solvent conditions, with no binding in less polar $CDCl_3$ solution. After the synthesis[382] of the chiral A_2B_2 mini receptor (242) the allylic side provided the links to immobilise the compound onto hplc silica, and showed that it could preferentially bind the L-forms of amino acids with enantioselectivity values up to 3.0 kcal/mol.

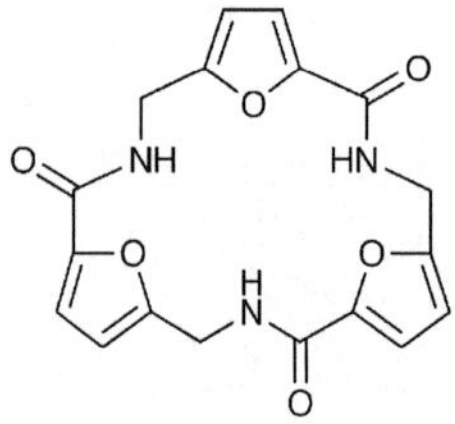

(240)

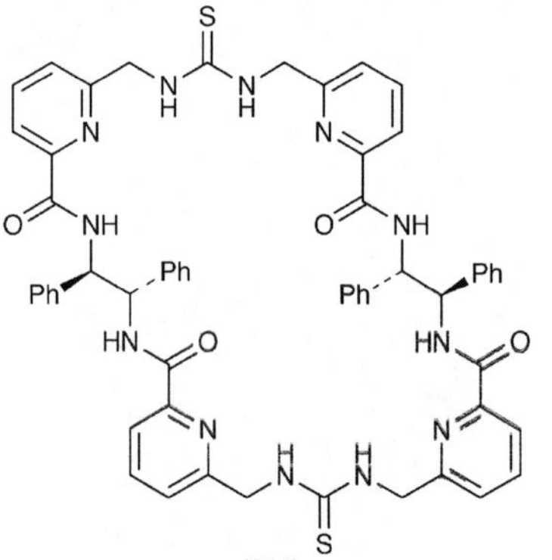

(241)

In model membranes it has been shown[383] that composites containing cysteine or serine ether and adamantane structures as in (243) exhibited 57–60% of the ion transport activity of gramicidin A. The macrocyclic peptide (244) and a trimeric equivalent have been formed[384] during the lactamisation of 1,1′-ferrocenylbis(alanine). NMR data confirmed that they were C_2 and C_3 symmetric molecules with electrochemical properties equal to ferrocene.

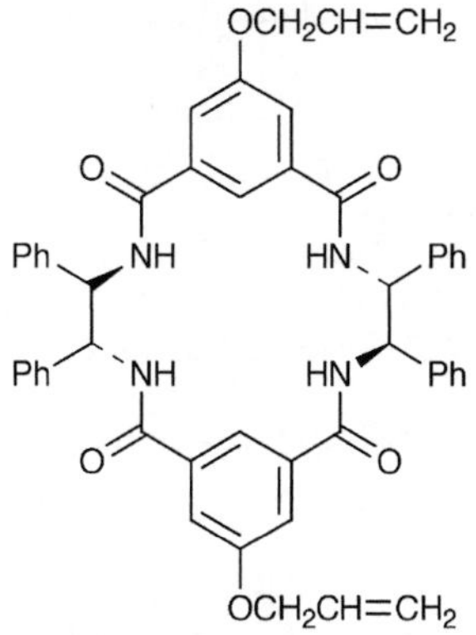

(242)

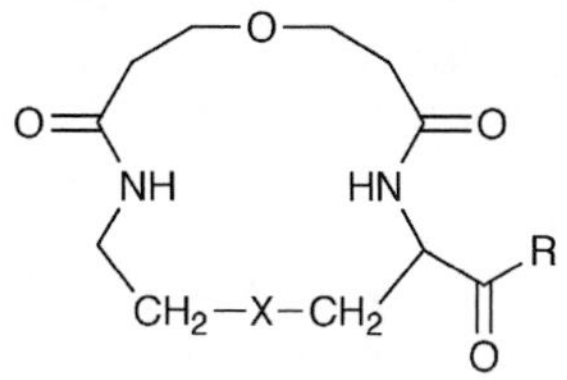

(243) R = OMe or LeuOMe or NH–

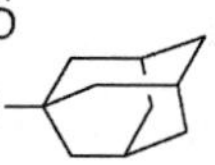

X = S–S or –OCO

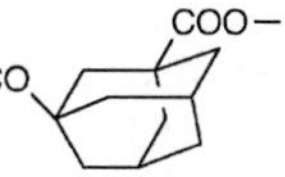

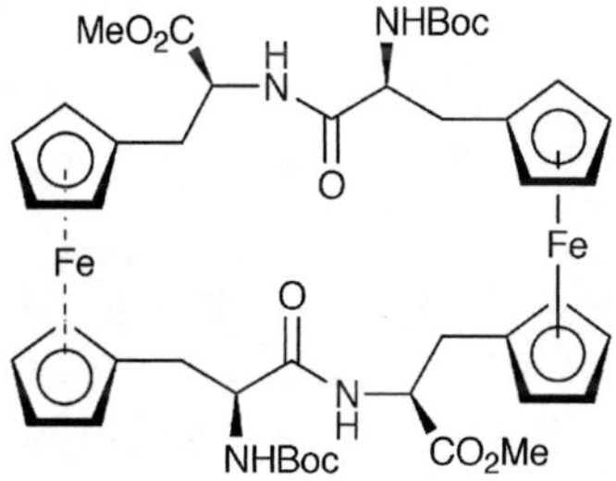

(244)

References

1. C.A. Selects on *Amino Acids, Peptides and Proteins*, published by the American Chemical Society and Chemical Abstracts Service, Columbus, Ohio.
2. *'The ISI Web of Knowledge Service for UK Education'* on http://wok.mimas.ac.uk.
3. *'Peptides 2002'*, Proceedings of the 27th European Peptide Symposium, Sorrento. eds. E. Benedetti and C. Pedone, Editioni Ziino, Castellamare di Stabia,(Na) Italy, 2002,1057pp.

4. *'Peptides: The Wave of the Future'*, Proceedings of the 17th American Peptide Symposium and 2nd International Peptide Symposium, San Diego, Ca. eds. M. Lebl and R. A. Houghten, American Peptide Society, 2001, 1111pp.
5. *'Peptide Science 2002'*, Proceedings of the 39th Japanese Peptide Symposium, ed. T. Yamada, Japanese Peptide Society, 2003, 444pp.
6. Synthesis of Peptides and Peptidomimetics, in *'Houben-Weyl Methods of Organic Chemistry'* 4th Edition, eds. M. Goodman, A. Felix, L. Moroder and C. Toniolo, Georg Thieme Verlag, Stuttgart and New York, 2001-2003, Vols. E22 a-e.
7. R. Rocchi and M. Gobbo, *Chimica Oggi/ Chem. Today*, 2002, **20**, 26.
8. A. Loffet, *J. Pept. Sci.*, 2002, **8**, 1.
9. S. Cappelletti, P. Annoni, G. DiGregorio, O. Storace and M. Pinori, *Chim. Oggi/ Chem. Today*, 2002, **20**, 47.
10. M. Verlander, *Chim. Oggi/Chem. Today*, 2002, **20**, 62.
11. M. Mizhiritskii and Y. Shpernat, *Chim. Oggi/Chem. Today*, 2002, **20**, 43.
12. M.J. Bursavich and D.H. Rich, *J. Med. Chem.*, 2002, **45**, 541.
13. A. McCluskey, A.T.R. Sim and J.A. Sakoff, *J. Med. Chem.*, 2002, **45**, 1151.
14. J.R. Lewis, *Nat. Prod. Rep.*, 2002, **19**, 223.
15. D.A. Horton, G.T. Bourne and M.L. Smythe, *J. Compt-Aided Mol. Design*, 2002, **16**, 415.
16. G. Jung, *Lett. Pept. Sci.*, 2001, **8**, 259.
17. P. Li, P. P. Roller and J. Xu, *Curr. Org. Chem.*, 2002, **6**, 411.
18. (*a*) T. Wang, Q. Yan, Zh. Li, J. You, X. Yu, Q. Yan and R. Xie, *Huaxue Yanjiu Yu Yingyong*, 2002, **14**, 3; (*b*) W. Zhenyu, Z. Yunhong, W. Hui and M. Fengqi, *Huaxue Jinzhan*, 2002, **14**, 113.
19. Y. Yang, D.-C. Yang, Q. Xiao and Y.-G. Zhong, *Youji Hauxue*, 2002, **22**, 239.
20. M. Mizhiritskii and Y. Shpernat, *Chimica Oggi/Chemistry Today*, 2002, **20**, 10.
21. J. Kim, I.-K. Hong, H.-J. Kim, H.-J. Jeong, M.-J. Choi, C.-N. Yoon and J.-H. Jeong, *Arch. Pharmacal. Res.*, 2002, **25**, 801.
22. J.T. Lundquist and J.C. Pelletier, *Org. Lett.*, 2002, **4**, 3219.
23. B. Thern, J. Rudolph and G. Jung, *Tetrahedron Lett.*, 2002, **43**, 5013.
24. R. Quaderer and D. Hilvert, *Chem. Comm. (Cambridge)*, 2002, 2620.
25. T.T. Tran, D. Bryant and M. Smythe, *Comput. Chem.*, 2002, **26**, 113.
26. D. Gottschling, J. Boer, A. Schuster, B. Holzmann and H. Kessler, *Angew. Chem. Int. Ed.*, 2002, **41**, 3007.
27. H. Zhou and W.A. van der Donk, *Org. Lett.*, 2002, **4**, 1335.
28. A.K. Galande and A.F. Spatola, *Lett. Pept. Sci.*, 2001, **8**, 247.
29. Y. Rew, S. Malkmus, C. Svensson, T.L. Yaksh, N.N. Chung, P.W. Schiller, J.A. Cassel, R.N. DeHaven, J.P. Taulane and M. Goodman, *J. Med. Chem.*, 2002, **45**, 3746.
30. V. Swali, M. Matteucci, R. Elliott and M. Bradley, *Tetrahedron*, 2002, **58**, 9101.
31. M.F. Mohd Mustapa, R. Harris, J. Mould, N.A.L. Chubb, D. Schultz, P.C. Driscoll and A.B. Tabor, *Tetrahedron Lett.*, 2002, **43**, 8359.
32. M.F. Mohd Mustapa, R. Harris, J. Mould, N.A.L. Chubb, D. Schultz, P.C. Driscoll and A.B. Tabor, *Tetrahedron Lett.*, 2002, **43**, 8363.
33. (*a*) A.G. Griesbeck, T. Heinrich, M. Oelgemoeller, J. Lex and A. Molis, *J. Amer. Chem. Soc.*, 2002, **124**, 10972; (*b*) A.G. Griesbeck, T. Heinrich, M. Oelgemoeller, J. Lex, A. Molis and A. Heidtmann, *Helv. Chim. Acta*, 2002, **85**, 4561.
34. T.M. Kinsella, C.T. Ohashi, A.G. Harder, G.C. Yam, W.Q. Li, B. Peelle, E.S. Pali, M.K. Bennett, S.M. Molineaux, D.A. Anderson, E.S. Masuda and D.G. Payan, *J. Biol. Chem.*, 2002, **277**, 37512.

35. K.F. Medzihradszky, N.P. Ambulos, A. Khatri, G. Osapay, H.A. Remmer, A. Somogyi and S.A. Kates, *Lett. Pept. Sci.*, 2002, **8**, 1.
36. M. Macht, M. Pelzing, P. Palloch, V. Sauerland and J. Volz, *Acta Biochim. Pol.*, 2001, **48**, 1109.
37. P.M. Fischer, *J. Pept. Sci.*, 2003, **9**, 9.
38. C.J. Dinsmore and D.C. Beshore, *Tetrahedron*, 2002, **58**, 3297.
39. H. Kanzaki, S. Yanagisawa, K. Kanoh and T. Nitoda, *J. Antibiotic.*, 2002, **55**, 1042.
40. A. Smelcerovic, M. Schiebel and S. Dordevic, *J. Serb. Chem. Soc.*, 2002, **67**, 27.
41. G. Degrassi, C. Aguilar, M. Bosco, S. Zahariev, S. Pongor and V. Venturi, *Curr. Microb.*, 2002, **45**, 250.
42. B.A. Fry and R. Loria, *Physiol. Mol. Plant Path.*, 2002, **60**, 1.
43. A. Moyama and R. Ogasawara, *Origins Life and Evol. of Biosphere*, 2002, **32**, 165.
44. W.-R. Li and J.H. Yang, *J. Comb. Chem.*, 2002, **4**, 106.
45. D.-X. Wang, M.-T. Liang, G.-J. Tian, H. Lin and H.-Q. Liu, *Tetrahedron Lett.*, 2002, **43**, 865.
46. T. Carbonell, I. Masip, F. Sanchez-Baeza, M. Delgado, E. Araya, O. Llorens, F. Corcho, J.J. Perez, E. Perez-Paya and A. Messeguer, *Mol. Diversity*, 2002, **5**, 131.
47. A. Golebiowski, J. Jozwik, S.R. Klopfenstein, A.-O. Colson, A.L. Grieb, A.F. Russell, V.L. Rastogi, C.F. Diven, D.E. Portlock and J.J. Chen, *J. Comb. Chem.*, 2002, **4**, 584.
48. J.J. Chen, A. Golebiowski, S.R. Klopfenstein and L. West, *Tetrahedron Lett.*, 2002, **43**, 4083.
49. A.L. Kennedy, A.M. Fryer and J.A. Josey, *Org. Lett.*, 2002, **4**, 1167.
50. M. Weigl and B. Wunsch, *Tetrahedron*, 2002, **58**, 1173.
51. I.L. Rodionov, L.N. Rodionova, L.K. Baidakova, A.M. Romashko, T.A. Balashova and V.T. Ivanov, *Tetrahedron*, 2002, **58**, 8515.
52. K.R.C. Prakash, Y. Tang, A.P. Kozikowski, J.L. Flippen-Anderson, S.M. Knoblach and A.I. Faden, *Bioorg. Med. Chem.*, 2002, **10**, 3043.
53. A. Hatzelt, S. Laschat, P.G. Jones and J. Grunenberg, *Eur. J. Org. Chem.*, 2002, 3936.
54. K.G. Poullennec, A.T. Kelly and D. Romo, *Org. Lett.*, 2002, **4**, 2645.
55. S. Jin and J. Liebscher, *Zeits. für Naturforsch.B: Chem. Sci.*, 2002, **57**, 377.
56. S.D. Bull, S.G. Davies, G.A. Christopher, M.D. O'Shea, E.D. Savory and E.J. Snow, *J. Chem. Soc. Perk. Trans 1*, 2002, 2442.
57. T. Mancilla, L. Carrillo, L.S. Zamudio-Rivera, H.I. Beltran and N. Farfan, *Org. Prep. Proc. Int.*, 2002, **34**, 87.
58. N.M. Khan, M. Cano and S. Balasubramanian, *Tetrahedron Lett.*, 2002, **43**, 2439.
59. N. Schaschke, A. Dominik, G. Matschiner and C.P. Sommerhoff, *Biorg. Med. Chem. Lett.*, 2002, **12**, 985.
60. D.A. Parrish and L.J. Mathias, *J. Org. Chem.*, 2002, **67**, 1820.
61. R. Butz, A.F. Garcia, T. Schneider, J. Starke, B. Tauscher and B. Trierweiler, *High Press. Res.*, 2002, **22**, 697.
62. Y. Ishida and T. Aida, *J. Amer. Chem. Soc.*, 2002, **124**, 14017.
63. M. Belohradsky, I. Cisarova, P. Holy, J. Pastor and J. Zavada, *Tetrahedron*, 2002, **58**, 8811.
64. W. Guo, J. Wang, C. Wang, J.Q. He, X.W. He and J.P. Cheng, *Tetrahedron Lett.*, 2002, **43**, 5665.
65. M. Conza and H. Wennemers, *J. Org. Chem.*, 2002, **67**, 2696.
66. M.R. Caira, E. Buyukbingol, A. Adejare and W.R. Millington, *Anal. Sci.*, 2002, **18**, 1175.

67. P. Bour, V. Sychrovsky, P. Malon, J. Hanzlikova, V. Baumruk, J. Pospisek and M. Budesinsky, *J. Phys. Chem. A*, 2002, **106**, 7321.
68. T.-J.M. Luo and G.T.R. Palmore, *Crystal Growth and Design*, 2002, **2**, 337.
69. L. Ciasullo, A. Casapullo, A. Cutignano, G. Bifulco, C. Debitus, J. Hooper, L. Gomez-Paloma and R. Riccio, *J. Nat. Prod.*, 2002, **65**, 407.
70. S. Lin and S.J. Danishefsky, *Angew. Chem. Int. Ed.*, 2002, **41**, 512.
71. M. Kaiser, M. Groll, C. Renner, R. Huber and L. Moroder, *Angew. Chem.Int. Ed.*, 2002, **41**, 780.
72. M. Kaiser, A.G. Milbradt and L. Moroder, *Lett. Pept. Sci.*, 2002, **9**, 65.
73. M. Kofod-Hansen, B. Peschke and H. Thogersen, *J. Org. Chem.*, 2002, **67**, 1227.
74. E.N. Prabhakaran, I.N. Rao, A. Boruah and J. Iqbal, *J. Org. Chem.*, 2002, **67**, 8247.
75. B. Wels, J.A. W. Kruitzer and R.M.J. Liskamp, *Org. Lett.*, 2002, **4**, 2173.
76. M. Haramura, A. Okamachi, K. Tsuzuki, K. Yogo, M. Ikuta, T. Kozono, H. Takanashi and E. Murayama, *J. Med. Chem.*, 2002, **45**, 670.
77. A.V.S. Sarma, M.H.V. Raman Rao, J.A.R.P. Sarma, R. Nagaraj, A.S. Dutta and A.C. Kunwar, *J. Biochem. Biophys. Meth.*, 2002, **51**, 27.
78. K. Haas and W. Beck, *Zeits. Anorg. Allgem. Chem.*, 2002, **628**, 788.
79. L. Yang, R.X. Tan, Q. Wang, W.Y. Huang and Y.X. Yin, *Tetrahedron Lett.*, 2002, **43**, 6545.
80. S.B. Singh, D.L. Zink, J.M. Liesch, R.T. Mosley, A.W. Dombrowski, G.F. Bills, S.J. Darkin-Rattray, D.M. Schmatz and M.A. Geotz, *J. Org. Chem.*, 2002, **67**, 815.
81. F. Berst, M. Ladlow and A.B. Holmes, *Chem. Commun. (Cambridge)*, 2002, 508.
82. H.K. Abbas, W.T. Shier, J.W. Gronwald and Y.W. Lee, *J. Nat. Toxins*, 2002, **11**, 173.
83. Z. Jiang, M.O. Barrett, K.G. Boyd, D.R. Adams, A.S.F. Boyd and J.G. Burgess, *Phytochemistry*, 2002, **60**, 33.
84. N. Loiseau, F. Cavalier, J.-P. Noel and J.-M. Gomis, *J. Pept. Sci.*, 2002, **8**, 335.
85. N. Loiseau, M. Delaforge, C. Minoletti, F. Andre, A. Garrigues, S. Orlowski and J.M. Gomis, *Biol. Reactive VI*, 2001, **500**, 343.
86. M. Royo, J. Farrera-Sinfreu. L. Sole and F. Albericio, *Tetrahedron Lett.*, 2002, **43**, 2029.
87. M. Liu, G.L. Tian and Y.H. Ye, *Chinese Chem. Lett.*, 2002, **13**, 1059.
88. (*a*) Y.-C. Tang, H.-B. Xie, G.-L. Tian and Y.-H. Ye, *J. Pept. Res.*, 2002, **60**, 95; (*b*) Y.-H. Ye, M. Liu, Y.-C. Tang and X. Jiang, *Chem. Commun.(Cambridge)*, 2002, 532.
89. X.-M. Gao, Y.-H. Ye, M. Bernd and B. Kutscher, *J. Pept. Sci.*, 2002, **8**, 418.
90. C.F. McCusker, P.J. Kocienski, F.T. Boyle and A.G. Schatzlein, *Bioorg. Med. Chem. Lett.*, 2002, **12**, 547.
91. S. Oishi, T. Kamano, A. Niida, Y. Odagaki, N. Hamanaka, M. Yamamoto, K. Ajito, H. Tamamura, A. Otaka and N. Fujii, *J. Org. Chem.*, 2002, **67**, 6162.
92. (*a*) F. Peri, R. Bassetti, E. Caneva, L. de Gioia, B. La Ferla, M. Presta, E. Tanghetti and F. Nicotra, *J. Chem. Soc. Perkin Trans 1*, 2002, 638; (*b*) F. Peri, L. Cipolla, E. Forni and F. Nicotra, *Monatsh. für Chem.*, 2002, **133**, 369.
93. M. Sukopp, L. Marinelli, M. Heller, T. Brandl, S.L. Goodman, R.W. Hoffmann and H. Kessler, *Helv. Chim. Acta*, 2002, **85**, 4442.
94. L. Belvisi, A. Caporale, M. Colombo, L. Manzoni, D. Potenza, C. Scolastico, M. Castorina, M. Cati, G. Giannini and C. Pisano, *Helv. Chim. Acta*, 2002, **85**, 4353.
95. S.L. Goodman, G. Holzemann, G.A.G. Sulyok and H. Kessler, *J. Med. Chem.*, 2002, **45**, 1045.
96. S. Rensing and T. Schrader, *Org. Lett.*, 2002, **4**, 2161.

97. B. Jeschke, J. Meyer, A. Jonczyk, H. Kessler, P. Adamietz, N.M. Meenen, M. Kantlehner, C. Goepfert and B. Nies, *Biomaterials*, 2002, **23**, 3455.
98. (*a*) H.H. Wasserman and R. Zhang, *Tetrahedron*, 2002, **58**, 6277; (*b*) H.H. Wasserman and R. Zhang, *Tetrahedron Lett.*, 2002, **43**, 3743.
99. M.J. Kavarana, D. Trivedi, M. Cai, J. Ying, M. Hammer, C. Cabello, P. Grieco, G. Han and V.J. Hruby, *J. Med. Chem.*, 2002, **45**, 2644.
100. D.R. Houston, K. Shiomi, N. Arai, S. Omura, M.G. Peter, A. Turberg, B. Synstad, V.G.H. Eijsink and D.M.F. van Aalten, *Proc. Natl. Acad. Sci. U.S.A.*, 2002, **99**, 9127.
101. H.Y. He, R.T. Williamson, B. Shen, E.I. Graziani, H.Y. Yang, S.M. Sakya, P.J. Petersen and G.T. Carter, *J. Amer. Chem. Soc.*, 2002, **124**, 9729.
102. T. Matsumoto, N. Tashiro, K. Nishimura and K. Takeya, *Heterocycles*, 2002, **57**, 477.
103. P. Grieco, A. Lavecchia, M.Y. Cai, D. Trivedi, D. Weinberg, T. MacNeil, L.H.T. Van der Ploeg and V.J. Hruby, *J. Med. Chem.*, 2002, **45**, 5287.
104. J. Aszodi, P. Fauveau, D. Melon-Manguer, E. Ehlers and L. Schio, *Tetrahedron Lett.*, 2002, **43**, 2953.
105. Y. Hitotsuyanagi, S. Motegi, H. Fukaya and K. Takeya, *J. Org. Chem.*, 2002, **67**, 3266.
106. P.J. Krenitsky and D.L. Boger, *Tetrahedron Lett.*, 2002, **43**, 407.
107. D. Yang, J. Qu, W. Li, Y.H. Zhang, Y. Ren, D.P. Wang and Y.D. Wu, *J. Am. Chem. Soc.*, 2002, **124**, 12410.
108. A. Katoh, Y. Inoue, H. Nagashima, Y. Nikita, J. Okhanda and R. Saito, *Heterocycles*, 2002, **58**, 371.
109. J. Strauss and J. Daub, *Org. Lett.*, 2002, **4**, 683.
110. P. Grieco, A. Carotenuto, R. Patacchini, C.A. Maggi, E. Novellino and P. Rovero, *Bioorg. Med. Chem.*, 2002, **10**, 3731.
111. K. Brzozwski, A. Legowska, S. Rodziewicz-Motowidlo, A. Liwo and K. Rolka, *Pol. J. Chem.*, 2002, **76**, 807.
112. G. Chen, S. Su and R. Liu, *J. Phys. Chem. B*, 2002, **106**, 1570.
113. S.B. Shu, C.Z. Cui, H.S. Son, J.S.U.Y. Won and K.S. Kim, *J. Phys. Chem. B*, 2002, **106**, 2061.
114. S. Kubik and R. Goddard, *Proc. Acad. Sci. U.S.A.*, 2002, **99**, 5127.
115. S. Kubik, R. Kirchner, D. Nolting and J. Seidel, *J. Am. Chem. Soc.*, 2002, **124**, 12752.
116. C. Cabrele, S. Fiori, S. Pegoraro and L. Moroder, *Chem. Biol.*, 2002, **9**, 731.
117. P.D. Adams, Y. Chen, K. Ma, M.G. Zagorski, F.D. Sonnichsen, M.L. McLaughlin and M.D. Barkley, *J. Am. Chem. Soc.*, 2002, **124**, 9278.
118. B. Dittrich, T. Koritsanszky, M. Grosche, W. Scherer, R. Flaig, A. Wagner, H.G. Krane, H. Kessler, C. Riemer, A.M.M. Schreurs and P. Luger, *Acta Crystall. Sec. B: Struct. Sci.*, 2002, **B58**, 721.
119. E. Politowska, P. Drabik, R. Kazmierkiewicz and J. Ciarkowski, *J. Receptor Signal Trans. Res.*, 2002, **22**, 393.
120. J. Malmstrom, A. Ryager, U. Anthoni and P.H. Nielsen, *Phytochemistry*, 2002, **60**, 869.
121. T. Matsumoto, K. Nishimura and K. Takeya, *Chem. Pharm. Bull.*, 2002, **50**, 857.
122. A. Wele, Y.J. Zhang, C. Caux, J.P. Brouard, L. Dubost, C. Guette, J.L. Pousset, M. Badiane and B. Bodo, *J. Chem. Soc. Perkin Trans. 1*, 2002, 2712.
123. A. Napolitano, I. Bruno, P. Rovero, R. Lucas, M.P. Peris, L.G. Paloma and R. Riccio, *J. Pept. Sci.*, 2002, **8**, 407.
124. M. Liu and Y.-H. Ye, *Chin. J. Chem.*, 2002, **20**, 1347.

125. H. Sugiyama, T. Shioiri and F. Yokokawa, *Tetrahedron Lett.*, 2002, **43**, 3489.
126. S. Schabbert, M.D. Pierschbacher, R.-H. Mattern and M. Goodman, *Bioorg. Med. Chem.*, 2002, **10**, 3331.
127. J.N. Tabudravu, M. Jaspars, L.A. Morris, J.J. Kettenes-van den Bosch and N. Smith, *J. Org. Chem.*, 2002, **67**, 8593.
128. P. Desai, M. Parchand, E. Coutinho, A. Saran, J. Bodi and H. Suli-Vargha, *Int. J. Biol. Macromol.*, 2002, **30**, 187.
129. J. Schripsema and D. Dagnino, *Mag. Res. Chem.*, 2002, **40**, 614.
130. C. Govaerts, J. Rozenski, J. Orwa, E. Roets, A. Van Schepdael and J. Hoogmartens, *Rapid Commun. Mass Spec.*, 2002, **16**, 823.
131. J.N. Tabudravu, N. Jioji, L.N. Morris, J.J. Kettens-van den Bosch and M. Jaspars, *Tetrahedron*, 2002, **58**, 7863.
132. P. Stefanowicz, *Acta Biochim. Pol.*, 2001, **48**, 1125.
133. P. Yang and Y. Song, *Prep. Biochem. Biotech.*, 2002, **32**, 381.
134. T. Matsumoto, A. Shishido, H. Morita, H. Itokawa and K. Takeya, *Tetrahedron*, 2002, **58**, 5135.
135. K. Kaczmarek, S. Jankowski, I.Z. Siemion, Z. Wieczorek, E. Benedetti, P. DiLello, C. Isernia, M. Saviano and J. Zabrocki, *Biopolymers*, 2002, **63**, 343.
136. M. Essler and E. Ruoslahti, *Proc. Natl. Acad. Sci. U.S.A.*, 2002, **99**, 2252.
137. X. Bu, X. Wu, G. Xie and Z. Guo, *Org. Lett.*, 2002, **4**, 2893.
138. R.M. Kohli, C.T. Walsh and M. Burkart, *Nature*, 2002, **418**, 658.
139. R.M. Kohli, J. Takagi and C.T. Walsh, *Proc. Natl. Acad. Sci. U.S.A.*, 2002, **99**, 1247.
140. L.S. Luo, R.M. Kohli, M. Onishi, U. Linne, M.A. Marahiel and C.T. Walsh, *Biochemistry*, 2002, **41**, 9184.
141. C.T. Walsh, *ChemBiochem*, 2002, **3**, 125.
142. J. Scherkenbeck, H.R. Chen and R.K. Haynes, *Eur. J. Org. Chem.*, 2002, 2350.
143. S.A. Krachkovskii, A.G. Sobol, T.V. Ovchinnikova, A.A. Tagaev, Z.A. Yakimenko, R.R. Azizbekyan, N.I. Kuznetsova, T.N. Shamshina and A.S. Arseniev, *Russ. J. Bioorg. Chem.*, 2002, **28**, 269.
144. K. Yamada, M. Unno, K. Kobayashi, H. Oku, H. Yamamura, S. Araki, H. Matsumoto, R. Katakai and M. Kawai, *J. Am. Chem. Soc.*, 2002, **124**, 12684.
145. S. Roy, H.G. Lombart, W.D. Lubell, R.E.W. Hancock and S.W. Farmer, *J. Pept. Res.*, 2002, **60**, 198.
146. H. Takiguchi, *Fukuoka Daigaku Rigaku Shuho*, 2002, **32**, 69.
147. M. Kiricsi, E.J. Prenner, M. Jelokhani-Niaraki, R.N.A.H. Lewis, R.S. Hodges and R.N. McElhaney, *Eur. J. Biochem.*, 2002, **269**, 5911.
148. A.C. Gibbs, T.C. Bjorndahl, R.S. Hodges and D.S. Wishart, *J. Am. Chem. Soc.*, 2002, **124**, 1203.
149. (*a*) M. Jelokhani-Niaraki, E.J. Prenner, C.M. Kay, R.N. McElhaney and R.S. Hodges, *J. Pept. Res.*, 2002, **60**, 23; (*b*) M. Kiricsi, E.J. Prenner, R.A.H. Lewis, M. Jelokhani-Niaraki, R.S. Hodges and R.N. McElhaney, *Biophys. J.*, 2002, **82**, 2727.
150. L.H. Kondejewski, D.L. Lee, S.W. Farmer, R.E.W. Hancock and R.S. Hodges, *J. Biol. Chem.*, 2002, **277**, 67.
151. T. Arai, K. Araki, K. Fukuma, T. Nakashima, T. Kato and N. Nishino, *Chem. Lett.*, 2002, 1110.
152. Y.J. Zang, T. Tanaka, Y. Betsumiya, R. Kusano, A. Matsuo, T. Uedo and I. Kouno, *Chem. Pharm. Bull.*, 2002, **50**, 258.

153. H. Morita, T. Niidome, H. Murakami, M. Kawazoe, T. Hatakeyama, Y. Kobasigawa, M. Matsusita, Y. Kumaki, M. Demura, K. Nitta and H. Aoyagi, *Pept. Sci.*, 2002, **38**, 343.
154. V.G. Shakottai, R. Sudha and P. Balaram, *J. Pept. Res.*, 2002, **60**, 112.
155. D. Altschuh, *J. Mol. Recog.*, 2002, **15**, 277.
156. J.-F. Guichou, L. Patiny and M. Mutter, *Tetrahedron Lett.*, 2002, **43**, 4389.
157. J.A. Smulik, S.T. Diver, F. Pan and J.O. Liu, *Org. Lett.*, 2002, **4**, 2051.
158. F. Loor, F.O. Tiberghien, T. Wenandy, A. Didier and R. Traber, *J. Med. Chem.*, 2002, **45**, 4598.
159. F. Loor, F.O. Tiberghien, T. Wenandy, A. Didier and R. Traber, *J. Med. Chem.*, 2002, **45**, 4613.
160. Q. Huai, H.Y. Kim, Y.D. Liu, Y.D. Zhao, A. Mondragon, J.O. Liu and H.M. Ke, *Proc. Natl. Acad. Sci. U.S.A.*, 2002, **99**, 12037.
161. L. Jin and S.C. Harrison, *Proc. Natl. Acad. Sci. U.S.A.*, 2002, **99**, 13522.
162. P. Sedmera, A. Jegorov, M. Buchta and L. Cvak, *J. Pept. Res.*, 2001, **58**, 229.
163. M. Kuzma, A. Jeorov, A. Hesso, J. Tornaeus, P. Sedmara and V. Havlicek, *J. Mass Spec.*, 2002, **37**, 292.
164. E. Ugazio, R. Cavalli and M.R. Gasco, *Int. J. Pharmaceutics*, 2002, **241**, 341.
165. Q. Wu, X. Dai, W. Huang, Y. Liu, X. Fan, M. Chen and P. Feng, *J. Chin. Pharm. Sci.*, 2002, **11**, 78.
166. B. Thern, J. Rudolph and G. Jung, *Angew. Chem. Int. Ed.*, 2002, **41**, 2307.
167. N. Sewald, *Angew. Chem. Int Ed.*, 2002, **41**, 4661.
168. D.J. Craik, M.A. Anderson, D.G. Barry, R.J. Clark, N.L. Daly, C.V. Jennings and J. Mulvenna, *Lett. Pept. Sci.*, 2001, **8**, 119.
169. A. Arrault, A. Witczak-Legrand, P. Gonzalez, N. Bontemps-Subielos and B. Banaigs, *Tetrahedron Lett.*, 2002, **43**, 4041.
170. J.C. Rodriguez, G.G. Holgado, R.I.S. Sanchez and L.M. Canedo, *J. Antibiot.*, 2002, **55**, 391.
171. S. Kehraus, G.M. Koenig, A.D. Wright and G. Woerheide, *J. Org. Chem.*, 2002, **67**, 4989.
172. R.A. Edrada, R. Ebel, A. Supriyano, V. Wray, P. Schupp, K. Steube, R. van Soest and P. Proksch, *J. Nat. Prod.*, 2002, **65**, 1168.
173. C.E. Salomon and D.J. Faulkner, *J. Nat. Prod.*, 2002, **65**, 689.
174. B. Fenet, F. Pierre, E. Cundliffe and M. Ciufolini, *Tetrahedron Lett.*, 2002, **43**, 2367.
175. K. Fujii, Y. Yahishi, T. Nakano, S. Imanishi, S.F. Baldia and K. Harada, *Tetrahedron*, 2002, **58**, 6873.
176. A. Bertram and G. Pattenden, *Heterocycles*, 2002, **58**, 521.
177. L. Vollbrecht, H. Steinmetz, G. Hofle, L. Oberer, G. Rihs, G. Bovermann and P. Von Matt, *J. Antibiot.*, 2002, **55**, 715.
178. (*a*) S.V. Ley and A. Priour, *Eur. J. Org. Chem.*, 2002, 3995; (*b*) S.V. Ley and A. Priour, *Org. Lett.*, 2002, **4**, 711.
179. T. Ritter and E.M. Carreira, *Angew. Chem. Int. Ed.*, 2002, **41**, 2489.
180. K. Nicolaou, M. Bella, D.Y.-K. Chen, X. Huang, T. Ling and S.A. Snyder, *Angew. Chem. Int. Ed.*, 2002, **41**, 3495.
181. K.S. Feldman, K.J. Eastman and G. Lessene, *Org. Lett.*, 2002, **4**, 3525.
182. (*a*) M. Doi and A. Asano, *Acta Crystallog. Section E: Structure Reports Online*, 2002, **E58**, o834; (*b*) M. Doi, A. Yumiba and A. Asano, *ibid.*, 2002, **E58**, o62.
183. F. Yokokawa, H. Sameshima, Y. In, K. Minoura, T. Ishida and T. Shioiri, *Tetrahedron*, 2002, **58**, 8127.

184. S. Deng and J. Taunton, *J. Am. Chem. Soc.*, 2002, **124**, 916.
185. N. Kutsumura, N.U. Sata and S. Nishiyama, *Bull. Chem. Soc. Japan*, 2002, **75**, 847.
186. C.J. Thomas, A.O. Chizhov, C.J. Leitheiser, M.J. Rishel, K. Konishi, Z.-F. Tao and S.M. Hecht, *J. Am. Chem. Soc.*, 2002, **124**, 12926.
187. Y. Zou, N.E. Fahmi, C. Vialas, G.M. Miller and S.M. Hecht, *J. Am. Chem. Soc.*, 2002, **124**, 9476.
188. H. Saito, T. Yamada, K. Okumura, Y. Yonezawa and C.-G. Shin, *Chem. Lett.*, 2002, 1098.
189. Y. Yonezawa, H. Saito, S. Suzuki and C. Shin, *Heterocycles*, 2002, **57**, 903.
190. S. Higashibayashi, K. Hashimoto and M. Nakata, *Tetrahedron Lett.*, 2002, **43**, 105.
191. S. Higashibayashi, T. Mori, K. Shinko, K. Hashimoto and M. Nakata, *Heterocycles*, 2002, **57**, 111.
192. G. Pattenden and T. Thompson, *Tetrahedron Letters*, 2002, **43**, 2459.
193. Y. Singh, M.J. Stoermer, A.J. Lucke, M.P. Glen and D.P. Fairlie, *Org. Lett.*, 2002, **4**, 3367.
194. G. Haberhauer and F. Rominger, *Tetrahedron Lett.*, 2002, **43**, 6335.
195. G. Jung, *Lett. Pept. Sci.*, 2001, **8**, 259.
196. A. Asano, K. Minoura, T. Yamada, A. Numata, T. Ishida, Y. Katsuya, Y. Mezaki, M. Sasaki, T. Taniguchi, M. Nakai, H. Hasegawa, A. Terashima and M. Doi, *J. Pept. Res.*, 2002, **60**, 10.
197. A. Asano, T. Yamada, A. Numata, Y. Katsuya, M. Sasaki, T. Taniguchi and M. Doi, *Biochem. Biophys. Res. Commun.*, 2002, **297**, 143.
198. L.A. Morris, B.F. Milne, G.S. Thompson and M. Jaspars, *J. Chem. Soc. Perkin I*, 2002, 1072.
199. B.F. Milne, L.A. Morris, M. Jaspars and G.S. Thompson, *J. Chem. Soc. Perkin Trans 2*, 2002, 1076.
200. P.V. Bernhardt, P. Comba, D.P. Fairlie, L.R. Gahan, G.R. Hanson and L. Lotzbeyer, *Chem.- A Eur. J.*, 2002, **8**, 1527.
201. R.M. Cusack, L. Grondahl, D.P. Fairlie, L.R. Gahan and G.R. Hanson, *J. Chem. Soc. Perkin Trans 2*, 2002, 556.
202. M. Soledade, C. Pedras, L.I. Zaharia and D.E. Ward, *Phytochemistry*, 2002, **59**, 579.
203. M. Hinaje, M. Ford, L. Banting, S. Arkle and B. Khambay, *Arch. Biochem. Biophys.*, 2002, **405**, 73.
204. J. Hong and L. Zhang, *Frontiers Biotech. Pharm.*, 2002, **3**, 193.
205. T. Hornbogen, M. Glinski and R. Zocher, *Eur. J. Plant. Path.*, 2002, **108**, 713.
206. A. Zampella, A. Randazzo, N. Borbone, S. Luciani, L. Trevisi, U. Debitus and M.V. D'Auria, *Tetrahedron Lett.*, 2002, **43**, 6163.
207. (*a*) V. Guerlavais, P.J. Carroll and M.M. Joullie, *Tetrahedron-Asymm.*, 2002, **13**, 675; (*b*) A. Zampella, M. Sorgente and M.V. D'Auria, *ibid.*, 2002, **13**, 681.
208. J.C. Thoen, A.I. Morales-Ramos and M.A. Lipton, *Org. Lett.*, 2002, **4**, 4455.
209. N.I. Kalinovskaya, T.A. Kuznetsova, E.P. Ivanova, L.A. Romanenko, V.G. Voinov, F. Huth and H. Laatsch, *Marine Tech.*, 2002, **4**, 179.
210. D. Sorensen, T.H. Nielsen, J. Sorensen and C. Christophersen, *Tetrahedron Lett.*, 2002, **43**, 4421.
211. H.S. Kim, M.H. Jung, S.D. Ahn, C.W. Lee, S.N. Kim and J.H. Ok, *J. Antibiot.*, 2002, **55**, 598.
212. H. Luesch, P.G. Williams, W.Y. Yoshida, R.E. Moore and V.J. Paul, *J. Nat. Prod.*, 2002, **65**, 996.
213. P.G. Williams, W.Y. Yoshida, R.E. Moore and V.J. Paul, *J. Nat. Prod.*, 2002, **65**, 29.

214. L.M. Nogle and W.H. Gerwick, *J. Nat. Prod.*, 2002, **65**, 21.
215. Y.C. Lin, J. Wang, X.Y. Wu, S.N. Zhou, L.L.P. Vrijmoed and E.B.G. Jones, *Aus. J. Chem.*, 2002, **55**, 225.
216. Y.J. Feng, J.W. Blunt, A.L.J. Cole and M.H.G. Munro, *Org. Lett.*, 2002, **4**, 2095.
217. K. Fujii, K. Sivonen, T. Nakano and K. Harada, *Tetrahedron*, 2002, **58**, 6863.
218. J. Kimura, Y. Takada, T. Inayoshi, Y. Nakao, G. Goetz, W.Y. Yoshida and P.J. Scheuer, *J. Org. Chem.*, 2002, **67**, 1760.
219. R.J. Capon, J. Ford, E. Lacey, J.H. Gill, K. Heiland and T. Friedel, *J.Nat. Prod.*, 2002, **65**, 358.
220. Y. Okada, S. Matsunaga, R.W.M. van Soest and N. Fusetani, *Org. Lett.*, 2002, **4**, 3039.
221. G.-Z. Yang and Y.-C. Li, *Helv. Chim. Acta*, 2002, **85**, 168.
222. K. Kurosawa, T. Nagase and N. Chida, *Chem. Commun. (Cambridge)*, 2002, 1280.
223. F. Sarabia, S. Chammaa and F.J. Lopez-Herrara, *Tetrahedron Lett.*, 2002, **43**, 2961.
224. D.W. Hoard, E.D. Moher, M.J. Martinelli and B.H. Norman, *Org. Lett.*, 2002, **4**, 1813.
225. M.J. Martinelli, R. Vaidyanathan, V.V. Khau and M.A. Staszak, *Tetrahedron Lett.*, 2002, **43**, 3365.
226. F. Yokokawa and T. Shioiri, *Tetrahedron Lett.*, 2002, **43**, 8673.
227. W. Jiang, J. Wanner, R.J. Lee, P.-Y. Bounaud and D.L. Boger, *J. Am. Chem. Soc.*, 2002, **124**, 5288.
228. Y. Chen, M. Bilban, C.A. Foster and D.L. Boger, *J. Am. Chem. Soc.*, 2002, **124**, 5431.
229. E.P. Schreiner, M. Kern and A. Steck, *J. Org. Chem.*, 2002, **67**, 8299.
230. L.A. Paquette, M. Duan, I. Konetzki and C. Kempmann, *J. Am. Chem. Soc.*, 2002, **124**, 4257.
231. M.K. Gurjar and S.R. Chaudhuri, *Tetrahedron Lett.*, 2002, **43**, 2435.
232. K.J. Hale and L. Lazarides, *Chem. Commun. (Cambridge)*, 2002, 1832.
233. K.J. Hale and L. Lazarides, *Org. Lett.*, 2002, **4**, 1903.
234. K. Makino, Y. Hemni and Y. Hamada, *Synlett*, 2002, **9**, 613.
235. B.H. Lee, F.E. Dutton, D.P. Thomson and E.M. Thomas, *Bioorg. Med. Chem. Lett.*, 2002, **12**, 353.
236. P. Jeschke, A. Harder, G. Von Samson-Himmelstjerna, W. Etzel, W. Gau, G. Thielking and G. Bonse, *Pest. Man. Sci.*, 2002, **58**, 1205.
237. A. Harder and G. Von Samson-Himmelstjerna, *Parasit. Res.*, 2002, **88**, 481.
238. S.D. Bruner, T. Weber, R.M. Kohli, D. Schwarzer, M.A. Marahiel, C.T. Walsh and M.T. Stubbs, *Structure*, 2002, **10**, 301.
239. C.C. Tseng, S.D. Bruner, R.M. Kohli, M.A. Marahiel, C.T. Walsh and S.A. Sieber, *Biochemistry*, 2002, **41**, 13350.
240. S. Ventkatraman, F.G. Njoroge, V.V. Girijavallabhan and A.T. McPhail, *J. Org. Chem.*, 2002, **67**, 3152.
241. M. Doi and A. Asano, *Acta Crystallog. Section E: Structure Reports Online*, 2002, **E58**, o935.
242. E. Stokvis, H. Rosing, L. Lopez-Lazaro, I. Rodriquez, I.M. Jimeno, J.G. Supko, J.H.M. Schellens and J.H. Beijnen, *J. Mass Spectrom.*, 2002, **37**, 992.
243. R.L. Bai, D.G. Covell, C.F. Liu, A.K. Ghosh and E. Hamel, *J. Biol. Chem.*, 2002, **277**, 32165.
244. K. Makino, T. Suzuki, S. Awane, O. Hara and Y. Hamada, *Tetrahedron Lett.*, 2002, **43**, 9391.

245. C. Nilanonta, M. Isaka, P. Kittakoop, S. Trakulnaleamsai, M. Tanticharoen and Y. Theotaranonth, *Tetrahedron*, 2002, **58**, 3355.
246. T. Ouchi, H. Miyazaki, H. Azimura, F. Tasaka, A. Hamada and Y. Ohya, *J. Polymer Sci, Pt A: Polymer Chem.*, 2002, **40**, 1218.
247. N. Masuoka, J. Zhang, L. Partoo, J. Ohta, K. Sasaki, H. Ebinuma and H. Kodama, *Arch. Biochem. Biophys.*, 2002, **407**, 184.
248. F. Hubalek, D.E. Edmondson and J. Pohl, *Anal. Biochem.*, 2002, **306**, 124.
249. M. Ruiz, V. Ojea, J.M. Quintela and J.J. Guillin, *Chem. Commun.(Cambridge)*, 2002, 1600.
250. D.-G. Liu, X.-Z. Wang, Y. Gao, B. Li, D. Yang and T.R. Burke, *Tetrahedron*, 2002, **58**, 10423.
251. K. Lee, M. Zhang, D. Yang and T.R. Burke, *Bioorg. Med. Chem. Lett.*, 2002, **12**, 3399.
252. S.W. Vetter and Z.Y. Zhang, *Curr. Prot. Pept. Sci.*, 2002, **3**, 365.
253. R.R. Flavell, M. Huse, M. Goger, M. Trester-Zedlitz, J. Kuriyan and T.W. Muir, *Org. Lett.*, 2002, **4**, 165.
254. W. Li, R.A. Boykins, P.S. Backlund, G.Y. Wang and H.C. Chen, *Anal. Chem.*, 2002, **74**, 5701.
255. J.F. Nemeth-Cawley, S. Karnik and J.C. Rouse, *J. Mass Spec.*, 2001, **36**, 1301.
256. C.S. Raska, C.E. Parker, Z. Dominski, W.F. Marzluff, G.L. Glish, R.M. Pope and C.H. Borchers, *Anal. Chem.*, 2002, **74**, 3429.
257. K.L. Bennett, A. Stensballe, A.V. Podtelejnikov, M. Moniatte and O.N. Jensen, *J. Mass Spec.*, 2002, **37**, 179.
258. S.C. Moyer, R.J. Cotter and A.S. Woods, *J. Am. Soc. Mass Spec.*, 2002, **13**, 274.
259. B.T. Ruotolo, G.F. Verbeck, L.M. Thomson, A.S. Woods, K.J. Killig and D.H. Russell, *J. Proteome Res.*, 2002, **1**, 303.
260. R. Ruijtenbeek, C. Versluis, A.J.R. Heck, F.A.M. Redegeld, F.P. Nijkamp and R.M.J. Liskamp, *J. Mass Spec.*, 2002, **37**, 47.
261. R.E. Bossio and A.G. Marshall, *Anal. Chem.*, 2002, **74**, 1674.
262. T. Young and L.L. Kiessling, *Angew. Chem. Int. Ed.*, 2002, **41**, 3449.
263. S.V. Campos, L.P. Miranda and M. Meldal, *J. Chem. Soc. Perkin Trans. 1*, 2002, 682.
264. M. Farzan, S. Chung, W.H. Li, N. Vasilieva, P.L. Wright, C.E. Schnitzler, R.J. Marchione, C. Gerard, N.P. Gerard, J. Sodroski and H. Choe, *J. Biol. Chem.*, 2002, **277**, 40397.
265. Y. Gao, *Nat. Prod. Rep.*, 2002, **19**, 100.
266. M. Takayanagi, *Yuki Gosei Kazaku Kyokaishi*, 2002, **60**, 240.
267. C. Lehmann, G. Bunkoczi, L. Vertesy and G.M. Sheldrick, *J. Mol. Biol.*, 2002, **318**, 723.
268. S. Weist, B. Bister, O. Puk, D. Bischoff, S. Pelzer, G.J. Nicholson, W. Wohlleben, G. Jung and R.D. Sussmuth, *Angew. Chem. Int. Ed.*, 2002, **41**, 3383.
269. S.S. Printsevskaya, A.Y. Pavlov, E.N. Olsufyeva, E.P. Mirchink, E.B. Isakova, M.I. Reznikova, R.C. Goldman, A.A. Branstrom, E.R. Baizman, C.B. Longley, F. Sztaricskai, G. Batta and M.N. Preobrazhenskaya, *J. Med. Chem.*, 2002, **45**, 1340.
270. S.S. Printsevskaya, E.N. Olsuf'eva, E.I. Lazhko and M.N. Preobrazhenskaya, *Russ. J. Bioorg. Chem.*, 2002, **28**, 65.
271. O. Yoshida, T. Yasukata, Y. Sumino, T. Munekage, Y. Narukawa and Y. Nishitana, *Bioorg. Med. Chem Lett.*, 2002, **12**, 3027.
272. T. Yasukata, H. Shindo, O. Yoshida, Y. Sumino, T. Munekage, Y. Narukawa and Y. Nishitani, *Bioorg. Med. Chem. Lett.*, 2002, **12**, 3033.

273. J. McAtee, S.L. Castle, Q. Jin and D.L. Boger, *Bioorg. Med. Chem. Lett.*, 2002, **12**, 1319.
274. K.C. Nicolaou and C.N.C. Boddy, *J. Am. Chem. Soc.*, 2002, **124**, 10451.
275. G. Bringmann, D. Menche, J. Muehlbacher, M. Reichert, N. Saito, S.S. Pfeiffer and B.H. Lipshutz, *Org. Lett.*, 2002, **4**, 2833.
276. A.J. Pearson, A.D. Velankar and J.-N. Heo, *ARKIVOC (online computer file)*, 2002, 49.
277. T.A. Blizzard, R.M. Kim, J.D. Morgan, J. Chang, J. Kohler, R. Kilburn, K. Chapman and M.L. Hamilton, *Bioorg. Med. Chem. Lett.*, 2002, **12**, 849.
278. N.E. Allen, D.L. LeTourneau, J.N. Hobbs and R.C. Thompson, *Antimicrob. Agents Chem.*, 2002, **46**, 2344.
279. B.Y. Sun, Z. Chen, U.S. Eggert, S.J. Shaw, J.V. LaTour and D. Kahne, *J. Am. Chem. Soc.*, 2001, **123**, 12722.
280. Z. Chen, U.S. Eggert, S.D. Dong, S.J. Shaw, B.Y. Sun, J.V. LaTour and D. Kahne, *Tetrahedron*, 2002, **58**, 6585.
281. T.J.D. Jorgensen, P. Hvelplund, J.U. Andersen and P. Roepstorff, *Int. J. Mass Spec.*, 2002, **219**, 659.
282. B. Bardsley, R. Zerella and D.H. Williams, *J. Chem. Soc. Perkin Trans 2*, 2002, 598.
283. C.F. Silverio, A. Plazas, J. Moran and F.A. Gomez, *J. Liq. Chrom. Rel. Technol.*, 2002, **25**, 1677.
284. S.J. Kim, L. Cegelski, R.D. Studelska, R.D. O'Connor, A.K. Mehta and J. Schaefer, *Biochemistry*, 2002, **41**, 6967.
285. H. Shiozawa, B.C.S. Chia, N.L. Davies, R. Zerella and D.H. Williams, *J. Am. Chem. Soc.*, 2002, **124**, 3914.
286. H.W. Chen, C.C. Tseng, R.K. Hubbard and C.T. Walsh, *Proc. Nat. Acad. Sci. U.S.A.*, 2001, **98**, 14901.
287. J. Pootoolal, M.G. Thomas, C.G. Marshall, J.M. Neu, B.K. Hubbard, C.T. Walsh and G.D. Wright, *Proc. Nat. Acad. Sci. U.S.A.*, 2002, **99**, 8962.
288. H.Y. He, R.T. Williamson, B. Shen, E.I. Graziani, H.Y. Yang, S.M. Sakya, P.J. Petersen and G.T. Carter, *J. Am. Chem. Soc.*, 2002, **124**, 9729.
289. H.Y. Aboul-Enein and M. Ali, *Farmaco*, 2002, **57**, 513.
290. H.Y. Aboul-Enein and I. Ali, *J. Liq. Chrom. Rel. Technol.*, 2002, **25**, 2337.
291. S. Alcaro, I. D'Acquarica, F. Gasparrini, D. Misiti, M. Pierini and C. Villani, *Tetrahedron-Asym.*, 2002, **13**, 69.
292. S.S. Chen, *J. Chin. Chem. Soc.*, 2002, **49**, 545.
293. S. Fanali, P. Catarcini and M.G. Quaglia, *Electrophoresis*, 2002, **23**, 477.
294. C. Brocke and H. Kunz, *Bioorg. Med. Chem.*, 2002, **10**, 3085.
295. K. Koeller and C.-H. Wong, *Glycoscience*, 2001, **3**, 2305.
296. Y. Mechref and M.V. Novotny, *Chem. Rev.*, 2002, **102**, 321.
297. M.R. Wormald, A.J. Petrescu, Y.-L. Pao, A. Glithero T. Elliott and R.A. Dwek, *Chem. Rev.*, 2002, **102**, 371.
298. R. Polt and S.A. Mitchell, *Glycoscience*, 2001, **3**, 2353.
299. M. Wagner and H. Kunz, *Angew. Chem. Int. Ed.*, 2001, **41**, 317.
300. M.-R. Lee, J. Lee and I. Shin, *Synlett.*, 2002, 1463.
301. J. Habermann, K. Stuber, T. Skripko, T. Reipen, R. Wieser and H. Kunz, *Angew. Chem. Int. Ed.*, 2002, **41**, 4249.
302. C.M. Taylor, C.A. Weir and C.G. Jorgensen, *Aus. J. Chem.*, 2002, **55**, 135.
303. J. Watabe, L. Singh, Y. Nakahara, Y. Ito, H. Hojo and Y. Nakahara, *Biosci. Biotech. Biochem.*, 2002, **66**, 1904.

304. S.D. Zhang, G. Liu, S.Q. Xia and P. Wu, *Chin. Chem. Lett.*, 2002, **13**, 17.
305. J. Tamura, A. Yamaguchi and J. Tanaka, *Bioorg. Med. Chem. Lett.*, 2002, **12**, 1901.
306. M. Cudic, H.C.J. Ertl and L. Otvos, *Bioorg. Med. Chem.*, 2002, **10**, 3859.
307. T. Ichiyanagi, M. Takatani, K. Sakamoto, Y. Hakahara, Y. Ito, H. Hojo and Y. Nakahara, *Tetrahedron Lett.*, 2002, **43**, 3297.
308. M. Gobbo, L. Biondi, F. Filira, R. Gennaro, M. Benincasa, B. Scolaro and R. Rocchi, *J. Med. Chem.*, 2002, **45**, 4494.
309. A. Takasu, T. Houjyou, Y. Inai and T. Hirabayashi, *Biomacromolecules*, 2002, **3**, 775.
310. J. Katajisto, T. Karskela, P. Heinonen and H. Loennberg, *J. Org. Chem.*, 2002, **67**, 7995.
311. K. Suzuki, I. Matsuo, M. Isomura and K. Ajisaka, *J. Carbohyd. Chem.*, 2002, **21**, 99.
312. O. Blixt, K. Allin, L. Pereira, A. Datta and J.C. Paulson, *J. Am. Chem. Soc.*, 2002, **124**, 5739.
313. S. Manabe, K. Sakamoto, Y. Nakahara, M. Sisido, T. Hohsaka and Y. Ito, *Bioorg. Med. Chem.*, 2002, **10**, 573.
314. G.I. Csonka, G.A. Schubert, A. Perczel, C.P. Sosa and I.G. Csizmadia, *Chem.- A Eur. J.*, 2002, **8**, 4718.
315. H. Moller, N. Serttas, H. Paulsen, J.M. Burchell, J. Taylor-Papadimitriou and B. Meyer, *Eur. J. Biochem.*, 2002, **269**, 1444.
316. L. Stella, M. Venanzi, M. Carafa, E. Maccaroni, M.E. Straccamore, G. Zanotti, A. Palleschi and B. Pispisa, *Biopolymers*, 2002, **64**, 44.
317. J. van Ameijde, H.B. Albada and R.M.J. Liskamp, *J. Chem. Soc. Perkin Trans 1*, 2002, 1042.
318. I. Shin, S. Park, K.S. Kim, J.W. Cho and D. Lim, *Bull. Korean. Chem. Soc.*, 2002, **23**, 15.
319. I. Saskiawan, M. Mizuno, T. Inazu, K. Haneda, S. Harashima, H. Kumagai and K. Yamamoto, *Arch. Biochem. Biophys.*, 2002, **406**, 127.
320. N. Wehofsky, R. Loser, A. Buchynskyy, P. Welzel and F. Bordusa, *Angew. Chem. Int. Ed.*, 2002, **41**, 2735.
321. L. Biondi, F. Filira, M. Gobbo and R. Rocchi, *J. Pept. Sci.*, 2002, **8**, 80.
322. J.-M. Beau, B. Vauzeilles and T. Skrydstrup, *Glycoscience*, 2001, **3**, 2679.
323. L. Colombo, M. diGiacomo and P. Ciceri, *Tetrahedron*, 2002, **58**, 9381.
324. X. Xu, G. Fakha and D. Sinou, *Tetrahedron*, 2002, **58**, 7539.
325. E. Brenna, C. Fuganti, P. Graselli, S. Serra and S. Zambotti, *Chem. – A Eur. J.*, 2002, **8**, 1872.
326. G.J. McGarvey, T.E. Benedum and F.W. Schmidtmann, *Org. Lett.*, 2002, **4**, 3591.
327. A. Dondoni, G. Mariotti and A. Marra, *J. Org. Chem.*, 2002, **67**, 4475.
328. E.G. Nolen, M.M. Watts and D.J. Fowler, *Org. Lett.*, 2002, **4**, 3963.
329. J.D. Parrish and R.D. Little, *Org. Lett.*, 2002, **4**, 1439.
330. F. Coutrot, M. Marraud, B. Maigret, C. Grison and P. Coutrot, *Lett. Pept. Sci.*, 2002, **8**, 107.
331. T. Nishikawa, K. Wada and M. Isobe, *Biosci. Biotech. Biochem.*, 2002, **66**, 2273.
332. P. Arya, A. Barkley and K.D. Randell, *J. Combinat. Chem.*, 2002, **4**, 193.
333. L. Cipolla, M. Rescigno, A. Leone, F. Peri, B. La Ferla and F. Nicotra, *Bioorg. Med. Chem.*, 2002, **10**, 1639.
334. J.P. Malkinson and R.A. Falconer, *Tetrahedron Lett.*, 2002, **43**, 9549.
335. D. Macmillan, A.M. Daines, M. Bayrhuber and S.L. Flitsch, *Org. Lett.*, 2002, **4**, 1467.
336. S.B. Cohen and R.L. Halcomb, *J. Am. Chem. Soc.*, 2002, **124**, 2534.
337. S. Knapp and D.S. Myers, *J. Org. Chem.*, 2002, **67**, 2995.

338. C. Grandjean, V. Santraine, J.-S. Fruchart, O. Melnyk and H. Gras-Masse, *Bioconj. Chem.*, 2002, **13**, 887.
339. R.A. Falconer, *Tetrahedron Lett.*, 2002, **43**, 8503.
340. F. Schweizer, *Angew. Chem. Int. Ed.*, 2001, **41**, 230.
341. M. Stoeckle, G. Voll, R. Guenther, E. Lohof, E. Locardi, S. Gruner and H. Kessler, *Org. Lett.*, 2002, **4**, 2501.
342. S.A.W. Gruner, V. Truffault, G. Voll, E. Locardi, M. Stockle and H. Kessler, *Chem.-A Eur. J.*, 2002, **8**, 4365.
343. A.N. Van Nhien, H. Ducatel, C. Len and D. Postel, *Tetrahedron Lett.*, 2002, **43**, 3805.
344. Y. Suhara, Y. Yamaguchi, B. Collins, R.L. Schnaar, M. Yanagishita, J.E.K. Hildreth, I. Shimada and Y. Ichikawa, *Bioorg. Med. Chem.*, 2002, **10**, 1999.
345. T. Ziegler, D. Roseling and L.R. Subramanian, *Tetrahedron-Asymmetry*, 2002, **13**, 911.
346. C. Peters, M. Wagner, M. Volkert and H. Waldmann, *Naturwissenschaften*, 2002, **89**, 381.
347. J.B. MacMillan, M.A. Ernst-Russell, J.S. de Ropp and T.F. Molinski, *J. Org. Chem.*, 2002, **67**, 8210.
348. A. Holtzel, D.G. Schmid, G.J. Nicholson, S. Stevanovic, J. Schimana, K. Gebbardt, H.-P. Fiedler and G. Jung, *J. Antibiot.*, 2002, **55**, 571.
349. S. Hiradate, S. Yoshida, H. Sugie, H. Yada and Y. Fujii, *Phytochemistry*, 2002, **61**, 693.
350. B.H. Williams, Y. Hathout and C. Fenselau, *J. Mass Spec.*, 2002, **37**, 259.
351. T.H. Nielsen, D. Sorensen, C. Tobiasen, J.B. Andersen, C. Christophersen, M. Givskov and J. Sorensen, *Appl. Env. Microb.*, 2002, **68**, 3416.
352. E. Peroni, G. Caminati, P. Baglioni, F. Nuti, M. Chelli and A.M. Papini, *Bioorg. Med. Chem. Lett.*, 2002, **12**, 1731.
353. B. Ludolph, F. Eisele and H. Waldmann, *J. Am. Chem. Soc.*, 2002, **124**, 5954.
354. F. Yokokawa, H. Sameshima, D. Katagari, T. Aoyama and T. Shioiri, *Tetrahedron*, 2002, **58**, 9445.
355. S.P. Creaser and B.R. Peterson, *J. Amer. Chem. Soc.*, 2002, **124**, 2444.
356. C. Moran, M. Rosa Infante and P. Clapes, *J. Chem. Soc. Perkin Trans 1*, 2002, 1124.
357. J.B. Bremner, J.A. Coates, D.R. Coghlan, D.M. David, P.A. Keller and S.G. Pyne, *New J. Chem.*, 2002, **26**, 1549.
358. C. Bijani, S. Varray, R. Lazaro, J. Martinez, F. Lamaty and N. Keiffer, *Tetrahedron Lett.*, 2002, **43**, 3765.
359. K. Akaji, *Yuki Gosei Kagaku Kyokaishi*, 2002, **60**, 500.
360. S. Sasmal, A. Geyer and M.E. Maier, *J. Org. Chem.*, 2002, **67**, 6260.
361. J.B. Bremner, J.A. Coates, P.A. Keller, S.G. Pyne and H.M. Witchard, *Synlett.*, 2002, 219.
362. V. Boucard, H. Sauriat-Dorizon and F. Guibe, *Tetrahedron*, 2002, **58**, 7275.
363. J. Efskind, H. Hope and K. Undheim, *Eur. J. Org. Chem.*, 2002, 464.
364. S. Hanessian and M. Angiolini, *Chem-A Eur J.*, 2002, **8**, 111.
365. M. Saviano, E. Benedetti, R.M. Vitale, B. Kaptein, Q. Broxterman, M. Crisma, F. Formaggio and C. Toniolo, *Macromolecules*, 2002, **35**, 4204.
366. U.-R. Inayat, M.A. Khan, G.A. Khan, L. Khan and V.U. Ahmad, *J. Chem. Soc. Pakistan*, 2001, **23**, 268.
367. T. Temal-Laieb, J. Chastanet and J. Zhu, *J. Am. Chem. Soc.*, 2002, **124**, 583.
368. B. Cao, H. Park and M.M. Joullie, *J. Am. Chem. Soc.*, 2002, **124**, 520.

369. A. Arasappan, K.X. Chen, F.G. Njoroge, T.N. Parekh and V. Girijavallabhan, *J. Org. Chem.*, 2002, **67**, 3923.
370. K.X. Chen, F.G. Njoroge, B. Vibulbhan, A. Buevich, T.-M. Chan and V. Girijavallabhan, *J. Org. Chem.*, 2002, **67**, 2730.
371. Y. Rew and M. Goodman, *J. Org. Chem.*, 2002, **67**, 8820.
372. R.P. Cheng, D.J. Suich, H. Cheng, H. Roder and W.F. DeGrado, *J. Am. Chem. Soc.*, 2001, **123**, 12710.
373. P. Johannesson, G. Lindeberg, A. Johansson, G.V. Nikiforovich, A. Gogoll, B. Synnergren, M. Le Greves, F. Nyberg, A. Karlen and A. Hallberg, *J. Med. Chem.*, 2002, **45**, 1767.
374. G. Fridkin and C. Gilon, *J. Pept. Res.*, 2002, **60**, 104.
375. P. Virta, J. Sinkkonen and H. Lonnberg, *Eur. J. Org. Chem.*, 2002, 3616.
376. X. Song, C.R. Xu, H.T. He and T.J. Siahaan, *Bioorg. Chem.*, 2002, **30**, 285.
377. H. Li and N. Sampson, *J. Pept. Res.*, 2002, **59**, 54.
378. L. Jiang and K. Burgess, *J. Am. Chem. Soc.*, 2002, **124**, 9028.
379. H. Huang, L. Mu, J. He and J.-P. Cheng, *Tetrahedron Lett.*, 2002, **43**, 2255.
380. T.K. Chakraborty, S. Tapadar and K.S. Kiran, *Tetrahedron Lett.*, 2002, **43**, 1317.
381. S. Rossi, G.M. Kyne, D.L. Turner, N.J. Wells and J.D. Kilburn, *Angew. Chem. Int. Ed.*, 2002, **41**, 4233.
382. F. Gasparrini, D. Misiti, M. Pierini and C. Villani, *Org. Lett.*, 2002, **4**, 3993.
383. D. Ranganathan, M.P. Samant, R. Nagaraj and E. Bikshapathy, *Tetrahedron Lett.*, 2002, **43**, 5145.
384. S. Maricic and T. Frejd, *J. Org. Chem.*, 2002, **67**, 7600.

Metal Complexes of Amino Acids and Peptides

BY E. FARKAS AND I. SÓVÁGÓ
Dept. Inorganic and Analytical Chemistry, University of Debrecen, POB 21, H-4010 Debrecen, Hungary

1 Introduction

This chapter deals with the synthesis, structures, kinetics and reactivity, solution equilibria and various applications of metal complexes of amino acids and peptides, including the most important synthetic derivatives and analogues of the naturally occurring ligands. The survey of the literature covers the papers published in the years 2001 and 2002. The major source of the articles reported here is CA Selects on Amino Acids, Peptides and Proteins,[1] but the title pages of the most common journals of inorganic, bioinorganic and coordination chemistry have also been surveyed. In addition to the hard copies of the abovementioned journals, the abstracts reported by the Web of Science Databases on the Internet[2] have also been searched. The number of publications dealing with the different aspects of the metal ion amino acid/peptide interactions is tremendously high, thus only papers that focus on the complexation reactions of these bioligands are included in this review. The results published on the metal complexes of amino acids and peptides are treated in separate sections, although there are many publications dealing with the coordination chemistry of both types of ligands. In addition to the naturally occurring amino acids, some synthetic analogues and derivatives of amino acids and peptides are also considered, provided they have a biological or theoretical significance or important applications in synthetic, analytical and medicinal chemistry.

The interest in the metal complexes of amino acids and other biologically related ligands arose some decades ago and the most important findings for the common transition elements with relatively small peptide molecules are now available in various review papers. As a consequence, the traditional structural, kinetic and thermodynamic characterization of complex formation reactions have recently been mainly focused on the less abundant elements with the derivatives of amino acids and longer oligopeptides consisting of 4 to 12 amino acids. The number of papers dealing especially with the synthetic, analytical and/or biomedical applications of these complexes has increased in the past few

Amino Acids, Peptides and Proteins, Volume 35

years. Among them, papers on the role of His and Cys residues in various metalloproteins, including a series of copper and zinc containing enzymes, zinc finger proteins, metallothioneins and prion proteins, were the focus of a great number of publications. Thus, in the case of peptides the papers published in the years 2001 and 2002 are classified into 5 subsections (a) synthesis and structural studies, (b) kinetics and reactivity (c) solution equilibria (d) complex formation with the peptides of His and/or Cys and (e) synthetic, analytical and biomedical applications. In the case of amino acids only 4 subsections are created and the results obtained for all amino acids are discussed together.

The most recent reviews published in 2001 and 2002 cover both the general features and some specific aspects of the coordination chemistry of amino acids and peptides. Volume 38 of the series on "Metal Ions in Biological Systems" (ed. H. Sigel) is devoted to the subject: "Probing of Proteins by Metal Ions and Their Low Molecular Weight Complexes". Chapter 1 of this volume summarizes the basic characteristics of the peptide or amide bonds emphasizing their capability to interact with protons and metal ions.[3] Discussions of the free energy of peptide bond hydrolysis and formation are also included. The metal ion promoted hydrolysis of the peptide amide bonds is reported in several other chapters. These reviews cover the role of platinum(II) and palladium(II),[4] copper(II) and nickel(II),[5,6] cobalt(III)[7] and lanthanide(III) ions[8] in the characterization of synthetic peptidases. The antitumor activity of cisplatin and related drugs has received increasing attention in the past two years and resulted in the publication of a great number of papers on the bioactivity and transport processes of platinum containing drugs. Various nitrogen and sulfur containing ligands are considered to be responsible both for the toxicity and transport of these platinum complexes. The interaction of the antitumoral platinum-group metallodrugs with albumin[9] and that of platinum(IV) with sulfur containing amino acids and peptides[10] have been reviewed recently.

The physiological effects of the metal complexes of aluminium(III), vanadium(IV/V) and dialkyltin(IV) ions and complexes have been a matter of debate in the past few years. The organotin(IV) complexes formed with biologically active ligands, including amino acids and peptides, have been reviewed and the possible biological implications discussed.[11,12] Vanadium complexes of different oxidation states are promising candidates for the treatment of diabetes and the metal ion is an important constituent of several metalloenzymes. The spectroscopic studies performed on the vanadium(IV) and vanadium(V) complexes of various bioligands have been reviewed.[13] Aluminium(III) ion is now mainly considered as a toxic metal ion for plants, animals and humans. The interaction of aluminium with small biomolecules[14] and its speciation in relation to the bioavailability, metabolism and toxicity of aluminium(III) have been surveyed recently.[15] The outstanding biological significance of calcium binding proteins is well-known and this type of interactions can be better understood by the investigation of synthetic calcium binding peptides.[16]

The other reviews published in the period 2001–2002 are devoted to the role of specific amino acid or side chain residues in the complex formation reactions with different metal ions or to the metal complexes of biologically active peptide molecules. The factors influencing the metal ion coordination of amide groups and imidazole side chains were evaluated via the examples of peptide complexes containing monodentate and/or chelating imidazole nitrogen donors.[17] The stability constants reported for the metal complexes of phosphonic acids have been critically evaluated.[18] The noncovalent interactions play a vital role in biological systems. Among them, hydrogen bonding is the most common, but the metal ion assisted weak interactions, including ring stacking and charge neutralization, are also very important and have been reviewed.[19,20] The metal binding ability and the structure-activity relationships were evaluated and reviewed for two important biomolecules, namely for metallothioneins[21] and the metalloantibiotic peptide bacitracin.[22] Modified ferrocenes and their analogues have long been exploited as redox probes, because these molecules are able to respond to the structural changes that will take place upon substrate binding. The widespread investigations on the synthesis, structure, electrochemistry and applications of the ferrocenyl derivatives of amino acids and peptides have been reviewed recently.[23,24]

2 Amino Acid Complexes

2.1 Synthesis and Structural Studies. – Although there has beeen continuous interest in metal complexation of amino acids for several decades, numerous papers have been published in this subject during the past two years. As usual, most of them relate to complexes of transition metal ions like vanadium, chromium, manganese, cobalt, nickel, copper, zinc, cadmium, rhenium, technetium, ruthenium. The most frequently used methods are IR-spectroscopy, X-Ray diffraction methods, calorimetry and various EPR, NMR and MS techniques.

Copper complexes of amino acids are of continuing interest.[25–33] Powder diffraction methods were used to establish the structures of four polymorf bis(glycinato)-copper(II) complexes, *cis*-$[Cu(Gly)_2]$ H_2O (**1a**), *trans*-$[Cu(Gly)_2]$ H_2O (**1b**), and their anhydrous modifications (**1c**) and (**1d**), respectively.[25]

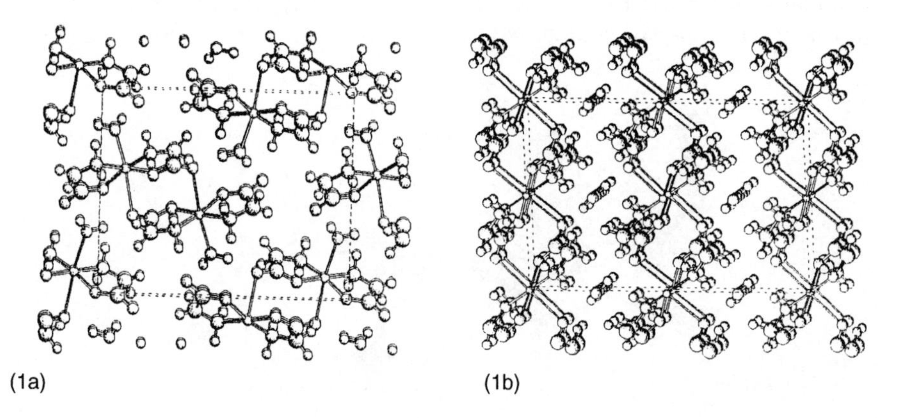

(1a) (1b)

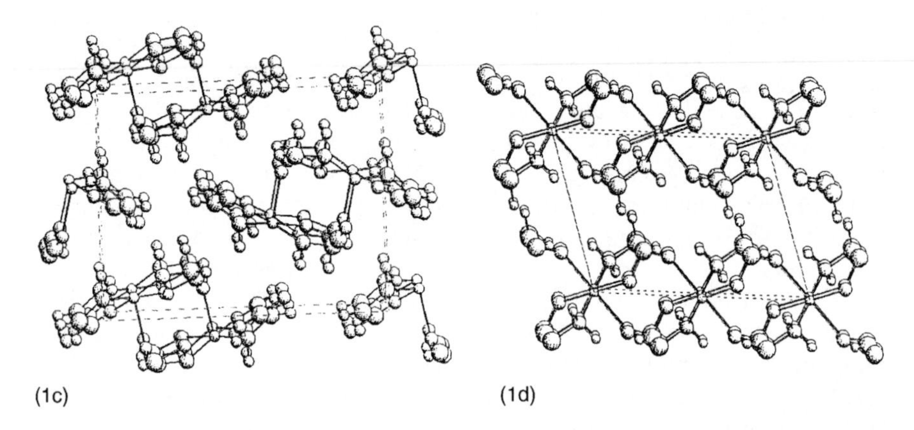

(1c) (1d)

The *cis*-complexes are orthorombic, the *trans*-complexes are monoclinic. In the hydrated complexes the *trans*-molecules are linked together by carbonyl oxygen atoms of neighbouring molecules, the *cis* ones by a combination of water molecules and carbonyl oxygen atoms. In the trans-complex the water molecule lies within a void of the lattice. All four complexes are stabilized by a network of hydrogen bonds.[25] When a 1 : 1 ratio of copper(II) to amino acid was used in another work, hexanuclear copper(II) complexes of Gly and Pro, $Na[NaCu_6(-Gly)_8(H_2O)_2](ClO_4)_6 \cdot 2H_2O$ and $[NaCu_6(Pro)_8(OH)](ClO_4)_4 \cdot H_2O$, respectively, were obtained.[26] There is no change in the coordination mode of *N,N*-dimethyl-L-α-Ala and *N,N*-dimethyl-L-Ile compared to their parent molecules, they both coordinate to a copper(II) ion via their amino and carboxylate groups. However, a different temperature dependence of the Brownian motion of the bis-complexes formed with these derivatives in dried deuterated methanol was found. While the EPR spectra of [Cu(N,N-dimethyl-L-α-Ala)$_2$] fitted well over the whole studied temperature range with characteristic parameters for the complex with one apical water molecule, with the Ile-derivative a more complex temperature dependence was obtained. In the latter case, the results provided acceptable fitting with the above model only below 300 K (**2a**). Above this temperature, the EPR spectra could be fitted with the parameters characteristic for a copper(II) complex which does not contain water molecules (**2b**).[27]

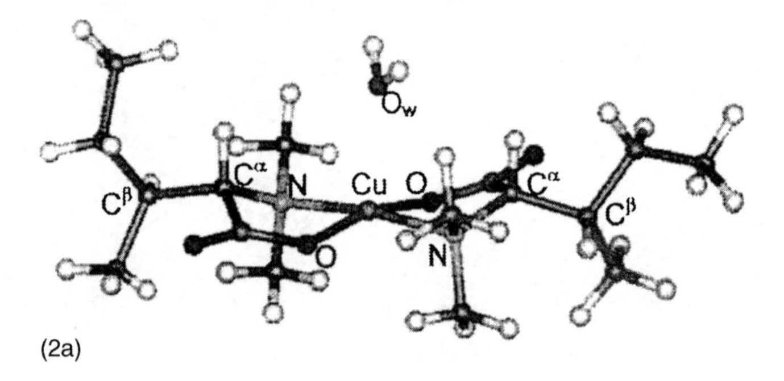

(2a)

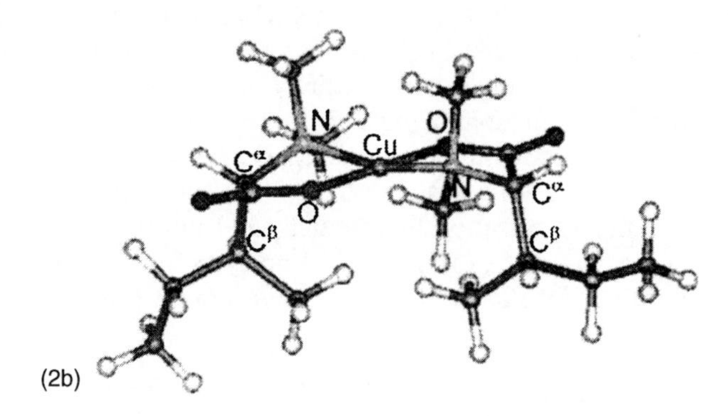

In another paper, structural and single crystal EPR results for the complex [Cu(Gln)$_2$] confirmed that this compound belongs to the family of copper(II)-amino acid complexes in which the side chain donor is not involved in the coordination. The complex shows a layered structure of copper(II) ions connected by equatorial–apical *syn–anti* carboxylate bridges.[28] Since copper–histidine coordination is highly abundant in many biological systems, numerous studies for investigation of Cu(II)-His complexes have already been performed. In a recent work W-band ^{1}H and ^{2}H ENDOR spectra of the Cu(II) complexes with L-His, DL-His-α-d,β-d$_2$ (deuterated at the α and β positions) and 1-methyl-His (MeHis) in frozen solutions at pH=7.3 were recorded. The hyperfine coupling of H_α was found to be highly sensitive to the coordination mode. Large isotropic hyperfine coupling was obtained for the complexes in which both nitrogens (amino and imidazole) of one His molecule are coordinated, but a significantly smaller coupling was found for the complex of MeHis, where the imidazole-N1 could not coordinate.[29] Although the usefulness of UV spectra for solving structural problems related to metal-ligand complex formation in aqueous solution is limited, UV and FT/IR methods were still found useful to study the Cu(II) complexes of some imidazole-containing ligands, L-His, *N*-acetyl-L-His, Hsm, L-His-OMe and β-Ala-His. The intensity of a shoulder due to the CT (charge transfer) between copper(II) and imidazole-N was found to be influenced by the coordination mode.[30]

For many reasons helicity and chirality are of intense current interest in chemistry, biochemistry, pharmaceutics, *etc.* Recently, a paper was published with the results of preparation and X-ray characterization of two new chiral coordination polymers, [Zn(SPA)(H$_2$O)$_2$]$_n$ and [Cu(SPA)(H$_2$O)$_2$]$_n$, where H$_2$SPA =4-sulfo-L-phenylalanine. The idea for choosing the sulfonate-containing ligand was that even if the metal–organic coordination polymers are multidimensional, they can still be soluble in water due to the strong hydrophilic capacity of the sulfonate group.[33]

β-Glutamic acid (β-Glu) and its metal complexes have received almost no attention because of their extremely low occurrence in biological systems. In the past two years one paper presenting results for Zn(II) and Li(I) complexation of β-Glu has been published. Neutralization of aqueous solution of β-Glu with zinc oxide or lithium hydroxyde resulted in the formation of two new complexes, $[Zn(\beta\text{-}GluH)_2(H_2O)_3]$ and $[Li(\beta\text{-}GluH)(H_2O)]$. Only the carboxylates play a role in the coordination in these complexes, the amino group remains protonated. This coordination mode is entirely different from the modes of α-Glu, where chelates are predominantly formed. No doubt this difference originates largely from the difference of the chelate ring size. Both complexes prepared show layer structure, but due to the different nature of the two metal ions, and their different solvation by residual water molecules, different structural patterns were obtained. Projection of monolayers for the Zn(II)- and Li(I)-containing complexes are shown in (**3a**) and (**3b**), respectively.[34]

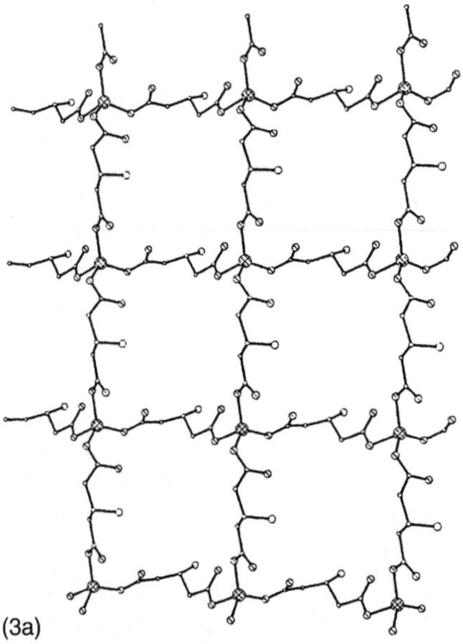

(3a) 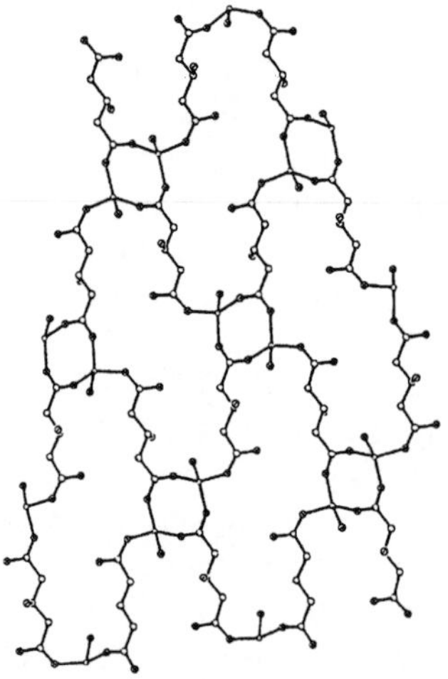 (3b)

The coordination chemistry of zinc(II) and the biological functions of this metal are in correlation with each other, and this correlation has initiated many investigations.[35–37] Recently, two new Zn(II)–amino acid complexes, $[Zn(Ile)_2(H_2O)_2]$ and $[Zn(Phe)_2]$, were prepared and structurally characterized. The former complex was found to show molecular structure (**4**), but the latter is a one-dimensional coordination polymer with η^2-carboxylate bridges (**5**).

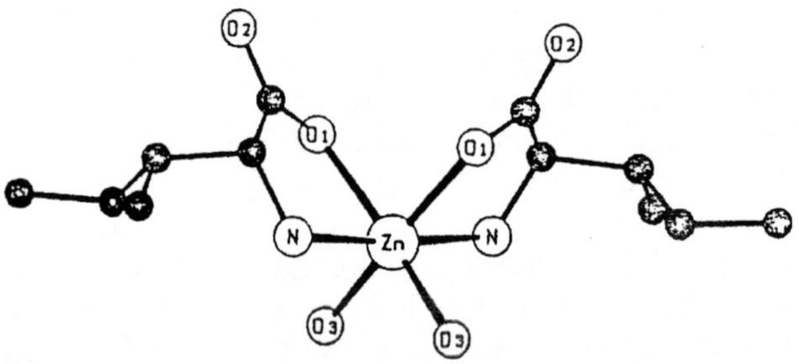

O2 O2 O1 O1 N Zn N O3 O3

(**4**)

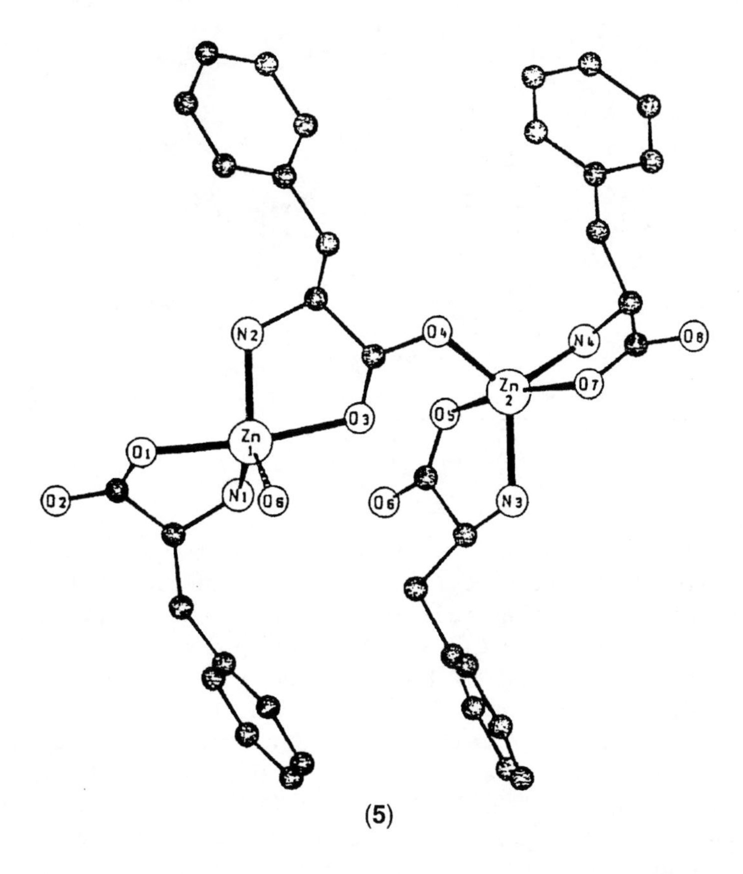

(5)

In the same work zinc(II)–pyrazolylborate–amino acid ternary complexes were also prepared. Monomeric and dinuclear structures and a large variety of the coordination modes were obtained.[35] Amino acids are among the supposed pre-biotic molecules. The conditions under which the complexes of amino acids with metal ions like zinc(II) and calcium(II) are formed, together with current ideas of the conditions on early Earth, support the assumption that the complexes of α-amino acids with zinc and calcium may have belonged to the pre-biotic chemical inventory on the early Earth. This paradigm initiated a study in which zinc and calcium salts, $ZnCl_2$, ZnO, $Zn(OH)_2$ and $Ca(OH)_2$, were reacted with racemic *i*-Val and Val. An unexpected compositional and structural diversity of the Zn(II)-*i*-Val complexes was found. Some of them were characterized by X-ray crystal structure analyses. Out of the complexes obtained the structure of the *cyclo*-$[Zn_6(i\text{-Val})_{12}]$ is shown in (**6**). Thermolyses of some of the solid products in a N_2 stream yielded a surprising number of organic products including larger molecules.[36]

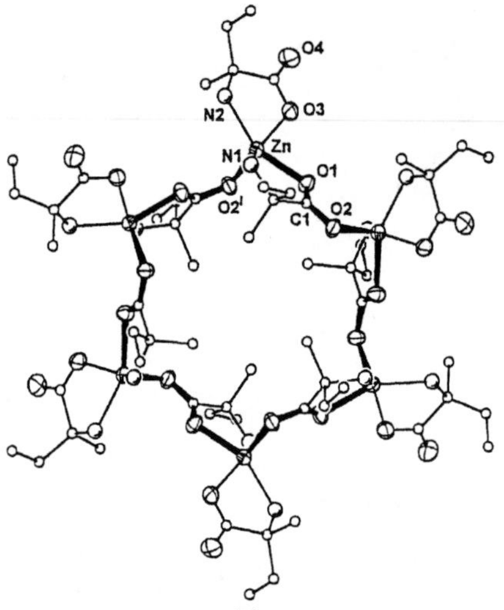

(6)

Due to the affinity of the toxic Cd(II) towards calcium binding sites which are rich in aspartyl and glutamyl residues, Cd(II)-Asp complexes have some interest. Two new complexes, $[Cd(AspH)NO_3]$ and [Cd(Asp)] were synthesized during the past two years. The former complex in its crystallized form is a two-dimensional polymer in which each of the cadmium ions is coordinated by seven oxygen atoms (four from carboxylates, three from nitrates) and the amino nitrogen is protonated and non-coordinated. In the latter complex, which was prepared at higher pH, the amino-N is also involved in the coordination.[38]

Owing to their biological and medical activity, there has been a growing interest in polyoxovanadates. In a recent work a new crown-shaped polyoxovanadium(V) cluster cation, $[V_6O_{12}(OH)_3(OOCCH_2CH_2NH_3)_3(SO_4)]^+$, involving a μ_6-sulfato anion, and zwitterionic μ-(β-Ala), was prepared. The single crystal obtained with sulfate counter ion was synthesized and characterized by X-ray diffraction methods (7).[39]

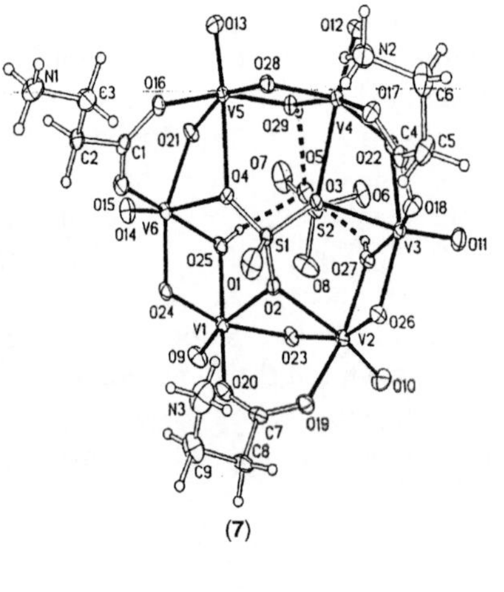

(7)

VO^{2+}–doped L-arginine phosphate monohydrate single crystal was also prepared and characterized. In this crystal, two types of magnetically inequivalent VO^{2+} sites occupying interstitial positions in the lattice with fixed orientations were identified.[40]

As model complexes for silver(I)-protein interactions, silver(I) complexes of Asp, Gly and Asn were prepared and characterized. It was found that the complex $[Ag_2(D\text{-}HAsp)(L\text{-}HAsp)] \cdot 1{,}5H_2O]$ contains only Ag–O bonds (**8a**), in the complex $[Ag(Gly)_2 \cdot H_2O]$ the O-Ag-O and N-Ag-N bonding units are alternately repeated (**8b**). N-Ag-O Units are repeated in the $[Ag(L\text{-}Asn)]_n$ (**8c**) and $[Ag(D\text{-}Asn)]_n$ complexes, and, all the above structures are different from that of Ag(I)-His, which involves only Ag-N bonds.[41]

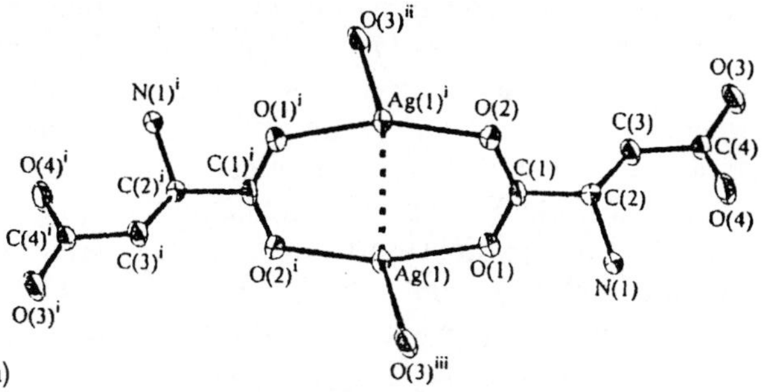

(8a)

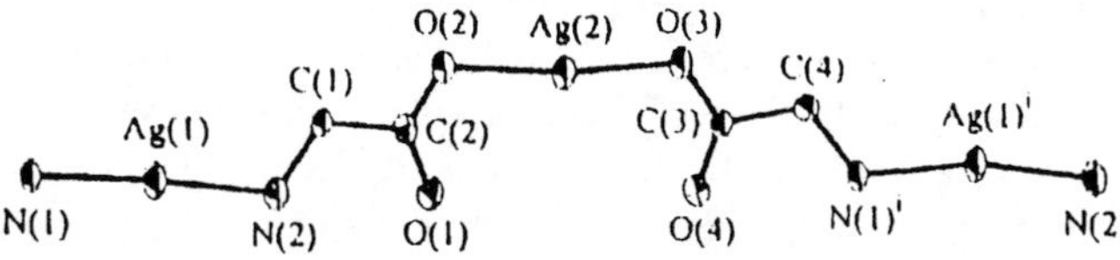

(8b)

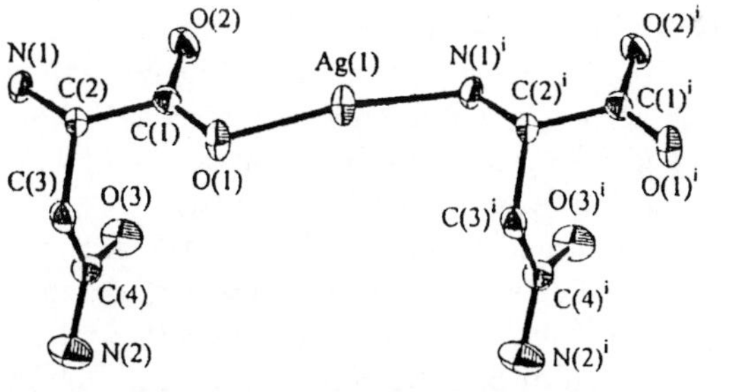

(8c)

Results of the first synthesis and characterization of Cr(V) complexes with non-sulfur-containing amino acids have been published.[42] The reduction of Cr(VI) in methanol in the presence of Gly, Ala and α-aminoisobutyric acid, Aib, showed that both Cr(V) and Cr(III) complexes were formed. A solid product isolated from the Ala-containing sample was EPR silent and was characterized as a dioxo-bridged dimeric species, $[Cr_2(\mu\text{-}O)_2(O)_2(Ala)_2(OCH_3)_2]^{2-}$. In the presence of excess Ala or Aib, tris-chelated Cr(III) complexes were formed.[42] Due to the low stability of the amino acid-Mn(II) complexes, such types of compounds are quite rarely studied. One paper, discussing the results of thermogravimetric and differential scanning calorimetric analysis of Mn(II)-Gly complexes was published during the period covered.[43]

Interactions of amino acids with metal[44–46] or metal oxide[47] surfaces are often studied as models for biomaterials formed by the adsorption of large biological molecules, such as proteins on metal surfaces. According to the results, the bonding of S-Pro to the Cu(110) surface occurs in an anionic form via the carboxylate oxygens and imino nitrogen. Both carboxylate oxygens are equidistant from the surface while the pyrrolidine ring is held at a small angle to the surface plane. The bonding of the proline layer to the surface is strong, creating a robust adlayer which is stable up to 450 K.[44] The interaction of basic and neutral amino acids with layered zirconium phosphates (α-ZrP and γ-ZrP) was also investigated. It was found that only the basic amino acids (Lys, Arg, His) could intercalate into the α-ZrP and γ-ZrP layers. When the intercalation compounds obtained in these cases were investigated by ^{31}P and ^{13}C solid-state NMR, the interaction of the P–OH group of the host phosphates with the ε-amino group of Lys and α-amino group of Arg and His was clearly observed.[48]

Among the elements, which can be used in nuclear medicine, technetium and rhenium are worth paying particular attention to. Many results for amino acid complexes of Re(I) and Tc(I), and Re(V) and Tc(V) complexes with sulfur-containing amino acids were previously published. During the past two years, the non-sulfur-containing His has been involved in several works.[49–51] When equivalent amounts of $ReOX_3(OPPh_3)(Me_2S)$ and L-His were reacted, the complex $[ReOX_2(L\text{-}His)]$ (where X=Cl, Br) was formed. It was found that His binds the tridentate way to the Re(V) center via its imidazole-N1, amino-N and carboxylate oxygen atoms in this mono-complex. By using 2 equivalents of L-His, the 2:1 complex $[ReO(L\text{-}His)_2]$ was obtained. This latter complex involves one N,N,O-tridentate and one N,N-bidentate anion. By refluxing the mono-complex in methanol, an oxo-bridged dinuclear complex $[ReOX_2(L\text{-}HisMe)_2O]$ containing N,N-bidentate histidine methyl ester (HisMe) is formed. Analogous reactions with Tc(V) were not succesful.[49] Ligand exchange reaction between $[Tc(CO)_3(thioether)]$, where thioether=1-carboxylato-3,6-dithiaheptane or 3,6-dithiaoctane, and His resulted in the formation of $[Tc(CO)_3His]$, in which the His is coordinated in N,N,O-tridentate mode.[51]

The idea that amino acid complexes of Sm(III) are suitable to transport this nuclide to abnormal cells initiated investigations on complexation of Sm(III) with different amino acids. Results obtained with Arg, Ser, Phe, Asp, His and β-Ala were published during the period reviewed. Only weak coordination of these amino acids via their carboxylate oxygens to the Sm(III) ion was found. Monomeric complexes predominate in solution, but polymeric structures exist in solid state. Based on the results it was concluded that the binary Sm(III)-amino acid complexes are not stable enough to transport this metal ion to tumor cells.[52] In addition to Sm(III), Nd(III) and Er(III) were also involved in another work. Only carboxylate coordinations of the studied amino acids (Gly, Ala, Glu and Val) were found in the clusters formed in all these systems. Of the X-Ray results, the ORTEP drawing of $[Sm_4(\mu_3\text{-}OH)_4\ (Gly)_5(H_2O)_{11}]^{8+}$ is shown in (**9**).[53]

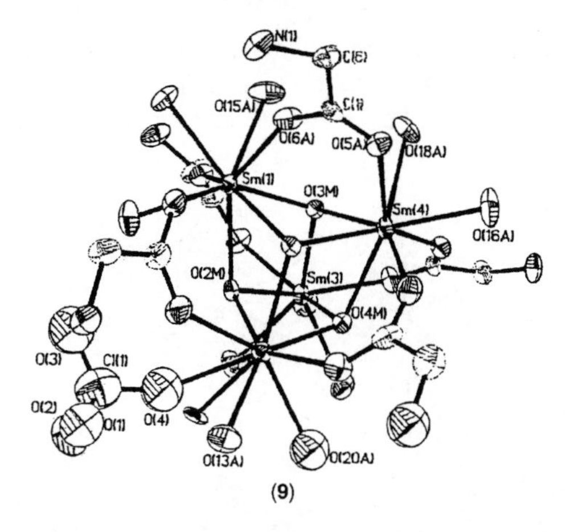

(9)

Since ternary complexes of the amino acids are often more relevant models for various biological systems than the binary ones, numerous studies have been performed during the past two years in this field.[54–64] It is known that interactions arising between π systems, between hydrophobic systems, or between oppositely charged side groups can contribute to the stability of the ternary complexes. Moreover, these so called "weak interactions" play a crucial role in the self-organization of molecules, supramolecular chemistry and molecular recognition. Because such interactions are generally more significant with aromatic ligands than with aliphatic ones, the metal complexes of aromatic amino acids are of great interest. Also the 2,2-bipyridine (bpy) or 4,4-bipyridine (4,4-bpy) and 1,10-phenanthroline (phen) are often chosen as "second" ligands. Three ternary copper(II) complexes, [Cu(L-Phe)(phen)(H_2O)]ClO_4, [Cu(L-Phe)(bpy)(H_2O)]ClO_4 and [Cu(L-His)(bpy)]$ClO_4 \cdot 1{,}5H_2O$, were synthesized and characterized by single crystal X-ray and spectroscopic (FT-IR, UV-Vis, EPR, ^{1}H NMR) techniques. While the former two complexes were found to form supramolecular networks via weak intra- and intermolecular π-π stacking and H-bonding interactions, the third complex exists as a weak dimer (**10**) in which the Cu ··· Cu separation is 3.811 Å.[54]

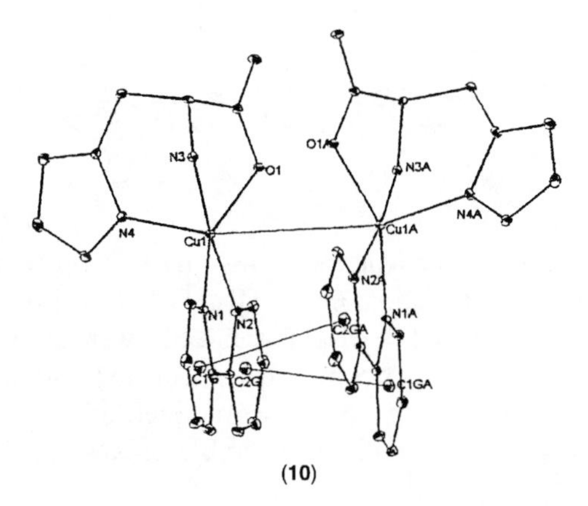

(10)

By using 4,4-bpy as a "second" ligand, a new type of Cu-Tyr complex, $[Cu_2(Tyr)_2(4,4\text{-bpy})\cdot 2H_2O]\cdot 2ClO_{4n}$ was prepared. The coordination mode of this complex is shown in (**11**), while a perspective view of its 2-D sheet is presented in (**12**).[55]

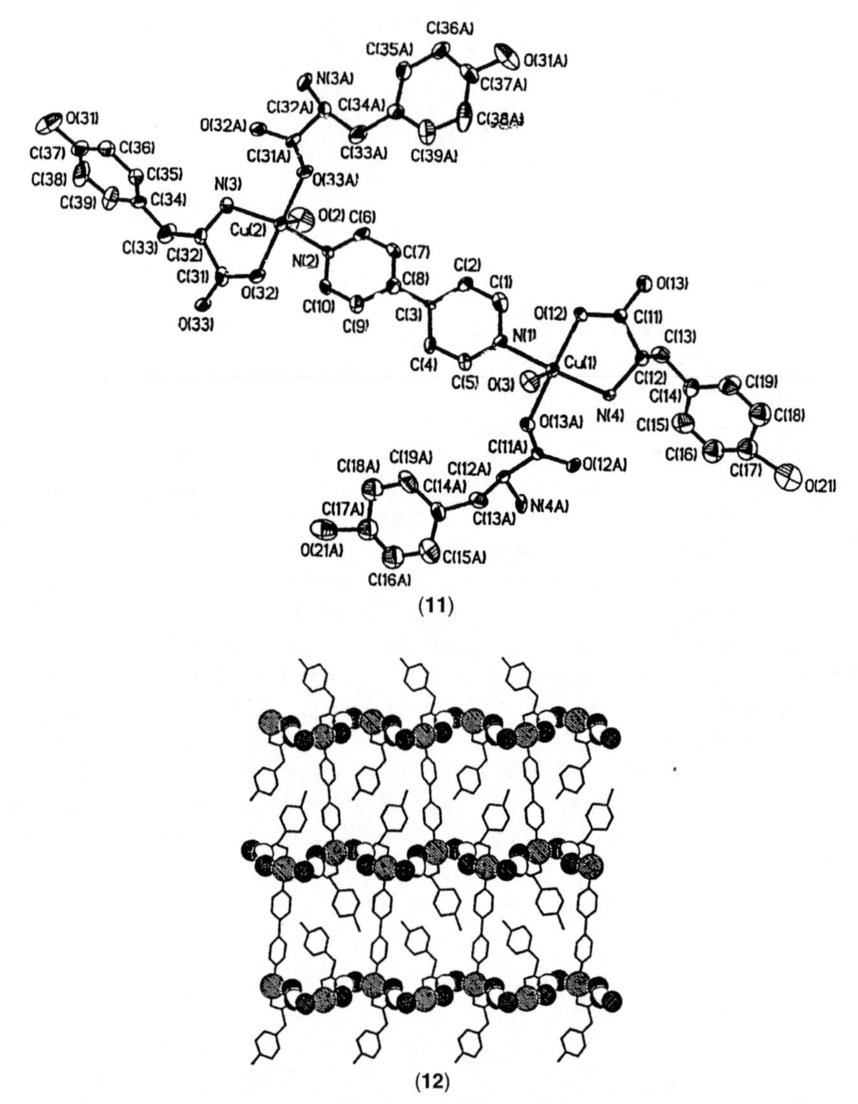

(**11**)

(**12**)

As models for metalloenzyme reactions, the ternary complexes of Cu(II), Ni(II) and Co(II) with cytidine and amino acids, L-Ala, L-Phe and L-Trp were synthesized and characterized.[57] Also Trp and Phe were chosen to prepare and investigate mixed-ligand copper complexes of cysteine thiolate. The complexes obtained were characterized by spectroscopic methods and voltametry.[58] To compare the homochiral-heterochiral assembly process of the chiral complex cations of nickel(II) with the well-studied cobalt(III) analogues, and to evaluate the effect of the metal ion and the steric effects of the side chain of an amino acid on the intermolecular H-bond pattern, the crystal and packing structure of three new racemic complexes with the formula [Ni(tren)(DL-amino acidato)Cl]·nH_2O were prepared. The amino acids involved in this work were DL-Val, DL-Phe and DL-Leu. Schematic representation of the homochiral-heterochiral helical chain generated by intermolecular H-bonds can be seen in (**13**).[59]

(**13**)

Several papers published during the past two years discuss results for Co(III) ternary complexes.[60–65] As is well known, NH-π interaction is one of the important factors in molecular recognition and in the stabilization of a molecular structure. The NH-π interaction was detected both in solid state and solution in the cobalt(III) ternary complexes formed with N,N-bis-(carboxymethyl)-(S)-phenylalanine and aromatic amino acids including (S)-Phe, (R)-Phe and (S)-Trp. To demonstrate this interaction, the structure of the complex formed with (S)-Trp is shown in (**14**).[60]

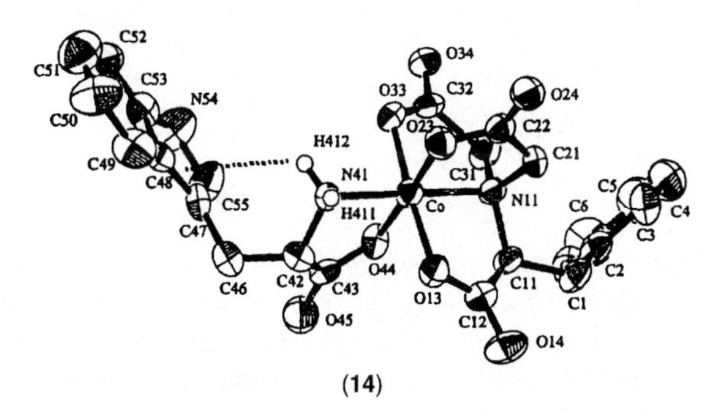

(**14**)

From ternary systems, where aqueous solution of MoO_3 with H_2O_2 and with amino acids such as Gly, Ser, Lys, His or Met was reacted, mononuclear [$MoO(O_2)_2L(H_2O)$] (L=Gly), dinuclear [$Mo_2O_4(O_2)_2L(H_2O)_4$] (L=Met or Ser) and trinuclear [$Mo_3O_7(O_2)_2L(H_2O)_6$] (L=Lys, His) complexes were isolated.[66]

There has been considerable current interest in the synthesis of rhodium and ruthenium ternary complexes involving amino acids or derivatives.[67–70] Some of the complexes were synthesized as possible DNA intercalators. By using five α-amino acids, Gly, Ala, Phe, Leu and Tyr, the ternary complexes $[Ru(bpy)_2(\text{amino acidate})]ClO_4$ were prepared and characterized. The complex with Tyr was also investigated by X-ray diffraction. The interaction of the synthesized complexes with DNA was monitored by fluorescence method. It was found that these stable complexes can effectively bind to DNA without causing any damage to its double helix.[67] In other work the interaction between various imidazole-containing agents like His or imidazole, as representatives of cellular donors, and the known "caged NO" reagent, $RuCl_3NO(H_2O)_2$ was investigated by NMR and IR.[68] Chiral cyclometallated rhodium complexes of amino acids such as Ala, Val, Phe, were successfully prepared. In addition to the two C,N-coordinated 2-phenylpyridine or benzo(h)chinoline molecules one amino acidate, coordinating via its amino nitrogen and carboxylate oxygen atoms, is also involved in the octahedral coordination sphere. It was found that the crystal of the complex formed with benzo(h)chinoline and Ala contains two diastereoisomers in the unit cell (**15**).[69]

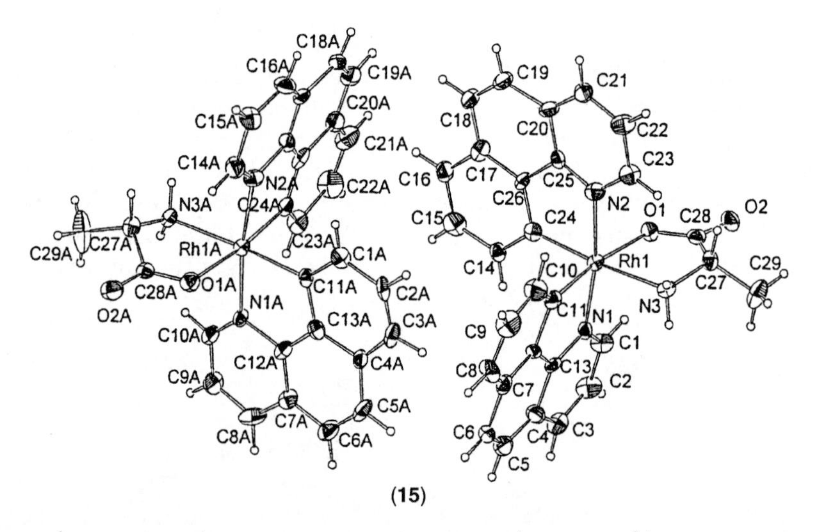

(**15**)

The ^{1}H NMR recognition probe, trans-$[Os(\eta^4\text{-}H_2)(en)_2]^{2+}$, was applied to study the coordination of amino acids to osmium. In the ternary complexes formed with α-amino acids, only carboxylate coordination was found in acidic solutions, but in alkaline conditions coordination of the amino nitrogen atom was also suggested.[71]

Two classes of lanthanide probe complexes for biomolecules may be defined. Coordinatively saturated complexes involving octa- or nonadentate ligands belong to the first class, while typically diaqua complexes, in which heptadentate ligands are coordinated to the lanthanide ion, belong to the second class. Displacement of the water molecule(s) in the latter type of complexes can lead to the formation of ternary complexes. Chiral diaqua complexes of Eu(III), Gd(III) and Yb(III) formed with heptadentate ligand, 1,4,7-tris[(R)-1-(1-phenyl)ethylcarbamoylmethyl]-1,4,7,10-tetraaza-cyclododecane, were reacted with Gly, Ser, Phe or His in aqueous media. Monoaqua ternary complexes were formed when Phe, His or Ser were reacted with Eu- or Gd-containing species at pH 6. However, both water molecules were displaced by Gly or Ser from the Yb complex, as confirmed by the crystal structures of these latter two adducts.[72] Results of a fluorescence lifetime study obtained for the Eu(III) complex of a heptadentate modified β-cyclodextrin are consistent with one water molecule coordinating to the metal ion. ^{1}H NMR results confirm that this complex of the β-cyclodextrin derivative form host-guest complexes with (R/S)-His, (R/S)-Tyr, (R/S)-Phe and (R/S)-Trp. The coordination of (R/S)-Tyr and (R/S)-His shows enantioselectivity.[73] Eight new ternary complexes of D-glucosamine and Gly with Cu(II), Ni(II), Co(III) and Fe(III) were synthesized and characterized by various spectroscopic methods including Raman scattering.[74] Gly was involved in a study in which several new Ni(II) ternary complexes were prepared and characterized as well as tested for their fungicidal actions.[75]

For many reasons, amino acids are versatile starting materials for the synthesis of different new ligands. Any structural modification may result in a completely different coordination behaviour of the new molecules compared to the parent amino acids. Numerous papers discussing metal complexes of amino acid derivatives have appeared during the past two years. The nitrogen atom of the amino group remains the main coordinating donor in amino acid esters. The interaction of such molecules with Co(III)-tetramethylchiroporphyrin chiral shift reagent was investigated by ^{1}H NMR and 1D TOCSY ^{1}H NMR techniques. It was found that in addition to the coordination of the amino nitrogen atom to the metal center in the apical position, hydrogen bonds and/or π-stacking also play a role in the interaction. The hydrogen-bonding capability of the porphyrin complex allows the conformation of coordinated amino acid esters to be uniquely defined within the chiral cavity on the time scale of NMR analysis, which yields a difference between the chemical shifts of the diastereomeric adducts.[76–77] Water soluble complexes of Zn(II) with porphyrin derivatives were used as receptors for various amino acids or their esters. The ligands were suggested to be recognized on the basis of carboxylate coordination. Chiral recognition evidence was not obtained, except in the cases of Gly-DL-Trp and Gly-L-Trp where significant hydrophobic interactions could also be supposed.[78–79] In another work, chirality induction by amino acid derivatives was observed in a bis-Zn(II)-porphyrin complex, in which two porphyrins are linked by a short ethane covalent bridge.[80]

Ni(II), Cu(II) and Zn(II) complexes of a derivative of Gly, (*N,N*-bispicolylamino)acylglycine ethyl ester, were synthesized and characterized by X-ray diffraction method. The structural flexibility of this tetradentate (N,N,N,O) ligand was demonstrated by the results. While the geometry of its Ni(II) complex was octahedral, a square-pyramidal complex was formed with Cu(II) and trigonal-bipyramidal with Zn(II).[81] Cu(II) complexes of (*N,N*-bispicolylamino)acylphenylalanine methyl ester were also synthesized by the same authors. When the structural, spectroscopic and electrochemical results obtained with the Gly and Phe derivatives were compared, the structure of the amino acid side chain had no effect on the properties of the copper(II) centre. This was in contrast to the previous results with zinc(II).[82] Sometimes the N,N,N donor set in zinc(II) complexes is created as a model for the active site of carbonic anhydrase. Although the new ligand, *N,N*-bis(3,5-dimethylpyrazolylmethyl)-L-alanine methyl ester involves three N donor atoms, only two of them (one from each pyrazole) was found to coordinate to a zinc(II) ion. The nitrogen atom of the L-alanine methyl ester was not coordinated as shown in (**16**).[83]

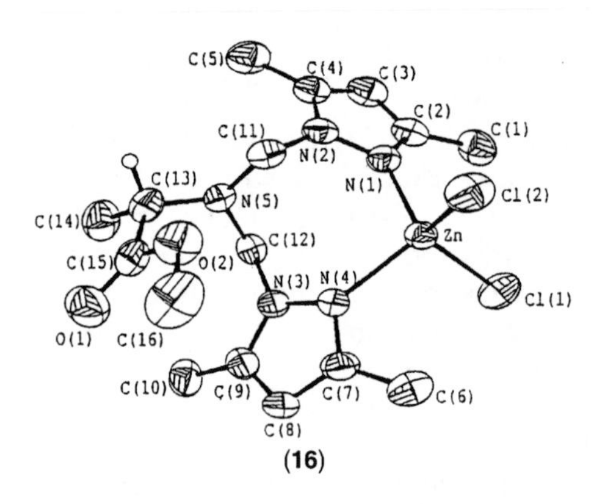

(**16**)

N,N-Bis(2-hydroxyethyl)glycine has an N-substituted amino, a carboxylate and two hydroxyl groups as coordinating functions in one molecule. Bis-complexes of this molecule with lanthanide metals, La(III), Ce(III), Nd(III), were synthesized and characterized. The crystal and molecular structure of the La(III) complex was determined by X-ray crystallography. The La(III) ion is nine-coordinated in this complex. In addition to the eight donor atoms coming from the two coordinated ligands, a carboxylato oxygen atom of a neighboring molecule is also coordinated to the metal center. Thus this complex forms a one-dimensional polymer chain in which the individual molecules are bridged by the carboxylato oxygen atoms.[84]

Significant current attention is given to metal complexes of different derivatives of amino acids as models for metalloenzymes. Very often metalloproteins possess dinuclear complex units as the active center. This is one reason why dinuclear complexes formed with bioligands like amino acid derivatives are the focus of many current studies.[85–90] The dinuclear Mn(II) complex of the tetradentate N-centered tripodal N,N-bis[(1-methylimidazol-2-y'l) methyl]glycinate (**17**) was found to be a good superoxide scavenger.[85]

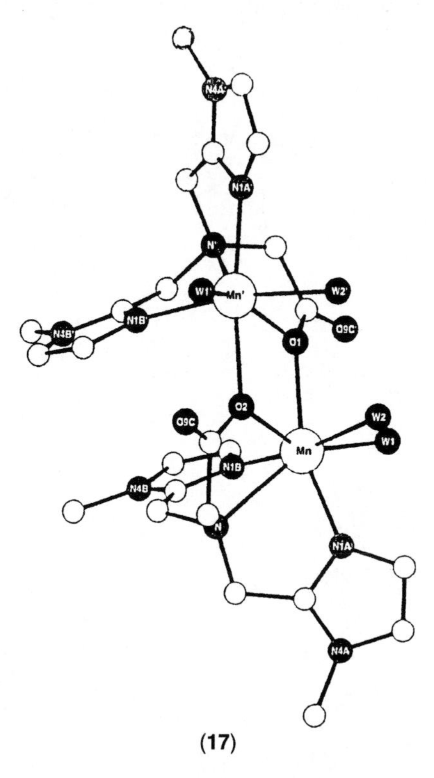

(17)

Dinuclear complexes of Ti(IV) with amino acid-bridged dicatechol ligands were also prepared and characterized.[87] Bidentate amino alcohols derived from L-Phe were synthesized and their Ti(IV) complexes were also prepared. The dimeric structure of the complexes was confirmed by X-ray crystallography.[88] Dithiocarbamate derivatives of selected natural amino acids, such as Asn, Gln, Ser, Thr, Tyr and Pro, were prepared. Reaction of these sulfur-containing molecules with Cu(II) was found to lead to the reduction of the metal ion and formation of dinuclear Cu(I) complexes. The only exception was the proline-based dithiocarbamate, which simply formed a bis-complex with Cu(II). In the anion formed, copper(II) is coordinated to four sulfur atoms.[90]

Many investigations have been performed on the metal complexation of different P-containing amino acid derivatives during the past two years.[91–97] A tridentate coordination mode of the commercial herbicide glyphosate (N-(phosphonomethyl)glycine) was found in its copper(II)[91] and both tridentate and monodentate in its uranium(VI)[92] complexes. The former complexes were characterized by IR and EXAFS, while those formed in the dioxouranium

(VI)-glyphosate-F^- binary and ternary systems were studied by multinuclear NMR. The structures of the binary complexes indentified in the uranium(VI)-glyphosate system are shown in (**18**).[92]

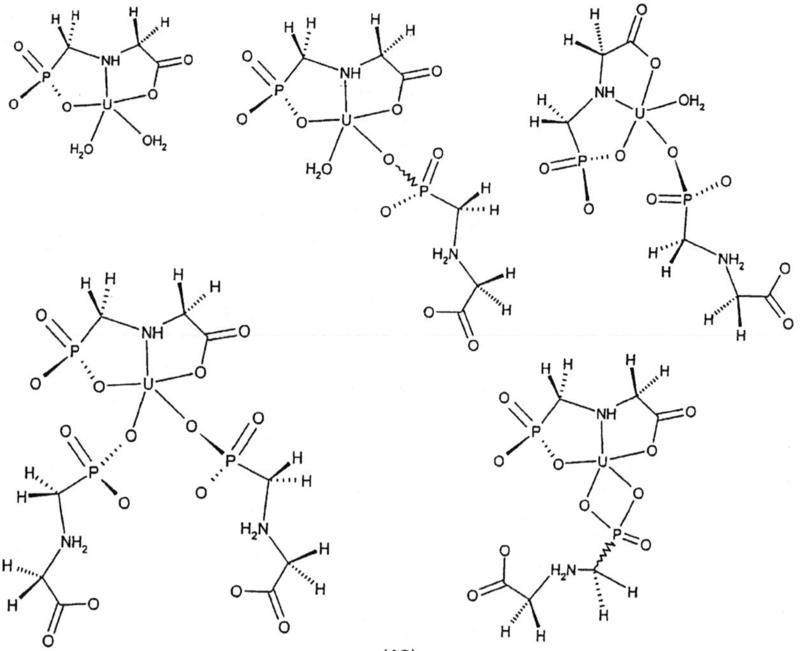

(**18**)

A new phenylalanine-based phosphine-type ligand was found to coordinate readily to Pd(II) and Pt(II), and form square planar complexes.[93] Pd, Pt, Rh and Au complexes of alanine-based phosphine-type ligands were also prepared and in several cases the structures were determined by X-ray diffraction method.[94] Surprisingly, five-coordinated ternary complexes were formed when a Pt(II)-aminophosphinate binary complex with thiol-containing molecules such as *N*-acetyl-L-Cys and glutathione was reacted.[95]

To develop new radiopharmaceuticals, Re(I),(III),(V) and Tc(I) complexes with selected amino acid derivatives were prepared.[98–100] A series of compounds were constructed from various potentially tridentate amino acid derivatives (**19**) by reacting them with different compounds containing $[Re(CO)_3]^+$ or $[ReCl_3]^+$ cores. Most of the complexes prepared were characterized by X-ray crystallography. The qualitative order suggested for the coordination preference of the different donors to the $[Re(CO)_3]^+$ core is the following: pyridyl nitrogen > carboxylate > halide > thioether.[98]

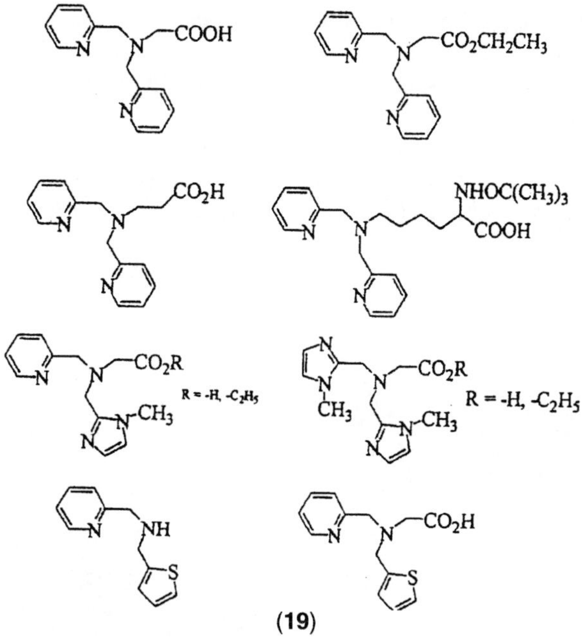

(19)

To find new compounds with insulin mimetic effects is a current challenge. Several new oxo peroxovanadium(V) complexes, including the compound with the amino acid derivative *N*-carboxymethylhistidine, were prepared. The complexes show a distorted pentagonal bipyramidal structure in which the peroxide ion binds in a side-on fashion to the V(V) centre. An insulin mimetic effect was obtained only with the compound formed with the derivative of His.[101] Interestingly, new Zn(II) complexes of some tetradentate derivatives of amino acids showed insulin mimetic effects. Of the complexes prepared, the X-ray structure analysis was performed for the complex $[Zn_3(N\text{-pyridylmethyl-His})_3(H_2O)_2]\cdot(ClO_4)_3\cdot 2H_2O$. Solution equilibrium studies were also performed, and the correlation between the stability of the complexes formed and their biological activity was evaluated.[102]

The known "*Pfeiffer Effect*" states that the interaction of an optically active probe with a labile racemic metal complex may induce a shift in the racemic equilibrium, resulting in an excess of one enantiomer over the other one. The effects of L-Pro and its derivatives, *N*-acetyl-L-Pro, L-Pro methyl ester, L-Pro benzyl ester on the equilibrium shift in the racemic $[Pr(ODA)_3]^{3-}$ complex, where ODA=2,2′-oxydiacetate, were studied by CD and ^1H NMR. The *Pfeiffer Effect* appeared with all studied ligands except *N*-acetyl-L-Pro. On the basis of the results, it was suggested that the chiral discriminatory interaction is first of all a combination of electrostatic and hydrophobic forces.[103]

Further derivatives, in which the amino acidato parts play a significant role in the metal complexation, should also be mentioned. For example, ternary complexes of N-phthaloylglycinate with Co(II), Ni(II), Cu(II), Zn(II), Cd(II) and with various mono-, bi- and tridentate amines as "second ligands" were prepared and investigated.[104] The X-ray structure of the [Cd(hippurate)$_2$(cytosine)(H$_2$O)]$_2$ complex (**20**) is the first crystal structure published for a cadmium(II)–amino acid derivative–nucleobase ternary compound.[105]

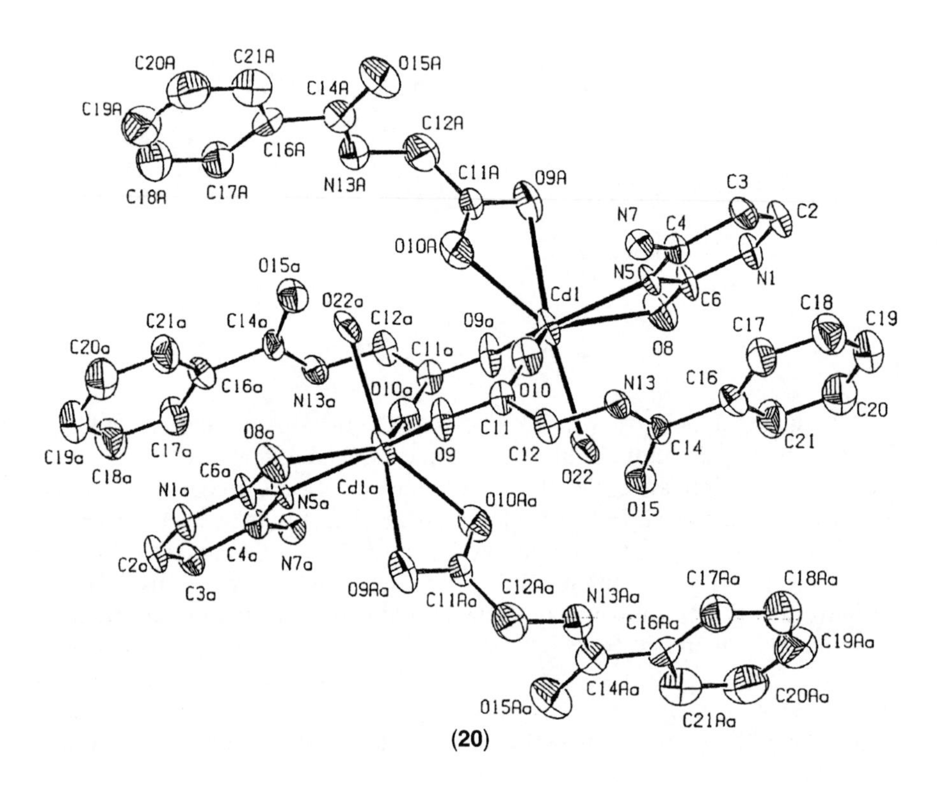

(**20**)

Thermal behavior of complexes of *S,S′*-methylenebis(cysteine) with Cu(II), Zn(II) and Cd(II) was also studied.[106] Calixarenes can be readily modified to give compounds with interesting new properties such as the inclusion of neutral molecules and ions. For example they can be functionalized as ionophores for metal ions. Three amino acid- substituted calix[4]arens, L^a, L^b and L^c (**21**), were synthesized and their complexes with Pd(II) by using electrospray mass spectrometry (ESMS) investigated.[107]

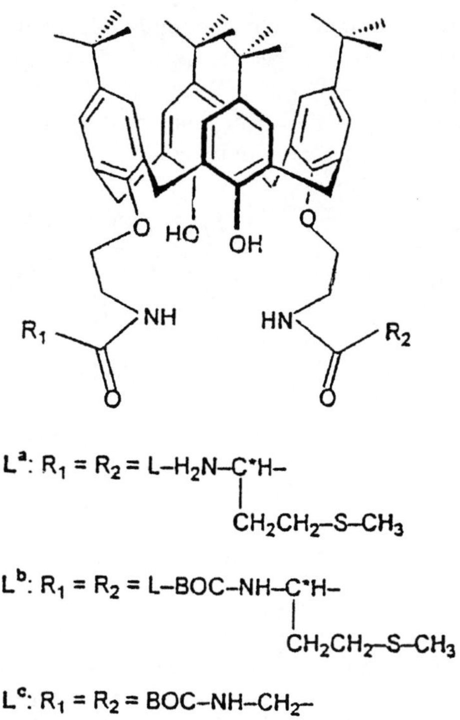

L^a: $R_1 = R_2 = L\text{–}H_2N\text{–}C^{*}H(CH_2CH_2\text{–}S\text{–}CH_3)\text{–}$

L^b: $R_1 = R_2 = L\text{–BOC–NH–}C^{*}H(CH_2CH_2\text{–}S\text{–}CH_3)\text{–}$

L^c: $R_1 = R_2 = \text{BOC–NH–}CH_2\text{–}$

(**21**)

Many complexes of amino acid based Schiff bases with various metal ions such as Ni(II),[108,109] Mn(III),[110] Ti(IV),[111,112] Zn(II),[113] Mg(II),[113] Ir(III),[114] Ru(III),[114] and Cu(II)[115] ions have been prepared during the past two years. Ni(II) complex of a new methionine-based Schiff base was prepared and characterized. The complex obtained was found to possess catalytic activity for polymerization of methyl methacrylate.[108] In another case derivatives of Gly and Ala were used for the preparation of chiral synthons of these amino acids.[109] Although Zn(II) complexes of Gly-based Schiff bases (derived from 3-formylsalicylic acid and Gly and salicylaldehyde and Gly) had vacant coordination sites, substitution of Zn(II) by Mg(II) happened when the complexes were reacted with Mg(II) in DMSO. The Mg(II) complexes obtained in this way were characterized by a X-ray diffraction method.[113] Cu(II) ternary complexes involving Gly or Ala or Phe-based Schiff base and bpy were synthesized and structurally characterized. The complexes showed catalytic activity in the oxidation reaction of ascorbic acid by dioxygen.[115] The condensation of 2-formylphenoxyacetic acid with Gly by template synthesis resulted in the formation of a new tetradentate ligand. The Cu(II) complex of this new amino acid derivative was prepared and characterized by X-ray analysis.[116]

A large number of organometallic compounds containing amino acids, peptides and derivatives have been studied.[117–122] Unexpectedly, when a C=C-bridged hippuric acid dimer was reacted with $[Cp^*IrCl_2]_2$ and $[Cp^*RhCl_2]_2$, ligand-bridged dinuclear complexes (**22a** and **22b**, respectively) with different bonding modes were obtained.[117]

(22a) (22b)

Several new metallocenyl type compounds of Val were synthesized with very good yields and with excellent diastereoselectivities.[118] To obtain unnatural, highly functionalized derivatives, amino acids with a "vinylstannate" moiety in their side chain were prepared.[120] In another paper the synthesis of chiral chloro-bridged η^6-phenylglycine complexes of ruthenium was published.[121] Organo tungsten complexes of amino acid derivatives were prepared and the structure of the complex (**23**) formed with the glycine derivative was confirmed by X-ray analysis.[122]

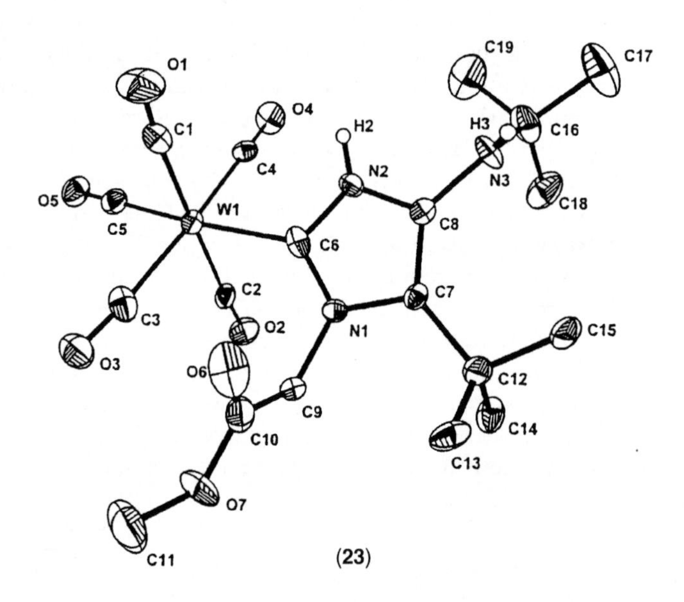

(23)

Sodium and potassium and, to a lesser extent, lithium cations have special importance from a biological point of view. They participate in numerous functions of living systems. Therefore, the understanding of the details of the local interactions between the alkaline metal ions and amino acids is a matter of considerable interest. Both experimental and theoretical investigations have been performed during the period of coverage.[123–124,129,130,132,133] Binding energies between alkali metal ions and seventeen different amino acids and their corresponding methyl esters were determined in the gas-phase by the kinetic method. It was found that the binding affinities of amino acids generally increase upon their conversion to methyl esters. Nevertheless, amino acids may bind more strongly to M^+, especially to larger metal ions, than their esters. This is the case with Pro and with the most basic amino acids, Lys, Arg.[123] The density functional theory was applied to study the interaction of α-Ala with Li^+, Na^+ and K^+. In the case of Li^+ and Na^+, the lowest energy structure was found to correspond to a N,O-coordinated α-Ala, while the complex with K^+[124] was O-monocoordinated.

The same theory was used to evaluate the structures, electron distributions and dissociation energies of gas phase ternary complexes of Cu(II) with bpy and Leu, Ile or Lys,[125] and also to study the interaction of Cu(I) and Cu(II) with α-Ala[126] and to characterize the stability of the Cu(II)-Phe-Val ternary complex in the gas phase.[127]

Theoretical calculations based on the structural data from the Cambridge Structural Database demonstrated that the cation-π interaction is ubiquitous in metal complexes. Examples were found for various metal ions and ligands.[128] The existence of alkali metal-π interactions with aromatic side chain of amino acids was proved several times.[129,130] In this context new results with transition metals were also obtained.[131,135]

Both theoretical calculations and experimental work were used to study the bonding mode of alkali metal and water to Val.[132] Without water it was suggested that Li^+ and Na^+ are coordinated to the nitrogen and carbonyl oxygen, whereas K^+ is coordinated to both oxygens of a valinate anion. The addition of a single water molecule did not significantly effect the relative energies calculated for the metal ion-Val clusters.[132] Between Gly and Na^+, the preference for chelate formation was also found in other work.[133] Quantum chemical calculations performed for the interaction of Gly with 15 different metal ions (M^+ or M^{2+}) in the gas phase provided the conclusion that the nature of the lowest energy isomer depends dramatically on the metal ion.[134] Electrospray ionization was used to generate and study complexes of Ag(I) with selected amino acids in the gas phase.[135] The investigation of the binding of sulfur-containing amino acid side chains to the toxic Al(III) and nontoxic Mg(II) in biological systems was the focus of theoretical calculations. Selected simple models were used to represent the side chains of Cys and Met. According to the results, Mg(II) binds to all of the neutral ligands studied, while Al(III) does not, and the binding of the anionic ligands is much stronger to both metal ions.[136] Additional theoretical results have also been published in selected papers.[137–139] For example, to search for different conformational and geometric isomers within a single simulation, a new searching algorithm for transition metal complexes was developed. The algorithm was used to study complexes of Tc with different ligands like D-penicillamine or its derivatives.[137] Chiral amino acid recognition by Zn(II)-porphyrin was analyzed by a molecular mechanics method based on the Tripos force field method.[138] In addition to semiempirical MO (PM3) and *ab initio* calculations, different experimental techniques were used to study the Cd(II) complexes of a new Trp derivative, *N*-deoxycholyl-1-tryptophan.[139]

Several $[Ru(bpy)_3]^{2+}$-containing new metalloamino acid derivatives have been synthesized and characterized during the past two years.[140–145] For example, a new functional model for the tyrosine and chlorophyll unit (P_{680}) of the natural photosystem II (PSII) was created. To achieve this goal, the ruthenium(II) complex was connected to one L-Tyr ethyl ester through an amide bond. Then this adduct was attached to the surface of nanocrystalline TiO_2 via the carboxylic acid groups linked to the bpy ligands. The reaction scheme proposed for the photo-induced electron transfer in the system is demonstrated in (**24**).[140]

(**24**)

To quantify the electronic interactions mediated by alkyl side chains, a series of compounds, in which the ω-amino groups of diaminopropionic acid, diaminobutyric acid, ornithine or lysine, were covalently attached to a $[Ru(bpy)]^{2+}$ fragment, were synthesized.[141] New $[Ru(bpy)_3]^{2+}$-type complexes in which L-Lys-containing side chains are attached to the bpy molecules were found to form nanometer scale fibrous assemblies in some organic solvents. The aggregates obtained are more efficient photosensitizers than the parent $[Ru(bpy)]^{2+}$.[142] In another work a $[Ru(bpy)]^{2+}$-based metalloamino acid was successfully incorporated into a water-soluble, alanine-based peptide. The electrochemical potential, absorption spectrum, emission and excitation spectra, and emission lifetime data for this new compound were determined.[143] A new copper-selective $[Ru(bpy)]^{2+}$-based emitter-receptor conjugate involving a Lys spacer (**25**) was developed during the period reviewed.[145]

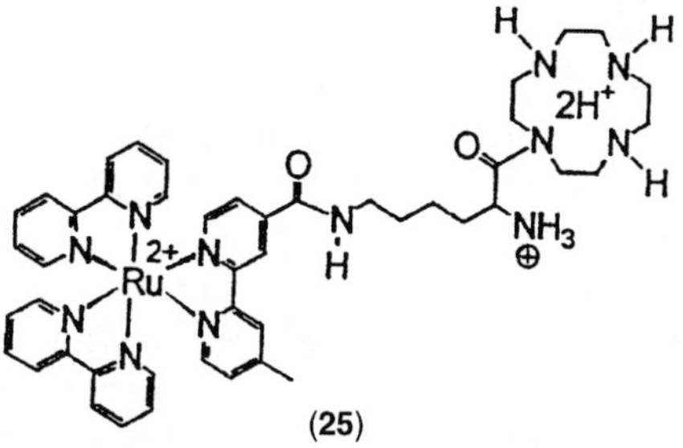

(25)

2.2 Solution Studies. – In spite of the enormous number of stability constants published over the past several decades for metal complexes of the natural amino acids, numerous new data are still being published. There is no doubt that the equilibrium systems, containing some 3d metal ions, especially Cu(II), Ni(II) or Zn(II), are the best characterized ones. In such cases, new data were obtained either by the reinvestigation of previously studied systems under different conditions (ionic strength, solvent), in the presence of an additional ligand or by using different techniques.[146–164] Several new data appeared for systems containing Co(II)[147,153,156,157,160] and Cd(II)[155,165–167] and, interestingly, the very rarely studied Hg(II) and Ag(I) were also involved in a few solution equilibrium studies.[168–172] Potentiometry was used to determine the stability constants for the binary and ternary complexes formed in the Hg(II)-L-Arg-canavanine system,[168] and for complexes of Ag(I) with Trp,[169] His,[170] Arg[171] and Glu.[172] With many metal ions, the strong competition between hydrolytic and complex formation processes makes the equilibrium systems very complicated. Fortunately, numerous new data, such as stability constants and a few thermodynamic data for both binary and ternary complexes of amino acids with metal ions, e.g. with V(IV),[173,174] V(V),[175–180] Mo(VI),[181] Fe(III),[182] Sc(III),[183] and with various rare earth metal ions, La(III),[184] Pr(III),[184] Sm(III),[184] Eu (III),[185] Ce(III),[186] Yb(III),[187] Gd(III),[187] Tb(III),[190] Dy(III)[190] and Th(IV)[188,189] were also obtained during the period of covered. Also p-block elements Al(III),[165,189,192] Tl(I)[193–194] and Pb(II)[165,189] have been investigated with amino acids. The influence of four essential amino acids Gly, Ser, Thr and His, on the bioavailability of aluminium was studied. Only one complex with $[Al_2LH_{-2}]$ stoichiometry was identified between the Al(III) ion and the four amino acids. Based on the very low formation constants determined for the $[Al_2LH_{-2}]$ complexes it was concluded that no risk is to be expected from any of the four amino acids investigated, in terms of increased aluminium bioavailability under normal conditions of alimentation and aluminium-based therapy.[192]

Although, the interaction between the amino acids and alkali and alkali earth metal ions is not very strong, they often coexist in biological fluids, so their weak interactions are of great significance. During the past two years, pH-metric investigations have been made on the Ca(II)- and Mg(II)-L-Gln binary[195] and Ca(II)- and Mg(II)-L-Gln-succinic acid ternary[196] systems in ethyleneglycol-water solvent. Solubility investigations were performed to estimate thermodynamic parameters for the interaction existing between L-Cys and sodium or magnesium chloride in aqueous solution.[197] A series of natural amino acids, Gly, α-Ala, β-Ala, L-Leu, L-Ile, DL-Ser, L-Thr, DL-Phe and DL-Met were involved in the study in which their interaction with the very toxic Be(II) ion was investigated.[198] In addition to the hydrolytic species of the Be(II) ion, the complexes $[Be(HL)]^{2+}$, $[Be_3(OH)_3(HL)]^{3+}$ and $[Be(OH)_2L]^-$ (L=deprotonated form of the amino acid), were found in measurable concentrations. The formation constants were determined by potentiometry.[198]

Molecular/chiral recognition in many cases occurs via the formation of ternary complexes, e.g. molecular recognition of amino acids and amino acid esters by Zn(II)-protoporphyrin,[199] of amino acid esters by the Zn(II) complex of threonine-modified porphyrin,[200] and by the Co(II)-salen complex,[201] or chiral recognition of L/D-amino acids by the Cu(II) complex of β-cyclodextrin functionalized by histamine (**26**). In the latter case, the inclusion of an aromatic side chain in the cyclodextrin cavity makes the *cis* arrangement in the ternary adducts of aromatic amino acids favoured.[202]

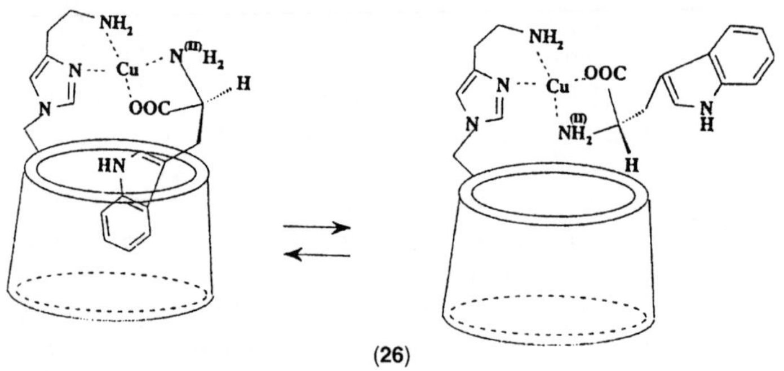

(**26**)

Solution equilibrium studies on metal complexation of numerous interesting amino acid-based ligands have been performed during the past two years.[203–219] Since imidazole nitrogen donor atoms of histidyl residues are among the most common metal binding sites of metalloproteins, model compounds are often synthesized to mimic the binding properties of imidazole nitrogen donor atoms in different environments. Cu(II), Ni(II) and Zn(II) complexes of N-phenylalanylbis(imidazol-2-yl)methylamine (Phe-BIMA) and N-histidylbis(imidazol-2-yl)methylamine (His-BIMA) (**27**) were studied by potentiometric, UV-VIS and EPR methods.[203]

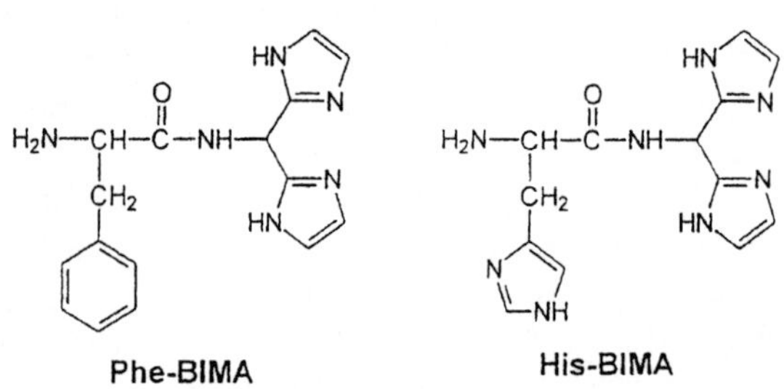

Phe-BIMA His-BIMA

(**27**)

The nitrogen donor atoms of the bis(imidazol-2-yl)methyl residues were described as the primary metal binding sites in all systems studied. Although, at higher pH, the copper(II) ion induces the deprotonation and coordination of the amide function in both ligands, nickel(II) can do it only in the case of His-BIMA. The histidyl side chain of His-BIMA also has a significant metal binding ability that provides a great versatility in the complex formation reactions of this ligand. As a consequence, this compound seems to be well-suited as a model for the active centre of various metalloenzymes. It forms different polynuclear complexes especially with Cu(II) ion. Of them, a trinuclear complex, $[Cu_3H_{-4}L_2]^{2+}$ (**28**), containing negatively charged imidazole bridges, seems a promising model for the SOD enzymes.[203] The metal binding ability of the bis(pyridin-2-yl)methyl derivatives was also studied and was found to be lower than that of the imidazole-based analogues.[204]

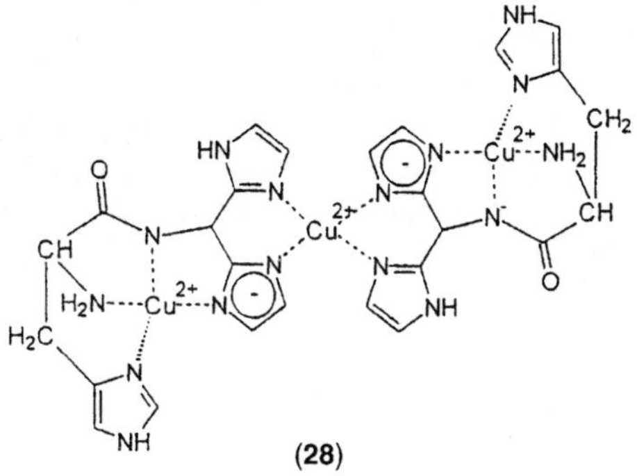

(28)

Potentiometry and spectrophotometry was used to investigate the Cu(II)-binding ability of a series of amino-phosphonates containing pyridine as a side chain. In some cases, the binding ability of these pyridine-containing ligands was found to be comparable to that of the powerful imidazole-containing analogues.[205] The complex formation between five *N*-D-gluconylamino acids (the derivatives of Gly, L-α-Ala, β-Ala, L-Ser and L-Met) and VO^{2+} ion was studied in aqueous solution by pH-metry, CD, UV-VIS and EPR methods. Various mono- and dinuclear complexes are formed in these systems. At pH *ca.* 5 a dinuclear spcies $[(VO)_2L_2H_{-4}]^{2-}$ predominates. In this species the deprotonation and coordination of the amido nitrogen and C(2)-OH is strongly suggested. The formation of a bis-complex, even at high excess of ligand, was not observed.[206] Equilibrium and spectroscopic results show that the C-terminal carboxylate is the primary anchor in the complex formation process between the dimethyltin(IV) and 2-hydroxy-hyppuric acid (Sal-Gly). But above pH 4, the dimethyltin(IV)-induced deprotonation and coordination of the amide group also occurs.[207]

Extensive efforts are currently made to develop new MRI agents (contrast enhancers in magnetic resonance imaging). New optically-active bis(amino acid) type ligands containing six and eight donor atoms were synthesized and their complexation with Gd(III) was studied. The hexadentate ligand formed a low stability complex with the Gd(III) ion. Although, the bis(amino acid)s with eight donor atoms obtained by the use of picolyl derivatives of His and Asp, formed more stable Gd(III) complexes, these species were found still to be insufficiently stable for use as MRI agents.[208] Europium(III) complexes of *N,N'*-ethylenebis(L-amino acid) ligands were found as useful chiral NMR shift reagents for some unprotected natural α-amino acids in a neutral aqueous solution. The stability constant of the Eu(III) complex of *N,N'*-ethylenebis(L-His) was determined by potentiometric titration.[209]

Numerous papers have been published during the past two years for metal complexes of aminohydroxamic acids.[210–218] These ambidentate type ligands are hydroxamic acid derivatives of amino acids. Depending on the character of the coordinating metal ion and also on the pH, these compounds are able to realize different coordination modes. Additional donor atoms in the side chain increase the number of possible coordination modes further.[210–212] The effects of side chain amino nitrogen donor atoms, and the side chain catecholate moiety, on the metal complexation of hydroxamic acid derivatives of 2,3-diaminopropionic acid, 2,4-diaminobutyric acid, ornithine, lysine and 2,3-dihydroxy-phenylalanine (Dopa) were studied by potentiometric, UV-VIS and EPR methods.[210,211] As a result of a combined potentiometric, spectrophotometric, CD and ESI-MS studies, the pentanuclear metallocrown complex (**29**) was found in the Cu(II)-S-tryptophanhydroxamic acid systems in methanol/water solution.[213]

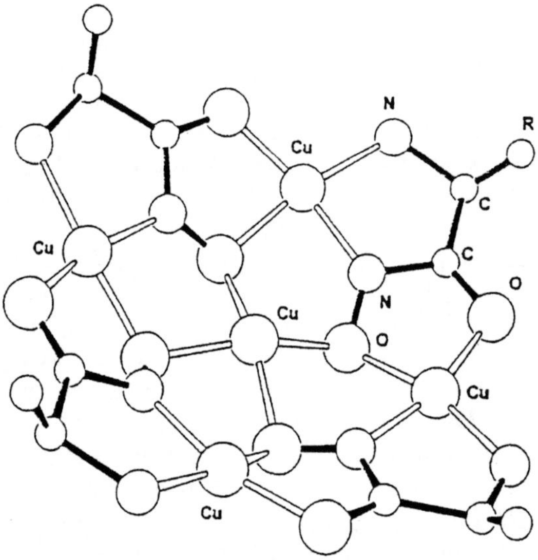

(**29**)

Results for ternary complexes of aminohydroxamic acids formed in Cu(II)- and Zn(II)-containing systems are discussed in other selected papers.[214–217] The redox potentials for a series of Fe(III)-aminohydroxamate complexes were determined. The values obtained at physiological pH are in the range of biological reducing agents, which makes these compounds potential siderophore models.[218]

^{1}H NMR, spectroscopic and potentiometric equilibrium results obtained for organometallic sandwich complexes of Dopa, dopamine and 5-hydroxytryptophan with (η^6-p-cymene)Ru(II) showed a dramatic decrease of the pK values relating to the first deprotonation of the catecholate functions and an even greater increase in the acidity of indole N1 and O5 donors in the Trp-containing compound. The deprotonation scheme determined for the Dopa-based complex by potentiometric and ^{1}H NMR titrations is shown in Scheme 1.[219]

2.3 Kinetic Studies. – Numerous papers published during the past two years deal with oxidation of amino acids by high oxidation state metal ions.[220–238] Frequently used oxidizing agents are permanganate (both under basic[220–223] and acidic[224] conditions), Mn(III),[225,226] Mn(IV),[227] Ir(IV),[228] V(V),[229] Ni(IV),[230] Co(III),[231] Fe(III),[232,233] Au(III),[234] Tl(III)[235] and Cr(VI).[236–238] There is no doubt that complex formation plays a crucial role in such reactions. Permanganate oxidant was used to oxidize L-hydroxyproline, L-Asn, L-Arg, L-Leu and L-Ile under basic conditions, while L-His was oxidized in sulfuric acid medium. The suggested structure of the catalytically active complex formed between the alkaline permanganate and L-hydroxyproline is shown

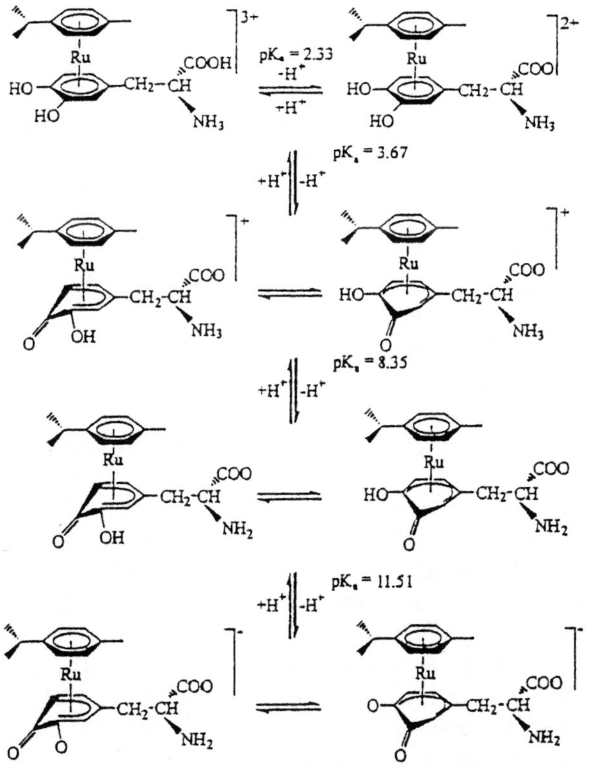

Scheme 1

in the structure (**30**).[220]

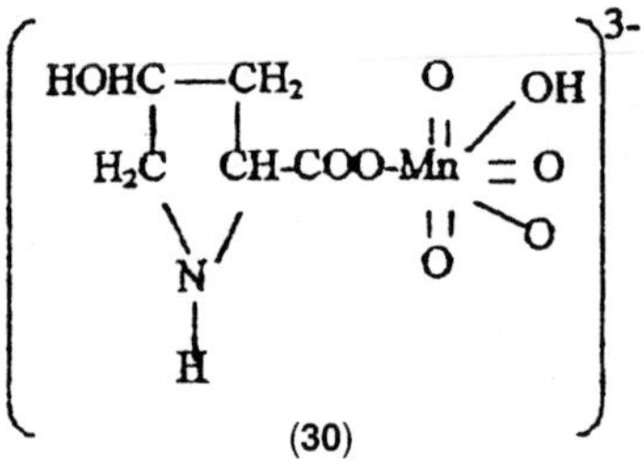

(**30**)

During the Ru(III)-catalysed oxidation of L-Arg by alkaline permanganate, spectral evidence for the formation of catalyst-substrate complex (**31**) was obtained.[222]

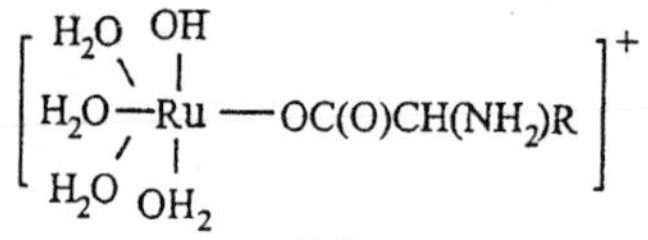

(**31**)

Ru(III)-catalysed oxidation of L-Leu and L-Ile by alkaline permanganate was also studied.[223] Ag(I) was found to catalyze the oxidation of L-His by Mn(VII) in acidic medium.[224] To obtain additional results for the mechanism of action of therapeutic gold compounds, the reaction of Au(III) with isotopically-labeled Glu, Ala and sarcosine was studied by NMR spectroscopy.[234] Oxidation reactions of dopamine, L-Dopa and derivatives by Mn(III), Mn(IV) were studied as models for the autooxidation process appearing naturally in neurons.[225,226] A new pathway in the reaction of dopamine and Dopa was observed when the oxidation by MnO_2 was performed in the presence of $S_2O_3^{2-}$ anion.[227] The oxidation reactions of sulfur-containing essential amino acids, Met, Cys, are frequently studied. During the period reviewed they were oxidized by various compounds of Ir(IV),[228] Co(III),[231] Fe(III),[232,233] Tl(III),[235] Cr(VI).[236,237] The Fe(II)-catalysed oxidative decarboxylation of amino acids by an organic agent chloramine-T in acidic medium was also investigated.[239]

Several papers discuss kinetic results for complexes of amino acids and derivatives with platinum[240–242] and palladium.[243–244] Spectrophotometry was used to study the kinetics of the interaction of D,L-penicillamine with $[Pt(en)(H_2O)_2]^{2+}$ (en=ethylenediamine). It was found that the reaction proceeds via rapid formation of an outer-sphere association complex, followed by two slow consecutive steps. The first is the formation of a sulfur-coordinated inner complex and the second is the chelation through the amino nitrogen.[240] In another work the reactivity of oxaliplatin and its metabolites (**32**) towards guanosine and L-Met in NaCl and Na_2HPO_4 solution at pH=7.4 was investigated. The aquated species (C) and especially the diaqua form (D) were always more reactive than the parent drug (A) itself. It was concluded that oxaliplatin, which is stable in water, has to be transformed in the presence of chloride to chloro-derivative (B), which is aquated to become active particularly versus guanosine.[241]

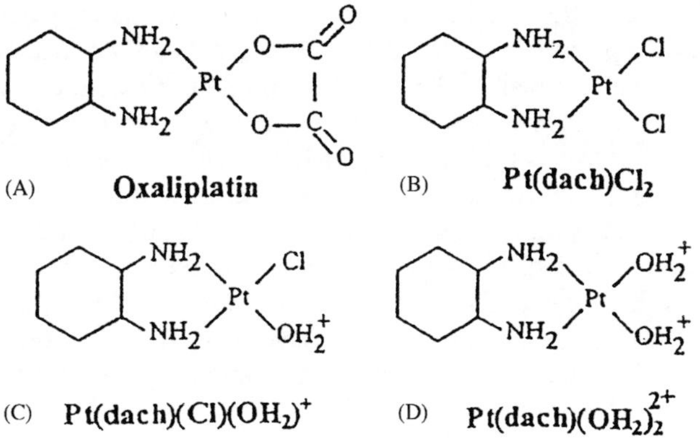

(A) **Oxaliplatin** (B) **Pt(dach)Cl$_2$**

(C) **Pt(dach)(Cl)(OH$_2$)$^+$** (D) **Pt(dach)(OH$_2$)$_2^{2+}$**

(32)

A detailed investigation was made on the kinetics and equilibria of the complex formation reactions of $[Pd(S\text{-methyl-L-Cys})(H_2O)_2]^+$ with inosine, inosine-5′-monophosphate, guanosine-5′-monophosphate, 1,1-cyclobutanedicarboxylate and L-Gly in aqueous solution, as a function of nucleophilic concentration, temperature and pressure. Two consecutive reaction steps, which both depend on the concentration of the nucleophile reactant, were observed under all conditions. An associative complex formation mechanism was supported by the negative entropies and volumes of activation obtained.[243] An associative mechanism was also suggested for the substitution reaction between DL-Met and $[Pd_2(bpy)_2(OH)_2]^{2+}$.[244]

The introduction of the C_p* function (C_p*=η^5-pentamethylcyclopentadienyl anion) into the cation $Ir(H_2O)_6{}^{3+}$ leads to a dramatic increase by 14 orders of magnitude in the rate of water exchange. By the use of $[Cp^*Ir(H_2O)_3]^{2+}$ and different bidentate ligands, *e.g.* DL-Pro, a series of half-sandwich monoaqua cations were prepared and characterized. It was found that the substitution process of the water molecule from the monoaqua cations by monodentate ligands obeys second-order kinetics and the magnitude of the rate constant depends on the nature of both the bidentate and monodentate ligands. The trinuclear complex crystallized from the DL-Pro-containing system was characterized by X-ray.[245]

A few results for the kinetics of formation reactions of metal ion – amino acid binary or ternary complexes appeared in the period of coverage.[246–250] Thermodynamic parameters (the activation enthalpies, the activation entropies and the activation free energies) and some kinetic parameters (the activation energies, the pre-exponential constant and the reaction order) were determined for the formation reaction of cobalt(II) complex with L-His in the temperature range of 25–50°C by microcalorimetry.[246] The rate constants for the reactions occurring between different Cu(II)-amino acid complexes and the tetraphenylporphine were determined and the relationship between the structure of the Cu(II)-amino acid complexes and the rate of their reactions with the macrocyclic ligand was evaluated.[247] The reactivity of the axial coordination sites of a vitamin B_{12} model Co(III) complex was investigated when this macrocyclic complex was reacted with five different amino acids and the rate constants were determined.[249] An *ab initio* computational study was performed in order to evaluate the associative versus dissociative mechanism in the formation of ternary complexes of Cu(II)-N-salicylidene-aminoacidate with 2-aminopyridine (or pyrimidine). Although the results indicate that the associative mechanism is more likely to occur, the dissociative mechanism cannot be completely discarded.[250] The factors affecting the ring-opening during the acid-catalysed hydrolysis of some carbonato complexes of Co(III), for example *fac*(N)-[Co(CO_3)(Gly)(en)] and *mer*(N)-[Co(CO_3)(Gly)(en)], were studied. The crystal structure of the above two complexes were also determined in the same work.[251]

The metal ion-promoted hydrolysis of α-amino acid esters has attracted considerable attention due to the relevance of these systems to the reactions of a variety of metalloenzymes. The kinetics of the hydrolysis of 4-nitrophenylglycine ester catalysed by the complexes $[Co(OH)(trien)(OH_2)]^{2+}$, $[Co(OH)(tren)(OH_2)]^{2+}$ and $[Co(OH)(en)_2(OH_2)]^{2+}$ was studied by spectrophotometry in weakly basic aqueous solution. The amines were used in these complexes to block four coordination sites of Co(III) and to leave only two sites for the interaction with the ester. The complexes were found to promote the hydrolysis of the ester significantly. The activation parameters for all three complex-promoted reactions were found to be comparable, thus suggesting a common mechanism for the studied reactions.[252] It is well-established that surfactants affect reaction rates by incorporating one or both of the reactants into, or at the surface of the micellar aggregates. The reaction of Cu(II)-Trp complex with ninhydrin was found to be catalyzed by cetyltrimethylammonium bromide cationic micelles. Not only the kinetics of the reaction was investigated, but also the binding constants of the reactions with the micelles were determined.[253]

Gold(I) is known to be a strong inhibitor for the catalytic activity of Se-glutathione peroxidase, which contains selenocysteine at its active binding site. This biological relevance initiated investigations of interaction of Au(I)-thiomalate $(AuStm)_n$ with two diselenides, selenocystine and selenocystamine. $(AuStm)_n$ exists as a polymer in both solid state and solution. The two diselenides were found to undergo very fast exchange reactions with $(AuStm)_n$. It was observed that the diselenides initially formed a selenylsulfide species which further underwent exchange reactions with $(AuStm)_n$, producing thiomalic disulfide.[254]

Mannich reactions of chelated amino acids have received considerable attention during the past two years.[255–258] To develop further the chemistry of this reaction type, reactions of metal complexes of β-amino acids[255,258] with formaldehyde and benzamide were investigated and interesting new compounds were obtained. In the case of α-amino acids, the effects of the α-carbon substituents on the preferred pathway of the reaction occurring between bis(α-amino acidato)copper(II) complexes and formaldehyde and benzamide were investigated.[256] In another investigation, cobalt(III) template chemistry was used to obtain some new compounds by simple procedures.[259,260] For example, the synthesis of 2-(nitromethyl)ornithine from ornithine was easily achieved,[259] or new amino acids and polyamine derivatives were obtained[260] in a few steps by template methodology.

2.4 Synthetic, Analytical and Biomedical Applications of Amino Acid Complexes. – A huge number of papers that came out during the period of coverage can be associated with these subjects. However, many of them are beyond the scope of this chapter or are discussed in other parts of this volume, e.g. in those dealing with different metal ion catalysed routes of synthesis to various amino acids/derivatives. Consequently, only papers that are related both to metal complexation and application of amino acids/derivatives are mentioned in this section.

A Ni(II) complex containing a glycine-based homochiral Schiff base (**33**) was successfully used for the synthesis of new, biologically active compounds, *syn*-(S)-β-hydroxy-α-amino acids[261] and 2-substituted 1-amino-cyclopropane carboxylic acids.[262] L-Val was attached covalently via its N- or C-termini to the Pd(II) complex of a *para*-functionalized "primer" ligand $C_6H_2(CH_2NMe_2)_2$-2,6. The obtained organopalladium(II) complexes were active as catalysts in selected reactions.[263]

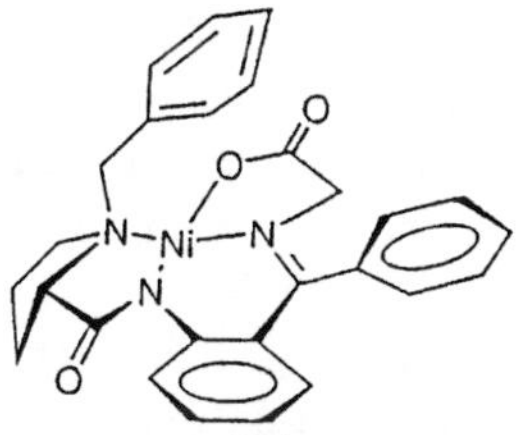

(33)

Primary and secondary iodo amino acids were used for the preparation of phosphine-containing amino acids. The developed route involves the formation of reactive organozinc/copper intermediates.[264]

Numerous investigations were performed to find optimum synthetic conditions for the preparation of various metal complexes of amino acids and derivatives.[265–270] In several cases, metal ion-amino acid interactions play a crucial role in newly developed separation procedures.[271–277] For example, the enantioseparation of amino acids by capillary electrophoresis was achieved by using an enantiomeric ternary complex, Cu(II)-(S)-3-aminopyrrolidine-L-His, as the chiral selector.[271] A new procedure for extraction of unprotected amino acids by selected ternary complexes of Ni(II) and Cu(II) was recently published.[274] In another work the influence of different chromatographic conditions on the retention and the separation of cyclic α-amino acid enantiomers on the Cu(II)-D-penicillamine chiral stationary phase was investigated. It was concluded that the Cu(II)-D-penicillamine complex was the primary docking site for the analytes, but the chiral discrimination was mainly based on hydrophobic interactions with the chiral stationary phase.[275] Teicoplanin is a macrocyclic glycopeptide that is highly effective as a chiral selector for LC enantiomeric separations of underivatized racemic amino acids. However, the chiral recognition of amino acids is dramatically decreased when Cu(II) is present in the mobile phase. There is no doubt that the complex formation between the Cu(II) ion and the amino acids is responsible for the loss of enantioselectivity.[276]

A new technology suggested for partitioning Hg(II) from actinides is based on the selective complexation of Hg(II) by L-Cys.[278] In a few cases the possibility for the determination of amino acids and the enantiomer composition of amino acid esters was provided by certain effects induced by their complex formation processes.[279–282] For example, the determination of Gly by fluorescence spectroscopy was made by evaluating the effect of Gly on the fluorescent characteristic of a selected Eu(III) complex.[279] The effect of rare earth metal ions on fluorescence spectra of Trp was also investigated.[280]

A new cysteine-selective electrode based on a Pb(II)-phthalocyanine complex as ionophore was developed. The potentiometric response of the electrode is believed to be based on the axial coordination of the cysteinate anion to the Pb(II) complex.[283] By taking the advantage of strong sulfur-gold interaction, a new type of gold electrode modified by L-Cys monolayers was created.[284] In another work a modified glassy carbon electrode was created in the following way: first, a polymer film of Trp was developed at a glassy carbon electrode by cyclic voltametry, then Ni(II) was incorporated into this film. After the incorporation of Ni(II), the catalytic activity of this electrode was greatly improved.[285]

Many of the investigations performed on metal ion-amino acid complexes during the past two years were initiated by the biological importance of such compounds. This is also true for many papers discussed in the former parts of this chapter. Thus, only some papers providing additional evidence for the biological activity of metal complexes of amino acids/derivatives are mentioned here. For example, new organotin(IV) complexes of *N*-methylglycine (sarcosine) showed *in vitro* cytotoxic activity against human adenocarcinoma HeLa cells.[286] A newly designed and synthesized complex of Gly with Au(III), chlorobischolylglycinatogold(III), was found to inhibit the growth of a variety of cell lines.[287] There are many reasons why 1,10-phenanthroline (phen) has been extensively used together with amino acids in the synthesis of metal complexes. Recently a great deal of attention has been paid to the recognition of DNA secondary structure by using this class of complex. Some new complexes have been designed and synthesized during the past two years in this area, for example: La(III) complex of a L-valine-based derivative of phen,[288] different Cu(II)-phen-amino acid ternary complexes and Cu(II)- dimethyl-phen-Gly ternary complex.[289] Other complexes such as Cu(II) complex of salicylideneglycinate Schiff base,[290] Ag(I) complexes of L-His[291] and some new mono- and polynuclear Cr(III) complexes of amino acids[292] were also tested. Some of these compounds demonstrated significant antimicrobal effects,[290,291] but the Cr(III) complexes tested did not show toxic effects.[292]

Three naturally occuring compounds, maltol, Gly and glucosamine, were used to synthesize a new effective chelating agent, 2-deoxy-2-(*N*–carbamoyl-methyl-[N′-2′-methyl-3′-hydroxypyrid-4′-one])-D-glucopyranose, for use as a Fe(III) and Al(III) sequestering agent under toxic overload conditions of these ions.[293] The vasodilator activity of CO released from the complex [Ru(CO)$_3$(L-Cys)] was described.[294] As functional models for cobalamine-independent methionine synthase, a series of new Zn(II)-pyrazolylborate-thiolate type ternary complexes, including two homocysteine derivative-containing ones, were prepared and structurally characterized.[295]

3 Peptide Complexes

3.1 Synthesis and Structural Studies on Peptide Complexes. – A huge number of peptide complexes containing a great variety of metal ions and peptide ligands have been synthesized and structurally characterized in the past two years. These studies were mainly performed by the X-ray diffraction analysis of the crystalline samples, but the usual spectroscopic techniques (UV-Vis, EPR, NMR, CD) were also extensively used in solution. The number of studies using mass spectrometric measurements[296,297] and theoretical calculations[298,299] has also increased and they provide useful information on the structural parameters and reactivity of peptide complexes in the gas phase, too.

The range of the central metal ions in the peptide complexes covers almost all major groups of the elements in the periodic table, e.g.: potassium,[298] magnesium,[300] calcium,[301] platinum,[296,302] vanadium,[303] cadmium,[301,304,305] lead,[301,306,307] zinc,[301] iridium,[308] nickel[305] and copper.[309–316] The crystal structure data on the platinum(IV) complexes of the most common dipeptides (Gly-Gly, Gly-Ala, Ala-Gly and Ala-Ala) represent the first report on the platinum(IV) induced deprotonation and coordination of the peptide amide group.[302] The biological activity of these platinum complexes has also been tested. The metal ion promoted deprotonation and coordination of the peptide amide functions have been described in the vanadium(III) and oxovanadium(IV) complexes of dipeptides (Gly-Ala, Gly-Val and Gly-Phe) in the mixed ligand complexes with phen.[303] Lead(II) and cadmium(II) complexes of Gly-Glu have been prepared in the solid state and structurally characterized.[301] The carboxylate functions were suggested as the exclusive metal binding sites for both metal ions, but the lead(II) compound shows a hemidirected coordination sphere (**34**) due to its stereochemically active lone pair, while a coordinatively saturated bis(ligand) complex (**35**) is formed with cadmium(II).

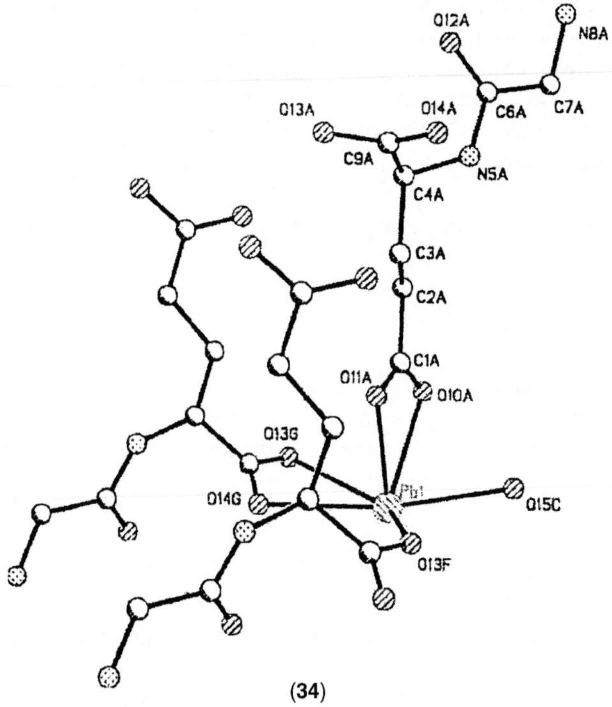

(34)

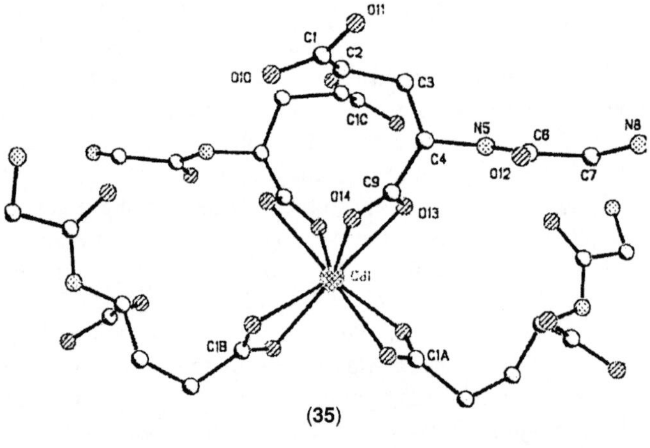

(35)

As usual, copper(II) complexes of peptides have been the most extensively studied. Crystal structure and EPR characteristics of the copper(II) complex of Trp-Gly have been reported[309] and the common distorted square planar geometry containing tridentately coordinated (NH_2,N^-,COO^-) ligands suggested. Copper(II) complexes of the dipeptide Aib-Aib (Aib=α-amino-isobutyric acid) have also been prepared.[310] Two complexes have been isolated with the same stoichiometry ($[CuH_{-1}L]$), but with different structures. One of them is described as a chain polymer (**36**), while the other is a 3D coordination polymer (**37**).

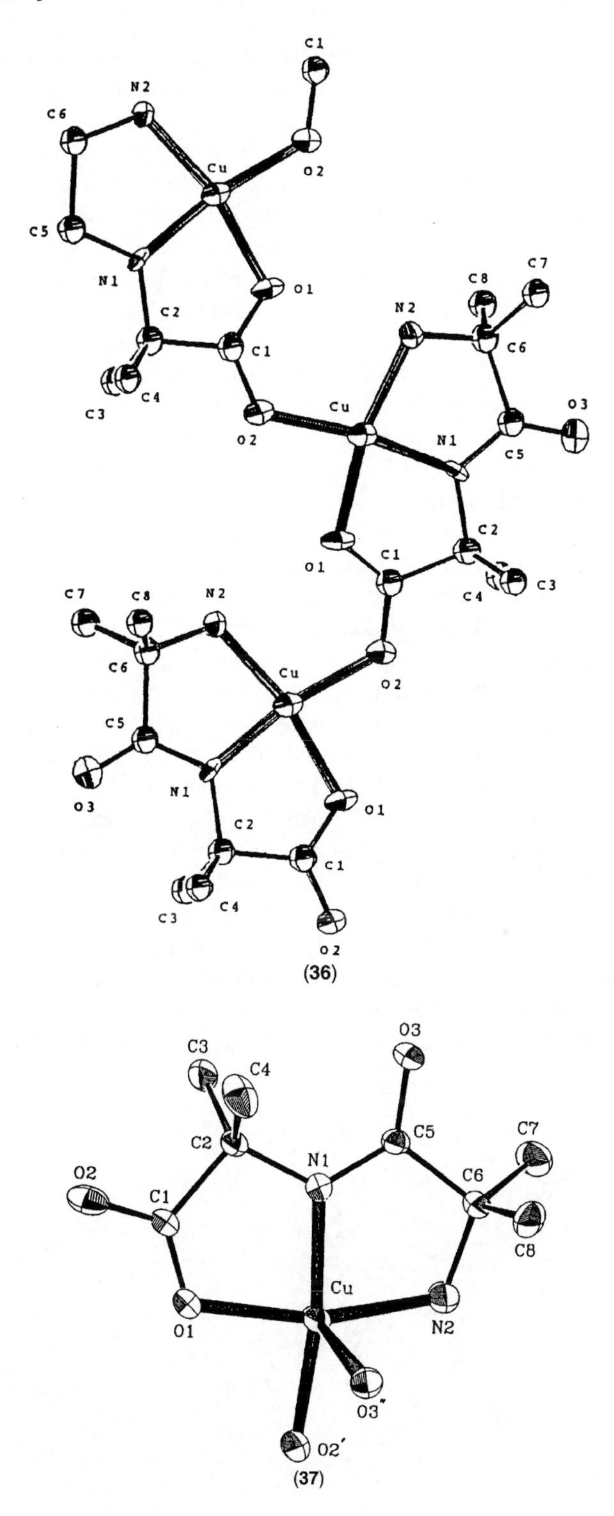
C1
N2
C6
Cu
O2
C5
N1
O1
C2
C1
C4
C3
C8
C7
N2
C6
O3
Cu
O2
N1
C5
C2
O1
C1
C4
C3
C7
C8
N2
C6
Cu
O2
C5
O3
N1
O1
C2
C1
C3
C4
O2
O3
C3
C4
C2
N1
C5
C7
C6
O2
C1
C8
Cu
O1
N2
O3″
O2′

(36)

(37)

The solid state structures of the copper(II) complexes of Ala-Ile, Ala-Thr and Ala-Tyr were also determined and an elongated square pyramidal coordination of the metal ions in a polymeric network containing common tridentate peptide ligands was suggested.[311] The superoxide dismutase activity of the complexes were also tested and it is remarkable that despite the very similar structures of the dipeptide complexes very different SOD-like activities were found. The tridentate coordination of the peptide ligands was described in the mixed ligand systems [Cu(II)(Tyr-Gly)-isocytosine and Cu(II)(Ala-Gly)-isocytosine] in the solid state.[312] Copper(II) complexes of β-casomorphin peptides were investigated by EPR and NMR spectroscopy in DMSO solution. These tetra- to hepta-peptides contain prolyl residues which are considered as break-points for amide coordination, prefer the formation of coordination isomers.[313]

Metal complexes of the various derivatives of the simple oligopeptides have also been extensively studied. Four synthetic patellamide-type cyclic peptides have been synthesized and their complexation with copper(II) studied.[317] Mono- and dinuclear complexes were detected in solution and their formations were described in slow equilibration reactions. The metal ion selectivity of another group of patellamide ligands containing both oxazole and thiazole rings was followed by MS and CD spectroscopic measurements.[318] The results revealed the enhanced copper(II) binding affinity of these cyclic peptide ligands. The factors influencing the calcium(II) binding affinity of cyclic peptides containing thiazole rings have also been studied[319] and a relationship between ligand flexibility and calcium(II) binding ability was suggested. The effect of calcium(II) binding on the conformation of a series of cyclo-[Pro-Phe-Phe-Ala-Xaa]$_2$ decapeptides was investigated in another report.[320] The synthesis of some novel cylindrical, conical and macrocyclic peptides via cyclooligomerization of functionalized thiazole amino acids have also been reported and their applications as molecular receptors discussed.[321] Haliclonamide A and B (**38**) were isolated from marine sponge and these molecules are also cyclic peptides containing oxazole rings.[322] Mass spectrometric and EPR measurements revealed that the haliclonamide ligands had an outstanding affinity for binding iron(III) and the molecules were suggested as potential non-siderophore metal binding peptides in natural oceans.

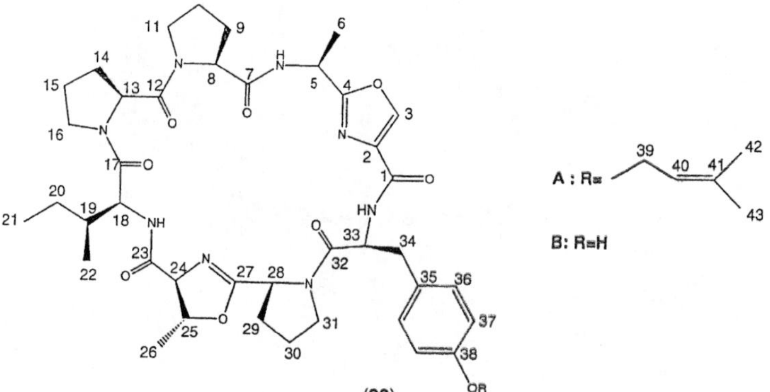

A : R=

B: R=H

(38)

The fascinating structural and catalytic properties of metalloenzymes have stimulated the design of synthetic polydentate ligands, which mimic the coordination environment of metal ions in proteins. These compounds are generally not built up from the natural amino acids, but contain the amide and/or side chain residues of the common peptide molecules. Most of these molecules contain the simple N-heterocyclic rings (pyrrole, pyridine and especially imidazole) and the donor groups are in chelating or tripodal arrangements. These molecules generally have outstanding metal binding affinity and the selectivity of metal binding can be finely tuned by the nature and location of donor atoms. The detailed description of the complex formation reactions of these ligands is beyond the scope of this compilation, but the structural variety of the complexes of these peptide models can be assessed from some selected references.[323–328]

Sulfonamides are also among the frequently studied derivatives of peptide molecules. The crystal structure of the nickel(II) complex of *N*-quinolin-8-yl-benzenesulfonamide revealed high nickel(II) binding affinity of the ligand in an octahedral coordination environment built up from 4N+2O donors.[329] Bleomycins are glycopeptide antitumor antibiotics and their biological activity is linked to their interaction with metal ions. The molecular modelling of Fe(II)-bleomycin and Co(II)-bleomycin systems led to the conclusion that both metal ion are present in a 6-coordinate structure and the Fe(II) and Co(II) adducts are isostructural.[330] The metal ion coordination of the nitrogen donor atoms of tertiary amide ligands is rather uncommon. The ligand *tert*-butoxy-carbonyl-(*S*)-alanine-*N,N*-bis(picolyl)amide (Boc-(S)-Ala-bpa) has been recently synthesized (**39**) and its interaction with copper(II) and cadmium(II) studied.[331] The binding of copper(II) resulted in the cleavage of the amide bond, while the formation of the unusual Cd–N(tertiary amide) bond was observed in the corresponding cadmium(II) complex. Possible implications for metal-induced conformational changes in proteins are also discussed.

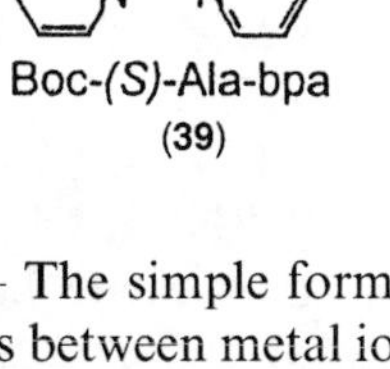

Boc-(*S*)-Ala-bpa

(**39**)

3.2 Kinetics and Reactivity. – The simple formation kinetics and mechanism of complex formation reactions between metal ions and peptide molecules have been scarcely studied in the past two years. On the other hand, a huge number of papers have been published on the metal ion catalyzed formation, hydrolysis and oxidation of peptide ligands and also on the chemical reactions modelling the catalytic activity and/or inhibition of metalloenzymes.

It has already been widely accepted that metal ions played a significant role in the formation of the first biopolymers under the conditions of the primitive earth. The model studies in this field started a few decades ago and have also continued in the past two years. The catalytic effect of alumina on the condensation of amino acids was investigated by HPLC measurements.[332] The results provide further support that alumina and related surfaces have played a crucial role in the prebiotic formation of the first peptide molecules. The readiness of amino acids to form dipeptides depended on the side chain residues and was found to decrease in the order: Gly > Ala > Leu > Val > Pro. The intermolecular condensation of L-alanine was also investigated and the effects of the different forms of alumina and the temperature dependence of the reaction discussed.[333] The synthesis of oligopeptides containing Gly and Tyr residues was investigated on the surfaces of clay minerals montmorillonite and Cu^{2+} exchanged hectorite.[334] Montmorillonite proved to be a more effective catalyst than hectorite and the formation of the sequence Gly-Tyr was favoured over the others.

The role of metal ions in some synthetic procedures of peptides has also been studied by several groups.[335–337] It was concluded that the univalent metal ions such as the alkaline metal ions can enhance both cyclization yield and rate from precursor linear peptides, while the di- or tri-valent metal ions may prevent cyclization.[335,336] The enthalpies and free energies coupled to the formation of the peptide bond in the gas phase have also been evaluated.[338]

Metal ion promoted cleavage of the peptide bonds has been the subject of numerous studies for several decades. The most important observations obtained in this field were reviewed in 2001 and the role of several different metal ions, including palladium(II) and platinum(II),[4] copper(II) and nickel(II),[5,6] cobalt(III)[7] and lanthanide ions,[8] were treated separately. The most recent studies on the metal ion assisted amide hydrolysis mainly devoted to the selectivity of this process. It has already been reported that palladium(II) complexes promote the hydrolysis of oligopeptides with high regioselectivity especially in the vicinity of His and Met residues. The hydrolysis of the neurotransmitter peptide methionine enkephalin (Ac-Tyr-Gly-Gly-Phe-Met) and some of its His and Met derivatives was studied recently by the use of NMR, HPLC and MALDI-MS measurements.[339] In the case of enkephalin only the Gly(3)–Phe(4) bond was cleaved and similar data were obtained for the other peptides. In conclusion, the site of cleavage is the bond involving the amino group of the residue that precedes the Met or His side chains. Another important observation of this study is that the rate of cleavage is a function of the stoichiometry of the palladium(II) complex. The cationic complexes with labile ligands $[Pd(H_2O)_4]^{2+}$ and $[Pd(NH_3)_4]^{2+}$ were more reactive than those containing anionic ($[PdCl_4]^{2-}$) or bidentate ligands ($[Pd(en)(H_2O)_2]^{2+}$). The relationship between the palladium(II) anchoring and cleavage sites of peptides was systematically investigated in another study.[340] It was concluded that the substrate lacking an anchoring residue cannot be cleaved. When the anchoring residue is present in position 1, 2 or 3 from the N-terminus, the amide nitrogen atoms upstream from the anchor become deprotonated and form chelates with palladium(II), thus "locking" this ion and preventing its approach to the scissile amide bond. When the anchoring residue is present in a position beyond

3 in the sequence and no other coordinating site interferes, the cleavage occurs at the second amide bond upstream from the anchor.[340] Similar observations were reported for the palladium(II) catalyzed hydrolysis of various forms of albumin.[341] In exceptional cases, such as one of the cleavage sites of cytochrome c [His(18)–Thr(19)], the cleavage was observed at the first peptide bond downstream from the anchor.[340,342] The results obtained for the metal ion interaction of N-protected peptides of tryptophan revealed that palladium(II) and platinum(II) promote the cleavage of the peptide bonds next to Trp residues.[343,344] The overall mechanism for the platinum(II) catalyzed hydrolysis of tryptophanyl peptides is shown in Scheme 2.

In addition to the above mentioned examples the catalytic activity of palladium(II) complexes was demonstrated in the hydrolytic reaction of acetoxime to acetone.[345] On the other hand, theoretical studies have been performed for the understanding of catalytic mechanism of matrix metalloproteinases and on the mechanism of peptide hydrolysis by thermolysin.[346,347]

The hydrolysis of dipeptides (Gly-Gly, Gly-Leu, Leu-Gly and Gly-Ser) promoted by the copper(II) complex of the ligand *cis,cis*-1,3,5-triaminocyclohexane

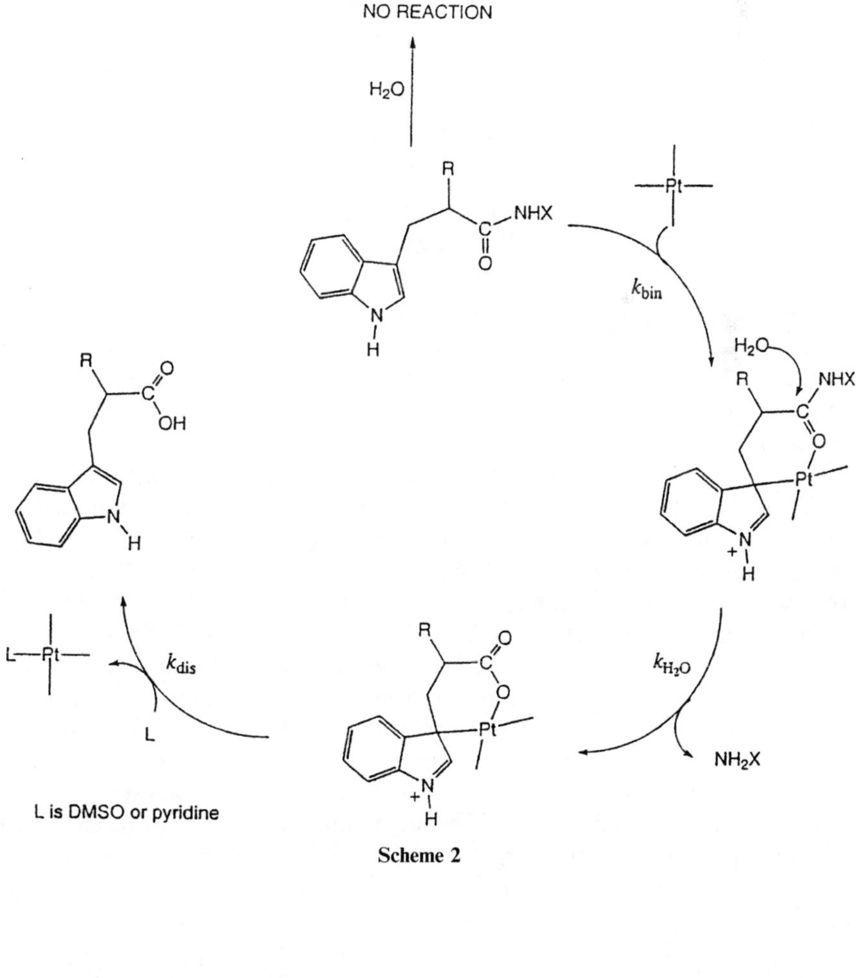

NO REACTION

H_2O

k_{bin}

k_{H_2O}

NH_2X

k_{dis}

L is DMSO or pyridine

Scheme 2

was investigated by HPLC method.[348] The kinetic parameters of the hydrolytic reaction, solution equilibria of the systems and solid state structure of the major species were determined. The amide hydrolysis of carboxyl-containing N-acyl amino acids was followed by the application of an artificial peptidase comprising a copper centre and a proximal guanidinium ion.[349] A dinuclear zinc(II) complex (**40**) containing the hydrogen bonded $(H_3O_2)^-$-bridge and the ligand [$Tp^{Me,Ph}$=hydrotris-(5,3-methylphenyl-pyrazolyl)borate] was isolated and its catalytic activity in peptide bond hydrolysis evaluated.[350]

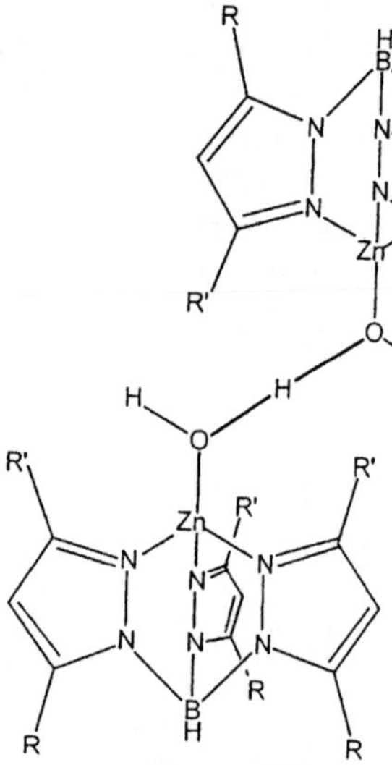

(40)

The hydrolysis of L-leucine anilide analogues containing either a carbonyl or thiocarbonyl group was studied in the presence of zinc and cadmium aminopeptidases. The enzyme containing zinc(II) was able to cleave both carbonyl and thiocarbonyl residues, while the cadmium analogue was effective only for the thiocarbonyl substrate.[351]

The catalytic role of different metal ions in these reactions makes it necessary to develop adequate experimental techniques for monitoring the hydrolytic reactions. A positional scanning approach was reported for the rapid screening of the sequence specificity of metal ion-assisted peptide hydrolysis.[352] The fragmentation of peptide complexes under mass spectrometric conditions is a useful technique for the determination of peptide sequences. A combination of mass spectrometric measurements with *ab initio* calculations was used to study the mechanism of the C-terminal residue cleavage in gas phase peptide-alkali metal ion complexes.[353] Fragmentation reactions of the dipeptides bonded in the $[Pt(L_3)M]^{2+}$-type complexes (where L_3 stands for triamines and M is dipeptide) were followed by electrospray ionization tandem mass spectrometric measurements. Different types of fragmentation reactions were observed depending on the nature and charge of coordinated ligands.[296]

The reactions of dipeptide complexes with dioxygen and/or the oxidation of peptides by inorganic oxidants are the subject of continuous interest. Raman and IR spectroscopic studies have been carried out on the cobalt(II)-carnosine (β-Ala-His) system and the various species formed in the presence of molecular oxygen structurally characterized.[354] Two dinuclear complexes containing a simple μ-peroxo bridge (**41**) and dibridged form (μ-peroxo,-μ-hydroxo) (**42**) have been identified as the major species at pH 9.

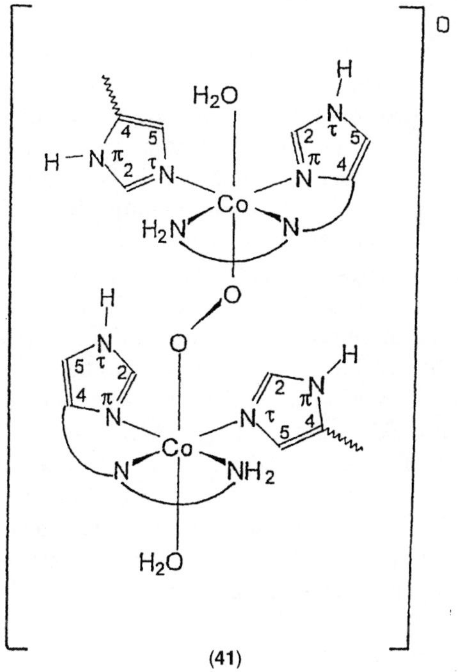

(**41**)

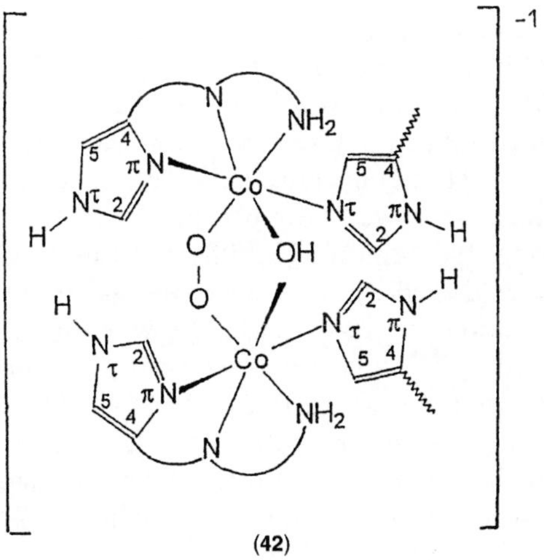

(**42**)

Several papers have been published on the oxidation of peptide molecules by manganese(III) salts.[355–358] The ligands involved several di-, tri- and penta-peptide analogues of elastine. Kinetic parameters of the reactions were determined and the oxidation products characterized. Chromium(V) complexes are considered as the active agents (or intermediates) in the Cr(VI)-induced carcinogenesis. Long-lived Cr(V)-complexes were obtained by the reduction of Cr(VI) with methanol in the presence of peptides. EPR measurements indicated that several different Cr(V)-peptide complexes were formed depending on the reaction conditions.[359]

The enzyme Cu,Zn-SOD had been treated with hydrogen-peroxide and the oxidized products fragmented with endoproteinases. Analysis of the resulting peptide mixture revealed that three out of the four Cu-coordinated His residues were specifically oxidized. The results demonstrate that copper-catalyzed oxidation of amino acid residues of proteins could be a versatile tool for the mass spectrometric determination of the copper binding sites of metalloproteins.[360]

The therapeutic effects and biological transport processes of platinum-containing anticancer drugs were extensively studied in the last three decades and this work was continued in 2001 and 2002. Most of these studies deal with the interaction of platinum compounds with nucleobases and derivatives, but several results on the peptide complexes have also been reported. The pH- and time-dependent reaction of *cis*[$PtCl_2(NH_3)_2$] with methionine- and histidine-containing peptides (Gly-Met, Gly-Gly-Met, Ac-His-Gly-Met and Ac-His-$(Ala)_3$-Met) was followed by HPLC and NMR techniques.[361] Thioether sulfur atoms were described as the primary metal binding sites, but the strong *trans*-effect of sulfur atoms resulted in the substitution of NH_3 by the tri- or tetra-dentate peptide ligands. The metal ion-induced amide hydrolysis and the formation of metastable macrochelates was also detected. In another study the reaction of [Pt(dien)Cl]Cl, [Pt(en)Cl_2] and cisplatin with hybrid molecules containing both N and S donor atoms was studied by various spectroscopic measurements.[362] Monofunctional Pt–S adducts were formed first and then transformed into Pt–N bonded complexes, in which guanine-N(7) was preferred. The reaction between monofunctional platinum(II) complexes ([Pt(Me_4-dien)Cl] and [Pt(Et_4dien)Cl] and the peptides Ac-Met-His and glutathione was investigated by the use of UV-Vis and NMR spectroscopies.[363] Interestingly, the binding of thioethers or imidazole-N donor functions was not detected in the first 24 hours of reaction, but coordination of imidazole-N(3) was observed on a longer time scale. However, both platinum complexes reacted easily with the thiol sulfur atom of glutathione. The differences in the kinetic parameters of the reactions were explained by steric effects of substituted dien ligands. The kinetics of the reaction of [Pt(digly)(H_2O)] with imidazole, pyrazole and triazole were studied by UV-Vis spectroscopy and the products were characterized.[364] The reaction of two different types of palladium(II) complexes ($[Pd(dien)Cl]^+$ and [Pd(Gly-Met)Cl]) with L-cysteine, DL-penicillamine and glutathione were studied by variable-temperature stopped-flow spectrophotometry. The reactivity of the thiol nucleophiles followed the sequence: DL-penicillamine < L-cysteine < glutathione.[365] The reactions of bifunctional palladium(II) complexes ($[Pd(en)(H_2O)_2]^{2+}$ and $[Pd(pic)(H_2O)_2]^{2+}$, pic=2-picolylamine) with N-alkyl-nucleobases and N-acetyl amino acids (Ac-Met, Ac-His) were followed by potentiometric and NMR measurements.[366] It was found that $[Pd(pic)(H_2O)_2]^{2+}$ had an enhanced tendency for thioether binding as compared to ($[Pd(en)(H_2O)_2]^{2+}$. On the other hand, the results revealed that monodentate binding of thioether ligands was kinetically preferred over the nitrogen donors, but the formation of [S(thioether),N(amide)] chelates was favored both kinetically and thermodynamically. The high affinity of palladium(II) for binding to imidazole-N of His residues made the application of the $[Pd(en)]^{2+}$ complex possible as a sequence-selective molecular pinch for α-helical peptides.[367] The $[Pd(en)]^{2+}$ complex was found to stabilize selectively the α-helix conformation of peptides having two His residues at *i* and $i + 3$ (or *4*) positions, while destabilization was observed with His in other positions. A new method for the facile preparation of monofunctionally *trans*-Pt^{II}-modified peptide nucleic acids (PNA) has been reported.[368] It gives the chance for the design of a new class of useful chemical peptide nucleic acid probes and antisense reagents.

An important application of chelating and other complexing agents is that they are potential enzyme inhibitors. A series of tripeptides containing C-terminal thiol residues was prepared and were found as the competitive inhibitors of the zinc metalloproteinase, thermolysin. The inhibition was explained by the coordination of the thiolate group to the catalytic zinc(II) ion and the ligand Z-Pro-Leu-cysteamine showed the highest inhibition.[369] Other peptides of cysteine were described as reversible competitive inhibitors of *Bacillus cereus* zinc β-lactamase.[370] The inhibition was explained by the substitution of hydroxide ion by the thiol function at the active site zinc(II). The synthesis of silanol and silanediol peptide analogues have been described in another report and their application for inhibition of angiotensin converting enzyme (ACE) evaluated.[371] N-Formyl hydroxyl amine and its analogues with alternative metal binding groups have been reported as potent peptide deformylase (PDF) inhibitors. The enzyme inhibition and antibacterial activity data revealed that the bidentate hydroxamic acid and N-formyl hydroxylamine structural motifs represent the optimum chelating groups.[372]

3.3 Solution Equilibria-Metal Ion Speciation of Peptide Complexes. – Stability constants of peptide complexes are generally determined by potentiometric measurements. The improvement of the computational methods in the past few years, however, has made it possible to apply various spectroscopic techniques for the simultaneous determination of stability constants and spectral parameters of peptide complexes. The analysis of EPR spectra obtained at various pH and ligand to metal concentration ratios yields the formation constants and the EPR parameters of each species, offering a rich source of information on the coordination modes of copper(II) complexes.[373,374] This method has been applied to the equilibrium and structural elucidation of copper(II) complexes of peptides containing seryl[373] and histidyl residues.[374] The major advantage of this technique is that it can provide information on the speciation and coordination modes even in the cases when the traditional methods, such as potentiometry or spectrophotometry, fail to detect or distinguish certain species. By pH-potentiometry, for example, it is difficult to show complexes formed without proton uptake or loss, the isomeric forms of the same species and the ratio of mono- and dinuclear complexes, while EPR can provide useful information on these questions.

The application of CD spectroscopy is another possibility for the simultaneous treatment of solution equilibria and structures of peptide complexes.[375,376] In this case the sign and intensity of CD spectra are well-related to the chirality and distance of the chiral center(s) in the ligand from the metal ion chromophore and the stability of the metal complexes. CD spectroscopic measurements were used for the equilibrium and structural characterization of copper(II) complexes of tripeptides containing alanyl residues in different positions[375] and for the copper(II), nickel(II) and palladium(II) complexes of tripeptides of methionine.[376] The results revealed that circular dichroism spectroscopy was an especially sensitive tool for the identification of weak charge transfer transition from Cu–S(thioether) bonds and the method could be used in highly diluted samples as compared to NMR measurements.

Steady-state luminescence investigations have been used to determine the binding constants of Eu(III) and Tb(III) ions with peptides. The peptides correspond to the putative calcium binding region of plant thionins.[377] In the case of some zinc(II) and cadmium(II) complexes of peptides theoretical calculations have been performed to understand the stoichiometry and coordination geometry of peptide complexes.[378] The influence of dielectric constants of dioxane-water mixture on the formation constants of copper(II)-triglycine complexes was evaluated on the basis of potentiometric measurements.[379] Both thermodynamic and kinetic parameters and characteristics of the electronic absoption spectra of the same copper(II)-triglycine system have been reported in another publication.[380]

Deprotonation and coordination of the peptide amide functions and the metal binding of the side chain residues are the two major factors influencing the coordination chemistry of peptide molecules. In the absence of side chain donor atoms the complexation with copper(II) is described via the coordination of the peptide backbone built up from fused 5-membered chelate rings involving nitrogen donor atoms. The inclusion of β-alanine into the peptide sequence results in the formation of 6-membered chelate ring. Copper(II) complexes of tri- and tetra-peptides containing β-alanine in all possible locations have been studied by potentiometric, UV-Vis and EPR spectroscopic methods.[381] The influence of the 6-membered chelate rings on the thermodynamic stability and spectroscopic properties of copper(II) complexes has been evaluated and the following stability order has been obtained for the $[NH_2,N^-,N^-,COO^-]$ binding mode of tripeptides: (5,6,5) > (5,5,6) > (5,5,5) > (6,5,5) > (6,5,6) > (5,6,6) > (6,6,6). The insertion of tetrazole rings $[-\Psi(CN_4)-]$ into the tetraalanine sequence is another way for the modification of the peptide backbone.[382] The presence of tetrazole rings slightly enhances the metal binding ability of the

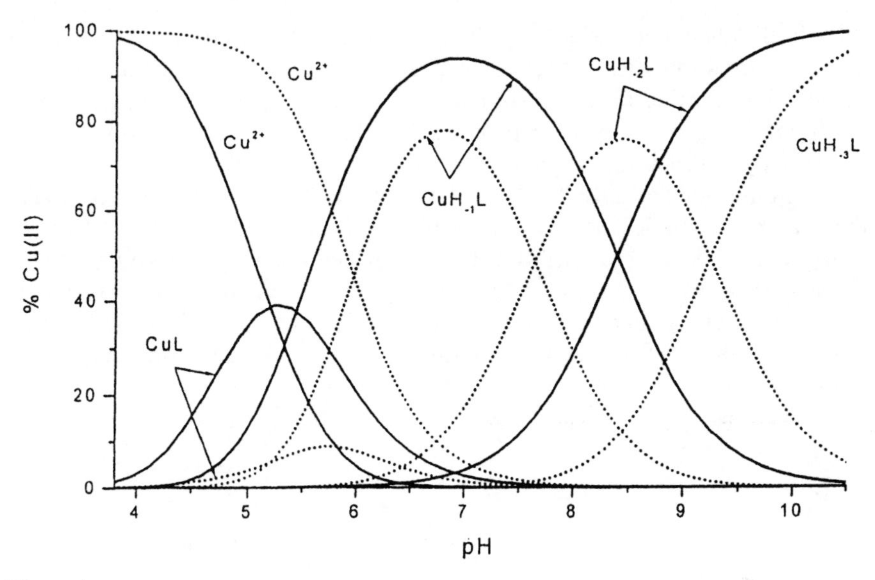

Figure 1

ligands as it is shown by Fig. 1 where the speciation of the Cu(II)-Ala-Ala [–Ψ(CN_4)–]Ala-Ala (solid line) and Cu(II)-tetraAla (dotted line) systems are compared.

The tetrazole moiety induces a specific peptide conformation which favours the formation of one or two stable chelating rings stabilizing a bent structure with 1N-coordination in [CuL] and 3N-coordination in [$CuH_{-2}L$] species.[382] α,β-Dehydro amino acids (Δ-amino acids) are the analogues of α-amino acids, with a double bond between the α- and β-carbon atoms. The studies on the binding ability of dehydro-tri- and tetra-peptides have shown that the α,β-double bond has a critical effect on the peptide coordination to copper(II) and nickel(II). The (Z)-isomer was found to be more effective in stabilizing the complexes formed and the stabilization was explained by steric and electronic effects.[383]

The effect of side chain residues on the stability of peptide complexes are the most studied with Cys and His residues, but peptides of these amino acids are discussed in the next paragraph. The role of alcoholic-OH (Ser and Thr), carboxylate (Asp and Glu), amino (Lys) and thioether (Met) functions have also been studied with various different metal ions. The interaction of Al(III) with a neurofilament heptapeptide fragment, Ac-Lys-Ser-Pro-Val-Val-Glu-Gly, was evaluated on the basis of potentiometric and NMR measurements and by molecular modelling calculations.[384] The data confirm the metal ion coordination of the carboxylate residues of Glu and Gly and suggest that aluminium(III) may play an important role in the aggregation of neuropeptides. The complexation of copper(II) with 2-(α-hydroxyethyl)thiamin pyrophosphate (HETPP) and the pentapeptide Asp-Asp-Asn-Lys-Ile has been studied in solution to obtain structural information on the active site of thiamin-dependent enzymes.[385,386] It has been suggested that the peptide offers three coordination sites (amino, Asp-COO^- and amide-N^-) at physiological pH and the coordination sphere of the metal ion is completed by two additional phosphate oxygen and nitrogen N(1′) of thiamin coenzyme. Potentiometric and spectroscopic studies have been performed on the dimethyltin(IV) complexes of 2-hydroxyhippuric acid (SalGly).[207] The formation of $[MH_{-1}L]$ was detected as the major species in the physiological pH range and a distorted trigonal bipyramidal geometry of tin(IV), containing tridentate (O^-,N^-,COO^-) bonded ligands with two equatorial methyl groups, was suggested (**43**). The data provide further argument for the dimethyltin(IV)-induced amide deprotonation in peptides and confirm that the C-terminal carboxylate is the primary anchor in this process.[93]

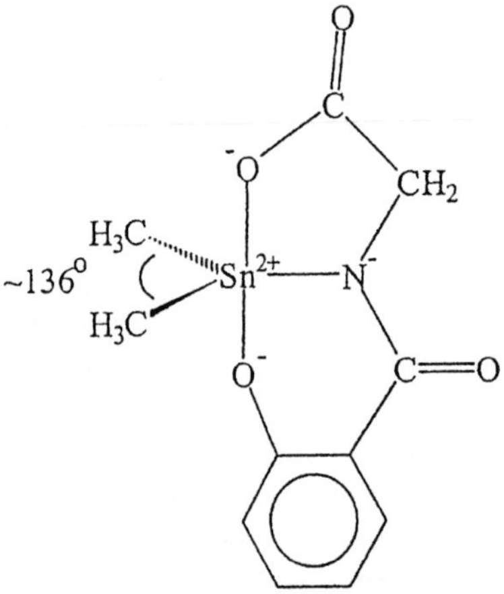

(43)

The acid-base properties and oxovanadium(IV) and copper(II) complexes of the peptide analogues, N-salicyl-Gly-Gly and N-salicyl-Gly-Gly-Gly, have been studied by potentiometric, UV-Vis and EPR measurements.[387] The amino-phenolate chelation was described as the primary anchor resulting in the formation of [ML] and $[ML_2]$ parent complexes. Deprotonation and metal ion coordination of the amide functions was also observed in all systems, but interestingly this process was especially favoured for oxovanadium(IV) with the dipeptide ligand. Alkylphosphinic acid analogues of Gly-Gly have been prepared and their complexation studied with Co(II), Ni(II), Cu(II) and Zn(II).[388] The formation of $[CuH_{-2}L_2]$ complexes with bis[NH_2,N^-(amide)] coordination was favoured for the phosphinic ligands. In the case of the other three metal ions the species [MHL], [ML] and $[ML_2]$ were detected. The metal ion coordination of the phosphinate group was suggested in the protonated complex, while (NH_2,CO) peptide-like coordination was reported for the parent complexes.

The ε-amino group of lysyl residue has a significant impact on the acid-base properties of the ligands, but its metal ion coordination is generally controversial. The nickel(II) complexes of dipeptides containing Lys residues have been studied by potentiometric, calorimetric, UV-Vis and CD spectroscopic techniques.[389] The participation of amide nitrogen in complex formation was suggested, while there was no evidence to support the metal binding of the ε-amino group. Potentiometric and spectroscopic measurements were performed on the copper(II), nickel(II) and zinc(II) complexes of two peptides, Gly-Lys(Gly) and Asp-ε-Lys, which contained the ε-amino group of lysine in the amide bond.[390] The stoichiometry of the major species formed in the copper(II)-Gly-Lys(Gly) system is $[CuH_{-1}L]$ and the EPR spectroscopic data indicated the existence of two isomeric forms of this complex (**44**).

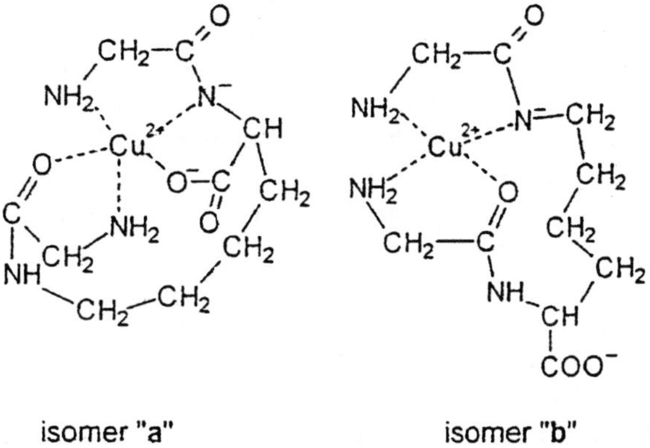

isomer "a" isomer "b"

(**44**)

The corresponding nickel(II) and zinc(II) complexes have been characterized by the formation of a remarkably stable mono(ligand) complex [ML] containing a bis(NH_2,CO) chelate and a "loop" around the metal ions (**45**).

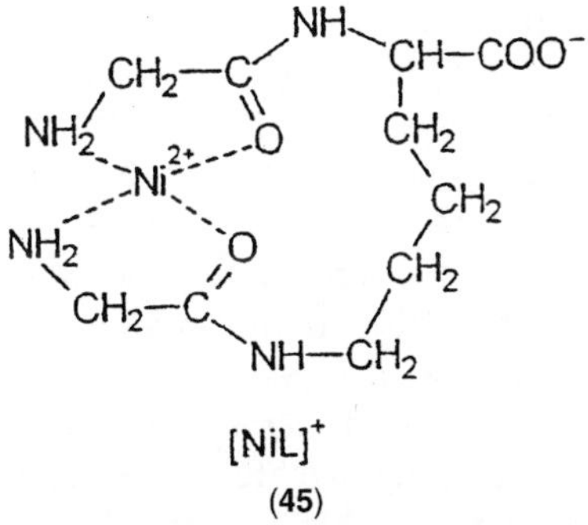

$[NiL]^+$

(**45**)

The metal binding sites of the ligand Asp-ε-Lys are well-separated and it results in the formation of various ligand-bridged dinuclear complexes.[390]

Complex formation between palladium(II) ions and the peptides Gly-Met, Gly-Met-Gly and Gly-Gly-Met was followed by potentiometric and ^1H NMR methods.[391] The complex $[PdH_{-1}L]$ was suggested as the major species for Gly-Met and Gly-Met-Gly containing [NH_2,N^-,S(thioether)]-coordination mode. The fourth coordination site of the metal ion in these complexes can be occupied by H_2O molecule, Cl^- or OH^- ions depending on the reaction conditions or by a second ligand molecule in the presence of excess of peptide. The second ligand binds monodentately via the thioether function in acidic media, but a migration to amino group coordination occurs in neutral solution. $[PdH_{-2}L]$ with $[NH_2,N^-,N^-$, S(thioether)]-coordination was obtained as the major species in a wide pH range and metal to ligand ratio in the Pd(II)-Gly-Gly-Met system. The formation of thioether bridged mixed metal complexes was also detected in the reaction of $[Pd(dien)]^{2+}$ with $[Cu(Gly\text{-}MetH_{-1})]$ or $[Ni(Gly\text{-}Met\text{-}GlyH_{-2})]$.[391]

Dihydroxamic acids are very effective chelating agents and the hydroxamate functional groups of these molecules are generally separated by carbon chains of different length including amide bonds. Three new dihydroxamic acids containing the peptide groups in different positions between the two functional groups have been synthesized and their complexation studied with Fe(III), Mo(VI) and V(V).[392] Although the termodynamic stability of the monochelated complexes formed with all three dihydroxamic acids were very similar, the bis-chelated complexes were favored for 2,5-DIHA, which supports the idea that this ligand has a proper preorganization for the coordination in octahedral complexes with metal ions having similar ionic radii to iron(III).

Peptide nucleic acids are synthetic analogues of DNA in which the natural phosphate-deoxyribose backbone is replaced by a peptide chain. In the case of chiral peptide nucleic acids (cPNA) the nucleobases are located in the side chains of peptide residues. Copper(II) and nickel(II) complexes of longer cPNA oligomers having 4, 6 and 8 thymines in the side chain were studied by potentiometric and spectroscopic techniques.[393] The result demonstrated that cPNA molecules were much more effective ligands for metal binding than the simple oligopeptides. The enhanced stability of the complexes was interpreted by the assumption that the nucleobase residues can interact with each other as well as with the peptide backbone.

Sulfonamides are also among the biologically active derivatives of peptides. Combined potentiometric and spectrophotometric measurements have been carried out on the copper(II) complexes of *N*-benzenesulfonyl derivatives of simple dipeptides built up from Gly, α- and β-Ala and Met residues. Metal ion coordination of the amide, sulfonamide and carboxylate functions was suggested and the differences in the equilibrium data of the complexes were explained by the different sizes of chelate rings and by metal to ligand and ligand-ligand interactions.[394] Mixed ligand complexes of the same sulfonamides have also been studied using acetate, 2-methyl-benzimidazole, glycinate and bipyridine as secondary ligands.[395] Equilibrium data and solid state structures on the mercury(II) complexes of *N*-carbonyl and *N*-sulfonyl amino acids have been reported.[396] All molecules behaved as simple carboxylate ligands at acidic pH, while a dianionic (N^-,O^-) chelating mode was suggested around neutrality with the involvement of amide nitrogen in metal binding. The molecular structure of the ternary complex $[Hg(bpy)_2(psgly\text{-}N,O)]$ (bpy=2,2′-bipyridine and psgly=*N*-2-nitrophenylsulphonylglycine) was determined by X-ray crystallography (**46**). In the complex, mercury(II) ion shows a distorted octahedral environment with N_5O donor set, in which 4 nitrogen atoms derive from bpy ligand, while the oxygen and the fifth nitrogen come from the psgly dianion.[396]

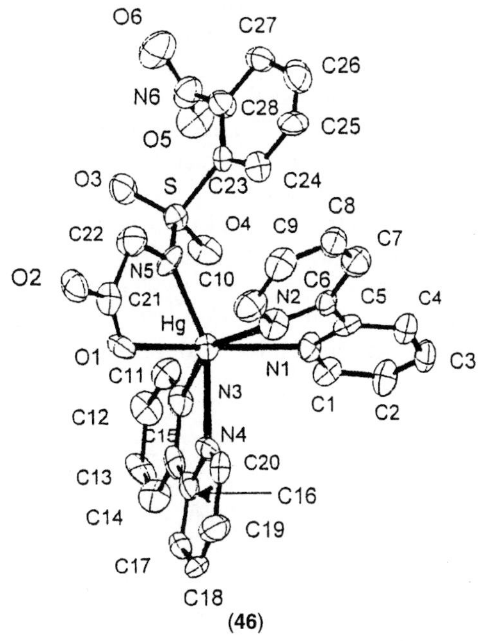

(46)

The factors influencing mixed ligand complex formation of peptides with biologically important mono- and bidentate ligands have also been extensively studied.[397–401] Solid state structures and thermodynamic stabilities of several ternary [Cu(II)(dipeptide)(diamine)] complex have been reported.[397] A distorted square pyramidal coordination geometry of the five-coordinate ternary complexes was suggested and the stabilization of the mixed ligand systems was explained by attractive ligand-ligand interactions involving the aromatic rings of the molecules.

Stability constants of the ternary complexes of copper(II) with picolylamine and amino acids or peptides have been determined and the effectiveness of these complexes in the hydrolysis of amino acid esters has also been tested.[398] Equilibrium data were reported on the mixed ligand mixed metal complexes of copper(II), nickel(II) and zinc(II) with Gly-Val and imidazole.[399] The formation of imidazolato bridged dinuclear complexes was suggested, which can be considered as promising models of the Cu,Zn-SOD enzyme. The stability constants of the complexes formed in the reaction of $[Pd(Me_2en)(H_2O)_2]^{2+}$ with various biologically relevant ligands, including peptides, have been determined and the binding modes of the major species discussed.[400]

3.4 Metal Complexes of Peptides Containing Histidyl and Cysteinyl Residues. – The imidazole nitrogen atom of histidyl and the thiol sulfur atom of cysteinyl residues are the most common binding sites in metalloproteins. The natural examples of these binding modes include various redox enzymes (such as Cu,Zn-SOD), electron transfer proteins, metalloproteins, zinc finger peptides and most recently the prion proteins and amyloid peptides. The outstanding biological significance of these biopolymers has attracted very wide and continuously increasing research in the field of metal complexes modelling the metal-imidazole and metal-thiol interactions. These studies generally cover all aspects of the coordination chemistry of these ligands, including the solution equilibria, kinetics and structural characterization of the complexes. As a consequence, all results obtained for the metal complexes of peptides containing histidyl and/or cysteinyl residues are discussed in this separate sub-section.

The metal complexes of simple di- and tri-peptides of histidine are still the subject of research, although a great deal of information has already been published in this field in the past few decades. The $H^+/H_2VO_4^-/H_2O_2$/Ala-His system was followed by potentiometric and ^{51}V NMR measurements.[402] Several mononuclear complexes containing both peroxo and peptide ligands were detected and a bis(peroxo) $[VO(O_2^{2-})(Ala\text{-}His)]^-$ complex (**47**) was found to predominate among them.

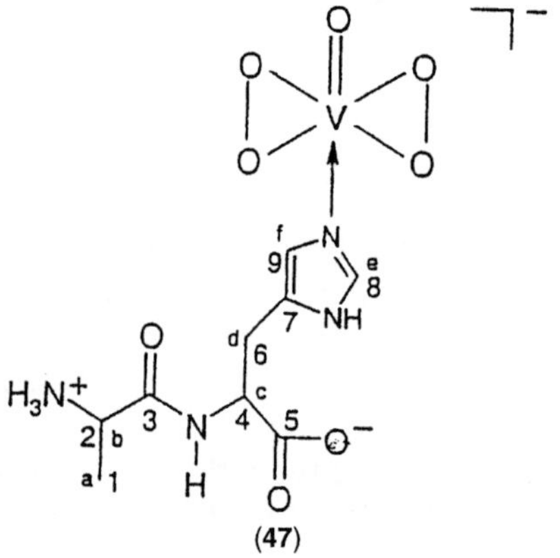

(**47**)

The complexes formed in the reaction of oxovanadium(IV) with carnosine (β-Ala-His) were studied by spectrophotometric and potentiometric techniques.[403] It was found that high excess of carnosine was able to suppress hydrolytic reactions of oxovanadium(IV) ions and the formation of $[VO(HL)_2]^{2+}$ was suggested as the major species at physiological pH. ESI-MS measurements were used for the elucidation of the speciation of supramolecular assemblies formed between equimolar amounts of carnosine and copper(II) or zinc(II) ions in diluted aqueous solution. The ESI-MS method was suggested as a very fast and effective tool to obtain stoichiometric information on the metal complexes formed at very low concentrations.[404] The copper(II) complexes of oligopeptides with protected amino groups (Z-His-OH, Z-His-Gly-OH, Z-Gly-His-OH and Z-Gly-Gly-His-OH) were studied by potentiometric and UV-Vis and EPR spectroscopic techniques.[405] The imidazole-N donor atoms were the exclusive metal binding sites in slightly acidic media. The deprotonation and coordination of the amide nitrogen atoms from the carbamate or peptide amide groups occured around pH 7 and resulted in chelation in the form of 2N (Z-His-OH and Z-His-Gly-OH), 3N (Z-Gly-His-OH) and 4N (Z-Gly-Gly-His-OH) complexes, supporting the anchoring ability of the imidazole ring. The tripeptide Gly-His-Lys has already been described as a liver cell growth factor and it was found to act synergistically with copper and iron. The copper(II) complexes of the tripeptide and its synthetic analogues were examined by various potentiometric and spectroscopic techniques and the biological activity of the natural and synthetic peptides was also tested.[406]

The active site of the enzyme Cu,Zn-superoxide dismutase (Cu,Zn-SOD) consists of copper(II) and zinc(II) ions bridged via a doubly deprotonated imidazolato ring. The tripeptide His-Val-His and the tetrapeptide His-Val-Gly-Asp can be considered as the simplest models for the copper(II) and zinc(II) binding sites, respectively. The interaction of copper(II) with the former and zinc(II) with the latter has been studied by potentiometric and spectroscopic methods.[407] Histamine-like coordination of the N-termini was suggested as the primary ligating mode in both systems, but in the case of the tripeptide the metal binding was supported by the coordination of the C-terminal His residue in the form of a macrochelate. The deprotonation and coordination of the amide functions was reported to occur in the Cu(II)-His-Val-His system and the major species formed by pH 8 was characterized by the [NH_2,N^-,N^-,N(Im)] albumin-like coordination. The terminally protected tetrapeptide Ac-His-Gly-His-Gly is another simple model for the active site of Cu,Zn-SOD. The interaction of copper(II) with the tetrapeptide was elucidated by the means of potentiometric and spectroscopic measurements.[408] The N(3) donor atoms of the histidyl side chains were described as the first anchoring sites of the ligand, while the deprotonation and coordination of the amide functions was suggested in slightly alkaline solution.

The N-terminal regions of human serum albumin (HSA, Asp-Ala-His-Lys-) and bovine serum albumin (BSA, Asp-Thr-His-Lys-) are known to provide specific binding sites for copper(II) and nickel(II) ions, with the third His residue thought to be mainly responsible for the specificity. It was found that the insertion of α-hydroxymethylserine (HmS) into these sequences further enhanced the metal binding ability of the ligands and tripeptide analogues HmS-HmS-His-OH/NH_2 were found to be the most effective albumin-like chelating agents.[409] To get further insight into the metal binding ability of serum albumin the copper(II) complexes of small model peptides (His-Lys, Ac-Thr-His-Lys, and Thr-His-Lys) have been studied.[410] The interaction of cobalt(II) ions with HSA and BSA was followed by UV-Vis spectrophotometry and a cooperativity in the cobalt binding affinity was observed.[411]

Secreted Protein, Acidic and Rich in Cysteine (SPARC) is a matricellular calcium binding protein consisting of 286 amino acids. All cysteines of the protein are present in disulfide form, but the molecule contains also two copper(II) binding sites, the strongest of which corresponds to the sequence

Lys-Gly-His-Lys ($SPARC_{120-123}$). Copper(II) and nickel(II) complexes of this tetrapeptide have been studied by potentiometric and spectroscopic measurements.[412] The complex [CuL] was described as the major species at physiological pH containing the albumin-like [NH_2,N^-,N^-,N(Im)] binding mode with two un-coordinated and protonated Lys residues. Copper(II) complexes of the terminally protected pentadecapeptide fragment ($SPARC_{114-128}$, Ac-TLEGTKKGHKLHLDY-NH_2) have also been studied.[413] The two histidyl residues were suggested as the primary metal binding sites of the peptide. The deprotonation and coordination of the amide nitrogens was observed around physiological pH and [2N(Im),$2N^-$] and [N(Im),$3N^-$] coordination modes were suggested in slightly and strongly alkaline solutions, respectively. The side chain donor atoms of Tyr and Lys residues were not involved in metal binding.

Histones are highly basic proteins that provide a scaffold for the DNA double helix in the cell nucleus. The carcinogenecity of nickel(II) compounds promoted the studies on the metal complexes of the oligopeptide segments of histones. Copper(II)[414] and nickel(II)[415] complexes of terminally protected hexapeptides (TESHHK, TASHHK, TEAHHK, TESAHK and TESHAK, modelling the ESHH motif of histone H2A) were studied by potentiometric, UV-Vis, CD, EPR and NMR spectroscopic measurements. The His imidazole-N donor atoms were identified as the primary ligating sites for both metal ions, but this type of coordination was followed by amide deprotonation and coordination by increasing pH. Metal ion coordination of the lysyl residues was ruled out. It is noteworthy that a nickel(II)-assisted hydrolysis of the peptides was observed in basic solution. The hydrolytic cleavage occurred at the Glu-Ser amide bond of TESHHK and required also the Ser(3) and His(5) residues in the hexapeptides.[415] In the copper(II)-TESHHK system [$CuH_{-1}L$] was reported as the major species at physiological pH, in which copper(II) was coordinated equatorially through imidazole of His(4) and amide functions of Ser(3) and His(4) and axially via imidazole of His(5). It was demonstrated that this complex reacted easily with H_2O_2 and the resulting reactive oxygen intermediates could have caused oxidative damage of 2′-deoxyguanosine.[416] The nickel(II) and copper(II) binding of the N-terminal tail of histone H4 (a 22-mer peptide, Ac-SGRGKGGKGLGKGGAKRHRKVL-NH_2) and its 7- and 11-amino acid derivatives with blocked lysyl residues (Ac-AK(Ac)RHRK(Ac)V-NH_2 and Ac-GK(Ac)GGAK(Ac)RHRK(Ac)V-NH_2) was also investigated.[417] Imidazole-N of His(18) was suggested to act as the anchoring site for both metal ions with all ligands. Amide deprotonation occured above pH 6 and the major species were described as 4N complexes. Although the nickel(II) binding ability of the 22-mer peptide was less effective than the copper(II) binding, the different conformation of the peptide complexes may suggest a specificity for nickel(II) binding. The tetradecapeptide (Ac-TRSRSHTSEGTRSR-NH_2) representing the C-terminal part of the nickel(II)-induced Cap43 protein was analyzed for nickel(II) and copper(II) binding.[418] The coordination modes of the two metal ions were rather similar, the imidazole-N of His being the primary metal binding site and followed by successive deprotonation of three amide functions. Interestingly, the species $[MH_{-4}L]$ was also detected and the extra proton loss was attributed to the guanidino moiety of arginyl residue. CMT-I (*Cucurbita maxima* trypsin inhibitor-I) is an oligopeptide consisting of 29 amino acids with 3 disulfide linkages, providing a high stability and rigidity of the molecule. The complexation of copper(II) with this squash trypsin inhibitor was monitored by potentiometric and spectroscopic techniques.[419] His(25) imidazole was suggested as the anchoring site and followed by the $[N(Im),N^{-}(His25),N^{-}(Glu24)]$ binding mode, which predominates at physiological pH.

Human prion diseases are characterized by the conversion of the normal prion protein (PrP^{C}) into a pathogenic isomorf (PrP^{Sc}). The polypeptide chains of PrP^{C} and PrP^{Sc} are identical in amino acid composition, differing only in their three-dimensional conformation. An increasing number of experimental data support the idea that prion proteins participate in the regulation of copper(II) and possibly modulates the concentration of reactive oxygen species. These observation gave a big impetus to the studies on the complex formation reactions of metal ions with various forms of prion proteins[420–425] and their oligopeptide segments.[426–429] MALDI-TOF MS measurements provided further evidence that the octarepeat domain in the N-terminal region (encompassing residues 60–91 in human PrP^{C}) was the major copper(II) binding site where the His residues of the 4 octapeptide units (PHGGGWGQ) were able to bind 4 copper ions.[420] Moreover, the binding of a fifth copper(II) was suggested at His(95). The metal binding capability of murine prion protein have been monitored by EPR spectroscopy. It was found that the N-terminal region binds copper(II) only at pH values above 5.0, whereas three different coordination types of copper(II) were observed in the C-terminal domain even at low pH values.[421,422] Other studies on the complexation of copper(II) with prion proteins in the presence of oxidants revealed that the proteins could be particularly susceptible to metal catalyzed oxidation.[423,424] It has also been demonstrated that a site-specific cleavage of the octarepeat region is induced by reactive oxygen species. The cleavage was both copper(II) and pH dependent, but was retarded in the presence of other divalent metal ions.[424] On the other hand, it has been demonstrated that the elevation of the level of manganese and, to a lesser extent of zinc, is accompanied with a significant decrease of copper bound to prion proteins. These results suggest that the altered metal ion occupancy of PrP plays a pivotal role in the pathogenesis of prion diseases.[425]

Crystal structure of the copper(II) complex (**48**) of Ac-His-Gly-Gly-Gly-Trp-NH_2 representing a pentapeptide fragment of the octarepeat has been determined recently.[426] Histidine imidazole, two deprotonated glycyl amide nitrogens and glycyl carbonyl oxygen were found as the equatorial binding sites with an axial water bridging to Trp indole residue. EPR measurements revealed that all forms of the protein (the pentapeptide, the octapeptide and the whole PrP itself) had the same binding mode in solution.

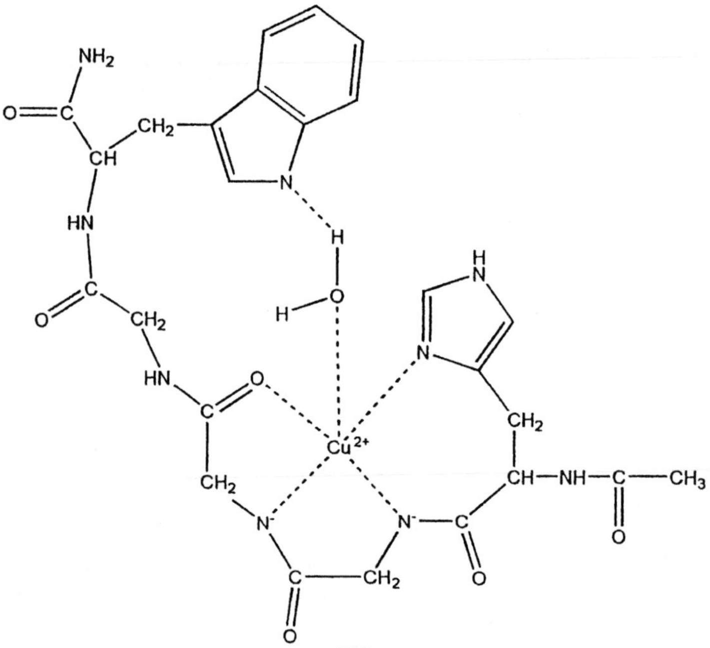

(48)

Similar observations were obtained from the potentiometric and spectroscopic measurements performed in the copper(II)-Ac-PHGGGWGQ-NH_2 system, but the equatorial binding of a fourth amide nitrogen atom was suggested in more basic solution.[427] One of the most interesting and unusual features of the metal binding centre of the octarepeat domains of PrP is that the binding of His imidazole promotes the ionization of amide functions on the C-terminal side of the His residue resulting in the formation of a 7-membered chelate ring. Copper(II) and nickel(II) complexes of the terminally free dodecapeptide consisting only of His and Gly residues (HG12, HGGGHGHGGGHG) have been studied by potentiometric and spectroscopic measurements.[428] Although the molecule doesn't have a natural amino acid sequence it can be a useful model either for prion proteins or Cu,Zn-SOD enzymes. The histamine-like coordination of the N-terminus was described as the primary ligating site followed by the amide deprotonation and coordination with increasing pH. The binding modes of the various species are shown on Scheme 3, where the species *a*, *b*, *c* and *d* were obtained at pH 5 and 10 in equimolar solution and 7 and 10 at 2 : 1 metal to ligand ratios, respectively.

The peptide PrP(106-126) and its analogues have been synthesized and their reactions studied with copper(II) and zinc(II). The results revealed that PrP(106–126), a peptide model for PrP^{Sc} toxicity, interacted with metal ions to form neurotoxic amyloidogenic structures. His(119) and Met(109) or (112) were suggested as the major metal binding sites of the peptides.[429]

(a) (b) (c) (d)

Scheme 3

Amyloid-*β* peptide (A*β*) is a normally soluble peptide found in all biological fluids, but it accumulates as the major constituent of the extracellular deposits that are pathologic hallmarks of Alzheimer's disease. These peptides also contain histidyl residues and their complex formation reactions with copper(II),[430–434] zinc(II)[430–431,435] and aluminium(III)[436,437] have been widely studied. It was found that A*β* easily binds copper(II) and zinc(II), inducing aggregation and giving rise to reactive oxygen species.[430] Metal ion coordination of His(6), His(13) and His(14) residues was suggested in both aqueous solution and lipid environments. Binding of a second copper(II) took place in a cooperative manner suggesting the involvement of bridging imidazole residue, similarly to the active site of Cu,Zn-SOD. In agreement with this finding the treatment with a quinoline-type copper(II)-zinc(II) chelator markedly and rapidly inhibited *β*-amyloid accumulation in Alzheimer's disease transgenic mice.[431] Copper(II) complexes of the N-terminal fragments (1-6, 1-9 and 1-10 amino acids) of human and mouse *β*-amyloid peptides have been studied by potentiometric and spectroscopic techniques.[432] The coordination modes of all peptides were rather similar, the N-terminal Asp being the primary ligating site supported by the coordination of imidazole-N of His(6) in the form of a macrochelate in a wide pH range (4–10). The metal ion coordination of the amide functions from His(6) and Ala(2) was suggested for the mouse and human fragments, respectively. This difference in the binding mode resulted in the enhanced metal binding ability of the mouse peptide as shown by Figure 2, where the distribution of copper(II) between the mouse (solid line) and human (dotted line) decapeptides are plotted.[432]

The results obtained for the copper(II) complexes of the 11–20 and 11–28 fragments of human and mouse *β*-amyloid peptides have also been reported.[433] The fragments contain histidyl residue in position-3 (His-13 in the native peptide) which corresponds to the effective albumin-like binding site

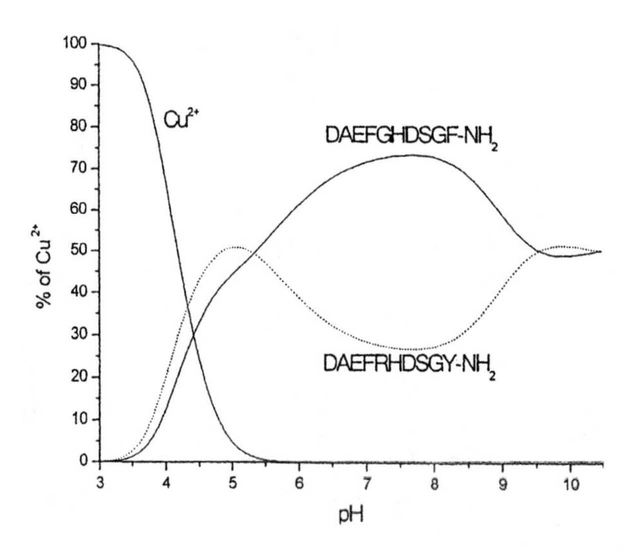

Figure 2

determining the coordination ability of the ligands. The pentapeptide fragment (Ac-His-Leu-His-Trp-His-NH_2) of βA4 amyloid precursor protein was synthesized and its complexation with copper(II) monitored. Previous studies on the protein suggest that it binds copper(II) very efficiently and then reduces them to copper(I) producing hydrogen peroxide. The results obtained for the copper(II) complex of the pentapeptide fragment situated in the cysteine-rich region (Cys144 and Cys158) of the protein indicate that the peptide also has a high affinity for copper(II) binding. The formation of a 3-imidazole coordinated species (**49**) was suggested at physiologically relevant pH and these data confirm that the His residues of the cysteine-rich region are responsible for metal binding.[434]

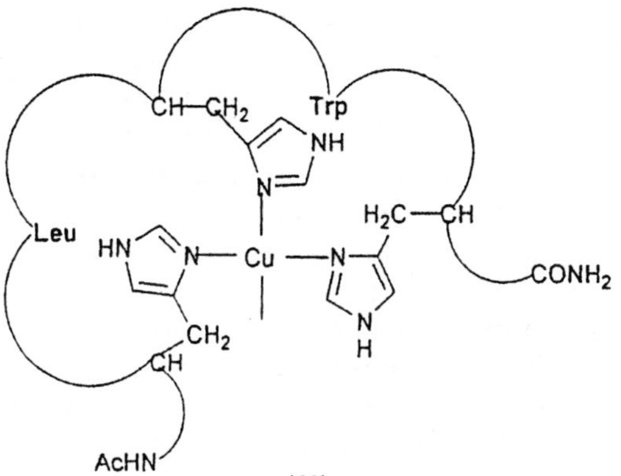

(**49**)

Aluminium(III) is often considered as a potential risk factor for Alzheimer's disease. The biological functions of the pancreatic polypeptide neurohormones, neuropeptide Y (NPY) and peptide YY (PYY) have been reviewed and it was suggested that the peptides could have a role in protecting the central nervous system from the effects of aluminium(III) toxicity.[436] In another study the promotion and formation of amyloid fibrils by ATP and, in particular, by its aluminium(III) complex has been reported.[437]

Both catalytic and structural zinc(II) ions in proteins are always coordinated by at least one His or Cys residues and this is especially true for the zinc finger peptides containing only these residues as metal binding sites. As a consequence, the zinc(II) complexes of His/Cys containing peptides are the subject of continuous interest. The stability constants formed in the reaction of Zn(II) with His-His in both free and terminally protected forms were determined by potentiometric measurements.[438] This work completed previous studies on similar peptides and the comparison of equilibrium data revealed some systematic trends. Among them it is important to note that the complex stability provided by exclusive Zn(II)–N(His-imidazole) coordination is usually low. The situation becomes better when additional donors, especially the terminal amino group or a Cys residue, of the peptides are involved in coordination. The cyclic dipeptides c-His-His, c-Gly-Cys, c-His-Lys and c-Cys-Cys have been prepared and their complexation studied.[439] Contrary to expectations the preorganization of the peptide donors does not enhance the complex stabilities in comparison to those of the non-cyclic peptides. The interaction of zinc(II) with dipeptides including Cys-Gly, His-Ala and His-Lys has been followed by potentiometric and ^{1}H NMR measurements.[440] It was found that the dipeptides acted as bidentate ligands in both the binary and ternary systems. A series of zinc finger mutant peptides corresponding to the second zinc finger domain of the human transcription factor (Sp1) have been synthesized and their complexation with zinc(II) and cobalt(II) studied.[441] Fluorescence emission studies revealed that the mutant peptides were capable of binding zinc despite removing one ligation site, while CD results clearly showed the induction of an α-helix by zinc(II) binding. A new 22-mer model peptide (P22), containing His residues at positions 10 and 14, was designed for maximal α-helicity in non-polar media. The peptide formed 1 : 1 adducts with zinc(II) halides and the histidyl residues were suggested as the metal binding sites.[442]

The copper(II) and cobalt(II) binding properties of two peptides, designed on the basis of the active site sequence of plastocyanin, have been explored.[443] The peptides contained three and four binding sites of the natural blue copper proteins, respectively. It was found that both peptides form 1 : 1 complexes with copper(II) indicating that the plastocyanin ligand loop could act as a metal binding site even in the absence of the remainder of folded protein but, by itself, could not stabilize the blue copper site. A mixed valence copper complex comprising a bridging thiolate/N donor ligand that models the CuHis–CysCu motif, found in nitrite reductase and multicopper oxidases, have been prepared and structurally characterized. The retention of dicopper structure was confirmed in solution providing sufficient base for studies on electron transfer or catalytic activity.[444]

Zinc(II) complexes of di- and tripeptides containing two Cys residues have been studied by potentiometric methods in solutions and prepared in the solid state.[445] ZnL and ZnL_2 were obtained as the major species with 2S and 4S coordination modes respectively, but Zn_2L_4 and Zn_3L_4 polynuclear complexes were also detected in some cases. 2-Mercaptopropionylglycine (MPG) is the simplest ligand containing both thiol and amide functions and widely used to study the anchoring ability of the thiol group. In the case of VO(IV)-MPG system the thiolate group proved to be an efficient anchor function which binds oxovanadium(IV) strongly and promotes the deprotonation and coordination of neighbouring amide group.[446] Ternary complexes of oxovanadium and MPG have also been studied with various N- and O-donor ligands. Mixed ligand complex formation was reported to be favoured with chelating (N,N) and disfavoured with (O,O) ligands.

Several papers have been published on the VO(IV)-glutathione system.[447–449] Two papers deal with the speciation of the system and plausible structures of the various species were suggested from spectroscopic measurements.[447,448] The results obtained for the VO(IV)-oxidized glutathione system revealed that this ligand binds VO^{2+} ions with an average efficiency and only glycine-type coordination was observed.[449] Stability constants and thermodynamic parameters have been determined for the glutathione complexes of a series of divalent metal ions[450] and the following stability order was obtained: Cu(II) > Pb(II) > Ni(II) > Zn(II) > Co(II) > Cd(II) > Mn(II) > Mg(II) > Ca(II). Glutathionyl cobalamin is a natural product which functions as an intermediate in the biosynthesis of the active B_{12} coenzymes adenosylcobalamin and methylcobalamin. To understand the outstanding stability of the glutathionyl adduct, two dipeptide derivatives (γ-Glu-Cys and Cys-Gly) of cobalamin were prepared. γ-Glu-Cys formed a stable adduct, while the other peptide, similarly to cysteine itself, formed an unstable adduct with cobalamine suggesting the role of the γ-amide bond in the stabilization.[451] The structures of phytochelatins are related to glutathione, but they are produced by plant, algae and fungi. Various cysteine-rich peptides ($(X\text{-}Cys)_7$-Gly, X=Glu, Asp, Lys, Gly, Ser and Gln) were synthesized and their metal binding ability and detoxification effect towards cadmium(II) analyzed.[452] It was concluded that the presence of thiol and carboxylate functions was essential for the formation of tight cadmium(II)-peptide complexes. The complexation between cadmium(II) and $(\gamma\text{-Glu-Cys})_2$-Gly was followed by a differential pulse voltammetric study, too.[453]

The hexapeptide fragment of mouse liver metallothionein (56-61, Lys-Cys-Thr-Cys-Cys-Ala) has been prepared and its complexation with cadmium(II) studied by electrochemical and EXAFS methods.[454,455] The formation of complexes with different stoichiometry was proposed and the applicability of the various experimental techniques discussed. ESI-MS results were reported for the binding of copper(I) to recombinant human metellothionein.[456] The data provide the first evidence for Cu_9S_9 and $Cu_{11}S_{11}$ diagonal clustering in the β- and α-domains respectively. The amino acid sequence CXXC is present in many metal binding proteins of the bacterial mercury detoxification system, MerP and MerA. Three 18-residue peptide fragment of MerP, containing CAAC, CACA and CCAA metal binding sites have been synthesized and their complexation studied.[457] It was found that the position of Cys residues largely affects the specificity of metal binding, in particular, the peptide with vicinal cysteines binds only mercury.

The thiol-disulfide redox couple has an outstanding biological significance. Oxidized glutathione (GSSG) is able to oxidize metallothionein (MT) with concomitant release of zinc(II), while glutathione (GSH) can reduce the oxidized protein to thionein, which then binds zinc(II). It has been shown that selenium compounds catalyze the oxidation of MT even under overall reducing conditions.[458] In this manner, the binding and release of zinc(II) is linked to redox catalysis by selenium compunds as shown by Scheme 4. The design and synthesis of agents that can abstract zinc from zinc finger modules (CCXX boxes) via thiol-disulfide exchange mechanism have also been discussed.[459] The platinum(IV) complex *trans*-$[Pt(en)_2Cl_2]^{2+}$ has been described as a disulfide forming agent, which rapidly and quantitatively converts fully reduced conotoxins to their disulfide bond-containing regioisomers. The reaction is a good base for the efficient synthesis of various disulfide ligands.[460]

Albumin is a well-known copper(II) and nickel(II) binding protein and its metal binding selectivity is related to the His(3) residue at the N-terminus.

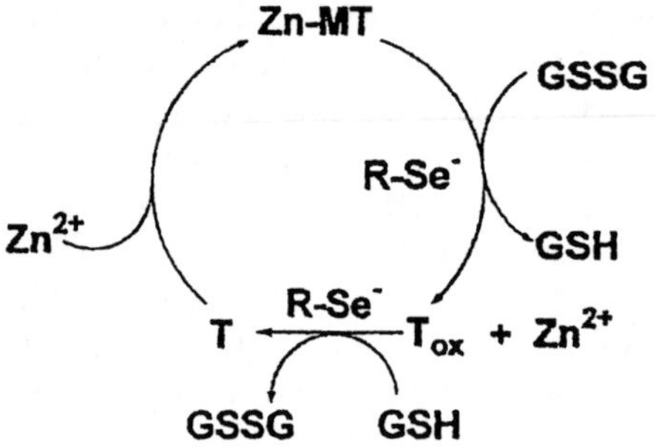

Scheme 4

Albumin, however, contains 35 Cys residues, 34 of which are involved in disulfide bond and only Cys(34) is a free thiol. Thermodynamic and spectroscopic measurements have been performed to understand the effect of Cys(34) on the copper(II) binding to bovine serum albumin.[461] Small metallopeptide aggregates may provide versatile options to understand the factors that lead to stable metalloproteins. The synthesis of a series of cysteine-containing peptides has been reported and their cadmium(II) and mercury(II) complexes studied. The metallated peptide aggregates exhibited pH dependent behaviour. At high pH cadmium(II) was bound in a trigonal planar geometry to 3 sulfurs of the three-stranded α-helical coiled coils.[462] Thiol groups of cysteine are the most common binding sites in the polynuclear metal sites of proteins (e.g. iron sulfur proteins). Several papers have been published on the synthesis of peptides (containing HC_4H_2, HC_4HC and HC_5H motifs) as scaffolds for the stabilization of a $[Ni^{II}\text{-X-}Fe_4S_4]$ bridged assembly.[463–465]

3.5 Synthetic, Analytical and Biomedical Applications of Peptide Complexes. – Metal ion intoxication is one of the major issues in bioinorganic chemistry and the peptides and related substances are promising agents in this field. The polyhydroxamate ligands are among the best known iron chelators. Their metal binding ability can be finely tuned by the distance between the hydroxamate functions. Two tripodal peptide hydroxamic acids, containing three [Ala-Ala-β-(OH)Ala] or three $[\text{Ala-Ala-}\beta\text{-(OH)Ala}]_2$ residues linked via tris(alanylaminoethyl)-amine have been prepared and their complexation with iron(III) studied.[466] Another iron/aluminium chelating compound (Feralex-G) containing amide bond and oxygen donors was synthesized from maltol, glycine and glucosamine.[293] Linear and cyclic hexapeptides containing the 3-hydroxy-4(1H)-pyridinone binding sites have also been prepared and the chemical properties and biological activity of the iron(III) complexes tested. Linear hexapeptides were found to remove efficiently iron(III) from human transferrin.[467] Another new metal ion chelator was obtained with the conjugation of EDTA to ethylphenylalaninate. The data obtained in the presence of Fe(II), H_2O_2 and the chelating agent demonstrated that the ligand could act as a radical scavanging and iron chelating antioxidant under physiologically relevant conditions.[468] The synthesis of the first peptide-oligonucleotide conjugate, designed to coordinate chromium(III), has been performed. The results obtained for the metal complexes of the peptide conjugate may help to understand the chromium-DNA interaction at a molecular level, which is supposed to be a risk factor for the development of cancer.[469] Functionalized biopolymers represent a new class of chelating agents. The synthesis of lysyl and imidazolyl conjugates of bovine serum albumin has been reported and the binding properties against divalent transition metal ions studied.[470] Poly(L-cysteine), a short synthetic biopolymer containing ~50 repeating Cys residues, has been shown to be an effective metal chelator both in solution and in immobilized form. The metal complexation of the immobilized synthetic biopolymer and its pH dependence was studied by tapping mode liquid cell atomic force microscopy. Conformational changes of the biopolymer are shown in Figure 3 under three different conditions: (a) reducing environment, (b) metal rich environment and (c) acidic environment.[471] The selective recovery of uranium and thorium ions from dilute aqueous solutions was examined by using animal biopolymers obtained from eggshell and silk proteins.[472]

The applications of peptides and related ligands as effective chelating and transporting agents of radionuclides is a matter of continuous interest. ^{99m}Tc is probably the most common isotope used in medical imaging, while the chemically related $^{186/188}$Re isotopes are promising candidates for internal radiotherapy in the treatment of cancer. As a consequence, a huge number of papers have been published on the coordination chemistry of technetium and renium revealing both the similarities and differences in the complex formation reactions of the metal ions. Peptides and their thiol or phosphonic derivatives seem to be the best ligands for binding the two metal ions and for the development of target specific radiopharmaceuticals.[473–484] Mercaptoacetyltriglycine (MAG_3) has become the standard ligand for kidney studies. It binds the oxotechnetium(V) core (TcO^{3+}) via three deprotonated amide and the thiolate functions (N_3S). The studies on the oxotechnetium(V) complexes of triglycine and various derivatives help to understand the role of thiol function in metal

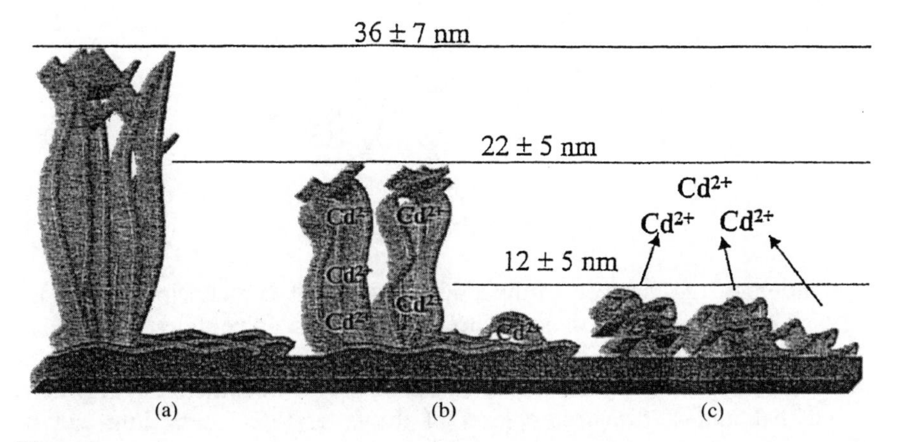

Figure 3

binding.[473] The solid state structures of ReO(V) complexes of some derivatives of MAG_3 containing a secondary amine instead of one of the amide functions have been determined. In addition to the preferred (N_3S) binding mode, the (N_2OS) binding was also observed and striking differences in the solution and solid state structures demonstrated.[474] The TcO(V) complex of the peptide Me_2Gly-Ser-Cys(Acm)-Gly-Thr-Lys-Pro-Pro-Arg was designed as an inflammation imaging agent targeting the tuftsin receptors,[477] whereas neuropeptide Y (NPY) analogues were described as potential tumor imaging agents.[484]

The investigations on the physico-chemical properties and applications of ferrocenoyl-peptides and other metallocene derivatives represent a rapidly growing area of bioorganometallic chemistry.[485–498] Modified ferrocenes and their analogues have long been exploited as redox probes and are able to respond to the structural changes that will take place upon substrate binding. The strategy for the incorporation of ferrocene (Fc) into a peptide framework under mild conditions has already been clarified and the redox potential of the ferrocenoyl groups attached to oligopeptides is significantly influenced by the secondary structures of the peptides. The solid state structure of Fc-Gly-Gly-OEt (**50**) has been determined by X-ray diffraction method and it is obvious that two adjacent molecules can interact in a head-to-head fashion engaging in H-bonding, resembling the interaction found in parallel β-sheets.[485] Several other Fc-dipeptides containing aspartyl[486] or prolyl[487] residues have also been structurally characterized and the presence of extensive H-bonding network is suggested.

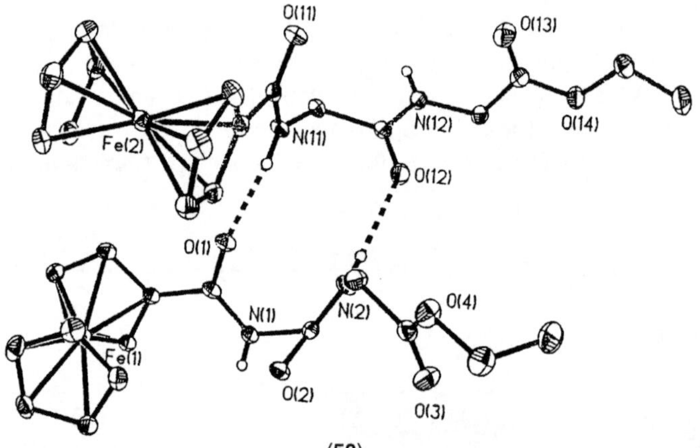

O(11)
O(13)
N(12)
O(14)
Fe(2)
N(11)
O(12)
O(1)
O(4)
N(1)
N(2)
Fe(1)
O(2)
O(3)

(50)

In addition to the ferrocenoyl-dipeptides several other metal ions were also involved in the synthesis of metallocene conjugates of peptides. The recent examples include tha cationic (peptide)titanocene complexes,[488] iridium(III),[489,490] ruthenium(II)[496] and rhodium(III)[497] derivatives. It has been shown that half-sandwich complexes of the latter three metal ions can be effectively used in peptide synthesis.[496,497]

Peptides and/or peptide complexes are often linked to other complexing agents and the resulting peptide conjugates can be widely used as sensors, markers or chelating agents for biomedical or analytical purposes.[144,499–511] The procedure for the preparation of a ruthenium(II) tris(bipyridyl) amino acid was described along with its incorporation into a helix forming peptide using standard solid-phase peptide synthesis. The studies on the physical properties of the conjugate indicated that the metalloamino acid behaved very much like $Ru^{II}(bipy)_3$.[144] Metal complexation studies have been carried out with Langmuir monolayers of histidyl peptide lipids. The peptide part of the system represented the Gly-His-Gly metal binding motif and was found to be effective towards copper(II) and zinc(II).[499] There is a considerable interest in the labelling of biomolecules with organometallic fragments that possess carbonyl ligands, because the corresponding bioconjugates can be detected in picomolar quantities. It was reported that Mo^0, Mo^I and Mo^{II} complexes with the [Mo(bis(2-picolyl)amine)$(CO)_3$] unit could be used for labelling simple amino acids or dipeptides at the N-termini. The attached biomolecule had only a slight influence on the spectroscopic properties of the metal complex and it is a well-suited marker for biomolecules.[500] The neuropeptide Leu^5-enkephalin was also labelled with organometallic molybdenum carbonyl complexes suggesting the possible application of these markers in neurochemistry.[501] Organoplatinum(II) complexes of the type [PtX(NCN)] have been developed for the labelling of amino acids and peptides at either their N- or C-termini or at the α-carbon atom. The ability of the molecule to bind SO_2 with concomitant change in the spectroscopic properties allows the applications in biochemistry.[502] Zinc(II) dipicolylamine-based dinuclear complexes as artificial receptors were found to bind and stabilize selectively the α-helix conformation of peptides having two His residues at specific positions.[503,504] The recognition of peptides based on metal ligand interaction offers several advantages over traditional (e.g. H-bonding, electrostatic and hydrophobic interactions, *etc.*) approaches. Specific binding to oligohistidine peptide deivatives has been achieved employing copper(II)- and nickel(II)-histidine interactions in aqueous medium.[508–511] When the pattern of copper(II) ions on a complex matched with the pattern of histidines on a peptide, a strong and selective binding was observed.

An important application of the simple copper(II)-peptide interaction is that it can be used for analytical purposes. Chromophore-labelled peptides are useful probes of the reactivity and physical properties of biomolecules and can act as sensors for monitoring the concentration of free metal ions in complex aqueous medium.[145,512–514] The development of a zinc(II)-selective sensor was based upon a 8-hydroxyquinoline complexing agent built into a model heptapeptide.[512] Dansylated polyamines containing amine and sulfonamide donor functions were also suggested as fluorescence sensors of metal ions.[514] Post-column complexation reactions of peptides with copper(II) were applied for the electrochemical detection of peptides at low concentrations. To understand and improve this procedure, the kinetics of the copper(II)-peptide interactions has been studied with various N-formylmethionyl chemoattractant and chemotactic peptides.[515] The application of copper(II) complexes of diaminodiamido-type ligands has been suggested in the chiral separation of amino acids.[273]

The use of metal complexes as catalysts is probably the most common application of coordination compounds. It is confirmed by more and more publications that various peptide complexes of transition elements can also be important catalysts. A 20-mer peptide, comprising an EF-hand calcium-binding motif was shown to bind lanthanide(III) ions and the Eu(III) complex promoted phosphate ester cleavage in a pH dependent manner.[516] Similar observations were reported for other lanthanide(III)-peptide systems.[517]

Zinc(II) model compounds with a histidine-containing pseudopeptide were also found to be active in the hydrolysis of natural and artificial phosphoesters.[518] A metallopeptide conjugate based upon the albumin-like binding sites has been characterized and the phosphate esters hydrolyzing activity demonstrated.[519] Other examples on the metallopeptide-based catalysts include the results obtained for catalytic effects of titanium,[520] copper(II),[521] palladium[263] and cobalt(III)[522] complexes.

References

1. C.A. Selects on Amino Acids, Peptides and Proteins, published by the American Chemical Society and Chemical Absracts Service, Columbus, Ohio.
2. The ISI Web of Science for Hungary on http://www.eisz.hu.
3. R.B. Martin, in *Metal Ions Biol. Syst.*, (ed. H. Sigel), 2001, **38**, 2.
4. N.M. Milović and N.M. Kostić, in *Metal Ions Biol. Syst.*, (ed. H. Sigel), 2001, **38**, 146.
5. G.M. Polzin and J.N. Burstyn, in *Metal Ions Biol. Syst.*, (ed. H. Sigel), 2001, **38**, 104.
6. G. Allen, in *Metal Ions Biol. Syst.*, (ed. H. Sigel), 2001, **38**, 197.
7. D.A. Buckingham and C.R. Clark, in *Metal Ions Biol. Syst.*, (ed. H. Sigel), 2001, **38**, 43.
8. M. Komiyama, in *Metal Ions Biol. Syst.*, (ed. H. Sigel), 2001, **38**, 25.
9. B.P. Espósito and R. Najjar, *Coord. Chem. Rev.*, 2002, **232**, 137.
10. M.D. Hall and T.W. Hambley, *Coord. Chem. Rev.*, 2002, **232**, 49.
11. M. Nath, S. Pokharia and R. Yadav, *Coord. Chem. Rev.*, 2001, **215**, 99.
12. L. Pellerito and L. Nagy, *Coord. Chem. Rev.*, 2002, **224**, 111.
13. E.J. Baran, *J. Coord. Chem.*, 2001, **54**, 215.
14. G. Berthon, *Coord. Chem. Rev.*, 2002, **228**, 319.
15. P. Rubini, A. Lakatos, D. Champmartin and T. Kiss, *Coord. Chem. Rev.*, 2002, **228**, 137.
16. G.S. Shaw, *Methods Mol. Biol.*, 2002, **173**, 175.
17. I. Sóvágó, K. Várnagy and K. Ösz, *Comm. Inorg. Chem.*, 2002, **23**, 149.
18. K. Popov, H. Rönkkömäki and L.H.J. Lajunen, *Pure Appl. Chem.*, 2001, **73**, 1641.
19. O. Yamauchi, A. Odani and S. Hirota, *Bull. Chem. Soc. Japan*, 2001, **74**, 1525.
20. O. Yamauchi, A. Odani and M. Takani, *J. Chem. Soc., Dalton Trans.*, 2002, 3411.
21. J. Chan, Z. Huang, M.E. Merrifield, M.T. Salgado and M.J. Stillman, *Coord. Chem. Rev.*, 2002, **233–234**, 319.
22. L.J. Ming and J.D. Epperson, *J. Inorg. Biochem.*, 2002, **91**, 46.

23. H.B. Kraatz and M. Galka, in *Metal Ions Biol. Syst.*, (ed. H. Sigel), 2001, **38**, 385.
24. D. Carmona, M.P. Lamata and L.A. Oro, *Eur. J. Inorg. Chem.*, 2002, **2239**.
25. S.M. Mousssa, R.R. Fenton, B.A. Hunter and B.J. Kennedy, *Aust. J. Chem.*, 2002, **55**, 331.
26. S. Hu, W. Du, J. Dai, L. Wu, C. Cui, Z. Fu and X. Wu, *J. Chem. Soc., Dalton Trans.*, 2001, 2963.
27. V. Noething-Laslo, N. Paulić, R. Basosi and R. Pogni, *Polyhedron*, 2002, **21**, 1643.
28. J.M. Schveigkardt, A.C. Rizzi, O.E. Piro, E.E. Castellano, R. Costa de Santana, R. Calvo and C.D. Brondino, *Eur. J. Inorg. Chem.*, 2002, 2913.
29. P. Manikandan, B. Epel and D. Goldfarb, *Inorg. Chem.*, 2001, **40**, 781.
30. E. Prenesti and S. Berto, *J. Inorg. Biochem.*, 2002, **88**, 37.
31. R.F. de Farias, *Trans. Met. Chem.*, 2002, **27**, 594.
32. A.C. Ukwueze and A.O. Fadario, *Int. J. Chem.*, 2002, **12**, 297.
33. Y.R. Xie, R.G. Xiong, X. Xue, X.T. Chen, Z. Xue and X.Z. You, *Inorg. Chem.*, 2002, **41**, 3323.
34. F. Wiesbrock and H. Schmidbaur, *J. Chem. Soc., Dalton Trans.*, 2002, 3201.
35. M. Rombach, M. Gelinsky and H. Vahrenkamp, *Inorg. Chim. Acta*, 2002, **334**, 25.
36. H. Strasdeit, I. Büsching, S. Behrends, W. Saak and W. Barklage, *Chem. Eur. J.*, 2001, **7**, 1133.
37. S. Chen, X. Yang, Z. Ju, H. Li and S. Gao, *Chem. Pap.*, 2001, **55**, 239.
38. L. Gasque, S. Bernès, R. Ferrari and G. Mendoza-Díaz, *Polyhedron*, 2002, **21**, 935.
39. C.H. Ng, C.W. Lim, S.G. Teoh, H.-K. Fun, A. Usman and S.W. Ng, *Inorg. Chem.*, 2002, **41**, 2.
40. P.A. Angeli Mary and S. Dhanuskodi, *Spectrochimica Acta, Part A*, 2001, **57**, 2345.
41. K. Nomiya and H. Yokoyama, *J. Chem. Soc., Dalton Trans.*, 2002, 2483.
42. H.A. Headlam, C.L. Weeks, P. Turner, T.W. Hambley and P.A. Lay, *Inorg. Chem.*, 2001, **40**, 5097.
43. R.F. de Farias, *Trans. Met. Chem.*, 2002, **27**, 594.
44. E.M. Marti, S.M. Barlow, S. Haq and R. Raval, *Surf. Sci.*, 2002, **501**, 191.
45. S.P. Ge, X.Y. Zhao, Z. Gai, R.G. Zao and W.S. Yang, *Chinese Phys.*, 2002, **11**, 839.
46. X. Zhao, R.G. Zhao and W.S. Yang, *Langmuir*, 2002, **18**, 443.
47. G. Martra, S. Horikoshi, M. Anpo, S. Coluccia and H. Hidaka, *Res. Chem. Intermed.*, 2002, **28**, 359.
48. A. Hayashi, S. Saito, Y. Nakatani, A. Nishiyama, Y. Matsumura, H. Nakayama and M. Tsuhako, *Phosph. Res. Bull.*, 2001, **12**, 129.
49. C. Tessier, F.D. Rochon and A.L. Beauchamp, *Inorg. Chem.*, 2002, **41**, 6527.
50. C. Tessier, A.L. Beauchamp and F.D. Rochon, *J. Inorg. Biochem.*, 2001, **85**, 77.
51. S. Seifert, J.-U. Künstler, A. Gupta, H. Funke, T. Reich, H.-J. Pietzsch, R. Alberto and B. Johannsen, *Inorg. Chim. Acta*, 2001, **322**, 79.
52. J. Torres, C. Kremer, E. Kremer, H. Pardo, L. Suescun, Á. Mombrú, S. Domíngez, A. Mederos, R. Herbst-Irmer and J.M. Arrieta, *J. Chem. Soc. Dalton Trans.*, 2002, 4035.
53. R. Wang, H. Liu, M.D. Carducci, T. Jin, C. Zheng and Z. Zheng, *Inorg. Chem.*, 2001, **40**, 2743.
54. P.S. Subramanian, E. Suresh, P. Dastidar, S. Waghmode and D. Srinivas, *Inorg. Chem.*, 2001, **40**, 4291.
55. J. Weng, M. Hong, Q. Shi, R. Cao and A.S.C. Chan, *Eur. J. Inorg. Chem.*, 2002, 2553.
56. H. Ishida, D. Hesek and Y. Inoue, *Jpn. Kokai Tokyo Koho JP*, 2001, **39**, 995.

57. P.R. Reddy and A.M. Reddy, *Indian J. Chem.*, 2002, **41A**, 2083.
58. S. Çakir, E. Biçer and A. Eleman, *Trans. Met. Chem.*, 2001, **26**, 89.
59. J. Cai, X. Hu, I. Bernal and L.-N. Ji, *Polyhedron*, 2002, **21**, 817.
60. H. Kumita, T. Kato, K. Jitsukawa, H. Einaga and H. Masuda, *Inorg. Chem.*, 2001, **40**, 3936.
61. J. Cai, X. Hu, X. Feng, L. Ji and I. Bernal, *Acta Crystallogr., Sect. B: Struct. Sci.*, 2001, **B57**, 45.
62. X. Hu, J. Cai, C. Chen, X.-M. Chen and L.-N. Ji, *Cryst. Eng.*, 2001, **4**, 141.
63. A.I. El-Said, A.S.A. Zidan, M.S. El-Meligy, A.A.M. Aly and O.F. Mohammed, *Synth. React. Inorg. Met.-Org. Chem.*, 2001, **31**, 633.
64. V.M. Dinović, S.R. Grgurić, X. Xing-You and T.J. Sabo, *J. Coord. Chem.*, 2001, **53**, 355.
65. J.M. Harrowfield, S.H. Jeong, M.K. Lee, Y. Kim, E. Rukmini, B.W. Skelton and A.H. White, *Austr. J. Chem.*, 2001, **54**, 63.
66. K. Serdiuk, R. Gancarz and M. Cieślak-Golonka, *Trans. Met. Chem.*, 2001, **26**, 538.
67. K. Majumder, R.J. Butcher and S. Bhattacharya, *Inorg. Chem.*, 2002, **41**, 4605.
68. J.M. Slocik, M.S. Ward, K.V. Somayajula and R.E. Shepherd, *Trans. Met. Chem.*, 2001, **26**, 351.
69. A. Böhm, K. Polborn and W. Beck, *Z. Naturforsch.*, 2001, **56b**, 293.
70. W. Ponikwar, P. Meyer and W. Beck, *Z. Naturforsch. B, Chem. Sci.*, 2002, **57**, 810.
71. X. Meng, P. Yang and H. Chen, *J. Inorg. Biochem.*, 2002, **92**, 28.
72. R.S. Dickins, S. Aime, A.S. Batsanov, A. Beeby, M. Botta, J.I. Bruce, J.A.K. Howard, C.S. Love, D. Parker, R.D. Peacock and H. Puschmann, *J. Am. Chem. Soc.*, 2002, **124**, 12697.
73. S.D. Kean, C.J. Easton, S.F. Lincoln and D. Parker, *Austr. J. Chem.*, 2001, **54**, 535.
74. Y. Ye, J.M. Hu and Y.E. Zeng, *J. Raman Spectr.*, 2001, **32**, 1018.
75. G. Seth and N.K. Mourya, *Asian J. Chem.*, 2002, **14**, 283.
76. M. Claeys-Bruno, D. Toronto, J. Pécaut, M. Bardet and J.-C. Marchon, *J. Am. Chem. Soc.*, 2001, **123**, 11067.
77. M. Claeys-Bruno, M. Bardet and J.-C. Marchon, *Magn. Res. Chem.*, 2002, **40**, 647.
78. H. Imai, K. Misawa, H. Munakata and Y. Uemori, *Chem. Lett.*, 2001, 688.
79. X. Peng, S. Xiao and W. Fu, *Huaxue Yanjiu*, 2001, **12**, 10.
80. V.V. Borovkov, N. Yamamoto, J.M. Lintuluoto, T. Tanaka and Y. Inoue, *Chirality*, 2001, **13**, 329.
81. N. Niklas, S. Wolf, G. Liehr, C.E. Anson, A.K. Powell and R. Alsfasser, *Inorg. Chim. Acta*, 2001, **314**, 126.
82. N. Niklas, F. Hampel, O. Walter, G. Liehr and R. Alsfasser, *Eur. J. Inorg. Chem.*, 2002, 1839.
83. S.-G. Roh, Y.-C. Park, D.-K. Park, T.-J. Kim and J.H. Jeong, *Polyhedron*, 2001, **20**, 1961.
84. Y. Inomata, T. Takei and F.S. Howell, *Inorg. Chim. Acta*, 2001, **318**, 201.
85. C. Policar, S. Durot, F. Lambert, M. Cesario, F. Ramiandrasoa and I. Morgenstern-Badarau, *Eur. J. Inorg. Chem.*, 2001, 1807.
86. S. Bihari, P.A. Smith, S. Parsons and P.J. Sadler, *Inorg. Chim. Acta*, 2002, **331**, 310.
87. M. Albrecht, M. Napp, M. Schneider, P. Weis and R. Fröhlich, *Chem. Eur. J.*, 2001, **7**, 3966.
88. C.K. Ho, A.D. Schuler, C.B. Yoo, S.R. Herron, K.A. Kantardjieff and A.R. Johnson, *Inorg. Chim. Acta*, 2002, **431**, 71.

89. I. Sougandi, R. Mahalakshmi, R. Kannappan, T.M. Rajendiram, R. Venkatesen and P.S. Rao, *Trans. Met. Chem.*, 2002, **27**, 512.
90. B. Macías, M.V. Villa, E. Chicote, S. Martín-Velasco, A. Castiñeiras and J. Borrás, *Polyhedron*, 2002, **21**, 1899.
91. J. Sheals, P. Persson and B. Hedman, *Inorg. Chem.*, 2001, **40**, 4302.
92. Z. Szabó, *J. Chem. Soc., Dalton Trans.*, 2002, 4242.
93. V. Iyengar, P.V. Galka, A. Pletsch and H.-B. Kraatz, *Can. J. Chem.*, 2002, **80**, 1562.
94. A.M.Z. Slawin, J.D. Woollins and Q. Zhang, *J. Chem. Soc., Dalton Trans.*, 2001, 621.
95. M.I. García-Seijo, A. Habtemariam, P. del Socorro Murdoch, R.O. Gould and M.E. García-Fernández, *Inorg. Chim. Acta*, 2002, **335**, 52.
96. A. Paladini, C. Calcagni, T. Di Palma, M. Speranza, A. Lagana, G. Fago, A. Filippi, M. Satta and A.G. Guidoni, *Chirality*, 2001, **13**, 707.
97. P.V. Galka and H.-B. Kraatz, *Chem. Ind.*, 2001, **82**, 589.
98. S.R. Banerjee, M.K. Levadala, N. Lazarova, L. Wei, J.F. Valliant, K.A. Stephenson, J.W. Babich, K.P. Maresca and J. Zubieta, *Inorg. Chem.*, 2002, **41**, 6417.
99. S.R. Banerjee, L. Wei, M.K. Levadala, N. Lazarova, V.O. Golub, C.J. O'Connor, K.A. Stephenson, J.F. Valliant, J.W. Babich and J. Zubieta, *Inorg. Chem.*, 2002, **41**, 5795.
100. M. Lipowska, L. Hansen, R. Cini, X. Xu, H. Choi, A.T. Taylor and L.G. Marzilli, *Inorg. Chim. Acta*, 2002, **339**, 327.
101. K. Kanamori, K. Nishida, N. Miyata, K. Okamoto, Y. Miyoshi, A. Tamura and H. Sakurai, *J. Inorg. Biochem.*, 2001, **86**, 649.
102. Y. Yoshikawa, K. Kawabe, M. Tadokoro, Y. Suzuji, N. Yanagihara, A. Nakayama, H. Sakuri and Y. Kojima, *Bull. Chem. Soc. Jpn.*, 2002, **75**, 2423.
103. T.N. Parac-Vogt, K. Binnemans and C. Görller-Walrand, *J. Chem., Soc. Dalton Trans.*, 2002, 1602.
104. L.H. Abdel-Rahman, *Trans. Met. Chem.*, 2001, **26**, 412.
105. M.C. Capllonch, A. García-Raso, A. Terrón, M.C. Apella, E. Espinosa and E. Molins, *J. Inorg. Biochem.*, 2001, **85**, 173.
106. M. Cavicchioli, P.P. Corbi, P. Melnikov and A.C. Massabni, *J. Coord. Chem.*, 2002, **55**, 951.
107. W. He, F. Liu, Y. Mei, Z. Guo and L. Zhu, *New J. Chem.*, 2001, **25**, 1330.
108. Z.-Y. Wu, D.-J. Xu and Z.-X. Feng, *Polyhedron*, 2001, **20**, 281.
109. A. Popkov, A. Gee, M. Nádvorník and A. Lyčka, *Trans. Met. Chem.*, 2002, **27**, 884.
110. I. Sakiyan, N. Gunduz and T. Gunduz, *Synth. React. Inorg. Met.-Org. Chem.*, 2001, **31**, 1175.
111. C. Hu, W. Zhang, Y. Xu, H. Zhu, X. Ren, C. Lu, Q. Meng and H. Wang, *Trans. Met. Chem.*, 2001, **26**, 700.
112. A.S. Saghiyan, H.H. Hambarcumyan, S.A. Dadayan, S.R. Haroutunyan, A.Ch. Hovhannesyan, A.M. Hovhannesyan, A.A. Avetisyan, V.L. Tararov, V.L. Maleev, Yu.N. Belokon and M. North, *Khim. Zh. Armenii*, 2002, **55**, 73.
113. A. Erxleben and D. Schumacher, *Eur. J. Inorg. Chem.*, 2001.
114. D. Koch, W. Hoffmüller, K. Polborn and W. Beck, *Z. Naturforsch.*, 2001, **56b**, 403.
115. P.A.N. Reddy, M. Nethaji and A.R. Chakravarty, *Inorg. Chim. Acta*, 2002, **337**, 450.
116. S. Shova, G. Novitchi, M. Gdaniec, Y.A. Simonov and C. Turta, *Russ. J. Inorg. Chem.*, 2001, **46**, 1685.

117. H. Dialer, S. Schumann, K. Polborn, W. Steglich and W. Beck, *Eur. J. Inorg. Chem.*, 2001, 1675.
118. H. Dialer, K. Polborn, W. Ponikwar, K. Sünkel and W. Beck, *Chem. Eur. J.*, 2002, **8**, 691.
119. D. Koch and W. Beck, *Z. Naturforsch.*, 2001, **56b**, 1271.
120. U. Kazmaier, D. Schauβ, S. Raddatz and M. Pohlman, *Chem. Eur. J.*, 2001, **7**, 456.
121. H. Dialer, P. Mayer, K. Polborn and W. Beck, *Eur. J. Inorg. Chem.*, 2001, 1051.
122. U. Kernbach, M. Mühl, K. Polborn, W.P. Fehlhammer and G. Jaouen, *Inorg. Chim. Acta*, 2002, **334**, 45.
123. J.M. Talley, B.A. Cerda, G. Ohanessian and C. Wesdemiotis, *Chem. Eur. J.*, 2002, **8**, 1377.
124. T. Marino, N. Russo and M. Toscano, *Inorg. Chem.*, 2001, **40**, 6439.
125. J.L. Seymour and F. Turecek, *J. Mass Spectrom.*, 2002, **37**, 533.
126. T. Marino, N. Russo and M. Toscano, *J. Mass Spectrom.*, 2002, **37**, 786.
127. E.N. Nikolaev and Yu.A. Borisov, *Khim. Fiz.*, 2002, **21**, 42.
128. M. Milčić and S.D. Zarić, *Eur. J. Inorg. Chem.*, 2001, 2143.
129. A. Gapeev and R.C. Dunbar, *J. Am. Chem. Soc.*, 2001, **123**, 8360.
130. J. Hu, L.J. Barbour and G.W. Gokel, *Proceed. Natl. Acad. Sci. USA*, 2002, **99**, 5121.
131. T. Shoeib, A. Cunje, A.C. Hopkinson and K.W.M. Siu, *J. Am. Soc. Mass Spectrom.*, 2002, **13**, 408.
132. R.A. Jockush, A.S. Lemoff and E.R. Williams, *J. Am. Chem. Soc.*, 2001, **123**, 12255.
133. R.M. Moision and P.B. Armentrout, *J. Phys. Chem. A*, 2002, **106**, 10350.
134. S. Hoyau, J.-P. Pelicier, F. Rogalewicz, Y. Hoppilliard and G. Gilles, *Eur. J. Mass Spectrom.*, 2001, **7**, 303.
135. B.A. Perera, M.P. Ince, E.R. Talaty and M.J. Van Stipdonk, *Rapid Commun. Mass Spectrom.*, 2001, **15**, 615.
136. J.M. Mercero, A. Irigoras, X. Lopez, J.E. Fowler and J.M. Ugalde, *J. Phys. Chem. A*, 2001, **105**, 7446.
137. C. Buda, S.K. Burt, T.R. Cundari and P.S. Shenkin, *Inorg. Chem.*, 2002, **41**, 2060.
138. C. Wang, Z. Zhu, Y. Li, R. Chen, X. Wen, F. Miao and A.S.C. Chan, *Gaodeng Xuexiao Huaxue Xuebao*, 2001, **22**, 262.
139. E. Virtanen, J. Tamminen, J. Linnanto, P. Maetttaeri, P. Vainiotalo and E. Kolehmainen, *J. Incl. Phenom. Macrocyclic Chem.*, 2002, **43**, 319.
140. R. Ghanem, Y. Xu, J. Pan, T. Hoffmann, J. Andersson, T. Polívka, T. Pascher, S. Styring, L. Sun and V. Sundström, *Inorg. Chem.*, 2002, **41**, 6258.
141. B. Geiβer, T. Skrivanek, U. Zimmermann, D.J. Stufkens and R. Alsfasser, *Eur. J. Inorg. Chem.*, 2001, 439.
142. M. Suzuki, C.C. Waraksa, T.E. Mallouk, H. Nakayama and K. Hanabusa, *J. Phys. Chem. B*, 2002, **106**, 4227.
143. A. Khatyr and R. Ziessel, *Synthesis*, 2001, **11**, 1665.
144. K.J. Kise, Jr and B.E. Bowler, *Inorg. Chem.*, 2002, **41**, 379.
145. B. Geiβer, B. König and R. Alsfasser, *Eur. J. Inorg. Chem.*, 2001, 1543.
146. O.Y. Zelenin, L.A. Kochergina, V.V. Chernikov and T.E. Zelenina, *Zh. Neorg. Khim.*, 2001, **46**, 160.
147. R.K.P. Singh and S. Aziz, *Res. J. Chem. Environ.*, 2001, **5**, 23.
148. O.Y. Zelenin, L.A. Kochergina and V.V. Chernikov, *Zh. Fiz. Khim.*, 2001, **75**, 583.
149. J. Fan, X. Shen and J. Wang, *Electroanalysis*, 2001, **13**, 1115.

150. G. Sheng-Li, L.J. Guo, F.X. Zhang and Y. Ma, *Wuli Huaxue Xuebao*, 2001, **17**, 573.
151. A. Doğan, F. Köseoğlu and E. Kilic, *Anal. Biochem.*, 2001, **295**, 237.
152. T.I. Lezhava, N.Sh. Ananiashvili, M.P. Kikabidze and N.O. Berdzenshvili, *Russ. J. Electrochem..*, 2001, **37**, 1305.
153. M.S. Babu, G.N. Rao, K.V. Ramana and M.S.P. Rao, *J. Indian Chem. Soc.*, 2001, **78**, 280.
154. M.S. Nair and M.A. Neelakantan, *Indian J. Chem.*, 2002, **41A**, 2088.
155. M.A. Kabir and M.R. Ullah, *J. Bangladesh Acad. Sci.*, 2002, **26**, 175.
156. A.A.A. Boraei and N.F.A. Mohamed, *Bull. Fac. Sci., Assiut Univ. B: Chemistry*, 2002, **31**, 17.
157. A.K. Cucu, M. Pekin, H.D. Demir and H.Y. Aboul-Enein, *Toxicol. Environ. Chem.*, 2001, **80**, 165.
158. R.N. Patel, N. Singh, R.P. Shrivastava, K.K. Shukla and P.K. Singh, *Proc.-Indian Acad. Sci. Chem. Sci.*, 2002, **114**, 115.
159. A.K. Molodkin, N.Y. Esina and N.K. Tinaeva, *Zh. Neorg. Khim.*, 2002, **47**, 953.
160. A.B. Patil and T.H. Mhaske, *Asian J. Chem.*, 2001, **13**, 1544.
161. N.E. Knyazeva, *Zh. Neorg. Khim.*, 2002, **47**, 809.
162. P.G. Rohankar and A.S. Aswar, *Indian J. Chem.*, 2001, **40A**, 1086.
163. A.V. Astapov, A.N. Amelin and Y.S. Peregudov, *Zh. Neorg. Khim.*, 2002, **47**, 1130.
164. A. Pérez-Cadenas, L. Godino-Salido, R. López-Garzón, P. Arranz-Mascrós, D. Gutiérrez-Valero and R. Cuesta-Martos, *Trans. Met. Chem.*, 2001, **26**, 581.
165. A. Sharma, K.D. Gupta and K.K. Saxena, *Ultra Sci. Phys. Sci.*, 2002, **14**, 111.
166. S.K. Singh and C.P.S. Chandel, *Oriental J. Chem.*, 2001, **17**, 239.
167. S.K. Singh and C.P.S. Chandel, *Orient J. Chem.*, 2001, **17**, 449.
168. A. Albourine, A. Assabbane, M. Elamine, Y. Ait-Ichiu and M. Petit-Ramel, *Bull. Electrochem.*, 2002, **18**, 203.
169. E.V. Legler, V.I. Kazbanov and A.S. Kazachenko, *Zh. Neorg. Khim.*, 2002, **47**, 341.
170. E.V. Legler, V.I. Kazbanov and A.S. Kazachenko, *Zh. Neorg. Khim.*, 2002, **47**, 158.
171. E.V. Legler, E.N. Drivol'skaya, V.I. Kazbanov, G.L. Pashkov and A.S. Kazachenko, *Zh. Neorg. Khim.*, 2001, **46**, 1404.
172. E.V. Legler, V.I. Kazbanov and A.S. Kazachenko, *Zh. Neorg. Khim.*, 2001, **46**, 1401.
173. R. Bapna, R. Gupta, P.C. Vyas and M. Arora, *J. Electrochem. Soc. India*, 2001, **50**, 15.
174. D.-Y. Chen, W.-L. Shi, S.-M. Chen and X.-M. Yan, *Chin. J. Chem.*, 2001, **19**, 449.
175. F. Gharib, H. Aghaei and A. Shamel, *Phys. Chem. Liquids*, 2002, **40**, 637.
176. M. Vadi and F. Gharib, *Int. J. Chem.*, 2002, **12**, 129.
177. F. Gharib and M. Vadi, *Zh. Neorg. Khim.*, 2002, **47**, 2091.
178. F. Gharib, M. Vadi, S. Momeni and H. Jalali, *Int. J. Chem.*, 2002, **12**, 55.
179. R.N. Patel, S. Sharma, K.K. Shukla, V.K. Soni, N. Singh and K.B. Pandeya, *J. Indian Chem. Soc.*, 2002, **79**, 831.
180. F. Gharib, M. Monajjemi, S. Ketabi and F. Zoroufi, *Zh. Neorg. Khim.*, 2001, **46**, 423.
181. F. Gharib, K. Zare, A. Taghvamanesh and M. Monajjemi, *J. Chem. Eng. Data*, 2001, **46**, 1140.
182. B.B. Tewari, *J. Chromatography A*, 2002, **962**, 233.
183. B.U. Khan, *Chem. Environ. Res.*, 2001, **10**, 219.

184. G. Ye, C. Wang, Z. Wang, Y. Liu and S. Qu, Wuhan Daxue Xuebao, *Ziran Kexueban*, 2001, **47**, 145.
185. H.B. Silber, T. Chang and E. Mendoza, *J. Alloys Compd.*, 2001, **323**, 190.
186. C.S. Kim and M.H. Chon, *Choson Minj. Inmin Kongh. Kwah. Tongbo*, 2001, 34.
187. J.-L. Wang and B.-S. Yang, *Wuji Huaxue Zuebao*, 2002, **18**, 577.
188. J. Wang, *Shenzhen Daxue, Xuebao Ligongban*, 2002, **19**, 54.
189. B.B. Tevari, *Bull. Kor. Chem. Soc.*, 2002, **23**, 705.
190. A. Asthana and K. Dwivedi, *Orient. J. Chem.*, 2002, **18**, 117.
191. G. Usha Rani and D. Das Manwal, *J. Electrochem. Soc. India*, 2001, **50**, 27.
192. S. Daydé, D. Champmartin, P. Rubini and G. Berthon, *Inorg. Chim. Acta*, 2002, **339**, 513.
193. F. Gharib, K. Zare, M. Habibi and A. Taghvamanesh, *Main Group Metal Chem.*, 2002, **25**, 283.
194. F. Gharib, K. Zare, A. Taghvamanesh, A. Samel and G. Shafiee, *Main Group Metal Chem.*, 2002, **25**, 647.
195. G. Nageswara Rao and S.B. Ronald, *J. Indian Chem. Soc.*, 2002, **79**, 796.
196. S.B. Ronald and G. Nageswara Rao, *J. Indian Chem. Soc.*, 2002, **79**, 799.
197. D. Li, Q. Yu, Y. Wang and D. Sun, *Indian J. Chem.*, 2002, **41A**, 1126.
198. V. Lubes, F. Brito M.L. Araujo, A. Vacca, S. Midollini and A. Mederos, *Ciencia*, 2002, **10**, 404.
199. A. Tong, Q. Yang, H. Dong, L. Li and C.W. Huie, *Anal. Sci.*, 2001, **17**, a207.
200. X. Peng, L. Liang, G. Yuan and S. Liu, *Huaxue Tongbao*, 2002, **65**, 126.
201. T. Liu, W.-J. Ruan, Y. Li, D.-Q. Jiang, Z.-A. Zhu, Y.-T. Chen and A.S.C. Chan, *Xuexiao Huaxue Xuebao*, 2001, **22**, 159.
202. R.P. Bonomo, V. Cucinotta, G. Maccarrone, E. Rizzarelli and G. Vecchio, *J. Chem. Soc., Dalton Trans.*, 2001, 1366.
203. K. Ösz, K. Várnagy, H. Süli-Vargha, D. Sanna, G. Micera and I. Sóvágó, *Inorg. Chim. Acta*, 2002, **339**, 373.
204. K. Ösz, K. Várnagy, I. Sóvágó, L. Lennert, H. Süli-Vargha, D. Sanna and G. Micera, *New. J. Chem.*, 2001, **25**, 700.
205. R. Lipinski, L. Chruscinski, P. Mlynarz, B. Boduszek and H. Kozlowski, *Inorg. Chim. Acta*, 2001, **322**, 157.
206. B. Gyurcsik, T. Jakusch and T. Kiss, *J. Chem. Soc., Dalton Trans.*, 2001, 1053.
207. A. Jancsó, T. Gajda, A. Szorcsik, T. Kiss, B. Henry, Gy. Vankó and P. Rubini, *J. Inorg. Biochem.*, 2001, **83**, 187.
208. H. Miyake, M. Watanabe, M. Takemura, T. Hasegawa, Y. Kojima, M.B. Inoue and Q. Fernando, *J. Chem. Soc., Dalton Trans.*, 2002, 1119.
209. M. Takemura, K. Yamato, M. Doe, M. Watanabe, H. Miyake, T. Kikunaga, N. Yanagihara and Y. Kojima, *Bull. Chem. Soc. Jpn.*, 2001, **74**, 707.
210. É.A. Enyedy, H. Csóka, I. Lázár, G. Micera, E. Garriba and E. Farkas, *J. Chem. Soc., Dalton Trans.*, 2002, 2632.
211. E. Farkas and H. Csóka, *J. Inorg. Biochem.*, 2002, **89**, 219.
212. N.M. Suhaib, H.M. Marafie, H.B. Youngo, F.M. Al-Sogair and M.S. El-Ezaby, *J. Coord. Chem.*, 2002, **55**, 933.
213. F. Dallavalle and M. Tegoni, *Polyhedron*, 2001, **20**, 2697.
214. D. Kroczewska, K. Bogusz, B. Kurzak and J. Jezierska, *Polyhedron*, 2002, **21**, 295.
215. B. Kurzak, K. Bogusz, D. Kroczewska and J. Jezierska, *Polyhedron*, 2001, **20**, 2627.
216. F. Dallavalle, G. Folesani, A. Sabatini, M. Tegoni and A. Vacca, *Polyhedron*, 2001, **20**, 103.

217. D. Kroczewska, B. Kurzak and E. Matczak-Jon, *Polyhedron*, 2002, **21**, 2183.
218. B. Nigović and N. Kujundžić, *Polyhedron*, 2002, **21**, 1661.
219. A. Schlüter, K. Bieber and W.S. Sheldrick, *Inorg. Chim. Acta*, 2002, **340**, 35.
220. S.A. Farokhi, A.K. Kini and S.T. Nandibewoor, *Inorg. React. Mechan.*, 2002, **4**, 67.
221. M.D.S.R. Kembhavi, A.L. Harihar and S.T. Nandibewoor, *Inorg. React. Mechan.*, 2001, **3**, 39.
222. N.N. Halligudi, S.M. Desai and S.T. Nandibewoor, *Trans. Met. Chem.*, 2001, **26**, 28.
223. A.K. Kini, S.A. Farokhi and S.T. Nandibewoor, *Trans. Met. Chem.*, 2002, **27**, 532.
224. H. Iloukhani and M. Moazenzadeh, *Phys. Chim. Liq.*, 2001, **39**, 429.
225. M. Farooqui, A. Zaheer, P.M.A. Khan, M. Ubale and S. Ubale, *Asian J. Chem.*, 2002, **14**, 1056.
226. M. Monajjemi, E. Moniri and H.A. Panahi, *J. Chem. Eng. Data*, 2001, **46**, 1249.
227. W.J. Barreto, S.R.G. Barreto, M.A. Santos, R. Schimidt, F.M.M. Paschoal, A.S. Mangrich and L.F.C. deOliveira, *J. Inorg. Biochem.*, 2001, **84**, 89.
228. O.A. Oyetunji and J.J. Tore, *Indian J. Chem. Sect. A, Inorg. Bio-inorg. Phys. Theor. Anal. Chem.*, 2002, **41A**, 1855.
229. S. Dubey and A. Pandey, *Bull. Pol. Acad. Sci. Chem.*, 2001, **49**, 183.
230. J.-H. Shan, H.-Y. Wei, L. Wang, S.-G. Shen, B.-S. Liu and H.-W. Sun, *Hebeisheng Kexueyuan Xuebao*, 2001, **18**, 93.
231. H.M. Abdel-Halim, *Int. J. Chem.*, 2001, **11**, 131.
232. P. Vani, K.K. Kishore, R. Rambabu and L.S.A. Dikshitulu, *Proc. Indian Acad. Sci., Chem. Sci.*, 2001, **113**, 351.
233. O. Nekrassova, G.D. Allen, N.S. Lawrence, L. Jiang and T.G.J. Jones, *Electroanalysis*, 2002, **14**, 1464.
234. J. Zou, J.A. Parkinson and P.J. Sadler, *J. Chinese Chem. Soc.*, 2002, **49**, 499.
235. P. Vani, T. Raja Rajeswari and L.S.A. Dikshitulu, *J. Indian Chem. Soc.*, 2001, **78**, 44.
236. M.A. Mansour, *Trans. Met. Chem.*, 2002, **27**, 818.
237. M. Verma and D. Doss, *Asian J. Chem.*, 2001, **13**, 1451.
238. N. Nalwaja, A. Jain and B.L. Hiran, *J. Indian Chem. Soc.*, 2002, **79**, 587.
239. S.D. Quine and B.T. Gowda, *Oxid. Commun.*, 2001, **24**, 450.
240. P.S. Sengupta, R. Sinha and G.S. De, *Trans. Met. Chem.*, 2001, **26**, 638.
241. S. Verstraete, O. Heudi, A. Cailleux and P. Allain, *J. Inorg. Biochem.*, 2001, **84**, 129.
242. A. Habtemariam, J.A. Parkinson, N. Margiotta, T.W. Hambley, S. Parsons and P.J. Sadler, *J. Chem. Soc., Dalton Trans.*, 2001, 362.
243. Ž.D. Bugarčić, M.M. Shoukry and R. van Eldik, *J. Chem. Soc., Dalton Trans.*, 2002, 3945.
244. S.C. Moi, A.K. Gosh and G.S. De, *Indian J. Chem. Sect. A, Inorg. Bioinorg. Phys. Theor. Anal. Chem.*, 2001, **40**, 1187.
245. T. Poth, H. Paulus, H. Elias, C. Dücker-Benfer and R. van Eldik, *Eur. J. Inorg. Chem.*, 2001, 1361.
246. G. Shengli, J. Mian, C. Sanping, H. Rongzu and S. Qizhen, *J. Therm. Anal. Calor.*, 2001, **66**, 423.
247. B.D. Berezin and G. Mamardashvili, *Russ. J. Coord. Chem.*, 2002, **28**, 771.
248. M. Thamae and T. Nyokong, *J. Porph. Phthaloc.*, 2001, **5**, 839.
249. X. Yan, G. Wang, W. Zhu, X. Liu, S. Zhu, H. Lin, Z. Zhu and R. Chen, *Zhongguo Yaowu Huaxue Zazhi*, 2001, **11**, 21.

250. Á. García-Raso, J.J. Fiol, F. Bádenas and F. Muñoz, *Polyhedron*, 2001, **20**, 2609.
251. Y. Kitamura, K. Hyodoh, Y. Nagawo, Y. Sasaki and N. Azuma, *Inorg. React. Mechan.*, 2002, **4**, 221.
252. A.A. Phulambrikar and C. Chatterjee, *Bull. Chem. Soc. Japan*, 2002, **75**, 1515.
253. K.-Ud- Din, M. Akram and Z. Khan, *Inorg. React. Mechan.*, 2002, **4**, 77.
254. S. Ahmad, A.A. Isab and M.I.M. Wazeer, *Inorg. React. Mechan.*, 2002, **4**, 95.
255. C.-H. Ng, S.-B. Teo, S.-G. Teoh, J.-P. Declercq and S.W. Ng, *J. Coord. Chem.*, 2002, **55**, 909.
256. C.H. Ng, T.S. Chong, S.G. Teoh, F. Adams and S.W. Ng, *J. Chem. Soc., Dalton Trans.*, 2002, 3361.
257. S. Kobayashi, R. Matsubara and H. Kitagawa, *Org. Lett.*, 2002, **4**, 143.
258. C.-H. Ng, S.-B. Teo, S.-G. Teoh, J.-P. Declercq and S.W. Ng, *Australian J. Chem.*, 2001, **54**, 743.
259. P.A. Butler, C.G. Crane, B.T. Golding, A. Hammersshoi, D.C. Hockless, T.B. Petersen, A.M. Sargeson and D.C. Ware, *Inorg. Chim. Acta*, 2002, **331**, 318.
260. G. Laval, W. Clegg, C.G. Crane, A. Hammershoi, A.M. Sargeson and B.T. Golding, *Chem. Commun.*, 2002, 1874.
261. Y.N. Belokon, K.A. Kochetkov, N.S. Ikonnikov, T.V. Strelkova, S.R. Harutyunyan and A.S. Saghiyan, *Teterahedron Assym.*, 2001, **12**, 481.
262. A. Debache, S. Collet, P. Bauchat, D. Danion, L. Euzenat, A. Hercouet and B. Carboni, *Teterahedron Assym.*, 2001, **12**, 761.
263. G. Guillena, G. Rodríguez and G. van Koten, *Teterahedron Lett.*, 2002, **43**, 3895.
264. S.J. Greenfield and S.R. Gilbertson, *Synthesis*, 2001, **15**, 2337.
265. S. Li, Y. Yi, L. Mo, H. Cheng, X. Guan and G. Dongye, *Guangdong Weiliang Yuansu Kexue*, 2001, **8**, 54.
266. Y. Kojima, K. Kabuto and Y. Sasaki, 2002, Jpn. Kokai Tokkyo Koho JP 2002, 20,358.
267. H.D. Ashmead and S.D. Ashmead, 2002, PCT. Int. Appl. WO 02 30,947.
268. S.D. Ashmead, D.C. Wheelwright, C. Ericson and M. Pedersen, 2002, PCT. Int. Appl. WO 02 30,948.
269. D.J.U. Miodragović, M.J. Malinar, S.M. Milosavljević, S.D. Zarić, D. Vučelić and M.B. Čelap, *J. Coord. Chem.*, 2002, **55**, 517.
270. X. Wei, W. Liang and Q. Feng, 2002, Faming Zhuanli Shenqing Gongkai Shuomingshu CN 1,343,660.
271. S. Zhao and Y.-M. Liu, *Anal. Chim. Acta*, 2001, **426**, 65.
272. P.E.M. Overdevest, T.J.M. de Bruin, E.J.R. Sudhoelter, K. van't Riet, J.T.F. Keurentjes and A. van der Padt, *Ind. Eng. Chem. Res.*, 2001, **40**, 5991.
273. G. Galaverna, R. Corradini, F. Dallavelle, G. Folesani, A. Dossena and R. Marchelli, *J. Chromatogr., A*, 2001, **922**, 151.
274. Y. Ihara, S. Kurose and T. Koyama, *Monatsh. Chem.*, 2001, **132**, 1433.
275. M. Schlauch and A.W. Frahm, *Anal. Chem.*, 2001, **73**, 262.
276. A. Berthod, A. Valleix, V. Tizon, E. Leonce, C. Caussignac and D.W. Armstrong, *Anal. Chem.*, 2001, **73**, 5499.
277. E.M. van der Ent, K. van't Riet, J.T.F. Keurentjes and A. van der Padt, *J. Membr. Sci.*, 2001, **185**, 207.
278. S.K. Fiskum, B.M. Rapko, M. Brian and G.J. Lumetta, *Solvent Extr. Ion Exch.*, 2001, **19**, 643.
279. K.-W. Cha, C.-I. Park and K.-W. Park, *Bull. Kor. Chem. Soc.*, 2002, **23**, 402.
280. Z. Fan, L. Du, X. Ji and H. Xie, *Guangpuxue Yu Guangpu Fenxi*, 2001, **21**, 682.

281. N. Teshima, T. Nobuta, T. Sakai and T. Kawashima, *Bunseki Kagaku*, 2001, **50**, 47.
282. M. Claeys-Bruno, M. Bardet and J.-C. Marchon, *Compt. Rend. Chim.*, 2002, **5**, 21.
283. S. Shahrokhian, *Anal. Chem.*, 2001, **73**, 5972.
284. S. Wang and D. Du, *Sensors*, 2002, **2**, 41.
285. M. Yurong, Y. Qiuxia, H. Wei and W. Xuelin, *J. Nat. Gas Chem.*, 2001, **10**, 147.
286. L. Ronconi, C. Marzano, U. Russo, S. Sitran, R. Graziani and D. Fregona, *J. Inorg. Biochem.*, 2002, **91**, 413.
287. J. Carrasco, J.J. Criado, R.I.R. Macías, J.L. Manzano, J.J.G. Marín, M. Medarde and E. Rodríguez, *J. Inorg. Biochem.*, 2001, **84**, 287.
288. Z.-M. Wang, H.-K. Lin, S.-R. Zhu, T.-F. Liu and Y.-T. Chen, *J. Inorg. Biochem.*, 2002, **89**, 97.
289. M. Chikira, Y. Tomizawa, D. Fukita, T. Sugizaki, N. Sugawara, T. Yamazaki, A. Sasano, H. Shindo, M. Palaniandavar and W.E. Antholine, *J. Inorg. Biochem.*, 2002, **89**, 163.
290. A. Valent, M. Melnik, D. Hudecová, B. Dudová, R. Kivekäs and M.R. Sundberg, *Inorg. Chim. Acta*, 2002, **340**, 15.
291. K. Nomiya and M. Oda, 2001, Jpn. Kokai Tokkyo Koho JP 2001 **335**, 405.
292. G. Maciejewska, M. Cieślak-Golonka, Z. Staszak and A. Szeląg, *Trans. Met. Chem.*, 2002, **27**, 473.
293. T.P.A. Kruck and T.E. Burrow, *J. Inorg. Biochem.*, 2002, **88**, 19.
294. R.A. Motterlini and B.E. Mann, 2002, PCT Int. Appl. WO 02 92,075.
295. U. Brand, M. Rombach, J. Seebacher and H. Vahrenkamp, *Inorg. Chem.*, 2001 **40**, 6.
296. M.L. Styles, R.A.J. O'Hair and W.D. McFadyen, *Eur. J. Mass Spectrom.*, 2001, **7**, 69.
297. V. Anbalagan, B.A. Perera, A.T.M. Silva, A.L. Gallardo, M. Barber, J.M. Barr, S.M. Terkarli, E.R. Talaty and M.J. Van Stipdonk, *J. Mass Spectrom.*, 2002, **37**, 910.
298. C.H.S. Wong, N.L. Ma and C.W. Tsang, *Chem. Eur. J.*, 2002, **8**, 4909.
299. B.K. Bluhm, S.J. Shields, C.A. Bayse, M.B. Hall and D.H. Russel, *Int. J. Mass Spectrom.*, 2001, **204**, 31.
300. J. Slaninova, L. Maletinska, J. Vondrasek and Z. Prochazka, *J. Peptid Sci.*, 2001, **7**, 413.
301. R. Ferrari, S. Bernés, C.R. de Barbarín, G. Mendoza-Díaz and L. Gasque, *Inorg. Chim. Acta*, 2002, **339**, 193.
302. M. Watabe, M. Kai, S. Asanuma, M. Yoshikane, A. Horiuchi, A. Ogasawara, T. Watanabe, T. Mikami and T. Matsumoto, *Inorg. Chem.*, 2001, **40**, 1496.
303. A.J. Tasiopoulos, E.J. Tolis, J.M. Tsangaris, A. Evangelou, J.D. Woollins, A.M.Z. Slawin, J.C. Pessoa, I. Correia and T.A. Kabanos, *J. Biol. Inorg. Chem.*, 2002, **7**, 363.
304. A. Asano, C.M. Sullivan, A. Yanagisawa, H. Kimoto and T. Kurotsu, *Anal. Bioanal. Chem.*, 2002, **374**, 1250.
305. A. Asano, C.M. Sullivan, A. Yanagisawa, C. Yato, H. Kimoto and T. Kurotsu, *Anal. Sci.*, 2001, **17**, a159.
306. C. Teljón, R. Olmo, J.M. Blanco, A. Romero and J.M. Teijón, *Polym. Int.*, 2001, **50**, 822.
307. B.T. Farrer and V.L. Pecoraro, *Curr. Op. Drug Disc. Developm.*, 2002, **5**, 937.
308. K. Polborn, W. Hoffmuller and W. Beck, *New Cryst. Sruct.*, 2001, **216**, 231.

309. O.R. Nascimento, A.J. Costa-Filho, D.I. De Morais, J. Ellena and L.F. Delboni, *Inorg. Chim. Acta*, 2001, **312**, 133.
310. M. Tiliakos, D. Raptis, A. Terzis, C.P. Raptopoulou, P. Cordopatis and E. Manessi-Zoupa, *Polyhedron*, 2002, **21**, 229.
311. G. Facchin, M.H. Torre, E. Kremer, O.E. Piro, E.E. Castellano and E.J. Baran, *J. Inorg. Biochem.*, 2002, **89**, 174.
312. A. García-Raso, J.J. Fiol, B. Adrover, A. Caubet, E. Espinosa, I. Mata and E. Molins, *Polyhedron*, 2002, **21**, 1197.
313. R. Basosi, N. D'Amelio, E. Gaggelli, R. Pogni and G. Valensin, *J. Chem. Soc., Perkin Trans. 2*, 2001, 252.
314. W.A. Tao, L. Wu and R.G. Cooks, *J. Am. Soc. Mass Spectrom.*, 2001, **12**, 490.
315. S.M. Barlow, S. Haq and R. Raval, *Langmuir*, 2001, **17**, 3292.
316. T. Yamada, T. Ichino, R. Yanagihara and T. Miyazawa, *Pept. Sci.*, 2001, **37**, 305.
317. P.V. Bernhardt, P. Comba, D.P. Fairlie, L.G. Gahan, G.R. Hanson and L. Lötzbeyer, *Chem. Eur. J.*, 2002, **8**, 1527.
318. L.A. Morris, M. Jaspars, J.J. Kettenes-van den Bosch, K. Versluis, A.J.R. Heck, S.M. Kelly and N.C. Price, *Tetrahedron*, 2001, **57**, 3185.
319. R.M. Cusack, M. Rodney, L. Grondahl, D.P. Fairlie, L.R. Gahan and G.R. Hanson, *J. Chem. Soc., Perkin Trans. 2*, 2002, 556.
320. G. Saviano, F. Rossi, E. Benedetti, C. Pedone, D.F. Mierke, A. Maione, G. Zanotti, T. Tancredi and M. Saviano, *Chem. Eur. J.*, 2001, **7**, 1176.
321. Y. Singh, N. Sokolenko, M.J. Kelso, L.R. Gahan, G. Abbenante and D.P. Fairlie, *J. Am. Chem. Soc.*, 2001, **123**, 333.
322. L.L. Guan, Y. Sera, K. Adachi, F. Nishida and Y. Shizuri, *Biochem. Biophys. Res. Commun.*, 2001, **283**, 976.
323. C.L. Weeks, P. Turner, R.R. Fenton and P.A. Lay, *J. Chem. Soc., Dalton Trans.*, 2002, 931.
324. N. Niklas, O. Walter, F. Hampel and R. Alsfasser, *J. Chem. Soc., Dalton Trans.*, 2002, 3367.
325. H. Kurosaki, R.K. Sharma, S. Aoki, T. Inoue, Y. Okamoto, Y. Sugiura, M. Doi, T. Ishida, M. Otsuka and M. Goto, *J. Chem. Soc., Dalton Trans.*, 2001, 441.
326. M. Gelinsky, R. Vogler and H. Vahrenkamp, *Inorg. Chem.*, 2002, **41**, 2560.
327. E. Kimura, T. Gotoh, S. Aoki and M. Shiro, *Inorg. Chem.*, 2002, **41**, 3239.
328. S. Chang, V.V. Karambelkar, R.D. Sommer, A.L. Rheingold and D.P. Goldberg, *Inorg. Chem.*, 2002, **41**, 239.
329. B. Macías, I. García, M.V. Villa, J. Borrás, A. Castiñeiras and F. Sanz, *Polyhedron*, 2002, **21**, 1229.
330. T.E. Lehmann, *J. Biol. Inorg. Chem.*, 2002, **7**, 305.
331. N. Niklas, F. Hampel, G. Liehr, A. Zahl and R. Alsfasser, *Chem. Eur. J.*, 2001, **7**, 5135.
332. J. Bujdák and B.M. Rode, *J. Inorg. Biochem.*, 2002, **90**, 1.
333. V.A. Basiuk and J. Sainz-Rojas, *Adv. Space Res.*, 2001, **27**, 225.
334. T.L. Porter, M.P. Eastman, E. Bain and S. Begay, *Biophys. Chem.*, 2001, **91**, 115.
335. Y. Ye, M. Liu, Y. Tang and X. Jiang, *Chem. Commun.*, 2002, 532.
336. M. Liu and Y. Ye, *Chin. J. Chem.*, 2002, **20**, 1347.
337. J. Cheng and T.J. Deming, *J. Am. Chem. Soc.*, 2001, **123**, 9457.
338. M. Remko and B.M. Rode, *Phys. Chem. Chem. Phys.*, 2001, **3**, 4667.
339. N.M. Milović and N.M. Kostić, *J. Am. Chem. Soc.*, 2002, **124**, 4759.
340. N.M. Milović and N.M. Kostić, *Inorg. Chem.*, 2002, **41**, 7053.
341. L. Zhu and N.M. Kostić, *Inorg. Chim. Acta*, 2002, **339**, 104.

342. X. Sun, L. Zhang, G. Yang, Z. Guo and L. Zhu, *New J. Chem.*, 2003, **27**, 818.
343. N.V. Kaminskaia and N.M. Kostić, *Inorg. Chem.*, 2001, **40**, 2368.
344. T. Matsubara and K. Hirao, *Organometallics*, 2001, **20**, 5056.
345. N.V. Kaminskaia and N.M. Kostić, *J. Chem. Soc., Dalton Trans.*, 2001, 1083.
346. V. Pelmenschikov and P.E.M. Siegbahn, *Inorg. Chem.*, 2002, **41**, 5659.
347. V. Pelmenschikov, M.R.A. Blomberg and P.E.M. Siegbahn, *J. Biol. Inorg. Chem.*, 2002, **7**, 284.
348. Y. Fujii, T. Kiss, T. Gajda, X.S. Tan, T. Sato, Y. Nakano, Y. Hayashi and M. Yashiro, *J. Biol. Inorg. Chem.*, 2002, **7**, 843.
349. J. Sun and S. Moon, *Inorg. Chem.*, 2001, **40**, 4890.
350. D.T. Puerta and S.M. Cohen, *Inorg. Chim. Acta*, 2002, **337**, 459.
351. D.L. Bienvenue, D. Gilner and R.C. Holz, *Biochemistry*, 2002, **41**, 3712.
352. T. Bantan-Polak and K.B. Grant, *Chem. Commun.*, 2002, 1444.
353. W.Y. Feng, S. Gronert, K.A. Fletcher, A. Warres and C.B. Lebrilla, *Int. J. Mass Spectrom.*, 2003, **222**, 117.
354. A. Torreggiani, P. Taddei and G. Fini, *Biopolymers*, 2002, **67**, 70.
355. M.N. Kumara, D.C. Gowda and K.S. Rangappa, *React. Kinet. Catal. Lett.*, 2001, **72**, 331.
356. B.K.K. Gowda, K.S. Rangappa and D.C. Gowda, *Ind. J. Chem.*, 2002, **41B**, 1039.
357. D.C. Gowda, B.K.K. Gowda and K.S. Rangappa, *Synth. React. Inorg. Met,-Org. Chem.*, 2001, **31**, 1109.
358. D.C. Gowda, B.K.K. Gowda and K.S. Rangappa, *J. Phys. Org. Chem.*, 2001, **14**, 716.
359. H.A. Headlam and P.A. Lay, *Inorg. Chem.*, 2001, **40**, 78.
360. T. Kurahashi, A. Miyazaki, S. Suwan and M. Isobe, *J. Am. Chem. Soc.*, 2001, **123**, 9268.
361. M. Hahn, M. Kleine and W.S. Sheldrick, *J. Biol. Inorg. Chem.*, 2001, **6**, 556.
362. V. Marchán, V. Moreno, E. Pedroso and A. Grandas, *Chem. Eur. J.*, 2001, **7**, 808.
363. M.I. Djuran, D.P. Dimitrijević, S.U. Milinković and Z.D. Bugarčić, *Trans. Met. Chem.*, 2002, **27**, 155.
364. S. Dey and P. Banerjee, *Inorg. React. Mechan.*, 2002, **4**, 159.
365. Z.D. Bugarcic, D. Ilic and M.I. Djuran, *Aust. J. Chem.*, 2001, **54**, 237.
366. Z. Nagy and I. Sóvágó, *J. Chem. Soc., Dalton Trans.*, 2001, 2467.
367. I. Hamachi, N. Kasagi, S. Kiyonaka, T. Nagase, Y. Mito-oka and S. Shinkai, *Chem. Lett.*, 2001, 16.
368. K.S. Schmidt, M. Boudvillain, A. Schwartz, G.A. van der Marel, J.H. van Boom, J. Reedijk and B. Lippert, *Chem. Eur. J.*, 2002, **8**, 5566.
369. K. Peters, G. Jahreis and E.M. Kötters, *J. Enzyme Inhibit.*, 2001, **16**, 339.
370. S. Bounaga, M. Galleni, A.P. Laws and M.I. Page, *Bioorg. Med. Chem.*, 2001, **9**, 503.
371. M. wa Mutahi, T. Nittoli, L. Guo and S. McN. Sieburth, *J. Am. Chem. Soc.*, 2002, **124**, 7363.
372. H.K. Smith, R.P. Beckett, J.M. Clements, S. Doel, S.P. East, S.B. Launchbury, L.M. Pratt, Z.M. Spavold, W. Thomas, R.S. Todd and M. Whittaker, *Bioorg. Med. Chem. Lett.*, 2002, **12**, 3595.
373. T. Szabó-Planka, Zs. Árkosi, A. Rockenbauer and L. Korecz, *Polyhedron*, 2001, **20**, 995.
374. T. Szabó-Planka, N.V. Nagy, A. Rockenbauer and L. Korecz, *Inorg. Chem.*, 2002, **41**, 3483.
375. B. Gyurcsik, I. Vosekalna and E. Larsen, *J. Inorg. Biochem.*, 2001, **85**, 89.

376. K. Ösz, B. Bóka, K. Várnagy, I. Sóvágó, T. Kurtán and S. Antus, *Polyhedron*, 2002, **21**, 2149.
377. M.P. Bemquerer, C. Bloch Jr., H.F. Brito, E.E.S. Teotonio and M.T.M. Miranda, *J. Inorg. Biochem.*, 2002, **91**, 363.
378. T.C. Ramalho and J.D. Figueroa-Villar, *Theochem.*, 2002, **580**, 217.
379. S.D. Chachere, P.J. Sondawale and M.L. Narwade, *Asian J. Chem.*, 2001, **13**, 671.
380. V.S. Ilakin, V.G. Shtyrlin, A.V. Zakharov and A.L. Kon'kin, *Russ. J. Gen. Chem.*, 2002, **72**, 349.
381. D. Sanna, Cs.G. Ágoston, G. Micera and I. Sóvágó, *Polyhedron*, 2001, **20**, 3079.
382. E. Lodyga-Chruscinska, G. Micera, D. Sanna, J. Olczak and J. Zabrocki, *Polyhedron*, 2001, **20**, 1915.
383. J. Świątek-Kozlowska, J. Brasuń, M. Luczkowski and M. Makowski, *J. Inorg. Biochem.*, 2002, **90**, 106.
384. T. Kiss, M. Kilyén, A. Lakatos, F. Evanics, T. Körtvélyesi, Gy. Dombi, Zs. Majer and M. Hollósi, *Coord. Chem. Rev.*, 2002, **228**, 227.
385. G. Malandrinos, M. Louloudi, Y. Deligiannakis and N. Hadjiliadis, *Inorg. Chem.*, 2001, **40**, 4588.
386. G. Malandrinos, M. Louloudi, Y. Deligiannakis and N. Hadjiliadis, *J. Phys. Chem. B*, 2001, **105**, 7323.
387. J. Costa Pessoa, I. Correia, T. Kiss, T. Jakusch, M.M.C.A. Castro and C.F.G.C. Geraldes, *J. Chem. Soc., Dalton Trans.*, 2002, 4440.
388. M. Lukáš, M. K□vala, P. Hermann, I. Lukeš, D. Sanna and G. Micera, *J. Chem. Soc., Dalton Trans.*, 2001, 2850.
389. C. Conato, S. Ferrari, H. Kozlowski, F. Pukidori and M. Remelli, *Polyhedron*, 2001, **20**, 615.
390. Cs. Kállay, K. Várnagy, I. Sóvágó, D. Sanna and G. Micera, *J. Chem. Soc., Dalton Trans.*, 2002, 92.
391. B. Bóka, Z. Nagy, K. Várnagy and I. Sóvágó, *J. Inorg. Biochem.*, 2001, **83**, 77.
392. E. Farkas, P. Buglyó, É.A. Enyedy, V.A. Gerlei and A.M. Santos, *Inorg. Chim. Acta*, 2002, **339**, 215.
393. J. Brasun, S. Ołdziej, M. Taddei and H. Kozlowski, *J. Inorg. Biochem.*, 2001, **85**, 79.
394. G.N. Mukherjee and P.K. Chakraborty, *J. Indian Chem. Soc.*, 2002, **79**, 137.
395. G.N. Mukherjee and P.K. Chakraborty, *J. Indian Chem. Soc.*, 2001, **78**, 565.
396. M. Saladini, L. Menabue, E. Ferrari and D. Iacopino, *J. Chem. Soc., Dalton Trans.*, 2001, 1513.
397. T. Yajima, M. Okajima, A. Odani and O. Yamauchi, *Inorg. Chim. Acta*, 2002, **339**, 445.
398. M.M. Shoukry, E.M. Khairy and A.A. El-Sherif, *Trans. Met. Chem.*, 2002, **27**, 656.
399. R.N. Patel, R.P. Shrivastava, N. Singh and S. Kumar, *Indian J. Chem.*, 2001, **40A**, 361.
400. M.M.A. Mohamed and M.M. Shoukry, *Polyhedron*, 2001, **20**, 343.
401. E.M. Shoukry, *Ann. Chim. (Rome, Italy)*, 2001, **91**, 627.
402. H. Schmidt, I. Andersson, D. Rehder and L. Pettersson, *Chem. Eur. J.*, 2001, **7**, 251.
403. Ma.T. Armas, A. Mederos, P. Gili, S. Domíngez, R. Hernández-Molina, P. Lorenzo, S. Vaz-Júnior, E.J. Baran, M.L. Araujo, V. Lubes and F. Brito, *Polyhedron*, 2002, **21**, 1513.
404. P. Mineo, D. Vitalini, D. La Mendola, E. Rizzarelli, E. Scamporrino and G. Vecchio, *Rapid Commun. Mass Spectrom.*, 2002, **16**, 722.

405. D. Sanna, Cs.G. Ágoston, I. Sóvágó and G. Micera, *Polyhedron*, 2001, **20**, 937.
406. C. Conato, R. Gavioli, R. Guerrini, H. Kozlowski, P. Młynarz, C. Pasti, F. Pulidori and M. Remelli, *Biochim. Biophys. Acta*, 2001, **1526**, 199.
407. A. Myari, G. Malandrinos, Y. Deligiannakis, J.C. Plakatouras, N. Hadjiliadis, Z. Nagy and I. Sóvágó, *J. Inorg. Biochem.*, 2001, **85**, 253.
408. M. Casolaro, M. Chelli, M. Ginanneschi, F. Laschi, L. Messori, M. Muniz-Miranda, A.M. Papini, T. Kowalik-Jankowska and H. Kozlowski, *J. Inorg. Biochem.*, 2002, **89**, 181.
409. P. Młynarz, D. Valensin, K. Kociolek, J. Zabrocki, J. Olejnik and H. Kozlowski, *New J. Chem.*, 2002, **26**, 264.
410. E.K. Quagraine, H.-B. Kraatz and R.S. Reid, *J. Inorg. Biochem.*, 2001, **85**, 23.
411. H. Liang, J. Huang, C.-Q. Tu, M. Zang, Y.-Q. Zhou and P.-W. Shen, *J. Inorg. Biochem.*, 2001, **85**, 167.
412. C. Conato, H. Kozlowski, P. Młynarz, F. Pulidori and M. Remelli, *Polyhedron*, 2002, **21**, 1469.
413. C. Conato, W. Kamysz, H. Kozlowski, M. Łuczkowski, Z. Mackiewicz, P. Młinarz, M. Remelli, D. Valensin and G. Valensin, *J. Chem. Soc., Dalton Trans.*, 2002, 3939.
414. M. Mylonas, J.C. Plakatouras, N. Hadjiliadis, A. Krężel and W. Bal, *Inorg. Chim. Acta*, 2002, **339**, 60.
415. M. Mylonas, A. Krężel, J. Plakatouras, N. Hadjiliadis and W. Bal, *J. Chem. Soc., Dalton Trans.*, 2002, 4296.
416. M. Mylonas, G. Malandrinos, J. Plakatouras, N. Hadjiliadis, K.S. Kasprzak, A. Krężel and W. Bal, *Chem. Res. Toxicol.*, 2001, **14**, 1177.
417. M.A. Zoroddu, M. Peana, T. Kowalik-Jankowska, H. Kozlowski and M. Costa, *J. Chem. Soc., Dalton Trans.*, 2002, 458.
418. M.A. Zoroddu, T. Kowalik-Jankowska, H. Kozlowski, K. Salnikow and M. Costa, *J. Inorg. Biochem.*, 2001, **85**, 47.
419. P. Młinarz, D. Valensin, H. Kozlowski, T. Kowalik-Jankowska, J. Otlewski, G. Valensin and N. Gaggelli, *J. Chem. Soc., Dalton Trans.*, 2001, 645.
420. K. Qin, Y. Yang, P. Mastrangelo and D. Westaway, *J. Biol. Chem.*, 2002, **277**, 1981.
421. G.M. Cereghetti, A. Schweiger, R. Glockshuber and S. Van Doorslaer, *Biophys. J.*, 2001, **81**, 516.
422. S. Van Doorslaer, G.M. Cereghetti, R. Glockshuber and A. Schweiger, *J. Phys. Chem. B*, 2001, **105**, 1631.
423. J.R. Requena, D. Groth, G. Legname, E.R. Stadtman, S.B. Prusiner and R.L. Levine, *Proc. Natl. Acad. Sci.*, 2001, **98**, 7170.
424. H.E.M. McMahon, A. Mangé, N. Nishida, C. Créminon, D. Casanova and S. Lehmann, *J. Biol. Chem.*, 2001, **276**, 2286.
425. B.-S. Wong, S.G. Chen, M. Colucci, Z. Xie, T. Pan, T. Liu, R. Li, P. Gambetti, M.-S. Sy and D.R. Brown, *J. Neurochem.*, 2001, **78**, 1400.
426. C.S. Burns, E. Aronoff-Spencer, C.M. Dunham, P. Lario, N.I. Avdievich, W.E. Antholine, M.M. Olmstead, A. Vrielink, G.J. Gerfen, J. Peisach, W.G. Scott and G.L. Millhauser, *Biochem.*, 2002, **41**, 3991.
427. M. Łuczkowski, H. Kozlowski, M. Stawikowski, K. Rolka, E. Gaggelli, D. Valensin and G. Valensin, *J. Chem. Soc., Dalton Trans.*, 2002, 2269.
428. G. Pappalardo, G. Impellizzeri, R.P. Bonomo, T. Campagna, G. Grasso and M.G. Saita, *New J. Chem.*, 2002, **26**, 593.
429. M.F. Jobling, X. Huang, L.R. Stewart, K.J. Barnham, C. Curtain, I. Volitakis, M. Perugini, A.R. White, R.C. Cherny, C.L. Masters, C.J. Barrow, S.J. Collins, A.I. Bush and R. Cappai, *Biochem.*, 2001, **40**, 8073.

430. C.C. Curtain, F. Ali, I. Volitakis, R.A. Cherny, R.S. Norton, K. Beyreuther, C.J. Barrow, C.L. Masters, A.I. Bush and K.J. Barnham, *J. Biol. Chem.*, 2001, **276**, 20466.
431. R.A. Cherny, C.S. Atwood, M.E. Xilinas, D.N. Gray, W.D. Jones, C.A. McLean, K.J. Barnham, I. Volitakis, F.W. Fraser, Y.-S. Kim, X. Huang, L.E. Goldstein, R.D. Moir, J.T. Kim, K. Beyreuther, H. Zheng, R.E. Tanzi, C.L. Masters and A.I. Bush, *Neuron*, 2001, **30**, 665.
432. T. Kowalik-Jankowska, M. Ruta-Dolejsz, K. Wiśniewska and L. Łankiewicz, *J. Inorg. Biochem.*, 2001, **86**, 535.
433. T. Kowalik-Jankowska, M. Ruta-Dolejsz, K. Wiśniewska and L. Łankiewicz, *Biochem.*, 2002, **92**, 1.
434. M. Łuczkowski, K. Wiśniewska, L. Łankiewicz and H. Kozlowski, *J. Chem. Soc., Dalton Trans.*, 2002, 2266.
435. D.M. Morgan, J. Dong, J. Jacob, K. Lu, R.P. Apkarian, P. Thiyagarajan and D.G. Lynn, *J. Am. Chem. Soc.*, 2002, **124**, 12644.
436. J. Croom and I.L. Taylor, *J. Inorg. Biochem.*, 2001, **87**, 51.
437. C. Exley and O.V. Korchazhkina, *J. Inorg. Biochem.*, 2001, **84**, 215.
438. R. Vogler and H. Vahrenkamp, *Eur. J. Inorg. Chem.*, 2002, 761.
439. P. Gockel, R. Vogler, M. Gelinsky, A. Meissner, H. Albrich and H. Vahrenkamp, *Inorg. Chim. Acta*, 2001, **323**, 16.
440. P.R. Raddy and S.K. Mohan, *Indian J. Chem.*, 2002, **41A**, 1816.
441. A. Nomura and Y. Sugiura, *Inorg. Chem.*, 2002, **41**, 3693.
442. M. Gelinsky and H. Vahrenkamp, *Eur. J. Inorg. Chem.*, 2002, **2458**.
443. R.G. Daugherty, T. Wasowicz, B.R. Gibney and V.J. DeRose, *Inorg. Chem.*, 2002, **41**, 2623.
444. W.-Z. Lee and W.B. Tolman, *Inorg. Chem.*, 2002, **41**, 5656.
445. R. Vogler, M. Gelinsky. L.F. Guo and H. Vahrenkamp, *Inorg. Chim. Acta*, 2002, **339**, 1.
446. T. Jakusch, P. Buglyó, A.I. Tomaz, J.C. Pessoa and T. Kiss, *Inorg. Chim. Acta*, 2002, **339**, 119.
447. M.T. Armas, A. Mederos, P. Gili, S. Domíngez, R. Hernández-Molina, P. Lorenzo, E.J. Baran, M.L. Araujo and F. Brito, *Polyhedron*, 2001, **20**, 799.
448. J.C. Pessoa, I. Tomaz, T. Kiss, E. Kiss and P. Buglyó, *J. Biol. Inorg. Chem.*, 2002, **7**, 225.
449. J.C. Pessoa, I. Tomaz, T. Kiss and P. Buglyó, *J. Inorg. Biochem.*, 2001, **84**, 259.
450. P.K. Singh, B.S. Garg, D.N. Kumar and B.K. Singh, *Indian J. Chem. A*, 2001, **40A**, 1339.
451. R.K. Suto, N.E. Brasch, O.P. Anderson and R.G. Finke, *Inorg. Chem.*, 2001, **40**, 2686.
452. H. Satofuka, T. Fukui, M. Takagi, H. Atomi and T. Imanaka, *J. Inorg. Biochem.*, 2001, **86**, 595.
453. B.H. Cruz, J.M. Díaz-Cruz, I. Šestáková, J. Velek, C. Ariño and M. Esteban, *J. Electroanal. Chem.*, 2002, **520**, 111.
454. M. Erk and B. Raspor, *J. Electroanal. Chem.*, 2001, **502**, 174.
455. M.S. Díaz-Cruz, J.M. Díaz-Cruz and M. Esteban, *Electroanalysis*, 2002, **14**, 899.
456. M.E. Merrifield, Z. Huang, P. Kille and M.J. Stillman, *J. Inorg. Biochem.*, 2002, **88**, 153.
457. T.M. DeSilva, G. Veglia, F. Porcelli, A.M. Prantner and S.J. Opella, *Biopolymers*, 2002, **64**, 189.
458. Y. Chen and W. Maret, *Eur. J. Biochem.*, 2001, **268**, 3346.

459. S. Ranganathan, K.M. Muraleedharan, P. Bharadwaj, D. Chatterji and I. Karle, *Tetrahedron*, 2002, **58**, 2861.
460. T. Shi and D.L. Rabenstein, *Tetrahedron Lett.*, 2001, **42**, 7203.
461. Y. Zang and D.E. Wilcox, *J. Biol. Inorg. Chem.*, 2002, **7**, 327.
462. M. Matzapetakis, B.T. Farrer, T.-C. Weng, L. Hemmingsen, J.E. Penner-Hahn and V.L. Pecoraro, *J. Am. Chem. Soc.*, 2002, **124**, 8042.
463. C.E. Laplaza and R.H. Holm, *J. Am. Chem. Soc.*, 2001, **123**, 10255.
464. K.B. Musgrave, C.E. Laplaza, R.H. Holm, B. Hedman and K.O. Hodgson, *J. Am. Chem. Soc.*, 2002, **124**, 3083.
465. C.E. Laplaza and R.H. Holm, *J. Biol. Inorg. Chem.*, 2002, **7**, 451.
466. Y. Hara and M. Akiyama, *J. Am. Chem. Soc.*, 2001, **123**, 7247.
467. A. Katoh, Y. Hikita, M. Harata, J. Ohkanda, T. Tsubomura, A. Higuchi, R. Saito and K. Harada, *Heterocycles*, 2001, **55**, 2171.
468. D.P. Naughton and M. Grootveld, *Bioorg. Med. Chem. Lett.*, 2001, **11**, 2573.
469. E.R. Civitello, R.G. Leniek, K.A. Hossler, K. Haebe and D.M. Stearns, *Bioconjugate Chem.*, 2001, **12**, 459.
470. G. Giraudi, C. Baggiani, C. Giovannoli, L. Anfossi, C. Tozzi and A. Vanni, *Annali Chim.*, 2001, **91**, 1.
471. T.C. Miller, E.-S. Kwak, M.E. Howard, D.A. Venden Bout and J.A. Holcombe, *Anal. Chem.*, 2001, **73**, 4087.
472. S.-I. Ishikawa, K. Suyama, K. Arihara and M. Itoh, *Biol. Trace Elem. Res.*, 2002, **86**, 227.
473. H.P. Vanbilloen, R. Busson and A.M. Verbruggen, *J. Labelled Cpd. Radiopharm.*, 2001, **44**, 99.
474. M. Lipowska, L. Hansen, X. Xu, P.A. Marzilli, A. Taylor, Jr. and L.G. Marzilli, *Inorg. Chem.*, 2002, **41**, 3032.
475. G. Ferro-Flores, F.DeM. Ramirez, M.G. Martinez-Mendoza, C. Artega De Murphy, M. Pedraza-Lopez and L. Garcia-Salinas, *J. Radioanal. Nucl. Chem.*, 2002, **251**, 7.
476. C. Qi, L. Yang, H. Zhang, X. Guo, S. Feng and B. Li, *Med. Chem. Res.*, 2002, **11**, 345.
477. E. Wong, S. Bennett, B. Lawrence, T. Fauconnier, L.F.L. Lu, R.A. Bell, J.R. Thornback and D. Eshima, *Inorg. Chem.*, 2001, **40**, 5695.
478. H. Gali, T.J. Hoffman, G.L. Sieckman, N.K. Owen, K.V. Katti and W.A. Volkert, *Bioconjugate Chem.*, 2001, **12**, 354.
479. A. Boschi, C. Bolzati, E. Benini, E. Malagó, L. Uccelli, A. Duatti, A. Piffanelli, F. Refosco and F. Tisato, *Bioconjugate Chem.*, 2001, **12**, 1035.
480. A. Lazzaro, G. Vertuani, P. Bergamini, N. Mantovani, A. Marchi, L. Marvelli, R. Rossi, V. Bertolasi and V. Ferretti, *J. Chem. Soc., Dalton Trans.*, 2002, 2843.
481. J.F. Valliant, R.W. Riddoch, D.W. Hughes, D.G. Roe, T.K. Fauconnier and J.R. Thornback, *Inorg. Chim. Acta*, 2001, **325**, 155.
482. J.K. Amartey, R.S. Parhar and I. Al-Jammaz, *Nucl. Med. Biol.*, 2001, **28**, 225.
483. S. Liu and D.S. Edwards, *Bioconjugate Chem.*, 2001, **12**, 554.
484. M. Langer, R. La Bella, E. Garcia-Garayoa and A.G. Beck-Sickinger, *Bioconjugate Chem.*, 2001, **12**, 1028.
485. P. Saweczko, G.D. Enright and H.-B. Kraatz, *Inorg. Chem.*, 2001, **40**, 4409.
486. M.V. Baker, H.-B. Kraatz and J.W. Quail, *New J. Chem.*, 2001, **25**, 427.
487. T. Moriuchi, K. Yoshida and T. Hirao, *J. Organomet. Chem.*, 2001, **637–639**, 75.
488. D. Harmsen, G. Erker, R. Fröhlich and G. Kehr, *Eur. J. Inorg. Chem.*, 2002, 3156.
489. W. Ponikwar, P. Mayer and W. Beck, *Eur. J. Inorg. Chem.*, 2002, **1932**.

490. D. Herebian and W.S. Sheldrick, *J. Chem. Soc., Dalton Trans.*, 2002, 966.
491. T. Moriuchi, K. Yoshida and T. Hirao, *Organometallics*, 2001, **20**, 3101.
492. A. Wieckowska, R. Bilewicz, A. Misicka, M. Pietraszkiewicz, K. Bajdor and L. Piela, *Chem. Phys. Lett.*, 2001, **350**, 447.
493. J. Sehnert, A. Hess and N. Metzler-Nolte, *J. Organomet. Chem.*, 2001, **637–639**, 349.
494. Y. Xu, P. Saweczko and H.-B. Kraatz, *J. Organomet. Chem.*, 2001, **637–639**, 335.
495. S. Maricic and T. Frejd, *J. Org. Chem.*, 2002, **67**, 7600.
496. K. Haas and W. Beck, *Eur. J. Inorg. Chem.*, 2001, 2485.
497. K. Haas and W. Beck, *Z. Anorg. Allg. Chem.*, 2002, **628**, 788.
498. H. Dialer, W. Steglich and W. Beck, *Z. Naturforsch. B*, 2001, **56**, 1084.
499. Q. Huo, G. Sui, Y. Zheng, P. Kele, R.M. Leblanc, T. Hasegawa, J. Nishijo and J. Umemura, *Chem. Eur. J.*, 2001, **7**, 4796.
500. D.R. van Staveren, E. Bothe, T. Weyhermüller and N. Metzler-Nolte, *Eur. J. Inorg. Chem.*, 2002, **1518**.
501. D.R. van Staveren and N. Metzler-Nolte, *Chem. Commun.*, 2002, 1406.
502. G. Guillena, G. Rodríguez, M. Albrecht and G. van Koten, *Chem. Eur. J.*, 2002, **8**, 5368.
503. Y. Mito-oka, S. Tsukiji, T. Hiraoka, N. Kasagi, S. Shinkai and I. Hamachi, *Tetrahedron Lett.*, 2001, **42**, 7059.
504. A. Ojida, Y. Mito-oka, M. Inoue and I. Hamachi, *J. Am. Chem. Soc.*, 2002, **124**, 6256.
505. M. Gochin, V. Khorosheva and M.A. Case, *J. Am. Chem. Soc.*, 2002, **124**, 11018.
506. A. Fedorova and M.Y. Ogawa, *Bioconjugate Chem.*, 2002, **13**, 150.
507. P. Scrimin, P. Tecilla, U. Tonellato, A. Veronese, M. Crisma, F. Formaggio and C. Toniolo, *Chem. Eur. J.*, 2002, **8**, 2753.
508. S. Sun, Md.A. Fazal, B.C. Roy, B. Chandra and S. Mallik, *Inorg. Chem.*, 2002, **41**, 1584.
509. Md.A. Fazal, B.C. Roy, S. Sun, S. Mallik and K.R. Rodgers, *J. Am. Chem. Soc.*, 2001, **123**, 6283.
510. A.N. Kapanidis, Y.W. Ebright and R.H. Ebright, *J. Am. Chem. Soc.*, 2001, **123**, 12123.
511. B.R. Hart and K.J. Shea, *J. Am. Chem. Soc.*, 2001, **123**, 2072.
512. N. Jotterand, D.A. Pearce and B. Imperiali, *J. Org. Chem.*, 2001, **66**, 3224.
513. N. Egashira, J. Piao, M. Sumihiro, E. Hifumi and T. Uda, *Chem. Sens.*, 2001, **17**, 145.
514. L. Prodi, M. Montalti and N. Zaccheroni, *Helv. Chim. Acta*, 2001, **84**, 690.
515. F. Shi, S.J. Woltman and S.G. Weber, *Anal. Chim. Acta*, 2002, **474**, 1.
516. Y. Kim and S.J. Franklin, *Inorg. Chim. Acta*, 2002, **341**, 107.
517. M. Sirish and S.J. Franklin, *J. Inorg. Biochem.*, 2002, **91**, 253.
518. K. Ichikawa, M. Tarnai, M.K. Uddin, K. Nakata and S. Sato, *J. Inorg. Biochem.*, 2002, **91**, 437.
519. C. Madhavaiah and S. Verma, *Bioconjugate Chem.*, 2001, **12**, 855.
520. N.S. Josephsohn, K.W. Kuntz, M.L. Snapper and A.H. Hoveyda, *J. Am. Chem. Soc.*, 2001, **123**, 11594.
521. Q. Xiang, X. Yan, N. Zhaodong, X. Zeng and J. Xie, *J. Dispersion Sci. Technol.*, 2001, **22**, 103.
522. K. Jitsukawa, T. Irisa, H. Einaga and H. Masuda, *Chem. Lett.*, 2001, 30.